Nanocrystal
Edited by Yoshitake Masuda

Published by InTech
Janeza Trdine 9, 51000 Rijeka, Croatia

Publishing Process Manager Silvia Vlase

Technical Editor Teodora Smiljanic

Cover Designer InTech Design Team

Additional hard copies can be obtained from orders@intechopen.com

Nanocrystal, Edited by Yoshitake Masuda
p. cm.
ISBN 978-953-307-199-2

For EU product safety concerns:
IN TECH d.o.o., Prolaz Marije Krucifikse Kozulić 3, 51000 Rijeka, Croatia, info@intechopen.com or visit our website at intechopen.com.

Meet the editor

Dr. Yoshitake Masuda is a senior research scientist at the National Institute of Advanced Industrial Science and Technology (AIST), Japan. He graduated from Tsukuba University in 1994, and received his Master of Engineering degree from Tsukuba University in 1996. He was an engineer at NGK Spark Plug Co. Ltd., from 1996 to 1998, and an assistant professor at Nagoya University from 2000 to 2006. Dr. Masuda received his Doctor of Engineering degree from Nagoya University in 2004, and his PhD thesis was on the Patterning of TiO_2 Thin Films and Particles using Self-assembled Monolayers.

Contents

Preface

Nanocrystals have been attracting much attention for future science and technology. Chapters are sought that address innovative solutions to the design, synthesis, crystallization, morphology control, self-assembly, nano/micro-structure formation, patterning, novel property and device application of Nanocrystals. The book involves Nanocrystals of metal oxides, semiconductors, compounds, noble metals, metals, inorganic materials, carbon materials, organic molecules, polymers, bio materials, hybrid materials, composites, etc. It features remarkable breakthrough on Nanocrystals and provides latest scientific knowledge and leading-edge technology. They offer research agenda and accelerate the research, development and diffusion of Nanocrystals.

In closing, I wish to express my sincere sense of gratitude to the authors, publishing process manager Ms. Silvia Vlase, and the publishing staff. I dedicate this book to my parents, Mr. Toshio Masuda and Ms. Nobuko Masuda, my sisters, Ms. Shinobu Horita and Ms. Satoe Amaya, my children, Ms. Yuuka Masuda, Ms. Arisa Masuda and Mr. Ikuto Masuda, and my wife, Ms. Yumi Masuda.

Yoshitake Masuda
National Institute of Advanced Industrial Science and Technology (AIST),
Japan

Part 1

Synthesis and Design of Nanocrystals

1

Morphology Control of Metal Oxide Nanocrystals

Yoshitake Masuda
National Institute of Advanced Industrial Science and Technology (AIST)
Anagahora, Shimoshidami, Moriyama-ku, Nagoya
Japan

1. Introduction

Metal oxides have been widely used in electro devices, optical devices, etc. Recently, liquid phase syntheses of them attract much attention as future technology and novel academic field. Especially, liquid phase syntheses of anisotropic particles or films are expected for next generation metal oxide devices. This section describes liquid phase morphology control of anisotropic metal oxide. They were realized by precise control of nucleation and crystal growth. They showed high performance of solution systems for future metal oxide devices. Liquid phase morphology control of anisotropic metal oxide particles and films would contribute to development of metal oxide science and technology.

2. Morphology control of acicular $BaTiO_3$ particles[1]

Acicular $BaTiO_3$ particles were developed using solution systems. The morphology of BaC_2O_4 $0.5H_2O$ was controlled to acicular shape. Its phase transition to $BaTiO_3$ was realized by introducing Ti ions from the coprecipitated amorphous phase. Acicular $BaTiO_3$ particles have an aspect ratio as high as 18 and the particle size can be controlled by varying the growth period of $BaC_2O_4 \cdot 0.5H_2O$ which governs the size of $BaC_2O_4 \cdot 0.5H_2O$ particles. Acicular particles of crystalline $BaTiO_3$ can be used for ultra-thin multilayer ceramic capacitors.
Multilayer ceramic capacitors (MLCC) are indispensable electronic components for advanced electronic technology[2-12], but larger capacity and smaller size are needed for future electronic devices. To meet these needs, $BaTiO_3$ particles were downsized, but ferroelectric ceramics lose their ferroelectricity when their particle size is decreased and lose ferroelectricity entirely at a critical size[2-11]. This is known as the size effect and it impedes the progress of MLCC, so a novel solution has been eagerly anticipated.
Here, we propose MLCC using acicular $BaTiO_3$ particles[13]. An ultra-thin ferroelectric layer and high capacity can be realized by acicular particles having a high aspect ratio. The short side provides an ultra-thin ferroelectric layer and the large volume caused by the long side avoids the loss of ferroelectricity at the critical size. Anisotropic $BaTiO_3$ particles are thus a candidate for MLCC. $BaTiO_3$ has, however, an isotropic cubic or tetragonal structure, and its morphology is extremely difficult to control due to its isotropic crystal faces. We focused on triclinic $BaC_2O_4 \cdot 0.5H_2O$ which has an anisotropic crystal structure and controlled the morphology of these particles by precisely controlling crystal growth. We also achieved

phase transition of $BaC_2O_4 \cdot 0.5H_2O$ to crystalline $BaTiO_3$ by introducing Ti ions from the coprecipitated amorphous phase. Having developed several key technologies, we were successfully able to produce anisotropic acicular $BaTiO_3$ particles.

Morphology control of $BaTiO_3$ to rod-shape was reported previously. Additionally, metal oxalates (MC_2O_4) have been used for synthesis of rod-shaped oxides or hydroxides. Y. Hayashi *et al.* reported preparation of rod-shaped $BaTiO_3$ from rod-shaped TiO_2-nH_2O and $BaCO_3$ in molten chloride at high temperature[14]. Li *et al.* reported preparation of nanoflakes and nanorods of $Ni(OH)_2$, $Co(OH)_2$ and Fe_3O_4 by hydrothermal conversion at 160 °C for 12 h from $MC_2O_4 \cdot 2H_2O$ in NaOH solutions[15]. Sun *et al.* prepared flowerlike SnC_2O_4 submicrotubes in ethanol solutions containing $SnCl_2$ and oxalic acid. They were annealed at 500 °C for 2 h in ambient atmosphere to obtain flowerlike SnO_2 submicrotubes[16].

Oxalic acid (252 mg) was dissolved into isopropyl alcohol (4 ml) (Fig. 1)[1]. Butyl titanate monomer (0.122 ml) was mixed with the oxalic acid solution, and the solution was then mixed with distilled water (100 ml). The pH of the solution was increased to pH = 7 by adding NaOH (1 M) and distilled water, while the volume of the solution was adjusted to 150 ml by these additions. The aqueous solution (50 ml) with barium acetate (39.3 mg) was mixed with the oxalic acid solution. The mixed solution containing barium acetate (0.77 mM), butyl titanate monomer (2 mM) and oxalic acid (10 mM) was kept at room temperature for several hours with no stirring, and the solution gradually became cloudy. Stirring causes the collision of homogeneously nucleated particles and destruction of large grown particles, and so was avoided in this process. The size of the precipitate was easily controlled from nanometer order to micrometer order by changing the growth period. Large particles were grown by immersion for several hours to evaluate the morphology and crystallinity in detail.

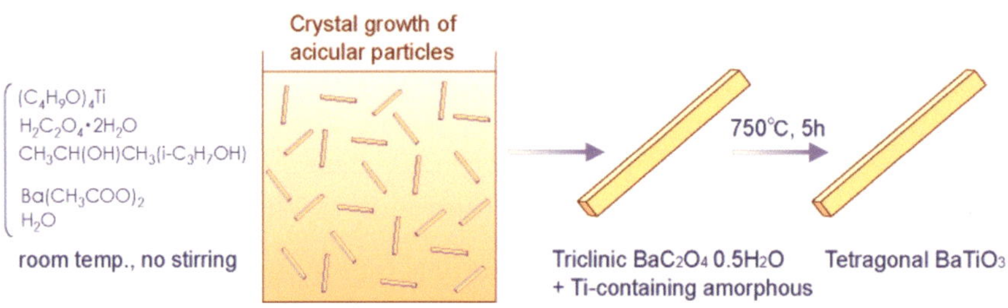

Fig. 1. Conceptual process for fabricating acicular $BaTiO_3$ particles. Morphology control of $BaC_2O_4 \bullet 0.5H_2O$ particles and phase transition to $BaTiO_3$. Reprinted with permission from Ref.[1], Masuda, Y., Yamada, T. and Koumoto, K., 2008, *Cryst. Growth Des.*, 8, 169. Copyright @American Chemical Society

Oxalate ions ($C_2O_4^{2-}$) react with barium ions (Ba^{2+}) to form barium oxalate ($BaC_2O_4 \cdot 0.5H_2O$). $BaC_2O_4 \cdot 0.5H_2O$ is dissolved in weak acetate acid provided by barium acetate ($(CH_3COO)_2Ba$), however, it can be deposited at pH 7 which is adjusted by adding NaOH. $BaC_2O_4 \cdot 0.5H_2O$ was thus successfully precipitated from the solution.

Acicular particles were homogeneously nucleated and precipitated from the solution. They were on average 23 μm (ranging from 19 to 27 μm) in width and 167 μm (ranging from 144 to 189 μm) in length, giving a high aspect ratio of 7.2 (Fig. 2). They had sharp edges and clear crystal faces, indicating high crystallinity. A gel-like solid was also coprecipitated from the solution as a second phase.

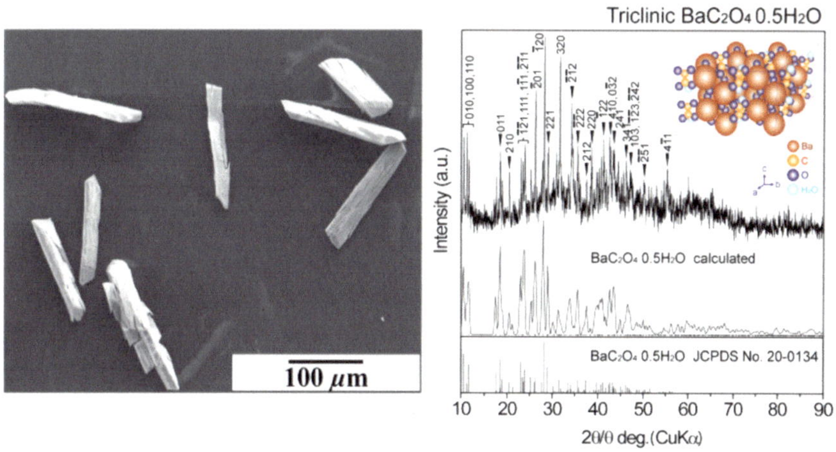

Fig. 2. SEM micrograph and XRD diffraction pattern of acicular $BaC_2O_4 \bullet 0.5H_2O$ particles precipitated from an aqueous solution at pH = 7. XRD diffraction measurement data (first step), XRD pattern calculated from crystal structure data16 (second step) and XRD pattern of JCPDS No. 20-134 (third step) are shown for triclinic $BaC_2O_4 \bullet 0.5H_2O$. Reprinted with permission from Ref.[1], Masuda, Y., Yamada, T. and Koumoto, K., 2008, *Cryst. Growth Des.*, 8, 169. Copyright @American Chemical Society

XRD diffraction patterns for the mixture of acicular particles and gel-like solid showed sharp diffraction peaks of crystalline $BaC_2O_4 \cdot 0.5H_2O$ with no additional phase. Acicular particles were crystalline $BaC_2O_4 \cdot 0.5H_2O$ and the gel-like solid would be an amorphous phase.

Fortunately, $BaC_2O_4 \cdot 0.5H_2O$ has a triclinic crystal structure as shown by the model calculated from structure data[17] (Fig. 2 XRD first step) and thus anisotropic crystal growth was allowed to proceed to produce an acicular shape. Each crystal face has a different surface energy and surface nature such as zeta potential and surface groups. Anisotropic crystal growth is induced by minimizing the total surface energy in ideal crystal growth. Additionally, site-selective adsorption of ions or molecules on specific crystal faces suppresses crystal growth perpendicular to the faces and so induces anisotropic crystal growth. These factors would cause anisotropic crystal growth of $BaC_2O_4 \cdot 0.5H_2O$ and hence allow us to control morphology and fabricate acicular $BaC_2O_4 \cdot 0.5H_2O$ particles. The positions of diffraction peaks corresponded with that of JCPDS No. 20-0134 (Fig. 2 XRD third step) and that calculated from crystal structure data[17] (Fig. 2 XRD second step), however, several diffraction peaks, especially 320 and 201, were enhanced strongly compared to their relative intensity. The enhancement of diffraction intensity from specific crystal faces would be related to anisotropic crystal growth; a large crystal size in a specific crystal orientation increases the x-ray diffraction intensity for the crystal face perpendicular to the crystal orientation.

EDX elemental analysis indicated the chemical ratio of the precipitate, which included acicular particles and gel-like solid, to be about Ba / Ti = 1 to 1.5. The chemical ratio indicated that the coprecipitated amorphous gel contained Ti ions. Additional Ba ions can be transformed into $BaCO_3$ by annealing and removed by HCl treatment in the next step. The ratio was thus controlled to slightly above Ba / Ti = 1 by adjusting the volume ratio of acicular particles and gel-like solid. Consequently, acicular particles of crystalline BaC_2O_4 ·

$0.5H_2O$ with Ti-containing gel-like solid were successfully fabricated in an aqueous solution process.

In comparison, isotropic particles of barium titanyl oxalate ($BaTiO(C_2O_4)_2 \cdot 4H_2O$) were precipitated at pH 2. $TiOC_2O_4$ was formed by the following reaction in which the reaction of oxalic acid ($H_2C_2O_4 \cdot 2H_2O$) with butyl titanate monomer ($(C_4H_9O)_4Ti$) and hydrolysis can take place simultaneously[18].

$$(C_4H_9O)_4 Ti + H_2C_2O_4 \cdot 2H_2O \rightarrow TiOC_2O_4 + 4C_4H_9OH + H_2O \qquad (a)$$

$TiO(C_2O_4)$ was then converted to oxalotitanic acid ($H_2TiO(C_2O_4)_2$) by the reaction:

$$TiO(C_2O_4) + H_2C_2O_4 \cdot 2H_2O \rightarrow H_2TiO(C_2O_4)_2 + 2H_2O \qquad (b)$$

Alcoholic solution containing oxalotitanic acid ($H_2TiO(C_2O_4)_2$) formed by reaction (b) was subjected to the following cation exchange reaction by rapidly adding an aqueous solution of barium acetate at room temperature:

$$H_2TiO(C_2O_4)_2 + Ba(CH_3COO)_2 \rightarrow BaTiO(C_2O_4)_2 \downarrow + 2CH_3COOH \qquad (c)$$

$BaTiO(C_2O_4)_2$ isotropic particles were formed by reaction (c).

On the other hand, neither $BaC_2O_4 \cdot 0.5H_2O$ nor $BaTiO(C_2O_4)_2$ was precipitated at pH 3 to pH 6. Gel-like solid was formed in the solution and their XRD spectra showed no diffraction peaks. The amorphous gel that precipitated at pH = 3 to 6 would be the same as the amorphous gel coprecipitated at pH 7.

These comparisons show that the crystal growth and morphology control of $BaC_2O_4 \cdot 0.5H_2O$ are sensitive to the solution conditions.

The precipitate was annealed at 750 °C for 5 h in air. Acicular $BaC_2O_4 \cdot 0.5H_2O$ particles were reacted with Ti-containing amorphous gel to introduce Ti ions to transform into crystalline $BaTiO_3$. X-ray diffraction of the annealed precipitate showed crystalline $BaTiO_3$ and an additional barium carbonate phase ($BaCO_3$). Excess precipitation of $BaC_2O_4 \cdot 0.5H_2O$ caused the generation of barium carbonate phase ($BaCO_3$) as expected.

The annealed precipitate was further immersed in HCl solution (1 M) to dissolve barium carbonate ($BaCO_3$). Acicular particles of crystalline $BaTiO_3$ were successfully fabricated with no additional phase. Particles showed acicular shape with 2.8×10×50 μm and x-ray diffraction of single-phase crystalline $BaTiO_3$ (Fig. 3). The high aspect ratio of the particles (17.8 = 50 / 2.8) would be provided by that of $BaC_2O_4 \cdot 0.5H_2O$ particles. The particle size of acicular $BaTiO_3$ can be easily controlled by the growth period and solution concentration for $BaC_2O_4 \cdot 0.5H_2O$ precipitation which decides the particle size of $BaC_2O_4 \cdot 0.5H_2O$.

$BaTiO_3$ has a cubic crystal structure at high temperature above phase transition and has a tetragonal crystal structure at room temperature. The cubic crystal structure is completely isotropic and the tetragonal crystal structure results from stretching a cubic lattice along one of its lattice vectors. For both of the crystal structures it is difficult to control anisotropic crystal growth, however, with our newly developed process we could successfully control the morphology and fabricate acicular particles. This was achieved by controlling the morphology of triclinic $BaC_2O_4 \cdot 0.5H_2O$ to acicular shape and the phase transition to $BaTiO_3$ by introducing Ti ions from the coprecipitated amorphous phase. The novel concept can be applied to a wide variety of morphology control and crystal growth control for advanced electronic devices composed of crystalline materials.

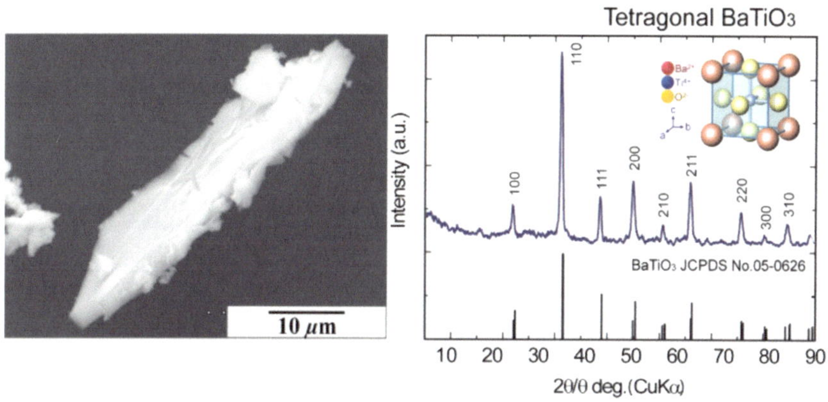

Fig. 3. SEM micrograph and XRD diffraction pattern of acicular $BaTiO_3$ particles after annealing at 750 °C for 5 h and HCl treatment. XRD diffraction measurement data (first step) and XRD pattern of JCPDS No. 05-0626 (second step) are shown for tetragonal $BaTiO_3$. Reprinted with permission from Ref.[1], Masuda, Y., Yamada, T. and Koumoto, K., 2008, *Cryst. Growth Des.*, 8, 169. Copyright @American Chemical Society

3. Morphology control of ZnO particles[19,20]

Ethylenediamine (H_2N-CH_2CH_2-NH_2, 15–45 mM, Sigma-Aldrich) was added to the zinc acetate aqueous solution ($Zn(CH_3COO)_2$, 15 mM, Kishida Chemical Co., Ltd.) to promote deposition of ZnO[19]. Zinc chelate ($(Zn(H_2N\text{-}CH_2CH_2\text{-}NH_2)^{2+})$) was formed from zinc acetate and ethylenediamine in reaction (a). ZnO was crystallized from zinc chelate and hydroxide ion (OH^-) in reaction (c).

$$Zn(CH_3COO)_2 + 2H_2NCH_2CH_2NH_2 \rightleftarrows Zn(H_2NCH_2CH_2NH_2)_2^{2+} + 2CH_3COO^- \ldots \quad (a)$$

$$H_2O \rightleftarrows OH^- + H^+ \ldots \quad (b)$$

$$Zn(H_2NCH_2CH_2NH_2)_2^{2+} + 2OH^- \longrightarrow ZnO + 2H_2NCH_2CH_2NH_2 + H_2O \ldots \quad (c)$$

The solution became turbid shortly after adding ethylenediamine. The molar ratio of ethylenediamine to Zn was [ethylenediamine] / [Zn] = (a) 1 : 1, (b) 2 : 1 or (c) 3 : 1. pH of the solutions were (a) pH=7.3, (b) pH=8.0 or (c) pH=8.7, respectively. Crystal growth rate and deposition of ZnO were attemped to control to change particle morphology. Si substrate (Newwingo Co., Ltd.) was immersed to evaluate deposited ZnO particles and particulate films. The solution in a glass beaker was kept at 60 °C for 3 h using a water bath. The silicon substrate was cleaned before immersion as described in references. The substrate was rinsed with distilled water after immersion.

ZnO particles having hexagonal cylinder shape were homogeneously nucleated and deposited in the aqueous solution containing 15 mM ethylenediamine ([ethylenediamine] / [Zn] = (a) 1 : 1) (Fig. 4a). X-ray diffraction patterns showed the deposition to be well crystallized ZnO (Fig. 5a). The relative intensity of (10-10) and (0002) is similar to that of randomly deposited ZnO particles, indicating the random orientation of deposited ZnO hexagonal cylinders, which is consistent with SEM observations. Crystals showed hexagonal facets of about 100-200 nm in diameter and about 500 nm in length. ZnO has a hexagonal crystal structure and thus hexagonal cylinders can be obtained by sufficiently slow crystal growth. A low crystal growth rate allows enough ions to diffuse to form a complete crystal structure.

Fig. 4. SEM micrographs of (a) ZnO hexagonal cylinder particles, (b) ZnO long ellipse particles, (c1) ZnO hexagonal symmetry radial whiskers and (c2) magnified area of (c1). Reprinted with permission from Ref.[19], Y. Masuda, N. Kinoshita and K. Koumoto, *J. Nanosci. Nanotechnol.*, in press. Copyright @American Scientific Publisher

ZnO having hexagonal cylinder shape showed strong photoluminescence intensity in the UV region at about 370-400 nm and weak intensity in the visible light region at about 530-550 nm by 350 nm excitation light which appears in spectra (Fig. 6a). ZnO crystals were reported to show UV luminescence (around 390 nm) attributed to band-edge luminescence and visible-light luminescence caused from oxygen vacancy (450-600 nm)[21,22]. Oxygen vacancies would be generated in ZnO during crystallization to show visible-light luminescence.

Concentration of ethylenediamine was increased twice to [ethylenediamine] / [Zn] = (b) 2 : 1. ZnO particles with long ellipse shape were deposited homogeneously from the solution (Fig. 4b). ZnO particles were about 100-200 nm in diameter and about 500 nm in length, and were similar to those of hexagonal cylinder shape. XRD showed the deposition to be well crystallized ZnO (Fig. 5b). Relative intensity of (10-10) diffraction is much stronger than that of (0002), indicating that mainly ZnO particles with long ellipse shape were laid on the silicon substrate. This was also observed in SEM micrographs (Fig. 4b). Deposition speed of ZnO with long ellipse shape was slightly faster than that of ZnO with hexagonal cylinder shape because of the high concentration of ethylenediamine. Ethylenediamine accelerates the crystallization of ZnO. In other words, supersaturation degree of the solution was increased by increasing the ethylenediamine concentration. As a result, ZnO grew slightly faster and formed not sharp hexagonal facets but rounded hexagonal cylinders, i.e., long ellipse shape. The photoluminescence spectrum of ZnO with long ellipse shape (Fig. 6b) was

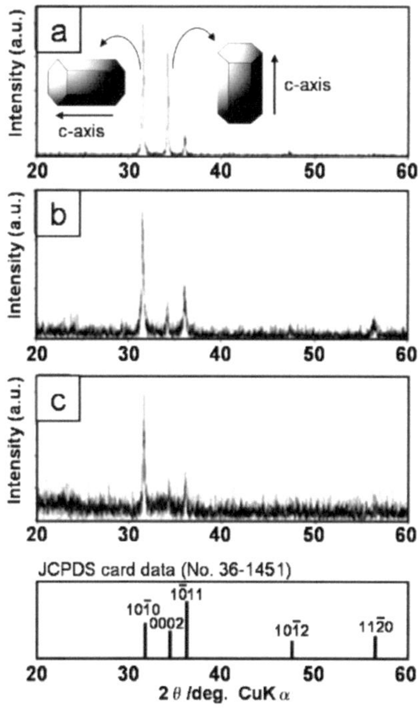

Fig. 5. XRD patterns of (a) ZnO hexagonal cylinder particles, (b) ZnO long ellipse particles, (c) ZnO hexagonal symmetry radial whiskers and ZnO (JCPDS No. 36-1451).
Reprinted with permission from Ref.[19], Y. Masuda, N. Kinoshita and K. Koumoto, *J. Nanosci. Nanotechnol.*, in press. Copyright @American Scientific Publisher

similar to that of ZnO with hexagonal cylinder shape (Fig. 6a). The luminescence property was clearly shown to be stable and not influenced by the synthesis conditions in the range of [ethylenediamine] / [Zn] = 1 : 1 to 2 : 1. High repeatability and stability of the photoluminescence property without being affected by the deposition conditions are major advantages of this system for large-scale production.

Further control the morphology was attempted by increasing the supersaturation degree. Ethylenediamine was added to the solution to be [ethylenediamine] / [Zn] = (c) 3 : 1 in the deposition process. ZnO whiskers were successfully grown and deposited from the solution. The morphology was drastically changed by the precise investigation and control of solution conditions. The whiskers were about 10–100 nm in diameter and about 1000–2000 nm in length (Fig. 4c1). A high aspect ratio was realized by high crystal growth rate of end faces to elongate the whiskers. Details of morphology were further evaluated. They were dispersed in water and dropped on a silicon substrate. Whiskers were connected to form hexagonal symmetry radial whiskers (Fig. 4c2). Six whiskers connected to form one particle. They had hexagonal symmetry. Tips of whiskers were finer than that at center of the particles. XRD showed the whiskers to be well crystallized ZnO (Fig. 5c). The relative intensity of (10-10) diffraction is much stronger than that of (0002), showing that mainly ZnO whiskers were laid on the substrate. This was also observed in SEM micrographs (Fig. 4c). Deposition of ZnO whiskers having high aspect ratio on the substrate would provide a ZnO network film having a high specific surface area. These whisker films can be applied to gas sensors[23,24]or solar cells[25] which require high specific surface area. The whisker films

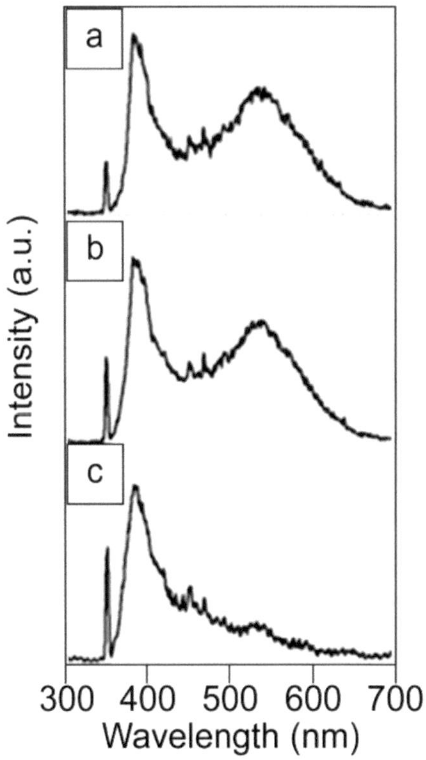

Fig. 6. Photoluminescence spectra of (a) ZnO hexagonal cylinder particles, (b) ZnO long ellipse particles and (c) ZnO hexagonal symmetry radial whiskers.
Reprinted with permission from Ref.[19], Y. Masuda, N. Kinoshita and K. Koumoto, *J. Nanosci. Nanotechnol.*, in press. Copyright @American Scientific Publisher

also have high conductivity per unit volume[26] compared to conventional particulate films or mesoporous materials because the whiskers carry an electric current for a long distance without grain boundaries. Photoluminescence intensity in the visible light region was quite different from that of ZnO with hexagonal cylinder shape or long ellipse shape (Fig. 5c). The large change of morphology indicates that the crystal growth mechanism differs greatly between ZnO particles having hexagonal cylinder or long ellipse shape and ZnO whiskers. Basically, ethylenediamine increased crystal growth rate to generate oxygen vacancies. The oxygen vacancies increased photoluminescence intensity in visible light region. However, hexagonal symmetry radial whiskers prepared in the solution of [ethylenediamine] / [Zn] = (c) 3 : 1 showed very weak photoluminescence intensity in visible light region compared to hexagonal cylinders or long ellipse particles. Excess ethylenediamine would decrease photoluminescence intensity in visible light region. Ion concentration of the solution of [ethylenediamine] / [Zn] = (c) 3 : 1 would drastically change at the initial stage. The solutions of [ethylenediamine] / [Zn] = (a) 1 : 1 and (b) 2 : 1 became turbid shortly after adding ethylenediamine and they changed to transparent gradually. The solutions of [ethylenediamine] / [Zn] = (c) 3 : 1 also became turbid shortly after adding ethylenediamine, however, they changed to transparent rapidly compared to that of [ethylenediamine] / [Zn] = (a) 1 : 1 and (b) 2 : 1. It indicated that ions were consumed to form particles rapidly in the solution of [ethylenediamine] / [Zn] = (c) 3 : 1 and ion concentration would decrease drastically with color change. It can be assumed that high concentration of ethylenediamine increased crystal growth rate to form ZnO particles at the first stage. The particles generated in the first stage would not be whiskers but small ZnO particles. Ion concentration decreased rapidly to make solution transparent by formation of small particles because Zn ions were consumed in formation of the particles. ZnO whiskers would grow slowly on the small particles in dilute solutions at second stage. Consequently, hexagonal symmetry radial whiskers were formed in the solution of [ethylenediamine] / [Zn] = (c) 3 : 1. This phenomenon was consistent with reported ZnO whiskers which had high crystallinity, high photoluminescence intensity in UV region and low photoluminescence intensity in visible light region. Novel properties such as unique morphology, high specific surface area, high conductivity per unit volume, low photoluminescence intensity in visible light region and high photoluminescence intensity in UV region may pave the way to a new age of ZnO devices. Furthermore, they can be fabricated on low heat resistant materials such as polymers, paper or organic materials for flexible devices.

4. Morphology control of multi-needle ZnO particles and their particulate films [27]

ZnO has attracted much attention as a next-generation gas sensor for CO[28-30], NH_3[31], NO_2[32], H_2S[33], H_2[29,34], ethanol[34,35], SF_6[34], C_4H_{10}[34] or gasoline[34] and dye-sensitized solar cells[26,36-39]. Sensitivity directly depends on the specific surface area of the sensing material. ZnO particles, particulate films or mesoporous material having high specific surface area were thus strongly required.

ZnO has been crystallized to a hexagonal cylinder shape for gas sensors or solar cells in many studies[26] by using the hexagonal crystal structure of ZnO at low supersaturation degree. However, strategic morphology design and precise morphology control for high specific surface area should be developed to improve the properties. ZnO particles should be controlled to have multi-needles or high surface asperity to increase the specific surface area by the crystallization at high supersaturation degree.

Recently, morphology control[20,40-42] and nano/micro manufacturing[43-46] of oxide materials were proposed in solution systems. Solution systems have the advantage of adjustment of supersaturation degree and high uniformity in the system for particle morphology control. However, many factors affect the system compared to gas phase systems or solid state reactions. Solution chemistry for oxide materials is therefore being developed and many areas remain to be explored.

Morphology control of ZnO particles to hexagonal cylinder shape, ellipse shape and multi-needle shape was recently developed[20]. Photoluminescence property was improved by changing the morphology and oxygen vacancy volume in this system. Morphology control of ZnO has also been proposed based on control of crystal growth[47-50]. W. Q. Peng et al. reported flower-like bunches synthesized on indium-doped tin oxide glass substrates through a chemical bath deposition process[47]. Y. R. Lin et al. fabricated nanowires on a ZnO-buffered silicon substrate by a hydrothermal method[48]. H. Zhang et al. prepared flowerlike, disklike, and dumbbell-like ZnO microcrystals by a capping-molecule-assisted hydrothermal process[49]. K. H. Liu et al. reported a hierarchical polygon prismatic Zn-ZnO core-shell structure grown on silicon by combining liquid-solution colloidals together with the vapor-gas growth process[50]. These studies showed high morphology controllability of

ZnO, however, morphology should be optimized to have high specific surface area to apply to solar cells or gas sensors.

In this study, the morphology design of ZnO particles was proposed for solar cells or gas sensors in which high specific surface area, high electrical conductivity and high mechanical strength are required. Multi-needle ZnO particles having ultrafine surface relief structure, as well as particulate films constructed from multi-needle particles and thin sheets were fabricated[27]. Morphology control was realized based on a new idea inspired from the morphology change in our former study[20]. High supersaturation degree of the solution was used for fast crystal growth which induces the formation of multi-needle particles and low super saturation was used for the formation of ZnO thin sheets.

Zinc nitrate hexahydrate ($Zn(NO_3)_2 \cdot 6H_2O$, > 99.0%, MW 297.49, Kanto Chemical Co., Inc.) and ethylenediamine ($H_2NCH_2CH_2NH_2$, > 99.0%, MW 60.10, Kanto Chemical Co., Inc.) were used as received. Glass (S-1225, Matsunami Glass Ind., Ltd.) was used as a substrate. Zinc nitrate hexahydrate (15 mM) was dissolved in distilled water at 60 °C and ethylenediamine (15 mM) was added to the solution to induce the formation of ZnO. Glass substrate was immersed in the middle of the solution at an angle and the solution was kept at 60 °C using a water bath for 80 min with no stirring. The solution became clouded shortly after the addition of ethylenediamine. Ethylenediamine plays an essential role in the formation of crystalline ZnO. ZnO was homogeneously nucleated and grown to form a large amount of particles to make the solution clouded. ZnO particles were gradually deposited and further grown on a substrate. Homogeneously nucleated particles precipitated gradually and the solution became light white after 80 min. The supersaturation degree of the solution was high at the initial stage of the reaction for the first 1 h and decreased as the color of the solution changed.

ZnO particulate films constructed from ZnO particles and thin sheets were fabricated by immersion for 48 h. The glass substrate was immersed in the middle of the solution at an angle and the solution was kept at 60 °C using a water bath for 6 h with no stirring. The solution was then left to cool for 42 h in the bath. The solution became clouded shortly after the addition of ethylenediamine and clear after 6 h. The bottom of the solution was covered with white precipitate after 6 h. The supersaturation degree of the solution was high at the initial stage of the reaction for the first 1 h and then decreased as the color of the solution changed.

Morphology of ZnO particles and particulate films was observed by a field emission scanning electron microscope (FE-SEM; JSM-6335FM, JEOL Ltd.) after heating at 150 °C for 30 min in vacuum for drying of carbon paste (Vacuum oven, VOS-201SD, EYELA, Tokyo Rikakikai. Co., Ltd.) and Pt coating for 3 nm (Quick cool coater, SC-701MCY, Sanyu Electronic Company). Crystal phases were evaluated by an X-ray diffractometer (XRD; RINT-2100V, Rigaku) with CuKα radiation (40 kV, 40 mA).

Morphology design of ZnO particles and particulate films. The morphology of ZnO particles was designed to increase the specific surface area, electrical conductivity and mechanical strength of the base material of solar cells and sensors (Fig. 7). Typical ZnO particles grown at low supersaturation degree are shown in Fig. 7a[20]. The particles show edged hexagonal faces and elongate parallel to the *c*-axis. ZnO particles grow to have a hexagonal cylinder shape by slow crystal growth due to the hexagonal crystal structure of ZnO. The morphology of ZnO particles was controlled to have a multi-needle shape in an aqueous solution (Fig. 7b)[20]. Multi-needle particles have a high specific surface area compared to hexagonal cylinder particles, but particulate films constructed from small particles have many grain boundaries which reduce the electrical conductivity (Fig. 7c). Particles should thus have large grain size to decrease grain boundaries and increase electrical conductivity (Fig. 7d). Furthermore, the specific surface area of ZnO particles should be increased to improve the sensing performance of sensors or generating efficiency of solar cells (Fig. 7e). An ultrafine surface relief structure on ZnO particles is a candidate morphology for increasing the specific surface area. Additionally, particles should be connected to each other and with a substrate by a combined member such as ZnO thin sheets (Fig. 7f). The thin sheets increase the mechanical strength of particulate films and help raise the electrical conductivity and specific surface area. Morphology control of ZnO particles was attempted based on the strategic morphology design for application to sensors.

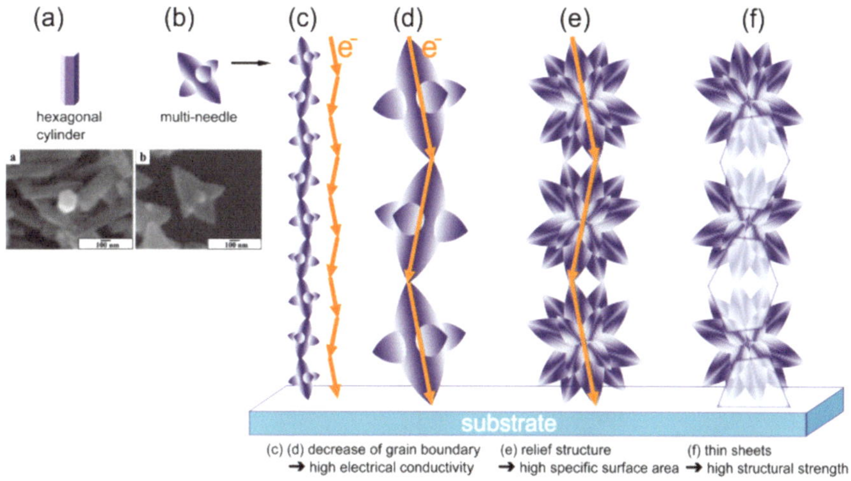

Fig. 7. Design and morphology control of ZnO particles and particulate films for high specific surface area, high electrical conductivity and high mechanical strength. Reprinted with permission from Ref.[27], Masuda, Y. and Kato, K., 2008, *Cryst. Growth Des.*, 8, 2633. Copyright @American Chemical Society

Morphology control of ZnO particles. After having been immersed in the solution for 80 min, the substrate covered with ZnO particles was evaluated by SEM and XRD. ZnO particles were shown to be multi-needle shape in which many needles were grown from the center of the particles (Fig. 8). The particles have more needles compared with the particles previously reported which were constructed from two large needles and several small needles[20]. The size of particles was in the range from 1 - 5 μm which is larger than the particles prepared previously[20] (Fig. 7b). Needles were constructed from an assembly of narrow acicular crystals and thus the side surfaces of needles were covered with arrays of pleats. The tips of the needles were rounded V-shape with many asperities. Edged hexagonal shapes were observed at the tips of needles, thus clearly showing high crystallinity and the direction of the *c*-axis. The *c*-axis would be the long direction of multi-needles and narrow acicular crystals. Elongation of the *c*-axis observed by SEM is consistent with high diffraction intensity of 0002 (Fig. 9). The 0002 diffraction intensity of multi-needle ZnO particles was much stronger than $10\bar{1}0$ or $10\bar{1}1$ peaks though 0002 diffraction is weaker than $10\bar{1}0$ or $10\bar{1}1$ diffractions in randomly orientated ZnO particles (JCPDS No. 36-1451). High diffraction intensity from (0002) planes which are perpendicular to the *c*-axis would be caused from the crystalline ZnO particles which grew to elongate the *c*-axis. The particles have more stacks of (0002) crystal planes compared to that of $(10\bar{1}0)$ planes which are parallel to the *c*-axis or $(10\bar{1}1)$ planes and hence the intensity from (0002) planes was stronger than that from $(10\bar{1}0)$ or $(10\bar{1}1)$ planes.

ZnO grows to a hexagonal cylinder shape at low supersaturation degree because of its hexagonal crystal structure, however, ZnO grows to a multi-needle shape at high supersaturation degree which induces fast crystal growth. ZnO was thus grown to a multi-needle shape in our solution in spite of its hexagonal crystal structure. The growth of ZnO was halted by a rapid decrease of supersaturation degree and removal of particles from the solution to obtain ZnO multi-needle particles having an ultrafine surface relief structure. The morphology of the ZnO particles was controlled by the fast crystal growth due to high supersaturation degree and by the suppression of crystal growth due to the rapid decrease of supersaturation degree and removal of particles from the solution.

Fig. 8. SEM micrograph (a-d) of multi-needle ZnO particles having ultrafine surface relief structure.

Reprinted with permission from Ref.[27], Masuda, Y. and Kato, K., 2008, *Cryst. Growth Des.*, 8, 2633. Copyright @American Chemical Society

Morphology control of ZnO particulate films. ZnO particulate films showed a multi-needle shape and were connected to each other by thin sheets (Fig. 10a-c). The morphology of the particles was similar to that of the particles prepared by immersion for 80 min to have a high specific surface area. Thin sheets had a thickness of 10 - 50 nm and width of 1 - 10 μm and were connected to particles closely with no clearance. The particulate films had continuous open pores ranging from several nm to 10 μm in diameter. The particulate films showed x-ray diffraction patterns of ZnO crystal with no additional phase (Fig. 11). Diffraction peaks were very sharp, showing high crystallinity of the particulate films. The high intensity of 0002 would be caused by elongation of multi-needle particles in the *c*-axis direction which increases the stacks of (0002) crystal planes.

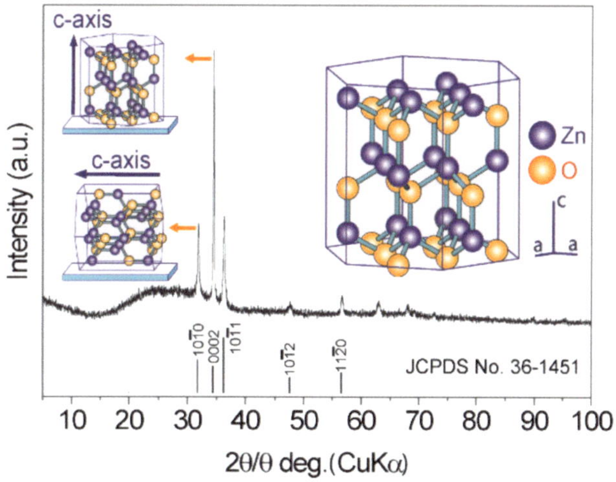

Fig. 9. XRD diffraction pattern of multi-needle ZnO particles having ultrafine surface relief structure.
Reprinted with permission from Ref.[27], Masuda, Y. and Kato, K., 2008, *Cryst. Growth Des.*, 8, 2633. Copyright @American Chemical Society

ZnO multi-needle particles having an ultrafine surface relief structure were prepared at 60 °C in the white solution during the initial 80 min (Fig. 7e). The supersaturation degree was high at the initial stage of the reaction due to the high concentration of ions. ZnO particles were then precipitated, making the bottom of the solution white and the solution itself clear. Ions were consumed to form ZnO particles and thus the ion concentration of the solution decreased rapidly. Thin sheets were formed at 25 °C in the clear solution after the formation of multi-needle particles (Fig. 7f). Solution temperature and supersaturation degree would influence on precipitates. Consequently, the particulate films constructed from multi-needle particles and thin sheets were successfully fabricated by the two-step growth (Fig. 7f).

For comparison, assemblies of thin sheets were prepared at air-liquid interfaces of the same solution we used in this study[51]. XRD patterns of the sheets were assigned to ZnO. The sheets had *c*-axis orientation parallel to the sheets, i.e., in-plane *c*-axis orientation. TEM observations showed the sheets were dense polycrystals consisted of nano-sized ZnO crystals. Electron diffraction pattern showed strong isotropic diffraction ring from (0002) planes. It suggested in-plane *c*-axis orientation of ZnO crystals which was consistent with XRD evaluations. Mechanical strength and electrical property would be affected by crystal orientation and microstructures. The sheets would have stronger mechanical strength compared to porous sheets because of their dense structure. They would have different electrical properties from randomly oriented sheets due to in-plane *c*-axis orientation, because ZnO has anisotropic electrical properties caused from anisotropic hexagonal crystal structure. The sheets prepared at air-liquid interfaces[51] would be similar to that prepared in the solutions (Fig. 10) because both of them were prepared from the same solution and showed XRD patterns assigned to ZnO[51] (Fig. 11). The sheets prepared in this study (Fig. 4, 5) would have similar mechanical and electrical properties to the sheets prepared at air-liquid interfaces[51].

Thin sheets were transformed to particles and porous particulate films by annealing at 500 °C for 1 h in air. The sheets did not maintain their thin sheet shape due to high slimness and/or phase transformation. Thin sheets would be inorganic films containing Zn ions such

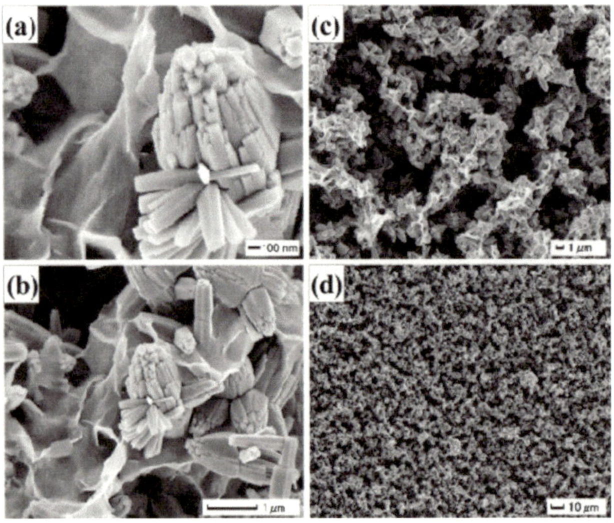

Fig. 10. SEM micrograph (a-d) of ZnO particulate films constructed from ZnO multi-needle particles and thin sheets.
Reprinted with permission from Ref.[27], Masuda, Y. and Kato, K., 2008, *Cryst. Growth Des.*, 8, 2633. Copyright @American Chemical Society

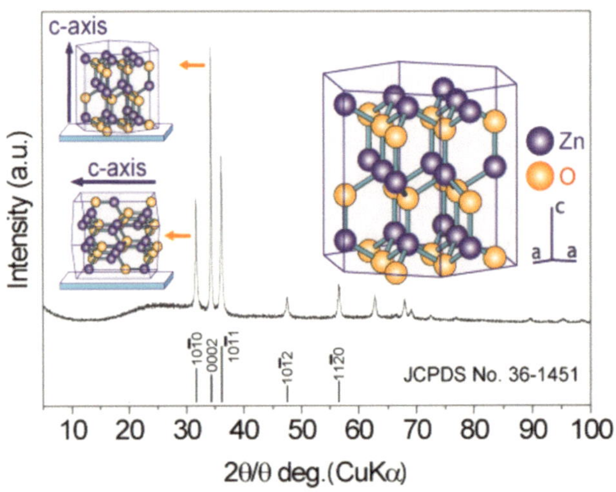

Fig. 11. XRD diffraction pattern of ZnO particulate films constructed from ZnO multi-needle particles and thin sheets.
Reprinted with permission from Ref.[27], Masuda, Y. and Kato, K., 2008, *Cryst. Growth Des.*, 8, 2633. Copyright @American Chemical Society

as crystalline ZnO, amorphous ZnO or zinc hydroxide and were transformed to porous ZnO particulate films by the annealing. Further investigation of the thin sheets would contribute to more precise morphology control of ZnO structure and further improvement of specific surface area. Additionally, precise evaluation of mechanical strength and electrical properties should be performed to clarify the potential of ZnO particulate films for sensors or solar cells and to produce guidelines for improving their properties.

Multi-needle ZnO particles having an ultrafine surface relief structure were successfully fabricated by the precise control of crystal growth in an aqueous solution. The morphology of ZnO was further controlled for ZnO particulate films constructed from ZnO multi-needle particles and thin sheets. The thin sheets connected particles to each other and with a substrate. The morphology design and morphology control described here will facilitate the progress of crystal science for developing future advanced materials and devices.

5. Morphology control of high *c*-axis oriented stand-alone ZnO self-assembled film[51]

Recently, crystalline ZnO[20,26,47-50,52-59] have been synthesized to utilize the high potential of the solution process for future devices and to realize green chemistry for a sustainable society. For instance, ZnO nanowire arrays have been synthesized using seed layers in aqueous solutions for dye-sensitized solar cell[26,52]. Full sun efficiency of 1.5% is demonstrated in this study. O'Brien et al. prepared specular ZnO films consisting of clumps of elongated triangular crystals[53], small ZnO spherical clumps consisting of particles of ca. 100 nm[53], ZnO films consisting of randomly rod-shaped particles of up to 1000 nm in length[53], ZnO films consisting of very thin random rod-shaped particles of ca. 1000 nm[53] and specular films consisting of flowers with well-formed triangle features[59] in aqueous solutions.

However, ZnO films have usually been prepared on substrates[54,57,60-76], and in particular crystalline ZnO films having high *c*-axis orientation require expensive substrates such as single crystals or highly-functional substrates. A simple and low-cost process for self-supporting crystalline ZnO films is expected to be used for a wide range of applications such as windows of optical devices or low-value-added products. Self-supporting crystalline ZnO films can also be applied by being pasting on a desired substrate such as low heat-resistant polymer films, glasses, metals or papers.

In this section, high *c*-axis oriented stand-alone ZnO self-assembled films was fabricated using an air-liquid interface[51]. ZnO was crystallized from an aqueous solution without heat treatment or catalyst. The ZnO film was fabricated at the air-liquid interface without using ammonia vapor.

Zinc nitrate hexahydrate ($Zn(NO_3)_2 \cdot 6H_2O$, > 99.0%, MW 297.49, Kanto Chemical Co., Inc.) and ethylenediamine ($H_2NCH_2CH_2NH_2$, > 99.0%, MW 60.10, Kanto Chemical Co., Inc.) were used as received. Zinc nitrate hexahydrate (15 mM) was dissolved in distilled water at 60°C and ethylenediamine (15 mM) was added to the solution to induce the formation of ZnO. The solution was kept at 60°C using a water bath for 6 h with no stirring. The solution was then left to cool for 42 h in the bath. Polyethylene terephthalate (PET) film, glass (S-1225, Matsunami Glass Ind., Ltd.) and an Si wafer (p-type Si [100], NK Platz Co., Ltd.) were used as substrates.

Morphology of ZnO film was observed by a field emission scanning electron microscope (FE-SEM; JSM-6335FM, JEOL Ltd.) and a transmission electron microscope (TEM; H-9000UHR, 300 kV, Hitachi). Crystal phase was evaluated by an X-ray diffractometer (XRD; RINT-2100V, Rigaku) with CuKα radiation (40 kV, 40 mA). Si wafer was used as a substrate for XRD evaluation. The crystal structure model and diffraction pattern of ZnO were calculated from ICSD (Inorganic Crystal Structure Database) data No. 26170 (FIZ Karlsruhe, Germany and NIST, USA) using FindIt and ATOMS (Hulinks Inc.).

The solution became clouded shortly after the addition of ethylenediamine by the homogeneous nucleation and growth of ZnO particles. ZnO particles were gradually

deposited to cover the bottom of the vessel, and the solution became light white after 1 h and clear after 6 h. The supersaturation degree of the solution was high at the initial stage of the reaction for the first 1 h and decreased as the color of the solution changed.
White films were formed at the air-liquid interface and they grew to large films. The films had sufficiently high strength to be obtained as stand-alone films. Additionally, a film was scooped to past onto a desired substrate such PET film, Si wafer, glass plate or paper, and the pasted ZnO film was then dried to bond it to the substrate. Both sides of the film can be pasted on substrate. The film physically adhered to the substrate. The film maintained its adhesion during immersion in lightly ultrasonicated water, however, it can be easily peeled off again by strong ultrasonication. The film can be handled easily from substrate to other substrate. It also can be attached strongly to substrate by annealing or addition of chemical regents such as silane coupling agent to form chemical bonds between the film and the substrate
The film grew to a thickness of about 5 μm after 48 h, i.e., 60°C for 6 h, and was left to cool for 42 h.
The air side of the stand-alone film had a smooth surface over a wide area due to the flat air-liquid interface (Fig. 12-a1), whereas the liquid side of the film had a rough surface (Fig. 12-b1). The films consisted of ZnO nano-sheets were clearly observed from the liquid side (Fig. 12-b2) and the fracture edge-on profile of the film (Fig. 12-c1, 12-c2). The nano-sheets had a thickness of 5-10 nm and were 1-5 μm in size. They mainly grew forward to the bottom of the solution, i.e., perpendicular to the air-liquid interface, such that the sheets stood perpendicular to the air-liquid interface. Thus, the liquid side of the film had many ultra-fine spaces surrounded by nano-sheet and had a high specific surface area. The air side of the film, on the other hand, had a flat surface that followed the flat shape of the air-liquid interface. The air-liquid interface was thus effectively utilized to form the flat surface of the film. This flatness would contribute to the strong adhesion strength to substrates for pasting of the film. The air-side surface prepared for 48 h had holes of 100-500 nm in diameter (Fig. 12-a2), and were hexagonal, rounded hexagonal or round in shape. The air-side surface prepared for 6 h, in contrast, had no holes on the surface. The air-side surface was well crystallized to form a dense surface and ZnO crystals would partially grow to a hexagonal shape because of the hexagonal crystal structure. Well-crystallized ZnO hexagons were then etched to form holes on the surface by decrease in pH. The growth face of the film would be liquid side. ZnO nano-sheets would grow to form a large ZnO film by Zn ion supply from the aqueous solution. Further investigation of the formation mechanism would contribute to the development of crystallography in the solution system and the creation of novel ZnO fine structures.
The film showed a very strong 0002 x-ray diffraction peak of hexagonal ZnO at $2\theta = 34.04°$ and weak 0004 diffraction peak at $2\theta = 72.16°$ with no other diffractions of ZnO (Fig. 13). (0002) planes and (0004) planes were perpendicular to the *c*-axis, and the diffraction peak only from (0002) and (0004) planes indicates high *c*-axis orientation of ZnO film. The inset figure shows that the crystal structure of hexagonal ZnO stands on a substrate to make the *c*-axis perpendicular to the substrate. Crystallite size parallel to (0002) planes was estimated from the half-maximum full-width of the 0002 peak to 43 nm. This is similar to the threshold limit value of our XRD equipment and thus the crystallite size parallel to (0002) planes is estimated to be greater than or equal to 43 nm. Diffraction peaks from a silicon substrate were observed at $2\theta = 68.9°$ and $2\theta = 32.43°$. Weak diffractions at $2\theta = 12.5°$, 24.0°, 27.6°, 30.5°, 32.4° and 57.6° were assigned to co-precipitated zinc carbonate hydroxide ($Zn_5(CO_3)_2(OH)_6$, JCPDS No. 19-1458).

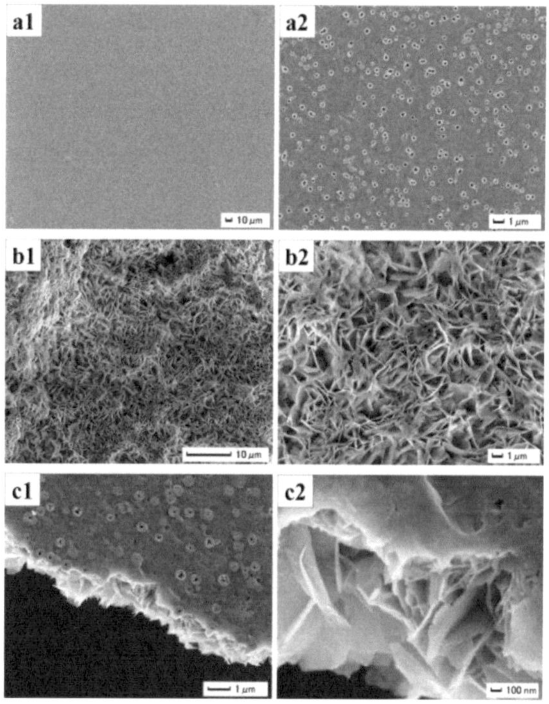

Fig. 12. SEM micrographs of high *c*-axis oriented stand-alone ZnO self-assembled film. (a1) Air-side surface of ZnO film. (a2) Magnified area of (a1). (b1) Liquid-side surface of ZnO film. (b2) Magnified area of (b2). (c1) Fracture cross section of ZnO film from air side. (c2) Magnified area of (c1).

Stand-alone ZnO film was further evaluated by TEM and electron diffraction. The film was crushed to sheets and dispersed in an acetone. The sheets at the air-liquid interface were skimmed by a cupper grid with a carbon supporting film. The sheets were shown to have uniform thickness (Fig. 14a). They were dense polycrystalline films constructed of ZnO nanoparticles (Fig. 14b). Lattice image was clearly observed to show high crystallinity of the particles. The film was shown to be single phase of ZnO by electron diffraction pattern. These observations were consistent with XRD and SEM evaluations.
The film pasted on a silicon wafer was annealed at 500°C for 1 h in air to evaluate the details of the films. ZnO film maintained its structure during the annealing (Fig. 15). The air side of the film showed a smooth surface (Fig. 15-a1) and the liquid side showed a relief structure having a high specific surface area (Fig. 15-b1, 15-b2). The air side showed the film consisted of dense packing of small ZnO nanosheets and the size of sheets increased toward the liquid-side surface (Fig. 15-a2). ZnO sheets would grow from the air side to the liquid side, i.e., the sheets would nucleate at the liquid-air interface and grow down toward the bottom of the solution by the supply of Zn ions from the solution. Annealed film showed X-ray diffractions of ZnO and Si substrate with no additional phases. As-deposited ZnO nano-sheets were shown to be crystalline ZnO because the sheets maintained their fine structure during the annealing without any phase transition. High *c*-axis orientation was also maintained during the annealing, showing a very strong 0002 diffraction peak.

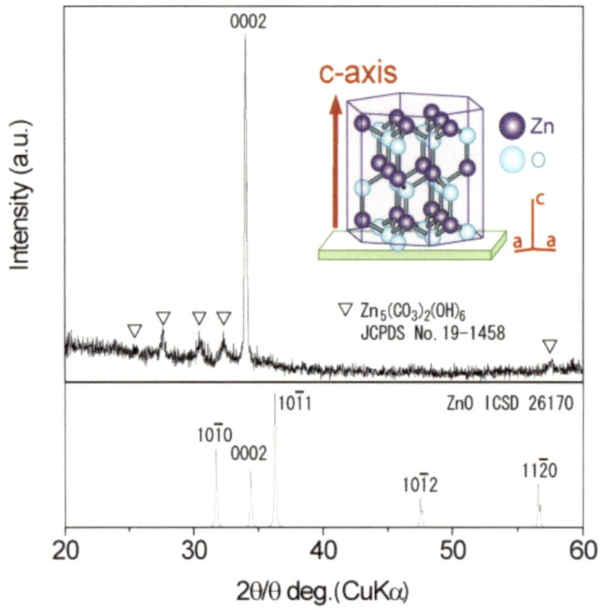

Fig. 13. XRD diffraction pattern of high *c*-axis oriented stand-alone ZnO self-assembled film. Reprinted with permission from Ref.[51], Masuda, Y. and Kato, K., 2008, *Cryst. Growth Des.*, 8, 275. Copyright @American Chemical Society

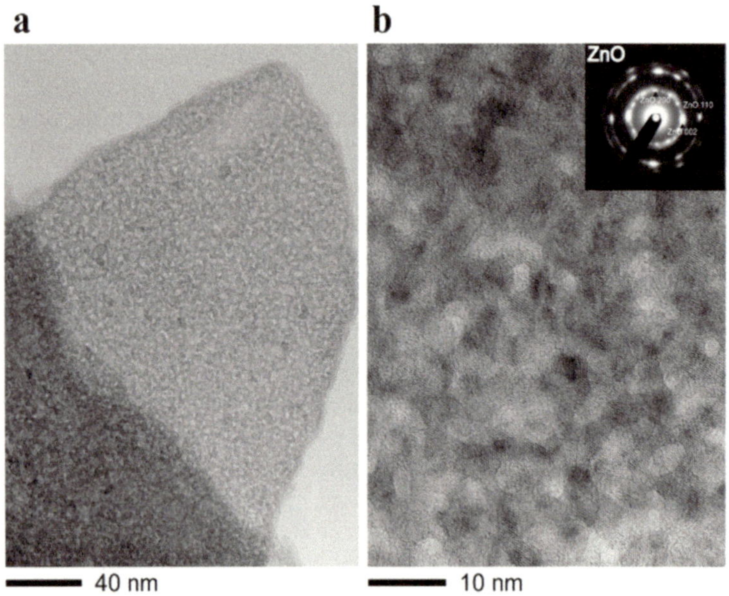

Fig. 14. (a) TEM micrograph of ZnO nano-sheets. (b) Magnified area of (a). (Insertion) Electron diffraction pattern of ZnO. Reprinted with permission from Ref.[51], Masuda, Y. and Kato, K., 2008, *Cryst. Growth Des.*, 8, 275. Copyright @American Chemical Society

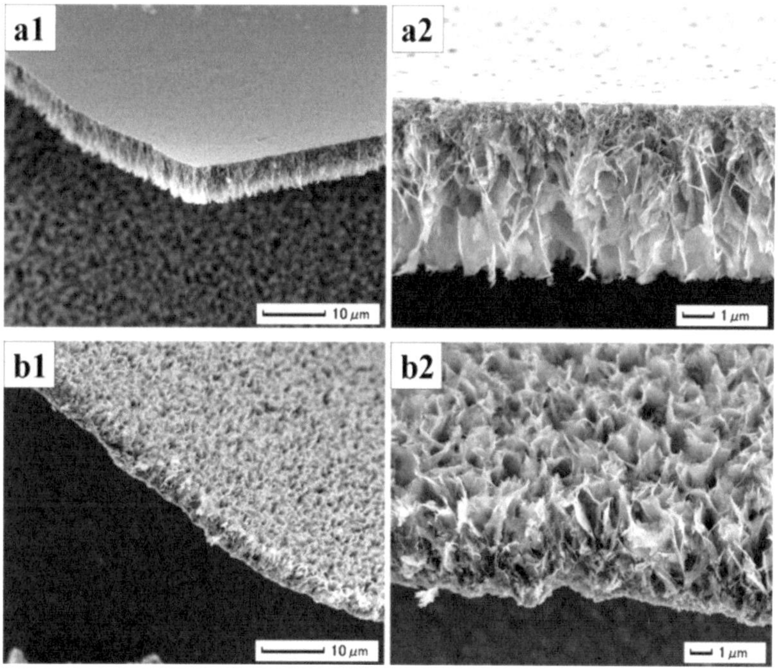

Fig. 15. SEM micrographs of high *c*-axis oriented stand-alone ZnO self-assembled film annealed at 500°C for 1 h in air. (a1) Fracture edge-on profile of ZnO film from air side. (a2) Cross-section profile of ZnO film from air side. (b1) Fracture edge-on profile of ZnO film from liquid side. (b2) Cross-section profile of ZnO film from liquid side.
Reprinted with permission from Ref.51, Masuda, Y. and Kato, K., 2008, *Cryst. Growth Des.*, 8, 275. Copyright @American Chemical Society

The solution was further kept at 25°C for 1 month to evaluate the details of the deposition mechanism. The film prepared at the air-liquid interface for 1 month was not hexagonal ZnO. The film showed strong X-ray diffractions of zinc carbonate hydroxide single phase. ZnO would be dissolved by decrease in pH. ZnO would be crystallized at the initial reaction stage for the first 48 h. ZnO was then gradually etched and dissolved by nitric acid and zinc carbonate hydroxide was crystallized using Zn ions which were supplied by the dissolution of crystalline ZnO.

High *c*-axis oriented stand-alone ZnO self-assembled film was fabricated using a simple solution process. The film consisted of ZnO nanosheets was crystallized at air-liquid interface. The nanosheets grew perpendicular to the film. The film had high *c*-axis orientation and showed a strong 0002 diffraction peak and weak 0004 peak. The air side of the film had a flat surface, whereas the liquid side had a rough surface having many ultra-fine spaces surrounded by ZnO nano-sheets. The rough surface of the liquid side can be utilized for sensors or dye-sensitized solar cells. The ZnO film was also pasted on a desired substrate such as PET films, Si substrate or glass plates. The surface of low heat-resistant flexible polymer film was modified with high *c*-axis oriented crystalline ZnO film without heat treatment. This low-cost, low-temperature technique can be used for a wide range of applications including sensors, solar cells, electrical devices and optical devices using the various properties of high *c*-axis oriented crystalline ZnO.

6. Morphology control of nanocrystal assembled TiO_2 particles[77]

Anatase TiO_2 particles, 100–200 nm in diameter, were developed in aqueous solution at 50°C. The particles were assemblies of nano TiO_2 crystals covered with nanorelief surface structures. The crystals grew anisotropically along the *c*-axis to form acicular crystals. The particles showed *c*-axis orientation due to high-intensity X-ray diffraction from the (004) planes. The particles had a BET specific surface area of 270 m^2/g. BJH and DFT/Monte-Carlo analysis of adsorption isotherm indicated the existence of pores ~3 nm and ~1 nm in diameter. Crystallization and self-assembly of nano TiO_2 were effectively utilized to fabricate nanocrystal assembled TiO_2 particles having high surface area and nanorelief surface structure.

Nanoporous TiO_2 architecture with micropores (<2 nm), mesopores (2–50 nm) and/or macropores (>50 nm) is of considerable interest for both scientific and technical applications. The latter include cosmetics, catalysts[78], photocatalysts[79-82], gas sensors[83,84], lithium batteries[85-87], biomolecular sensors[88] and dye-sensitized solar cells[89,90]. Crystalline anatase generally exhibits higher properties than rutile in photocatalysts, biomolecular sensors and dye-sensitized solar cells. Electrons are obtained from dyes adsorbed on TiO_2 electrodes in sensors and solar cells. Photoelectric conversion efficiency strongly depends on dye adsorption volume and surface area of TiO_2. High surface area is required to achieve high efficiency and sensitivity of the devices. Additionally, the surface of TiO_2 should be covered with nano/microrelief structures to adsorb large amounts of dye, molecules and DNA for biomolecular sensors and dye-sensitized solar cells.

TiO_2 nanoparticles have been prepared by flame synthesis[91,92], ultrasonic irradiation[93,94], chemical vapor synthesis[95], sol-gel methods[79,96-100], sonochemical method[101] and liquid phase deposition of amorphous TiO_2[102-105]. High temperature in the treatment processes, however, causes aggregation of nanoparticles and decreased surface area. Formation of nanorelief structures on the surface of the particles is difficult to achieve in these processes.

Highly porous materials have been prepared via template-based methods, including soft templates (surfactants, chelating agents, block polymers, etc.)[106-109] and hard templates (porous anionic alumina, porous silica, polystyrene spheres, carbon nanotubes, etc.)[110,111]. However, the nanostructures of these materials usually change due to amorphous-phase crystallization to anatase TiO_2 during annealing. This decreases the surface area and damages the surface nanostructures.

In this section, porous anatase TiO_2 particles were developed in aqueous solution. Nano TiO_2 was crystallized in the solution to assemble into particles 100–200 nm in diameter. The surface of the particles was covered with nanorelief structures. The particles showed *c*-axis orientation due to anisotropic crystal growth of TiO_2 along the *c*-axis. BET surface area of the particles was estimated to be 270 m^2/g[112]. BJH and DFT/Monte-Carlo analysis of adsorption isotherm indicated the existence of pores ~2.8 nm and ~3.6 nm in diameter, respectively. The existence of micropores ~1 nm was also indicated. Crystallization and self-assembly of acicular TiO_2 were effectively utilized to fabricate nanocrystal assembled TiO_2 particles having high surface area and nanorelief surface structure.

Ammonium hexafluorotitanate ($[NH_4]_2TiF_6$) (Morita Chemical Industries Co., Ltd., FW: 197.95, purity 96.0%) and boric acid (H_3BO_3) (Kishida Chemical Co., Ltd., FW: 61.83, purity 99.5%) were used as received. Ammonium hexafluorotitanate (12.372 g) and boric acid (11.1852 g) were separately dissolved in deionized water (600 mL) at 50°C. Boric acid solution was added to ammonium hexafluorotitanate solution at a concentration of 0.15 and 0.05 M, respectively.

The solution was kept at 50°C for 30 min using a water bath with no stirring, after which it was centrifuged at 4000 rpm for 10 min (Model 8920, Kubota Corp.). Preparation for centrifugation, centrifugation at 4000 rpm, deceleration from 4000 to 0 rpm and preparation for removal of supernatant solution took 4 min, 10 min, 10 min and 8 min, respectively. The particles contacted with the solution, the temperature of which was gradually lowered for 32 min after maintaining 50°C for 30 min. Precipitates were dried at 60°C for 12 h after removal of supernatant solution.

The crystal phase of the particles was evaluated by X-ray diffractometer (XRD; RINT-2100V, Rigaku) with CuKα radiation (40 kV, 30 mA). Diffraction patterns were evaluated using JCPDS, ICSD (Inorganic Crystal Structure Database) data (FIZ Karlsruhe, Germany and NIST, USA) and FindIt. Morphology of TiO_2 was observed by transmission electron microscopy (TEM; JEM2010, 200 kV, JEOL). Zeta potential and particle size distribution were measured by electrophoretic light-scattering spectrophotometer (ELS-Z2, Otsuka Electronics Co., Ltd.) with automatic pH titrator. Samples of 0.01 g were dispersed in distilled water (100 g) and ultrasonicated for 30 min prior to measurement. The pH of colloidal solutions was controlled by the addition of HCl (0.1 M) or NaOH (0.1 M). Zeta potential and particle size distribution were evaluated at 25°C and integrated 5 and 70 times, respectively.

Nitrogen adsorption-desorption isotherms were obtained using Autosorb-1 (Quantachrome Instruments) and samples of 0.137 g were outgassed at 110°C under 10^{-2} mmHg for 6 h prior to measurement. Specific surface area was calculated by BET (Brunauer-Emmett-Teller) method using adsorption isotherms. Pore size distribution was calculated by BJH (Barrett-Joyner-Halenda) method using adsorption isotherms because an artificial peak was observed from BJH size distribution calculated from desorption branches. Pore size distribution was further calculated by DFT/Monte-Carlo method (N_2 at 77 K on silica (cylinder/sphere, pore, NLDFT ads. model), adsorbent: oxygen) using adsorption branches. Total pore volume (V) and average pore diameter (4V/A) were estimated from pores smaller than 230 nm (diameter) at P/Po = 0.99 through the use of BET surface area (A). However, adsorption volume increased drastically at high relative pressure in isotherms, indicating a large data error at high relative pressure. Therefore, total pore volume and average pore diameter estimated from pores smaller than 11 nm at P/Po = 0.80 would provide useful information. Relative pressure P/Po = 0.80 was selected for estimations because it was lower than the drastic adsorption increase and higher than adsorption hysteresis. The data can be compared to that of TiO_2 having similar morphology with small errors. Total pore volume (cumulative pore volume) was also estimated by BJH method from pores smaller than 154 nm. This usually has a small error compared to that estimated from isotherm data including high relative pressure such as P/Po = 0.99 because estimation by BJH method is not effected by adsorption volume errors at high relative pressure.

Liquid phase crystal deposition of anatase TiO_2. The solution became clouded about 10 min after mixing ammonium hexafluorotitanate solution and boric acid solution. The particles were homogeneously nucleated in the solution, turning the solution white.

Deposition of anatase TiO_2 proceeds by the following mechanisms[43,113,114]:

$$TiF_6^{2-} + 2H_2O \rightleftarrows TiO_2 + 4H^+ + 6F^- \ldots \quad \text{(a)}$$

$$BO_3^{3-} + 4F^- + 6H^+ \longrightarrow BF_{4-} + 3H_2O \ldots \quad \text{(b)}$$

Equation (a) is described in detail by the following two equations:

$$TiF_6^{2-} \xrightarrow{nOH^-} TiF_{6-n}(OH)_n^{2-} + nF^- \xrightarrow{(6-n)OH^-} Ti(OH)_6^{2-} + 6F^- \ldots \quad (c)$$

$$Ti(OH)_6^{2-} \longrightarrow TiO_2 + 2H_2O + 2OH^- \ldots \quad (d)$$

Fluorinated titanium complex ions gradually change into titanium hydroxide complex ions in an aqueous solution, as shown in Eq. (c). The increase of F^- concentration displaces Eqs. (a) and (c) to the left; however, the produced F^- can be scavenged by H_3BO_3 (BO_3^{3-}) as shown in Eq. (b) to displace Eqs. (a) and (c) to the right. Anatase TiO_2 is formed from titanium hydroxide complex ions ($Ti(OH)_6^{2-}$) in Eq. (d).

Crystal phase of TiO_2 particles. X-ray diffraction peaks for the particles were observed at 2θ = 25.1, 37.9, 47.6, 54.2, 62.4, 69.3, 75.1, 82.5 and 94.0° after evaluation of N_2 adsorption. They were assigned to the 101, 004, 200, 105 + 211, 204, 116 + 220, 215, 303 + 224 + 312 and 305 + 321 diffraction peaks of anatase TiO_2 (JCPSD No. 21-1272, ICSD No. 9852) (Fig. 16).

The 004 diffraction intensity of randomly oriented particles is usually 0.2 times the 101 diffraction intensity as shown in JCPDS data (No. 21-1272). However, the 004 diffraction intensity of the particles deposited in our process was 0.36 times the 101 diffraction intensity. Additionally, the integral intensity of the 004 diffraction was 0.18 times the 101 diffraction intensity, indicating the *c*-axis orientation of the particles. Particles were not oriented on the glass holder for XRD measurement. Therefore, TiO_2 crystals would be an anisotropic shape in which the crystals were elongated along the *c*-axis. The crystals would have a large number of stacks of c planes such as (001) planes compared to stacks of (101) planes. The diffraction intensity from the (004) planes would be enhanced compared to that from the (101) planes.

Crystallite size perpendicular to the (101) or (004) planes was estimated from the full-width half-maximum of the 101 or 004 peak to be 3.9 nm or 6.3 nm, respectively. Elongation of crystals in the *c*-axis direction was also suggested by the difference in crystallite size.

TEM observation of TiO_2 particles. The particles were shown to be assemblies of nano TiO_2 crystals (Fig. 17a). Particle diameter was estimated to be 100–200 nm. Relief structures had formed on the surfaces and open pores had formed inside because the particles were porous assemblies of nanocrystals.

Nanocrystals were shown to have acicular shapes (Fig. 17b). They were about 5–10 nm in length. The longer direction of acicular TiO_2 is indicated by the black arrow. The inserted FFT image shows the 101 and 004 diffractions of anatase TiO_2. Nanocrystals are assigned to the single phase of anatase TiO_2. It is notable that the diffraction from the (101) planes has a ring shape due to random orientation but that from the (004) planes was observed only in the upper right region and lower left region in the FFT image. Anisotropic 004 diffractions indicated the direction of the *c*-axis, which was perpendicular to the (004) planes, as shown by the white arrow. It was roughly parallel to the longer direction of acicular TiO_2. These results suggest that acicular TiO_2 grew along the *c*-axis to enhance the diffraction intensity from the (004) planes. Crystal growth of anatase TiO_2 along the *c*-axis was previously observed in TiO_2 films[113]. Anisotropic crystal growth is one of the features of liquid phase crystal deposition.

Acicular nanocrystals showed lattice images of anatase TiO_2 (Fig. 17c). They were constructed of anatase TiO_2 crystals without amorphous or additional phases. Anatase crystals were not covered with amorphous or additional phases even at the tips. Bare anatase crystal with nanosized structure is important to achieve high performance for catalysts and devices.

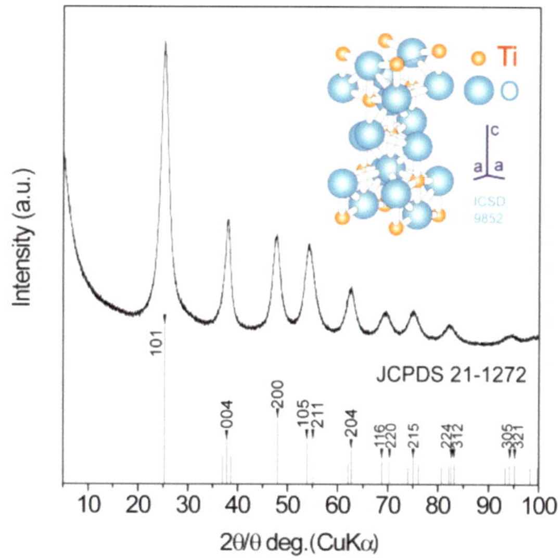

Fig. 16. XRD diffraction pattern of anatase TiO_2 particles.
Reprinted with permission from Ref.[77], Masuda, Y. and Kato, K., 2008, *Cryst. Growth Des.*, 8, 3213. Copyright @American Chemical Society

Crystallization of TiO_2 was effectively utilized to form assemblies of acicular nanocrystals in the process. Open pores and surface relief structures were successfully formed on the particles.

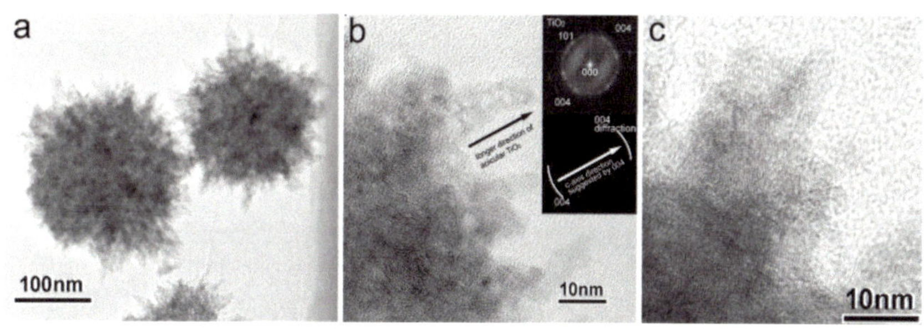

Fig. 17. (a): TEM micrograph of anatase TiO_2 particles. (b): Magnified area of (a) showing morphology of acicular crystals. Insertion in (b): FFT image of (b) anatase TiO_2. (c): Magnified area of (a) showing lattice images of anatase TiO_2.
Reprinted with permission from Ref.77, Masuda, Y. and Kato, K., 2008, *Cryst. Growth Des.*, 8, 3213. Copyright @American Chemical Society

Zeta potential and particle size distribution. The dried particles were dispersed in water to evaluate zeta potential and particle size distribution after evaluation of N_2 adsorption. The particles had positive zeta potential of 30.2 mV at pH 3.1, which decreased to 5.0, −0.6, −11.3 and −36.3 mV at pH 5.0, 7.0, 9.0 and 11.1, respectively (Fig. 18). The isoelectric point was estimated to be pH 6.7, slightly higher than that of anatase TiO_2 (pH 2.7-6.0)[115]. Zeta potential is very sensitive to the particle surface conditions, ions adsorbed on the particle surfaces, and the kind and concentration of ions in the solution. The variations in zeta potential were likely caused by the difference in the surface conditions of TiO_2 particles, affected by the interaction between particles and ions in the solution.

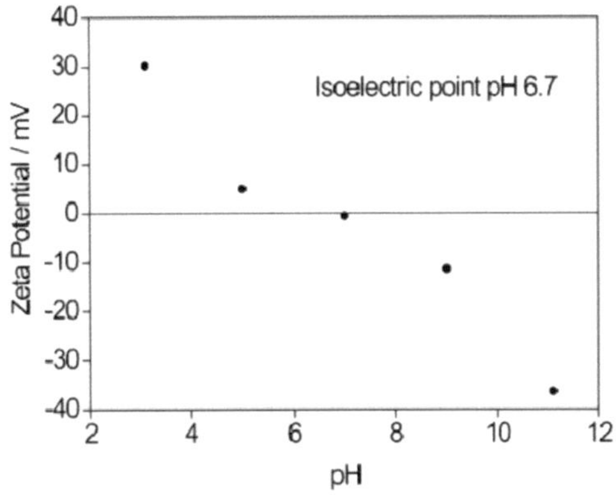

Fig. 18. Zeta potential of anatase TiO_2 particles as a function of pH.
Reprinted with permission from Ref.[77], Masuda, Y. and Kato, K., 2008, *Cryst. Growth Des.*, 8, 3213. Copyright @American Chemical Society

Mean particle size was estimated to be ~550 nm in diameter with a standard deviation (STD) of 220 nm at pH 3.1 (Fig. 19a). This was larger than that observed by TEM. Slight aggregation occurred at pH 3 because the particles were dried completely prior to measurement. Particle size increased with pH and showed a maximum of near the isoelectric point (550 nm at pH 3.1, 3150 nm at pH 5, 4300 nm at pH 7, 5500 nm at pH 9 or 2400 nm at pH 11.1) (Fig. 19b). Strong aggregation resulted from the lack of repulsion force between particles near the isoelectric point.

The particles were generated in the solution at pH 3.8 in this study. It would be suitable to obtain repulsion force between particles for crystallization without strong aggregation.

N_2 adsorption characteristics of TiO_2 particles. TiO_2 particles exhibited N_2 adsorption-desorption isotherms of Type IV (Fig. 20a). The desorption isotherm differed from adsorption isotherm in the relative pressure (P/P_0) range from 0.4 to 0.7, showing mesopores in the particles. BET surface area of the particles was estimated to be 270 m^2/g (Fig. 20b). This is higher than that of TiO_2 nanoparticles such as Aeroxide P25 (BET 50 m^2/g, 21 nm in diameter, anatase 80% + rutile 20%, Degussa), Aeroxide P90 (BET 90–100 m^2/g, 14 nm in diameter, anatase 90% + rutile 10%, Degussa), MT-01 (BET 60 m^2/g, 10 nm in diameter, rutile, Tayca Corp.) and Altair TiNano (BET 50 m^2/g, 30–50 nm in diameter, Altair Nanotechnologies Inc.)[116]. A high BET surface area cannot be obtained from particles having a smooth surface even if the particle size is less than 100 nm. A high BET surface area would be realized by the unique morphology of TiO_2 particles constructed of nanocrystal assemblies.

Total pore volume and average pore diameter were estimated from pores smaller than 230 nm at P/Po = 0.99–0.431 cc/g and 6.4 nm, respectively. They were estimated to be 0.212 cc/g and 3.1 nm, respectively, from pores smaller than 11 nm at P/Po = 0.80. Total pore volume was also estimated by the BJH method from pores smaller than 154 nm to be 0.428 cc/g. Average pore diameter was estimated to be 6.3 nm using BET surface area.

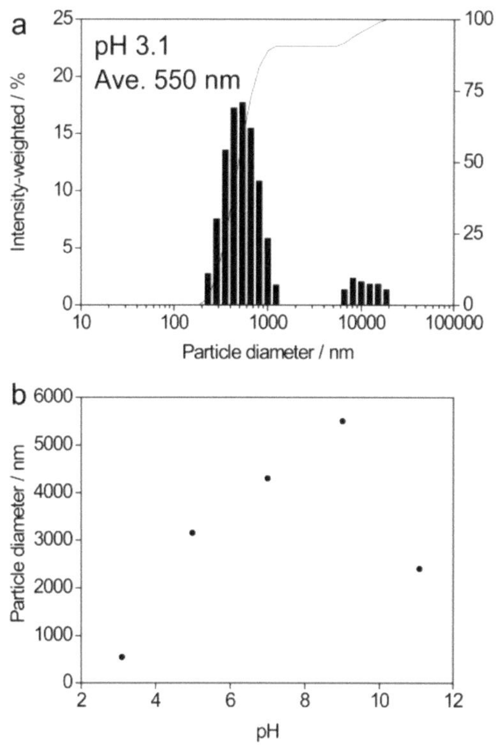

Fig. 19. (a): Particle size distribution of anatase TiO_2 particles at pH 3.1. (b): Particle size of anatase TiO_2 particles as a function of pH.
Reprinted with permission from Ref.77, Masuda, Y. and Kato, K., 2008, *Cryst. Growth Des.*, 8, 3213. Copyright @American Chemical Society

Pore size distribution was calculated by the BJH method using adsorption isotherms (Fig. 20c). It showed a pore size distribution curve having a peak at ~2.8 nm and pores larger than 10 nm. TiO_2 particles would have mesopores of ~2.8 nm surrounded by nanocrystals. Pores larger than 10 nm are considered to be interparticle spaces. The pore size distribution also suggested the existence of micropores smaller than 1 nm.
Pore size distribution was further calculated by the DFT/Monte-Carlo method. The model was in fair agreement with adsorption isotherms (Fig. 20d). Pore size distribution showed a peak at ~3.6 nm that indicated the existence of mesopores of ~3.6 nm (Fig. 20e). The pore size calculated by the DFT/Monte-Carlo method was slightly larger than that calculated from the BJH method because the latter method is considered to have produced an underestimation[117-119]. The pore size distribution also suggested the existence of micropores of ~1 nm, probably resulting from microspaces surrounded by nanocrystals and the uneven surface structure of nanocrystals.
The particles were shown to have a large surface area as well as micropores of ~1 nm, mesopores of ~2.8-3.6 nm and pores larger than 10 nm, by N_2 adsorption characteristics. Assembly of acicular nanocrystals resulted in unique features and high surface area.
TiO_2 particles were generated in the solutions at 90°C for 1h using an oil bath with no stirring for comparison. The solutions became clouded after the addition of boric acid solutions into ammonium hexafluorotitanate solutions. High temperature accelerated

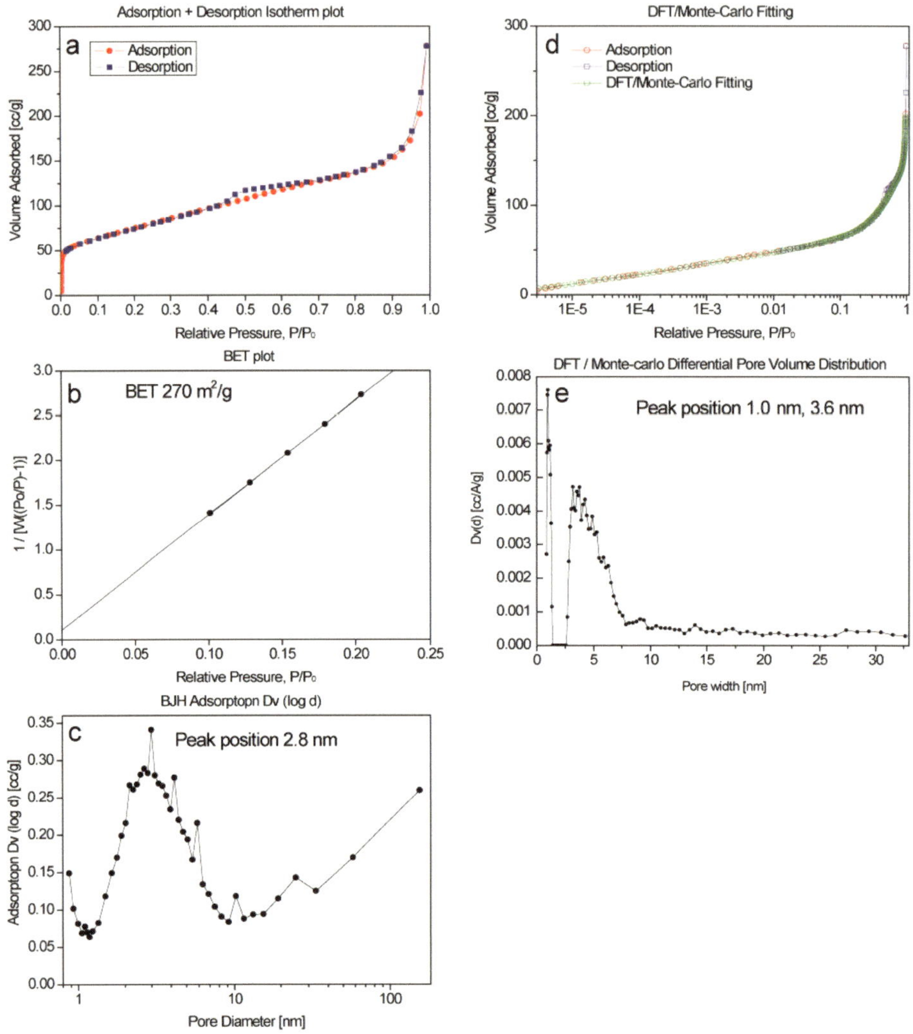

Fig 20. (a): N_2 adsorption-desorption isotherm of anatase TiO_2 particles. (b): BET surface area of anatase TiO_2 particles. (c): Pore size distribution calculated from N_2 adsorption data of anatase TiO_2 particles using BJH equation. (d): N_2 adsorption-desorption isotherm and DFT/Monte-Carlo fitting curve of anatase TiO_2 particles. (e): Pore size distribution calculated from N_2 adsorption data of anatase TiO_2 particles using DFT/Monte-Carlo equation.

Reprinted with permission from Ref.77, Masuda, Y. and Kato, K., 2008, *Cryst. Growth Des.*, 8, 3213. Copyright @American Chemical Society

crystal growth of TiO_2. Hydrogen chloride of 0.6 ml was added into the solutions of 200ml to decrease crystallization speed of TiO_2. The pH of the solutions was 2.4 one hour after mixing the solutions. BET surface area of the particles was estimated to 18 m^2/g. This is much lower than that of the particles prepared at 50°C and slightly lower than that prepared at 90°C for 8 min in our previous work (44 m^2/g)[120]. Formation of TiO_2 was accelerated at high temperature and it decreased surface area. The particles grew in the solutions to decrease surface area as function of time. Crystallization of TiO_2 was shown to be strongly affected by growth conditions such as solution temperature and growth time.

Anatase TiO_2 particles, 100–200 nm in diameter, were successfully fabricated in aqueous solution. They were assemblies of nanocrystals 5–10 nm that grew anisotropically along the *c*-axis to form acicular shapes. The particles thus had nanorelief surface structures constructed of acicular crystals. They showed *c*-axis orientation due to high-intensity X-ray diffraction from the (004) crystal planes. The particles had a high BET surface area of 270 m^2/g. Total pore volume and average pore diameter were estimated from pores smaller than 230 nm at P/Po = 0.99–0.43 cc/g and 6.4 nm, respectively. They were also estimated from pores smaller than 11 nm at P/Po = 0.80–0.21 cc/g and 3.1 nm, respectively. BJH and DFT/Monte-Carlo analysis of adsorption isotherm indicated the existence of pores ~2.8 and ~3.6 nm, respectively. Additionally, the analyses suggested the existence of micropores of ~1 nm. Crystallization and self-assembly of nano TiO_2 were effectively utilized to fabricate nanocrystal assembled TiO_2 particles having high surface area and nanorelief surface structure.

7. Morphology control of multi-needle TiO_2 particles[121]

Flower-like multi-needle anatase TiO_2 particles were developed in aqueous solutions. They were pure anatase TiO_2 crystals containing no cores, organic binders or solvents. Furthermore, micro-structured silicon wafers were covered with the TiO_2 particles uniformly in the solutions. Their unique crystals growth and physicochemical profiles were precisely evaluated and discussed.

Micro-structured silicon wafers and their surface modification to super-hydrophilic surfaces

Silicon wafers were modified to have micro-structures on the surfaces with cutting work. They were cut using a precise diamond cutter under running water. Width and height of salient lines were 200 μm and 150 μm, respectively. They were formed at 500 μm intervals. They were blown by air to remove dust and were exposed to vacuum-ultraviolet light (VUV light, low-pressure mercury lamp PL16-110, air flow, 100 V, 200 W, SEN Lights Co., 14 mW/cm^2 for 184.9 nm at a distance of 10 mm from the lamp, 18 mW/cm^2 for 253.7 nm at a distance of 10 mm from the lamp) for 10 min in air. Bare silicon surfaces were covered with small amount of surface contamination. The VUV irradiation modified them to clean surfaces that showed super hydrophilic surfaces of water contact angle about 0-1 °.

Morphology control of anatase TiO_2

Ammonium hexafluorotitanate (206.20 mg) and boric acid (186.42 mg) were separately dissolved in deionized hot water (100 mL) at 50°C. Boric acid solution was added to ammonium hexafluorotitanate solution at concentrations of 15 mM and 5 mM, respectively. The silicon wafers having patterned surfaces were immersed in the middle of the solutions with the bottom up at an angle. They were tilted at 15 degrees to the upright. The solutions were kept at 50°C for 19 hours or 7 days using a drying oven (Yamato Scientific Co., Ltd., DKN402) with no stirring. The substrates were washed with running water and dried by air blow. The solutions were centrifuged at 4000 rpm for 10 min (Model 8920, Kubota Corp.). Precipitated particles were dried at 60°C for 12 h after removal of supernatant solutions. The particles were dispersed in distilled water. They were centrifuged and dried again for purification.

(1) Morphology control of multi-needle TiO_2 particles TEM observation of multi-needle TiO_2 particles

Flower-like multi-needle TiO_2 particles were successfully formed in aqueous solutions (Fig. 21a, b). Needle shaped crystals grew from the center of the particles. Especially, the needles grew parallel to TEM observation direction from the center of the particles as shown in red circles. They clearly showed that needles radiated in all directions to form flower-like morphology. Each particle had about 6-10 taper needles. Width and length of them were about 200 nm and 100 nm, respectively. Aspect ratio was about 2 (200 nm / 100 nm). Width of the needles became smaller as growth direction to have tips. The particles had no core or pore at the center of their bodies. Nucleation and crystal growth were well controlled to have flower-like multi-needle morphology.
Electron diffraction pattern showed that the particles were single phase of anatase TiO_2 crystals (Fig. 21c). Interplanar spacing of (004), (200), and (204) planes were estimated to 0.243 nm, 0.201 nm and 0.151 nm. Diffractions from (004), (200), and (204) planes were clear single spots (Fig. 21c). Needle shaped crystal was shown to single crystal of anatase TiO_2. Long direction of the needle shaped crystals was perpendicular to (004) crystal faces (Fig. 21c). It indicated that anatase TiO_2 crystals grew along *c*-axis to form needle shape morphology. They were thus surrounded by *a*-faces of anatase TiO_2 crystals.
Surfaces of the needle crystals were observed carefully (Fig. 21c, d). There were no amorphous layers or second phase layers on the surfaces. The particles had pure and bare anatase TiO_2 surfaces.

XRD analysis of multi-needle TiO_2 particles

XRD analysis showed that the particles were single phase of anatse TiO_2. X-ray diffraction peaks were observed at 2θ = 25.12, 36.8, 37.7, 47.7, 53.7, 54.7, 62.4, 68.5, 69.8, 74.7, 82.2, 93.7 and 94.4 °. Pure anatase TiO_2 particles were obtained from the aqueous solutions in this study. Anatase phase have been prepared using high temperature annealing in many reports. It caused deformation of nano/micro-structures, aggregation of the particles and decrease of surface area. However, crystallization of anatase TiO_2 was realized at 50 °C in this study to avoid degradation of the properties.
Crystallite size perpendicular to the (101), (004) or (200) planes was estimated from the full-width half-maximum of the 101, 004 or 200 peak to be 18.2, 29.8 or 19.2 nm, respectively. Crystallite size perpendicular to (004) was larger than that of others. Difference in crystallite size indicated anisotropic crystal growth along *c*-axis.
For comparison, nanocrystal assembled TiO_2 particles were prepared in aqueous solutions at 50˚C for 30 min in previous report[77]. Crystallite size perpendicular to the (101), (004) or (200) planes was estimated to be 3.9, 6.3 or 4.9 nm, respectively. The crystalline degree of the multi-needle TiO_2 nanostructures was much improved compared with that of the previous nanocrystal[77]. The key to achieve highly crystalline degree even at low-temperature, i.e., 50˚C, was slow and long-term growth. It was achieved by low super saturation degree of the solutions. Concentration of ammonium hexafluorotitanate and boric acid were one-tenth of the previous nanocrystal[77]. Synthesis parameters of multi-needle TiO_2 nanostructures were developed based on previous reports as follows. Acicular nanocrystals were homogeneously formed immediately after mixing of two solutions in previous report[77]. They aggregated into nanocrystal assembled particles. The particles were removed from the solutions 30 minutes after the mixing to prevent further crystal growth. The crystallite size was thus small and it contributed to high specific surface area. On the other hand, TiO_2 films were

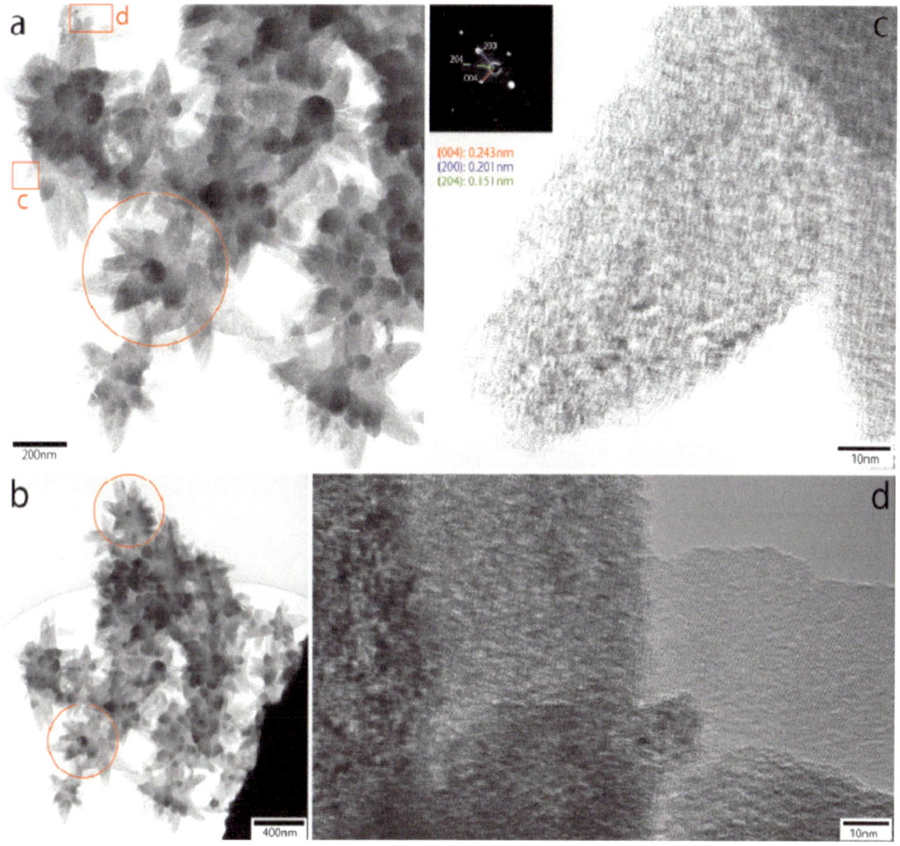

Fig. 21. Transmission electron microscope images of the multi-needle TiO_2 particles. (a, b) Low magnification images of the particles. (c) High resolution image of the needles grew from the center of the particle showing taper shape of the needle. (d) High resolution image of surface of the needles showing bare TiO_2 surface. (c) and (d) are magnified area of (a) and (b). Insert in (c) is electron diffraction pattern indicating crystal phase and interplanar spacing. Red circles indicate typical images of the multi-needle TiO_2 particles.
Reprinted with permission from Ref.[121], Masuda, Y., Ohji, T. and Kato, K., 2010, *Cryst. Growth Des.*, 10, 913. Copyright @American Chemical Society

prepared in the same solutions at 50˚C for several hours[114,122,123]. The films were consisted of two layers. Under layer was consisted of small nanocrystals. They were formed at an early stage of immersion period in high ion concentration solutions. Ions were consumed gradually to form the crystals. Upper layer of acicular crystal assembly was then formed. They grew in the solutions with low ion concentrations. It indicated that low ion concentration and low super saturation degree realized formation of acicular TiO_2 crystals. Additionally, size of acicular crystals was much larger than initially deposited nanocrystals. We tried to form multi-needle particles consisted of large acicular crystals on the basis of these results. Crystal growth in the solution with low ion concentration and low super saturation degree was utilized for anisotropic crystal growth. They were grown to large crystals with long period such as 7 days. Slow growth rate caused formation of euhedral crystals that were affected by crystal structure of tetragonal anatase.

Raman spectroscopy of multi-needle TiO_2 particles

The particles showed Raman peaks at 157 cm^{-1}, 412 cm^{-1}, 506 cm^{-1} and 628 cm^{-1}. It was typical of anatase TiO_2 phase. They were assigned to Eg (v6) mode (157 cm^{-1}), B1g mode (412 cm^{-1}), doublet of the A1g and B1g modes (506 cm^{-1}), and Eg (v1) mode (628 cm^{-1}), respectively. Additional peaks indicating rutile TiO_2 or other phases were not observed. It was notable that Eg (v6) mode shifted with respect to those of bulk crystals or sintered powders. For the single crystal, Ohsaka et al. determined following allowed bands: 144 cm^{-1} (Eg), 197 cm^{-1} (Eg), 399 cm^{-1} (B1g), 513 cm^{-1} (A1g), 519 cm^{-1} (B1g) and 639 cm^{-1} (Eg)[124]. TiO_2 particles (P25, Degussa) showed peaks at 143.3 cm^{-1} (Eg), 196 cm^{-1} (Eg), 396 cm^{-1} (B1g), 516 cm^{-1} (A1g+B1g) and 638 cm^{-1} (Eg). Raman peak of Eg (v6) mode (144 cm^{-1}) has been reported to shift due to several factors such as effect of crystalline size (quantum size confinement effect)[125,126], temperature[125] or pressure[127]. Additionally, nitrogen doped anatase TiO_2 particles of 8.40-9.80 nm in diameter were reported to have Eg (v6) mode at 151 cm^{-1}[126]. The multi-needle particles showed large peak shift of Eg (v6) mode from 144 cm^{-1} to 157 cm^{-1} because of fluorine doping effect and size effect. The large shift of Eg (v6) mode was one of the characteristics of the multi-needle TiO_2 particles.

FT-IR absorbance of multi-needle TiO_2 particles

The particles showed absorption spectra in infrared light region. The absorption bands related to TiO_2 were observed at 907 cm^{-1} and 773 cm^{-1}. They were assigned to stretching vibrations of Ti=O and -Ti-O-Ti-, respectively. They were consistent with TEM, XRD and Raman analyses. The bands observed in the range of 1400-1750 cm^{-1} were attributed to bending vibrations of O-H[128]. Absorption band in frequency range 600-450 cm^{-1} were reported to be attributed to stretching vibrations of Ti-F bonds in TiO_2 lattice[129,130]. The bands were observed at 513, 532, 540, 558 or 568 cm^{-1} from $xTiOF_2 \cdot yBaF_2 \cdot zMnF_2$ glasses[129]. They indicated that F ions were partially replaced to O ions in the lattice[129]. Actually, absorption bands were observed at 455, 488, 505, 520, 552 and 567 cm^{-1} in the spectra of the multi-needle TiO_2 particles. Additionally, absorption band at 1080 cm^{-1} was reported to be assigned to surface fluorinated Ti-F species[131]. Absorption band was observed at about 1058 cm^{-1} from the multi-needle TiO_2 particles. These analyses suggested that F ions were partially doped into TiO_2 crystals to replace O ions.

UV-Vis spectroscopy and band gap of multi-needle TiO_2 particles

Optical property of the particles was evaluated with UV-Vis spectroscopy. Transparency gradually decreased as decrease of wavelength. The bulk band gap structures are known as direct-transition for anatase and indirect-transition for rutile in titania, respectively. Band gap of the particles was estimated to 3.20 eV (388 nm) by assuming the direct-transition. It was similar to that of anatase TiO_2 nanostructures such as anatase TiO_2 nanorods (3.2 eV (388 nm))[132], anatase TiO_2 nanowalls (3.2 eV (388 nm))[132], anatase TiO_2 nanotubes (3.2 eV (388 nm))[133], single-layered TiO_2 nano-sheet with a thickness less than 1 nm (3.15 eV (394 nm))[134], stacked TiO_2 nano-sheets (3.15 eV (394 nm))[134] and anatase TiO_2 films (3.2 eV (388 nm))[135]. It was higher than that of rutile TiO_2 films (2.9eV (428 nm))[135], rutile TiO_2 nanorods (3.0 eV (414 nm))[132], rutile TiO_2 single crystal (3.0 eV (414 nm), SHINKOSHA Co., Ltd.), and lower than that of amorphous TiO_2 films (3.5 eV (355 nm))[135].

N_2 adsorption characteristics of multi-needle TiO_2 particles

BET surface area of the particles was estimated to be 178 m^2/g from an adsorption branch in the range of $P/Po = 0.1-0.29$. Average pore diameter was estimated to be 11.1 nm using BET surface area. They were estimated to 166 m^2/g and 11.9 nm, 151 m^2/g and 17.6 nm from P/Po=0.05-0.1 or P/Po=0.02-0.07, respectively. BET surface area was higher than that of TiO_2 nanoparticles such as Aeroxide P25 (BET 50 m^2/g, 21 nm in diameter, anatase 80% + rutile 20%, Degussa), Aeroxide P90 (BET 90–100 m^2/g, 14 nm in diameter, anatase 90% + rutile 10%, Degussa), MT-01 (BET 60 m^2/g, 10 nm in diameter, rutile, Tayca Corp.) and Altair TiNano (BET 50 m^2/g, 30–50 nm in diameter, Altair Nanotechnologies Inc.)[116]. Pore size distribution calculated by BJH method indicated that spaces of 2-3 nm were existed in the particles. DFT/Monte-Carlo analysis showed several types of mesopores in the range of 3-10 nm (3.5 nm, 4.9 nm, 6.3 nm and 10 nm) and micropores of ~0.8 nm. Total pore volume and average pore diameter were estimated from pores smaller than 241 nm at P/Po = 0.99 to be 0.493 cc/g and 13 nm, respectively. They were estimated to be 0.158 cc/g and 4.2 nm, respectively, from pores smaller than 11 nm at P/Po = 0.81. Total pore volume was also estimated by the BJH method from pores smaller than 156 nm to be 0.505 cc/g using desorption isotherm. Total pore volume was also estimated by the BJH method from pores smaller than 160 nm to be 0.628 cc/g using adsorption isotherm.

(2) Surface coating of micro-structured substrates with multi-needle TiO_2 particles SEM observation of the surface coatings

Micro-structured silicon wafers were immersed in the aqueous solutions at 50°C for 7 days. The substrates were successfully covered with multi-needle TiO_2 particles (Fig. 22-a). Both of salient regions and concave regions were modified with the particles uniformly (Fig. 22-b1, c1). The original micro-structure of silicon wafers were well maintained because the TiO_2 surface coatings were uniform thin layers of about 200-600 nm in thickness (Fig. 22-b2, c2). The coating layers were consisted of flower-like multi-needle TiO_2 particles of about 200-400 nm in diameter (Fig. 22-b3, c3, red circles). The particles had several needles which grew from the center of the particles. The needles had taper shape along growth direction. The particle in red circle of Fig. 22-b3 clearly indicated that tetragonal crystal phase of anatase TiO_2 caused fourfold symmetry crystal growth of four needles parallel to the substrate. The *a*-axis and *c*-axis of the particle were parallel and perpendicular to the substrate, respectively. This shape was basic morphology of the particles in this system. Other needles also grew from the center to form multi-needle shape as shown in red circles in Fig. 22-c3. Salient regions were slightly trapezoidal geometry (Fig. 22-a, b1). TiO_2 particles were deposited on walls of them. The area between the salient and concave regions was thus white in SEM images.

For comparison, micro-structured silicon wafers were immersed in the aqueous solutions at 50°C for 19 hours. The multi-needle TiO_2 particles were formed on both of salient regions and concave regions (Fig. 23-a). The micro-structures of the substrates were maintained because the surface coatings were uniform thin layers (Fig. 23-b1, c1). However, coverage of the surfaces was different from that immersed for 7 d. The TiO_2 particles covered approximately half of the salient regions and one third of concave regions (Fig. 23-b2, c2). Coverage of the salient regions was higher than that of concave regions because concave regions set back far from the salient region surfaces. The particles and ions were not well supplied to concave regions. The particles had several needles grown from the center (Fig. 23-b3, c3). The morphology of them was similar to that immersed for 7d, in contrast, their size was slightly smaller to be about 100-300 nm. These observations indicated growth mechanism as follows. The particles were homogeneously nucleated from ions in the solutions. They were then adhered on the substrates to form surface coatings. The particles further grew to increase number and size of the needles.

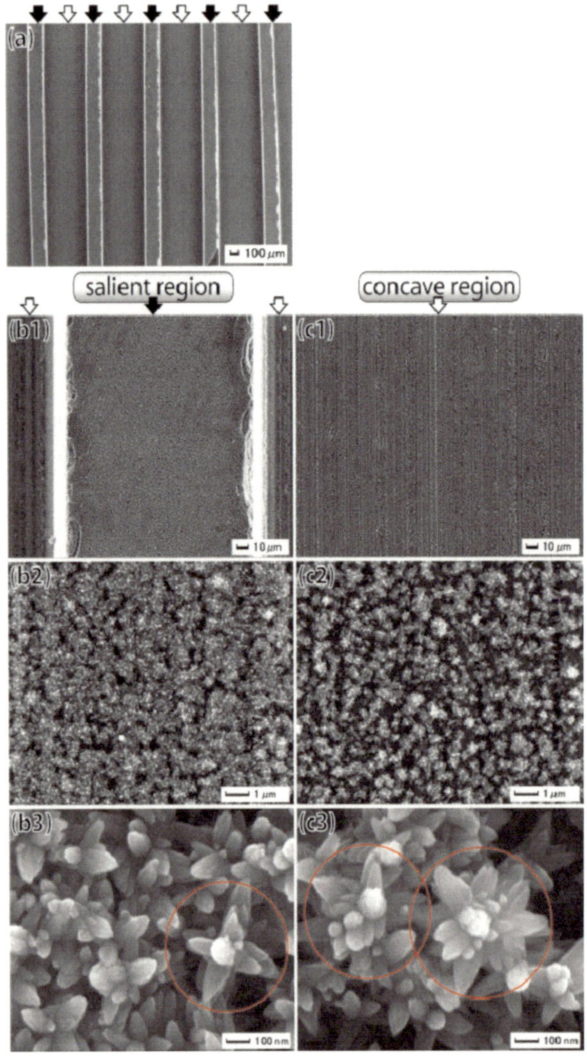

Fig. 22. (a) SEM micrograph of a micro-structured substrate coated with the multi-needle TiO_2 particles deposited for 7 d. (b1) Salient region coated with the multi-needle TiO_2 particles. (b2) Magnified area of (b1) showing uniform surface coating. (b3) Magnified area of (b2) showing morphology of the multi-needle TiO_2 particles. (c1) Concave region coated with the multi-needle TiO_2 particles. (c2) Magnified area of (c1) showing uniform surface coating. (c3) Magnified area of (c2) showing morphology of the multi-needle TiO_2 particles. Black arrows show salient regions. White arrows show concave regions. Red circles indicate typical images of the multi-needle TiO_2 particles.
Reprinted with permission from Ref.[121], Masuda, Y., Ohji, T. and Kato, K., 2010, *Cryst. Growth Des.*, 10, 913. Copyright @American Chemical Society

XPS analysis of the surface coatings

Surfaces of the TiO_2 coatings were analyzed with XPS. Titanium, oxygen, carbon, silicon and fluorine were observed from the surface before (Fig. 24-1a) and after (Fig. 24-1b) Ar^+

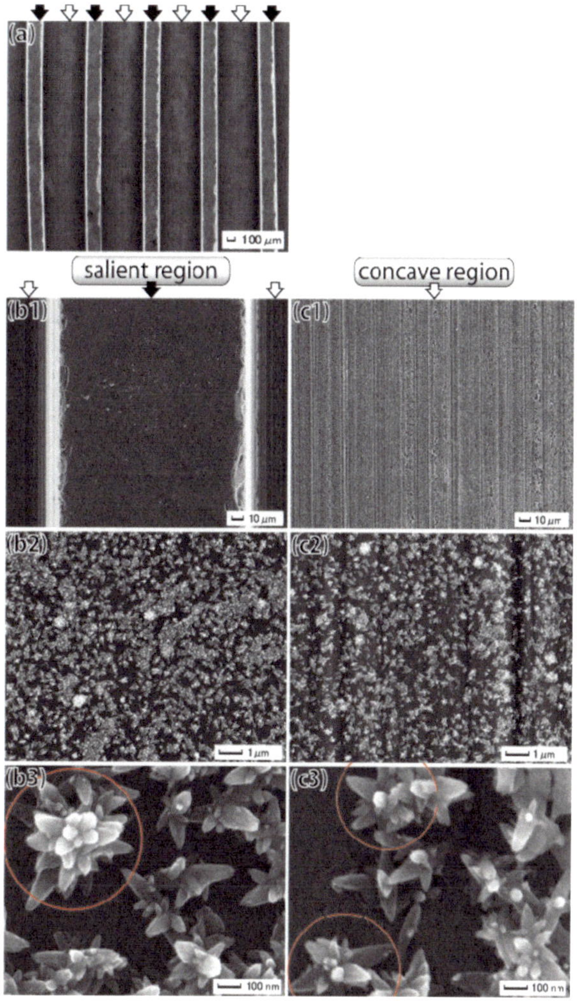

Fig. 23. (a) SEM micrograph of a micro-structured substrate coated with the multi-needle TiO_2 particles deposited for 19 h. (b1) Salient region coated with the multi-needle TiO_2 particles. (b2) Magnified area of (b1) showing uniform surface coating. (b3) Magnified area of (b2) showing morphology of the multi-needle TiO_2 particles. (c1) Concave region coated with the multi-needle TiO_2 particles. (c2) Magnified area of (c1) showing uniform surface coating. (c3) Magnified area of (c2) showing morphology of the multi-needle TiO_2 particles. Black arrows show salient regions. White arrows show concave regions. Red circles indicate typical images of the multi-needle TiO_2 particles.
Reprinted with permission from Ref.[121], Masuda, Y., Ohji, T. and Kato, K., 2010, *Cryst. Growth Des.*, 10, 913. Copyright @American Chemical Society

sputtering. Ti $2p_{3/2}$ spectrum was observed at 458.6 eV (Fig. 24-2a). The binding energy was higher than that of Ti metal (454.0 eV), TiC (454.6 eV), TiO (455.0 eV), TiN (455.7 eV) and Ti_2O_3 (456.7 eV), and similar to that of TiO_2 (458.4-458.7 eV)[136-138]. This suggested that the titanium atoms in the particles were positively charged relative to that of titanium metal by formation of direct bonds with oxygen. Ti 2p spectrum changed its shape by the sputtering due to the decrease of cation valence (Fig. 24-2b). The phenomenon was often observed in sputtering of titanium oxide[136]. The spectrum after sputtering was not suitable for estimation of chemical ratio because the ratio was slightly changed by decrease of titanium cation valence. The chemical ratio of Ti to O was thus estimated from the spectra before the sputtering.

O 1s spectrum was decomposed into two Gaussian curves after removal of background (Fig. 24-3a). The integral intensity ratio of O 1s peak at 532.12 eV to O 1s peak at 529.85 eV was 0.662 : 0.338. The binding energy of O 1s peak at 529.85 eV was similar to that of TiO_2 (529.9 eV[138], 530.1 eV[136,137]) showing that oxygen was negatively charged compared to neutral oxygen molecules (531.0 eV) through the formation of direct bonds with Ti. High binding energy component at 532.12 eV was assigned to oxygen atoms combined to carbon atoms as C-O or C=O (532.8 eV). It was decreased by the sputtering for 10 sec (Fig. 24-3b). It was included in surface contaminations. Drastic decreasing of C 1s spectrum by the sputtering supported this ascription (Fig. 24-4a, 4b). Chemical ratio of Ti to O was estimated to 1 : 1.76 using Ti 2p at 458.6 eV (Fig. 24-2a) and O 1s at 529.85 eV (Fig. 24-3a). It was slightly smaller than that expected from TiO_2. Oxygen vacancy and doping of fluorine ions would decrease oxygen volume. Fluorine was, in fact, observed from the surfaces at 684.5 eV (Fig. 24-5a). Chemical ratio was estimated to Ti : O : F = 1 : 1.76 : 0.18. Sum of oxygen and fluorine was 1.94 which was similar to 2 expected from TiO_2. Fluorine spectrum intensity was not decreased by the sputtering to be Ti : F = 1 : 0.16 (Fig. 24-5b). These indicated that fluorine would not be included in surface contaminations but in the TiO_2 particles. Fluorine has been reported to improve properties of TiO_2. Fluorination of TiO_2 increased surface acidity due to the strongest electronegativity of fluorine[139]. Fluorine doping into TiO_2 increased the surface OH radicals that were suitable for photocatalytic reactions and improved photocorrosion resistance[140-145]. These properties expand application area of TiO_2.

Si $2p_{3/2}$ was observed at 98.6 eV and 102.7 eV (Fig. 24-6a). It was detected from silicon regions uncovered with TiO_2. The binding energies were similar to those in silicon wafers (99.6 eV) and SiO_2 (103.4 eV). The latter component was decreased by the sputtering because surface native oxide layer of amorphous SiO_2 was removed (Fig. 24-6b).

In this section, Multi-needle TiO_2 nanostructures having high surface area, fluorine doping and novel physicochemical characteristics were successfully fabricated. Unique aqueous synthesis realized their distinct morphologies and properties. The needle crystals grew along *c*-axis from the center of the particles. They were surrounded by *a*-faces of anatase TiO_2 crystals. Width and length of needles were about 200 nm and 100 nm. Aspect ratio was about 2 (200 nm / 100 nm). Diffraction patterns showed that the particles were single phase of anatase TiO_2. Interparticle spaces and micro/meso pores of 1-10 nm allowed us to realize high surface area of 178 m^2/g. Large Raman peak shift of Eg (v6) mode suggested fluorine doping and size effect. Band gap was estimated to 3.20 eV (388 nm) with UV-Vis. These were characteristics of the multi-needle TiO_2 particles. Furthermore, self-assembly surface coating of micro-structured substrates was successfully realized. The coating layers were consisted of multi-needle TiO_2 particles of about 200-400 in diameter. XPS analyses indicated chemical bonds between Ti and O. Chemical ratio was estimated to Ti : O: F = 1 : 1.76 : 0.18 suggesting fluorine doping in TiO_2. The multi-needle particles and the surface coating with the particles having nano/micro structures would contribute to future metal oxide devices of solar cells, photo catalysts and high sensitive sensors.

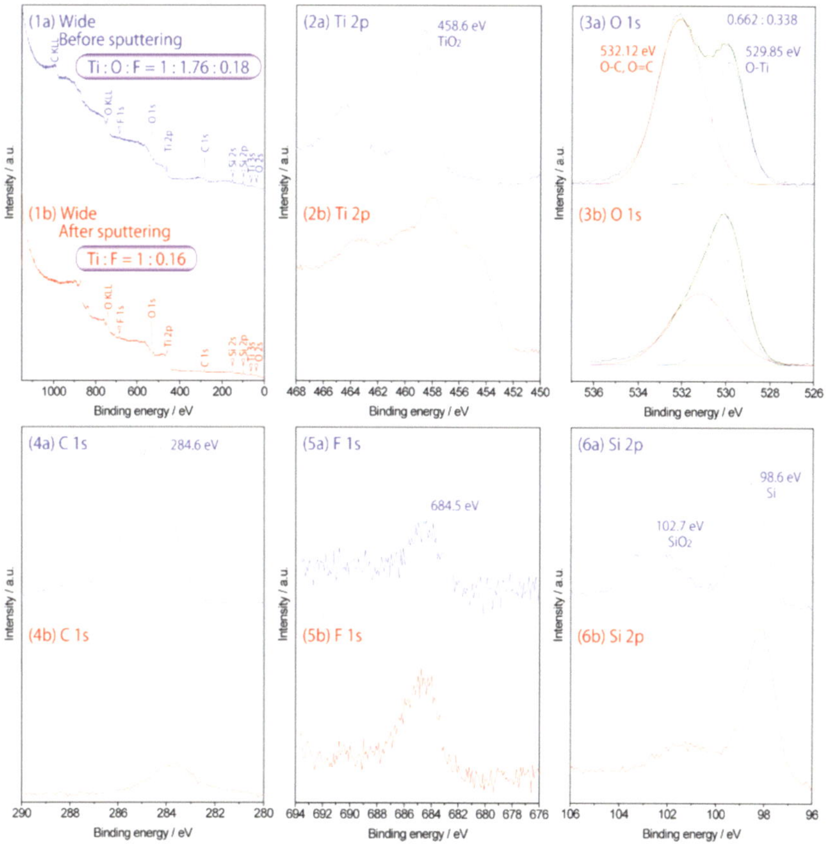

Fig. 24. XPS spectra of (1) wide scan, (2) Ti 2p, (3) O 1s, (4) C 1s, (5) F 1s and (6) Si 2p for a micro-structured substrate coated with the multi-needle TiO_2 particles deposited for 7d (a) before and (b) after Ar^+ sputtering. (3a) O1s spectra decomposed to two peaks after removal of background.

Reprinted with permission from Ref.[121], Masuda, Y., Ohji, T. and Kato, K., 2010, *Cryst. Growth Des.*, 10, 913. Copyright @American Chemical Society

8. Morphology control of nano-sheet assembled tin oxide[146]

Tin oxide particles were synthesized in aqueous solutions. They were consisted of nanosheets of tin oxide crystals. The sheets were about 50-100 nm in size and 5-10 nm thickness. X-ray diffraction analysis revealed that the particles were crystals of SnO_2 and SnO. The particles had BET surface area of 85 m^2/g estimated with N_2 adsorption characteristics. BJH analysis indicated that mesopores of 3.9 nm in size contributed to increase surface area.

Tin dioxide (SnO_2), with a rutile-type crystalline structure, is an n-type wide band gap (3.5 eV) semiconductor. It is an oxide of great interest for gas sensors[1-3], optical device[147], lithium batteries[148-151], white pigments for conducting coatings, transparent conducting coatings for furnaces and electrodes [9], surge arrestors (varistors)[152,153], catalysts[154,155], opto-conducting coatings for solar cells[156], dye-sensitized molecular sensors, etc. Transparency, semiconductivity and surfaced properties of tin oxide are suitable for these applications.

Nanoparticles of tin dioxide have been synthesized by several methods such as precipitation[151,157], hydrothermal synthesis[158,159], sol-gel[160,161], hydrolytic[162], carbothermal reduction[163] and polymeric precursor[164] methods. A variety of SnO_2 nanostructures including nanowires, nanobelts[165], nanotubes[150,166,167], nanorods[168,169], spirals, nanorings[170], zigzag nanobelts[171], grains[172], flakes[172], plates[172], meshes[173] and columnar thin film[149] were synthesized.

Tin chlorides ($SnCl_2$ or $SnCl_4$) were commonly used in many reports. However, chlorine ions were difficult to remove from the systems and it seriously altered superficial and electrical properties. For instance, chlorine ions caused sensitivity degradation of gas sensors[174], aggregation of particles[175] and increasing of sintering temperatures[176]. Chloride problems can be avoided through usage of organic tin compounds, such as alkoxides. However, these reagents were costly which makes industrial syntheses implementation hardly attainable. Additionally, sol-gel processes using alkoxides formed amorphous phases. High temperature annealing which increased cost, energy consumption and CO_2 emission was necessary to obtain tin oxide crystals.

Recently, aqueous syntheses of metal oxide crystals including tin oxide were developed[1,20,177]. Morphology control of metal oxide crystals were realized by anisotropic crystal growth. They were induced by crystal growth control or organic additives. Thermodynamically stable crystal faces, for instance, depend on crystal growth conditions. Organic molecules adsorbed on typical crystal faces suppress crystal growth perpendicular to the faces. Morphology control realized improvement of properties of metal oxide devices.

In this study, tin oxide particles were prepared in aqueous solutions[146]. They were assembly of nanosheets. Morphology, crystal phases and N_2 adsorption characteristics were evaluated. Specific surface area and size distribution of pores were analyzed using isotherms.

Aqueous synthesis of nano-sheet assembled tin oxide particles. SnF_2 (Wako Pure Chemical Industries, Ltd., No. 202-05485, FW: 156.71, purity 90.0%) was used as received. Distilled water in polypropylene vessels (200 mL) were capped with polymer films and kept at 90°C. SnF_2 (870.6 mg) was dissolved in the distilled water at 90°C to be 5 mM. The solutions were kept at 90°C for 30 min using a drying oven (Yamato Scientific Co., Ltd., DKN402) with no stirring. The solutions became clouded shortly after the addition of SnF_2. The bottoms of the vessels were covered with white precipitates. The solutions were centrifuged at 4000 rpm for 10 min (Model 8920, Kubota Corp.). Precipitated particles were dried at 60°C for 12 h after removal of supernatant solutions.

The solutions became clouded shortly after the addition of SnF_2 because of the homogeneous nucleation and growth of tin oxide particles. SnO_2 and SnO are formed in the aqueous solutions as follows:

$$SnF_2 \xrightarrow{OH^-} SnF(OH) + F^- \xrightarrow{OH^-} Sn(OH)_2 + F^- \cdots \quad \text{(a)}$$

$$Sn(OH)_2 \longrightarrow SnO + H_2O \cdots \quad \text{(b)}$$

$$Sn(OH)_2 \longrightarrow SnO_2 + H_2 \cdots \quad \text{(c)}$$

$$SnO + H_2O \longrightarrow SnO_2 + H_2 \cdots \quad \text{(d)}$$ [178,179]

The particles were gradually deposited to cover the bottom of the vessels. The supersaturation degree of the solutions was high at the initial stage and decreased as the color of the solutions changed.

Precipitated particles showed broad X-ray diffraction peaks at 2θ = 27, 34, 38, 52 and 57.5. They were assigned to SnO_2 (JDPDS No. 41-1445. The peaks had large width due to small crystallite size of SnO_2.

Morphology of nano-sheet assembled tin oxide particles. Nano-sheet assembled tin oxide particles were successfully fabricated. Crystallization of tin oxide nanosheets in aqueous solutions allowed us to obtain unique morphology of tin oxide. The particles with 300-800 nm in diameter were observed with FE-SEM (Fig. 25). They were not dense particles but assemblies of tin oxide nanosheets. The sheets were about 50-100 nm in size and 5-10 nm thickness. Aspect ratio was estimated to about 10. They were randomly aggregated to form the particles. The particles had thus continuous open pores inside. It contributed to increase surface area of the particles.

The nanosheets would be generated in the solutions homogeneously. They were aggregated to form particles which made solutions cloudy. They were then precipitated to cover bottoms of vessels. These were consistent with color change observation of solutions.

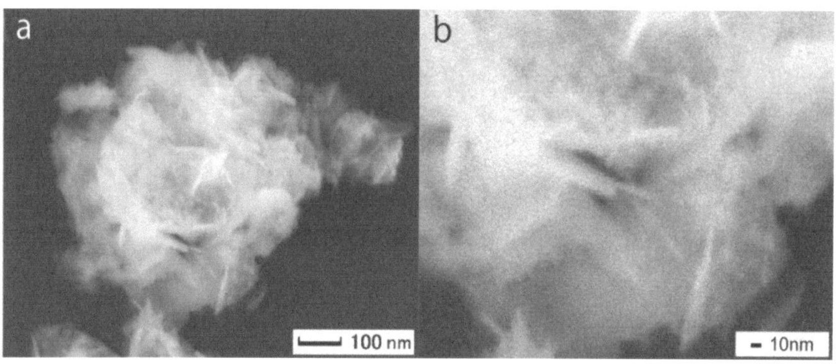

Fig. 25. SEM micrographs of (a) tin oxide particles and (b) magnified area of (a).
Reprinted with permission from Ref.[146], Masuda, Y. and Kato, K., 2009, *J. Cryst. Growth*, 311, 593. Copyright @Elsevier

N_2 adsorption characteristics of nano-sheet assembled tin oxide particles. The particles exhibited N_2 adsorption-desorption isotherms of Type IV. The desorption isotherm differed from adsorption isotherm in the relative pressure (P/P_0) range from 0.45 to 0.97, showing mesopores in the particles. BET surface area of the particles was estimated to be 85 m^2/g. This is higher than that of SnO_2 nanoparticles such as SnO_2 (BET 47.2 m^2/g, 18.3nm in diameter, No. 549657-25G, Aldrich), SnO_2 (BET 25.9 m^2/g, 34nm in diameter, Yamanaka & Co., Ltd.), SnO_2 (BET 23 m^2/g, 26nm in diameter, No. 37314-13, NanoTek, C. I. Kasei Co.,Ltd.) and In_2O_3-SnO_2 (BET 3-6m^2/g, 100-300nm in diameter, Sumitomo Chemical Co., Ltd.). The particles were assembly of nanosheets. Unique morphologies of the sheets contributed to increase surface area of the particles.

Total pore volume and average pore diameter were estimated to 0.343 cc/g and 16.1 nm, respectively from pores smaller than 259 nm at P/Po = 0.9925. They were estimated to be 0.088 cc/g and 4.1 nm, respectively, from pores smaller than 10.6 nm at P/Po = 0.7994. Total pore volume was also estimated by the BJH method from pores smaller than 174 nm to be 0.354 cc/g.

Pore size distribution was calculated by the BJH method using desorption braches. It showed a pore size distribution curve having a strong peak at ~3.9 nm and a broad peak from 25 nm to 175 nm. The particles would have a large amount of mesopores with ~3.9 nm in diameter and mesopores with 25-175 nm in diameter. The pores were spaces surrounded by nanosheets and interparticle spaces. Micropores smaller than 2 nm were not suggested by BJH pore size distribution.

In this section, nanosheet assembled tin oxide particles were fabricated by aqueous solution synthesis. Anisotropic crystal growth of tin oxide was effectively utilized to form nanosheet assembled structure. They were crystallized at ordinary temperature without annealing. It allowed us to avoid aggregation of the particles and decrease of surface area. The particles were 300-800 nm in diameter and were crystals of SnO_2 and SnO. The sheets were about 50-100 nm in size and 5-10 nm thickness. The particles had BET surface area of 85 m^2/g estimated with N_2 adsorption characteristics. BJH analysis indicated that mesopores of 3.9 nm in size contributed to increase surface area. The particles were prepared by environmentally friendly process. The system had advantages on low cost, low energy consumption and low CO_2 emission.

9. Enhancement of surface area of tin oxide[180]

Tin oxide nano-crystals with high surface area were firstly synthesized in aqueous solutions at 50°C. BET surface area was successfully reached to 194 m^2/g[180]. It was much higher than that of SnO_2 (BET 47.2 m^2/g, Aldrich), SnO_2 (BET 25.9 m^2/g, Yamanaka & Co., Ltd.), SnO_2 (BET 23 m^2/g, C. I. Kasei Co.,Ltd.) and In_2O_3-SnO_2 (BET 3-6m^2/g, Sumitomo Chemical Co., Ltd.). N_2 adsorption characteristics revealed that they had pores of 1-3 nm which contributed high surface area. TEM, ED and XRD indicated morphology, crystal structure and chemical composition of nano-crystals. Novel process allowed us to avoid sintering and deformation of the crystals and hence, realized high surface area and unique morphology.

In this study, tin oxide nanosheets were formed in the solutions. BET surface area was successfully reached to 194 m^2/g. Origin of high surface area was discussed with pore size distribution and morphology observations.

Synthesis of tin oxide nanosheets. SnF_2 (870.6 mg) was dissolved in distilled water (200 mL) of 50°C to be 5 mM. The solutions were kept at 50°C for 20 min and then at 28°C for 3 days without stirring. The nanocrystals precipitated to cover bottom of the vessels. For comparison, the solutions were centrifuged at 4000 rpm for 10 minafter keeping at 50°C for 20 min. Precipitated particles were dried at 60°C for 12 h after removal of supernatant solutions.

Morphology and crystal phase of tin oxide nanosheets.

The nanosheets synthesized at 50°C for 20min and at 28°C for 3 days were mixture of SnO_2 main phase and SnO additional phase (Fig. 26a). X-ray diffraction peaks at 2θ = 26.5, 33, 51.4, 62, 64.5, 80 and 89 were assigned to 110, 101, 211, etc. of SnO_2 (JDPDS No. 41-1445). 101, 110 and 002 diffraction peaks of SnO (JDPDS No. 06−0395) were over lapped to peaks of SnO_2. For comparison, diffraction pattern of the nanosheets synthesized at 50°C for 20min was shown in Fig. 26b. Half maximum full-width of the peaks was smaller than that in Fig. 26a. Lower SnO content resulted sharp peaks of SnO_2 in Fig. 26b. These observations indicated that SnO was mainly formed at 28°C rather than at 50°C.

The nanosheets were well dispersed in ethanol. The nanosheets in supernatant solutions were skimmed with Cu grids for TEM observations. They were 20-50 nm in diameter having

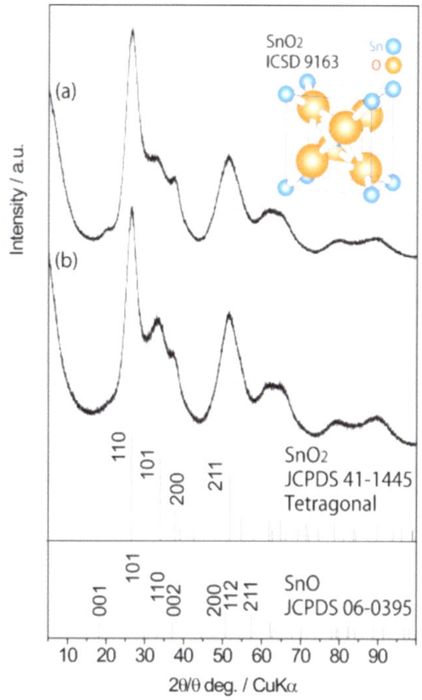

Fig. 26. XRD patterns of (a) tin oxide nanosheets fabricated at 50°C for 20min and at 28°C for 3 days, and (b) tin oxide nanosheets fabricated at 50°C for 20min.
Reprinted with permission from Ref.[180], Masuda, Y., Ohji, T. and Kato, K., 2010, *J. Am. Ceram. Soc.*, 93(8), 2140. Copyright @Wiley

uniform thickness (Fig. 27A). Similar structures were observed from many areas. We observed that they curled up during long-term observation. It indicated that they had sheet structure. Electron beams damaged them to transform the structures. Some of them tightly connected each other (Fig. 27B). They had clear interfaces without pores or small grains. The nanosheet showed electron diffractions (Fig. 27B, Insert). Lattice spacing calculated from spots indicated with a red line and a yellow line were 0.283 nm and 0.154 nm, respectively. They can be assigned to SnO_2 and/or SnO. The former one can be assigned to101 crystal plane of SnO_2 (0.264 nm) or 110 crystal plane of SnO (0.269 nm). The later one can be assigned to 310 crystal plane of SnO_2 (0.149 nm), 221 crystal plane of SnO_2 (0.148 nm), 202 crystal plane of SnO (0.149 nm) or 103 crystal plane of SnO (0.148 nm). Additionally, diffraction spots related to lattice spacing of 0.283 nm were observed at upper part. They were shown with two white circles and a red line in Fig. 27B Insert. The double spots indicated that the area shown in a white circle in Fig. 27B was consisted of two crystals, i.e., stacked two nanosheets. High magnification image was also obtained from other observation area (Fig. 27C). The structure was thinner than that in Fig.27A and 27B. Clear image was not obtained in low magnification images due to low contrast. However, it showed clear high magnification image and lattice fringe (Fig. 27C). Electron diffraction patterns showed lattice spacing of 0.277 nm (red line), 0.279 nm (green line) and 0.196 nm (yellow line). They were assigned to 110, 1-10 and 200 crystal planes of SnO, respectively. Lattice spacing and their angles were well matched to that of SnO. Chemical composition was estimated from several points of tightly-packed area. The area included nanosheets and spherical crystals. Chemical composition varied among each positions in the range of Sn : O = 1 : 1.7-2.7 which was similar to that of SnO_2 rather than SnO. These observations indicated that the crystals were mixture of SnO_2 and SnO.

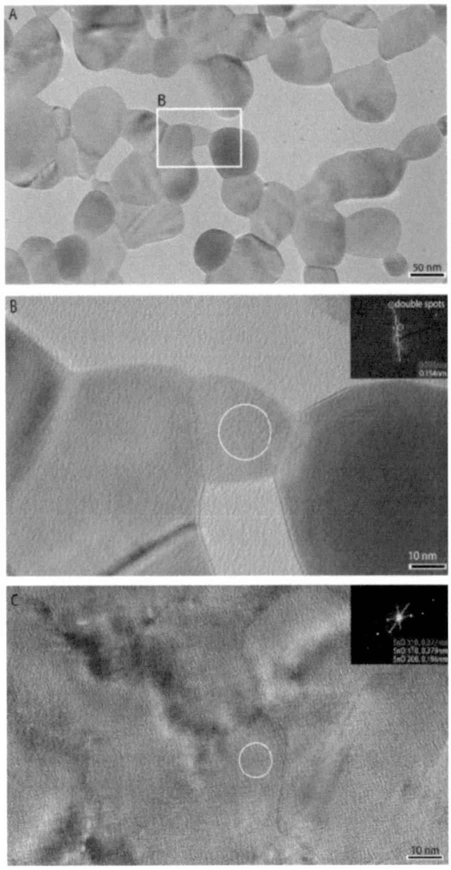

Fig. 27. TEM micrographs and electron diffraction patterns of tin oxide nanosheets. (b): high magnification image of (a). (c): high magnification image of other area.
Reprinted with permission from Ref.[180], Masuda, Y., Ohji, T. and Kato, K., 2010, *J. Am. Ceram. Soc.*, 93(8), 2140. Copyright @Wiley

N_2 adsorption characteristics of tin oxide nanosheets.

BET surface area of the nanosheets was estimated from N_2 adsorption isotherm (Fig. 28a). The nanosheets had successfully high surface area of 194 m^2/g (Fig. 28b). It was much higher than that of SnO_2 nanoparticles such as SnO_2 (BET 47.2 m^2/g, 18.3nm in diameter, No. 549657-25G, Aldrich), SnO_2 (BET 25.9 m^2/g, 34nm in diameter, Yamanaka & Co., Ltd., Osaka, Japan), SnO_2 (BET 23 m^2/g, 26nm in diameter, No. 37314-13, NanoTek, C. I. Kasei Co.,Ltd., Tokyo, Japan) and In_2O_3-SnO_2 (BET 3-6m^2/g, 100-300nm in diameter, Sumitomo Chemical Co., Ltd.). Additionally, it was more than double higher than that of previous report (85 m^2/g)[181]. Pore size distribution was analyzed with BJH method using adsorption isotherm (Fig. 28c). It indicated that the nanosheets included pores of 1-2 nm. Micropore analysis was performed with DFT/Monte-Carlo Fitting which was completely-consistent with isotherms (Fig. 28d). Pores of 1-3 nm were shown to be in the nanosheets (Fig. 28e). The micro and meso-pores contributed high surface area of 194 m^2/g in this system. For comparison, BET surface area of the nanosheets synthesized at 50°C for 20min was estimated to 146 m^2/g. It was also higher than that of nanoparticles in former studies.

In this section, nanosheets of tin oxides were fabricated in aqueous solutions at ordinary temperature. BET surface area successfully reached to 194 m^2/g. It was much higher than that of nanoparticles in former studies. 2-dimensional sheet structure was one of ideal structures for high surface area per unit weight. Nano sized thickness directly contributed high surface area. Crystalline nanosheets were prepared without high temperature annealing which degraded surface area and nanostructures. High surface area and unique nanostructures of the sheets can be applied to gas sensors, dye-sensitized solar cells and molecular sensors.

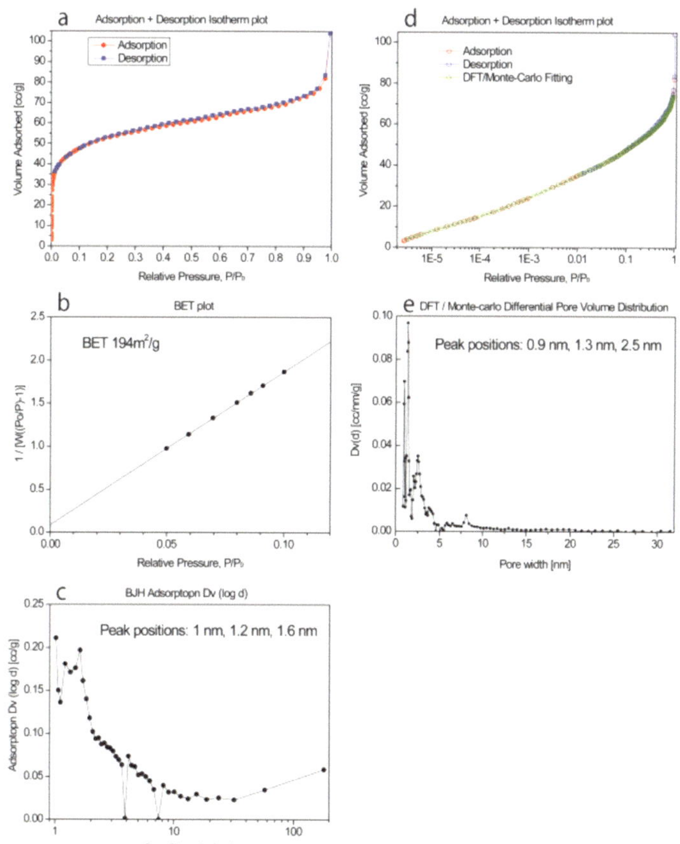

Fig. 28. (a): N_2 adsorption-desorption isotherm of tin oxide nanosheets. (b): BET surface area of tin oxide nanosheets. (c): Pore size distribution calculated from N_2 adsorption data of tin oxide nanosheets using BJH equation. (d): N_2 adsorption-desorption isotherm and DFT/Monte-Carlo fitting curve of tin oxide nanosheets. (e): Pore size distribution calculated from N_2 adsorption data of tin oxide nanosheets using DFT/Monte-Carlo equation. Reprinted with permission from Ref.[180], Masuda, Y., Ohji, T. and Kato, K., 2010, *J. Am. Ceram. Soc.*, 93(8), 2140. Copyright @Wiley

10. Tin oxide coating on polytetrafluoroethylene films in aqueous solutions[182]

Tin Polytetrafluoroethylene(PTFE) films were successfully coated with tin oxide in aqueous solutions. Tin oxide was crystallized in the solution and formed nanocrystal coatings on the polymer films. The coatings were consisted of SnO_2 and SnO crystals. They were assemblies of tin oxide nanosheet of about 10 nm to 50 nm in size and about 5 nm in thickness. The nanocrystal films can be exfoliated from the PTFE substrates. Tin oxide nanocrystal films had a rough liquid-surface and a dense substrate-side surface. Transparency of PTFE films coated with tin oxide was same to that of bare PTFE films in the range from 400 nm to 800 nm. Tin oxide decreased transparency about 25 % at 320 nm. The PTFE films coated with tin oxide nanocrystals can be pasted on desired substrates.

Organic-inorganic hybrid materials such as metal oxide electronics on polymer flexible films have received considerable attention in recent years for light-weight flexible sensors, displays, dye-sensitized solar cells, etc. The polymer films offer the advantages of flexibility, light-weight, low-cost and impact resistance.

Tin oxide (SnO_2) is an important semiconductor with a wide band gap of 3.6 eV at room temperature. They have been widely used in gas sensors[167,183], optical devices[147], lithium batteries[148-150], etc. A novel type of bio-sensor was proposed to detect environmental toxins such as bisphenol-A or dioxin[88,184]. Tin oxide is a candidate material for the sensor because of its suitable band gap, surface characteristics, and high transparency.

Metal oxides including tin oxide have been synthesized with high temperature processes at several hundred degrees for many years. Recently, the aqueous syntheses of metal oxides have attracted much attention as a next generation science and technology[1,51,114,177]. Aqueous systems are environmentally friendly and have advantages of low energy consumption, low cost and an organic solvent free process.

In recent years, metal oxide films and their microstructures were fabricated on organic surfaces such as polyethylene terephthalate (PET) films[185,186] or self-assembled monolayers[20,113,187,188]. However, metal oxide formation on polytetrafluoroethylene (PTFE) was difficult compared to PET films. PTFE films are widely used in electronic applications because of low chemical reactivity, low coefficient of friction, high melting point (327 °C), high-corrosion resistance, a high dielectric strength over many different frequencies, a low dissipation factor, and a high surface resistivity. They were selected as substrates in this study.

In this section, tin oxide nanocrystals were prepared on PTFE films in aqueous solutions[182,184]. They were crystallized in the solution containing tin ions to form films consisted of nanosheets. The process realized tin oxide film formation without high temperature annealing and unique morphology of tin oxide crystals.

(1) Aqueous synthesis of tin oxide nanocrystals on PTFE films

SnF_2 (870.6 mg) was dissolved in the distilled water (200 mL) at 90°C to be 5 mM. PTFE films (thickness: 50 μm, ASF-110, Chukoh) with silicone adhesive (thickness: 30 μm) were pasted on quarts substrates (25 × 50 × 1 mm). They were immersed in the middle of the solutions with the bottom up at an angle or with the top up at an angle. They were tilted at 15 degrees to the upright. The solutions were kept at 90°C using a drying oven (Yamato Scientific Co., Ltd., DKN402) for 2 h with no stirring. The solutions became lightly clouded after 2 h. The as-deposited nanocrystals on substrates were rinsed under running water and dried by a strong air spray. Additionally, the solutions kept at 90°C for 2 h were centrifuged at 4000 rpm for 10 min (Model 8920, Kubota Corp.). Precipitated particles were dried at 20°C for 12 h after the removal of supernatant solutions.

(2) Morphology of tin oxide nanocrystals on PTFE films

Bare PTFE films pasted on quarts substrates had cracks of about 50 nm to 200 nm in length (Fig. 29). Longer directions of cracks were perpendicular to the extensional direction. They were formed during adhesive processes. The quarts substrates and the PTFE films were transparent or slightly white, respectively. The PTFE films can be pasted on desired substrates such as quarts, metals and polymers with silicone adhesive.

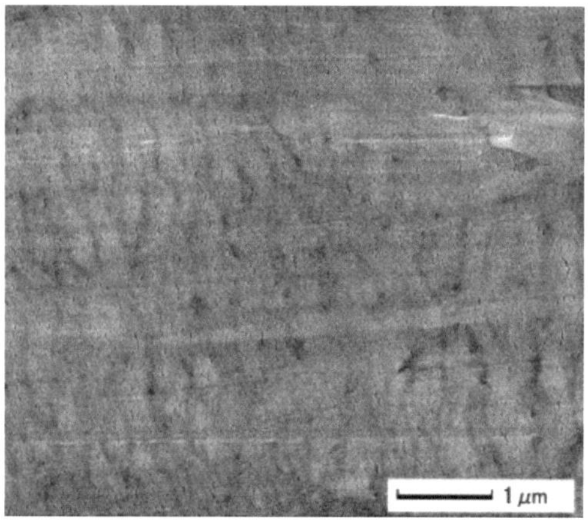

Fig. 29. SEM micrograph of a bare PTFE film.
Reprinted with permission from Ref.[182], Masuda, Y. and Kato, K., 2010, *Polym. Adv. Technol.*, 21(3), 211. Copyright @Wiley

The surface of PTFE films was completely covered with assemblies of nanosheets (Fig. 30a). Uniform formation of tin oxide coatings is one of the advantages of solution processes. Large sheets were also observed from the surfaces (Fig. 30b). They were about 200 nm to 300 nm in size and about 10 nm in thickness (Fig. 30b). Some of them stood perpendicular to the PTFE films at an angle. They had an angular outline, which was connected by straight lines. They were caused from anisotropic crystal growth of tin oxide reflected in their crystal structure. The large sheets connected in a cross shape were also observed at the lower right (Fig. 30b). Nanosheets were roughly estimated to about 10 nm to 50 nm in size and about 5 nm in thickness (Fig. 30c). They connected to each other to form continuous films on the PTFE surfaces. The tin oxide films were exfoliated from PTFE films by scratch using a metal spatula. Exfoliated tin oxide films were placed on a substrate for SEM observation. Three exfoliated tin oxide films were observed to be partially-overlapping (Fig. 30d). The top film and bottom film showed their liquid-side surfaces. They were similar to that observed in Fig. 30a-c. The middle film showed a substrate-side surface, which was contacted with PTFE film during the immersion period. Substrate-side surface of the middle film was indicated by a white arrow. It had a dense surface, which consisted of nanocrystals of about 5 nm to 10 nm in size. These observations indicated that dense films consisted of nanocrystals of about 5 nm to 10 nm in size were formed on the PTFE films at first stage of tin oxide film formation. Tin oxide then grew to sheet-shapes to form assemblies of nanosheets. Additionally, large sheet crystals grew on the tin oxide dense films.

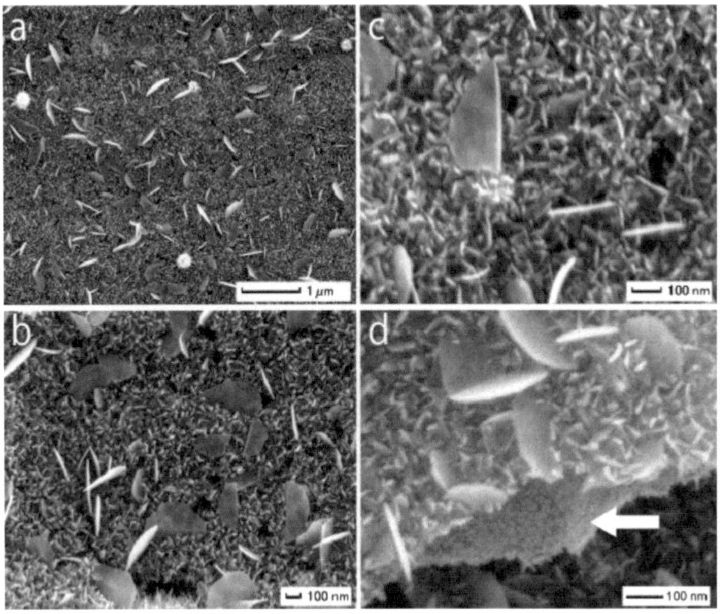

Fig. 30. SEM micrographs of (a) tin oxide nanocrystals on a PTFE film and (b, c) magnified area of (a) showing morphology of nanocrystals and (d) exfoliated tin oxide nanocrystal films showing liquid-side surface and substrate-side surface. Substrate-side surface is indicated by a white arrow.

Reprinted with permission from Ref.[182], Masuda, Y. and Kato, K., 2010, *Polym. Adv. Technol.*, 21(3), 211. Copyright @Wiley

(3) Crystal phase of tin oxide nanocrystals on PTFE films

X-ray diffraction pattern obtained from a PTFE film coated with tin oxide was similar to that from a bare PTFE film. Diffraction peaks assigned to tin oxide were not observed because of thin film thickness.

Precipitated particles were evaluated after drying. They showed broad X-ray diffraction peaks at 2θ = 27, 34, 38 and 52. They were assigned to 110, 101, 200 or 211 diffraction peaks from SnO_2 (JDPDS No. 41-1445) (Fig. 3c). The peaks were wide in width due to the small crystallite size of SnO_2. Peaks were also observed 2θ = 29.5, 32.0, 37.4, 48.0, 50.4 and 57.4. They were assigned to 101, 110, 002, 200, 112 or 211 diffraction peaks from SnO (JDPDS No. 06-0395).

XRD analysis indicated that the particles obtained from the solutions consisted of SnO_2 crystals and SnO crystals.

(4) XPS analysis of tin oxide nanocrystals on PTFE films

Carbon, fluorine and oxygen were detected from bare PTFE films (Fig. 31b). Chemical ratio of them was estimated to C : F : O = 1 : 2.05 : 0.02. It was consistent with chemical composition of polytetrafluoroethylene (PTFE, C : F = 1 : 2). Small amount of oxygen was detected from surface contamination. PTFE films coated with tin oxide showed spectra of tin, oxygen, fluorine and carbon (Fig. 31a). Spectral peak corresponding to Sn $3d_{5/2}$ was observed at 487.2 eV (Fig. 31c1). The binding energy was similar to that of SnO_2 (486.3 eV[189], 486.5 eV[190], 486.6 eV[191], 487.3 eV) and higher than that of Sn metal (484.8 eV, 484.85 eV,

484.87 eV, 484.9 eV, 485.0 eV), which suggested tin atoms in surface coatings were positively charged by forming direct bonds with oxygen. Binding energy of O 1s centered at about 531.2 eV corresponds to that of SnO_2 (Fig. 31c2). Chemical ratio of the coatings was estimated to Sn : O : F : C = 1 : 1.88 : 7.25 : 3.82 and C : F = 1 : 1.90. It was indicated that the surface coatings consisted mainly of SnO_2 (Sn : 1 : 2). They were formed on polytetrafluoroethylene films (PTFE, C : F = 1 : 2). Difference and similarity between XPS analyses and XRD analyses suggested deposition mechanism of tin oxides. Crystal phases and chemical compositions of the surface coatings were different from those of the precipitated particles. SnO_2 crystallized on the films to form surface coatings, on the other hand, SnO_2 and SnO homogeneously crystallized to form particles in the solutions. Pure SnO_2 coatings were thus successfully formed on the PTFE films.

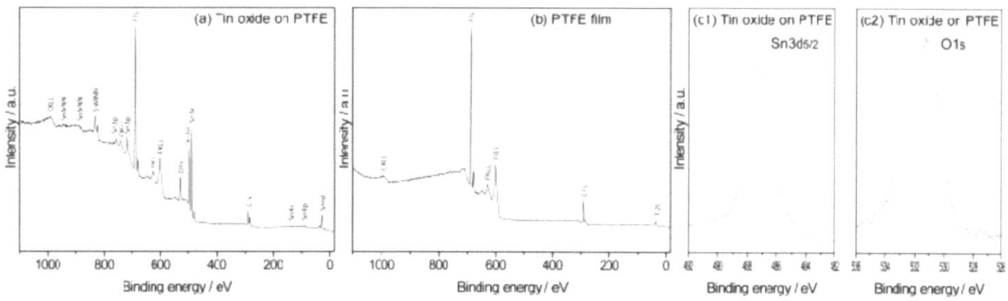

Fig. 31. XPS spectra of (a) tin oxide nanocrystals on a PTFE film and (b) a bare PTFE film. (c1) Sn $3d_{5/2}$ spectrum and (c2) O 1s spectrum of tin oxide nanocrystals on a PTFE film. Reprinted with permission from Ref.[182], Masuda, Y. and Kato, K., 2010, *Polym. Adv. Technol.*, 21(3), 211. Copyright @Wiley

(5) Optical property of tin oxide nanocrystals on PTFE films

Quarts substrates had high transparency in the range from 200 nm to 850 nm (Fig. 32, black line). PTFE films pasted on quarts substrates were visually observed to be slight white. Their transparency was lower than that of quarts (Fig. 32, red line). Especially, it decreased as decrease of wavelength below 350 nm. PTFE films coated with tin oxide showed the same transparency as bare PTFE films in the range from 400 nm to 850 nm (Fig. 32, blue line). They decreased transparency about 25 % at 320 nm. Tin oxide particles precipitated from the solution were evaluated for comparison. They had absorption peak centered at 320 nm. This absorption was caused by tin oxide. These analyses indicated that tin oxide was deposited on PTFE films immersed in the solutions and they decrease transparency at 320 nm.

11. Conclusion

Polytetrafluoroethylene (PTFE) films were successfully coated with tin oxide nanocrystals. Tin oxide was crystallized in aqueous solutions to form nanosheet-assembled films. They were about 10 nm to 50 nm in size and about 5 nm in thickness. Large sheets of about 200 nm to 300 nm in size and about 10 nm in thickness were also crystallized on the surfaces. X-ray diffraction analysis indicated that tin oxide was a mixture of SnO_2 and SnO. PTFE films coated with tin oxide were transparent in the range from 400 nm to 850 nm. Tin oxide on the films had absorption centered at 320 nm. Tin oxide coated PTFE films can be pasted on desired substrates. Hybrid tin oxide-PTFE composites may be useful for increasing the potential application of tin oxide film as flexible electronics.

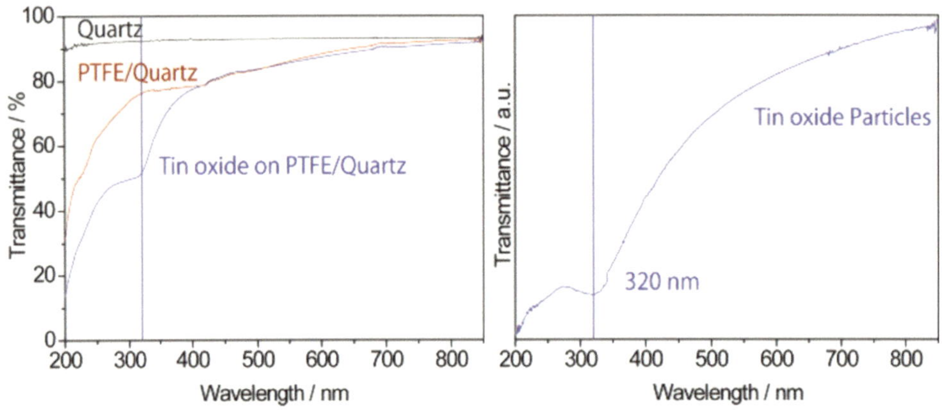

Fig. 32. UV-Vis spectra of (a) a quarts substrate (black line), (b) a PTFE film pasted on a quarts substrate (red line) and (c) tin oxide nanocrystals on a PTFE film pasted on a quarts substrate (blue line). UV-Vis spectrum of tin oxide particles.
Reprinted with permission from Ref.[182], Masuda, Y. and Kato, K., 2010, *Polym. Adv. Technol.*, 21(3), 211. Copyright @Wiley

12. Summary

Liquid phase morphology control of metal oxide was developed by precise control of crystal growth in this study. Various nano/micro-structures of them were fabricated in solution systems. These would contribute to development of future metal oxide devices and develop new academic fields.

13. References

[1] Masuda, Y.; Yamada, T.; Koumoto, K. *Cryst. Growth Des.* 2008, *8*, 169-171.

[2] Tanaka, K.; Suzuki, K.; Fu, D. S.; Nishizawa, K.; Miki, T.; Kato, K. *Jpn. J. Appl. Phys.* 2004, *43*, 6525-6529.

[3] Aoyagi, S.; Kuroiwa, Y.; Sawada, A.; Kawaji, H.; Atake, T. *J. Therm. Anal. Cal.* 2005, *81*, 627-630.

[4] Luan, W. L.; Gao, L.; Guo, J. K. *Ceram. Inter.* 1999, *25*, 727-729.

[5] Wang, X. H.; Chen, R. Z.; Gui, Z. L.; Li, L. T. *Mater. Sci. Eng. B* 2003, *99*, 199-202.

[6] Sakabe, Y.; Yamashita, Y.; Yamamoto, H. *J. Euro. Ceram. Soc.* 2005, *25*, 2739-2742.

[7] Polotai, A. V.; Ragulya, A. V.; Randall, C. A. *Ferroelectrics* 2003, *288*, 93-102.

[8] Yashima, M.; Hoshina, T.; Ishimura, D.; Kobayashi, S.; Nakamura, W.; Tsurumi, T.; Wada, S. *J. Appl. Phys.* 2005, *98*.

[9] Frey, M. H.; Xu, Z.; Han, P.; Payne, D. A. *Ferroelectrics* 1998, *206*, 337-353.

[10] Frey, M. H.; Payne, D. A. *Phys. Rev. B* 1996, *54*, 3158-3168.

[11] Wada, S.; Suzuki, T.; Noma, T. *J. Ceram. Soc. Japan* 1996, *104*, 383-392.

[12] Masuda, Y.; Koumura, T.; Okawa, T.; Koumoto, K. *J. Colloid Interface Sci.* 2003, *263*, 190-195.

[13] Masuda, Y.; Koumoto, K.; Ueyama, R. Japanese Patent Application Number: P2007-113579, 2007.

[14] Hayashi, Y.; Kimura, T.; Yamaguchi, T. *J. Mater. Sci.* 1986, *21*, 757-762.

[15] Li, X. L.; Liu, J. F.; Li, Y. D. *Mater. Chem. Phys.* 2003, *80*, 222-227.
[16] Sun, H.; Kang, S. Z.; Mu, J. *Mater. Lett.* 2007, *61*, 4121-4123.
[17] Mutin, J. C.; Dusausoy, Y.; Protas, J. *J. Solid State Chem.* 1981, *36*, 356-364.
[18] Potdar, H. S.; Deshpande, S. B.; Date, S. K. *J. Am. Ceram. Soc.* 1996, *79*, 2795-2797.
[19] Masuda, Y.; Kinoshita, N.; Koumoto, K. *J. Nanosci. Nanotechnol.*, in press.
[20] Masuda, Y.; Kinoshita, N.; Sato, F.; Koumoto, K. *Cryst. Growth Des.* 2006, *6*, 75-78.
[21] Wu, X. L.; Siu, G. G.; Fu, C. L.; Ong, H. C. *Appl. Phys. Lett.* 2001, *78*, 2285-2287.
[22] Kang, J. S.; Kang, H. S.; Pang, S. S.; Shim, E. S.; Lee, S. Y. *Thin Solid Films* 2003, *443*, 5-8.
[23] Golego, N.; Studenikin, S. A.; Cocivera, M. J. *J. Electrochem. Soc.* 2000, *147*, 1592-1594.
[24] Sberveglieri, G. *Sens. Actuators B: Chem.* 1995, *23*, 103-109.
[25] Pauporte, T.; Lincot, D. *Electrochem. Acta.* 2000, *45*, 3345-3353.
[26] Law, M.; Greene, L. E.; Johnson, J. C.; Saykally, R.; Yang, P. D. *Nature Mater.* 2005, *4*, 455-459.
[27] Masuda, Y.; Kato, K. *Cryst. Growth Des.* 2008, *8*, 2633-2637.
[28] Gong, H.; Hu, J. Q.; Wang, J. H.; Ong, C. H.; Zhu, F. R. *Sens. Actuators B: Chem.* 2006, *115*, 247-251.
[29] Moon, W. J.; Yu, J. H.; Choi, G. M. *Sens. Actuators B: Chem.* 2002, *87*, 464-470.
[30] Chang, J. F.; Kuo, H. H.; Leu, I. C.; Hon, M. H. *Sens. Actuators B: Chem.* 2002, *84*, 258-264.
[31] Tang, H. X.; Yan, M.; Zhang, H.; Li, S. H.; Ma, X. F.; Wang, M.; Yang, D. R. *Sens. Actuators B: Chem.* 2006, *114*, 910-915.
[32] Shishiyanu, S. T.; Shishiyanu, T. S.; Lupan, O. I. *Sens. Actuators B: Chem.* 2005, *107*, 379-386.
[33] Wagh, M. S.; Patil, L. A.; Seth, T.; Amalnerkar, D. P. *Mater. Chem. Phys.* 2004, *84*, 228-233.
[34] Xu, J. Q.; Pan, Q. Y.; Shun, Y. A.; Tian, Z. Z. *Sens. Actuators B: Chem.* 2000, *66*, 277-279.
[35] Paraguay, D. F.; Miki-Yoshida, M.; Morales, J.; Solis, J.; Estrada, L. W. *Thin Solid Films* 2000, *373*, 137-140.
[36] Baxter, J. B.; Aydil, E. S. *Appl. Phys. Lett.* 2005, *86*, 53114.
[37] Katoh, R.; Furube, A.; Hara, K.; Murata, S.; Sugihara, H.; Arakawa, H.; Tachiya, M. *J. Phys. Chem. B* 2002, *106*, 12957-12964.
[38] Karuppuchamy, S.; Nonomura, K.; Yoshida, T.; Sugiura, T.; Minoura, H. *Solid State Ionics* 2002, *151*, 19-27.
[39] Keis, K.; Bauer, C.; Boschloo, G.; Hagfeldt, A.; Westermark, K.; Rensmo, H.; Siegbahn, H. *J. Photochem. Photobiol. A* 2002, *148*, 57-64.
[40] Kasuga, T.; Hiramatsu, M.; Hosono, A.; Sekino, T.; Niihara, K. Adv. Mater. 1999, 11, 1307-+.
[41] Oaki, Y.; Imai, H. *J. Am. Chem. Soc.* 2004, *126*, 9271-9275.
[42] Xia, Y. N.; Yang, P. D.; Sun, Y. G.; Wu, Y. Y.; Mayers, B.; Gates, B.; Yin, Y. D.; Kim, F.; Yan, Y. Q. *Adv. Mater.* 2003, *15*, 353-389.
[43] Masuda, Y.; Saito, N.; Hoffmann, R.; De Guire, M. R.; Koumoto, K. *Sci. Tech. Adv. Mater.* 2003, *4*, 461-467.
[44] Masuda, Y.; Ieda, S.; Koumoto, K. *Langmuir* 2003, *19*, 4415-4419.
[45] Masuda, Y.; Jinbo, Y.; Yonezawa, T.; Koumoto, K. *Chem. Mater.* 2002, *14*, 1236-1241.
[46] Nakanishi, T.; Masuda, Y.; Koumoto, K. *Chem. Mater.* 2004, *16*, 3484-3488.
[47] Peng, W. Q.; Qu, S. C.; Cong, G. W.; Wang, Z. G. *Cryst. Growth Des.* 2006, *6*, 1518-1522.
[48] Lin, Y. R.; Yang, S. S.; Tsai, S. Y.; Hsu, H. C.; Wu, S. T.; Chen, I. C. *Cryst. Growth Des.* 2006, *6*, 1951-1955.

[49] Zhang, H.; Yang, D. R.; Li, D. S.; Ma, X. Y.; Li, S. Z.; Que, D. L. *Cryst. Growth Des.* 2005, *5*, 547-550.

[50] Liu, K. H.; Lin, C. C.; Chen, S. Y. *Cryst. Growth Des.* 2005, *5*, 483-487.

[51] Masuda, Y.; Kato, K. *Cryst. Growth Des.* 2008, *8*, 275-279.

[52] Greene, L. E.; Law, M.; Tan, D. H.; Montano, M.; Goldberger, J.; Somorjai, G.; Yang, P. D. *Nano Lett.* 2005, *5*, 1231-1236.

[53] O'Brien, P.; Saeed, T.; Knowles, J. *J. Mater. Chem.* 1996, *6*, 1135-1139.

[54] Yamabi, S.; Imai, H. *J. Mater. Chem.* 2002, *12*, 3773-3778.

[55] Vayssieres, L.; Keis, K.; Lindquist, S. E.; Hagfeldt, A. *J. Phys. Chem. B* 2001, *105*, 3350-3352.

[56] Vayssieres, L. *Adv. Mater.* 2003, *15*, 464-466.

[57] Saito, N.; Haneda, H.; Sekiguchi, T.; Ohashi, N.; Sakaguchi, I.; Koumoto, K. *Adv. Mater.* 2002, *14*, 418-421.

[58] Saito, N.; Haneda, H.; Seo, W. S.; Koumoto, K. *Langmuir* 2001, *17*, 1461-1469.

[59] Saeed, T.; O'Brien, P. *Thin Solid Films* 1995, *271*, 35-38.

[60] Kakiuchi, K.; Hosono, E.; Kimura, T.; Imai, H.; Fujihara, S. *J. Sol-Gel Sci. Technol.* 2006, *39*, 63-72.

[61] Schwenzer, B.; Gomm, J. R.; Morse, D. E. *Langmuir* 2006, *22*, 9829-9831.

[62] Yu, H. D.; Zhang, Z. P.; Han, M. Y.; Hao, X. T.; Zhu, F. R. *J. Am. Chem. Soc.* 2005, *127*, 2378-2379.

[63] Choi, K. S.; Lichtenegger, H. C.; Stucky, G. D.; McFarland, E. W. *J. Am. Chem. Soc.* 2002, *124*, 12402-12403.

[64] Yin, M.; Gu, Y.; Kuskovsky, I. L.; Andelman, T.; Zhu, Y.; Neumark, G. F.; O'Brien, S. *J. Am. Chem. Soc.* 2004, *126*, 6206-6207.

[65] Feng, X. J.; Feng, L.; Jin, M. H.; Zhai, J.; Jiang, L.; Zhu, D. B. *J. Am. Chem. Soc.* 2004, *126*, 62-63.

[66] Yoshida, T.; Tochimoto, M.; Schlettwein, D.; Wohrle, D.; Sugiura, T.; Minoura, H. *Chem. Mater.* 1999, *11*, 2657-2667.

[67] Lee, J. Y.; Yin, D. H.; Horiuchi, S. *Chem. Mater.* 2005, *17*, 5498-5503.

[68] Kopalko, K.; Godlewski, M.; Domagala, J. Z.; Lusakowska, E.; Minikayev, R.; Paszkowicz, W.; Szczerbakow, A. *Chem. Mater.* 2004, *16*, 1447-1450.

[69] Turgeman, R.; Gershevitz, O.; Deutsch, M.; Ocko, B. M.; Gedanken, A.; Sukenik, C. N. *Chem. Mater.* 2005, *17*, 5048-5056.

[70] Turgeman, R.; Gershevitz, O.; Palchik, O.; Deutsch, M.; Ocko, B. M.; Gedanken, A.; Sukenik, C. N. *Cryst. Growth Des.* 2004, *4*, 169-175.

[71] Mirica, E.; Kowach, G.; Evans, P.; Du, H. *Cryst. Growth Des.* 2004, *4*, 147-156.

[72] Lin, Y. R.; Tseng, Y. K.; Yang, S. S.; Wu, S. T.; Hsu, C. L.; Chang, S. J. *Cryst. Growth Des.* 2005, *5*, 579-583.

[73] Gao, Y. F.; Nagai, M. *Langmuir* 2006, *22*, 3936-3940.

[74] Wu, X. D.; Zheng, L. J.; Wu, D. *Langmuir* 2005, *21*, 2665-2667.

[75] Peterson, R. B.; Fields, C. L.; Gregg, B. A. *Langmuir* 2004, *20*, 5114-5118.

[76] Liu, T. Y.; Liao, H. C.; Lin, C. C.; Hu, S. H.; Chen, S. Y. *Langmuir* 2006, *22*, 5804-5809.

[77] Masuda, Y.; Kato, K. *Cryst. Growth Des.* 2008, *8*, 3213-3218.

[78] Carlson, T.; Giffin, G. L. *J. Phys. Chem.* 1986, *90*, 5896-5900.

[79] Zhang, Z. B.; Wang, C. C.; Zakaria, R.; Ying, J. Y. *J. Phys. Chem. B* 1998, *102*, 10871-10878.

[80] Wang, R.; Hashimoto, K.; Fujishima, A. *Nature* 1997, *388*, 431-432.

[81] Choi, W. Y.; Termin, A.; Hoffmann, M. R. *J. Phys. Chem.* 1994, *98*, 13669-13679.
[82] Sung, Y. M.; Lee, J. K.; Chae, W. S. *Cryst. Growth Des.* 2006, *6*, 805-808.
[83] Kumazawa, N.; Islam, M. R.; Takeuchi, M. *J. Electroanal. Chem.* 1999, *472*, 137-141.
[84] Ferroni, M.; Carotta, M. C.; Guidi, V.; Martinelli, G.; Ronconi, F.; Sacerdoti, M.; Traversa, E. *Sens. Actuators B: Chem.* 2001, *77*, 163-166.
[85] Wagemaker, M.; Kentgens, A. P. M.; Mulder, F. M. *Nature* 2002, *418*, 397-399.
[86] Arico, A. S.; Bruce, P.; Scrosati, B.; Tarascon, J. M.; Van Schalkwijk, W. *Nature Mater.* 2005, *4*, 366-377.
[87] Guo, Y. G.; Hu, Y. S.; Maier, J. *Chem. Commun.* 2006, *26*, 2783-2785.
[88] Tokudome, H.; Yamada, Y.; Sonezaki, S.; Ishikawa, H.; Bekki, M.; Kanehira, K.; Miyauchi, M. *Appl. Phys. Lett.* 2005, *87*, 213901-213903.
[89] Nazeeruddin, M. K.; De Angelis, F.; Fantacci, S.; Selloni, A.; Viscardi, G.; Liska, P.; Ito, S.; Takeru, B.; Gratzel, M. G. *J. Am. Chem. Soc.* 2005, *127*, 16835-16847.
[90] Wang, P.; Zakeeruddin, S. M.; Moser, J. E.; Humphry-Baker, R.; Comte, P.; Aranyos, V.; Hagfeldt, A.; Nazeeruddin, M. K.; Gratzel, M. *Adv. Mater.* 2004, *16*, 1806-1811.
[91] Morrison, P. W.; Raghavan, R.; Timpone, A. J.; Artelt, C. P.; Pratsinis, S. E. *Chem. Mater.* 1997, *9*, 2702-2708.
[92] Yang, G. X.; Zhuang, H. R.; Biswas, P. *Nanostruct. Mater.* 1996, *7*, 675-689.
[93] Yu, J. C.; Yu, J. G.; Ho, W. K.; Zhang, L. Z. *Chem. Commun.* 2001, *19*, 1942-1943.
[94] Huang, W. P.; Tang, X. H.; Wang, Y. Q.; Koltypin, Y.; Gedanken, A. *Chem. Commun.* 2000, *15*, 1415-1416.
[95] Seifried, S.; Winterer, M.; Hahn, H. *Chem. Vap. Deposition* 2000, *6*, 239-244.
[96] Scolan, E.; Sanchez, C. *Chem. Mater.* 1998, *10*, 3217-3223.
[97] Wang, C. C.; Ying, J. Y. *Chem. Mater.* 1999, *11*, 3113-3120.
[98] Burnside, S. D.; Shklover, V.; Barbe, C.; Comte, P.; Arendse, F.; Brooks, K.; Gratzel, M. *Chem. Mater.* 1998, *10*, 2419-2425.
[99] Zhang, H. Z.; Finnegan, M.; Banfield, J. F. *Nano Lett.* 2001, *1*, 81-85.
[100] Sung, Y. M.; Lee, J. K. *Cryst. Growth Des.* 2004, *4*, 737-742.
[101] Perkas, N.; Pol, V.; Pol, S.; Gedanken, A. *Cryst. Growth Des.* 2006, *6*, 293-296.
[102] Deki, S.; Aoi, Y.; Hiroi, O.; Kajinami, A. *Chem. Lett.* 1996, *6*, 433-434.
[103] Deki, S.; Aoi, Y.; Yanagimoto, H.; Ishii, K.; Akamatsu, K.; Mizuhata, M.; Kajinami, A. *J. Mater. Chem.* 1996, *6*, 1879-1882.
[104] Deki, S.; Aoi, Y.; Asaoka, Y.; Kajinami, A.; Mizuhata, M. *J. Mater. Chem.* 1997, *7*, 733-736.
[105] Kishimoto, H.; Takahama, K.; Hashimoto, N.; Aoi, Y.; Deki, S. *J. Mater. Chem.* 1998, *8*, 2019-2024.
[106] Choi, S. Y.; Mamak, M.; Coombs, N.; Chopra, N.; Ozin, G. A. *Adv. Func. Mater.* 2004, *14*, 335-344.
[107] Antonelli, D. M.; Ying, J. Y. *Angew. Chem., Int. Ed.* 1995, *34*, 2014-2017.
[108] Shibata, H.; Ogura, T.; Mukai, T.; Ohkubo, T.; Sakai, H.; Abe, M. *J. Am. Chem. Soc.* 2005, *127*, 16396-16397.
[109] Shibata, H.; Mihara, H.; Mlikai, T.; Ogura, T.; Kohno, H.; Ohkubo, T.; Sakait, H.; Abe, M. *Chem. Mater.* 2006, *18*, 2256-2260.
[110] Tian, B. Z.; Liu, X. Y.; Yang, H. F.; Xie, S. H.; Yu, C. Z.; Tu, B.; Zhao, D. Y. *Adv. Mater.* 2003, *15*, 1370-+.
[111] Ryoo, R.; Joo, S. H.; Kruk, M.; Jaroniec, M. *Adv. Mater.* 2001, *13*, 677-681.

[112] Masuda, Y.; Kato, K. Japan, 2007.
[113] Masuda, Y.; Sugiyama, T.; Seo, W. S.; Koumoto, K. *Chem. Mater.* 2003, *15*, 2469-2476.
[114] Masuda, Y.; Kato, K. *Chem. Mater.* 2008, *20*, 1057-1063.
[115] Furlong, D. N.; Parfitt, G. D. *J. Colloid Interface Sci.* 1978, *65*, 548-554.
[116] Wahi, R. K.; Liu, Y. P.; Falkner, J. C.; Colvin, V. L. *J. Colloid Interface Sci.* 2006, *302*, 530-536.
[117] Kruk, M.; Jaroniec, M. *Chem. Mater.* 2001, *13*, 3169-3183.
[118] Ravikovitch, P. I.; Odomhnaill, S. C.; Neimark, A. V.; Schuth, F.; Unger, K. K. *Langmuir* 1995, *11*, 4765-4772.
[119] Lastoskie, C.; Gubbins, K. E.; Quirke, N. *J. Phys. Chem.* 1993, *97*, 4786-4796.
[120] Katagiri, K.; Ohno, K.; Masuda, Y.; Koumoto, K. *J. Ceram. Soc. Japan* 2007, *115*, 831-834.
[121] Masuda, Y.; Ohji, T.; Kato, K. *Crystal Growth & Design* 2010, *10*, 913-922.
[122] Masuda, Y.; Kato, K. *Thin Solid Films* 2008, *516*, 2547-2552.
[123] Masuda, Y.; Kato, K. *J. Cryst. Growth* 2009, *311*, 512-517.
[124] Ohsaka, T.; Izumi, F.; Fujiki, Y. *J. Raman Spectrosc.* 1978, *7*, 321-324.
[125] Gao, K. *Physica Status Solidi B-Basic Solid State Physics* 2007, *244*, 2597-2604.
[126] Cong, Y.; Zhang, J.; Chen, F.; Anpo, M. *J. Phys. Chem. C* 2007, *111*, 6976-6982.
[127] Ohsaka, T.; Yamaoka, S.; Shimomura *Solid State Communications* 1979, *30*, 345-347.
[128] Yu, J. G.; Yu, H. G.; Cheng, B.; Zhao, X. J.; Yu, J. C.; Ho, W. K. *Journal of Physical Chemistry B* 2003, *107*, 13871-13879.
[129] Ignat'eva, L. N.; Polishchuk, S. A.; Antokhina, T. F.; Buznik, V. T. *Glass Physics and Chemistry* 2004, *30*, 139-141.
[130] Nakamoto, K. *Infrared Spectra of Inorganic and Coordination Compounds*; Wiley: New York, 1970.
[131] Padmanabhan, S. C.; Pillai, S. C.; Colreavy, J.; Balakrishnan, S.; McCormack, D. E.; Perova, T. S.; Gun'ko, Y.; Hinder, S. J.; Kelly, J. M. *Chemistry of Materials* 2007, *19*, 4474-4481.
[132] Wu, J. J.; Yu, C. C. *Journal of Physical Chemistry B* 2004, *108*, 3377-3379.
[133] Perathoner, S.; Passalacqua, R.; Centi, G.; S. Su, D.; Weinberg, G. *Catalysis Today* 2007, *122*, 3-13.
[134] Sato, H.; Ono, K.; Sasaki, T.; Yamagishi, A. *Journal of Physical Chemistry B* 2003, *107*, 9824-9828.
[135] Naik, V. M.; Haddad, D.; Naik, R.; Benci, J.; Auner, G. W. In *Symposium on Solid-State Chemistry of Inorganic Materials IV held at the 2002 MRS Fall Meeting*; AlarioFranco, M. A., Greenblatt, M., Rohrer, G., Whittingham, M. S., Eds.; Materials Research Society: Boston, Ma, 2002, p 413-418.
[136] Masuda, Y.; Sugiyama, T.; Lin, H.; Seo, W. S.; Koumoto, K. *Thin Solid Films* 2001, *382*, 153-157.
[137] Huang, D.; Xiao, Z.-D.; Gu, J.-H.; Huang, N.-P.; Yuan, C.-W. *Thin Solid Films* 1997, *305*, 110-115.
[138] Zhang, F.; Mao, Y.; Zheng, Z.; Chen, Y.; Liu, X.; Jin, S. *Thin Solid Films* 1997, *310*, 29-33.
[139] Li, D.; Haneda, H.; Hishita, S.; Ohashi, N.; Labhsetwar, N. K. *Journal of Fluorine Chemistry* 2005, *126*, 69-77.
[140] Wang, C. M.; Mallouk, T. E. *Journal of Physical Chemistry* 1990, *94*, 423-428.
[141] Minero, C.; Mariella, G.; Maurino, V.; Pelizzetti, E. *Langmuir* 2000, *16*, 2632-2641.

[142] Minero, C.; Mariella, G.; Maurino, V.; Vione, D.; Pelizzetti, E. *Langmuir* 2000, *16*, 8964-8972.
[143] Vohra, M. S.; Kim, S.; Choi, W. In *International Symposium on Photochemistry at Interfaces*; Elsevier Science Sa: Sapporo, Japan, 2002, p 55-60.
[144] Fu, H. B.; Zhang, L. W.; Zhang, S. C.; Zhu, Y. F.; Zhao, J. C. *Journal of Physical Chemistry B* 2006, *110*, 3061-3065.
[145] Irie, H.; Watanabe, Y.; Hashimoto, K. *Journal of Physical Chemistry B* 2003, *107*, 5483-5486.
[146] Masuda, Y.; Kato, K. *J. Cryst. Growth* 2009, *311*, 593-596.
[147] Ginley, D. S.; Bright, C. *MRS Bull.* 2000, *25*, 15-21.
[148] Idota, Y.; Kubota, T.; Matsufuji, A.; Maekawa, Y.; Miyasaka, T. *Science* 1997, *276*, 1395 - 1397.
[149] Zhang, Y. L.; Liu, Y.; Liu, M. L. *Chem. Mater.* 2006, *18*, 4643-4646.
[150] Wang, Y.; Lee, J. Y.; Zeng, H. C. *Chem. Mater.* 2005, *17*, 3899-3903.
[151] Bose, A. C.; Kalpana, D.; Thangadurai, P.; Ramasamy, S. *Journal of Power Sources* 2002, *107*, 138-141.
[152] Pianaro, S. A.; Bueno, P. R.; Longo, E.; Varela, J., A. *J. Mat. Sci. Lett.* 1995, *14*, 692-694.
[153] Bueno, P. R.; de Cassia-Santos, M. R.; Leite, E. R.; Longo, E.; Bisquert, J.; Garcia-Belmonte, G.; Fabregat-Santiago, F. *Journal of Applied Physics* 2000, *88*, 6545-6548.
[154] Tagawa, T.; Kataoka, S.; Hattori, T.; Murakami, Y. *Applied Catalysis* 1982, *4*, 1-4.
[155] Park, P. W.; Kung, H. H.; Kim, D. W.; Kung, M. C. *Journal of Catalysis* 1999, *184*, 440-454.
[156] Chopra, K. L.; Major, S.; Pandya, D. K. *Thin Solid Films* 1983, *102*, 1-46.
[157] Sergent, N.; Gelin, P.; Perier-Camby, L.; Praliaud, H.; Thomas, G. *Sensors and Actuators B-Chemical* 2002, *84*, 176-188.
[158] Baik, N. S.; Sakai, G.; Miura, N.; Yamazoe, N. *J. Am. Ceram. Soc.* 2000, *83*, 2983-2987.
[159] Ristic, M.; Ivanda, M.; Popovic, S.; Music, S. *Journal of Non-Crystalline Solids* 2002, *303*, 270-280.
[160] Broussous, L.; Santilli, C. V.; Pulcinelli, S. H.; Craievich, A. F. *Journal of Physical Chemistry B* 2002, *106*, 2855-2860.
[161] Zhang, J. R.; Gao, L. *Journal of Solid State Chemistry* 2004, *177*, 1425-1430.
[162] Deng, Z. X.; Wang, C.; Li, Y. D. *Journal of The American Ceramic Society* 2002, *85*, 2837-2839.
[163] Leite, E. R.; Gomes, J. W.; Oliveira, M. M.; Lee, E. J. H.; Longo, E.; Varela, J. A.; Paskocimas, C. A.; Boschi, T. M.; Lanciotti, F.; Pizani, P. S.; Soares, P. C. *Journal of Nanoscience and Nanotechnology* 2002, *2*, 125-128.
[164] Leite, E. R.; Maciel, A. P.; Weber, I. T.; Lisboa, P. N.; Longo, E.; Paiva-Santos, C. O.; Andrade, A. V. C.; Pakoscimas, C. A.; Maniette, Y.; Schreiner, W. H. *Advanced Materials* 2002, *14*, 905-908.
[165] Dai, Z. R.; Gole, J. L.; Stout, J. D.; Wang, Z. L. *J. Phys. Chem. B* 2002, *106*, 1274 -1279.
[166] Liu, Y.; Liu, M. L. *Adv. Funct. Mater.* 2005, *15*, 57 - 62.
[167] Huang, J.; Matsunaga, N.; Shimanoe, K.; Yamazoe, N.; Kunitake, T. *Chem. Mater.* 2005, *17*, 3513-3518.
[168] Liu, Y. K.; Zheng, C. L.; Wang, W. Z.; Yin, C. R.; Wang, G. H. *Adv. Mater.* 2001, *13*, 1883 - 1887.

[169] Sun, J. Q.; Wang, J. S.; Wu, X. C.; Zhang, G. S.; Wei, J. Y.; Zhang, S. Q.; Li, H.; Chen, D. R. *Cryst. Growth Des.* 2006, *6*, 1584-1587.
[170] Yang, R. S.; Wang, Z. L. *J. Am. Chem. Soc.* 2006, *128*, 1466-1467.
[171] Duan, J. H.; Yang, S. G.; Liu, H. W.; Gong, J. F.; Huang, H. B.; Zhao, X. N.; Zhang, R.; Du, Y. W. *J. Am. Chem. Soc.* 2005, *127*, 6180 -6181.
[172] Ohgi, H.; Maeda, T.; Hosono, E.; Fujihara, S.; Imai, H. *Cryst. Growth Des.* 2005, *5*, 1079-1083.
[173] Uchiyama, H.; Imai, H. *Cryst. Growth Des.* 2007, *7*, 841-843.
[174] Niesz, D. E.; Bennett, R. B.; Snyder, M. J. *Am. Ceram. Soc. Bull.* 1972, *51*, 677-680.
[175] Roosen, A.; Hausener, H. *Adv. Ceram. Mater.* 1988, *3*, 131.
[176] Vasylkiv, O.; Sakka, Y. *Journal of The American Ceramic Society* 2001, *84*, 2489-2494.
[177] Niesen, T. P.; DeGuire, M. R. *J. Electroceramics* 2001, *6*, 169-207.
[178] Baes, C. F.; Mesiner, R. E. *The Hydrolysis of Cations*; John Wiley & Sons, Inc., Wiley-Interscience: New York, 1976.
[179] Ararat Ibarguena, C.; Mosqueraa, A.; Parrab, R.; Castrob, M. S.; Rodríguez-Páeza, J. E. *Mater. Chem. Phys.* 2007, *101*, 433-440.
[180] Masuda, Y.; Ohji, T.; Kato, K. *Journal of The American Ceramic Society* 2010, *93*, 2140-2143.
[181] Masuda, Y.; Kato, K. *Journal of Crystal Growth* 2009, *311*, 593-596.
[182] Masuda, Y.; Kato, K. *Polymers for Advanced Technologies* 2010, *21*, 211-215.
[183] Shukla, S.; Seal, S.; Ludwig, L.; Parish, C. *Sens. Actuators B: Chem.* 2004, *97*, 256-265.
[184] Masuda, Y.; Kato, K.; Sonezaki, S.; Ajimi, M.; Bekki, M. Japan, 2008, p P 2008-227389.
[185] Xiang, J. H.; Masuda, Y.; Koumoto, K. *Adv. Mater.* 2004, *16*, 1461-1464.
[186] Xiang, J. H.; Zhu, P. X.; Masuda, Y.; Koumoto, K. *Langmuir* 2004, *20*, 3278-3283.
[187] Masuda, Y.; Yamagishi, M.; Koumoto, K. *Chem. Mater.* 2007, *19*, 1002-1008.
[188] Masuda, Y.; Wakamatsu, S.; Koumoto, K. *J. Euro. Ceram. Soc.* 2004, *24*, 301-307.
[189] Themlin, J. M.; Chtaib, M.; Henrard, L.; Lambin, P.; Darville, J.; Gilles, J. M. *Phys. Rev. B* 1992 *46*, 2460-2466.
[190] Yan, L.; Pan, J. S.; Ong, C. K. *Mater. Sci. Eng. B* 2006, *128*, 34-36.
[191] Wagner, C. D. *Practical Surface Analysis*; 2 ed.; John Wiley, 1990; Vol. 1.

2

Size- and Shape-Controlled Synthesis of Monodisperse Metal Oxide and Mixed Oxide Nanocrystals

Thanh-Dinh Nguyen and Trong-On Do
Department of Chemical Engineering, Laval University, Quebec
Canada

1. Introduction

A nanocrystal or nanoparticle (not fully crystalline) is defined as a particle with size in range of 1 to 100 nm (10^2 to 10^7 atoms) from zero (0D) to three dimensions (3D), which exhibits the unique physiochemical properties due to the quantum size effect that cannot be anticipated from bulk counterparts. Strictly speaking, the name of "nanocrystal" is only used for crystalline nanoparticle, and is however a more general term which can refer to both crystalline and non-crystalline nanoparticles. Accordingly, their particle size is intermediate between the size of molecule and bulk solid (Rao, Müller and Cheetham 2005, Sorensen 2009). Nanocrystals can be formed in a variety of shapes including dot, sphere, cube, rod, triangle, hexagon and many others. In this size range, they possess an immense surface area per unit volume, a very large percentage of atoms in the surface. As a result, their unexpected properties can be obtained as compared to those of both individual atoms/molecules and bulk counterpart of the same chemical composition.

Size- and shape-dependent properties of the nanocrystals can be tuned by changing the dimension and designing the shape (Rao et al. 2005). Due to the materials at the nanoscale, low coordination number, surfaced edge and corner atoms are usually chemically reactive, catalytically active and polarisable surface, contributing to their high chemical potential. For example, the high surface area is of particular importance regarding heterogeneous catalytic reactions, because of the increase of interaction of reactive molecules and active sites on the catalyst surface (Abbet and Heiz 2005). Furthermore, the particle size not only affects their surface area, but also arise new properties, due to the quantum-size effect (e.g., electron confinement and surface effect) (Kroes et al. 2002, Kamat et al. 2010).

Considerable efforts have recently been devoted to the preparation of metal oxide and mixed oxide nanomaterials due to both their unique properties and their technological applications (Seshadri 2005, Burda et al. 2005, Mao et al. 2007, Yin and Alivisatos 2005). Metal oxides including the transition metals and rare earths, display a wide variety of complex structures and interesting electronic and magnetic properties associated with the changes in electronic structure and bonding and in the presence of ordered defect complexes or extended defects. The nanostructured mixed oxides can greatly generate new synergetic properties and improve the overall application performance, that is not available from single metal oxide species, due to the appropriate combination of individual oxide

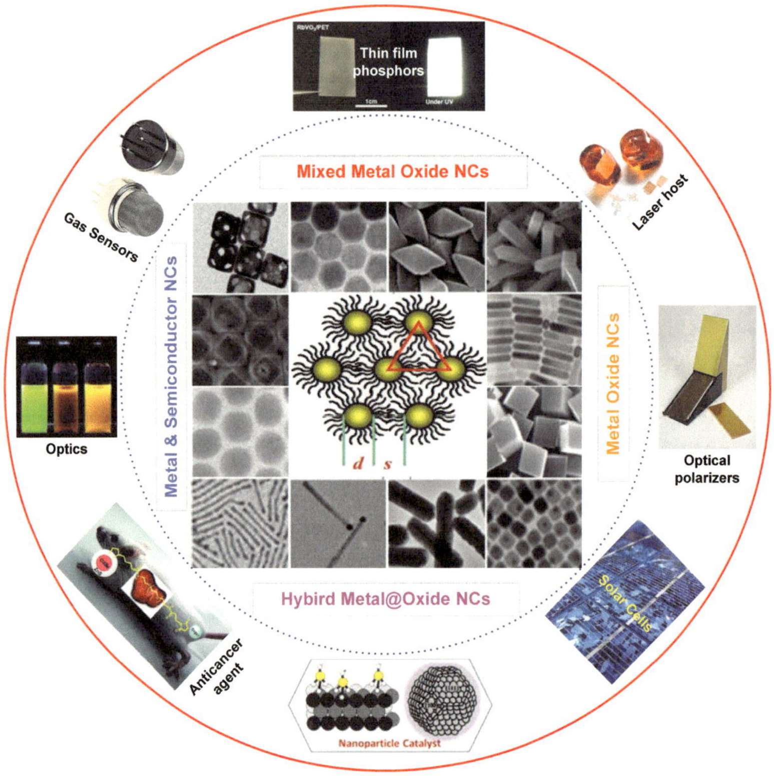

Fig. 1. Representative shapes of inorganic nanocrystals developed to date. These kinds of nanocrystals utilized as key building blocks for the fabrication of novel nano-systems for catalytic, electronic, and biomedical technologies.

components (Redl et al. 2003). Furthermore, the size- and shape-dependent properties of nanomaterials raising expectations for a better performance generally are a consequence of quantum confinement within the particle (Alivisatos 1996). The precise controlled synthesis of the size, shape, chemical composition, crystal structure, and surface chemistry of nanomaterials allows to obtain their unique properties, which have become one of the most challenging issues faced by nanomaterial researchers. The properties of nanocrystals are drastically altered in the shape and size change, making nanocrystals as ideal candidates for many applications, such as in catalysis, energy storages, optoelectronics, sensing, magnetic resonance imaging, biomedicine. (Jun, Choi and Cheon 2006b, Hao et al. 2010, Kinge, Crego-Calama and Reinhoudt 2008, Ying 2000, Na, Song and Hyeon 2009, Jun, Choi and Cheon 2006a). A general picture in Figure 1 is schematically illustrated for the features and new phenomena of the nanoscale materials.

The size- and shape-dependent properties of colloidal metal oxide and mixed oxide structures at the nanoscale makes great demands to the synthetic methodology. Therefore, it is a great challenge to develop a "synthetic chemistry" of nanocrystals that is as precise as that used to make building units. This allows scientists to study the effect of these synthesis parameters which impart to the unique collective properties of monodisperse nanocrystals. There are two different approaches to synthesize nanocrystals: the"top-down" approach, which utilizes physical methods, and the "bottom-up" approach, which employs solution-

phase colloidal chemistry. Using the top-down approach, the production of a large quantity of nanocrystals can be achieved, however, uniform-sized nanocrystals and their size control is very difficult to obtain. In contrast, using the "bottom-up" approach, the colloidal chemical methods can be used to synthesize uniform nanocrystals with controlled particle size through chemical nucleation and growth process in bulk solution, although generally only subgram quantities are produced (Schmid 2005). The use of atoms or molecules as building blocks has the advantage, that provides control over crystallite size and shape with a precision well beyond that of top-down lithography (Park et al. 2007). Indeed, bottom-up assembly of well-defined nanoscale building blocks into nanocrystals with controlled size and shape represents a powerful tool to fabricate novel multi-component materials and devices (Nagarajan 2008). In order to control the crystal growth, the capping agents are often used to decrease the surface energies of crystals. The surfactants bind selectively to the different crystallographic faces, so that shape of nanocrytals can be controlled by the nonselective or selective surfactants (2008, Cushing, Kolesnichenko and O'Connor 2004). The use of surfactant molecules, and consequently, result in oxide nanocrystals comprising an inorganic core coated with a layer of organic surfactant molecules. This organic capping provides electronic and chemical passivation of the surface dangling bonds, prevents uncontrolled growth and agglomeration of the nanoparticles, and permits chemical manipulations of the nanoparticles similarly to large molecules having their solubility and reactivity determined by the nature of the surface ligands. The most commonly used ones in colloidal syntheses include alkyl- thiols, long chain amines, carboxylic and phosphonic acids, phosphine oxides, phosphine, phosphates, phosphonates, and various coordinating (e.g., ethers, THF, DMF) or non-coordinating solvents (e.g., alkanes, alkenes).
Understanding growth behavior and morphology evolution is crucial for an efficient synthesis and a good control of inorganic nanocrystals. In the bottom-up syntheses, for the growth process of nuclei, the behavior was described by the classical Ostwald ripening mechanism, in which the growth of larger particles at the expense of smaller ones driven by surface energy reduction. This phenomenon was extensively used to explain the formation of thermodynamically stable nanocrystals with nearly spherical morphologies. For the controlled self-assembly of nanoparticles into well-defined anisotropic nanostructures, organic capping reagents usually play critical roles in reducing the activity of the nanocrystal surface to promote or tune the ordered self-assembly (Kinge et al. 2008, Sellinger et al. 1998, Malenfant Patrick et al. 2007, Yin and Alivisatos 2005). An oriented attachment mechanism could offer as an additional tool to design advanced materials with anisotropic properties and could be used for the synthesis of more complex crystalline one-dimensional structures. In addition, the sterically diffusive kinetics and selective binding or nonbinding of surfactant molecules to different faces of the growing nanocrystal can also control the product's morphology due to the possibility of breaking the limitations of crystal growth dynamically. In some cases, the formation of the intrinsic anisotropic nanocrystals is found to be a highly kinetics-driven process, which occurs far away from the thermodynamic equilibrium, and must be overdriven by high precursor monomer concentrations.
The reaction medium is crucial in the solution-based approaches. To date, noble metal nanocrystals obtained from solution-based methods that mainly used the organic reagent as solvent medium such as toluene, diphenyl ether, oleic acid/oleylamine, etc. In organic solvent systems, expensive organometallic precursors, toxic and environmentally unfriendly organic solvents are often not compatible with biomedical applications. Water, as an

environmentally friendly solvent with the most abundant resource, and most metal nitrates and chloride salts were used as starting materials, can overcome these barriers. Further, due to the high solubility of metal salt precursors in aqueous media, the aqueous-based routes can be used for the synthesis of pure products in high yield. Therefore, the development of general synthetic strategies to produce the size- and shape-controlled metal oxide, mixed oxide nanocrystals in terms of low cost, environmentally benign reagents, mild synthesis conditions, and potential for large-scale production are needed among the important research topics of the advanced materials chemistry.

In the present chapter, we provide a brief account of our own recent results to synthesize different types of monodisperse colloidal metal oxide and mixed oxide nanocrystals, focusing on one-phase and two-phase solvo-hydrothermal surfactant-assisted approaches. Based on our approaches, a variety of metal oxide and mixed oxide nanocrystals with different sizes, shapes, and phases are obtained using simple chemical reactions (e.g., solvo-hydrothermal reactions), choosing appropriate reaction systems (precursor, surfactant.), and controlling reaction parameters (monomer concentration, temperature and time). The chapter is organized as follows: Introduction in Section 1; Essential concepts in the nucleation and crystal growth process of size- and shape-controlled nanocrystals in Section 2; subsequently, we present the main results in our laboratory along with the results from other research groups related to metal oxide nanocrystals in Section 3, followed by mixed metal oxide nanocrystals in Section 4. The synthetic procedures, the formation mechanisms, and the controlled growth of nuclei based on kinetic and thermodynamic conditions as well as the selection of capping agents will also be discussed to control the size and shape, and conclusion in Section 5.

2. General strategies for surfactant-assisted synthesis of colloidal metal oxide and mixed oxide nanocrystals

In a typical synthesis of inorganic nanocrystals, the precursor compound in bulk solution is decomposed to generate atoms followed by the precipitation starting from dissolved atoms as building blocks to form the nanocrystals. A understanding of the process and parameters controlling the precipitation helps to improve the engineering of the growth of nanocrystals to the desired size and shape. The precipitation process then basically consists of a nucleation step followed by crystal growth stages. Generally, there are three kinds of nucleation processes: homogeneous nucleation, heterogeneous nucleation, and secondary nucleation. For the chemical colloidal nanocrystal synthesis, homogeneous nucleation occurs in the absence of a solid interface by combining solute molecules to produce nuclei. Homogeneous nucleation occurs due to the driving force of the thermodynamics because the supersaturated solution is not stable in energy. Seed formation proceeds according to the LaMer model are shown in Figure 2. This mechanism reported in the early 50's on the basis of the crystallization study of the solution-phase synthesis of monodisperse sulfur colloids in ethanol (LaMer and Dinegar 1950). According to LaMer plot for the crystal nucleation process, in which the concentration of atoms steadily increases with time as the precursor is decomposed by heating, colloidal nanocrystal formation comprises the following three steps: (i) The atoms start to aggregate into nuclei via self-nucleation as increasing the monomer concentration in the solution to supersaturation levels; (ii) Then monomers continuously aggregate on the pre-existing nuclei or seed which leads to gradual decrease in the monomer concentration. As long as the concentration of reactants is kept below the critical level, further nucleation is discouraged; (iii) With a continuous supply of atoms via ongoing precursor decomposition, the nuclei will grow into nanocrystals of increasingly larger size until an equilibrium state is reached between the atoms on the surface of the nanocrystal and the atoms in the solution (Watzky and Finke 1997).

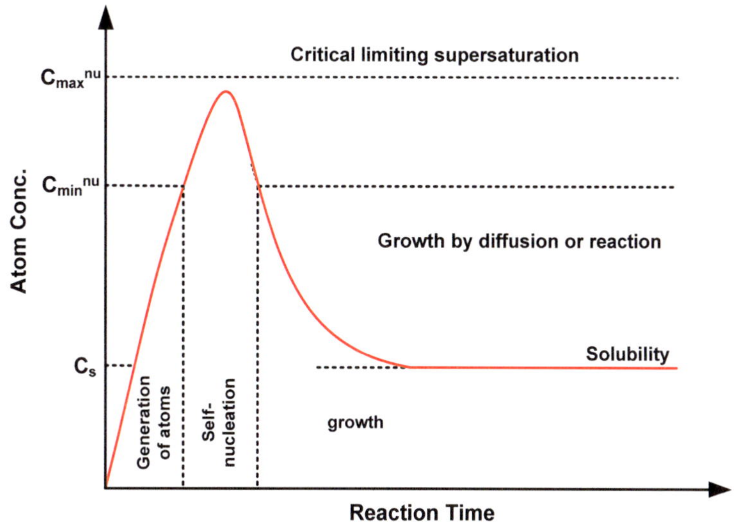

Fig. 2. Plot of La Mer model for the generation of atoms, nucleation, and subsequent growth of colloidal synthesis (LaMer and Dinegar 1950)

After the formation of nuclei, the subsequent growth stages also strongly govern the final morphology of the nanocrystals. Generally, the nanocrystal growth can occur under two different regimes, either in a thermodynamically controlled or kinetically controlled growth regime. The manipulation between thermodynamic and kinetic growth regimes is thus a critical factor in determining nanoparticle shape (Kwon and Hyeon 2009) The final nanoparticle morphology can be controlled by dictating the shape of nuclei and directing the growth of the nuclei and/or nanocrystals. Nuclei can take on a variety of shapes determined by the chemical potentials of the different crystallographic faces, which are in turn highly dependent on the reaction environment such as temperature and solute concentration. The nuclei shape can have a strong effect on the final nanocrystal shape, for example, through selected growth of high-energy crystal faces of the nuclei (Skrabalak and Xia 2009, Pileni 2007, Searcy 1983). In the present of surfactant in bulk solution, the products are capped by surfactant molecules, resulting in the restriction of the particle growth as well as the good dispersibility of the product in reaction solvent. This is particularly important for shape control: to obtain a highly shape-monodisperse yield of nanocrystals, nucleation must occur rapidly and instantaneously.

The surfactant-assisted synthetic methods provide convenient and powerful pathway for the reproducible controlled synthesis of nanocrystals because these methods allow for the metal oxide and mixed oxide nanocrystals to be precisely adjusted in terms of their size, shape, composition, and phase structure on the nanometer scale. Nanocrystals obtained by the surfactant-assisted route, in general, exhibit excellent crystallinity and monodispersity. In the following section, we develop the simple one-phase and two-phase surfactant-assisted routes for the shape-and size-controlled synthesis of colloidal monodisperse metal oxide and mixed metal oxide nanocrystals.

Critical reaction parameters that have strong effect on the growth of the metal oxide and mixed oxide nanocrystals including precursor and surfactant concentration, the molar ratio of precursor to capping ligand, and reaction temperature and time, are precisely adjusted to control over their sizes and shapes in the crystal nucleation-growth stages. In addition, to understand the formation process of nanocrystals in the bulk solution, we discuss the possible mechanism for the shape and size control of the nanocrystals obtained from our surfactant-assisted approaches. In all cases, there are some advantages for these routes because of the use of inorganic salt precursors, instead of expensive metal alkoxides, and quite mild synthetic conditions. Particularly, these synthesis methods are scalable to multigram in a single run using the same synthetic conditions.

3. Metal oxide nanocrystals

Within the broad family of functional materials, metal oxides play a very important role in many scientific and technological areas (Lu, Chang and Fan 2006) For decades they have been extensively investigated their physiochemical properties and useful applications by solid-state chemists. Metal oxides including the transition metals and rare earths are able to form a large diversity of oxide compounds, giving the inspiration for designing new materials. The crystal structures ranging from simple rock salt to complex oxide are often built by the metal-oxygen bonds varying nearly ionic to covalent or metallic. The oxidic materials exhibit fascinating electronic and magnetic properties associating with the changes in electronic structure and bonding (Gariglio, Gabay and Triscone 2010). Additionally, metal oxides having multivalent oxidation states have attracted much attention among specialists because they often exhibit superior catalytic reaction performance (Antonini and et al. 1987) Many progresses, such as hot injection, co-precipitation, microemulsion, nonhydrolytic sol-gel process, and so on, have been devoted to fabricate metal oxide nanocrystals.

Vanadium oxides (V_2O_{5-x}) are of interest due to their versatile redox activity and layered structures (Shah et al. 2008) They are a key technological material widely used in various fields such as chemical sensing (Livage 1991), actuators (Gu et al. 2003), high-energy lithium batteries (Poizot et al. 2000), and electric field-effect transistors (Muster et al. 2000) As a target for the shape-controlled NC synthesis, recent studies in vanadium oxide NCs have focused on the development of synthetic approaches toward nanotubes, nanobelts, nanofibers, nanowires, nanorods, and so on, as well as their shape-dependent properties. Recently, our group demonstrated a simple modified solvothermal method for the multigram scale synthesis of uniform vanadium oxide NCs using vanadium(V) diperoxo alkylammonium complexes in toluene or toluene/water medium in the presence of aliphatic amines as capping agent (Nguyen and Do 2009b) The V(V) diperoxo tetraoctylammonium complexes, $VO(O_2)_2(TOA)$, were prepared from the two-phase system of V(V) diperoxo aqueous solution and toluene containing tetraoctylammoniumligands (TOA^+) (Figure 3A). Under the solvothermal treatments, $VO(O_2)_2(TOA)$ complex precursors are decomposed and generate vanadium monomers and then grew into vanadium nanocrystals. They are capped by oleylamine molecules and easily dispersed in organic medium. The XRD results revealed that the as-made vanadium oxide nanocrystal samples corresponds to monoclinic rutile-type VO_2 structure. However, the XRD pattern of the calcined sample exhibits the orthorhombic V_2O_{5-x} structure. Furthermore, their color changed from blue-black to yellow after calcination. This indicates the transformation from the monoclinic rutile phase to the orthorhombic phase of these samples.

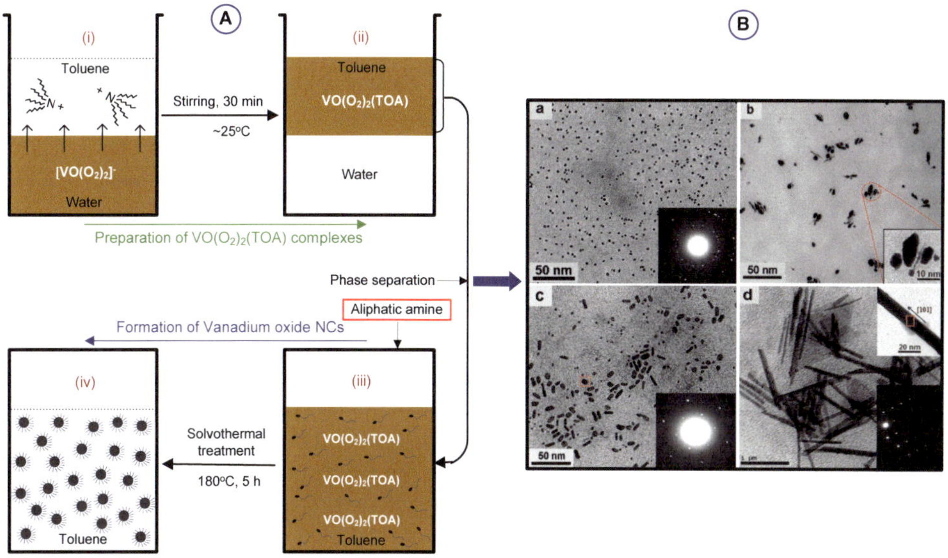

Fig. 3. (A) Schematic illustration of the preparation of vanadium(V) diperoxo tetraoctylammonium ($VO(O)_2$(TOA)) complexes, followed by the formation of alkyl amines-capped vanadium oxide nanocrystals; (B) Effect of water content in the synthesis mixture on the shape transformation of nanospheres into nanorods. TEM images and corresponding SAED patterns of the as-made VO_2 nanocrystals synthesized from V(V) diperoxo tetraoctylammonium complexes (0.04 mol/L) in oleylamine (5 mL) at 180 °C for 5 h with various water/toluene solvent ratios in volume (W/T): a) W/T = 0:40, b) W/T = 2:40, c) W/T = 8:40, and d) W/T = 20:40 (Nguyen and Do 2009b)

The monodisperse vanadium oxide NCs with different sizes and shapes including nanospheres, nanocubes, nanorices, and nanorods can be achieved by the control of various reaction parameters, such as types of V(V) diperoxo alkylammonium complexes, and alkyl chain length of capping agents in synthesis mixture. Remarkably, a significant effect of water content on the size and shape of vanadium oxide NCs has been observed. Figure 3B shows representative TEM images of these samples synthesized at different water/toluene volume ratio (W/T) varing from 0:40 to 2:40, 8:40, and 20:40. Only uniform quasispherical NCs with an average size of 4 nm were obtained in the absence of water in the synthesis medium. While the W/T value was as low as 2:40, aggregated nanoparticles beside some un-uniformed small nanorods were formed. When the W/T value increased to 8:40, mostly short vanadium oxide nanorods were generated. However, when the W/T value increased to 20:40, the vanadium product is composed of uniformly sized and shaped rods with 20 nm in width and 150-300 nm in length. The SAED pattern taken from a single vanadium oxide nanorod reveals the single crystal nature of the nanorod, and it further confirmed that the nanorods' elongation axis was along the [101] direction. The shape elongation of vanadium oxide nanocrystals in the increase of water could be explained by a lateral aggregation of individual nanorods along the longitudinal axis and further their fusion to form aligned nanorods at the high water content in the synthesis mixture.

Some research groups reported the preparation of size-turnable VO_x nanotubes via the aging and hydrothermal process of various vanadium sources including bulk V_2O_5 powders, vanadium(V) peroxo gels, and vanadium(V) triisopropoxides using aliphatic amines as structure-directing templates (Tenne 2004, Spahr et al. 1998, Corr et al. 2008) Schlogl et al. (Pinna et al. 2003) presented a reverse micelle technique to prepare V_2O_5 nanorods and nanowires from a colloidal self-assembly made of sodium bis(ethyl-2-hexyl)sulfosuccinate Na(AOT)/isooctane/H_2O. Park et al. (Guiton et al. 2004) synthesized single-crystalline VO_2 nanowires with rectangular cross sections using a vapor transport method. Zhang et al. (Li et al. 2007) showed that the belt-, olive-, petal-shaped VO_2 NCs could be synthesized with high concentrations of the reducing oxalic acid agent through the hydrothermal route. Baughman et al. (Gu et al. 2003) reported the synthesis of V_2O_5 nanofibers at room temperature from ammonium metavanadate and acidic ion-exchange resin in water. The resulting V_2O_5 nanofibers could deliver dramatically higher specific discharge capacities than micrometer-sized V_2O_5 fibers. The Whittingham's group has also developed a method to produce vanadium oxide nanofibers with dimensions of less than 140 nm by coating oxides on polylactide fibers (Lutta et al. 2005).

Erbium-compound nanomaterials consisting of hexagonal $Er(OH)_3$, monoclinic ErOOH, or cubic Er_2O_3 are particularly attractive among the rare earth oxides due to their remarkably electrical and optical properties. These unique properties orginate from the intra Er^{3+} $4f$ shell transition from its first excited state ($^4I_{3/2}$) to the ground state ($^4I_{5/2}$) is related to the emission band of around 1.54 μm, which is one of the standard telecommunication wavelengths. As a consequence, this minimum absorption has become ideal candidates for use in lasers and optical amplifiers for sensing applications. However, little work has been reported concerning 3D erbium compound materials with controllable size, from micro- to nanostructures, and shapes such as spheres, wrinkle-surfaced spheres, and flowers. Recently, we successfully synthesized the erbium-compound micro- and nanostructures consisting of $Er(OH)_3$, ErOOH, and Er_2O_3 from the reaction of erbium nitrate in basic solution containing ethanol/decanoic acid via ligand-assisted hydrothermal route (Figure 4A) (Nguyen, Dinh and Do 2010). The reactions take place in a water/ethanol solution, a "one polar phase" system at the relatively low temperature of ≤180 °C. The capping products were precipitated at the bottom of a Teflon cup instead of becoming dispersed in the toluene phase as described in the previous two-phase methods. A central feature of this work is the generation of products in the morphology, composition, and phase structure control, which is simple, economical, versatile, and using water as an environmentally benign solvent. As shown Figure 4B, by only tuning the temperature in the reaction system, monoclinic ErOOH and cubic Er_2O_3 phases can be obtained. Furthermore, various particle sizes in the range of thousands to tens of nanometers and a variety of shapes can be achieved simply by varying the synthetic conditions including the concentration of decanoic acid and erbium precursor and the amount of water. The crystalline phase- and particle size-dependent luminescence results indicated that the luminescence properties depend not only on the crystalline phase but also on the particle size of products. The luminescence intensity increases with a decrease of particle size.

Li et al. (Wang et al. 2005b) reported a general hydrothermal method for the synthesis of a variety of nanocrystals by a liquid-solid-solution reaction. The system consists of metal salt, sodium linoleate (solid), ethanol-linoleic acid liquid phase and water-ethanol solution at different reaction temperatures under hydrothermal conditions. TiO_2 nanoparticles and nanorods can be also obtained by solvothermal reaction of titanium butoxide, linoleic acid, triethylamine, and cyclohexane (Li et al. 2006). The decomposition of NH_4HCO_3 which provide H_2O for the hydrolyzation reaction is found to be an important factor to shape evolution of particles. In the presence of NH_4HCO_3, the fast hydrolyzation of precursors with the water leads to the formation of nanoparticles. In the absence of NH_4HCO_3, in contrast, the slow nonhydrolytic condensation of precursors produces titania nanorods with uniform diameters of 3.3 nm, and a length of up to 25 nm. TiO_2 nanowires could be also produced by solvothermal treatment of a mixture containing titanium tetra-isopropoxide, ethylenediamine, and ethylene glycol (Xie and Shang 2007). The diameter of nanowires was controlled by changing the amount of ethylenediamine.

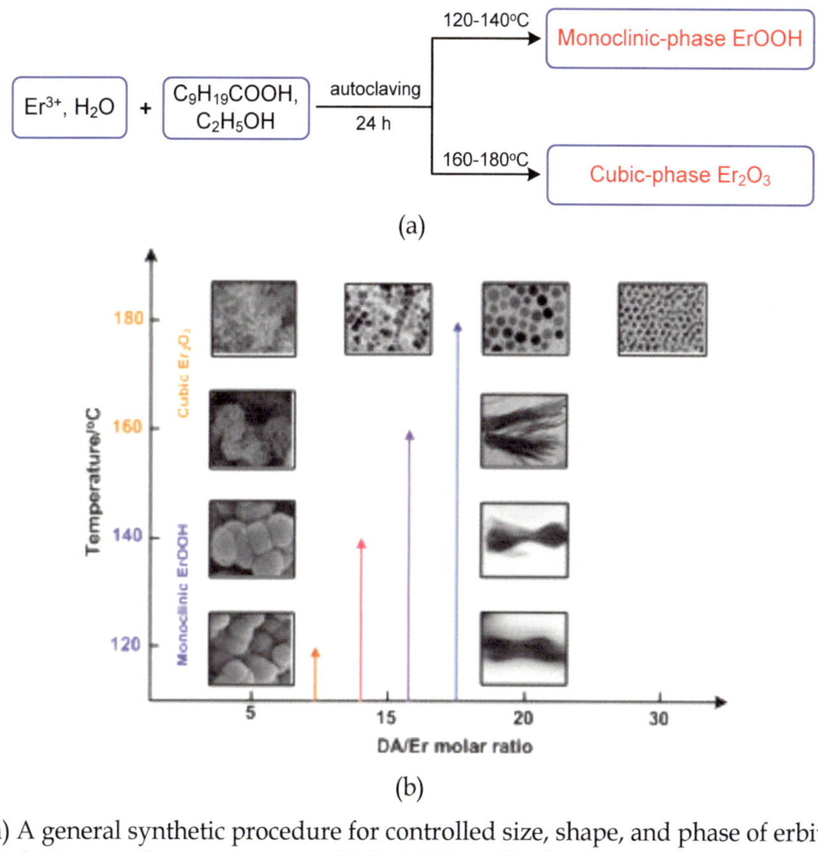

Fig. 4. (a) A general synthetic procedure for controlled size, shape, and phase of erbium-compound micro- and nanostructures; (b) ErOOH and Er_2O_3 micro- and nanostructures with different sizes, shapes, and phases obtained as a function of reaction temperature and decanoic acid/erbium molar ratio (Nguyen et al. 2010)

Our group recently also synthesized the TiO_2 nanocrystals with well-controlled shapes on the basis of solvothermal technique using both acid oleic acid (OA) and oleyamine (OM) as two capping surfactants, and water vapor as hydrolysis agent (Dinh et al. 2009). It is demonstrated that, the presence of water vapor along with the desired OA:OM molar ratio plays crucial roles in controlling size and shape of TiO_2 nanocrystals. In particular, the shape of TiO_2 changed from rhombic to truncated rhombic and to sphere as the OA:OM ratio increased from 4:6 to 5:5 and to 6:4, respectively. Increasing the amount of titanium butoxide

(TB) led to the formation of elongated particles. For example, when the TB:OA:OM molar ratio changed from 1:6:4 to 2:6:4, the shape of TiO_2 evolved from spheres to dog bone-like particles with uniform size. The solvothermal reaction of $Mn(NO_3)_2$/oleylamine/dodecanol recently flourished by Li et al. (Li et al. 2010b) was a successful way for shape control of highly monodisperse Mn_3O_4 nanocrystals with dot, rod, wire shapes. Moreover, the as-prepared hydrophobic spherical or elongated nanoparticles were used as building blocks to be rationally assembled into three-dimensional (3D) Mn_3O_4 colloidal spheres with a facile ultrasonication strategy. The as-prepared colloidal spheres were chemically converted to $LiMn_2O_4$ nanomaterials in a simple solid-state reaction. Such materials showed distinct electrochemical performance, mainly depending on their crystallinity and particle size.

The Niederberger's group developed a nonaqueous sol-gel approach for the synthesis a variety of metal oxide nanocrystals involving the solvothermal treatment of metallic alkoxide precursors in benzyl alcohol solvent (Garnweitner and Niederberger 2008). As the presence of halide impurities in the final oxides obtained by these routes may be a drawback, alternative halide-free methods have been developed as well.

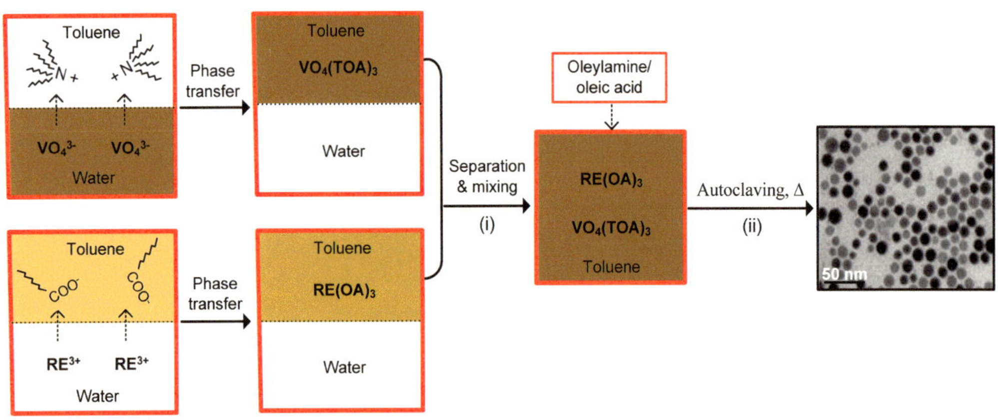

Fig. 7. Schematic illustration for the formation of the $RE(OA)_3$ and $VO_4(TOA)_3$ complexes (i) and the $REVO_4$ nanocrystals (ii) (Nguyen et al. 2009a, Nguyen et al. 2009b)

The shape of $SmVO_4$ and $CeVO_4$ NCs can be controlled by the synthesis temperature. TEM images of the samples synthesized solvothermally at 150 °C and 180 °C for 16 h are shown in Figures 8a,b (Nguyen et al. 2009a). Nearly cubic-shaped $SmVO_4$ and round-shaped $CeVO_4$ nanocrystals with an average diameter of 15 nm were found at 150 °C. When the synthesis temperature increased to 180 °C for 16 h, both uniform $SmVO_4$ and $CeVO_4$ nanospheres were observed, however, the diameter is unchanged. The transformation of both cubic-shaped $SmVO_4$ and round-shaped $CeVO_4$ NCs into uniform nanospheres, while preserving the particle size by increasing synthesis temperature from 150 °C to 180 °C can be explained by Wuff facets theory. The XPS results for characterization of these nanomaterials exhibite that only one oxidation state of samarium, cerium, and vanadium for each metal (e.g., Sm^{3+}, Ce^{3+}, V^{5+}) was observed on the particle surface at the nanoscale, even after calcination, while the existence of two oxidation states of these metals was found (e.g., Sm^{3+}/Sm^{2+}, Ce^{4+}/Ce^{3+}, V^{5+}/V^{4+}) in the corresponding single metal oxide nanocrystals (Nguyen et al. 2009a).

For this approach, $SmVO_4$ was selected as a typical example for discussion on the experimental results. It was found that their size and shape were controlled by the nature

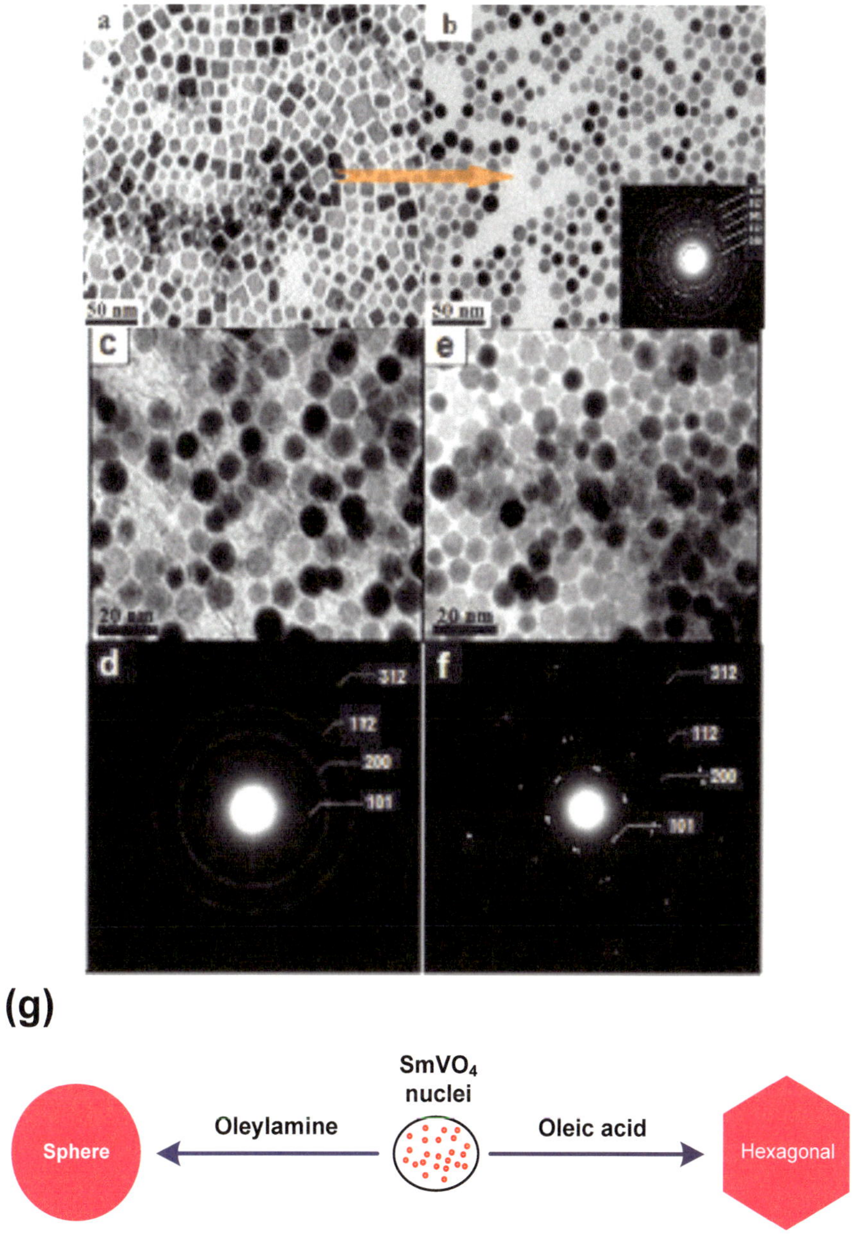

Fig. 8. Effects of the reaction temperature and surfactants with different functional groups on the shape control. TEM images of 15 nm-sized $SmVO_4$ nanocrystals synthesized at different temperatures for 16 h: (a) nanocubes at 150 °C and (b) nanospheres at 180 °C;(Nguyen et al. 2009a) TEM images and SAED patterns of the $SmVO_4$ nanocrystals synthesized from a 1:1 mixture of $Sm(OA)_3$ and $VO_4(TOA)_3$ in toluene using various surfactants: (c,d) oleylamine, 17 nm nanospheres, (e,f) oleic acid, 17 nm nanohexagons, and (g) Schematic illustration for the shape control of the $SmVO_4$ nanocrystals (Nguyen et al. 2009b).

and amount of capping surfactant as well as the metal complex precursor concentration (Nguyen et al. 2009b) Figure 8c-f displays TEM images of the $SmVO_4$ nanocrystals obtained using two different surfactants (e.g., oleylamine and oleic acid), and the corresponding SAED patterns. When only oleylamine was used, nanospheres with ~17 nm in diameter were obtained (Figure 8c,d). However, using oleic acid as capping surfactant instead of oleylamine under the same synthesis conditions, hexagonal-like $SmVO_4$ nanocrystals with no significant change in particle size (~17 nm in diameter) were formed (Figures 8e,f). The SAED results revealed that both the samples are indexed to a tetragonal $SmVO_4$ single crystal. The selective-shape formation of the $SmVO_4$ nanocrystals nanospheres and nanohexagonons can be resulted in the nonselective and selective absorption of oleylamine or oleic acid, respectively.

The effect of oleylamine (OM) and metal complex precursor concentrations on the growth of $SmVO_4$ nanoparticles was also studied, as shown in Figure 9 (Nguyen et al. 2009b). In the absence of OM, only irregular nanocrysqtals with aggregated pearl-chain-like structures often formed. When the OM concentration increased from 0.025 to 0.060, 0.129, 0.230, 0.034, and 0.43 M, the particle size of $SmVO_4$ nanoparticles decreased from ~30 to 3 nm. The reason for this behaviour may be due to the high degree of surfactant protection and stabilization of nanocrystals with increasing the OM concentrations in the bulk reaction solution. The larger and irregular sizes of the nanocrystals at low OM concentrations as compared to those obtained at high OM concentrations may result in insufficient coverage to the nanocrystal surface and induce their aggregation. Furthermore, in all the cases, the spherical nanocrystals were produced. The formation of spherical NPs could be due to the nonselective surfactant character of oleylamine (OM). On the other hands, it was found that the shape of $SmVO_4$ nanocrystals elongated from ~3 nm cores into ~3 nm x 200 nm wires as increasing the metal complex precursor concentrations from 0.065 to 0.130, 0.195, 0.260 M. These results reveal that the shape evolution from nanocores to nanowires can be controlled by increasing the precursor monomer concentration, which is strongly associated with the increase of chemical potential in the bulk solution as well as the dominant oriented attachment for the formation of nanowires (Nguyen et al. 2009b)

Due to their remarkable luminescence properties, Li's group(Liu and Li 2007) has also developed a general oleic acid-assisted hydrothermal method for the synthesis of a series of colloidal rare earth orthovanadate nanocrystals through the hydrothermal reaction of metal nitrate, NaOH, NH_4VO_3, oleic acid, ethanol/water mixture. The morphologies of all products were mainly square sheetlike shape with average diameters of ~20-40 nm. The products were formed by capping of oleic acid to Ln^{3+} first, then oleic acid attached rare earth ion (Ln^{3+}) reacted with VO_4^{3-} to form $LnVO_4$ nuclei at the water-oleic acid interface. This was followed by crystal growth until the nanocrystals were large enough to fall to the bottom of the vessel. Further, Eu^{3+}-doped $LnVO_4$ nanocrystals emitted intense red light. The author also used this route to synthesize colloidal uniform rare earth floride nanocrystals (Li, Peng and Li 2009, Wang et al. 2006). Lin et al. (Xu et al. 2010) synthesized and studied the luminescence properties of Ln^{3+} (Ln = Eu, Dy, Sm, Er)-doped YVO_4 nanocrystals via the trisodium citrate-assisted hydrothermal process. Haase et al (Sun et al. 2006c) also demonstrated an increase of visible emission intensity of Er^{3+}-doped YVO_4 nanocrystals due to photoadsorption and energy transfer of Er^{3+} ions to the host YVO_4 . An another general ultrasonic irradiation route for the lanthanide orthovanadate $LnVO_4$ (Ln = La-Lu) nanocrystals from the aqueous solution of $Ln(NO_3)_3$ and NH_4VO_3 without any surfactant were also reported by Lin et al. (Yu et al. 2008). The resulting $LnVO_4$ nanocrystals had

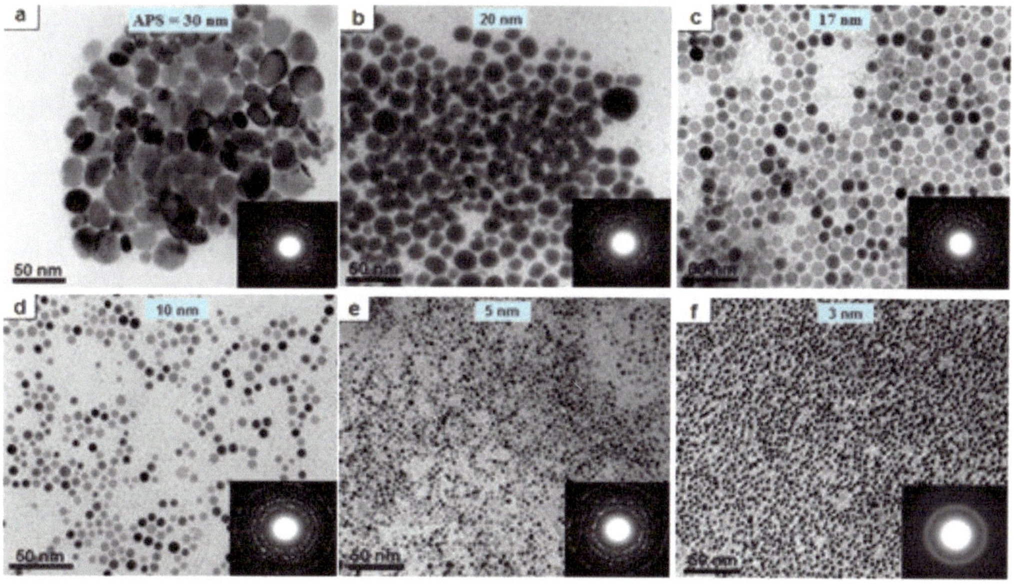

Fig. 9. Effect of the oleylamine concentration in the bulk solution on the particle size. TEM images and corresponding inset SAED patterns of the $SmVO_4$ nanocrystals synthesized using the different molar ratios of oleylamine:precursor (*OM:P*): (a) *OM:P* = 0.8, 30 nm roundlike nanocrystals, (b) *OM:P* = 2, 20 nm nanospheres, (c) *OM:P* = 4, 17 nm nanospheres, (d) *OM:P* = 8, 10 nm nanospheres, (e) *OM:P* = 12, 5 nm nanospheres, and (f) *OM:P* = 17, 3 nm nanocores. Average particle size = APS (Nguyen et al. 2009b).

spindle-like shape with the equatorial diameter of 30-70 nm and the length of 100-200 nm, which were the aggregates of small particles of 10-20 nm. Furthermore, the Eu^{3+} and Dy^{3+}-doped $LnVO_4$ (Ln = La, Gd, Lu) samples showed the characteristic dominant emissions of Eu^{3+} at 613 nm and Dy^{3+} at 572 nm, respectively, as a result of an energy transfer from VO_4^{3-} to Eu^{3+} or Dy^{3+}.

Manganese tungstate ($MnWO_4$) is one of the most promising mixed metal oxide nanomaterials, which exhibits high sensitivity to humidity change and unique magnetic property (Qu, Wlodarski and Meyer 2000, Arkenbout et al. 2006, Heyer and et al. 2006). Hence, it has attracted considerable research interest for potential applications such as photocatalysts, humidity sensors, optical fibers, photoluminescence and scintillator materials (Xing et al. 2008, Qu et al. 2000, Bharati, Singh and Wanklyn 1982). Several efforts have been devoted to the synthesis of $MnWO_4$ nanoparticles and especially focused on the shape and dimensional control (Zhou et al. 2008b, Zhang et al. 2008b). We recently developed a new approach for the aqueous-phase "one-step" synthesis of uniform single-crystalline $MnWO_4$ nanoparticles with controlled shape and the self-assembled mesocrystal microspheres/microapples with high yield using $Mn(NO_3)_2$ and Na_2WO_4 as precursors and bifunctional amino acid biomolecules as capping agent. $MnWO_4$ nuclei were early formed by the combination of Mn^{2+} cations and WO_4^{2-} anions and then grew into nanocrystals (Nguyen et al. 2011b). The nanoparticle products were capped by the amino head groups of 6-aminohexanoic acid (AHA) biomolecules and their surface become hydrophilic owing to other end of the uncoordinated carboxylic groups. Because of the hydrophilic surface character, the final products can be suspended in water medium.

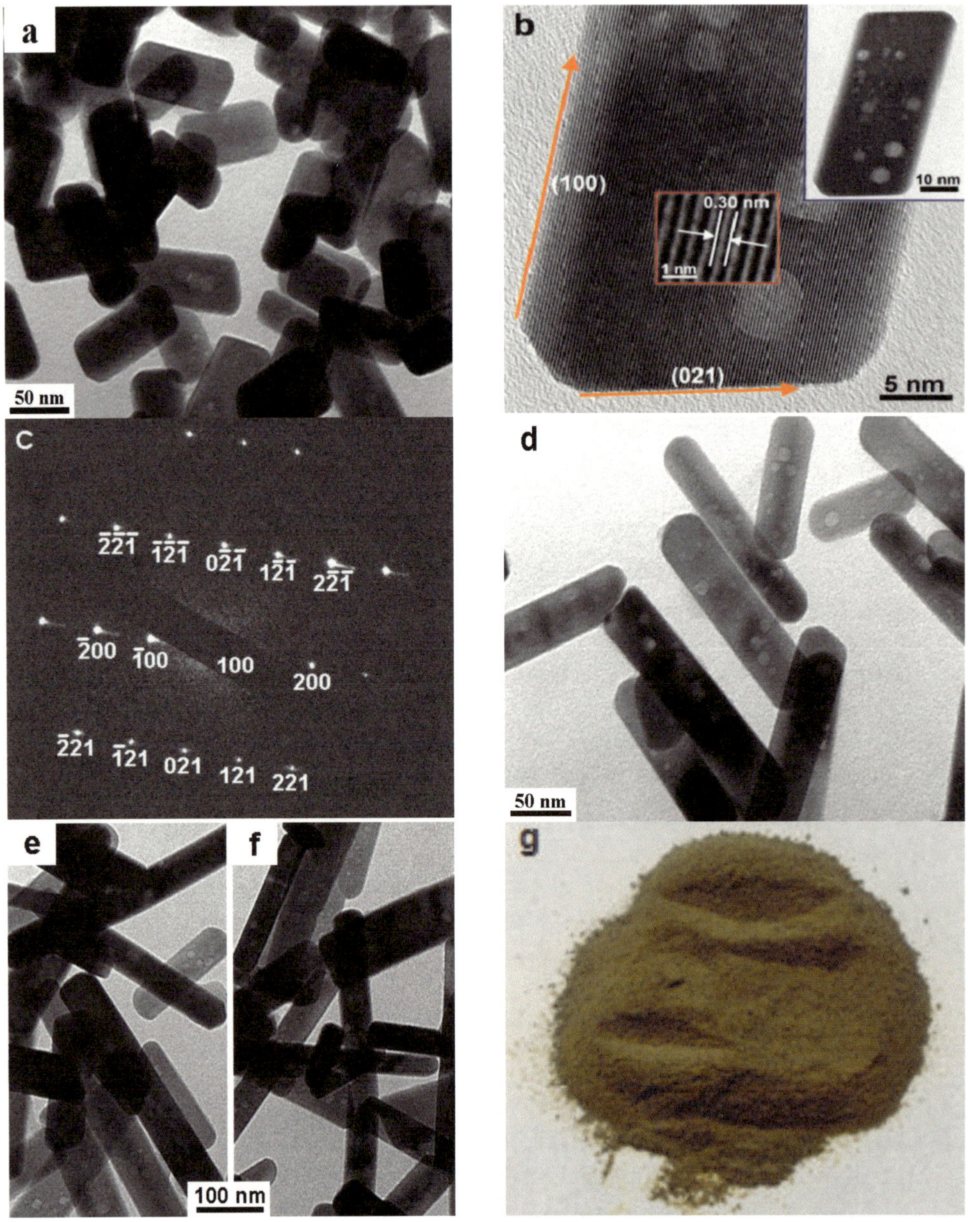

Fig. 10. TEM images of the $MnWO_4$ nanocrystals synthesized from an aqueous solution of 0.015 M $Mn(NO_3)_2$ and 0.015 M Na_2WO_4, pH = 9, 180 °C for 20 h, using the different 6-aminohexanoic acid/(Mn+W) molar ratios (*AHA/P*): (a) 25 nm x 50 nm nanobars, *AHA/P* = 2:1; (b) HRTEM image of an individual nanobar; (c) SAED pattern of a single bar taking along [100] zone axis; (d) 25 nm x 100 nm nanorods, *AHA/P* = 10:1; (e) 25 nm x 150 nm nanorods, *AHA/P* = 20:1; (f) 25 nm x 150 nm nanorods, *AHA/P* = 30:1; (g) one photo of ~16 grams of 6-aminohexanoic acid-capped $MnWO_4$ nanobar powders synthesized using $[Mn^{2+}]$ = $[WO_4^{2-}]$ of 0.122 M, 0.243 M of AHA, pH = 9, 180 °C for 20 h, in a 700 mL-sized autoclave (Nguyen et al. 2011b).

The experimental results revealed that the shape elongation of $MnWO_4$ nanoparticles is affected by capping 6-aminohexanoic acid concentration in the aqueous solution. Figure 10 shows TEM/HRTEM images of the $MnWO_4$ nanoparticles synthesized using AHA concentration ranging from o.031 to 0.305, 0.610, 0.915 M, corresponding to the AHA/(Mn+W) molar ratio ranging from 2:1 to 10:1, 20:1, 30:1 (Nguyen et al. 2011b). The elongation of nanoparticles from 25 nm x 50 nm-sized nanobar to 25 nm x 150 nm-sized nanorods with increasing 6-aminohexanoic acid (AHA) concentration from 0.031 to 0.610 M was clearly observed. The high-resolution TEM results (Figure 10b) of a $MnWO_4$ nanobar suggest that the nanobar is single crystal with an interplanar spacing of 0.30 nm, which corresponds to the separation between the (200) lattice planes of monoclinic $MnWO_4$. The side surfaces of the nanobar were bounded by (100) plane and the ends of the nanobar were enclosed by the (021) plane. The (100) planes are oriented parallel to the nanobars' growth axis, suggesting that the growth direction of the single-crystalline nanobar occurs preferentially along the [100] direction (*c*-axis), in good agreement with the SAED data (Figure 10c). The shape evolution could suggest that the (021) faces of nanocrystals could be selectively adsorbed and stabilized by AHA molecules, while the (100) faces were uncovered. The crystal grew anisotropically along the [100] direction due to their higher surface energies resulting in the nanorod product. Because AHA adsorbed onto only specific (021) faces, the nanorods with high aspect ratios produced at high AHA concentration could be due to the oriented attachment of nanocrystals predominantly during the synthesis.

In this synthesis approach, water is adopted as the continuous solution phase and inorganic salts were used as starting materials. Due to the high solubility of the salts in aqueous solution, it is applicable to synthesize the nanoparticles in scale up by using the high precursor monomer concentrations. This aqueous-based method is thus a promising way in the academic laboratory as well as can be expanded to the industrial scale in a simple way. In fact, we obtained as much as 16 grams of ~25 nm x 50 nm $MnWO_4$ nanobars per single run in a 700 mL-sized autoclave when the high precursor concentration of 0.122 M was used (Figure 10g) (Nguyen et al. 2011b).

The self-assembly of tailored nanobuilding units into three-dimensional (3D) mixed metal oxide microarchitectures has recently received considerable interest. Many novel and fascinating properties of these materials and useful applications are predicted depending not only on the complex morphology but also on the order degree of single-crystalline nanoparticles in microarchitectures. Sacrificial organic surfactants can act as structure-directing agents or soft templates and are widely used to design nano/microstructures with peculiar morphologies. The amino acid molecules are considered as fine assemblied agents. In this work, we also found that the formation of self-assemblied $MnWO_4$ 3D hierarchical microspheres from nanobars is favorable at relatively low precursor concentration (0.0076 M) of Mn^{2+} cation and WO_4^{2-} anion precursors in the initial synthesis solution in the presence of 6-aminohexanoic acid surfactant (Nguyen et al. 2011b) SEM/TEM images in Figure 11A shows that self-assemblied products is quite monodisperse microspheres with two size populations, 3-5 μm and 8-16 μm. The peripheral surface of the microsphere is rough and each microsphere is composed of numerous disordered nanobars suggesting that the formation of microspheres is likely driven by the interaction of inter-nanobars.

The FTIR results clarified that the assembly mechanism of the formed microspheres caused by generating the peptide chains in bulk solution during synthesis. Only the amino ($-NH_2$) group of AHA molecules capped on the surface of $MnWO_4$ nanoparticles and the free carboxylic (-COOH) terminus was oriented outward. The polypeptide chains were formed

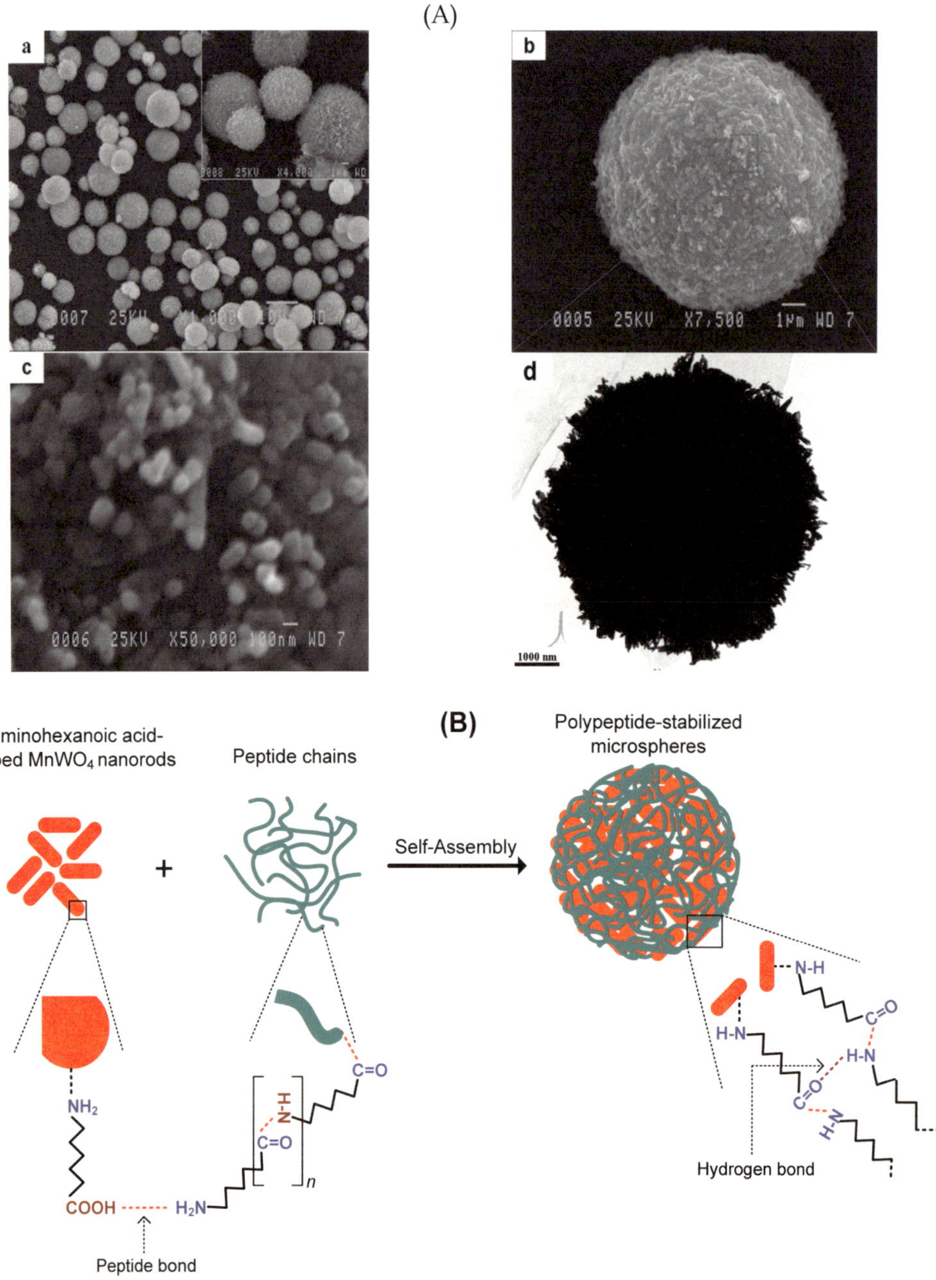

Fig. 11. (A) Different-magnification SEM (a-c) and TEM (d,e) images, inset SAED pattern of the self-assembled $MnWO_4$ hierarchical microspheres synthesized using $[Mn^{2+}] = [WO_4^{2-}]$ of 0.0076 M, 0.0305 M of AHA, pH = 9, 180 °C for 20 h. (f) SEM image of the broken $MnWO_4$ microspheres achieving ultrasonic treatment. (B) A possible proposed mechanism for the construction of the 3D hierarchical $MnWO_4$ microspheres from self-assembly of nanobars using low precursor monomer concentration (0.0076 M) (Nguyen et al. 2011b).

through the interaction between this uncoordinated carboxylic group on the nanobar surface and the amino group of residual 6-aminohexanoic acid in aqueous solution. The self-assembly of the peptide structure led to produce the assembled $MnWO_4$ microspheres. Further, the microsphere morphology can thus be controlled by adjusting the concentration of the peptide, that is, the concentration of 6-aminohexanoic acid (Nguyen et al. 2011b).

This suggests that with a low concentration of Mn^{2+} and WO_4^{2-} precursors (0.0076 M, (Mn+W)/AHA molar ratio of 0.25:1) and a low nuclei of $MnWO_4$ was formed and a small amount of $MnWO_4$ nanoparticles was produced. Due to the relatively high surfactant concentration (0.031 M), an excess amount of free 6-aminohexanoic acid exists in the aqueous synthesis solution, the dipeptide/polypeptide process of both $-NH_2$ and -COOH groups of free 6-aminohexanoic acid occurred on the basis of nucleophile mechanism, resulting in the formation of polypeptide chains. Because only small amount of $MnWO_4$ nanobars was yielded in bulk solution and because of the high AHA concentration, the peptide reaction continued and the reaction between the amino groups of the formed amino acid sequence of protein and the uncoordinated carboxylic groups on the nanoparticle surface to generate the polypeptide-stabilized $MnWO_4$ nanobars. Subsequently, highly oriented backbone-backbone intermolecular hydrogen-bonding interactions of numerous amphiphilic polypeptide chains via either antiparallel or parallel arrangements were proceeded to spontaneous-assemble into polypeptide-stabilized $MnWO_4$ microspheres. These interactions are thermodynamically favorable due to the reduction of the particle surface energy when the interface is eliminated. On the contrary, a large amount of monodisperse nanobars was yielded using a higher precursor monomer concentration (0.015 M, (Mn+W)/AHA molar ratio of 0.48:1) with a faster nucleation of $MnWO_4$. The excess amount of free 6-aminohexanoic acid in synthesis solution decreased strongly because they were consumed more for the capping on nanobar surface. A possible proposed mechanism of the "one-step" formation of self-assemblied $MnWO_4$ hierarchical microspheres is illustrated in Figure 11B. The photoluminescence results indicated that the PL emission intensity of the $MnWO_4$ nanobars is higher than that of the $MnWO_4$ microspheres indicating the decrease in the luminescence efficiency of the microspheres due to nanobars inside of microspheres (Nguyen et al. 2011b).

$MnWO_4$ nanomaterials have been synthesized and revealed several morphologies such as flower-like clusters, nanowires, nanoplates, and nanorods (Zhang et al. 2007b, Zhou et al. 2008b, Thongtem, Wannapop and Thongtem 2009, Chen et al. 2003, Zhou et al. 2008a). Additionally, the kinetic control of $MnWO_4$ nanoparticles grown for tailored structural properties were studied by Li et al. (Tong et al. 2010). To extend for these tungstates, Wong et al. (Zhang et al. 2008a) developed a room-temperature template-directed method for the synthesis of single-crystalline alkaline-earth-metal tungstate AWO_4 (A = Ca, Sr, Ba) nanorods with unique optoelectronic properties. The 2D self-organization of $BaWO_4$ nanorods at the water-air interface was performed on the basis of a Langmuir-Blodgett technique (Kim et al. 2001). By using the catanionic reversed micelle templating method, single-crystalline $BaWO_4$ nanorods with high-aspect-ratio were produced (Shi et al. 2003b). The penniform $BaWO_4$ nanostructures could be generated in reverse micellar system containing double-hydrophilic block copolymers (Shi et al. 2003a). Also used a microemulsion with solvothermal association, the $SrWO_4$ nanorods of 100 nm in diameter and 500-1500 nm in length were yielded (Sun et al. 2006b). In the solution-based self-assembly process, through properly employing organic additives to the reaction system, the growth of inorganic nanocrystals can be rationally directed to obtain products with desirable morphologies and/or hierarchical structures. For example, Cascales et al. (Esteban-Betegón, Zaldo and Cascales 2010) have produced the tetragonal scheelite-phase Yb^{3+}-doped $NaGd(WO_4)_2$ micro/nanostructures with preserved photoluminescence properties from hydrothermal preparations using metal nitrate and chloride reagents. Shen et al. (Liu et al. 2004) synthesized the single crystalline $ZnWO_4$ nanorods via the direct aggregation of the amorphous nanoparticles, and promote the crystallization process of the amorphous $ZnWO_4$ particles derived from the precipitation reaction in a nanorod/amorphous nanoparticle coexisting system. Yu et al. (Zhang et al. 2007c) have used the refluxing method to fabricate the uniform core-shell heterostructured $ZnWO_4$@MWO_4 (M = Mn, Fe) nanorods with optical and antiferromagnetic characters. The direct crystallization of the MWO_4 nuclei occurred on the backbone of $ZnWO_4$ nanorods without surfactant. The formation of the hybrid $ZnWO_4$@MWO_4 (M = Mn, Fe) nanorods based on the oriented aggregation. The shell thickness of MWO_4 could be tuned by changing the molar ratio of these raw materials.

Many other research groups have also used the solvo-hydrothemal method to synthesize the mixed metal oxide nanocrystals. For example, Yu et al. (Zhou et al. 2009) reported an effective ethylene glycol (EG)-assisted solvothermal method to synthesize hierarchical $FeWO_4$ microcrystals using $FeCl_3 \cdot 6H_2O$ and $Na_2WO_4 \cdot 2H_2O$ as precursors. It was found that the organic solvent EG played a critical role as both a reducing agent and a structure-directing agent in driving such architectures assembled by oriented attachment of primary nanoparticles. Moreover, a certain amount of CH_3COONa was necessary for the formation of such unique platelike $FeWO_4$ microcrystals. The photocatalytic property of as-synthesized hierarchical $FeWO_4$ microcrystals was also studied, which shows excellent photocatalytic activity for the degradation of rhodamine B. Cascales et al. (Esteban-Betegón et al. 2010) has produced the tetragonal scheelite-phase Yb^{3+}-doped $NaGd(WO_4)_2$ micro/nanostructures with preserved photoluminescence properties from hydrothermal preparations using metal nitrate and chloride reagents. The Li's group has successfully prepared a novel hollow CeO_2-ZrO_2 nanocages with controlled shapes, sizes, and compositions by adding zirconium(IV) into the glycol solution containing CeO_2 nanospheres under solvothermal treatment (Liang et al. 2008). The formation of cage-structured CeO_2-ZrO_2 nanomaterials was explained by the Kirkendall effect. When Zr^{4+} ions were added into the system, they readily doped into ceria to form a solid solution of the type $Ce_{1-x}Zr_xO_2$; meanwhile, the diffusion rate of the special secondary nanostructure of the resulting clusters was obviously much faster than that of the single-element nanostructure, leading to the formation of hollow nanostructures of the $Ce_{1-x}Zr_xO_2$ type. Yu and coworkers(Gu et al. 2003) noted that hydrothermal heating at 180 °C of an ammoniacal $FeCl_2$ solution containing metallic Zn leads to the formation of octahedrally-shaped $ZnFe_2O_4$ nanoparticles with an average size of 300 nm. Using liquid-solid-solution approach, Li et al. has been successfully employed multicomponent nanocrystals not only with mixed metal oxide nanocrystals ($LnVO_4$ and $LnPO_4$) but also with lanthanide doped $NaYF_4$ nanocrystals (Liu and Li 2007).

Related progress using this two-phase approach was also applied for the synthesis of a variety of mixed inorganic nanomaterials under solvo/hydrothermal treatment. A modified two-phase approach for the synthesis of the CdS nanocrystals at the water/toluene interface was carried out by mixing cadimium myristate toluene solution and thioure aqueous solution (Wang et al. 2005a). Additionally, these authors used a seeding-growth technique to change the CdS nanocrystal sizes. They also synthesized the CdSe and CdSe/CdS core/shell nanocrystals using selenourea as a selenium precursor by this two-phase approach. The CdSe/CdS core/shell nanocrystals exhibited quantum yields up to 60-80%, the photoluminescence of these nanocrystals can be clearly observed even without UV lamp irradiation (Pan et al. 2005a). By an alternant growth technique, the authors also synthesized two multi-shell nanocrystals of CdS/CdSe/CdS/CdSe/CdS and CdSe/CdS/CdSe/CdS/CdSe/CdS nanocrystals (Pan et al. 2006). In all cases, the obtained products had the narrow size distributions due to the use of low reaction precursors. Xu et al. (Wang et al. 2009) have recently achieved rare-earth doped $NaYF_4$ up conversion fluorescent nanocrystals using water/ethanol/rare-earth stearate/oleic acid system. Kaskel et al. (Du et al. 2007) also used this two phase solvothermal method to synthesize of $BaTiO_3$ nanocrystals using mixed titanium(IV)-*n*-butoxide and barium acetate precursors.

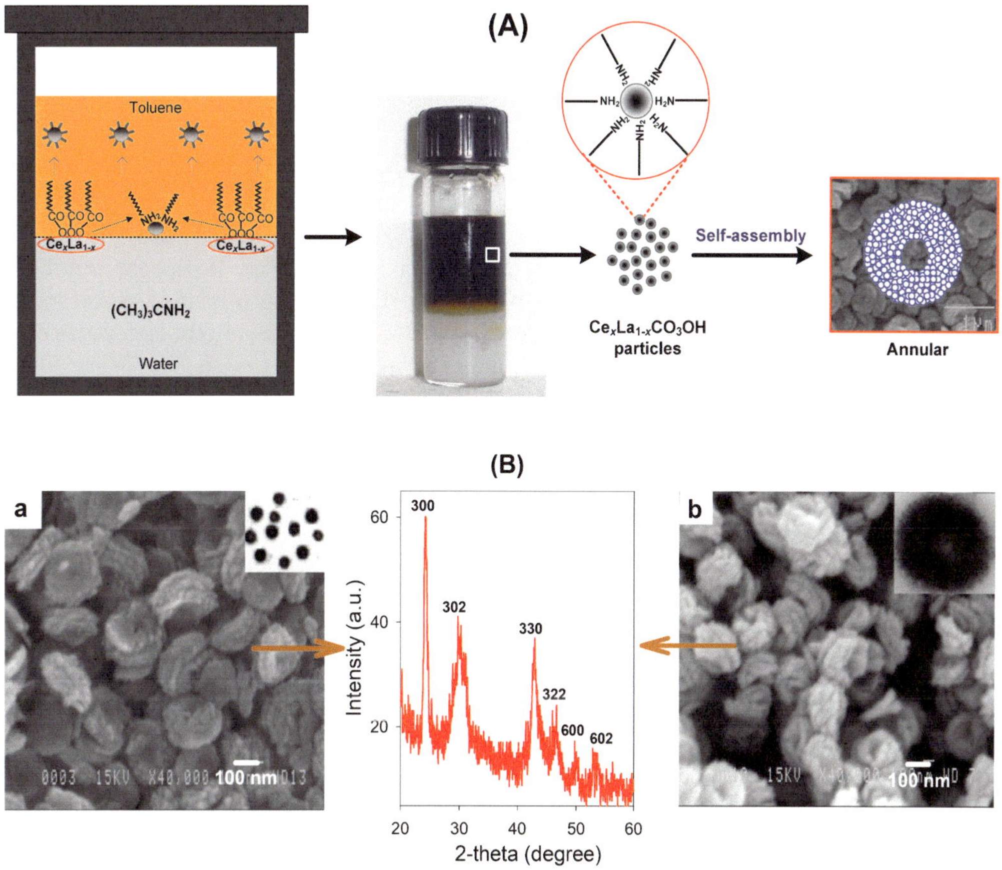

Fig. 12. (A) Two-phase protocol for the synthesis of the undoped and cerium doped $LaCO_3OH$ annular-shaped nanoarchitectures; (B) SEM/TEM images of undoped $LaCO_3OH$ (a) and cerium doped $LaCO_3OH$ (b) samples having the same pure hexagonal-phase and annular shape (Nguyen, Dinh and Do 2011a).

Our group recently reported a modified two-phase method for the synthesis of alkyl chain-capped metal particles (e.g., Cu and Au) and metal oxide nanoparticles (TiO_2 and ZrO_2), followed by their cooperative assemblage into unusual hybrid metal/metal oxide nanocrystal mesostructured materials (Mrabet, Zahedi-Niaki and Do 2008). The catalytic activity of these hybrid products for the CO oxidation was also employed. Indeed, these hybrid metal/metal oxide nanocatalysts exhibited high surface areas, narrow pore size distributions, and exceptional catalytic properties in the oxidation of CO, even surpassing the performance of commercial noble metal catalysts.

Cerium doped lanthanum(III) carbonate hydroxide ($Ce_xLa_{1-x}CO_3OH$) has drawn a great deal of interest as a promising luminescent material because of the empty 4f shell of La^{3+} and the lack of electronic f-f transitions (Binnemans 2009, Bünzli 2010). This is mainly due to their unique electronic and optical properties arising from the 4f electrons of cerium (Su et al. 2009). Cerium incorporation in the $LaCO_3OH$ structure can improve the reactive performance, because of the generation of crystalline defects (Meiser, Cortez and Caruso 2004, Ding et al. 2001). Thus its physical and chemical properties can be controlled by atomic-scale precision, and this type of materials can be tailored to possess specific properties. These make these cerium doped $LaCO_3OH$ materials to have potential applications in catalysis,(Sun et al. 2006a) high-quality phosphors,(Mai et al. 2006) up-conversion materials,(Auzel 2003) oxygen-ion conducting electrolytes (Etsell and Flengas 1970).

We recently reported the fabrication of undoped and cerium doped $LaCO_3OH$ annular-shaped nanoarchitectures with high specific surface area via the thermolysis of the binary source precursor, $Ce_xLa_{1-x}(oleate)_3$ complex (x = 0 - 20 mol%), in water-toluene system containing *tert*-butylamine/oleylamine (Nguyen et al. 2011a). The two-phase synthetic procedure consists of two steps, as shown in Figure 12A: (i) the preparation of $Ce_xLa_{1-x}(oleate)_3$ complex from the reaction between respective lanthanide nitrate and potassium oleate in a water-toluene solution; (ii) the formation of mesoporous $Ce_xLa_{1-x}CO_3OH$ annular-shaped nanoarchitectures in an autoclave containing a water-toluene mixture composed of $Ce_xLa_{1-x}(oleate)_3$/*tert*-butylamine/oleylamine at 180 °C for 24 h. The solid-solution cerium-lanthanum oxide particles were produced from the decarbonation and dehydration of $Ce_xLa_{1-x}CO_3OH$ upon annealing. The $Ce_xLa_{1-x}CO_3OH$ nanoarchitectures were capped by the amine groups of oleylamine molecules, the exposed hydrophobic alkyl groups were well-immersed in toluene, and guaranted the good dispersibility of the product in toluene phase.

The XRD and SEM/TEM results (Figure 12B) revealed that the formed product without cerium doping exhibited the pure hexagonal-phase $LaCO_3OH$ structure with annular shape (Nguyen et al. 2011a). The monodisperse $LaCO_3OH$ annular-shaped nanoarchitectures revealed rough surface, narrow size distribution, average particle diameter of <400 nm, high specific surface area (~100 $m^2.g^{-1}$), and are composed of numerous small 3-5 nm particle assemblies. The phase structure and morphology of $Ce_xLa_{1-x}CO_3OH$ is unchanged as cerium doping concentration ranging from 5, to 10, 15, 20 mol%. This suggested that cerium ions tend to incorporate into the $LaCO_3OH$ lattice leading to a homogeneous $Ce_xLa_{1-x}CO_3OH$ structure. No segregation of cerium species on the surface of the annular-shaped nanoarchitecture was observed, even up to 20 mol% of cerium. This can be explained by) no significant difference in ionic radii of Ce^{3+} (1.150 Å) and La^{3+} (1.172 Å), which allows the replacement of La^{3+} by trivalent Ce^{3+} ions in $LaCO_3OH$. In contrast, by only 5.0 mol% copper in the $LaCO_3OH$ structure, a separate phase: hexagonal $LaCO_3OH$, monoclinic CuO, and cubic Cu_2O phases was observed by its XRD pattern. Because the radius of La^{3+} ion is much larger than that of Cu^+ ion ($r(La^{3+})$ = 1.172 Å, $r(Cu^{2+})$ = 0.87 Å, $r(Cu^+)$ = 0.91 Å), The copper ions cannot enter the lattice of $LaCO_3OH$ by occupying the La^{3+} ion sites (Nguyen et al. 2011a).

The photoluminescence (PL) emission spectra (Figure 13) of the colloidal $Ce_xLa_{1-x}CO_3OH$ nanoarchitectures with various doping levels (x = 0-20 mol%) in toluene were recorded on exciting at 360 nm (Nguyen et al. 2011a). Using the same particle concentrations, the colloidal solutions of these $Ce_xLa_{1-x}CO_3OH$ samples show the same spectral peak positions at 424, 448, 486, 529, 560 nm, which can be attributed to the charge-transfer transition in $Ce_xLa_{1-x}CO_3OH$ structure. The broad backgrounds of the luminescence spectra of these doped samples can be due to the self-assembly of small particles (3-5 nm) for the formation of the aggregated $Ce_xLa_{1-x}CO_3OH$ structure with annular shape. However, the effect of variation in emission can be correlated to the cerium doping concentration. The intensity was found to increase with increase in cerium doping level. The emission intensity of $Ce_xLa_{1-x}CO_3OH$ sample increases with increase in cerium concentration to 20 mol% and then exhibits a gradual decrease upon further increase in cerium doping content. The initial increase in emission intensity can be associated with the increase in relative concentration of the defects in the crystal structure. The subsequent decrease in emission intensity can be primarily attributed to the increase in particle size of the formed Ce doped $LaCO_3OH$ heterogeneous structure.

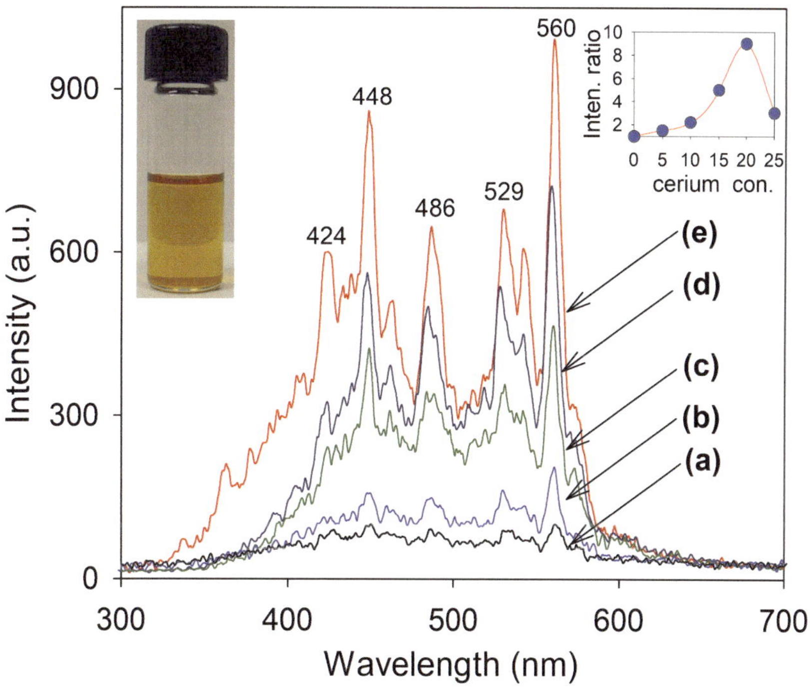

Fig. 13. Photoluminescence emission spectra (under excitation at 360 nm) of the as-synthesized $Ce_xLa_{1-x}CO_3OH$ nanoannular samples with various doping levels (mol%): (a) x = 0; (b) x = 5; (c) x = 10; (d) x = 15; (e) x = 20. Inset: one photo of transparent toluene solution containing colloidal 10 mol% cerium doped $LaCO_3OH$ nanoannulars and the relationship between the cerium concentration and the PL intensity ratio (Nguyen et al. 2011a).

Xie et al. (Xie et al. 2009) fabricated the 1D layer-by-layer hexagonal-phased $La(OH)CO_3$ hierarchical nanomaterials with a diameter of ~700 nm and lengths in the range of 6-8 μm from the self-assembly of the single $La(OH)CO_3$ nanoplates via the hydrothermal reaction of bulk La_2O_3 powders with NH_2CH_2COOH. Under the thermal procedure, the obtained 1D $La(OH)CO_3$ nanostructures could be converted to porous $La(OH)_3$ and La_2O_3 nanorods. The $La(OH)CO_3$ was first transformed into La_2O_3 through calcination at 900 °C and then further change into hexagonal-phased $La(OH)_3$ by a hydrolysis process. The doping ion obviously plays a crucial role in modifying the interesting optical properties of the produced mixed oxides. The orthorhombic-phase Eu^{3+}/Tb^{3+} doped La_2OCO_3/La_2O_3 nano/microcrystals with multiform morphologies were synthesized via a homogeneous precipitation (Li et al. 2010a). The multiform $LaCO_3OH$ products such as flake, flower, rhombuse, two-double hexagram, sandwichlike spindle, peach-nucleus-shaped nanocrystals obtained by changing the carbon sources [$CO(NH_2)_2$, Na_2CO_3, $NaHCO_3$, $(NH_4)_2CO_3$, and NH_4HCO_3], NH_4^+, Na^+ ions, and pH values of the initial solution. $LaCO_3OH$ was easily converted to $La_2O_2CO_3$ and La_2O_3 under annealing at suitable temperatures. The excitation and selective emission spectroscopy were applied to the microstructures and difference in $^5D_0 \rightarrow {}^7F_2$ transition of Eu^{3+} ions in $La_2O_2CO_3$ and La_2O_3 host lattices. In addition, the optical luminescence properties of 1-5 mol% Eu^{3+}/Tb^{3+}-doped La_2OCO_3/La_2O_3 phosphors were strongly dependent on their morphologies and sizes. Reddy et al. (Reddy, Katta and Thrimurthulu 2009) synthesized the novel nanocrystalline $Ce_{1-x}La_xO_{2-\delta}$ (x = 0.2) solid solutions via a modified coprecipitation. The mixed $Ce_{1-x}La_xO_{2-\delta}$ nanocomposites well matched with the standard fluorite type cubic phase of CeO_2. Variation in the lattice parameter of CeO_2 was ascribed to the partial substitution of Ce^{4+} with La^{3+}. The incorporation of La^{3+} into CeO_2 lattice led to the lattice expansion in unit cell volume because the ionic radius of La^{3+} (0.11 nm) is larger than Ce^{4+} (0.097 nm). The two oxidation states (Ce^{3+} and Ce^{4+}) coexisted in the segregation of La^{3+} in the $Ce_{1-x}La_xO_{2-\delta}$ structure. These $Ce_{1-x}La_xO_{2-\delta}$ catalysts were evaluated for OSC and CO oxidation activity. The nanosized Ce-La solid exhibited superior catalytic performance and thermal stability in comparison to heterostructured Ce-Zr solid. The excellent catalytic activity of the $Ce_{1-x}La_xO_{2-\delta}$ samples are originated from the features of structure, redox behavior, bulk oxygen mobility.

A variety of nanostructured metal oxide and mixed oxide materials have been synthesized by various methods, covering a wide range of compositions and tunable size/shape, especially over the past decade. In comparison, microemulsions can also be used to synthesize monodisperse mixed oxide nanocrystals with various morphologies; however, this method requires a large amount of solvent and small-scale production. In terms of size and shape control of the magnetic nanocrystals, thermal decomposition seems an improved method has been developed to date; however, this techniques often require the use of expensive organometallic or metal alkoxide precursors and performed at high temperature under argon or nitrogen atmosphere. In terms of simplicity of the synthesis, one-phase and two-phase solvo-hydrothermal surfactant-assisted methods are almost the preferred pathways for the large-scale synthesis of high-quality inorganic nanocrystals with controlled size and shape at relatively mild reaction temperature.

Eventually, the obtained metal oxide and mixed oxide nanomaterials are of great vitality and offers immense opportunities for chemistry, physics, biology, materials and engineering. Interaction among scientists with different backgrounds will undoubtedly create new science, and in particular new materials, with unforeseen technological possibilities. What is noteworthy is that such nanomaterials are likely to benefit not only the catalytic industry, but also to contribute to most aspects of the electronic and biotechnological industries.

5. Summary and outlook

In summary, this chapter provides an overview of some recent progresses related to solution-based syntheses of monodisperse metal oxide and mixed oxide colloidal nanocrystals. Compared to the traditional methods such as sonochemistry, solid-phase reaction, gas-phase reaction, coprecipitation, the one-phase and two-phase solvo-hydrothermal surfactant-assisted pathways are much more easily controlled in the size, shape, composition, and phase structure. Several current approaches developed in our group have some conveniences including nontoxic and inexpensive reagents (e.g., inorganic metal salts as starting precursors), water as environmentally benign medium, high-yield, and large-scale products.

It is pointed out that the major interest of such metal oxide and mixed oxide nanocrystals is systematically studied their catalytic properties in our laboratory for the next time. The ability to manipulate precisely the size/shape and the surface of nanocrystals has opened up a number of potential applications for the new further materials: nanoelectronics, optics, solar cells, magnetic resonance imaging (MRI), biomedicine, etc. Considering the astonishing rate at which progress is being made on several areas, we may expect that the impact of this field of nanotechnology on our daily lives will grow markedly in the near future. Finally, we expect that this chapter would be useful to readerships and could also form the basis of a course on the nanoscale shape-controlled subject.

6. Acknowledgment

This work was supported by the Natural Sciences and Engineering Research Council of Canada (NSERC) through a strategic grant.

7. References

Abbet, S. & U. Heiz. 2005. *Nanocatalysis*. Wiley-VCH Verlag GmbH & Co. KGaA.

Alivisatos, A. P. (1996) Semiconductor Clusters, Nanocrystals, and Quantum Dots. *Science*, 271, 933-937.

Antonini, G. M. & et al. (1987) Atomic Local Coordinations and Multivalent States in $YBa_2Cu_3O_{9-\delta}$ Superconductors. *EPL (Europhysics Letters)*, 4, 851.

Arkenbout, A. H., T. T. M. Palstra, T. Siegrist & T. Kimura (2006) Ferroelectricity in the cycloidal spiral magnetic phase of $MnWO_4$. *Physical Review B*, 74, 184431.

Auzel, F. (2003) Upconversion and Anti-Stokes Processes with f and d Ions in Solids. *Chemical Reviews*, 104, 139-174.

Bharati, R., R. A. Singh & B. M. Wanklyn (1982) Electrical conduction in manganese tungstate. *Journal of Physics and Chemistry of Solids*, 43, 641-644.

Binnemans, K. (2009) Lanthanide-Based Luminescent Hybrid Materials. *Chemical Reviews*, 109, 4283-4374.

Brust, M., M. Walker, D. Bethell, D. J. Schiffrin & R. Whyman (1994) SYNTHESIS OF THIOL-DERIVATIZED GOLD NANOPARTICLES IN A 2-PHASE LIQUID-LIQUID SYSTEM. *Chem. Commun.*, 7, 801-802.

Bünzli, J.-C. G. (2010) Lanthanide Luminescence for Biomedical Analyses and Imaging. *Chemical Reviews*, 110, 2729-2755.

Burda, C., X. Chen, R. Narayanan & M. A. El-Sayed (2005) Chemistry and Properties of Nanocrystals of Different Shapes. *Chemical Reviews,* 105, 1025-1102.

Chen, S.-J., X.-T. Chen, Z. Xue, J.-H. Zhou, J. Li, J.-M. Hong & X.-Z. You (2003) Morphology control of $MnWO_4$ nanocrystals by a solvothermal route. *Journal of Materials Chemistry,* 13, 1132-1135.

Corr, S. A., M. Grossman, J. D. Furman, B. C. Melot, A. K. Cheetham, K. R. Heier & R. Seshadri (2008) Controlled Reduction of Vanadium Oxide Nanoscrolls: Crystal Structure, Morphology, and Electrical Properties. *Chemistry of Materials,* 20, 6396-6404.

Cushing, B. L., V. L. Kolesnichenko & C. J. O'Connor (2004) Recent Advances in the Liquid-Phase Syntheses of Inorganic Nanoparticles. *Chemical Reviews,* 104, 3893-3946.

Ding, S.-J., D. W. Zhang, P.-F. Wang & J.-T. Wang (2001) Preparation and photoluminescence of the Ce-, Tb- and Gd-doped lanthanum borophosphate phosphor. *Materials Chemistry and Physics,* 68, 98-104.

Dinh, C. T., Nguyen, T.D., Kleitz, F. and Do. T. O. (2009). Shape-controlled synthesis of highly crystalline titania nanocrystals, *ACS nano,* 3, pp. 3737-3743.

Djerdj, I., D. Arčon, Z. Jagličić & M. Niederberger (2007) Nonaqueous Synthesis of Manganese Oxide Nanoparticles, Structural Characterization, and Magnetic Properties. *The Journal of Physical Chemistry C,* 111, 3614-3623.

Du, H., S. Wohlrab, Wei & S. Kaskel (2007) Preparation of $BaTiO_3$ nanocrystals using a two-phase solvothermal method. *Journal of Materials Chemistry,* 17, 4605-4610.

Esteban-Betegón, F. t., C. Zaldo & C. n. Cascales (2010) Hydrothermal Yb^{3+}-Doped $NaGd(WO_4)_2$ Nano- and Micrometer-Sized Crystals with Preserved Photoluminescence Properties. *Chemistry of Materials,* 22, 2315-2324.

Etsell, T. H. & S. N. Flengas (1970) Electrical properties of solid oxide electrolytes. *Chemical Reviews,* 70, 339-376.

Fu, Q. & T. Wagner (2007) Interaction of nanostructured metal overlayers with oxide surfaces. *Surface Science Reports,* 62, 431-498.

Gariglio, S., M. Gabay & J.-M. Triscone (2010) Oxide materials: Superconductivity on the other side. *Nat Nano,* 5, 13-14.

Garnweitner, G., M. Antonietti & M. Niederberger (2005) Nonaqueous synthesis of crystalline anatase nanoparticles in simple ketones and aldehydes as oxygen-supplying agents. *Chemical Communications,* 397-399.

Garnweitner, G. & M. Niederberger (2008) Organic chemistry in inorganic nanomaterials synthesis. *Journal of Materials Chemistry,* 18, 1171-1182.

Ghoshal, T., S. Biswas, P. M. G. Nambissan, G. Majumdar & S. K. De (2009) Cadmium Oxide Octahedrons and Nanowires on the Micro-Octahedrons: A Simple Solvothermal Synthesis. *Crystal Growth & Design,* 9, 1287-1292.

Gu, G., M. Schmid, P.-W. Chiu, A. Minett, J. Fraysse, G.-T. Kim, S. Roth, M. Kozlov, E. Munoz & R. H. Baughman (2003) V_2O_5 nanofibre sheet actuators. *Nat Mater,* 2, 316-319.

Guiton, B. S., Q. Gu, A. L. Prieto, M. S. Gudiksen & H. Park (2004) Single-Crystalline Vanadium Dioxide Nanowires with Rectangular Cross Sections. *Journal of the American Chemical Society,* 127, 498-499.

Hao, R., R. Xing, Z. Xu, Y. Hou, S. Gao & S. Sun (2010) Synthesis, Functionalization, and Biomedical Applications of Multifunctional Magnetic Nanoparticles. *Advanced Materials,* 22, 2729-2742.

Heyer, O. & et al. (2006) A new multiferroic material: $MnWO_4$. *Journal of Physics: Condensed Matter,* 18, L471.

Jun, Y.-w., J.-s. Choi & J. Cheon (2006a) Shape Control of Semiconductor and Metal Oxide Nanocrystals through Nonhydrolytic Colloidal Routes. *Angewandte Chemie International Edition,* 45, 3414-3439.

Kalai Selvan, R., A. Gedanken, P. Anilkumar, G. Manikandan & C. Karunakaran (2009) Synthesis and Characterization of Rare Earth Orthovanadate (RVO_4; R = La, Ce, Nd, Sm, Eu, Gd) Nanorods/Nanocrystals/Nanospindles by a Facile Sonochemical Method and Their Catalytic Properties. *Journal of Cluster Science,* 20, 291-305.

Kamat, P. V., K. Tvrdy, D. R. Baker & J. G. Radich (2010) Beyond Photovoltaics: Semiconductor Nanoarchitectures for Liquid-Junction Solar Cells. *Chemical Reviews,* 110, 6664-6688.

Kim, F., S. Kwan, J. Akana & P. Yang (2001) Langmuir−Blodgett Nanorod Assembly. *Journal of the American Chemical Society,* 123, 4360-4361.

Kinge, S., M. Crego-Calama & D. N. Reinhoudt (2008) Self-Assembling Nanoparticles at Surfaces and Interfaces. *ChemPhysChem,* 9, 20-42.

Kroes, G.-J., A. Gross, E.-J. Baerends, M. Scheffler & D. A. McCormack (2002) Quantum Theory of Dissociative Chemisorption on Metal Surfaces. *Accounts of Chemical Research,* 35, 193-200.

Kwon, S. G. & T. Hyeon. 2009. *Kinetics of Colloidal Chemical Synthesis of Monodisperse Spherical Nanocrystals.* John Wiley & Sons, Inc.

LaMer, V. K. & R. H. Dinegar (1950) Theory, Production and Mechanism of Formation of Monodispersed Hydrosols. *Journal of the American Chemical Society,* 72, 4847-4854.

Li, G., K. Chao, H. Peng, K. Chen & Z. Zhang (2007) Low-Valent Vanadium Oxide Nanostructures with Controlled Crystal Structures and Morphologies. *Inorganic Chemistry,* 46, 5787-5790.

Li, G., C. Peng, C. Zhang, Z. Xu, M. Shang, D. Yang, X. Kang, W. Wang, C. Li, Z. Cheng & J. Lin (2010a) Eu^{3+}/Tb^{3+}-Doped $La_2O_2CO_3/La_2O_3$ Nano/Microcrystals with Multiform Morphologies: Facile Synthesis, Growth Mechanism, and Luminescence Properties. *Inorganic Chemistry,* 49, 10522-10535.

Li, P., C. Nan, Z. Wei, J. Lu, Q. Peng & Y. Li (2010b) Mn_3O_4 Nanocrystals: Facile Synthesis, Controlled Assembly, and Application. *Chemistry of Materials,* 22, 4232-4236.

Li, P., Q. Peng & Y. Li (2009) Dual-Mode Luminescent Colloidal Spheres from Monodisperse Rare-Earth Fluoride Nanocrystals. *Advanced Materials,* 21, 1945-1948.

Li, X.-L., Q. Peng, J.-X. Yi, X. Wang & Y. Li (2006) Near Monodisperse TiO_2 Nanoparticles and Nanorods. *Chemistry – A European Journal,* 12, 2383-2391.

Liang, X., X. Wang, Y. Zhuang, B. Xu, S. Kuang & Y. Li (2008) Formation of CeO_2-ZrO_2 Solid Solution Nanocages with Controllable Structures via Kirkendall Effect. *Journal of the American Chemical Society,* 130, 2736-2737.

Liu, B., S.-H. Yu, L. Li, F. Zhang, Q. Zhang, M. Yoshimura & P. Shen (2004) Nanorod-Direct Oriented Attachment Growth and Promoted Crystallization Processes Evidenced in Case of ZnWO4. *The Journal of Physical Chemistry B,* 108, 2788-2792.

Liu, J. & Y. Li (2007) General synthesis of colloidal rare earth orthovanadate nanocrystals. *Journal of Materials Chemistry,* 17, 1797-1803.

Livage, J. (1991) Vanadium pentoxide gels. *Chemistry of Materials,* 3, 578-593.

Lu, J. G., P. Chang & Z. Fan (2006) Quasi-one-dimensional metal oxide materials--Synthesis, properties and applications. *Materials Science and Engineering: R: Reports,* 52, 49-91.

Lutta, S. T., H. Dong, P. Y. Zavalij & M. S. Whittingham (2005) Synthesis of vanadium oxide nanofibers and tubes using polylactide fibers as template. *Materials Research Bulletin,* 40, 383-393.

Mai, H.-X., Y.-W. Zhang, R. Si, Z.-G. Yan, L.-d. Sun, L.-P. You & C.-H. Yan (2006) High-Quality Sodium Rare-Earth Fluoride Nanocrystals:□ Controlled Synthesis and Optical Properties. *Journal of the American Chemical Society,* 128, 6426-6436.

MalenfantPatrick, R. L., J. Wan, S. T. Taylor & M. Manoharan (2007) Self-assembly of an organic-inorganic block copolymer for nano-ordered ceramics. *Nat Nano,* 2, 43-46.

Mao, Y., T.-J. Park & S. S. Wong (2005) Synthesis of classes of ternary metal oxide nanostructures. *Chemical Communications,* 5721-5735.

Mao, Y., T.-J. Park, F. Zhang, H. Zhou & S. S. Wong (2007) Environmentally Friendly Methodologies of Nanostructure Synthesis. *Small,* 3, 1122-1139.

Meiser, F., C. Cortez & F. Caruso (2004) Biofunctionalization of Fluorescent Rare-Earth-Doped Lanthanum Phosphate Colloidal Nanoparticles. *Angewandte Chemie International Edition,* 43, 5954-5957.

Mrabet, D., M. H. Zahedi-Niaki & T.-O. Do (2008) Synthesis of Nanoporous Network Materials with High Surface Areas from the Cooperative Assemblage of Alkyl-Chain-Capped Metal/Metal Oxide Nanoparticles. *The Journal of Physical Chemistry C,* 112, 7124-7129.

Muster, J., G. T. Kim, V. Krstić, J. G. Park, Y. W. Park, S. Roth & M. Burghard (2000) Electrical Transport Through Individual Vanadium Pentoxide Nanowires. *Advanced Materials,* 12, 420-424.

Na, H. B., I. C. Song & T. Hyeon (2009) Inorganic Nanoparticles for MRI Contrast Agents. *Advanced Materials,* 21, 2133-2148.

Nagarajan, R. 2008. Nanoparticles: Building Blocks for Nanotechnology. In *Nanoparticles: Synthesis, Stabilization, Passivation, and Functionalization,* 2-14. American Chemical Society.

Nguyen, T.-D., C.-T. Dinh & T.-O. Do (2009a) Monodisperse Samarium and Cerium Orthovanadate Nanocrystals and Metal Oxidation States on the Nanocrystal Surface. *Langmuir,* 25, 11142-11148.

Nguyen, T.-D., C.-T. Dinh & T.-O. Do (2010) Shape- and Size-Controlled Synthesis of Monoclinic ErOOH and Cubic Er_2O_3 from Micro- to Nanostructures and Their Upconversion Luminescence. *ACS Nano,* 4, 2263-2273.

Nguyen, T.-D., C.-T. Dinh & T.-O. Do (2011a) Two-Phase Synthesis of Colloidal Annular-Shaped $CexLa_{1-x}CO_3OH$ Nanoarchitectures Assemblied from Small Particles and Their Thermal Conversion to Derived Mixed Oxides. *Inorganic Chemistry,* 2011, 50, 1309-1320.

Nguyen, T.-D., C.-T. Dinh, D.-T. Nguyen & T.-O. Do (2009b) A Novel Approach for Monodisperse Samarium Orthovanadate Nanocrystals: Controlled Synthesis and Characterization. *The Journal of Physical Chemistry C,* 113, 18584-18595.

Nguyen, T.-D. & T.-O. Do (2009a) General Two-Phase Routes to Synthesize Colloidal Metal Oxide Nanocrystals: Simple Synthesis and Ordered Self-Assembly Structures. *The Journal of Physical Chemistry C,* 113, 11204-11214.

Nguyen, T.-D. & T.-O. Do (2009b) Solvo-Hydrothermal Approach for the Shape-Selective Synthesis of Vanadium Oxide Nanocrystals and Their Characterization. *Langmuir,* 25, 5322-5332.

Nguyen, T.-D., D. Mrabet & T.-O. Do (2008) Controlled Self-Assembly of Sm_2O_3 Nanoparticles into Nanorods: Simple and Large Scale Synthesis using Bulk Sm_2O_3 Powders. *The Journal of Physical Chemistry C,* 112, 15226-15235.

Nguyen, T.-D., D. Mrabet, T.-T.-D. Vu, C.-T. Dinh & T.-O. Do (2011b) Biomolecule-assisted route for shape-controlled synthesis of single-crystalline $MnWO_4$ nanoparticles and spontaneous assembly of polypeptide-stabilized mesocrystal microspheres. *CrystEngComm,* 2011, 13, 1450-1460.

Norris, D. J., A. L. Efros & S. C. Erwin (2008) Doped Nanocrystals. *Science,* 319, 1776-1779.

Pan, D., Q. Wang, S. Jiang, X. Ji & L. An (2005a) Synthesis of Extremely Small CdSe and Highly Luminescent CdSe/CdS Core–Shell Nanocrystals via a Novel Two-Phase Thermal Approach. *Advanced Materials,* 17, 176-179.

Pan, D., Q. Wang, J. Pang, S. Jiang, X. Ji & L. An (2006) Semiconductor "Nano-Onions" with Multifold Alternating CdS/CdSe or CdSe/CdS Structure. *Chemistry of Materials,* 18, 4253-4258.

Pan, D., N. Zhao, Q. Wang, S. Jiang, X. Ji & L. An (2005b) Facile Synthesis and Characterization of Luminescent TiO_2 Nanocrystals. *Advanced Materials,* 17, 1991-1995.

Park, J., J. Joo, S. G. Kwon, Y. Jang & T. Hyeon (2007) Synthesis of Monodisperse Spherical Nanocrystals. *Angewandte Chemie International Edition,* 46, 4630-4660.

Pileni, M. P. (2007) Control of the Size and Shape of Inorganic Nanocrystals at Various Scales from Nano to Macrodomains. *The Journal of Physical Chemistry C,* 111, 9019-9038.

Pinna, N., M. Willinger, K. Weiss, J. Urban & R. Schlögl (2003) Local Structure of Nanoscopic Materials:□ V_2O_5 Nanorods and Nanowires. *Nano Letters,* 3, 1131-1134.

Poizot, P., S. Laruelle, S. Grugeon, L. Dupont & J. M. Tarascon (2000) Nano-sized transition-metal oxides as negative-electrode materials for lithium-ion batteries. *Nature,* 407, 496-499.

Qu, W., W. Wlodarski & J.-U. Meyer (2000) Comparative study on micromorphology and humidity sensitive properties of thin-film and thick-film humidity sensors based on semiconducting $MnWO_4$. *Sensors and Actuators B: Chemical,* 64, 76-82.

Rao, C. N. R., A. Müller & A. K. Cheetham. 2005. *Nanomaterials – An Introduction*. Wiley-VCH Verlag GmbH & Co. KGaA.

Reddy, B. M., L. Katta & G. Thrimurthulu (2009) Novel Nanocrystalline $Ce_{1-x}La_xO_{2-\delta}$ (x = 0.2) Solid Solutions: Structural Characteristics and Catalytic Performance. *Chemistry of Materials,* 22, 467-475.

Redl, F. X., K. S. Cho, C. B. Murray & S. O'Brien (2003) Three-dimensional binary superlattices of magnetic nanocrystals and semiconductor quantum dots. *Nature,* 423, 968-971.

Schmid, G. 2005. *General Introduction*. Wiley-VCH Verlag GmbH & Co. KGaA.

Searcy, A. W. (1983) The equilibrium shapes of crystals and of cavities in crystals. *Journal of Solid State Chemistry,* 48, 93-99.

Sellinger, A., P. M. Weiss, A. Nguyen, Y. Lu, R. A. Assink, W. Gong & C. J. Brinker (1998) Continuous self-assembly of organic-inorganic nanocomposite coatings that mimic nacre. *Nature,* 394, 256-260.

Seshadri, R. 2005. *Oxide Nanoparticles.* Wiley-VCH Verlag GmbH & Co. KGaA.

Shah, P. R., M. M. Khader, J. M. Vohs & R. J. Gorte (2008) A Comparison of the Redox Properties of Vanadia-Based Mixed Oxides. *The Journal of Physical Chemistry C,* 112, 2613-2617.

Shi, H., L. Qi, J. Ma & H. Cheng (2003a) Polymer-Directed Synthesis of Penniform $BaWO_4$ Nanostructures in Reverse Micelles. *Journal of the American Chemical Society,* 125, 3450-3451.

Shi, H., L. Qi, J. Ma, H. Cheng & B. Zhu (2003b) Synthesis of Hierarchical Superstructures Consisting of BaCrO4 Nanobelts in Catanionic Reverse Micelles. *Advanced Materials,* 15, 1647-1651.

Skrabalak, S. E. & Y. Xia (2009) Pushing Nanocrystal Synthesis toward Nanomanufacturing. *ACS Nano,* 3, 10-15.

Sorensen, C. M. 2009. *Particles as Molecules.* John Wiley & Sons, Inc.

Spahr, M. E., P. Bitterli, R. Nesper, M. Müller, F. Krumeich & H. U. Nissen (1998) Redox-Active Nanotubes of Vanadium Oxide. *Angewandte Chemie International Edition,* 37, 1263-1265.

Su, L. T., A. I. Y. Tok, Y. Zhao, N. Ng & F. Y. C. Boey (2009) Synthesis and Electron−Phonon Interactions of Ce3+-Doped YAG Nanoparticles. *The Journal of Physical Chemistry C,* 113, 5974-5979.

Sun, C., J. Sun, G. Xiao, H. Zhang, X. Qiu, H. Li & L. Chen (2006a) Mesoscale Organization of Nearly Monodisperse Flowerlike Ceria Microspheres. *The Journal of Physical Chemistry B,* 110, 13445-13452.

Sun, L., Q. Guo, X. Wu, S. Luo, W. Pan, K. Huang, J. Lu, L. Ren, M. Cao & C. Hu (2006b) Synthesis and Photoluminescent Properties of Strontium Tungstate Nanostructures. *The Journal of Physical Chemistry C,* 111, 532-537.

Sun, Y., H. Liu, X. Wang, X. Kong & H. Zhang (2006c) Optical Spectroscopy and Visible Upconversion Studies of $YVO_4:Er^{3+}$ Nanocrystals Synthesized by a Hydrothermal Process. *Chemistry of Materials,* 18, 2726-2732.

Tenne, R. (2004) Materials physics: Doping control for nanotubes. *Nature,* 431, 640-641.

Thongtem, S., S. Wannapop & T. Thongtem (2009) Characterization of $MnWO_4$ with flower-like clusters produced using spray pyrolysis. *Transactions of Nonferrous Metals Society of China,* 19, s100-s104.

Tong, W., L. Li, W. Hu, T. Yan, X. Guan & G. Li (2010) Kinetic Control of $MnWO_4$ Nanoparticles for Tailored Structural Properties. *The Journal of Physical Chemistry C,* 114, 15298-15305.

Wachs, I. E. (2005) Recent conceptual advances in the catalysis science of mixed metal oxide catalytic materials. *Catalysis Today,* 100, 79-94.

Wang, D.-S., T. Xie, Q. Peng, S.-Y. Zhang, J. Chen & Y.-D. Li (2008) Direct Thermal Decomposition of Metal Nitrates in Octadecylamine to Metal Oxide Nanocrystals. *Chemistry – A European Journal,* 14, 2507-2513.

Wang, M., J.-L. Liu, Y.-X. Zhang, W. Hou, X.-L. Wu & S.-K. Xu (2009) Two-phase solvothermal synthesis of rare-earth doped NaYF4 upconversion fluorescent nanocrystals. *Materials Letters,* 63, 325-327.

Wang, Q., D. Pan, S. Jiang, X. Ji, L. An & B. Jiang (2005a) A New Two-Phase Route to High-Quality CdS Nanocrystals. *Chemistry – A European Journal,* 11, 3843-3848.

Wang, X., J. Zhuang, Q. Peng & Y. Li (2005b) A general strategy for nanocrystal synthesis. *Nature,* 437, 121-124.

Wang, X., J. Zhuang, Q. Peng & Y. Li (2006) Hydrothermal Synthesis of Rare-Earth Fluoride Nanocrystals. *Inorganic Chemistry,* 45, 6661-6665.

Watzky, M. A. & R. G. Finke (1997) Transition Metal Nanocluster Formation Kinetic and Mechanistic Studies. A New Mechanism When Hydrogen Is the Reductant: Slow, Continuous Nucleation and Fast Autocatalytic Surface Growth. *Journal of the American Chemical Society,* 119, 10382-10400.

Xia, Y., Y. Xiong, B. Lim & S. E. Skrabalak (2009) Shape-Controlled Synthesis of Metal Nanocrystals: Simple Chemistry Meets Complex Physics?. Angewandte Chemie International Edition, 48, 60-103.

Xie, J., Q. Wu, D. Zhang & Y. Ding (2009) Biomolecular-Induced Synthesis of Self-Assembled Hierarchical La(OH)CO_3 One-Dimensional Nanostructures and Its Morphology-Held Conversion toward La_2O_3 and $La(OH)_3$. *Crystal Growth & Design,* 9, 3889-3897.

Xie, R.-C. & J. Shang (2007) Morphological control in solvothermal synthesis of titanium oxide. *Journal of Materials Science,* 42, 6583-6589.

Xing, Y., S. Song, J. Feng, Y. Lei, M. Li & H. Zhang (2008) Microemulsion-mediated solvothermal synthesis and photoluminescent property of 3D flowerlike $MnWO_4$ micro/nanocomposite structure. *Solid State Sciences,* 10, 1299-1304.

Xu, Z., X. Kang, C. Li, Z. Hou, C. Zhang, D. Yang, G. Li & J. Lin (2010) Ln^{3+} (Ln = Eu, Dy, Sm, and Er) Ion-Doped YVO_4 Nano/Microcrystals with Multiform Morphologies: Hydrothermal Synthesis, Growing Mechanism, and Luminescent Properties. *Inorganic Chemistry,* 49, 6706-6715.

Yan, T., X. Wang, J. Long, H. Lin, R. Yuan, W. Dai, Z. Li & X. Fu (2008) Controlled preparation of In_2O_3, InOOH and $In(OH)_3$via a one-pot aqueous solvothermal route. *New Journal of Chemistry,* 32, 1843-1846.

Yan, Z.-G. & C.-H. Yan (2008) Controlled synthesis of rare earth nanostructures. *Journal of Materials Chemistry,* 18, 5046-5059.

Yin, Y. & A. P. Alivisatos (2005) Colloidal nanocrystal synthesis and the organic-inorganic interface. *Nature,* 437, 664-670.

Ying, J. Y. (2000) Nanostructural tailoring: Opportunities for molecular engineering in catalysis. *AIChE Journal,* 46, 1902-1906.

Yu, C., M. Yu, C. Li, C. Zhang, P. Yang & J. Lin (2008) Spindle-like Lanthanide Orthovanadate Nanoparticles: Facile Synthesis by Ultrasonic Irradiation, Characterization, and Luminescent Properties. *Crystal Growth & Design,* 9, 783-791.

Zhang, F., M. Y. Sfeir, J. A. Misewich & S. S. Wong (2008a) Room-Temperature Preparation, Characterization, and Photoluminescence Measurements of Solid Solutions of Various Compositionally-Defined Single-Crystalline Alkaline-Earth-Metal Tungstate Nanorods. *Chemistry of Materials,* 20, 5500-5512.

Zhang, F., Y. Yiu, M. C. Aronson & S. S. Wong (2008b) Exploring the Room-Temperature Synthesis and Properties of Multifunctional Doped Tungstate Nanorods. *The Journal of Physical Chemistry C,* 112, 14816-14824.

Zhang, J., S. Ohara, M. Umetsu, T. Naka, Y. Hatakeyama & T. Adschiri (2007a) Colloidal Ceria Nanocrystals: A Tailor-Made Crystal Morphology in Supercritical Water. *Advanced Materials,* 19, 203-206.

Zhang, L., C. Lu, Y. Wang & Y. Cheng (2007b) Hydrothermal synthesis and characterization of $MnWO_4$ nanoplates and their ionic conductivity. *Materials Chemistry and Physics,* 103, 433-436.

Zhang, Q., X. Chen, Y. Zhou, G. Zhang & S.-H. Yu (2007c) Synthesis of ZnWO4@MWO4 (M = Mn, Fe) Core−Shell Nanorods with Optical and Antiferromagnetic Property by Oriented Attachment Mechanism. *The Journal of Physical Chemistry C,* 111, 3927-3933.

Zhao, N., W. Nie, X. Liu, S. Tian, Y. Zhang & X. Ji (2008) Shape- and Size-Controlled Synthesis and Dependent Magnetic Properties of Nearly Monodisperse Mn_3O_4 Nanocrystals. *Small,* 4, 77-81.

Zhao, N., W. Nie, J. Mao, M. Yang, D. Wang, Y. Lin, Y. Fan, Z. Zhao, H. Wei & X. Ji (2010) A General Synthesis of High-Quality Inorganic Nanocrystals via a Two-Phase Method. *Small,* 6, 2558-2565.

Zhao, N., D. Pan, W. Nie & X. Ji (2006) Two-Phase Synthesis of Shape-Controlled Colloidal Zirconia Nanocrystals and Their Characterization. *Journal of the American Chemical Society,* 128, 10118-10124.

Zhou, H., Y. Yiu, M. C. Aronson & S. S. Wong (2008a) Ambient template synthesis of multiferroic $MnWO_4$ nanowires and nanowire arrays. *Journal of Solid State Chemistry,* 181, 1539-1545.

Zhou, Y.-X., H.-B. Yao, Q. Zhang, J.-Y. Gong, S.-J. Liu & S.-H. Yu (2009) Hierarchical FeWO4 Microcrystals: Solvothermal Synthesis and Their Photocatalytic and Magnetic Properties. *Inorganic Chemistry,* 48, 1082-1090.

Zhou, Y.-X., Q. Zhang, J.-Y. Gong & S.-H. Yu (2008b) Surfactant-Assisted Hydrothermal Synthesis and Magnetic Properties of Urchin-like $MnWO_4$ Microspheres. *The Journal of Physical Chemistry C,* 112, 13383-13389.

3

Colloidal Hybrid Nanocrystals: Synthesis, Properties, and Perspectives

Jie Zeng[1,2], Xiaoping Wang[1] and J. G. Hou[1]
[1]*Hefei National Laboratory of Physical Science at Microscale, University of Science and Technology of China, Hefei*
[2]*Department of Biomedical Engineering, Washington University, St. Louis*
[1]*P. R. China*
[2]*USA*

1. Introduction

Colloidal nanocrystals are mesoscopic materials occupying the region between the atomistic and the macroscopic worlds.[1] The physical properties of these tiny crystals manifest the transition from molecular limit to the solid state providing benchmark systems for experimental and theoretical studies.[1] In recent years, there have been tremendous developments in the synthetic control of nanocrystal size, shape, and composition, thus allowing the tailoring of their properties.[2-5] These properties, alongside with the ability to manipulate them using the powerful scaffolds of chemical syntheses, also leads to potential applications of colloidal nanocrystals in diverse fields involving physics and biology.[6-11]

A tremendous amount of research in recent years has been directed towards the design and synthesis of multicomponent nanostructures.[5-13] These nanostructures combine two or more components into one solid structure without the use of organic linking molecules, therefore each component is in direct contact with another through one or more of its crystal facets.[14] Such complex structures have the potential to combine magnetic, plasmonic, semiconducting[15] and other physical or chemical properties into a single nanomaterial.[9, 16-17] Efforts to create these multicomponent nanostructures have largely been driven by their increased functionality.[18-19] This increase in function combined with the potential for enhanced, and often tunable, chemical and physical properties makes these nanostructures useful in applications otherwise inaccessible by their single component counterparts.[20] For example, such nanostructures have already found applications in areas such as multimodal biomedical imaging/sensing[21] and photocatalysis.[22-23]

To date, a variety of multicomponent nanostructures have already been successfully synthesized and can be sorted into two groups: *i*) Centrosymmetric structures that are symmetric about a center point, *i.e.* core-shell nanostructures, which most often combine two or more semiconducting materials in order to enhance their photoluminescence;[24] and *ii*) Non-centrosymmetric structures that are asymmetric relative to a center point, *i.e.* hybrid nanostructures.[11, 25-27] In recent years, many research groups have successfully synthesized various multicomponent nanomaterials including semiconductor-semiconductor,[28-29] metal-metal,[30-32] semiconductor-metal,[33-39] and insulator/metal systems[26,40-42] with either a centrosymmetric structure or non-centrosymmetric structure. This chapter will focus on the non-centrosymmetric hybrid nanostructures. The final structure of these multicomponent materials can take a variety of forms, the most common being that of a "dumbbell" or "matchstick". Another common structure is the heterodimer, which consists of two spherical nanocrystals in intimate contact with each other through selected crystal facets.[43]

Advances in the solution-based synthesis of single component nanostructures have given us the building blocks to create such intricate hybrid nanostructures. It is well known that the properties of single component nanostructures depend greatly on the size and shape of the nanomaterial,[44-45] and that a solution-based synthetic route allows us to access an enormous range of morphologies.[46-55] Likewise, the wet-chemical synthesis of hybrid nanostructures has the potential to create a wide variety of interesting structures.[56-60] Such syntheses will also provide us with the avenues to study the formation mechanisms involved which can in turn be exploited to create targeted multicomponent nanostructures with particular chemical and physical properties. The synthesis of single component nanostructures has been studied for many years and researchers can now manipulate the formation mechanisms involved to produce a desired shape in a high yield. The same formation mechanisms can be useful for the synthesis of some hybrid nanostructures. For example, Ostwald ripening, a well-known mechanism existing in the growth of a crystal, plays a more interesting role in hybrid nanostructure than in single component systems as seen in the synthesis of gold-tipped CdSe nanorods (see Section 2). Due to the increased complexity of hybrid nanostructures, there are many intricate growth mechanisms involved in their syntheses. In addition, the growth mechanisms involved in single-component systems do not always extrapolate to the hybrid nanostructures. In this case, mechanisms such as heterogeneous nucleation and core-shell versus hybrid growth come into play. Further development of the wet-chemical synthesis of hybrid nanostructures will enable us to gain a better understanding of the growth mechanisms involved and learn how to use them to design and synthesize targeted hybrid nanostructures.

The complex structure of hybrid nanostructures offers an interesting system to study how the chemical and physical properties of the individual nanomaterials are affected by the intimate interaction with another component. At the least, multicomponent hybrid nanostructures combine the properties of the individual components independent of each other, creating a multifunctional material.[61] On a higher level, they also have the potential to enhance the inherent properties and possibly even create new ones.[23] For example, Au NPs are usually considered to be catalytically inert. However, when they are combined with a metal oxide compound, *i.e.* Fe_3O_4, the hybrid structure shows high catalytic activity for CO oxidation.[62] This increase in activity arises from the electron transfer from the oxide support to the Fermi level of the adjacent Au NPs.[63-64] In addition, these hybrid nanostructures have been shown to add another parameter with which to finely tune the properties of the individual components.[65-68] For example, Yu and co-workers have demonstrated the ability to tune the UV-vis emission of ZnO-Au nanorods by simply changing the mole ratio between ZnO and the gold precursor, $HAuCl_4$.[67]

In this chapter, we will discuss the growth mechanisms involved in the synthesis of hybrid nanostructures. We will then present a case-study of various hybrid nanostructures that have already been successfully synthesized. Finally, we will detail their physical properties and some of the interesting applications of these multicomponent hybrid nanostructures, followed by perspectives.

2. Mechanisms of the formation of colloidal hybrid nanocrystals

Multiple factors can affect size and shape of nanocrystals produced in a solution through the nucleation and growth process. Taking metallic materials as an example, at the very initial stages, the zero-valence metal atoms form through either reduction of ions or bond breaking of compounds. These metal atoms collide to produce small clusters that are thermodynamically unstable and can dissolve before they reach a critical radius (r^*) or overcome a critical free energy barrier (ΔG^*) and become thermodynamically stable nuclei. These nuclei grow into nanoparticles at the consumption of free atoms in solution or unstable small clusters ($r < r^*$).69 In the classical nucleation theory, the nucleation process can be described according to Gibbs free energy. The excess free energy (ΔG_r), which contains two competing terms, *i.e.* the changes in surface and bulk free energies, reaches the maximum when clusters grow to the critical size. The excess free energy can be described by the following equations.

$$\Delta G_r = 4\pi r^2 \gamma + \frac{4}{3}\pi r^3 \Delta G_V = 4\pi r^2 \gamma - \frac{4}{3}\pi r^3 \cdot \frac{RTlnS}{V_m} \quad (1)$$

where r is radius of the clusters, γ surface free energy per unit area, R ideal gas constant, ΔG_V change of free energy between solute atoms in solution and bulk crystal per unit volume, T reaction temperature, V_m molar volume of bulk crystal, and S ratio between solute concentrations at saturation and equilibrium conditions. Based on the Eq. (1) and Boltzmann distribution, the number of clusters (N_r) as a function of radius r can be written as:

$$N_r = N_0 \times exp\left(-\frac{\Delta G_r}{RT}\right) \quad (2)$$

where N_0 is the total number of free solute atoms per unit volume. When solute is undersaturated, namely, $S \leq 1$, ΔG_r remains positive and increases with growth of clusters. Nucleation is not favored in such systems. Only when the solute is supersaturated, *i.e.* $S > 1$, ΔG_r can decrease with the increase of cluster radius, and clusters become stable with the growth. The relationship between ΔG_r or N_r and r under the supersaturated condition can be illustrated in Figure 1.

As illustrated in Figure 1A, there is a critical excess free energy, ΔG^*, associated with the critical radius of cluster, r^*. When the radius is smaller than the critical value, the system lowers its free energy by dissolving clusters, and at the same time new ones form due to spontaneous collisions. The expressions of critical radius r^* and maximum excess free energy ΔG^*can then be written as the followings.[69]

$$r^* = -\frac{2\gamma}{\Delta G_V} = -\frac{2\gamma V_m}{RTlnS} \quad (3)$$

$$\Delta G^* = \frac{16\pi\gamma^3 V_m^2}{3(RTlnS)^2} \quad (4)$$

The smaller the critical radius or the maximum excess free energy is, the easier the nuclei form, since the clusters need to incorporate few atoms or overcome a small energy barrier to

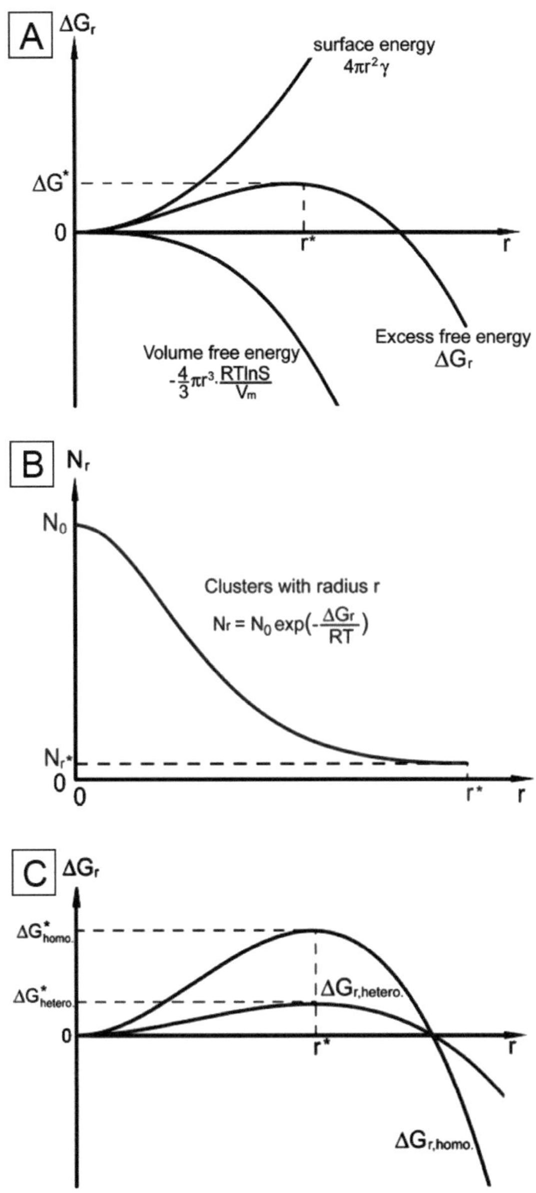

Fig. 1. Illustrations of (A) the generally overall excess free energy, (B) the average number of cluster, and (C) the specifically overall excess free energy for heterogeneous and homogeneous nucleation, as functions of cluster size, respectively.

become stable. Nuclei can hardly form if a reaction system has large values of critical radius or maximum excess free energy. However, when small nanocrystals are present in the system, they can release some of the surface energy through interaction with other species. This interaction changes the subsequent nucleation and growth of clusters. In essence, the critical radius, excess Gibbs free energy, number of nuclei, and rate for a second species formed on existing particles can be described using the following equations.[69-72]

$$r^*_{hetero} = -\frac{2\gamma V_m}{RTlnS} \quad (5)$$

$$\Delta G^*_{hetero} = \frac{16\pi\gamma^3 V_m^2}{3(RTlnS)^2} \cdot S(\theta) \quad (6)$$

where $S(\theta)$, a numerical value less than unity, is a shape factor describing the geometric relationship between heterogeneous nucleation sites per unit volume. The overall excess free energy (ΔGr) as a function of cluster size (r) for homogeneous and heterogeneous nucleation and growth can be schematically depicted in Figure 1C. The critical energy barrier is generally smaller for heterogeneous nucleation than that for homogeneous one by a factor of $S(\theta)$, while they should have similar critical nucleus radius.

In addition to the theories and equations mentioned above, we can also use the kinetic theory of nucleation, developed based on Gibb's formalism, to describe how the preference for homogeneous nucleation or heterogeneous nucleation is different in different synthetic environments. Using the Arrhenius reaction velocity equation commonly used to determine the rate of a thermally activated process, the rate of nucleation, J, can be expressed as:

$$J = A\exp(-\frac{\Delta G}{kT}) = A\exp\left\{-f\frac{16\pi\Omega^2\gamma_i^3}{3(kT)^3\left[\ln(1+\sigma)\right]^2}\right\},\ 0 < f \le 1 \quad (7)$$

where A is the pre-exponential factor, Ω is the molecular volume, γ_i is the interfacial free energy between the solid nuclei and the liquid phase, and σ is the relative supersaturation. The variable, f, is a parameter describing the influence of foreign bodies on the nucleation energy barrier. For homogeneous nucleation f is equal to 1. Since f can never be greater than 1, heterogeneous nucleation is always kinetically more favorable than homogeneous nucleation. Also, for a given synthetic system where the temperature, T, and the interfacial energy, γ_i, between the two materials are fixed, it can be derived that heterogeneous nucleation will dominate at low supersaturations while homogeneous nucleation is preferred at high supersaturations.[74] In other words, genuine homogeneous nucleation can be regarded as an upper-limit to heterogeneous nucleation in that it requires a very high degree of supersaturation, which may not be easily achieved under standard reaction conditions.[75] Although many models have been proposed over the last few decades to describe the nucleation and growth of small particles,[76-81] we still do not have a thorough understanding of the nature of homogeneous and heterogeneous nucleation, particularly when foreign particles are involved.

Many of the reports concerning the synthesis of multicomponent nanostructures recently have presented either a core-shell or hybrid structure, both of which originate from a heterogeneous nucleation process. Despite this similarity, the details of their epitaxial growth mechanisms are not the same and result in completely different structures.

To reveal the detailed mechanism, realizing the role of lattice mismatch (F) that exists between the deposited material and that in existing seed particles played in the seeded growth processes is definitely important. Note that the size of atoms and the crystallines of these two materials are usually different, and thus leading to specifically preferred arrangements for these materials. Considering material A nucleates on the surface of material B that acts as a seeding particle and provides heterogeneous nucleation sites, the lattice mismatch F can be defined as:[81]

$$F = (a_A - a_B) / a_B \tag{8}$$

where a_A and a_B are lattice parameters of growing planes for materials A and B, respectively. This mismatch induces a positive strain energy (γ_{strain}), which increases rapidly with the growth of nuclei. The larger the mismatch is, the more positive this strain energy becomes. Interfaces formed between metals A and B during the nucleation and growth, results in a new energy term. The interplay between the interfacial energy (γ_i) and surface energy of these two materials (γ_A and γ_B) may determine the outcome of growth modes. The overall excess energy ($\Delta\gamma$) is created to include the contribution of these four energy terms according to the following equation:[81]

$$\Delta\gamma = \gamma_A + \gamma_i + \gamma_{strain} - \gamma_B \tag{9}$$

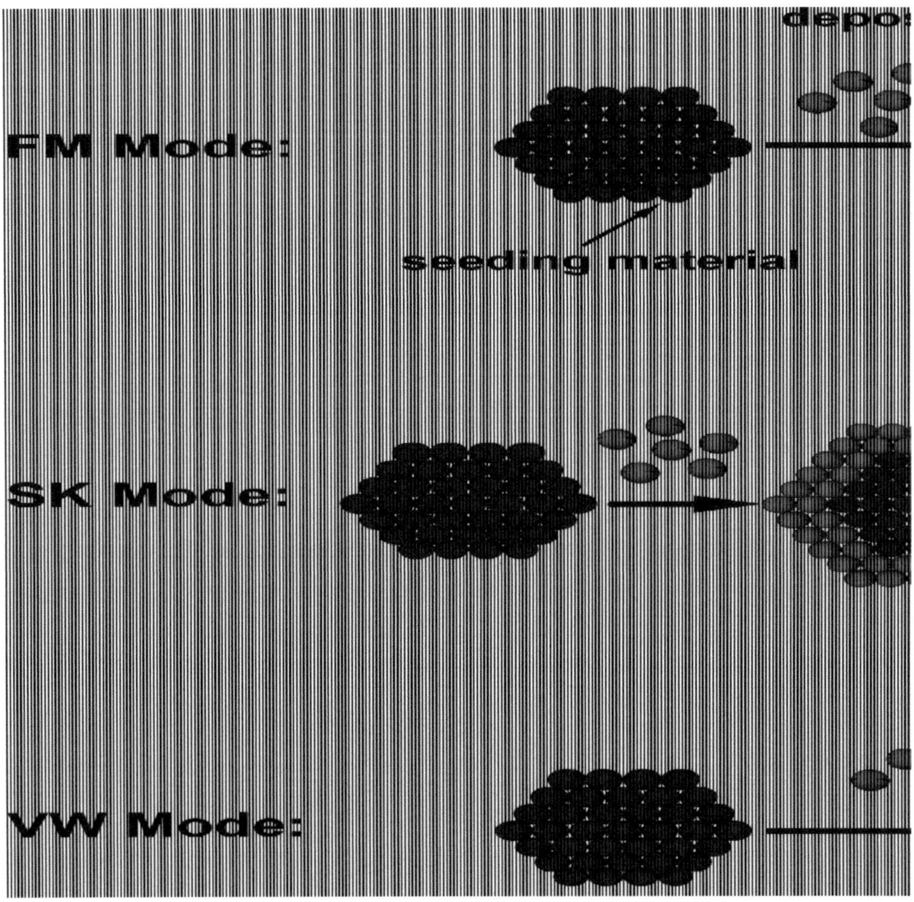

Fig. 2. Schematic illustrations of the FM Mode, the VW Mode, and the SK Mode for the overgrowth of a secondary material on a seed particle. *Courtesy: Nano Today, 2009, 4, 143-164.*

In general, three different types of growth modes can be observed according to the values of overall excess energy $\Delta\gamma$ (Fig. 2).[82] They are: *i*) Frank-van der Merwe (FM) or layer-by-layer Mode, *ii*) Volmer-Weber (VW) or 3D island Mode, and *iii*) Stranski-Krastanow (SK) or 3D island-on-wetting-layer Mode. In the FM Mode, layer-by-layer growth requires that the sum of the surface energy of the deposited material, the strain energy, and the interface energy be less than the surface energy of the seeding material in order for wetting to occur. This means that the value of the overall excess free energy should be negative. On the contrary, if the value of $\Delta\gamma$ is positive, which means the deposition will increase the surface energy of the structure, the layer can lower its total energy by forming isolated, thick 3D islands and can relax the strain by using interfacial misfit dislocations.[83] This type of growth is defined by the SK Mode. It has been demonstrated experimentally that the epitaxial growth of several semiconductor/semiconductor systems with large lattice-mismatches (>~2%) (e.g., Ge/Si,[84-85] InGaAs/GaAs,[86-87] and InGaAs/GaAs[88]) proceeds via the SK Mode.[89] It is found that during layer-by-layer growth the strain energy γ_{strain} increases linearly with the number of strained layers. As a result, when the thickness arrives at a certain point, the sum of γ_i, γ_A, and γ_{strain} exceeds γ_B. At this point the growth mode transforms from the FM Mode to the SK Mode leading to the formation of 3D islands on the 2D wetting layer. This particular type of growth is known as the VW Mode. Those three modes are very useful for predicting and tailoring the growth outcome of multicomponent nanostructures. For example, in order to get a hybrid nanostructure, the overgrowth should satisfy the conditions applied in either the SK Mode or the VW Mode. By doing so, the growth will form isolated 3D islands rather than layers resulting in a hybrid nanostructure instead of a core-shell structure.

In the epitaxial growth of multicomponent nanostructures, many other factors also have an important role in the overgrowth of the secondary phase. For example, a large lattice mismatch prevents the conformal overgrowth of the secondary material. If the lattice mismatch between the two materials is greater than ~5%, the strain energy term will have a large influence making the SK Mode or the VW Mode the preferred growth mode. In addition to the surface energy and lattice mismatch, other factors such as metal bond energy,[90] pH of growth solution,[91] and stirring rate[73] should also be considered in various overgrowth systems.

While these concepts are first developed to explain the film growth, there should not be a fundamental difference between film growth and colloidal synthesis in terms of heterogeneous nucleation. The three growth modes should be able to use, at least in some cases, in understanding the formation of a range of complex nanostructures synthesized in solution phase, such as core-shell and hybrid nanostructures.

Additional considerations need to be taken into consideration during growth of heterogeneous nuclei, which can determine the finial morphology. One key factor is the so-called Ostwald ripening.[92-94] The driving force for this effect arises from the overall energy of the particles involved.[95] In order to lower the total energy of the system, smaller crystallites, which have higher surface energy, dissolve into solution over time. The dissolved species then regrow onto the larger particles.[96-97] This general definition only requires that mass be transferred between particles, however, in other cases both mass and electron transfers are required in order for an Ostwald ripening process to occur.

Banin and co-workers have demonstrated an interesting example of this electrochemical Ostwald ripening process, in which electron transfer was involved across connected particles (Fig. 3).[98] They have developed a series of methods to selectively grow gold

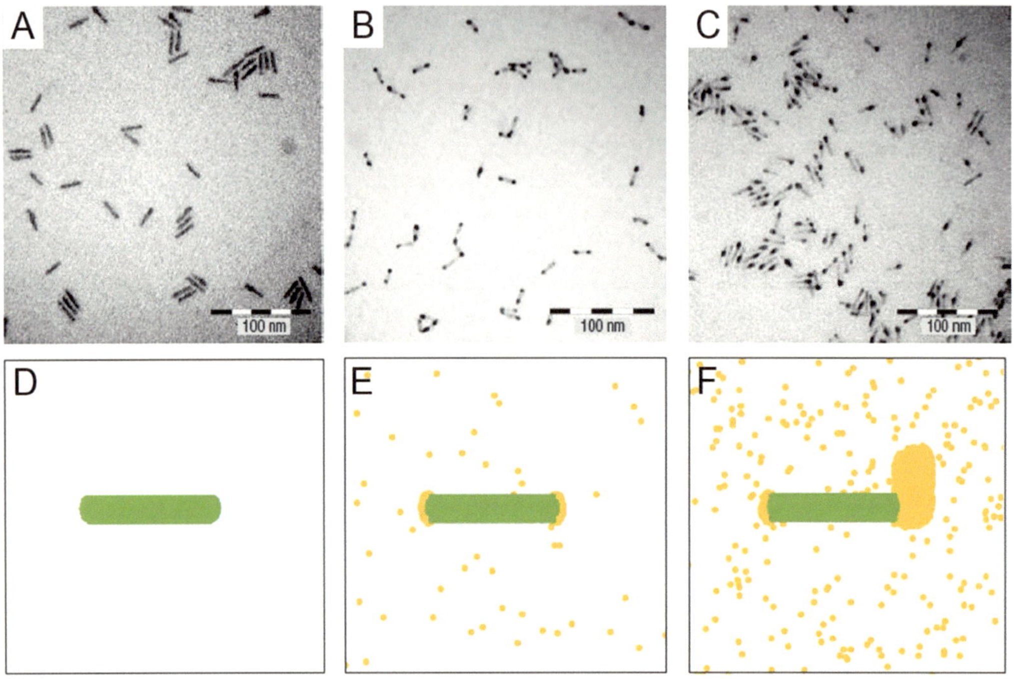

Fig. 3. Effect of increasing Au/rod molar ratio on the growth. Top panels show the experimental results. (A) Original rods of dimension 25×4 nm. (B) "Nano-dumbbells" (NDBs) after adding 1.3 μmol of gold precursors to (A). (C) "Nano-bell-tongues" (NBTs) after adding 1.3 μmol of gold precursors to (B). Lower panels show the theoretical results. (D) Presentation of the system on a two-dimensional lattice. (E) Snapshot of final morphology at low gold concentration (the two-dimensional gold density is 0.001) yielding two-sided growth. (F) Snapshot of final morphology at high gold density (0.01) showing one-sided growth. *Courtesy: Nat. Mater., 2005, 4, 855-863.*

nanoparticles onto the tips of colloidal CdSe nanorods and tetrapods. During the growth of Au-tipped CdSe nanorods, a structural transition from a two-sided "nano-dumbbell" (NDB) to a one-sided "nano-bell-tongue" (NBT) occurs as the gold concentration is increased.[98] They proposed such following process: as the gold precursor in the solution becomes depleted, the thermodynamically stable size for the gold islands increases. Gold islands smaller than this thermodynamically stable size will dissolve back into the growth medium. As gold atoms are oxidized to ions, which can complex with the surfactants in solution, electrons are released into the nanorod. These electrons can then diffuse along the surface of the nanorod to the larger gold particle on the opposite side. The electrons at the second metal tip can then reduce a gold ion from solution, resulting in the overall transport of material from one tip to the other. They observed that during the formation of NBTs no ripening occurs among different NBTs, *i.e.* all of the nanorods have at least one gold tip. They also noticed that no such ripening occurred during the growth of Au-tipped tetrapods. These observations suggest that the electrons are transferred along the semiconductor nanorod rather than through solution. The nanorods transport electrons from the oxidized end to the other so they can reduce the free metal ions in solution. In the case of tetrapods, the large amounts of crystal defects in the central region act as a barrier to the free movement of electrons, therefore no ripening can occur. Electrochemical Ostwald ripening has been shown previously to only occur on a conductive substrate which facilitates electron transfer from smaller particles to larger ones.[99] Therefore, some kind of conductive pathway is necessary in order for this type of Ostwald ripening to occur.

3. Designer hybrid nanocrystals

In this section, we will review some of the successful approaches that have been reported for the synthesis of hybrid nanoparticles.[100-112]

Sun and co-workers synthesized dumbbell-like Au-Fe_3O_4 nanoparticles by using controlled nucleation of Fe_3O_4 on Au nanoparticles without any pretreatment of the Au surface.[100] The decomposition of $Fe(CO)_5$, followed by room-temperature air oxidation leads to iron nucleation on the surface of the Au nanoparticles and the formation of Au-Fe_3O_4 hybrid nanoparticles. One advantage of this method is that each part of the dumbbell can be manipulated by regulating the synthetic conditions. The size of both the Au component and the Fe_3O_4 component can be controlled up to 20 nm in diameter by changing the ratio between $HAuCl_4$ and OLA, and the ratio between $Fe(CO)_5$ and Au independently. As shown in Figure 4, a Fe_3O_4 (111) plane grows onto an Au (111) plane, giving the dumbbell-like structure. When the solvent was changed from a nonpolar hydrocarbon to slightly polarized diphenyl ether, flower-like Au-Fe_3O_4 hybrid nanoparticles were obtained. Similar flower-like Au-Fe_3O_4 hybrid nanostructures were reported by Grzybowski and co-workers.[113] The Fe_3O_4 component can also be converted to γ-Fe_2O_3 and further to α-Fe_2O_3 under high-temperature annealing conditions. In addition, since Fe_3O_4 can be dissolved while Au stays intact in 0.5 M H_2SO_4 solution, the as-prepared Au-Fe_3O_4 nanoparticles could be etched to form single-component Au nanoparticles and Fe_3O_4 nanoparticles.[114] Similar processes were applied to synthesize Au-MnO nanoflowers reported by Tremel and co-workers[101] where the size and morphology of the nanostructure could be varied by changing the molar ratio of $Mn(acac)_2$ to $Au(Ac)_3$.

Alternatively, heterodimers can be formed by taking advantage of lattice mismatch and selective annealing at relatively low temperatures. Xu and co-workers reported a one-pot synthesic method for generating FePt-CdS hybrid nanostructures.[102] After the growth of the FePt seed particles, elemental S was added to the reaction. The high affinity between the FePt component and elemental S allows S to be deposited on the surface of FePt to form a FePt-S core-shell structure. The subsequent addition of $Cd(acac)_2$, HDD, and TOPO produced metastable FePt-CdS nanoparticles, where CdS was amorphous. Upon further annealing at 280 ^{o}C, the amorphous CdS crystallized, and the lattice mismatch between FePt and CdS crystals made the core/shell system metastable. This instability led to the formation of the FePt-CdS hybrid nanoparticles, as shown in Figure 5. The extension of this synthesis led to the formation of γ-Fe_2O_3-MS (M=Zn, Cd, Hg) hybrid nanoparticles where γ-Fe_2O_3 nanoparticles were made first and used as seeds.[115]

Recently, our group developed a simple, alternative method to prepare CdSe-Au hybrid nanoparticles as shown in Figure 6.[104] Wurtzite CdSe nanoparticles synthesized using the procedure developed by Peng[116] with some modifications[117-118] were used as seeds. Then a specific amount of a Au(I)-$SC_{12}H_{25}$ (-SR) stock solution mixed with a toluene solution containing the CdSe nanoparticles was reduced by ethylene glycol under agitation. The size of the Au components could be easily tuned by varying the volume ratio of seed solution to Au(I)-SR stock solution. The large lattice mismatch between CdSe and Au (~50%)[119] and the

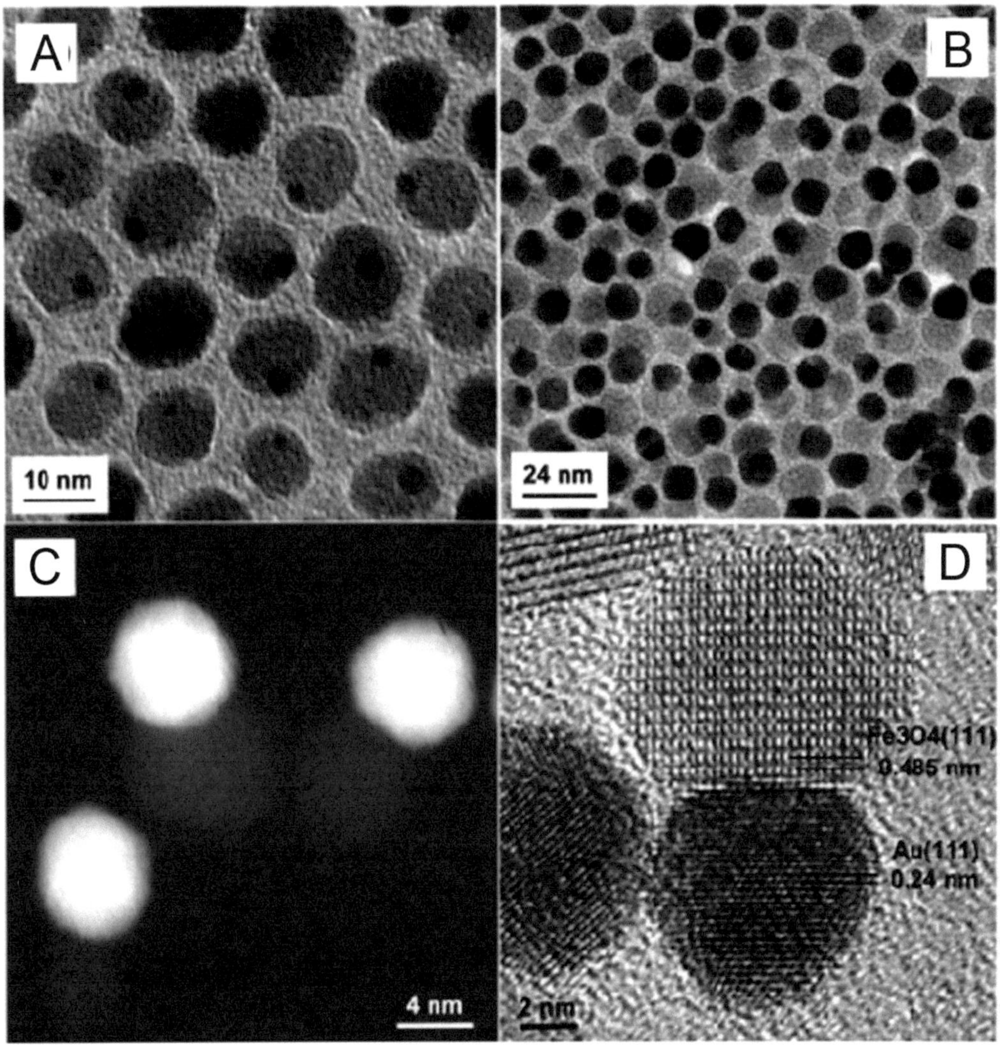

Fig. 4. TEM and STEM images of the dumbbell-like Au-Fe_3O_4 nanoparticles: (A) TEM image of the 3-14 nm Au-Fe_3O_4 particles; (B) TEM image of the 8-14 nm Au-Fe_3O_4 particles; (C) HAADF-STEM image of the 8-9 nm Au-Fe_3O_4 particles; and (D) HRTEM image of one 8-12 nm Au-Fe_3O_4 particle. *Courtesy: Nano Lett., 2005, 5, 379-382.*

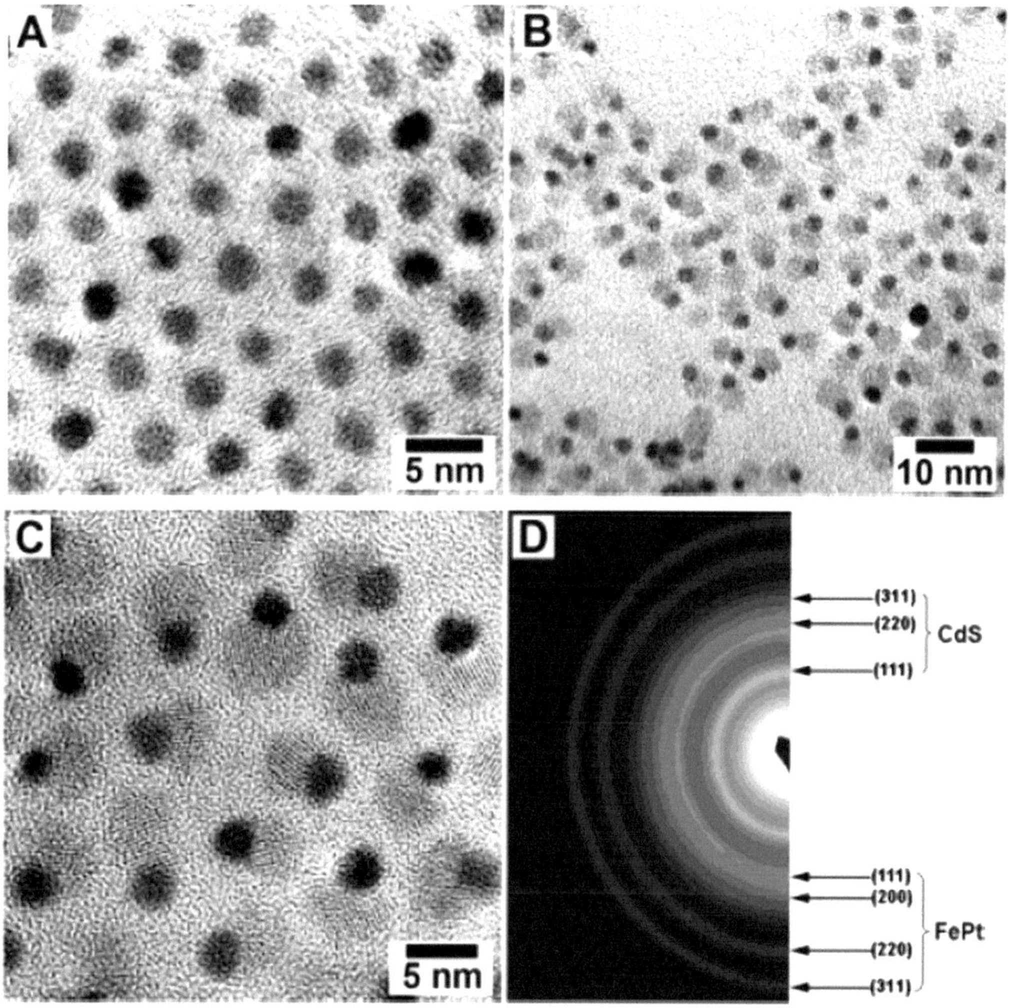

Fig. 5. (A) A TEM image of FePt nanoparticles. Schematic illustration of the synthesis of FePt-CdS hybrid nanoparticles. (B) TEM image of the as-prepared FePt seed particles. (C) TEM image and (D) High-resolution TEM image of the FePt-CdS hybrid nanoparticles. *Courtesy: J. Am. Chem. Soc., 2004, 126, 5664-5665.*

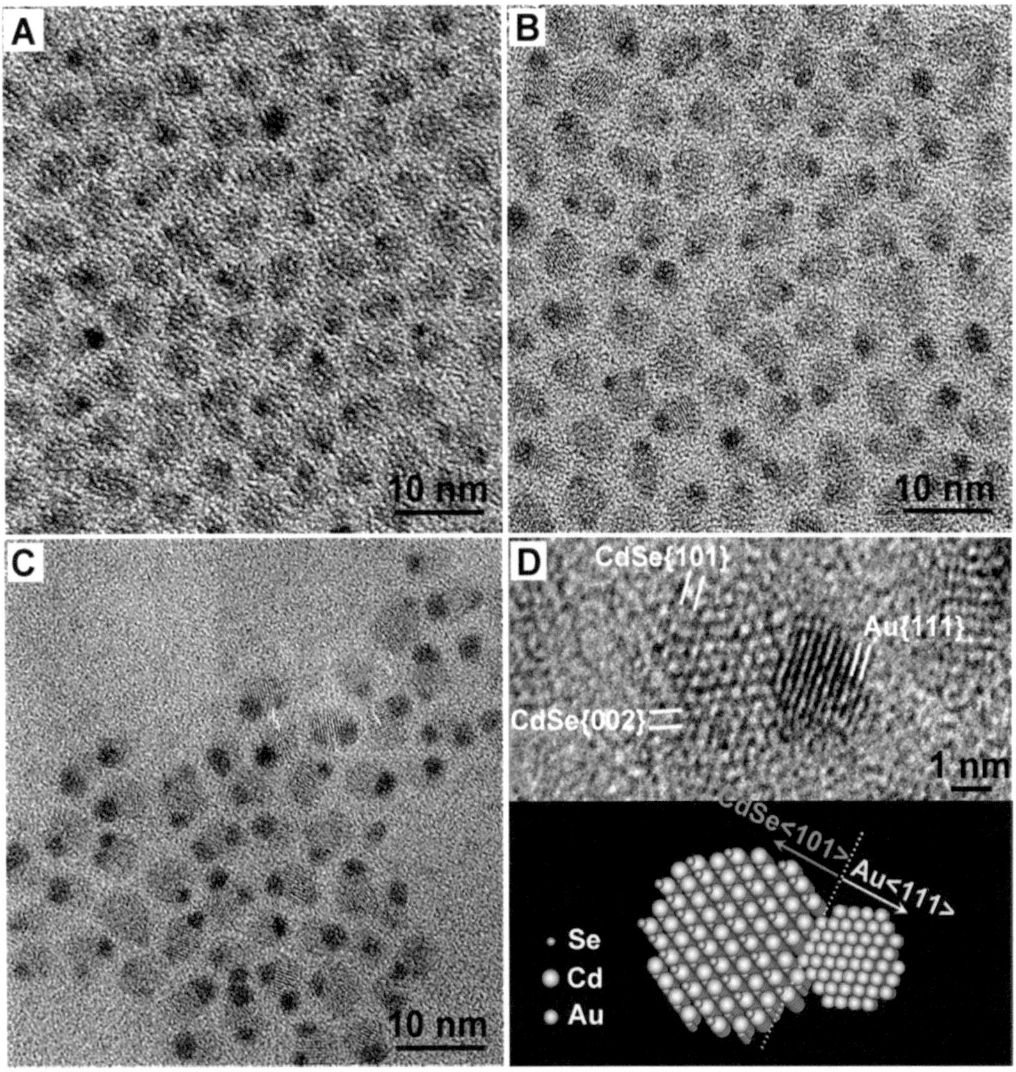

Fig. 6. (A-C) TEM images of CdSe-Au hybrid nanoparticles obtained by changing QAu from 2, 4 to 6 μmol, respectively. The nanoparticles are seen as dimeric units, all with a similar size of ~3.4 nm for the CdSe component. (D) High-resolution TEM image and a schematic drawing of the CdSe-Au hybrid nanoparticles. *Courtesy: Adv. Mater., 2010, 22, 1936-1940.*

self-catalytic reduction of Au(I) ions by the Au clusters prevented the formation of a complete Au shell on the CdSe seed. The surface of the CdSe nanoparticles provided a heterogeneous nucleation site for the formation of Au(0) clusters by simultaneously lowering the energy barrier and serving as a reasonable catalytic site for the reduction of the Au(I)-SR precursor. One important detail in the synthesis we want to emphasize here is the use of a Au(I)-SR complex. This complex has been demonstrated to decompose and serve as an effective source of elemental Au,[120-123] compared to a Au(III) organometallic complex.

Based on the method for the preparation of CdSe-Au hybrid nanoparticles,[104] our group has successfully synthesized PbSe-Au hybrid nanoparticles in which the Au component only covered one side of the PbSe particle like a cap and the overall shape of the hybrid nanoparticles was similar to a Chinese tumbler (Fig. 7, A and B). In a typical procedure, the Au(I) stock solution, ethylene glycol, and toluene containing the PbSe seed particles were mixed together and heated at 50 °C for 2 h. The color of the solution changed from green-brown to dark brown gradually, indicating the formation of PbSe-Au hybrid nanoparticles. Besides PbSe-Au, FePt-Au hybrid nanoparticles have also been obtained through the same method (Fig. 7, C and D).

Ying and co-workers reported the synthesis of PbS-Au hybrid nanostructures using a general protocol for transferring metal ions from an aqueous to an organic medium.[109] This process involves mixing an aqueous solution of $HAuCl_4$ with an ethanolic solution of DDA, and then extracting the Au(III) ions into an organic layer to form a Au(III)-DDA compound. Typically, 5 ml of PbS organosol in toluene were mixed with 5 ml of Au(III)-DDA compound in toluene and the mixture was aged for 1 h.

Compared to the growth of CdSe-Au or PbSe-Au nanoparticles, the growth rate of Au particles on Cu_2O seeds reported by our group was found to be much slower.[104] Prior studies have shown that the addition of Ag could facilitate the nucleation and growth of Au nanocrystals, and promote the self-catalyzed reduction of Au ions.[124-125] Therefore, in order to accelerate the reaction, we introduced both a Au(I) stock solution and a Ag(I) stock solution into the system to produce Cu_2O-AuAg hybrid nanoparticles.[104] Cu_2O seeds were prepared by a previously reported procedure.[126] For the synthesis of Cu_2O-AuAg hybrid nanoparticles, Au(I) stock solution and ethylene glycol were added to a light green suspension of Cu_2O seeds. After stirring for 15 min at the room temperature, the Ag(I) stock solution was added. After another 15 min, the solution became purple. The reaction was further continued in air for 2 h, and the products were precipitated with a copious amount of ethanol then collected by centrifugation. As shown in Figure 7, E and F, the shape of the resulting hybrid nanoparticles is between that of a Chinese tumbler and a gourd.

A technique based on performing seeded growth at a liquid/liquid interface under mild conditions was devised to synthesize heterodimers coupling a magnetic section and a noble metal domain.[110] Examples of hybrid nanocrystals synthesized by this biphasic strategy are shown in Figure 8. In the reported procedure, an aqueous metal salt solution was brought in contact with an immiscible organic solvent (such as dichlorobenzene, dichloromethane, hexane, or DOE) in which surfactant-capped γ-Fe_2O_3/Fe_3O_4 or FePt seeds were dissolved. Upon ultrasonic irradiation under inert atmosphere, an emulsion was formed that supposedly consisted of continuous aqueous phase containing "colloidosomes", namely organic microdroplets stabilized by the hydrophobic seeds self-assembled at the organic/water interfaces. The seeding nanocrystals provided catalytic sites onto which the Ag^+ or $AuCl_4^-$ ions were reduced to the respective Ag or Au upon sonication, respectively. As the seeds were only partially exposed to the aqueous phase, metal deposition was spatially restricted to a small surface region and proceeded self-catalytically, thus resulting in a single metal domain on each seed. These hybrid nanocrystals were proven to accommodate a site-differential surface distribution of biomolecules to enable multiple tasks in biomedicine applications.[110] In another recent study, Gu and co-workers have also synthesized similar hybrid nanostructures by mixing FeS-surface-modified Fe_3O_4 nanoparticles and Ag(Ac) in toluene using OA as the capping agent.[111]

Our group has recently reported that by using FePt-CdS heterodimers as the seeds, FePt-CdS-Au ternary hybrid nanoparticles could be obtained by applying the same methodology

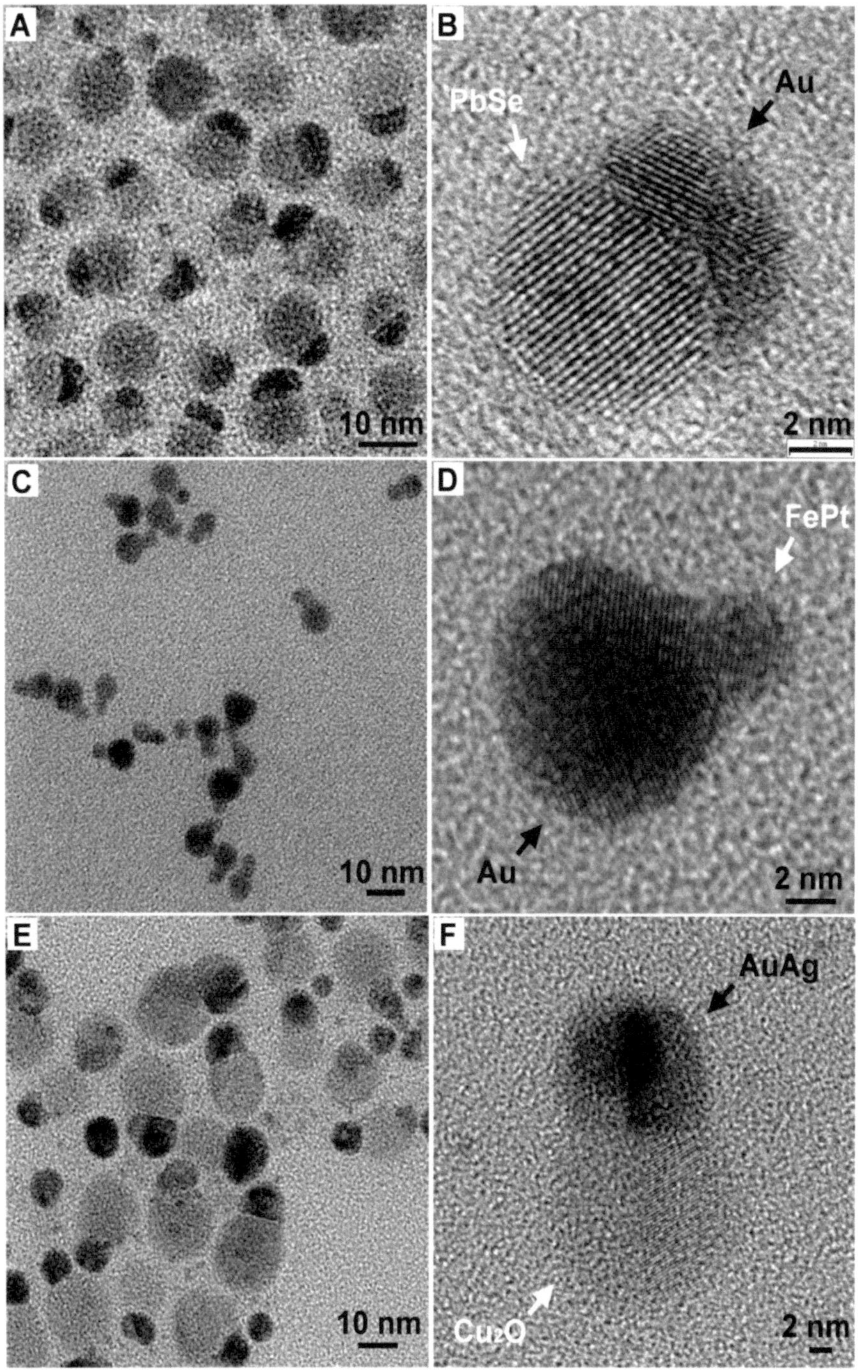

Fig. 7. (A) TEM and (B) high-resolution TEM images of PbSe-Au hybrid nanocrystals. (C) TEM and (D) high-resolution TEM images of FePt-Au hybrid nanocrystals. (E) TEM and (F) high-resolution TEM images of Cu_2O-AuAg hybrid nanocrystals. *Courtesy: Adv. Mater., 2010, 22, 1936-1940.*

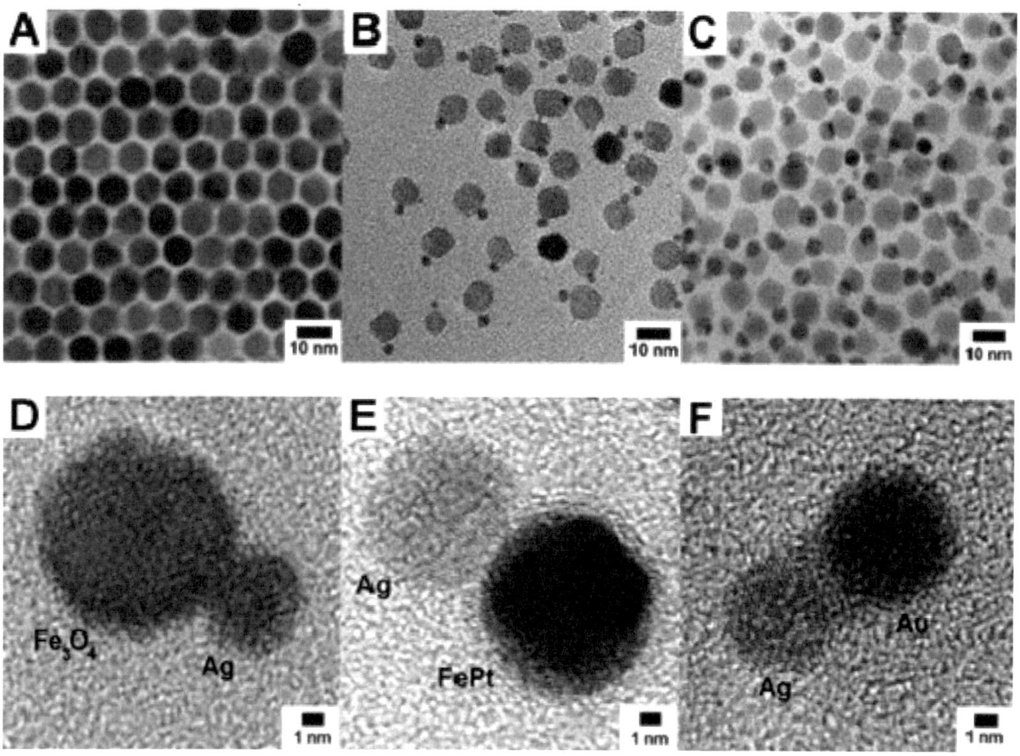

Fig. 8. TEM images of (A) the as-prepared Fe_3O_4 nanoparticles; the Fe_3O_4−Ag heterodimers after (B) 10 min reaction and (C) after reaction stopped at 30 min. HRTEM of (D) Fe_3O_4−Ag heterodimers, (E) FePt−Ag heterodimers, and (F) Au−Ag heterodimers. *Courtesy: J. Am. Chem. Soc., 2005, 127, 34-35.*

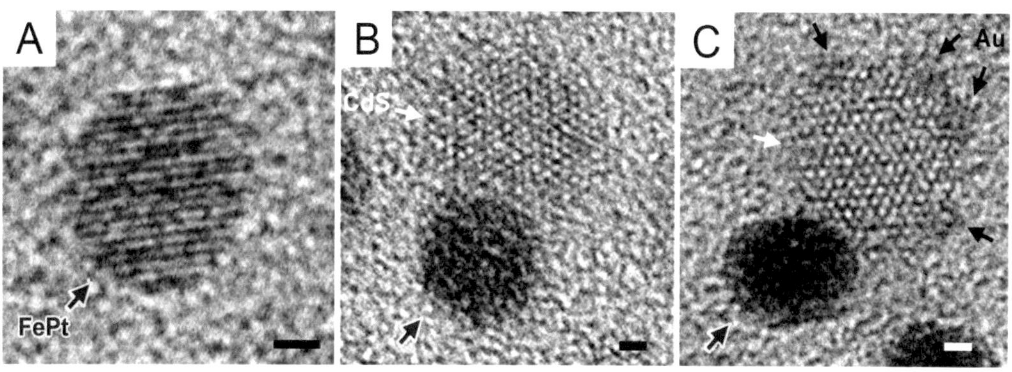

Fig. 9. High-resolution TEM images of (A) FePt, (B) FePt-CdS, and (C) FePt-CdS-Au nanocrystals. All the scale bars correspond to 1 nm. *Courtesy: Adv. Mater., 2010, 22, 1936-1940.*

used in binary systems.[104] In a typical process, two drops of ethylene glycol and a Au(I) stock solution were added to a diluted suspension of FePt-CdS seeds in toluene, and the mixture was slowly heated to 45 °C and kept for 2 h. As shown in Figure 9, even though both the FePt and the CdS components can act as a catalytic site for the heterogeneous nucleation of Au in a solution, the Au atoms preferred to deposit on the surface of the CdS portion rather than the FePt portion. This observation indicates that there exsits a strong coupling between the two portions of the hybrid seeds, and it significantly impacts on the heterogeneous nucleation process. This coupling effect probably exists widely in binary and multicomponent material systems.

We would like to point out that when spherical seeds (including CdSe, PbSe, and Cu_2O nanoparticles) were used, the hybrid dimeric nanostructures would be the most-commonly observed nanostructures, as shown in Figures, 6 and 7. However, when seed particles with a non-spherical shape were used, such as the CdS-FePt particles, some edge regions on the CdS surface might also provide additional nucleation sites to form nanoparticles with multiple components (Fig. 9C, where four patches of Au were formed on the same particle).

Zeng and co-workers reported a general approach to the synthesis of ternary hybrid nanostructures with a magnetic portion (Fe_3O_4), a metallic portion (Au), and a semiconductor portion (PbSe or PbS).[112] They used binary nanoparticles (specifically, peanut-like Fe_3O_4-Au nanoparticles with 12-nm Fe_3O_4 and 4-nm Au) as seeds. In a typical process, a Pb-oleate complex was formed by mixing PbO, OA and phenyl ether in a 100 mL three-necked flask while heating the mixture at 120 °C. Peanut-like Fe_3O_4-Au nanoparticles in hexane were then injected. After the hexane was removed by distillation, Se-TOP or S-TOP solution was rapidly injected into the reaction mixture at 160 °C. The reaction was quenched by injection of hexane after 1 min. The products are shown in Figure 10. The PbSe/PbS portions were selectively deposited on the surface of the Au portion rather than on the surface of the Fe_3O_4 portion, indicating the great impact of the coupling between two portions of one seed particle on the heterogeneous nucleation process.

The heterogeneous nucleation mechanisms involved in the systems mentioned above suggest that the well-defined structure of the seeds has a great impact on the nucleation and growth of a secondary material. It seems the site with the smallest curvature radius on the seed surface usually acts as the nucleation site for the secondary material because of the high activity at this site. For example, Au particles preferred to form on the tips of CdSe nanorods or tetrapods instead of on their lateral sides. Controlling the nucleation of the secondary material at specific sites on the seeds will help to control the chemical and physical properties of the hybrid nanostructures. However, only a few successful methods have been reported.[127] Our group exploited the use of Cu_2O nanocubes as seeds to accurately position the AuAg clusters during the heterogeneous nucleation process.[104] The experimental observation indicates that the AuAg clusters were formed at all corners of the Cu_2O nanocube.[104] This result further supports the heterogeneous nucleation mechanism described in Section 2. Another successful example has been reported by Yang and co-workers have demonstrated that during the early stages of Pt growth, or at a low Pt concentrations, Pt deposition appeared to occur preferentially only on one tip of the CdS nanorods.[128]

4. Properties and potential applications

One promising application for hybrid nanostructures is in the area of photocatalysis. We take the semiconductor-metal hybrid nanostructures as an example. When semiconductors are irradiated with light greater than their band-gap energy they excite an electron which

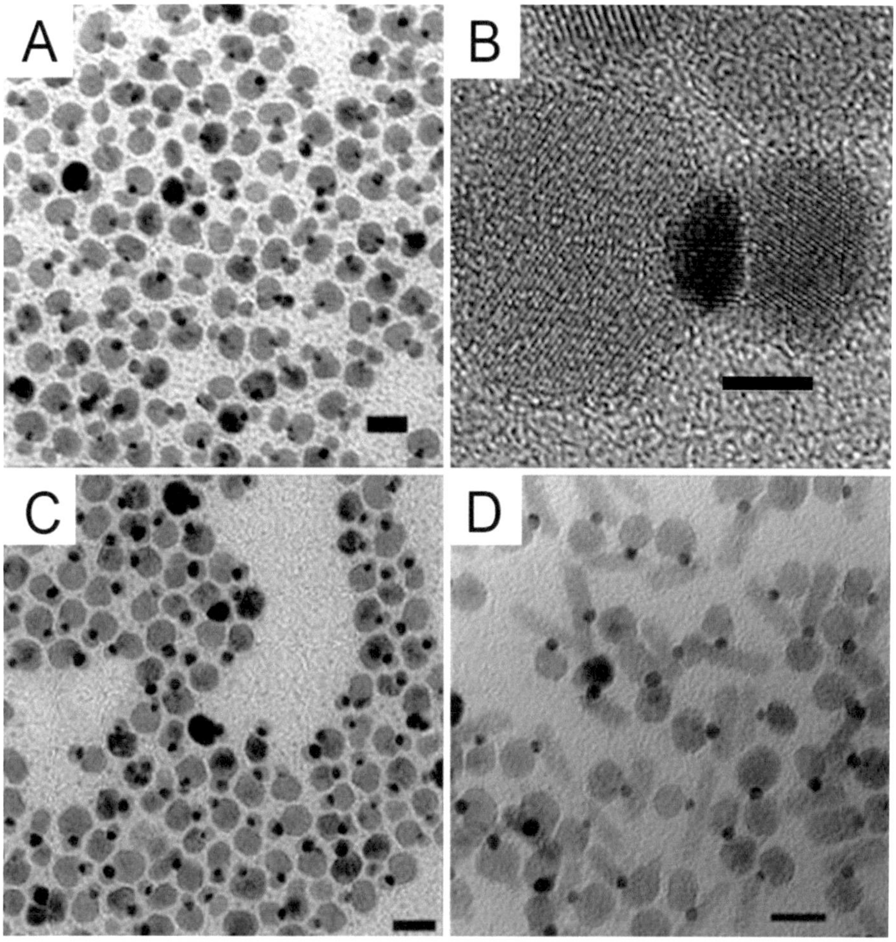

Fig. 10. (A) TEM and (B) High-resolution TEM images of Fe_3O_4-Au-PbSe ternary hybrid nanostructure. (C-D) TEM images of Fe_3O_4-Au-PbS ternary hybrid nanostructure. The scale bars are 20 nm in (A), (C), and (D) and 4 nm in (B). *Courtesy: Nano Lett., 2006, 6, 875-881.*

creates an electron and hole pair. In the case of small nanometer sized particles, the recombination of these pairs occurs on such a rapid time scale that the charges become useless for further redox reactions. To overcome this rapid recombination, researchers deposited a metallic component directly onto the semiconductor nanomaterial in order to aid in charge separation. The metal acts as a charge reservoir which enables the charges to be stored and utilized for a variety of redox reactions. A great example of these particles is the CdSe-Au nanodumbbells synthesized by Banin and co-workers.[23] These particles, when irradiated with visible light, can efficiently separate and store charges for immediate or long-term use in redox reactions (shown in Fig. 11).

In addition to being useful photocatalysts, many multicomponent nanostructures have found application as multimodal biomedical imaging and sensing agents. The increased complexity of these structures allows them to accomplish multiple tasks simultaneously. For example, they can be functionalized with multiple biomolecules by exploiting the different surface chemistries of the components and then used to sense multiple biomarkers

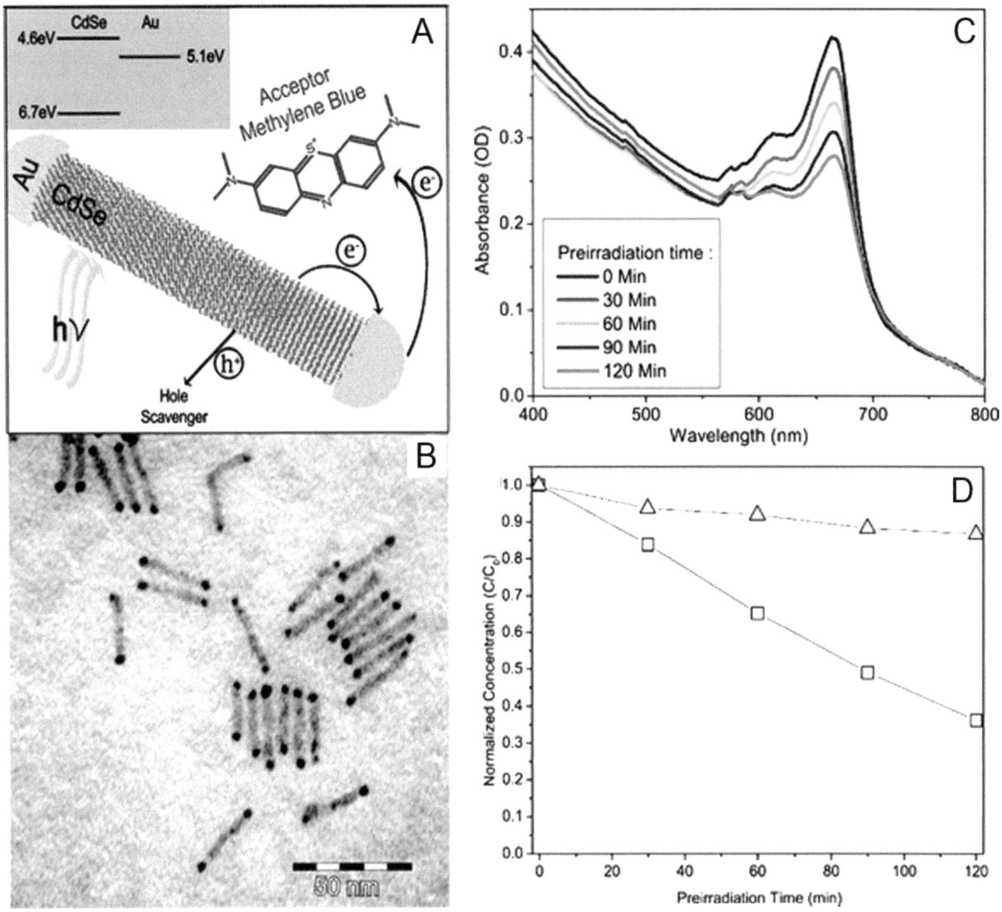

Fig. 11. (A) Scheme of a light-induced charge separation mechanism in a nanodumbbell in which the photogenerated electron–hole pair separates so that the electron resides at the gold tip and the hole at the CdSe nanorod. (B) TEM image of CdSe–Au hybrid nanodumbbells synthesized in aqueous solution. (C) Set of absorbance spectra of methylene blue–nanodumbbells solution in which the double peak absorbance feature of the methylene blue is noticeable, and each spectrum relates to a different pre-irradiation time at 532 nm of the nanodumbbell solution before the addition of the methylene blue. (D) Normalized concentration of methylene blue dye reduced by CdSe nanorods–gold nanoparticles mixture (open blue triangles) and by hybrid CdSe–Au nanodumbbells solution (open black squares) versus pre-irradiation time. *Courtesy: Nano Lett., 2008, 8, 637-641.*

simultaneously. Xu and co-workers have demonstrated this functionality using their Fe_3O_4-Ag hybrid nanostructures.[61] They were able to confirm that they had one molecule bound to the Fe_3O_4 component and another bound to the Ag component. The authors believe that these particles could be useful in areas such as protein binding, molecular imaging and pathogen detection. In addition to having increased imaging and sensing capabilities, the magnetic Fe_3O_4 component adds another degree of functionality to these particles.

Another potential use for hybrid nanocrystals involves the growth of metal tips on one-dimensional semiconductor nanostructures that will serve as integrated electrical contacts to external circuits. A number of strategies have evolved for integrating semiconductor nanorods and nanowires into electrical devices. Typically, nanoscale metal contacts are deposited using electron-beam lithography, focused ion beam deposition, or other methods, onto a nanostructure on a substrate. Metal tips on hybrid nanoparticles may be useful as integrated attachment points to external circuits. The conductance difference present in a metal–semiconductor hybrid nanocrystal is demonstrated in Figure 12, in which gold-tipped CdSe nanorods are characterized using scanning tunneling microscopy (STM).[15] The tip region exhibits lower resistance as compared to the CdSe rod region. In addition, higher tunneling current was observed through the tip, suggesting the tip as a potential electrical contact. The utility of the metal tips for electrical contacts was directly demonstrated by Sheldon and co-workers, who used a trapping method to localize Au-tipped CdSe nanorods between Au electrodes. Transport measurements showed that for CdSe nanorods with gold tips grown in solution, the conductance improved remarkably by five orders of magnitude in comparison to CdSe nanorods.[129] This establishes that hybrid nanostructures allow a path for highly improved electrical connectivity of the semiconductor part to the circuit.

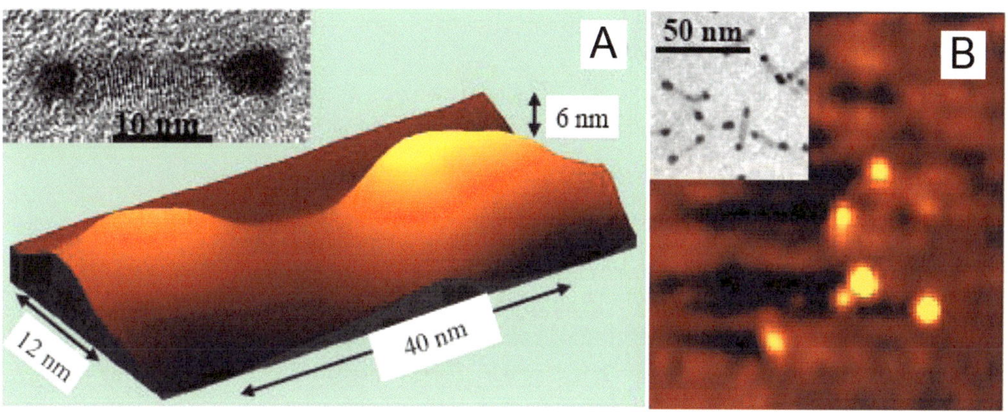

Fig. 12. (A) STM topography image of a single Au-tipped CdSe nanorod (nanodumbbell) taken with V_B = 2 eV. (B) STM current image (150 × 150 nm²) acquired at a subband gap bias (0.5 eV) showing the gold tips of three nanodumbbells. The insets show high resolution (A) and low resolution (B) TEM images of nanodumbbells. *Courtesy: Phys. Rev. Lett. 2005, 95, 056805.*

The hybrid nanocrystals represent a new set of building blocks as they combine more than one sections with different properties in a single particle. If coupled with biological molecules capable of molecular recognition, these nanocrystals could be chained together through their "anchoring points" (Fig. 13).[95] This could pave the way to assemblies that would behave as nano-machines, equipped, for instance, with magnets for navigation, fluorescent regions that could enable them to be tracked, anchoring regions bearing molecular receptors and chemical releasing agents, and so on. Researchers have already reported substrates bearing a repeating motif of nanocrystals organized in well-defined geometries, which can act as binding sites.[130] Anchoring nanocrystals to these substrates with a high selectivity and in a predictable orientation can be seen as the analogue of the lock-and-key mechanism that operates in biological systems, and that constitutes the very foundation of self-assembly as realized by nature.[95]

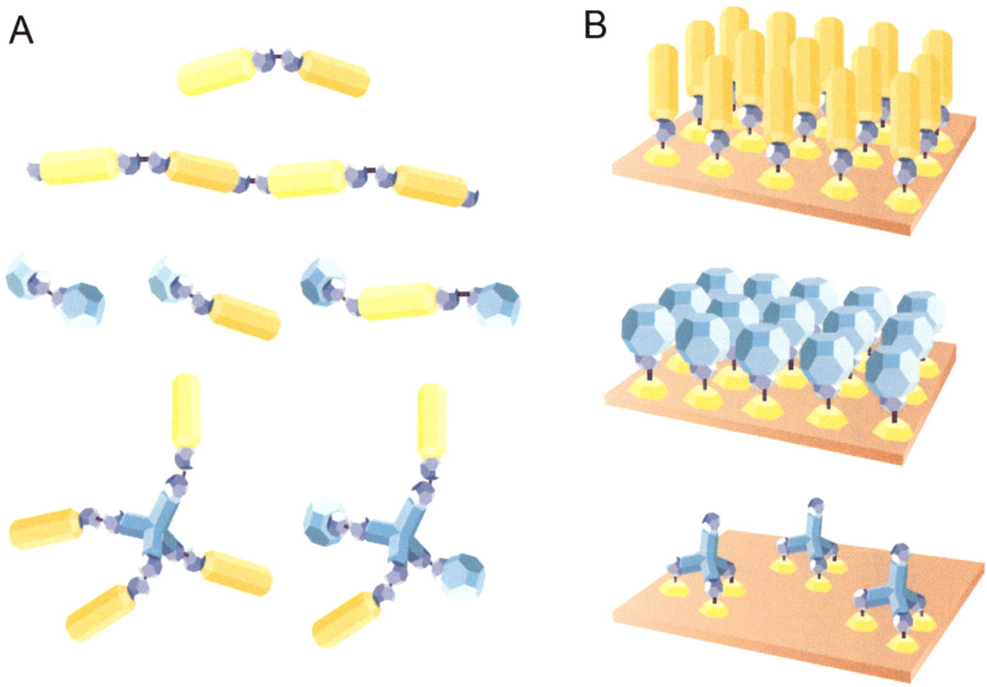

Fig. 13. Strategies for assembly. Nanocrystals carrying anchoring points might be used to build complex nanoscale assemblies, using, for instance, biomolecules as glue. (A) These assemblies could be realized directly in solution, forming, for instance, repeating chains or propeller-like structures. (B) Anchoring of gold-tipped nanocrystals on patterned surfaces is yet another possibility to arrange nanocrystals in programmed locations and orientations *Courtesy: Nat. Mater., 2005, 4, 801-802.*

Grzybowski and co-workers have demonstrated the elegant assembly of Fe_3O_4-Au hybrid nanostructures by extending the concepts of bond strength and steric hindrance to the nanometer regime.[131] In their system, they used a dithiol linker to "react" or aggregate the hybrids. They observed that as they increased the concentration of this linker, the number of hybrid particles per cluster increased. They equated this kind of control to the same kind of control one would have by varying bond strength. Holding the dithiol concentration constant and increasing the size of the bulky Fe_3O_4 domain resulted in a decrease in the number of particles per cluster. This type of control mimics the kind of control observed when varying steric hindrance on the molecular level. By varying these two controls relative to each other, they were able to precisely control the number of hybrid particles in each cluster (Fig. 14).

5. Perspectives

This chapter presents many of the diverse types of colloidal hybrid nanocrystals that have been achieved over the last several years. While the range of material combinations, geometries, and properties that have been presented is impressive, this area of study still remains challenges in industry applications. One challenge is to optimize the growth conditions and thus to obtain hybrid nanocrystals with well-defined structures and

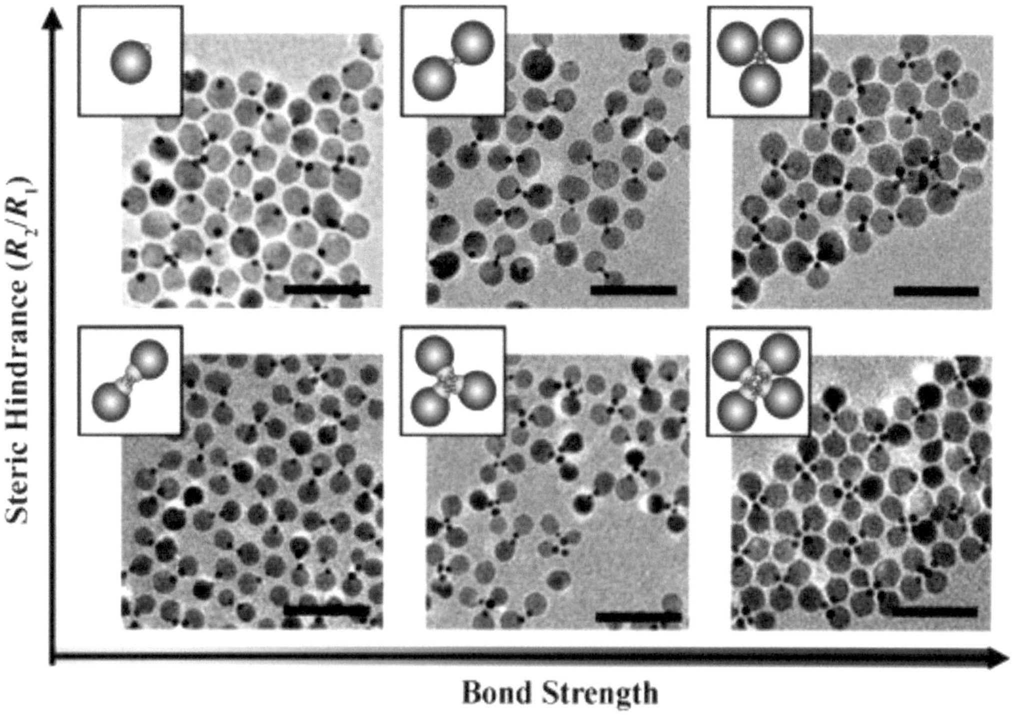

Fig. 14. TEM images of nanoclusters obtained for different bond strengths (determined by the dithiol concentration) and steric hindrance (quantified by the ratio, $s=R_2/R_1$). The insets provide a schematic illustration of the most probable nanoparticle cluster. The images correspond to dithiol concentrations of 0.3 M, 5 M, and 500 M (left to right) and s of 2.9 and 3.8 (bottom and top, respectively). All scale bars are 50 nm. *Courtesy: Angew. Chem. Int. Ed., 2009, 48, 9477-9480.*

controllable physical properties. Since sufficient strategies for fabricating hybrid nanocrystals have been developed, a coherent picture of how reaction conditions and material combinations would be tuned to influence the growth mechanism of the hybrid nanocrystals is yet to emerge. As these strategies become more robust, a diverse library of material combinations, arranged in different geometries, will provide insight into the interaction of different material systems at the nanometer scale.[132] Prospectively, such material combinations may lead to multiple functions, such as luminescence associated with catalysis. Contact points on the hybrids may allow further development of bottom-up assembly strategies for constructing complex electrical and optical devices, such as those for solar energy harvesting. Applications in other fields, such as spintronics or biological probes, need to be demonstrated upon development of specific hybrid systems. Although such explorations are still in an very early stage, the ability to tune the properties or impart multiple functionalities to hybrid nanocrystals shows high potential for their future inclusion in emerging modern technologies.

6. Acknowledgement

This work is supported by NSFC under Grant Nos. 50721091, 90921013 and 11074231 as well as by Chinese Academy of Sciences (CAS) and MOST of China (2011CB921403). J.Z. also acknowledges the financial support from the CAS (the startup fund of the outstanding doctoral dissertation award of CAS).

7. References

[1] P. D. Cozzoli, T. Pellegrino and L. Manna, *Chem. Soc. Rev.*, 2006, 35, 1195-1208.
[2] W. Shi, H. Zeng, Y. Sahoo, T. Y. Ohulchanskyy, Y. Ding, Z. L. Wang, M. Swihart and P. N. Prasad, *Nano Lett.*, 2006, 6, 875-881.
[3] H. Zeng and S. Sun, *Adv. Funct. Mater.*, 2008, 18, 391-400.
[4] R. Hao, R. Xing, Z. Xu, Y. Hou, S. Gao and S. Sun, *Adv. Mater.*, 2010, 22, 2729-2742.
[5] Y. Wu, J. Xiang, C. Yang, W. Lu and C. M. Lieber, *Nature*, 2004, 430, 61-65.
[6] D. Wang and Y. Li, *J. Am. Chem. Soc.*, 2010, 132, 6280-6281.
[7] R. D. Robinson, B. Sadtler, D. O. Demchenko, C. K. Erdonmez, L.-W. Wang and A. P. Alivisatos, *Science*, 2007, 317, 355-358.
[8] F. Wurm and A. F. M. Kilbinger, *Angew. Chem. Int. Ed.*, 2009, 48, 8412-8421.
[9] X. Peng, M. C. Schlamp, A. V. Kadavanich and A. P. Alivisatos, *J. Am. Chem. Soc.*, 1997, 119, 7019-7029.
[10] W. C. W. Chan and S. Nie, *Science*, 1998, 281, 2016-2018.
[11] P. D. Cozzoli, T. Pellegrino and L. Manna, *Chem. Soc. Rev.*, 2006, 35, 1195-1208.
[12] Y. Yin and A. P. Alivisatos, *Nature*, 2005, 437, 664-670.
[13] M. Casavola, V. Grillo, E. Carlino, C. Giannini, F. Gozzo, E. F. Pinel, M. A. Garcia, L. Manna, R. Cingolani and P. D. Cozzoli, *Nano Lett.*, 2007, 7, 1386-1395.
[14] S. Kinge, M. Crego-Calama and D. N. Reinhoudt, *ChemPhysChem*, 2008, 9, 20-42.
[15] D. Steiner, T. Mokari, U. Banin and O. Millo, *Phys. Rev. Lett.*, 2005, 95, 056805.
[16] I. L. Medintz, H. T. Uyeda, E. R. Goldman and H. Mattoussi, *Nat. Mater.*, 2005, 4, 435-446.
[17] V. Subramanian, E. E. Wolf and P. V. Kamat, *J. Phys. Chem. B*, 2003, 107, 7479-7485.
[18] M. Green, *Small*, 2005, 1, 684-686.
[19] H. Y. Lin, Y. F. Chen, J. G. Wu, D. I. Wang and C. C. Chen, *Appl. Phys. Lett.*, 2006, 88, 161911-161913.
[20] S. Deka, A. Falqui, G. Bertoni, C. Sangregorio, G. Poneti, G. Morello, M. D. Giorgi, C. Giannini, R. Cingolani, L. Manna and P. D. Cozzoli, *J. Am. Chem. Soc.*, 2009, 131, 12817-12828.
[21] J. Choi, Y. Jun, S. Yeon, H. C. Kim, J. Shin and J. Cheon, *J. Am. Chem. Soc.*, 2006, 128, 15982-15983.
[22] P. V. Kamat, *J. Phys. Chem. C*, 2007, 111, 2834-2860.
[23] R. Costi, A. E. Saunders, E. Elmalem, A. Salant and U. Banin, *Nano Lett.*, 2008, 8, 637-641.
[24] X. Wang, X. Ren, K. Kahen, M. A. Hahn, M. Rajeswaran, S. Maccagnano-Zacher, J. Silcox, G. E. Cragg, A. L. Efros and T. D. Krauss, *Nature*, 2009, 459, 686-689.
[25] C. Wang, C. Xu, H. Zeng and S. Sun, *Adv. Mater.*, 2009, 21, 3045-3052.
[26] A. Perro, S. Reculusa, S. Ravaine, E. B. Bourgeat-Lami and E. Duguet, *J. Mater. Chem.*, 2005, 15, 3745-3760.
[27] S. C. Glotzer and M. J. Solomon, *Nat. Mater.*, 2007, 6, 557-562.

[28] B. O. Dabbousi, J. RodriguezViejo, F. V. Mikulec, J. R. Heine, H. Mattoussi, R. Ober, K. F. Jensen and M. G. Bawendi, *J. Phys. Chem. B*, 1997, 101, 9463-9475.
[29] M. A. Hines and P. Guyot-Sionnest, *J. Phys. Chem.*, 1996, 100, 468-471.
[30] E. Cho, P. Camargo and Y. Xia, *Adv. Mater.*, 2010, 22, 744-748.
[31] F. Fan, D. Liu, Y. Wu, S. Duan, Z. Xie, Z. Jiang and Z. Tian, *J. Am. Chem. Soc.*, 2008, 130, 6949-6951.
[32] P. H. C. Camargo, Y. Xiong, L. Ji, J. M. Zuo and Y. Xia, *J. Am. Chem. Soc.*, 2007, 129, 15452-15453.
[33] V. Subramanian, E. Wolf and P. V. Kamat, *J. Phys. Chem. B*, 2001, 105, 11439-11446.
[34] T. Mokari, E. Rothenberg, I. Popov, R. Costi and U. Banin, *Science*, 2004, 304, 1787-1790.
[35] P. X. Gao, C. S. Lao, Y. Ding and Z. L. Wang, *Adv. Funct. Mater.*, 2006, 16, 53-62.
[36] Y. Lei and W. K. Chim, *J. Am. Chem. Soc*, 2005, 127, 1487-1492.
[37] Z. Sun, Z. Yang, J. Zhou, M. Yeung, W. Ni, H. Wu and J. Wang, *Angew. Chem. Int. Ed.*, 2009, 48, 2881-2885.
[38] Y. Wang, M. Li, H. Jia, W. Song, X. Han, J. Zhang, B. Yang, W. Xu and B. Zhao, *Spec. Acta A-Mol. Biomol. Spec.*, 2006, 64, 101-105.
[39] J. Yang, H. I. Elim, Q. B. Zhang, J. Y. Lee and W. Ji, *J. Am. Chem. Soc.*, 2006, 128, 11921-11926.
[40] M. Lahav, E. A. Weiss, Q. B. Xu and G. M. Whitesides, *Nano Lett.*, 2006, 6, 2166-2171.
[41] K. Fujimoto, K. Nakahama, M. Shidara and H. Kawaguchi, *Langmuir*, 1999, 15, 4630-4635.
[42] A. Ohnuma, E. C. Cho, P. H. C. Camargo, L. Au, B. Ohtani and Y. N. Xia, *J. Am. Chem. Soc.*, 2009, 131, 1352-1353.
[43] J. Yang, L. Levina, E. H. Sargent and S. O. Kelley, *J. Mater. Chem.*, 2006, 16, 4025-4028.
[44] C. Burda, X. Chen, R. Narayanan and M. A. El-Sayed, *Chem. Rev.*, 2005, 105, 1025-1102.
[45] M. P. Pileni, *Nat. Mater.*, 2003, 2, 145-150.
[46] B. Wiley, Y. Sun, B. Mayers and Y. Xia, *Chem. Eur. J.*, 2005, 11, 454-463.
[47] Y. Xia, Y. Xiong, B. Lim and S. Skrabalak, *Angew. Chem. Int. Ed.*, 2009, 48, 60-103.
[48] Y. Xiong and Y. Xia, *Adv. Mater.*, 2007, 19, 3385-3391.
[49] S. Kumar and T. Nann, *Small*, 2006, 2, 316-329.
[50] L. Manna, E. C. Scher and A. P. Alivisatos, *J. Am. Chem. Soc.*, 2000, 122, 12700-12706.
[51] N. Zhao and L. Qi, *Adv. Mater.*, 2006, 18, 359-362.
[52] X. Teng and H. Yang, *Nano Lett.*, 2005, 5, 885-891.
[53] C.-C. Chen, C.-Y. Chao and Z.-H. Lang, *Chem. Mater.*, 2000, 12, 1516-1518.
[54] L. Gou and C. J. Murphy, *Chem. Mater.*, 2005, 17, 3668-3672.
[55] J. Wang, Q. Chen, C. Zeng and B. Hou, *Adv. Mater.*, 2004, 16, 137-140.
[56] R. Buonsanti, V. Grillo, E. Carlino, C. Giannini, M. L. Curri, C. Innocenti, C. Sangregorio, K. Achterhold, F. G. Parak, A. Agostiano and P. D. Cozzoli, *J. Am. Chem. Soc.*, 2006, 128, 16953-16970.
[57] A. Figuerola, A. Fiore, R. Di Corato, A. Falqui, C. Giannini, E. Micotti, A. Lascialfari, M. Corti, R. Cingolani, T. Pellegrino, P. D. Cozzoli and L. Manna, *J. Am. Chem. Soc.*, 2008, 130, 1477-1487.
[58] T. Pellegrino, A. Fiore, E. Carlino, C. Giannini, P. D. Cozzoli, G. Ciccarella, M. Respaud, L. Palmirotta, R. Cingolani and L. Manna, *J. Am. Chem. Soc.*, 2006, 128, 6690-6698.
[59] L. Carbone, S. Kudera, C. Giannini, G. Ciccarella, R. Cingolani, P. D. Cozzoli and L. Manna, *J. Mater. Chem.*, 2006, 16, 3952-3956.

[60] J. Yang, J. Peng, Q. Zhang, F. Peng, H. Wang and H. Yu, *Angew. Chem.*, 2009, 121, 4051-4055.
[61] H. Gu, Z. Yang, J. Gao, C. K. Chang and B. Xu, *J. Am. Chem. Soc.*, 2005, 127, 34-35.
[62] M. Haruta, T. Kobayashi, H. Sano and N. Yamada, *Chem. Lett.*, 1987, 405-408.
[63] Z.-P. Liu, X.-Q. Gong, J. Kohanoff, C. Sanchez and P. Hu, *Phys. Rev. Lett.*, 2003, 91, 266102.
[64] A. Bulgac, M. M. Forbes and A. Schwenk, *Phys. Rev. Lett.*, 2006, 97, 020402.
[65] R. Paiva and R. D. Felice, *ACS Nano*, 2008, 2, 2225-2236.
[66] N. Liu, B. S. Prall and V. I. Klimov, *J. Am. Chem. Soc.*, 2006, 128, 15362-15363.
[67] W. Zhang, Y. Lu, T. Zhang, W. Xu, M. Zhang and S. Yu, *J. Phys. Chem. C*, 2008, 112, 19872-19877.
[68] M. Cheng, S. Liu, H. Zhou, Z. Hao and Q. Wang, *Opt. Lett.*, 2007, 32, 2125-2127.
[69] T. Young, *Philosophical Transactions of the Royal Society of London*, 1805, 95, 65-87.
[70] J. W. Mullin, *Crystallization*, Butterworth-Heinemann, Oxford, 2001.
[71] M. Volmer, *Z. Electrochem.*, 1929, 35, 555.
[72] M. Volmer, *Kinetic der Phasebildung*, Steinkopf, Dresden, 1939.
[73] E. V. Khamskii, *Crystallization from solutions*, Consultants bureau, New York, 1969.
[74] O. Söhnel and J. W. Mullin, *J. Cryst. Growth*, 1978, 44, 377-382.
[75] X. Y. Liu, *J. Chem. Phys.*, 2000, 112, 9949-9955.
[76] V. K. LaMer and R. H. Dinegar, *J. Am. Chem. Soc.*, 1950, 72, 4847-4854.
[77] V. P. Skripov, *Current Topics in Material Science*, North-Holland, Amsterdam, 1977.
[78] A. A. Chernov, *Modern Crystallography III - Crystal Growth*, Springer, Berlin, 1984.
[79] B. Mutaftschiev, *Handbook on Crystal Growth*, North-Holland, Amsterdam, 1993.
[80] P. S. R. Lacmann, *Current Topics in Materials Science*, North-Holland, Amsterdam, 1977.
[81] S. A. Chambers, *Surf. Sci. Rep.*, 2000, 39, 105-180.
[82] Z. Peng, H. Yang, *Nano Today*, 2009, 4, 143-164.
[83] D. J. Eaglesham and M. Cerullo, *Phys. Rev. Lett.*, 1990, 64, 1943.
[84] Y. W. Mo, D. E. Savage, B. S. Swartzentruber and M. G. Lagally, *Phys. Rev. Lett.*, 1990, 65, 1020.
[85] B. Voigtlander and A. Zinner, *Appl. Phys. Lett.*, 1993, 63, 3055-3057.
[86] S. Guha, A. Madhukar and K. C. Rajkumar, *Appl. Phys. Lett.*, 1990, 57, 2110-2112.
[87] D. Leonard, K. Pond and P. M. Petroff, *Phys. Rev. B*, 1994, 50, 11687-11692.
[88] M. Sopanen, H. Lipsanen and J. Ahopelto, *Appl. Phys. Lett.*, 1995, 67, 3768-3770.
[89] Y. Chen and J. Washburn, *Phys. Rev. Lett.*, 1996, 77, 4046.
[90] F.-R. Fan, D.-Y. Liu, Y.-F. Wu, S. Duan, Z.-X. Xie, Z.-Y. Jiang and Z.-Q. Tian, *J. Am. Chem. Soc.*, 2008, 130, 6949-6951.
[91] K.-C. Huang and S. H. Ehrman, *Langmuir*, 2006, 23, 1419-1426.
[92] W. Ostwald, *Z. Phys. Chem.*, 1897, 22, 289.
[93] W. Ostwald, *Z. Phys. Chem.*, 1900, 34, 495.
[94] *IUPAC Compendium of Chemical Terminology, 2nd Edition*, 1997.
[95] P. D. Cozzoli and L. Manna, *Nat. Mater.*, 2005, 4, 801-802.
[96] B. Liu and H. C. Zeng, *Small*, 2005, 1, 566-571.
[97] W. J. Plieth, *J. Electroanal. Chem.*, 1986, 204, 343-349.
[98] T. Mokari, C. G. Sztrum, A. Salant, E. Rabani and U. Banin, *Nat. Mater.*, 2005, 4, 855-863.
[99] M. Zinke-Allmang, L. C. Feldman and M. H. Grabow, *Surf. Sci. Rep.*, 1992, 16, 377-463.

[100] H. Yu, M. Chen, P. M. Rice, S. X. Wang, R. L. White and S. Sun, *Nano Lett.*, 2005, 5, 379-382.
[101] T. D. Schladt, M. I. Shukoor, K. Schneider, M. N. Tahir, F. Natalio, I. Ament, J. Becker, F. D. Jochum, S. Weber, O. Köhler, P. Theato, L. M. Schreiber, C. Sönnichsen, H. C. Schröder, W. E. G. Müller and W. Tremel, *Angew. Chem. Int. Ed.*, 2010, 49, 3976-3980.
[102] H. Gu, R. Zheng, X. Zhang and B. Xu, *J. Am. Chem. Soc.*, 2004, 126, 5664-5665.
[103] F. Wang and W. E. Buhro, *J. Am. Chem. Soc.*, 2007, 129, 14381-14387.
[104] J. Zeng, J. Huang, C. Liu, C. H. Wu, Y. Lin, X. Wang, S. Zhang, J. Hou and Y. Xia, *Adv. Mater.*, 2010, 22, 1936-1940.
[105] S. Chakrabortty, Jie A. Yang, Yee M. Tan, N. Mishra and Y. Chan, *Angew. Chem. Int. Ed.*, 2010, 49, 2888-2892.
[106] G. Menagen, J. E. Macdonald, Y. Shemesh, I. Popov and U. Banin, *J. Am. Chem. Soc.*, 2009, 131, 17406-17411.
[107] A. E. Saunders, I. Popov and U. Banin, *J. Phys. Chem. B*, 2006, 110, 25421-25429.
[108] T. Mokari, A. Aharoni, I. Popov and U. Banin, *Angew. Chem. Int. Ed.*, 2006, 45, 8001-8005.
[109] J. Yang, E. Sargent, S. Kelley and J. Y. Ying, *Nat. Mater.*, 2009, 8, 683-689.
[110] H. Gu, Z. Yang, J. Gao, C. K. Chang and B. Xu, *J. Am. Chem. Soc.*, 2005, 127, 34-35.
[111] L. Zhang, Y.-H. Dou and H.-C. Gu, *J. Colloid Interface Sci.*, 2006, 297, 660-664.
[112] W. Shi, H. Zeng, Y. Sahoo, T. Y. Ohulchanskyy, Y. Ding, Z. L. Wang, M. Swihart and P. N. Prasad, *Nano Lett.*, 2006, 6, 875-881.
[113] Y. Wei, R. Klajn, A. O. Pinchuk and B. A. Grzybowski, *Small*, 2008, 4, 1635-1639.
[114] Y. Lee, M. A. Garcia, N. A. F. Huls, S. Sun, *Angew. Chem. Int. Ed.*, 2010, 49, 1271-1274.
[115] K. W. Kwon and M. Shim, *J. Am. Chem. Soc.*, 2005, 127, 10269-10275.
[116] L. Qu and X. Peng, *J. Am. Chem. Soc.*, 2002, 124, 2049-2055.
[117] J. Zeng, W. Lu, X. Wang, B. Wang, G. Wang and J. G. Hou, *J. Colloid Interface Sci.*, 2006, 298, 685-688.
[118] J. Zeng, C. Liu, J. Huang, X. Wang, S. Zhang, G. Li and J. Hou, *Nano Lett.*, 2008, 8, 1318-1322.
[119] W. Lu, B. Wang, J. Zeng, X. Wang, S. Zhang and J. G. Hou, *Langmuir*, 2005, 21, 3684-3687.
[120] X. Lu, H.-Y. Tuan, Brian A. Korgel and Y. Xia, *Chem. Eur. J.*, 2008, 14, 1584-1591.
[121] Z. Li, J. Tao, X. Lu, Y. Zhu and Y. Xia, *Nano Lett.*, 2008, 8, 3052-3055.
[122] X. Lu, M. S. Yavuz, H.-Y. Tuan, B. A. Korgel and Y. Xia, *J. Am. Chem. Soc.*, 2008, 130, 8900-8901.
[123] Z. Li, W. Li, Pedro H. C. Camargo and Y. Xia, *Angew. Chem.*, 2008, 120, 9799-9802.
[124] T. K. Sau and C. J. Murphy, *Langmuir*, 2004, 20, 6414-6420.
[125] B. Nikoobakht and M. A. El-Sayed, *Chem. Mater.*, 2003, 15, 1957-1962.
[126] S. U. Son, I. K. Park, J. Park and T. Hyeon, *Chem. Comm.*, 2004, 778-779.
[127] S. Kudera, L. Carbone, M. F. Casula, R. Cingolani, A. Falqui, E. Snoeck, W. J. Parak and L. Manna, *Nano Lett.*, 2005, 5, 445-449.
[128] S. E. Habas, P. Yang and T. Mokari, *J. Am. Chem. Soc.*, 2008, 130, 3294-3295.
[129] M. T. Sheldon, P. E. Trudeau, T. Mokari, L. W. Wang, A. P. Alivisatos, *Nano Lett.*, 2009, 9, 3676-3682.
[130] R. Glass, M. Moller and J. P. Spatz, *Nanotech.*, 2003, 14, 1153-1160.

[131] Y. Wei, K. Bishop, J. Kim, S. Soh and B. Grzybowski, *Angew. Chem. Int. Ed.*, 2009, 48, 9477-9480.

[132] R. Costi, A. E. Saunders and U. Banin, *Angew. Chem. Int. Ed.*, 2010, 49, 4878-4897.

4

Nanocrystals in Metallic Glasses

Rainer J. Hebert
University of Connecticut
U.S.A

1. Introduction

Nanocrystals are considered as isolated, nanoscale particles in modern science or as grains of nanocrystalline material. The latter material is comprised of nanocrystals that form a 3-dimensional polycrystal made of nanocrystals and grain-boundaries. Historically, however, nanocrystals are firmly rooted in colloid science and for a long time nanocrystals were used unknowingly as components of composite materials. For example, the coloring of glasses with colloidal gold nanocrystals dates back to the Romans (Freestone et al., 2007). In light of the historical use of nanocrystals as components of macroscopic composite materials it is not surprising that modern materials science continues and expands the use of nanocrystals for composite materials. The ever increasing array of nanoscale objects, for example, nanowires, nanofibers, nanobelts, nanopillars, or nanotubes, along with improving synthesis and characterization options offers a broad range of possible macroscopic composites with nano-particle components. The range of applications is no longer limited to functional properties, but includes structural applications. While the nanocrystals are per definition crystalline and thus reveal a periodic arrangement of atoms, the surrounding matrix can be crystalline or amorphous. For metals-based nanocrystals, the modern era of composite research dawned with the discovery of Guinier-Preston zones and remained focused on nanocrystals embedded in crystalline matrices. For ceramic materials, an important application emerged with the dispersion of nanocrystals in amorphous or glassy ceramic matrices. For example, cook tops are widely available today that are made of ceramic glasses containing dispersions of oxide nanocrystals. On the metals side, bulk composite materials comprised of nanocrystals embedded in amorphous metallic matrices are still relatively novel materials by comparison with their ceramic counterparts.

The interest in nanocrystals for metallic glasses has had two related motivations. From a viewpoint of fundamental material science, metallic glasses offer a very convenient approach for studying crystallization reactions. For conventional metals or alloys, solidification occurs almost instantaneously and it is usually only possible to study the completely crystallized phase experimentally but not the process of crystallization. Crystallization of a metallic glass, by comparison, can be induced as a "slow-motion" process that enables detailed experimental studies of the crystallization process. The formation of nanocrystals from metallic glasses thus represents an ideal vehicle to test and validate crystallization theories. The second motivation for studying nanocrystals in metallic glasses is much more practical and concerns improvements in properties. Iron-based metallic glasses were among the earliest metals-based glasses and it was soon discovered that the crystallization of transition-metal nanocrystals

improved the soft-magnetic properties. A number of commercial products based on Fe-nanocrystals dispersed in melt-spun amorphous ribbons emerged from the early crystallization studies. With the emergence of bulk metallic glasses starting about 1992 the interest in the synthesis of nanocrystal/metallic glass composite materials shifted toward a much broader range of applications and in particular toward structural applications. The role of nanocrystals for fundamental crystallization studies and for applications is related, since only an understanding of the nanocrystal formation enables a controlled synthesis and thus controlled improvements in properties.
This chapter addresses mainly the role of nanocrystals for properties of metallic glasses and the synthesis of nanocrystals in metallic glasses. Although differences exist between metallic glasses and amorphous alloys, in this chapter both terms are used synonymously. For a discussion of the differentiation between amorphous alloys and metallic glasses the reader is referred to an overview article on metallic glasses (Greer, 1995).

2. Synthesis of nanocrystals in amorphous matrices

2.1 Thermal synthesis

The research program from which the first non-thin film metallic glass emerged focused on the formation of Cu-Ag non-equilibrium solid solutions (Duwez, 1981). The solid solutions were obtained from rapid quenching experiments, leaving the solidified products in a highly non-equilibrium condition. The key idea behind Duwez' ground-breaking experiments was to prevent atoms from re-arranging from nearly homogeneous liquid solutions into solid clusters with different compositions. The rapid quenching limited the time at which atoms had sufficient mobility for re-arrangements to such a short period that re-arrangements were effectively inhibited. While the re-arrangements into equilibrium configurations were inhibited, the quenched alloys retained a thermodynamic driving force for transitioning into the phases that would have formed at lower cooling rates. This driving force is fundamental for the synthesis of nanocrystals in metallic glasses, since the nanocrystals represent phases with lower free energies. The different modes of crystallization and thus of reducing the metastability of metallic glasses are shown in Fig. 1. Both diagrams show the free energy—usually Gibbs free energy, but other energy functions such as the Helmholtz free energy could be used as well—as a function of composition. In polymorphous crystallization the product phase has the same composition as the parent phase. In Fig. 1 a the amorphous parent phase transforms into a terminal solid solution with the same composition, c_{poly}. The crystallization product does not have to be a solid solution, however, and Fig. 1 b shows a polymorphous transition from the amorphous phase into a compound phase ("Compound I"). A second crystallization mode is primary crystallization. In this mode the amorphous phase transforms into a terminal solid solution and an amorphous phase with a higher solute content than the initial amorphous phase. The compositions of the terminal solid solution, $c_{p,s.s}$, and the new amorphous phase, $c_{p,a.}$, are obtained from a common tangent construction. Several important nanocrystal/metallic glass composite materials are obtained from primary crystallization reactions, for example, Al-based metallic glasses, or Fe-based metallic glasses. In both systems the crystallization products are nearly pure Al and Fe phases, respectively. A third crystallization mode—eutectic crystallization--is highlighted in Fig. 1 b. In this mode the parent amorphous phase crystallizes into two crystalline products with compositions that are again obtained from the common tangent construction.

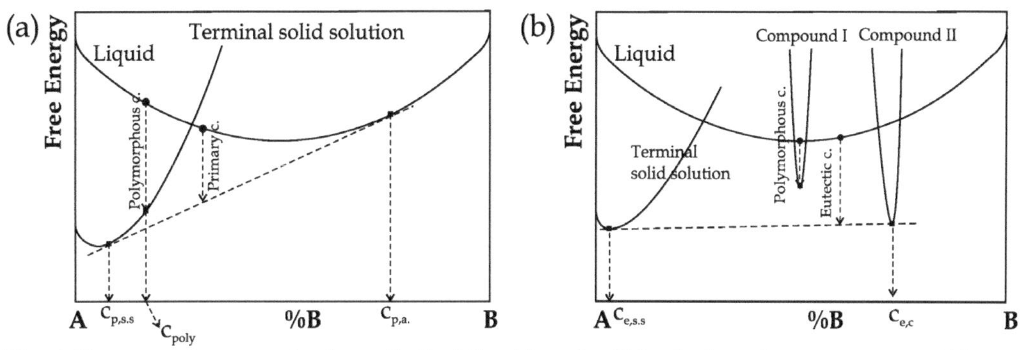

Fig. 1. Free energy curves for hypothetical liquid and solid solution phases vs. composition

Figure 1 shows the different modes of crystallization, but it does not indicate if nanocrystals can be obtained from all crystallization modes. To examine the relation between crystallization mode and the microstructure of the crystallization products, it is necessary to consider the mechanisms of crystallization. Two fundamentally different mechanisms exist: phase separation and nucleation and growth. Both mechanisms are different, but can occur in sequence. For phase separation to occur, the free energy has to reveal a specific composition dependence that is shown schematically in Fig. 2. In composition region II the free energy curve is concave, i.e., $d^2G/dc^2 < 0$. If the initial alloy composition lies within region II any variation in composition lowers the free energy. This scenario is indicated with arrows in Fig. 2. If the initial alloy with composition $c_{initial}$ separates, maybe due to local fluctuations, into two regions with composition c_a and c_b, the mixture of regions with composition c_a and c_b has a lower free energy than the initial region with composition $c_{initial}$. This transition is referred to as spinodal decomposition (Cahn, 1961). If the initial alloy composition lies within regions I and I′, however, a small fluctuation in composition will raise the overall free energy. To lower the overall free energy in region I and I′, compositional fluctuations have to be finite to lower the free energy.

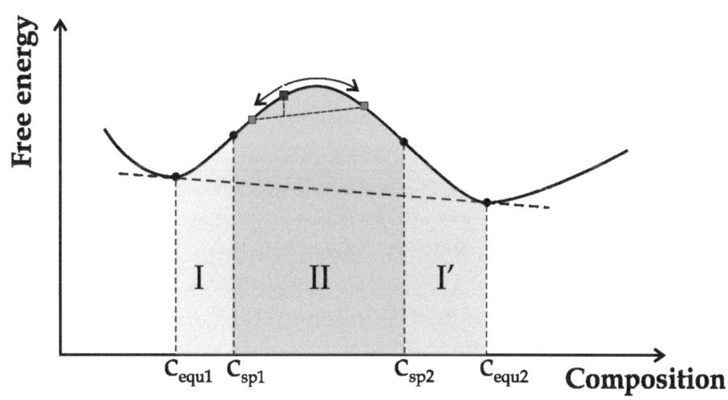

Fig. 2. Free energy vs. composition for a system that reveals spinodal decomposition

In spinodal decomposition transformations an initially single phase separates into two phases with different compositions but the same phase. The composition differences between the compositionally different regions can be minimal, in contrast to nucleation-based phase transformations. In the latter case, thermally induced fluctuations in a parent phase lead to clusters that are thermodynamically stable once their size exceeds a critical size. Unlike for spinodal decomposition reactions, however, the composition of the new phase is different by a finite amount from the parent phase.

The nucleation and spinodal decomposition theories have been applied successfully for nanocrystal crystallization reactions in metallic glasses. However, several experimental observations suggest that there are open issues related to nucleation or spinodal decomposition reactions that call for continued studies for metallic glasses. For example, several studies on Al-based metallic glasses demonstrated that the addition of minor amounts of some solute atoms—typically on the order of 1 %--can oftentimes dramatically impact the nanocrystal number density, size distribution, and the glass forming ability. The addition of small amounts of V and Ti to $Al_{88}Y_7Fe_5$ metallic glasses induced a tendency for local ordering in the vicinity of the V and Ti atoms, leading to clusters with about 1 nm in size that have a distorted fcc arrangement (Sadoc et al., 2007). These clusters are now recognized as medium-range ordered regions in the metallic glasses (Stratton et al., 2005) that can decisively influence not only the size distribution and particle number density of the nanocrystals (Perepezko et al., 2010) but furthermore the phase formation (Sadoc, Heckmann et al., 2007). The current understanding of microalloying in metallic glasses is very encouraging for designing metallic glasses with specific nanocrystal dispersions. Further work is necessary to understand the formation of inhomogeneities and their effect on the nucleation or phase separation reactions. The classical nucleation theory is based on the notion of random fluctuations in atomic configurations. This notion might have to be modified for atomic configurations that are locally neither fully amorphous nor crystalline.

2.1.1 Growth of crystallites in metallic glasses

Depending on whether partial processes act in a parallel or sequential manner, the fastest (growth mechanism acting parallel) or slowest (sequential) process controls the growth-rate (Kingery et al., 1976). For polymorphic phase transformations the composition of the parent and new phase is identical and long-range diffusion is not necessary. The growth rate of the new phase is limited by the speed at which atoms at the interface between the old and new phase can rearrange from the old to the new phase. This growth mode is referred to as interface-controlled. If the growing particle has a composition different from the matrix, solute diffusion in the matrix is necessary to accommodate the composition change across the phase boundary. In this latter case, one can distinguish between interface and diffusion-controlled growth. Aside from polymorphous transformations where long-range diffusion is not required, interface control is mainly observed for highly anisotropic interfaces and in discontinuous precipitation reactions. Due to the importance of this growth mode it has been treated in several publications (Hillert, 1968; Christian, 1975; Purdy, 1981; Ratke and Vorhees, 2001). For the formation of nanocrystals in metallic glasses, diffusional growth is of particular interest, since the the nanocrystals that have been synthesized in metallic glasses based on thermal processing are often primary nanocrystals and thus require a solute redistribution or diffusion during growth. Among the first and most influential investigations of diffusion-controlled growth is the work by Zener (Zener, 1949) who considered the growth of a spherical particle in a supersaturated matrix of infinite size and a stationary particle/matrix boundary. This growth problem has been treated in detail in two textbooks (Glicksman, 2000; Ratke and Vorhees, 2001).

A more recent example for growth in ternary systems with greatly differing values of the diffusion coefficients is the growth of primary Al in Al-RE-TM systems (RE: rare-earth, TM:

transition metal). For amorphous metallic metal-metal alloys it has been found that the diffusivity correlates with the atomic size of the diffusing species but not necessarily for metal-metalloid amorphous alloys (Sharma and Banerjee, 1989). With respect to the amorphous Al-RE-TM alloys, this observation suggests that the diffusivities of the rare-earth metals in the amorphous matrix should be low compared with the diffusion of the transition-metal atoms. Atom-probe field-ion microscopy (APFIM) experiments were conducted with partially crystalline amorphous Al-Ni-Ce, Al-Ni-Y and Al-Ni-Sm alloys and the concentration profiles of the individual elements were measured. For the amorphous $Al_{87}Ni_{10}Ce_3$ sample the analysis showed that Ce was enriched in a layer surrounding the primary Al nanocrystallites. The thickness of the layer was approximately 3nm (Hono et al., 1995). The Ni atoms did not show any segregation at the nanocrystal/amorphous matrix interface. A similar investigation with amorphous $Al_{88}Ni_8Sm_4$ (Zhang et al., 2002) and $Al_{88}Ni_4Sm_8$ (Gloriant et al., 2001) showed, however, that the Sm concentration at the interface was not enhanced although the size difference between Sm and Ce atoms is practically zero and a similar diffusion behavior in the amorphous matrix would therefore be expected. It must be noted, however, that the ribbons in (Hono, Zhang et al., 1995) were not annealed but were taken from a portion of the ribbons that partially crystalline already after the melt-spinning. The Al-Ni-Sm samples, by contrast, were annealed to induce the nanocrystallization and it was argued that in this latter case metastable equilibrium was established. Aside from the Al-based amorphous alloys, Fe-based systems and their crystallization behavior have also been investigated in detail. Solute accumulation at the interface between bcc-Fe nanocrystals and the amorphous matrix was observed by APFIM for an amorphous $Fe_{90}Zr_7B_3$ alloy after annealing at 723K for 60min (Zhang et al., 1996). In this case Zr atoms piled-up at the interfaces. The blocking effect of the solute rich layer on the growth of the nanocrystals is discussed in (Koester, 1993).

Similar to the situation for nucleation, recent experimental observations suggest that the classical growth theories might not capture the growth scenario for some metallic glasses accurately. The observation of new growth mechanisms seem to be related to the presence of local clusters with semi-crystalline atomic configurations. Liu and coworkers examined the growth process of nanocrystals in a deeply undercooled $Zr_{65}Ni_{25}Ti_{10}$ bulk metallic glass that contained nanoscale clusters with imperfect ordering in the as-cast state. A step-wise growth process was discovered that started with a transition of the clusters into one-dimensional, then two-dimensional, and eventually three-dimensional fcc Zr_2Ni nanocrystals (Liu et al., 2007). The initial semi-crystalline nanoscale clusters developed directly into nanocrystals, but it is not clear if a different growth mechanism would have unfolded at reduced undercooling levels. It is clear, however, that the key to the design of nanocrystal dispersions in metallic glasses is a better understanding of the nanoscale clusters with distorted crystalline atomic arrangements or icosahedral structure, their formation and role for nucleation and growth.

2.2 Mechanical processing – deformation-induced crystallization

The transformation mechanisms for crystallization that are highlighted in 2.1 require thermal motion of atoms to initiate the transformations. Other modes of crystallization have been observed, however, that are not initiated with externally applied heating of metallic glasses. These externally induced crystallization modes include irradiation and deformation of metallic glasses. Recently, evidence was found that nanocrystals developed on surfaces of Zr-based metallic glasses in corrosion (Paillier et. al., 2010) pits as a direct response to corrosion (Paillier et. al., 2010). Electrochemical interactions with metallic glasses can therefore be added to the list of external driving modes for crystallization. In the following, the experimental evidence for deformation-induced crystallization is surveyed. Mechanisms for external, driven crystallization are then summarized. This section is concluded with a summary of deformation-affected crystallization. In contrast to deformation-induced crystallization, deformation can impact crystallization during heat treatments that follow deformation. From a historical perspective, the impact of rolling or tensile loading on annealing-induced crystallization was studied early on, i.e., starting in the early 1970s. Deformation-induced crystallization was not observed until the early 1990s (Schulz et al., 1990).

2.2.1 Experimental findings for deformation-induced crystallization

The majority of deformation processes to induce crystallization are based on mechanical alloying. In this approach amorphous powder or amorphous ribbon pieces are deformed in ball mills. Containers filled with amorphous powder or pieces are shaken and contain steel or ceramic balls that collide, trapping powder particles or ribbon pieces in between. Examples of crystallization reactions that were induced with milling are highlighted in Table 1. In the examples highlighted in Table 1 the samples were sealed in containers that were filled with argon. Some ball milling experiments were conducted with the container being immersed in liquid nitrogen or other cooling media to maintain the samples at ambient temperatures. It was demonstrated that without cooling the temperature of the milled samples can increase by 60 to 120 K (Koch, 1989). The size of the crystallites was in all cases limited to less than about 20 nm. Indeed, Schulz and coworkers stated in what has been the first report on milling-induced crystallization of metallic glasses that the size of the nanocrystals in their milled $Fe_{78}Si_9B_{13}$ amorphous ribbons was less than 3 nm and that in order to thermally induce nanocrystals at this size the samples would have had to be annealed for months (Schulz et al., 1990). To achieve a high particle number density and a small size of crystallites that develop from a nucleation and growth process it is necessary to limit the growth process but to achieve a high nucleation frequency. With annealing this can only be achieved if the annealing temperature remains low compared to the glass transition range. Under these annealing conditions the time for nanocrystals to develop could be on the order of months as Schulz and coauthors stated. Ball milling, therefore, provides a faster processing route for nanocrystal formation at nanocrystal sizes of less than about 10 nm than standard heat treatments.

The formation of nanocrystals is not limited to ball milling as a deformation technique, but has furthermore been observed during bending of metallic glass samples. The bending experiments have been conducted mostly with melt-spun ribbons, although in one example micrometer-sized simple beams were machined and subsequently deformed in a simple bending mode (Ogura et al., 2001). During bending, large compressive and tensile stresses develop on the surfaces of the bent samples. Nanocrystals were observed to form on the compressive side of bent melt-spun ribbons, but not on the tensile side for bending experiments with Al-based amorphous alloys (Jiang and Atzmon, 2003). Atzmon's and coworker's observation of an asymmetric nanocrystallization process contrasts with bending experiments of Ni-P micro-beams; nanocrystals developed on the tensile side of the bent beam (Ogura, Tarumi et al., 2001). While the ball milling-induced experiments were conducted mostly with Fe-based metallic glasses, the majority of the bending experiments that started with Chen and coworkers' report in Nature (Chen et al., 1994) were conducted with Al-based amorphous melt-spun ribbons. The common thread between the

System	Deformation specific	Reference
Fe78Si9B13, Fe66Co18Si1B15	Milling in SPEX 8000 mill with steel balls, high-energy ball milling, formation of α-Fe(Si) nanocrystals	(Schulz et. al. 1990)
Fe78Si9B13	Cryogenic ball milling; formation of 2 nm Fe-nanocrystals	(Huang et. al. 1995)
Fe78Si9B13	Low-energy ball milling; Fe nanocrystals develop during milling at 250 oC, but not during annealing at the same temperature without milling.	(Xu and Atzmon, 1998)
Al90Fe5Gd5 Al90Fe5Ce5	Milling in SPEX 8000 with Cr-steel balls; formation of Al nano-crystals.	(He et. al. 1995)
Fe78Si9B13	Cryogenic ball milling, formation of 2-13 nm α-Fe(Si) and Fe2B nanocrystals	(Huang et. al. 1996)
Fe80B20	Formation of a-Fe and Fe3B nanocrystals, followed by formation of Fe2B nanocrystals and disappearance of Fe3B nanocrystals.	(Fan et. al. 1995)
Fe40Ni40P14B6	Nanocrystals develop initially during high-energy ball milling at the surfaces of the amorphous ribbons; at longer annealing times the bulk of the ribbons/powders crystallizes into γ-(Fe,Ni) and $(Fe,Ni)_3(P,B)$	(Fan et. al. 1999)

Table 1. Examples of ball-milling induced crystallization reactions

bending-induced nanocrystallization reactions is the crystallization process only taking place on one side of the bent samples and the tendency of the nanocrystallites to have at least a partial degree of orientation. Table 2 provides a survey of reports for bending-induced crystallization.

System	Deformation specific	Reference
Al90Fe5Gd5 Al87Ni8.7Y4.3 Al90Fe5Ce5 Al85Ni10Ce5	Bending of ribbons through 180° at room temperature; formation of Al nanocrystals , 7-10 nm in size, in shear bands except for the Al85Ni10Ce metallic glass.	(Chen et. al. 1994)
Ni-11.5 wt% P	Bending of micrometer-size cantilever induces Ni nanocrystals at the tension side of the cantileve surface. Nanocrystals had the same orientation.	(Ogura et. al. 2001)
Al90Fe5Gd5	Bending of melt-spun amorphous ribbons through 180° at -40 °C induces nanocrystals on the compressive side but not on the tensile side.	(Jiang et. al. 2003) (Jiang and Atzmon, 2003)

Table 2. Examples of bending-induced crystallization reactions

In addition to ball-milling and bending, cold-rolling and instrumented indentation have been used to induce nanocrystals in metallic glasses. Nanoindentation-induced crystallization was reported for bulk metallic $Zr_{52.5}Cu_{17.9}Ni_{14.6}Al_{10}Ti_5$ glass (Kim et al., 2002). The Zr_2Ni nanocrystallites had a size between 10 and 40 nm. Moreover, the crystallographic orientation of the nanocrystals was not perfectly random. Instrumented indentation studies on a $Al_{90}Fe_5Gd_5$ metallic glass revealed an isotropic arrangement of Al nanocrystals that developed during indentation at 100 nm/s and 10 nm/s (Jiang et. Al., 2003). The average size of the nanocrystals was 6.4 nm for the 100 nm/s indentation rate and 3.8 nm for the

lower rate. The density of the nanocrystals increased toward the bottom of the indent where the strain is the highest. Furthermore the nanocrystal density is higher for the lower indentation rate.
Crystallization reactions were furthermore induced with intense cold rolling and folding of melt-spun amorphous ribbons (Hebert and Perepezko, 2004; Hebert et al., 2004; Hebert et al., 2005; Park et al., 2005; Louzguine-Luzgin and Inoue, 2006; Hebert and Perepezko, 2007). Similarly to the milling- and bending-induced crystallization reactions the size of the nanocrystals was limited to about 15-20 nm. Nanocrystals developed in some metallic glasses during rolling and folding, but for some metallic glass compositions, notably the $Al_{85}Ni_{10}Ce_5$ metallic glass nanocrystals did not develop until the highest strains that could be imparted on the melt-spun ribbons with the rolling and folding approach.

2.2.2 Mechanisms for deformation-induced crystallization

The main challenge in understanding the mechanism of deformation-induced nanocrystal formation in metallic glasses is to discern between thermal and athermal effects. Under the intense deformation modes that have been used to induce nanocrystals it is inevitable that shear bands develop during the deformation process. Proposed mechanisms for deformation-induced crystallization either follow the argument of adiabatic heating in shear bands as source for crystallization or they follow the argument of enhanced kinetics in shear bands due to the dilatation of the metallic glass in shear bands. Several studies focused on estimates or experimental analyses of the transient temperatures that develop during the shear band formation and propagation (Lewandowski and Greer, 2006; Hongwen et al., 2007; Battezzati and Baldissin, 2008; Jiang et al., 2008). The shear strains within shear bands are estimated to be on the order of 100 % (Jiang et al., 2007). Reported temperature rises vary widely between tenths of a degree to on the order of 1,000 K. Indirect evidence for adiabatic heating as a source for deformation-induced crystallization was provided, for example, in Csontos' and Shiflet's work (Csontos and Shiflet, 1997). Composition profiles were measured around Al nanocrystals of a $Al_{90}Fe_5Gd_5$ metallic glass that was deformed at room temperature and annealed in a separate experiment. The composition profiles for both processing routes were practically identical, suggesting a heating effect as the same source for crystallization in both cases. Further indirect evidence for shear band-related heating as the mechanism for deformation-induced crystallization was obtained in Kim and coworkers' study (Kim et al., 2006). The high-temperature Laves phase that formed in the shear bands during room temperature cold rolling was identical with the Laves phase that formed in the same $Ti_{40}Zr_{29}Cu_9Ni_8Be_{14}$ metallic glass upon annealing. On the other hand, evidence was put forth that the deformation-induced crystallization could not be the result of only adiabatic heating. Shear bands formed on the compressive and the tensile sides of bent Al-based metallic glasses, but nanocrystals developed only on one side. With adiabatic heating as the source and mechanism for crystallization, nanocrystals should develop on both sides bar any major differences in the shear bands between the tensile and compressive sides. Observations of a partial to complete ordering of the deformation-induced nanocrystals (Ogura, Tarumi et al., 2001) are difficult to rationalize based only on a heating argument. With arguments for both sides it seems that the deformation-induced crystallization reactions have both a thermal and a deformation component. Deformation-induced crystallization was more recently observed during uniaxial compression of bulk $Cu_{50}Zr_{43}Al_7$ metallic glass at room temperature (Lee et al., 2006). Lee and coworkers suggested that the hydrostatic stress component lowered the energy barrier for nucleation while the shear stress component lowered the energy barrier for diffusion (Lee, Huh et al., 2006).

2.3 Mechanical processing – deformation prior and during thermally induced crystallization

Crystallization reactions can occur during deformation at room temperature as described in 2.2, but historically the first relation between deformation and crystallization was established for cold-rolling of melt-spun glassy ribbons followed by annealing. Since then the sequence of deformation and annealing has been studied not only for cold-rolling but furthermore for uniaxial compression. A third group of experiments included deformation during annealing. In this set of experiments the stress state can be selected; hydrostatic and uniaxial stress states were examined for their effect on thermally induced crystallization.

2.3.1 Deformation followed by thermally induced crystallization

The effect of cold-rolling on the crystallization behavior was studied starting in the 1970s, but initially the rolling experiments were coupled with annealing treatments after the cold-rolling (Masumoto and Maddin, 1975; Luborsky et al., 1976; Calvayrac et al., 1980; Noskova et al., 1989; Jin et al., 2001). These studies examined the effects that the cold-rolling had on the thermally induced crystallization process. Masumoto and Maddin determined that during the cold-rolling the atomic arrangement in the $Pd_{80}Si_{20}$ glass became more disordered with the formation of additional irregularities. The crystallization temperature increased with rolling deformation (Masumoto and Maddin, 1975). Luborsky and coworkers cold-rolled $Fe_{40}Ni_{40}P_{14}$ metallic glass ribbons at the same reduction levels than Masumoto of 40 % thickness reduction. Small-angle X-ray scattering experiments indicated the disappearance of scattering centers and thus an increase in randomness with rolling (Luborsky, Walter et al., 1976). Kulik and Matyja compressed stacks of $Pd_{100-x}Si_x$ metallic glass ribbons and observed that the crystallization temperature decreased for the compressed ribbons. The amount of compression was comparable to the roll reduction in Masumoto's and Luborsky's work. Moreover, the decrease in crystallization temperature with compression was a function of composition and with a higher Si content the difference between the crystallization temperatures of deformed and undeformed samples increased (Kulik and Matyja, 1980). Calvayrac and coworkers cold-rolled $Cu_{60}Zr_{40}$ metallic glass ribbons and found a less topologically defined atomic arrangement after cold-rolling. Changes in the crystallization behavior were not found, however (Calvayrac, Harmelin et al., 1980). Noskova and coworkers examined the effect of cold-rolling on the phase formation of $Fe_{81}Si_7B_{12}$ metallic glass ribbons. The main finding was an increase in the number density and a reduction in size of the nanocrystals during the initial primary crystallization reactions following cold-rolling (Noskova, Vil'danova et al., 1989). These results can be compared to the more recent findings of cold-rolling induced nanocrystal formation at room temperature. At rolling reductions of about 40 % a more disordered state seems to be achieved according to Masumoto's, Luborsky's, and Kulik's work. With further cold-rolling, however, nanocrystals develop in shear bands and thus at least locally the level of ordering increases at higher deformation levels. Several questions remain open, however: without doubt, shear bands develop at rolling reductions of 40 %. The Pd-based and Fe-based alloys with about 80 at % metal content and about 20 % metalloid content seem to be "inert" to crystallization during deformation while the primary-crystallizing Al-based metallic glasses reveal deformation-induced crystallization. The crystallization behavior of metallic glasses following deformation is thus strongly composition dependent. Further work is necessary to identify what aspects of the compositional differences are ultimately responsible for the differences in the response to deformation.

2.3.2 Deformation superposed on thermally induced annealing

Hydrostatic and uniaxial tensile stresses were superposed on thermally induced annealing. The impact of hydrostatic stress conditions on crystallization has been considered in several studies for melt-spun amorphous ribbons and bulk metallic glasses. Emmens and coworkers examined the crystallization behavior of $Pd_{75}Ag_5Si_{20}$ metallic glass under a hydrostatic stress of 600 MPa and observed a shift of the crystallization onset temperatures to higher temperatures (Emmens et al., 1975). They argued that an increase in hydrostatic pressure enhanced the nucleation rate due to a higher density of the crystallizing phase. At the same time, the growth rate was inhibited due to a reduced mobility under hydrostatic pressure (Emmens, Vrijen et al., 1975). Iwasaki and Masumoto examined the crystallization behavior of melt-spun $Pd_{80}Si_{20}$ metallic glass at a hydrostatic pressure of 10,000 MPa (Iwasaki and Masumoto, 1978). Their results indicated that the hydrostatic pressure retarded crystallization over the entire temperature range of crystallization. Iwasaki and Masumoto pointed out that the hydrostatic stress limited the mobility of atoms and thus reduced diffusion-controlled growth. Ye and Lu examined the primary crystallization of $Al_{89}La_6Ni_5$ metallic glass under hydrostatic pressure and found a lowering of the crystallization onset temperature under hydrostatic pressure (Ye and Lu, 1999). The thermodynamic argument based on a higher driving force for crystallization under hydrostatic pressure was tested with nanoindentation-based crystallization at different loading rates (Jiang et al., 2003). The indentation-induced nanocrystals were larger for the higher loading rate than for the lower rate. Jiang and coworkers argued that with the same hydrostatic stress surrounding the indent for both loading rates the slower rate exposes the metallic glass for a longer time to the hydrostatic pressure. Since the slower rate and therefore longer pressure exposure resulted in smaller nanocrystals than for the higher loading rate, the thermodynamic driving force argument could not hold to explain the effect of hydrostatic pressure on crystallization. It was suggested that pressure effect was predominantly kinetic in nature. The effects of hydrostatic pressure on crystallization were summarized in Suryanarayana's and Inoue's book (Suryanarayana and Inoue, 2011):

- The density increase with crystallization of metallic glasses promotes crystallization under hydrostatic pressure. For polymorphous transformation reactions, diffusion is not required and thus hydrostatic pressure should promote crystallization.
- Hydrostatic pressure reduces atomic mobility due to a decrease in the specific volume. For primary and eutectic crystallization reactions the kinetics of the crystallization process should be impeded.
- Hydrostatic pressure can change the crystallization pathway or the crystallization products.

A similar effort included the application of tensile stresses during thermal crystallization (Maddin and Masumoto, 1972; Patterson and Jones, 1979; Tiwari et al., 1982; Claus and von Heimendahl, 1983). The results of the tensile test experiments suggest that the tensile stress improves the volume diffusion capability and thus promotes crystallization for diffusion-controlled growth cases. For crystallization reactions that are interface controlled the tensile stress does not impact the crystallization behavior (Claus and von Heimendahl, 1983).

2.4 Mechanical processing of metallic glass/crystal composites

The synthesis approaches for metallic glass composites containing crystalline second phases fall in one of two general categories. Crystallites can develop "in-situ", i.e., during annealing of metallic glass precursors or during the quenching of the liquid alloys into glassy metals. This approach is very useful and efficient for the synthesis of metallic glasses containing nanocrystals. In addition to the annealing of metallic glasses and the formation of crystallites directly during quenching, metallic glass/crystal composites can be synthesized "ex-situ". The most important example of ex-situ processing is melt-infiltration (Dandliker et al., 1998). In this approach metallic glass is heated up to the supercooled liquid state and subsequently pushed into a form containing the second phase. The form containing the metallic glass and the second phase is then quenched. The second phases that have been used so far include metals, alloys, and nonmetallic materials. Different shapes have been used including fibers, rods, or particles. A major advantage of the melt-infiltration process is that the volume fraction of the second phase can be raised beyond the thermodynamic limits inherent to thermal processing. On the downside it is very difficult to achieve dispersions of nano-sized second phases without agglomeration of the second phase particles.

3. Properties of nanocrystals in amorphous matrices

Most studies of nanocrystal/metallic glass composites focus on the overall composite properties and the impact of the nanocrystals on the composite properties. From particle-matrix composite studies with crystalline materials it is known that not only the size distribution and particle number densities determine the composite properties, but furthermore the morphology of the particles, the defects that might exist in the particles, and the particle composition. This chapter highlights the current knowledge of defects in nanocrystals that are embedded in metallic glasses, their morphology and composition.

3.1 Defects in nanocrystals

Thermally induced nanocrystals in metallic glasses were initially considered to be defect free. This notion appears to be true for the vast majority of nanocrystals that develop during annealing of amorphous precursor alloys. Most investigations focused on Al-based amorphous alloys (Inoue, 1998; Abrosimova and Aronin, 2002) and revealed a defect free nature of the Al nanocrystals that grow to about 20 nm in size during controlled heat treatment (Abrosimova and Aronin, 2007). Aronin, however, reported in 2001 the presence of twins and dislocations in Al nanocrystals of annealed $Al_{86}Ni_{11}Yb_3$ melt-spun ribbons for a nanocrystal with a size of about 25 nm (Aronin, 2001). Louzguine-Luzgin and Inoue found evidence for dislocations in Al nanocrystals with sizes of less than 7 nm in as-spun $Al_{85}Y_4Ni_5Co_2Pd_4$ amorphous ribbons (Louzguine-Luzgin and Inoue, 2006; Louzguine-Luzgin and Inoue, 2007). The dislocations in Louzguine-Luzgin's work were observed for nanocrystals that impinged on each other due to extremely high particle number densities of 10^{24} m^{-3}. It was suggested that the dislocations developed due to microstrains in the nanocrystals and due to the formation of nanocrystals directly during quenching. The quenching, it was argued, inhibited the annihilation of dislocations due to the limited time at elevated temperatures during the quench process. From the reports of Al nanocrystals that were defect free after annealing and the reports of dislocations in nanocrystals that developed directly during quenching a conclusion could be drawn that defects could only

develop in quenched-in nanocrystals. The situation is more complex, though. For a $(Ni_{70}Mo_{30})_{90}B_{10}$ amorphous alloy, stacking faults were observed for thermally induced nanocrystals and for nanocrystal sizes exceeding about 5 nm (Abrosimova and Aronin, 2007). At smaller sizes, the nanocrystals that were comprised of fcc Ni-Mo solid solutions were defect free. The defect character was observed with high-resolution transmission electron microscopy and inferred from X-ray analysis (Abrosimova and Aronin, 2007). The same authors did not observe defects in fcc Al nanocrystals that developed in Al-Ni-RE (rare earth) melt-spun ribbons. The argument to rationalize the differences between the nanocrystal defect existence in the Ni-based and Al-based amorphous alloys was based on defect energies. Unlike the Al nanocrystals that are nearly pure Al, the Ni nanocrystals contain significant solute content. The alloying changes the electron concentration and thus the stacking fault energy. While the stacking fault energy is higher for Ni than for Al, the Ni-Mo solid solution has a much reduced stacking fault energy that is below the stacking fault energy of Al (Abrosimova and Aronin, 2007). The low stacking fault energy enables the formation of stacking faults. Microstrains due to, for example, differences in the thermal expansion coefficient, provide the necessary impetus for the formation of the stacking faults (Abrosimova and Aronin, 2007). The existence of defects in isolated nanoparticles or nanocrystals that are embedded in a matrix was considered by Gryaznov in light of dislocation concepts such as image forces and annihilation at interfaces (Gryaznov et al., 1991). The predictions obtained from Gryaznov's work did not, however, agree with the experimental observations in Abrosimova's work (Abrosimova and Aronin, 2002).
Nanocrystals that develop during intense deformation reactions at ambient deformation temperatures might be expected to contain defects due to the mechanical deformation underlying their genesis. A very limited number of studies focused on the defect content of deformation-induced nanocrystals. Studies on intensely cold-rolled and folded $Al_{88}Y_7Fe_5$ metallic glasses demonstrated that for nanocrystal sizes of less than about 5-10 nm the primary Al nanocrystals were defect free (Hebert et al., 2006). Some Al nanocrystals, however, clearly revealed the presence of defects as shown in Fig. 3.

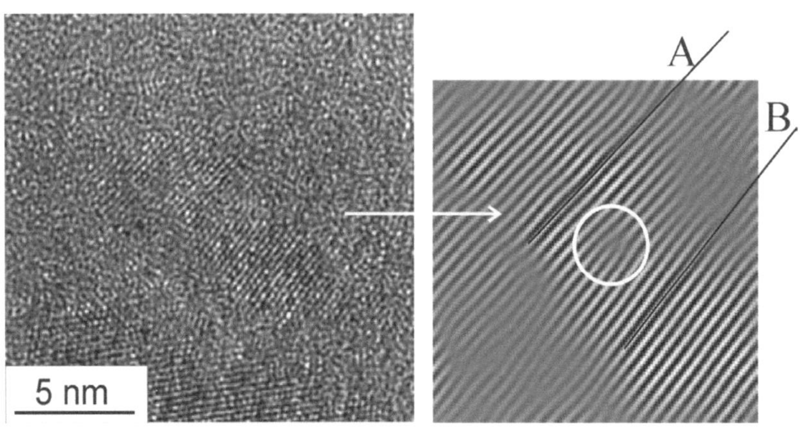

Fig. 3. High-resolution transmission electron microscopy image of an Al nanocrystal in a rolled and folded $Al_{88}Y_7Fe_5$ metallic glass ribbon (left). The Fourier transform (right) reveals the presence of a defect that appears to be a dislocation (Reprinted from (Hebert, Perepezko et al., 2006), with permission from Elsevier)

3.2 Morphology of nanocrystals

Transmission electron microscopy studies have repeatedly shown that nanocrystals developing in amorphous matrices have a spherical shape in the early growth stages, i.e., at a size of about 5-10 nm. An example of a spherical nanocrystal is shown in Fig.4 a.

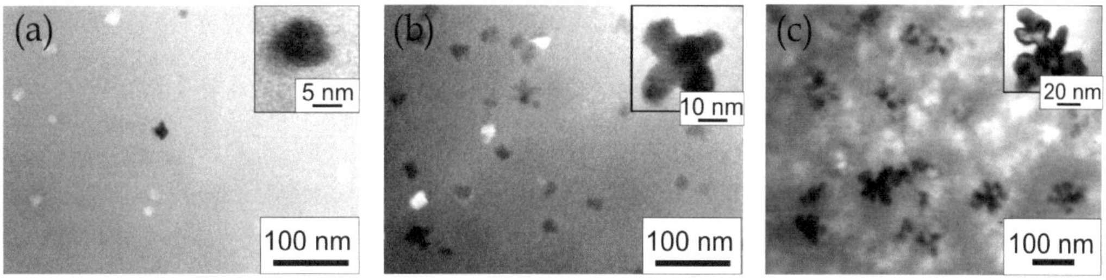

Fig. 4. Bright-field TEM images of $Al_{88}Y_7Fe_5$ melt-spun metallic glass ribbon. (a) after annealing at 245 °C for 10 min. (b) after annealing at 245 °C for 30 min. (c) at 245 °C for 100 min (adapted from (Perepezko and Hebert, 2002)).

The spherical shape of nanocrystals breaks down during further growth. For Al nanocrystals in amorphous Al-based alloys, a dendritic morphology started to develop at sizes of about 10-50 nm as shown in Fig. 4b. While the spherical shape of the nanocrystals during their early growth stage might be expected based on the largely isotropic nature of the metallic glass precursor, recent studies based on molecular-dynamics simulations and high-resolution electron microscopy suggest a morphology of the nanocrystals during the nucleation stage that deviates from the spherical symmetry. The step-wise growth process from one-dimensional ordered arrays into two- and three-dimensional nanocrystals for Zr-Ni-Fi metallic glasses is one example (Liu, Chen et al., 2007).

The transition of the nanocrystals from spherical to dendritic shape might be explained as interface instability. Mullins and Sekerka (Mullins and Sekerka, 1963) investigated the stability of infinitesimal undulations at the interface between a growing particle and the surrounding matrix under the condition that the concentration field can be calculated from the solution of the Laplace-equation. The velocity of each interface element can then be obtained from the mass-conservation equation:

$$v = \frac{D}{C - c_s} \frac{\partial c}{\partial n} \tag{1}$$

where C is the concentration inside the particle, c_s is the concentration in the matrix at the interface and $\partial c / \partial n$ is the concentration gradient at the interface normal to the interface. In order to "generate" undulations, the following ansatz was used for the description of the particle interface:

$$r(\theta,\varphi) = R + \delta Y_l^m(\theta,\varphi) \tag{2}$$

where R is not a function of the angular variables, Y_l^m are the orthonormal set of eigenfunctions for the solution of the Laplace equation for spherical coordinates (spherical harmonics) and δ is the amplitude of the spherical harmonics that is sufficiently small for higher order terms in δ ($\cong \delta^2, \delta^3, \ldots$) to be negligible. Mullins and Sekerka also included the capillarity effect on the concentration outside the particle at the interface. As a result, an expression for $\dot{\delta}_l$ was obtained, i.e. the change in time of the amplitudes of all spherical harmonic functions. Since this expression is a difference between a concentration gradient term (> 0) and a capillarity term (< 0), the amplitudes can grow (interface becoming unstable) or decay, depending on the relative magnitude of both terms. A key result in (Mullins and Sekerka, 1963) is that the critical size that marks the onset of instability for any of the spherical harmonics is 7R*, with R* the critical size for nucleation. The result for the stability analysis are illustrated schematically in Fig. 5 which is taken from Mullins and Sekerka's seminal paper on interface instability (Mullins and Sekerka, 1963). The solid line denotes the critical size as a function of the parameter $-\frac{c_\infty - c_0}{c_0}$, where c_0 is the concentration of the matrix at the interface for a planar interface (no capillarity effect) and c_∞ the concentration of the matrix at infinity. In the hatched area, the particle is stable, but above the solid line the morphology of the growing particle becomes unstable. The dashed line corresponds to the critical nucleus size. Two trajectories are highlighted that illustrate different growth paths. For an isolated particle in an infinite matrix, the ratio $-\frac{c_\infty - c_0}{c_0}$ is constant therefore the trajectory is vertical. Once the morphology becomes unstable, it continues to be unstable. In case of diffusion-field impingement, c_∞ is time-dependent, therefore the trajectory can be curved. This can lead to the situation depicted in path 2. After an intermediate stage where the particle is unstable it reaches the stable region again.

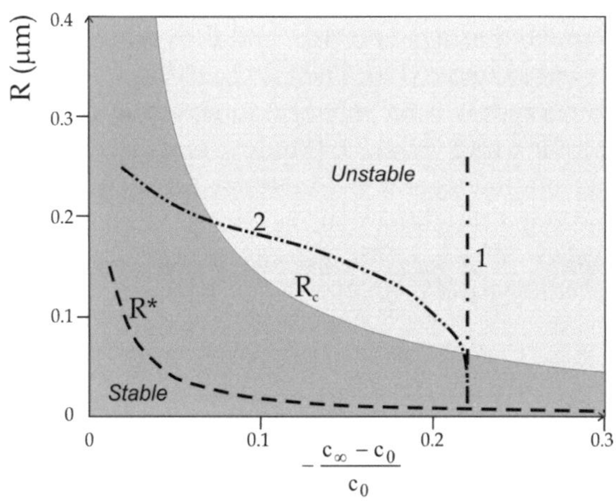

Fig. 5. Critical radius for a spherical particle to become unstable with respect to the spherical shape as a function of the parameter $\frac{c_\infty - c_0}{c_0}$ (adapted from (Mullins and Sekerka, 1963).

The morphology of deformation-induced nanocrystals is not necessarily spherical or dendritic. High-resolution transmission electron microscopy images of Al nanocrystals that developed during intense cold-rolling and folding of metallic glass $Al_{88}Y_7Fe_5$ ribbon revealed prolate shapes of the nanocrystals as shown in Fig. 6. A high-resolution transmission electron image of an Al nanocrystal in the bent region of a $Al_{90}Fe_5Gd_5$ metallic glass ribbon, however, revealed a nearly spherical shape (Jiang and Atzmon, 2003). Additional insight can be gained from molecular-dynamic simulations. An orientation relationship was observed for the shear- and crystallographic direction of deformation-induced Ni nanocrystals (Tarumi et al., 2000).

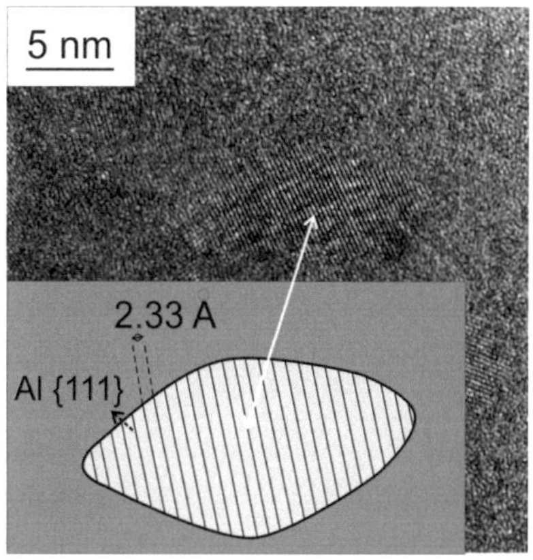

Fig. 6. High resolution transmission electron microscopy image of a Al nanocrystals that developed during repeated cold-rolling and folding of an array of melt-spun $Al_{88}Y_7Fe_5$ metallic glass ribbons (Reprinted from (Hebert, Perepezko et al., 2006), with permission from Elsevier).

4. Effect of nanocrystals on properties of nanocrystal/amorphous composites

4.1 Mechanical properties

Metallic glasses in general offer high strength, elastic resilience (Schuh et al., 2007), and in many cases additional advantageous properties such as high wear resistance (Greer et al., 2002). The substantial differences in the mechanical behavior between crystalline and glassy metals derive from the vastly different deformation mechanisms. Dislocations and stacking faults along with twins are defects that are responsible for the plastic deformation of crystalline materials. At room temperature most metallic glasses deform above a critical stress with the formation and propagation of shear bands. These shear bands are thin—about 20-50 nm—planar regions that deform heavily while the surrounding regions are not or very little affected. During deformation of monolithic metallic glasses a single shear band can propagate throughout the sample and in the process weaken the metallic glass along the shear band to the point of fracture if the deformation is unrestricted (Hays et al., 2000). The Achilles heel of most monolithic metallic glasses for structural applications is thus their limited ductility in tension due to the unrestricted shear band propagation. Many attempts have been made to improve plastic deformation in metallic glasses, including but not limited to specific selections of alloying elements for improved ductility (Gu et al., 2008) or the identification of bulk-glass forming systems that reveal modest plasticity even in the as-cast condition (Schroers and Johnson, 2004). Among the most successful strategies, though, is the dispersion of crystalline particles in amorphous matrices. For micrometer-sized crystalline particles, ductility can be gained but at the expense of a decrease in the overall composite strength (Bae et al., 2003). Figure 7 a, for example, highlights the dispersion of micrometer-sized brass particles in a warm-extruded $Ni_{59}Zr_{20}Ti_{16}Si_2Sn_3$ metallic glass. While the monolithic glass did not reveal plasticity, a 5 % plastic strain was achieved at 40 vol % brass, but the strength decreased by about 40 %.

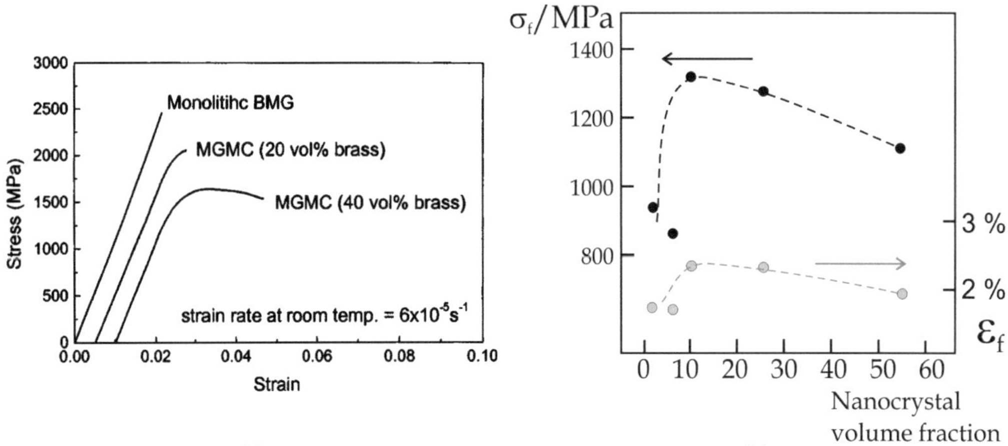

(a) (b)

Fig. 7. (a) The stress vs strain curves of the monolithic sample and MGMCs containing 20 vol % brass and 40 vol % brass tested under the uniaxial compressive condition at room temperature. The load is applied along the extrusion direction (Reprinted with permission from (Bae, Lee et al., 2003). Copyright 2003, American Institute of Physics). (b) Fracture strength σ_f and strain at fracture, e_f, for $Al_{88}Y_2Ni_9Fe_1$ metallic glass as a function of volume fraction of fcc Al nanocrystals (adapted from (Yeong-Hwan et al., 1991)).

Unlike for micrometer-sized crystals in metallic glasses the dispersion of nanocrystals can not only improve the strain to fracture, but can furthermore improve the fracture strength as shown in Fig. 7 b. The dispersion of Al nanocrystals in $Al_{88}Y_2Ni_9Fe_1$ metallic glass raised the fracture stress from about 950 MPa to over 1300 MPa while the strain to fracture increased slightly rather than decreased for a volume fraction of the nanocrystals between 0 and 10 % but decreased at higher volume fractions (Yeong-Hwan, Inoue et al., 1991). The decrease in strain to fracture or ductility with increasing volume fraction of nanocrystals appears to be a general trend (Bian et al., 2002).

Since the deformation behavior of metallic glasses at room temperature and stresses above the elastic limit is tied to the formation and propagation of shear bands, a key topic in understanding the role of nanocrystals and micrometer-sized crystals for the mechanical behavior of metallic glasses is the interaction between shear bands and nano-/micro crystals. This interaction involves not only the role of crystals for interacting with existing and propagating shear bands, but also the role of crystals for the formation of shear bands. It was indeed suggested early on in the development of crystal/metallic glass composites that the role of the crystalline second phase in the amorphous matrices was twofold: first, to serve as initiation or nucleation sites for shear bands and secondly to attract or pin shear bands (Hays, Kim et al., 2000). A parallel between shear bands and their interaction with second phase crystals and dislocations in crystalline materials and their interaction with crystalline second phases is the importance of elastic stress fields. Several studies focused on the nature of stress fields in the vicinity of crystals and shear bands. Donovan noted (Donovan and Stobbs, 1983) that interactions between shear bands and other sources for elastic stresses in the metallic glasses such as second phase crystals depend on the dilation of the specific volume in shear bands during shear band propagation (Spaepen and Turnbull, 1974; Argon et al., 1985). The dilation of the volume within shear bands induces compressive stresses in the surrounding matrix. The compressive stress field surrounding shear bands would indicate a repulsive interaction with crystals that are surrounded by compressive elastic stress fields and an attraction in case of tensile stress fields. Different sources contribute to elastic stress fields in the vicinity of crystals in metallic glasses. For in-situ formed crystals—either micrometer-sized dendrites or nanocrystals—the annealing induced crystallization reduces the specific volume of the crystalline phase for most metallic glass systems relative to the metallic glass matrix. The stress fields that could develop during the crystallization process, however, are considered to be alleviated due to the elevated temperatures of the metallic glasses during thermally induced crystallization. The elevated temperatures allow the atoms in the surrounding of the crystallites to relax during the crystallization process and therefore potential elastic stresses are relieved. Elastic stresses surrounding crystals furthermore develop during the cooling process from the crystallization process due to the thermal expansion mismatch between crystallites and metallic glasses.

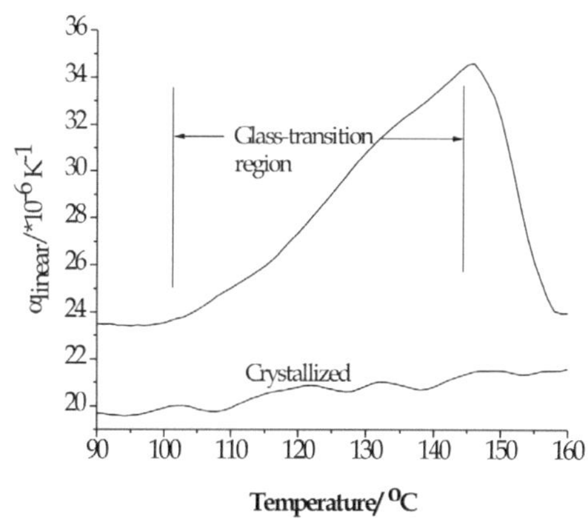

Fig. 8. Thermal expansion coefficient of crystallized and glassy $(Ce_{0.72}Cu_{0.28})_{78.5}Al_{10}Fe_{10}Si_{1.5}$ bulk metallic glass

Figure 8 highlights differences in the true thermal expansion coefficient between a crystallized $(Ce_{0.72}Cu_{0.28})_{78.5}Al_{10}Fe_{10}Si_{1.5}$ metallic glass and the same composition in a glassy state. The temperature range depicted in Fig. 8 includes the glass transition region; the steep drop in the thermal expansion coefficient for the metallic glass sample represents the onset and evolution of crystallization of the metallic glass (Mubarok and Hebert). Upon loading of metallic glasses containing crystals, additional stresses develop due to differences in the elastic and plastic behavior of crystals and glassy matrices. Ott and coworkers examined the stresses in and around micrometer-sized Ta crystals in Zr-Cu-Ni-Ta-Al metallic glasses (Ott et al., 2005). The micrometer-sized Ta crystals were softer than the metallic glass. Plastic misfit strains developed when the applied load induced plastic strain in the Ta particles while the glassy matrix still deformed elastically. The plastic misfit strains induced localized stress fields in the immediate vicinity of the Ta particles. Shear bands could thus develop in the vicinity of the particles, but could not propagate into the glassy matrix where the stresses remained below the critical stress for yielding. A further important finding in Ott's work (Ott, Sansoz et al., 2005) was that the stress in the glassy matrix was below the applied stress at some matrix locations. The stress fields surrounding the Ta particles were thus highly inhomogeneous and enabled shear band formation and propagation in some areas, but not in other regions until the applied stress increased the stress even in low-stress regions to the point of yielding. As a result, the shear band propagation could not be straight but would be expected to the wavy and possibly deviating from the planes of maximum shear stress. Ott's work was conducted for micro-meter sized crystals and the situation of a soft crystal embedded in a stronger glassy matrix could change for the situation of nanocrystals embedded in glassy matrices. One of the explanations for hardness increases of nanocrystal/metallic glass composites in fact invokes the idea of a hardness increase due to the formation of nanocrystals with hardness levels greater than those of the metallic glass matrices (Yeong-Hwan, Inoue et al., 1991). While the effects of elastic and plastic misfit stresses for the role of shear band propagation have been clearly recognized, further work is necessary to understand in particular the formation of stresses in metallic glasses containing nanocrystals and their effect on shear bands.

4.2 Wear properties

Wear properties of metallic glasses have been studied since 1979 (Boswell, 1979). Until the discovery of bulk metallic glasses, wear studies were conducted on Fe-based and later on Al-based metallic glass melt-spun ribbons (Klinger and Feller, 1983; Miyoshi and Buckley, 1983; Miyoshi and Buckley, 1983; Miyoshi and Buckley, 1984; Kishore et al., 1987; Lim and Ashby, 1987; Gloriant, 2003). Reviews on the wear behavior of melt-spun, ribbon-shaped metallic glasses were prepared by Lim and Ashby (Lim and Ashby, 1987) and Greer, Rutherford, and Hutchings (Greer et al., 2002). The vast majority of studies on the effect of crystallites in metallic glasses on wear properties indicate improvements in the wear behavior with the formation of crystallites. The wear resistance of $Al_{88}Ni_4Sm_8$ metallic glass, for example, increased after an initial phase before reaching a plateau (Gloriant and Greer, 1998). Improvements with partial devitrification of Al nanocrystals were found furthermore for partially crystallized Al-Ni-Y based amorphous alloys (Greer, Rutherford et al., 2002), annealed $Zr_{41}Ti_{14}Cu_{12.5}Ni_{10}Be_{22.5}$ bulk metallic glasses (Li et al., 2002), or partially crystallized bulk $Cu_{50}Hf_{41.5}Al_{8.5}$ metallic glasses (Maddala et al., 2010). The wear improvements at first glance seem to be expected, since the hardness increases in many cases during devitrification. Archard's empirical rule relates hardness directly to wear resistance (Archard, 1957). The hardness, however, continues to increase with further crystallization and often reaches a maximum for the fully crystallized condition (Maddala, Mubarok et al., 2010). The wear resistance does not continue to improve, but often reaches a maximum at a certain degree of crystallization (Boswell, 1979; Vom Wege et al., 1988; Maddala, Mubarok et al., 2010). A possible explanation for the wear changes with crystallization was given early on in Boswell's and Zum Gahr's work (Boswell, 1979; Zum Gahr and Noecker, 1981). With increasing crystallization the hardness increases, but the toughness of the material decreases, i.e., the brittleness of the metallic glass increases. The increasing brittleness of the metallic glass with annealing and crystallization causes a change in the wear mechanism

toward micro-fractures and delamination. Maddala and coworkers have recently confirmed the wear behavior changes with measurements of notch toughness and hardness at different crystallization stages (Maddala and Hebert, 2011).
Several results on the wear behavior of metallic glasses containing nanocrystals, however, seem to contradict the mechanism of a wear improvement with nanocrystal formation and eventual deterioration at higher nanocrystal number densities. Abrasive wear tests on a SiC paper, for example, showed that the wear resistance of the Co-based glass improved with increasing degree of crystallization when fine-grained SiC paper was used (Vom Wege, Skrotzki et al., 1988; Vom Wege et al., 1991). Upon testing with coarse-grained SiC paper, however, the wear resistance deteriorated with increasing crystallinity. The crystallization products were primary Co and Co_3B particles with sizes of 15-20 nm (Vom Wege, Skrotzki et al., 1988). Tam and Shek examined the wear behavior of Cu-Zr-Ti metallic glasses and observed that the hardness did not scale with the wear behavior (Tam and Shek, 2004). Upon closer examination, however, it appears that in Tam's work and Vom Wege's work the wear mechanism was based on micro-fractures. This clearly shows that for wear studies not only the sample need to be considered, but the wear load, condition, atmosphere, amongst others. It seems, however, that under conditions that cause micro-ploughing and micro-cutting the formation of nanocrystals has a positive impact on the wear behavior. Clearly, more work is necessary to identify the relations between the level of crystallization, external wear conditions, and wear behavior.

4.3 Corrosion properties

Metallic glasses in general compare favorably with their crystalline counterparts in their corrosion behavior, although exceptions are known (Hashimoto, 2002; Scully and Lucente, 2005; Scully et al., 2007). The corrosion behavior of single-phase metallic glasses benefits from the lack of structural or chemical inhomogeneities such as grain boundaries or inclusions. It was suggested that metallic glasses promote the formation of amorphous oxides (Scully, Gebert et al., 2007). Many metallic glasses reveal solute levels that far exceed those of their crystalline counterpart systems. The enhanced solute content in the metallic glass can increase the solute content in oxide layers and thus improve the corrosion behavior. The change in interatomic distances and coordination numbers with transition from crystalline to amorphous atomic arrangement furthermore affects electrochemical properties and corrosion properties (Scully, Gebert et al., 2007). Among the earliest corrosion resistance studies on metallic glasses were Fe-metalloid metallic glasses containing chromium. The early studies on Fe-Cr-P-C melt-spun ribbons were continued more recently when bulk metallic Fe-glasses were developed. Bulk metallic glasses in the Fe-Cr-Mo-C-B system revealed corrosion rates as low as 10^{-3} mm/year in concentrated hydrochloric acid.
While the corrosion advantages of fully amorphous metallic glasses are undisputed, the formation of nanocrystals might be expected to negatively impact their corrosion behavior. The nanocrystals, after all, represent both structural and chemical inhomogeneities and entail interfaces with the surrounding amorphous matrices. The chemical inhomogeneities are not limited to the nanocrystal/amorphous matrix interface, but extend into the amorphous matrix due to gradients in solute concentrations that exist at least during the grow stages of nanocrystals that form non-polymorphically. Studies on the corrosion behavior of nanocrystal-containing metallic glasses have been conducted so far for Fe-based,

Ni-based, Zr-based, and Al-based systems. An overall trend, however, does not appear to exist. Even within the same system the corrosion behavior improves for some compositions, but deteriorates for other compositions.

Lucente and Scully examined the corrosion behavior of Al-based amorphous-nanocrystalline alloys containing Al nanocrystals in detail. Four melt-spun amorphous alloys, $Al_{90}Fe_5Gd_5$, $Al_{85}Fe_7Gd_8$, $Al_{90}Co_3Ce_7$, and $Al_{87}Ni_7Gd_6$ were exposed to 0.6 M NaCl electrolyte in an electrolytic cell to test the resistance to micrometer-scale pitting (Lucente and Scully, 2008). The resistance to pitting can be determined quantitatively from measurements of the pitting and repassivation potentials in anodic polarization scans. The pitting potential reflects a resistance to pit initiation and stabilization (Lucente and Scully, 2008). For all four amorphous alloys the repassivation potential remained unchanged when the fully amorphous alloys were partially devitrified. The pitting potential increased compared to the fully amorphous alloy, except for the $Al_{90}Co_3Ce_7$ amorphous alloy that revealed a small decrease in pitting potential with nanocrystal formation. It is noteworthy that the pitting potential increased significantly for the $Al_{85}Fe_7Gd_8$ amorphous alloy despite a nanocrystal size of 100 nm. The increase in pitting potential with nanocrystallization relative to the fully amorphous state suggests that the devitrification inhibits pit initiation. In a second set of experiments, the growth and repassivation kinetics were determined for artificial pits that were induced electrolytically. The growth kinetics of the artificial pits was not affected for the amorphous alloys containing nanocrystals. The unchanged repassivation potential was interpreted as an overall growth kinetics of pits larger than 10 micrometers that were unaffected by the nanocrystal formation.

5. Conclusion

Nanocrystals can significantly improve properties of metallic glasses, for example, mechanical properties or magnetic properties. Some progress has been made to understand the mechanisms behind these improvements, but in many cases a clear understanding has not been achieved, yet. Further work is therefore not only necessary, but promises further advancements. Progress will have to come from a combination of smart experimentation, advanced diagnostics, and computer simulations. Arguably the most interesting aspect of nanocrystals in metallic glasses, though, is the longest-standing issue: their synthesis. Changes in local atomic configuration or changes in nanocrystal number densities by orders of magnitude, or changes in the phase formation with a 1 at % alloying addition offer exciting opportunities to design novel composite materials.

6. References

Abrosimova, G. E., &Aronin, A. S. (2002). The fine structure of FCC nanocrystals in Al- and Ni-based alloys. *Physics of The Solid State*, 44, 6, pp. 1003-1007.

Abrosimova, G. E., &Aronin, A. S. (2007). Size effect on the structure of Al- and Ni-based nanocrystals. *Physics of the Solid State*, 50, 1, pp. 159-163.

Archard, J.F. (1957). Elastic deformation and the laws of friction. *Proc. Roy. Soc. Lond.*, A243, pp. 190-205.

Argon, A. S., Megusar, J., &Grant, N. J. (1985). Shear band induced dilations in metallic glasses. *Scripta Met.*, 19, pp. 591-596.

Aronin, A. S. (2001). Formation and structure of nanocrystals in an $Al_{86}Ni_{11}Yb_3$ metallic glass. *Physics of the Solid State*, 43, 1, pp. 2003-2011.
Bae, D. H., Lee, M. H., Kim, D. H., &Sordelet, D. J. (2003). Plasticity in Ni59Zr20Ti16Si 2Sn3 metallic glass matrix composites containing brass fibers synthesized by warm extrusion of powders. *Applied Physics Letters*, 83, 12, pp. 2312-2314.
Battezzati, L., &Baldissin, D. (2008). Quantitative evaluation of lengthscales for temperature rise in shear bands and for failure of metallic glasses. *Scripta Materialia*, 59, 2, pp. 223-226.
Bian, Z., He, G., &Chen, G. L. (2002). Investigation of shear bands under compressive testing for Zr-base bulk metallic glasses containing nanocrystals. *Scripta Materialia*, 46, 6, pp. 407-412.
Boswell, P. G. (1979). Wear resistance of liquid quenched metallic glass. *Journal of Materials science*, 14, pp. 1505-1507.
Cahn, J.W. (1961). On spinodal decomposition, Acta Metall., 9, pp. 795-801
Calvayrac, Y., Harmelin, M., Quivy, A., Chevalier, J. P., &Bigot, J. (1980). COLD-ROLLING AND SUBSEQUENT ANNEALING OF AMORPHOUS Cu60Zr40. *Scripta metallurgica*, 14, 8, pp. 895-898.
Chen, H., He, Y., Shiflet, G. J., &Poon, S. J. (1994). Deformation-induced nanocrystal formation in shear bands of amorphous alloys. *Nature*, 367, 6463, pp. 541-543.
Christian, J.W. (1975). The theory of transformations in metals and alloys, Pergamon.
Claus, J. C., &von Heimendahl, M. (1983). Influence of superimposed tensile stress on metallic glass crystallization. *Zeitschrift fur Metallkunde*, 74, 11, pp. 744-750.
Csontos, A. A., &Shiflet, G. J. (1997). Composition profiles associated with nanocrystal formation in aluminum-rich metallic glasses, Orlando, FL, USA, TMS.
Dandliker, R. B., Conner, R. D., &Johnson, W. L. (1998). Melt infiltration casting of bulk metallic-glass matrix composites. *Journal of Materials Research*, 13, 10, pp. 2896-2901.
Donovan, P. E., &Stobbs, W. M. (1983). Shear band interactions with crystals in partially crystallised metallic glasses. *Journal of Non-Crystalline Solids*, 55, 1, pp. 61-76.
Duwez, P. (1981). Metallic glasses - historical background. Glassy Metals I. H.-J. Guentherodt and H. Beck. Berlin Heidelberg New-York, Springer-Verlag. I: 19-23.
Emmens, W.C., Vrijen, J., &Radelaar, S. (1975). Crystallization of amorphous $Pd_{0.75}Ag_{0.05}Si_{0.2}$ under hydrostatic stress. *J. Non-Cryst. Solids*, 18, pp. 299-302.
Freestone, I., Meeks, N., Sax, M., &Higgitt, C. (2007). The Lycurgus Cup - A Roman Nanotechnology. *Gold Bulletin*, 40, 4, pp. 270-276.
Glicksman, M.E. (2000). Diffusion in solids, John Wiley & Sons, Inc.
Gloriant, T. (2003). Microhardness and abrasive wear resistance of metallic glasses and nanostructured composite materials. *Journal of Non-Crystalline Solids*, 316, pp. 96-103.
Gloriant, T., &Greer, A. L. (1998). Al-based nanocrystalline composites by rapid solidification of Al-Ni-Sm alloys. *NanoStruct. Mater.*, 10, 3, pp. 389-396.
Gloriant, T., Ping, D.H., Hono, K., Greer, A.L., &Baro, M.D. (2001). Nanostructured Al88Ni4Sm8 alloys investigated by transmission electron and field-ion microscopies. *Mater. Sci. Engr. A*, A304-306, pp. 315-320.
Greer, A. L. (1995). Metallic Glasses. *Science*, 267, pp. 1947-1953.
Greer, A. L., Rutherford, K. L., &Hutchings, I. M. (2002). Wear resistance of amorphous alloys and related materials. *International Materials Reviews*, 47, 2, pp. 87-112.

Greer, A. L., Rutherford, K. L., &Hutchings, M. (2002). Wear resistance of amorphous alloys and related materials. *International Materials Reviews*, 47, 2, pp. 87-112.

Gryaznov, V. G., Polonsky, I. A., Romanov, A. E., &Trusov, L. I. (1991). Size effects of dislocation stability in nanocrystals. *Physical Review B (Condensed Matter)*, 44, 1, pp. 42-46.

Gu, X. J., Poon, S. J., Shiflet, G. J., &Widom, M. (2008). Ductility improvement of amorphous steels: roles of shear modulus and electronic structure. *Acta Materialia*, 56, 1, pp. 88-94.

Hashimoto, K. (2002). In pursuit of new corrosion-resistant alloys. *Corrosion*, 58, pp. 715.

Hays, C. C., Kim, C. P., &Johnson, W. L. (2000). Microstructure controlled shear band pattern formation and enhanced plasticity of bulk metallic glasses containing in situ formed ductile phase dendrite dispersions. *Physical Review Letters*, 84, 13, pp. 2901-2904.

Hebert, R. J., Boucharat, N., Perepezko, J. H., Rosner, H., &Wilde, G. (2005). Deformation-induced crystallization reactions in amorphous $Al_{88}Y_7Fe_5$ alloy, Phoenix, AZ, United States, Minerals, Metals and Materials Society, Warrendale, PA 15086, United States.

Hebert, R. J., &Perepezko, J. H. (2004). Effect of cold-rolling on the crystallization behavior of amorphous $Al_{88}Y_7Fe_5$ alloy, Oxford, UK, Elsevier.

Hebert, R. J., Perepezko, J. H., Rosner, H., &Wilde, G. (2006). Dislocation formation during deformation-induced synthesis of nanocrystals in amorphous and partially crystalline amorphous Al88Y7Fe5 alloy. *Scripta Materialia*, 54, 1, pp. 25-29.

Hebert, R.J., &Perepezko, J. H. (2007). Driven nanocrystal catalysis for amorphous Al-Y-Fe alloys. *Met. Trans. A*, to be published, pp.

Hebert, Rainer J., Rosner, Harald, Wilde, Gerhard, &Perepezko, John H. (2004). Deformation-induced devitrification of Al-based amorphous alloys. *JOM*, 56, 11, pp. 269.

Hillert, M. (1968). The role of interfaces in phase transformations. The mechanism of phase transformations in crystalline solids, Manchester, The Institute of Metals, London.

Hongwen, Zhang, Subhash, G., &Maiti, S. (2007). Local heating and viscosity drop during shear band evolution in bulk metallic glasses under quasistatic loading. *Journal of Applied Physics*, 102, 4, pp. 043519-043511.

Hono, K., Zhang, Y., Tsai, A.P., Inoue, A., &Sakurai, T. (1995). Solute partitioning in partially crystallized Al-Ni-Ce(-Cu) metallic glasses. *Scripta Met. Mat.*, 32, 2, pp. 191-196.

Inoue, A. (1998). Amorphous, nanoquasicrystalline and nanocrystalline alloys in Al-based systems. *Progr. Mater. Sci.*, 43, pp. 365-.

Iwasaki, H., &Masumoto, T. (1978). Effect of high pressure on the crystallization of an amorphous $Pd_{80}Si_{20}$ alloy. *J. Mater. Sci.*, 13, pp. 2171-2176.

Jiang, W. H., &Atzmon, M. (2003). The effect of compression and tension on shear-band structure and nanocrystallization in amorphous $Al_{90}Fe_5Gd_5$: A high-resolution transmission electron microscopy study. *Acta Materialia*, 51, 14, pp. 4095-4105.

Jiang, W. H., Liu, F. X., Liao, H. H., Choo, H., Liaw, P. K., Edwards, B. J., &Khomami, B. (2008). Temperature increases caused by shear banding in as-cast and relaxed Zr-based bulk metallic glasses under compression. *Journal of Materials Research*, 23, 11, pp. 2967-2974.

Jiang, W. H., Liu, F. X., Liaw, P. K., &Choo, H. (2007). Shear strain in a shear band of a bulk-metallic glass in compression. *Applied Physics Letters*, 90, 18, pp.

Jiang, W. H., Pinkerton, F. E., &Atzmon, M. (2003). Effect of strain rate on the formation of nanocrystallites in an Al-based amorphous alloy during nanoindentation. *Journal of Applied Physics*, 93, 11, pp. 9287-9290.

Jin, H. J., Zhou, F., Wang, L. B., &Lu, K. (2001). Effect of plastic deformation on thermal stability in metallic glasses. *Scripta Materialia*, 44, 7, pp. 1083-1087.

Kim, D. H., Chang, H. J., Kim, Y. M., Kim, Y. J., &Chattopadhyay, K. (2006). On the origin of nanocrystals in the shear band in a quasicrystal forming bulk metallic glass Ti40Zr29Cu9Ni8Be14. *Scripta Materialia*, 55, 6, pp. 509-512.

Kim, J. J., Choi, Y., Suresh, S., &Argon, A. S. (2002). Nanocrystallization during nanoindentation of a bulk amorphous metal alloy at room temperature. *Science*, 295, 5555, pp. 654-657.

Kingery, W.D., Bowen, H.K., &Uhlmann, D.R. (1976). Introduction to ceramics, John Wiley & Sons, Inc.

Kishore, Sudarsan, U., Chandran, N., &Chattopadhyay, K. (1987). On the wear mechanism of iron and nickel based transition metal-metalloid metallic glasses. *Acta Metallurgica*, 35, 7, pp. 1463-1473.

Klinger, R., &Feller, H. G. (1983). Sliding friction and wear resistance of the metallic glass $Fe_{40+}Ni_{40}B_{20}$ *Wear*, 86, pp. 287-297.

Koch, C. C. (1989). Materials synthesis by mechanical alloying. Annual review of material science. Vol.19. Palo Alto, CA, USA, Annual Reviews: 121-143.

Koester, U. (1993). Phase transformations in rapidly solidified alloys. *Key Engr. Mat.*, 81-83, pp. 647-662.

Kulik, T., &Matyja, H. (1980). The effect of plastic deformation of amorphous Pd-Si alloys on their thermal properties. *Journal of Materials Science*, 15, 12, pp. 3169-3172.

Lee, Seok-Woo, Huh, Moo-Young, Chae, Soo-Won, &Lee, Jae-Chul (2006). Mechanism of the deformation-induced nanocrystallization in a Cu-based bulk amorphous alloy under uniaxial compression. *Scripta Materialia*, 54, 8, pp. 1439-1444.

Lewandowski, J. J., &Greer, A. L. (2006). Temperature rise at shear bands in metallic glasses. *Nature Materials*, 5, 1, pp. 15-18.

Li, G., Wang, Y. Q., Wang, L. M., Gao, Y. P., Zhang, R. J., Zhan, Z. J., Sun, L. L., Zhang, J., &Wang, W. K. (2002). Wear behavior of bulk Zr41Ti14Cu12.5Ni10Be22.5 metallic glasses. *Journal of Materials Research*, 17, 8, pp. 1877-1880.

Lim, S.C., &Ashby, M.F. (1987). Overview No. 55 Wear mechanism maps. *Acta Metallurgica*, 35, 1, pp. 1-24.

Liu, X. J., Chen, G. L., Hui, X. D., Hou, H. Y., Yao, K. F., &Liu, C. T. (2007). Growth mechanism from nano-ordered clusters to nanocrystals in a deeply undercooled melt of Zr-Ni-Ti metallic glass. *Journal of Applied Physics*, 102, 6, pp.

Louzguine-Luzgin, D. V., &Inoue, A. (2006). Comparative study of the effect of cold rolling on the structure of Al-RE-Ni-Co (RE=rare-earth metals) amorphous and glassy alloys. *Journal of Non-Crystalline Solids*, 352, 36-37, pp. 3903-3909.

Louzguine-Luzgin, D. V., &Inoue, A. (2006). Observation of linear defects in Al particles below 7 nm in size. *Journal of Materials Research*, 21, 6, pp. 1347-1350.

Louzguine-Luzgin, D. V., &Inoue, A. (2007). Investigation of a rapidly solidified Al-based nanocomposite with extremely high number density of precipitates. *Materials*

Science & Engineering A (Structural Materials: Properties, Microstructure and Processing), 449-451, pp. 1026-1028.

Luborsky, F. E., Walter, J. R., &LeGrand, D. G. (1976). Cold-rolling and annealing of amorphous ribbons, USA.

Lucente, A., &Scully, J. R. (2008). Localized corrosion of Al-based amorphous-nanocrystalline alloys with solute-lean nanocrystals: pit stabilization. *J. Electrochem. Soc.*, 155, 5, pp. C234-C243.

Maddala, D., &Hebert, R.J. (2011). Effect of notch toughness and hardness on sliding wear of Cu50Hf41.5Al8.5 bulk metallic glass. *Scripta Mater., submitted*, pp.

Maddala, D., Mubarok, A., &Hebert, R.J. (2010). Sliding wear behavior of Cu50Hf41.5Al8.5 bulk metallic glass. *Wear*, 269, 7-8, pp. 572-580.

Maddin, R., &Masumoto, T. (1972). The deformation of amorphous palladium-20 at.% silicon. *Material Science and Engineering*, 9, 3, pp. 153-162.

Masumoto, T., &Maddin, R. (1975). Structural stability and mechanical properties of amorphous metals. *Material Science and Engineering*, 19, 1, pp. 1-24.

Miyoshi, Kazuhisa, &Buckley, Donald H. (1983). Friction and wear of some ferrous-base metallic glasses, Lewis Res. Cent.,Natl. Aeronaut. Space Adm.,Cleveland,OH,USA.: 24 pp.

Miyoshi, Kazuhisa, &Buckley, Donald H. (1983). Sliding-induced crystallization of metallic glass, Lewis Res. Cent.,Natl. Aeronaut. Space Adm.,Cleveland,OH,USA.: 23 pp.

Miyoshi, Kazuhisa, &Buckley, Donald H. (1984). Friction and wear of some ferrous-base metallic glasses. *ASLE Transactions*, 27, 4, pp. 295-304.

Mubarok, A., &Hebert, R.J. The Kauzmann temperature of a Ce69.5Al10Cu20Co0.5 bulk metallic glass based on calorimetric and volumetric analyses. *in preparation*, pp.

Mullins, W.W., &Sekerka, R.F. (1963). Morphological stability of a particle growing by diffusion or heat flow. *J. Appl. Phys.*, 34, 2, pp. 323-329.

Noskova, N. I., Vil'danova, N. F., Tagirov, R. I., Potapov, A. P., &Glazer, A. A. (1989). Influence of rolling reduction on the crystallization process of amorphous alloy $Fe_{81}Si_7B_{12}$ *Phys. Met. Metall.*, 67, 6, pp. 136-143.

Ogura, A., Tarumi, R., Shimojo, M., Takashima, K., &Higo, Y. (2001). Control of nanocrystalline orientation using application of a stress field in an amorphous alloy. *Appl. Phys. Lett.*, 79, 7, pp. 1042-1044.

Ott, R. T., Sansoz, F., Molinari, J. F., Almer, J., Ramesh, K. T., &Hufnagel, T. C. (2005). Micromechanics of deformation of metallic-glass-matrix composites from in situ synchrotron strain measurements and finite element modeling. *Acta Materialia*, 53, 7, pp. 1883-1893.

Paillier, J., Mickel, C., Gostin, P.F., Gebert, A. (2010). Characterization of corrosion phenomena of Zr-Ti-Cu-Al-Ni metallic glass by SEM and TEM, Mater. Charact., 61, pp. 1000-1008.

Park, J. S., Lim, H. K., Kim, J. H., Chang, H. J., Kim, W. T., Kim, D. H., &Fleury, E. (2005). In situ crystallization and enhanced mechanical properties of the Zr41.2Ti13.8Cu12.5Ni10Be22.5 alloy by cold rolling. *Journal of Non-Crystalline Solids*, 351, 24-26, pp. 2142-2146.

Patterson, J., &Jones, D.R.H. (1979). The effect of homogeneous deformation on the crystallization of a metallic glass. *Scripta Met.*, 13, pp. 947-949.

Perepezko, J. H., &Hebert, R. J. (2002). Amorphous aluminum alloys - synthesis and stability. *JOM*, 54, 3, pp. 34-39.

Perepezko, J. H., Imhoff, S. D., &Hebert, R. J. (2010). Nanostructure development during devitrification and deformation. *Journal of Alloys and Compounds*, 495, 2, pp. 360-364.

Purdy, G.R. (1981). Interface diffusion-controlled kinetics and mechanisms. Int. Conf. on solid-solid phase transformations, Pittsburgh, The Metallurgical Society AIME.

Ratke, L., &Vorhees, P.W. (2001). Growth and Coarsening, Springer.

Sadoc, Anne, Heckmann, Olivier, Nassif, Vivian, Proux, Olivier, Hazemann, Jean-Louis, Xing, L. Q., &Kelton, K. F. (2007). Local order and nanostructure induced by microalloying in Al-Y-Fe amorphous alloys. *Journal of Non-Crystalline Solids*, 353, 29, pp. 2758-2766.

Schroers, Jan, &Johnson, William L. (2004). Ductile bulk metallic glass. *Physical Review Letters*, 93, 25, pp. 255506-255501-255506-255504.

Schuh, C. A., Hufnagel, T. C., &Ramamurty, U. (2007). Overview No.144 - Mechanical behavior of amorphous alloys. *Acta Materialia*, 55, 12, pp. 4067-4109.

Schulz, R., Trudeau, M. L., Dussault, D., &Van Neste, A. (1990). Crystallization of amorphous alloys by high energy mechanical deformation. *Colloque de Physique C4*, 51, Suppl. No.14, pp. C-4 259 - C-254 264.

Schulz, R., Trudeau, M. L., Dussault, D., &Van Neste, A. (1990). Crystallization of amorphous alloys by high energy mechanical deformation, France.

Scully, J. R., Gebert, A., &Payer, J. H. (2007). Corrosion and related mechanical properties of bulk metallic glasses. *Journal of Materials Research*, 22, 2, pp. 302-313.

Scully, J. R., &Lucente, A. (2005). Corrosion of amorphous metals. *Met. Mat. Trans. B*, 13B, pp. 476.

Sharma, S.K., &Banerjee, S. (1989). Some correlations for diffusion in amorphous alloys. *J. Mater. Res.*, 4, 3, pp. 603-606.

Spaepen, Frans, &Turnbull, David (1974). Mechanism for the flow and fracture of metallic glasses. *Scripta Metallurgica*, 8, 5, pp. 563-568.

Stratton, W. G., Hamann, J., Perepezko, J. H., Voyles, P. M., Mao, X., &Khare, S. V. (2005). Aluminum nanoscale order in amorphous Al92Sm8 measured by fluctuation electron microscopy. *Applied Physics Letters*, 86, 14, pp. 141910-141913.

Suryanarayana, C., &Inoue, A. (2011). Effect of pressure during annealing. Bulk Metallic Glasses, CRC Press: 250-254.

Tam, C. Y., &Shek, C. H. (2004). Abrasive wear of Cu60Zr30Ti10 bulk metallic glass. *Materials Science and Engineering a-Structural Materials Properties Microstructure and Processing*, 384, 1-2, pp. 138-142.

Tarumi, Ryuichi, Ogura, Akio, Shimojo, Masayuki, Takashima, Kazuki, &Higo, Yakichi (2000). Molecular dynamics simulation of crystallization in an amorphous metal during shear deformation. *Japanese Journal of Applied Physics, Part 2: Letters*, 39, 6 B, pp. 611-613.

Tiwari, R. S., Claus, J. C., &von Heimendahl, M. (1982). The effect of tensile stress on the crystallization kinetics of Metglas 2826 $Fe_{40}Ni_{40}P_{14}B_6$. *Material Science and Engineering*, 55, 1, pp. 1-7.

Vom Wege, F., Kumpfert, J., &Hornbogen, E. (1991). Mikrokratzertests an einer schmelzgesponnenen und nachfolgend angelassenen Co-Basislegierung. *Z. Metallkde.*, 82, pp. 209-216.

Vom Wege, F., Skrotzki, B., &Hornbogen, E. (1988). Abrasivverschleiss einer schmelzgesponnenen und nachfolgend angelassenen Co-Basis Legierung. *Z. Metallkde.*, 79, pp. 492-498.

Yeong-Hwan, Kim, Inoue, A., &Masumoto, T. (1991). Ultrahigh mechanical strengths of $Al_{88}Y_2Ni_{10-x}M_x$ (M=Mn, Fe or Co) amorphous alloys containing nanoscale FCC-Al particles. *Materials Transactions, JIM*, 32, 7, pp. 599-608.

Zener, C. (1949). Theory of growth of spherical precipitates from solid solution. *J. Appl. Phys.*, 20, pp. 950-953.

Zhang, Y., Hono, K., Inoue, A., Makino, A., &Sakurai, T. (1996). Nanocrystalline structural evolution in Fe90Zr7B3 soft magnetic material. *Acta Mat.*, 44, 4, pp. 1497-1510.

Zhang, Y., Warren, P.J., &Cerezo, A. (2002). Effect of Cu addition on nanocrystallization of Al-Ni-Sm amorphous alloy. *Mater. Sci. Engr. A*, A327, pp. 109-115.

Zum Gahr, K.H., &Noecker, H. (1981). Abrasivverschleiss metallischer Glaeser. *Metall*, 35, 10, pp. 988-995.

Paillier, J., Mickel, C., Gostin, P.F., Gebert, A. (2010). Characterization of corrosion phenomena of Zr-Ti-Cu-Al-Ni metallic glass by SEM and TEM, *Mater. Charact.*, 61, pp. 1000-1008.

Ye, F., &Lu, K. (1999) Pressure effect on crystallization kinetics of an Al-La-Ni amorphous alloy, Acta Mater., 47,8, pp. 2449-2454

5

Preparation of Nano Al_5O_6N via Shock Wave Plasmas Technique

Zuoshan Wang
College of Chemistry, Chemical Engineering and Material Science, Soochow University, Suzhou, P.R.China

1. Introduction

1.1 Plasmas technique

Preparation of nano-materials can be divided into three types: gas-phase-, solid state- and liquid-methods. Material researchers have made some achievements in their own fields using different methods. Among them, the gas-phase evaporation method in vacuum is usually filled into the low-pressure inert gas interior (N_2, He, Ne, Ar, etc.). Heating sources such as resistance, plasma, electron beam, laser and high frequency induction are used to make raw material gasificated or to form into plasma, which will lose energy during the collision with the inert gas atoms and then condensed into fine particles by quencher [1]. The technique of plasma plays an important role in preparing nano-materials, especially the metal particle nano-materials, which are prepared by gas-phase methods.

The conception of "plasma" was first reported by Langmuir in 1927, which was firstly applied to investigate the electric discharge phenomena of mercury vapor under low pressure. Plasma is a highly ionized gas mixture, which is consists of electrons, ions, atoms and molecules. The whole system is neutral as positive and negative charges are equal. So, it has different properties from the common gases. The inner electrons, irons, and even neutral particles have higher energy, and all reactions are proceeding under high excited state. This is totally different from the typical chemical reactions, for example, in plasma the inert gas could be highly chemical active and form XeF_6 and O_2F_2 and so on.

According to the degree of gas dissociation, plasma can be divided into high-temperature plasma (high temperature can reach tens of thousands to tens of millions centigrade) and low-temperature plasma (hundreds to tens of thousands centigrade). Low temperature plasma can be divided into thermal plasma (thousands to tens of thousands centigrade) and cold plasma (hundreds centigrade). To prepare ultrafine particle by plasma, chemical reaction can take place (plasma physical method) or not (plasma chemical method).

1.1.1 Plasma physical method

In plasma physical method ultra-fining particles can be obtained by the phase transformation through the energy of plasma. Firstly, the metal powders melt and evaporate into gas using plasma. , Then, the metal powders condense and grow into ultra-fine particles when cooling down. This technique is the most effective to prepare metal and metal alloy nano-particles with high purity and small uniform size.

Wang et al.[2] have prepared the single crystalline phase nano-Tantalum ultrafine particles using the direct-current arc plasma (DC). They designed a Tantalum-to-Tantalum solution using Tantalum stick as the cathode and Tantalum, which could evaporate in a fixed water-cooled copper anode. Then the mixed argon and hydrogen were filled into the vacuum chamber. The Tantalum evaporated under high temperature of arc plasma produced by hydrogen arc, and then smoke-like Tantalum nano-particles precipitated on the low temperature surface. Finally, black nano-Tantalum metal particles were collected after purification. The single crystalline Tantalum nano-particles obtained using this method have a black color, homogeneous particle with an average diameter of 10 nm, and good dispersion. Until now, the nano-powders of Au, Fe, Ni, Ag and Zn have been prepared using the method of direct-current arc plasma (DC).[3-5]

1.1.2 Plasma chemical method

In Plasma chemical processing, some chemical reactions can happen to generate and refine new products. The vapors of volatile metal halide or hydride or metal organics are usually used as raw materials in this method. The nonmetal oxides such as carbides, nitrides and borides ultrafine powders could be synthesized through gas-phase thermal decomposition or other chemical reactions with the energy generated from plasma.

Han et al.[6] prepared silicon nitride nano-spheres with high purity, loose structure and size distribution of 20~30 nm by radio frequency plasma vapor deposition(R.F, CVD) technique, using $SiCl_4$ and NH_3 as raw materials. Waterfall flows of Titanium atoms generated from the Titanium anticathode and cathode by (O_2 + Ar)arc plasma, and were oxidized to form TiO_2 nano-crystals in the ionized oxygen plasma region, which has been reported before. The prepared TiO_2 nano-powders were highly dispersed sphere-like particles with an average diameter around 20 nm, and the grains were mixed structure of anatase and rutile.[7]

Radio frequency plasma would not produce the impurities of electrode materials in plasma because it dose not use electrode. It could use gas during the preparation process. At the same time, the residence time of the compounds would increase due to the larger plasma space, and the reactants could be completely reacted and heated. Shang Shuyong et al.[8] in Sichuan University have prepared the Aluminum nitride ultra-fine powders by DC arc thermal plasma using micro aluminum powders as raw material. During the preparation process, aluminum powders were inputted by nitrogen as carrier gas, so, the flow of powder and the distribution of temperature in the reaction tube were achieved by adjusting the flow rate of nitrogen gas. Nitrogen and Aluminum reacted in the reaction chamber, and after that, ammonia was added at the end of the chamber to form aluminum nitride powers. Then, the powers rapidly quenched, nucleated, grown up and finally were collected by bag collector. As-prepared Aluminum nitride was pure and the particles with an average diameter of 100nm were highly dispersed.

Studies have shown that high temperature plasma method is fast, easy to achieve mass production, and it can prepare almost any kind of nano-materials, especially for nano metal materials. The prepared nano-material is zero-dimensional and homogeneous.

1.1.3 Issues in typical plasma methods

Although plasma method has certain advantages in the preparation of nano-materials, there is also a drawback of exorbitant energy consumption, which limits its space of industrialization.

The plasma methods can be divided into the method of direct-current arc plasma (DC), radio frequency plasma (RF) and hybrid plasma (HP) according to the ways of generating plasma. As heat induced plasma, they all have the properties of high temperature, containing plenty of electrons, ions and other active particles, controllable atmosphere etc., which offers the many new possibilities for material processing and creates new opportunities for scientific research on material properties. However, further production has revealed the drawback of excessive energy consumption of thermally induced plasma.

It can reach a high core temperature but small scale high temperature zone via direct-current arc plasma (DC). The plasma generator or reactor always have huge temperature gradient (10^5~10^6 K/m) and velocity gradient (10^4~10^5 m/ s•m), so it is difficult to move the particles exactly to the high temperature zone of the generator or reactor of arc plasma and make the target materials gain proper and effective heat. It is more difficult to load the gas reactants into the high temperature zone of plasma reactor to mix and react completely.

Radio frequency plasma generator or reactor has a bigger high temperature zone, which is good for reaction. The lower velocity of plasma in reactor (usually 10 m/s scale), controllable atmosphere, and no electrode pollution are benefit to obtain clean products. However, Radio frequency plasma reactor itself has large energy consumption and low energy efficiency (less than 50%), so it is difficult to reduce product cost. Furthermore, the input particles or gas reactants need to be organized elaborately in consideration of the complex flow field in reactor.

Recently, the schemes of tandem connection of direct-current arc and radio frequency plasma, parallel connection of multi direct-current arc and direct-current arc using ring-like anode have been used to enlarge the high temperature zone and offer a better material producing surroundings or improve the yields of product. However, all above schemes inevitably make the system more complex and make investment larger. Therefore, it is necessary to explore novel plasma to make up for its shortages.

1.1.4 Design of novel plasma

Making a general survey of DC, RF and combination of the two methods, although the plasma is generated in different ways, they all prepare ultra fine powders in the same principle. Raw material particles (solid or liquid) are rapidly heated by the thermal plasma with high temperature, containing various electrons, ions and active particles in excited state. Then the particles are evaporated or cracked to complete the necessary reactions of gas reactant, and finally grow up to form the goaled product after quenching, nucleating. The process is shown in Fig.1.

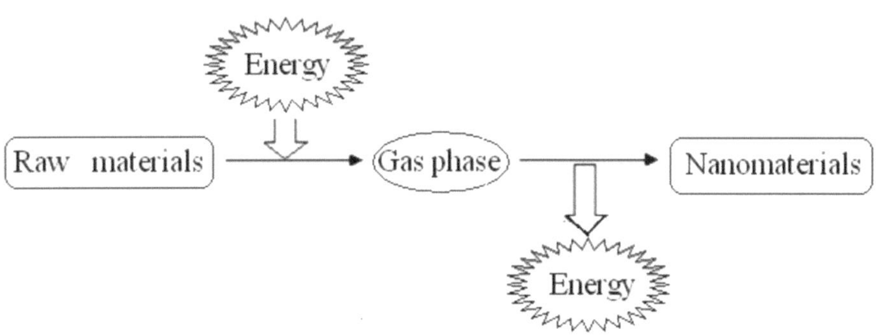

Fig. 1. A schematic illustration of preparing nano-materials using plasma

The process from solid to gas then nano-powders is determined by the preparation of goals with regard to the preparation craft, which can be seen from Fig.1. The exploration of space is very small, but the different energy ways to generate plasma and the following forms of energy release have left a wide thinking space to the researchers.

1.1.5 Shock wave plasmas technique

There are various methods and ways to generate plasma, which are related to lots of microcosmic processes, physical effects and experimental methods. The most common methods used to generate plasma in chemical fields are mainly listed as follows: [9]

(1) Gas discharge plasmas

The gas molecules are ionized through colliding with the accelerated charged particles especially the electrons in the electric field. Plasma forms by disruptive discharge of gas under the effect of cathode secondary emission and other mechanisms. The plasma currently actually used in laboratory and production is produced by gas discharge. In particular, the most widely used is radio frequency discharge.

(2) Photo ionization and laser emission ionization

The plasma is formed by ionizing the molecule through gathering the energy of incident photons. The precondition is that the photon energy should be larger than the first ionization energy. The laser emission belongs to photo ionization in itself, but it not only possesses single photon ionization but also multi photon ionization in ionized mechanism, which results in difference in ionization mechanism from normal photo ionization methods. For example, In multi photon ionization, the atoms or molecules ionized by absorbing numerous of photons at the same time. So argon plasma can be obtained by irradiating the argon gas using ruby laser. In recent years, application of laser plasma in chemical fields such as chemical deposition by laser plasma has gained an obvious ascended trend, because it is easy to achieve plasma with high temperature and high density.

(3) Irradiation

The plasma could also be generated by irradiating the gas with various rays or particles beams such as the α 、 β 、 γ rays produced by radioisotope, X-rays given off from X-ray tube and beams of electron or ion accelerated by accelerator, etc.. The plasma generated via α rays is equivalent to the gas molecules collided and ionized with nuclear power particles. β rays is a beam of election current, the generated ionization is equivalent to ionize by high speed election collision. While γ rays and X rays could be recognized as photo ionization. The collision ionization of gas molecules is obtained by accelerated particles such as beams of electrons and ions. The ion beams show lots of advantages due to the controllable accelerated energy, intensity of flow and impulse.

(4) Combustion method

This is a method already well known as thermal ionization, which generates ionization by the collision between the atoms and molecules with sufficient kinetic energy with the help of thermal motion, and the generated plasma called flame plasma.

(5) Shock wave methods

The plasma generated by the adiabatic compression of gas called shock wave during the process that shock waves pass through the sample gas. Actually, it belongs to thermal ionization.

1.2 Preparation of nanometer Al_5O_6N via shock wave plasmas technique

1.2.1 Introduction

It has drawn much attention from the society since the Al_5O_6N with spinel structure was found[10-17]. Compared with the normal Al_2O_3, the crystal lattice in Al_5O_6N recombines, because the atom "O" is displaced by the atom "N". It results that the Al_5O_6N not only keeps the excellences from the white gem, such as higher tenacity and rigidity, but also has the outstanding isotropic characteristics, which avoids briefringent effect from aeolotropism within the use of Al_2O_3 poly-crystal. As a result, it is hopeful for Al_5O_6N to replace the large cost white gem in the armed techno-field, and become the preferred material in the aspect of high-speed missile rectifier and the bulletproof glass of the new pattern armored-car [18]. At present, the transparent ceramics of Al_5O_6N has been produced for business in foreign countries, and it is also being studied step by step in China. From the literatures all over the world, it is not hard to see that the very point to prepare the Al_5O_6N transparent ceramics is getting ceramics powder with higher purity and agglutinating activity. It is easy to get Al_5O_6N ceramics powder with higher purity by the existing technic, but it is difficult to get better agglutinating activity. The study of nano-tech shows that when the ceramics come to the nano scale, the agglutinating activity will be improved effectively. The nano particles have very small size, big BET surface area and high rate of diffusion, which results in the speedup of densification in the process of agglutination and reducer of the agglutinating temperature and time greatly. For example, the common Al_2O_3 ceramics agglutinate under 1750°C . However, when doped with 0.1%~0.5% nano Al_5O_3 powder, the agglutinating temperature can be reduced to less than 1500°C , and at the same time, the intensity and tenacity of the 85 and 95 ceramics doped with nano Al_5O_3 can also be improved to 1.5 times[3]. Therefor, whether agglutinated immediately or used as the filler in the process, the preparation of high purity nano Al_5O_6N transparent ceramics is helpful for raising the agglutinating activity of Al_5O_6N.

It is difficult to prepare nano Al_5O_6N powder with high purity, because the high temperature becomes the huge bottleneck to avoid agglomeration. We use a new method named blast wave plasma to prepare the Al_5O_6N quickly by the detonation of the mixed explosives. However, as the reaction is so fast and complex, it is difficult to study its process in detail. Here, we designed some contrast experiments, analyzed the reaction mechanism of detonation for nano Al_5O_6N with the energetic material TNT and the metal Al powder as the starting materials in water, which will establish the theoretical basis for the development of the technology.

1.2.2 Design of the experiments

1.2.2.1 Construction of shock wave plasma field

Actually, the reported direct-current discharge, radio frequency discharge and microwave discharge are gas discharges generating plasma to convert the electric energy to thermal energy. The electron beams and ion beams plasma belong to irradiation method. From the ways to achieve energy, the gather of transient state super high energy from the detonation of energetic material provides the possibility of generating plasma, and the preparation of nano-materials using shock wave plasma could be designed based on this.

The energy of shock wave plasma is obtained from powerful thermal energy and strong shock wave instantaneously released by the detonation of material with high energy such as TNT, RDX, PETN, HMX etc.. The transient state production of solid powder materials was achieved by powerful shock wave, super high pressure and temperature in microenvironment formed by the strong plasma field. The preparation of nano-materials by shock wave plasma has brought in the extreme conditions of powerful shock wave, super high pressure and compulsory rapidly cooling besides the necessary high temperature, which is different from other plasma fields. The introduction of the extreme conditions is hopeful to achieve the chemical reaction or physical process, which is hard to achieve under normal conditions, and brings new energy to the techniques of nano-material preparation.

Using explosion of energetic materials to synthesize materials began with the preparation of nano-diamond.[19] Soviet scientists first reported the detonation synthesis of nano-diamond in 1984, and then the synthesis prelude using detonation was opened. They gained a trial-production in 1990 by detonating 140 kg explosive charge in the explosion chamber with a diameter of 12 m [20]. Since then, the U.S., Japan, Germany and other developed countries have also reported the synthesis of nano-diamond by detonating one after another[21,22]. Since 1992, Jin Zengshou in Lanzhou Institute of Chemical Physics, Xu Kang in Gansu Research Institute of Chemical Industry and Hui Shourong in Beijing Institute of Technology have also done the research [23,24]. However, the experimental scale and performance of the products were behind that of the developed countries. The related explosion synthesis mechanism was discussed on this basis. The detonation of energetic materials is a transient state process of high temperature and high pressure, and so, it is very difficult to measure directly. Many researchers have to find the relative evidence of the synthesis mechanism only from the detonation products.

Samara[25] supposed that the detonation of energetic materials was a fast reaction process that firstly generated the mixture of electriferous atoms or elementide called low temperature plasma, and then formed stable phase of diamond in thermodynamics. Ree et al.[26] thought that carbon phase in the detonation products was first separated from fluid phase to form semi-stable state graphite and changed to become diamond by phase transformation. Tinov et al.[27] considered that the molecules of explosive reaction decomposed to be various atoms, and a part of carbon atoms combined to form graphite in the reaction zone, then produced diamond through phase transformation. Yamada et al[28] observed extremely round spherical carbon particles in detonation products by electron microscope analysis. Accordingly, he presented the generate mechanism of diamond phase under high temperature and pressure was that the initial gas state carbon condensed to liquid carbon drop in the detonation reaction zone because of supersaturating, and the liquid carbon drop formed diamond by phase transformation. Hui Shourong got the similar results by electron microscope analysis.

From the mechanism studied by the former researchers, it is not difficult to find that the plasma can be obtained during the detonation process with a phase transformation from gas carbon to liquid carbon drop and even graphite. Therefore, it is completely possible to construct the transient state shock wave plasma field with the help of detonation process in energetic materials.

1.2.2.2 Selection of energetic materials

The main energetic materials here is TNT, which has the scientific name of 2,4,6-trinitrotoluene, and molecular formula of $C_7H_5N_3O_6$, relative molecular weight of 227.13, and -73.96% for the oxygen balance. Its structure is shown in Fig.2.

The pure TNT is colorless needle-like crystal. Its industrial products are yellowish flaps with physical properties of low hygroscopic, non-volatile at room temperature and difficult to dissolve in water, slightly dissolve in ethanol, CCl_4 and CS_2. TNT is neuter. It can't corrode metal or react with heavy metal oxide obviously. It performs satisfactorily for pressing and casting, and has good stability for experiment.

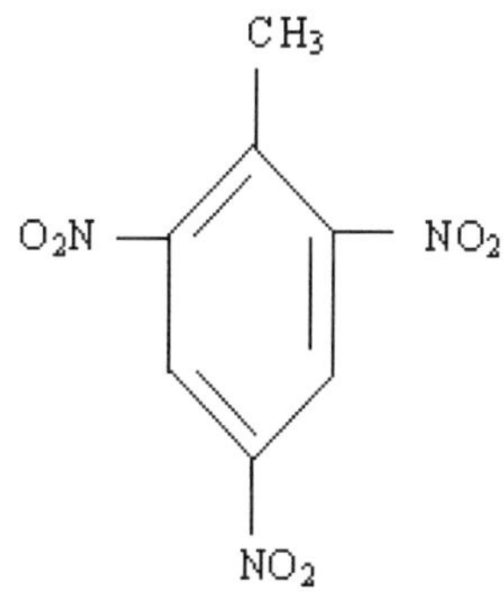

Fig. 2. Structure of TNT

TNT is a typical negative oxygen balanced explosive, which would produce large amount of carbons, and the reaction equation is as follows:

$$C_7H_5N_3O_6 = 2.5H_2O + 3.5CO + 1.5N_2 + 3.5C \quad (1\text{-}1)$$

Different from normal condition, the free carbon and nitrogen generated through the detonation of TNT have high activity, which creates beneficial material conditions for design and synthesis of nano-materials. In this sense, TNT is the energy source as well as important material source. The powerful shock, super high pressure and temperature have provided several effective ways to form the transient state of high thermal energy plasma field.

1.2.2.3 Selection of cooling medium in shock wave plasma field

Among the reported methods for plasma, gasified materials or gaseous products lose energy while colliding with the atoms of inert gas, and then they are cooled rapidly to form ultra-fine particles, which is a process of energy release. The fast energy release can make the particles in plasma field reach a thermodynamic stable state and avoid agglomeration.

The theoretical study of underwater ordnance shows that [29] the shock wave generated by explosive under water could be rapidly absorbed by aqueous medium. The spread velocity of shock wave in water is several times larger than that of sound (about 1520 m/s) in water. Therefore, the shock wave could be far away from the centre of explosion and the energy provided by shock wave in plasma field decreases fast to achieve the partly energy release, as the shock waves transmit fast in water. At the same time, impulse bubbles formed by explosion also help to the energy release rapidly. So, we choose water as the cooling medium in the research as it can not only cool down faster, but also possess the advantages of easy acquired, economy and environmental friendly.

Above all, with the general requirements in the area of material preparation encourages environmental protection and energy saving in nowadays, the dual features of transient state energy field and material field in shock wave plasma field make this method possess good development prospects. The reactions which are difficult to achieve in normal conditions are expected to be achieved when combining the chemical theory with the extremely preparation conditions. It is important and meaningful to the development of novel materials, and it enriches the theories of explosion physics and chemistry.

1.2.3 Experimental procedures

1.2.3.1 Preparation of Al_5O_6N nano-powders

All reagents were commercially available and used without further purification. In a typical experiment, the negative oxygen explosive TNT and the metal Al powder(average diameter 200 mu, 99.0%W) were mixed together with a mass ratio of 4:1, and the mixture was loaded into a mould and pressed into a column shape with a density of 1.60g/cm^3.
Then the column mixture was put into a 1 L stainless autoclave, which was special-designed according to the temperature and pressure during detonation. After filling 500 mL water, the autoclave was sealed and the column mixture was ignited through the detonator outside. The detonation resultant was collected and filtered with 230 mesh screen to remove the coarse impurity. The filtrate was filtered again after depositing for 5 h to obtain the dark precursor. After treatment in inorganic acid for 1 h at 80°C , dark solid was obtained by filtration, washing with absolute ethanol and distilled water, and drying at 60°C for 2 h. The carbon in the dark precursor was removed through calcination at 600–700°C for 1 h to obtain gray powder. The calcination temperature schedule of the precursor was decided through a differential thermal analysis/thermogravimetry (DTA/TG, model STA449C, NETZSCH, Germany company), using an ascending rate of 10 K/min from 23°C to 800°C and a velocity of air flow of 20 mL/min.

1.2.3.2 Instruments

Crystalline phase was determined by X-ray diffraction (XRD, model D/Max-IIIB, Shimadzu) instrument with Cu Ka radiation (λ = 0.15418 nm). The scanning rate of 0.05^0/S was applied to record the pattern in the 2θ range of 10–70^0. The crystal size was calculated with Scherer's formula.
Laser Common Focus Raman Spectrum Instrument (model Renishaw-1000, England) was used to analyze the degree of graphitization of carbon particles in the mixture at ambient temperature, employing an argon ion laser at an excitation wavelength of 514.5 nm. The scan area ranges from 100~2000 cm^{-1}. The capacity of the incident light is 5 mW. The slit breadth of the monochromator is 24 μm. The diameter of the light spot is 1~2 μm.
Morphological feature and particle size of the products were investigated by a high-resolution transmission electron microscope (HRTEM, model JEM-2100, Japan), using an accelerating voltage of 200 kV.
Analysis of element content was done with Thermo ESCALAB 250 XPS instrument (America) using Al Kα radiation (hυ = 1486.6 eV).

1.2.3.3 The antioxidancy of Al_5O_6N

Differential thermal analyzer (DTA, model STA449C, NETZSCH) is employed to analyze the temperature course of nano Al5O6N at the reaction of oxygenation and denitrify. The heating rate is 10K/min.The flow rate of air is 20mL/min. The sample' quality is 10.181 mg. The scan area ranges from 23~1360°C and the sample pond is made by Al_2O_3.

1.2.4 Results and discussion

1.2.4.1 XRD analysis of the precursor

The key idea of our method for the synthesis of Al_5O_6N is to carry out a shock wave plasma process including the detonation of explosive column, decomposition of chemistry substances, dispersion and rebuilding of free atoms or gas ions. The precursor including carbon and Al_5O_6N was produced during this process. Fig.3 displays the typical XRD pattern of the precursor.

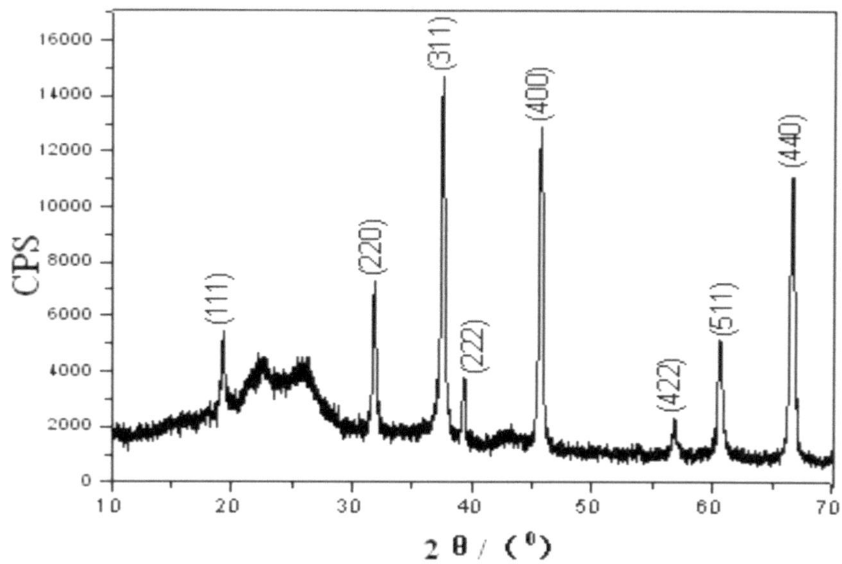

Fig. 3. XRD pattern of the precursor

It shows that the half band widths are 4.586, 2.808, 2.394, 2.288, 1.984, 1.621, 1.526, 1.402 and the position and intensity are according with JCPDS 48-0686, which confirms that the major constituent of the powder is Al_5O_6N.

The main peaks have an obvious broaden phenomena and the strongest diffraction appearing at 37.54^0 with the half band width of 0.440 can be indexed to (311) plane. According to Scherer equation, the average size of the particles is 28 nm. The whole spectrum has a higher diffraction mount and in $20\sim30^0$ exits some low and broad peaks, which indicates the product contains ultrafine imperfectly crystallize carbon in the detonation process, especially in the diffraction position at 25.5^0 where exists a low and broad graphite diffraction.

1.2.4.2 Raman analysis of the precursor

The following laser common focus Raman spectrum in Fig.4 approves the crystallographic structure of the carbon in the precursor. Fig.4 shows that there exists two wide peaks at 1344.81 cm^{-1} (D-band) and 1575.94 cm^{-1} (G-band). The D-peak at 1344.81 cm^{-1} is related to a resonant Raman process, generally at 1370 cm^{-1}, it represents a peak of disordered graphite with primary imperfections, and the decrease of its intensity indicates the increase of degree of graphitization. The G-peak at 1575.94 cm^{-1} is associated with an E_{2g} mode of graphite and related to the vibration of sp^2 -bonded carbon atoms in a two- dimensional hexagonal lattice, such as in a graphite layer, which is consistent with the reported data [30]. Comparing with the Raman spectrum of nature monocrystal graphite, in which there is only one acute peak at 1580 cm^{-1} (G-band), the Raman spectrum of the precursor is unsymmetrical and the peaks are widened obviously, which indicates that the carbon in the precursor consists of amorphous carbon and nanographite. The above analysis proves that the precursor is the composite of nanometer Al_5O_6N and incomplete crystalline nanographite.

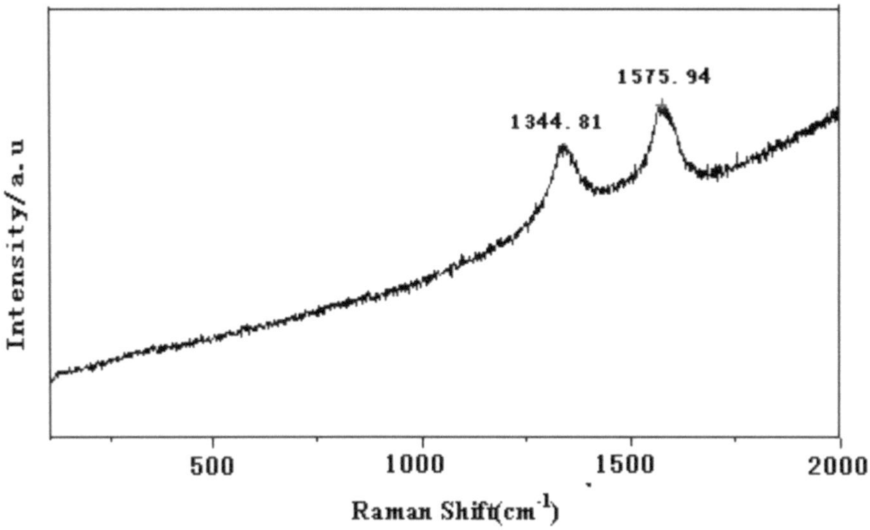

Fig. 4. Raman spectrum of the precursor

1.2.4.3 TEM images of the precursor

Fig.5 shows the TEM image of the sample prepared in the optimum technology. It is concluded that the sample is particle-like with a uniform diameter from 20~40 nm and has no obvious accumulation.

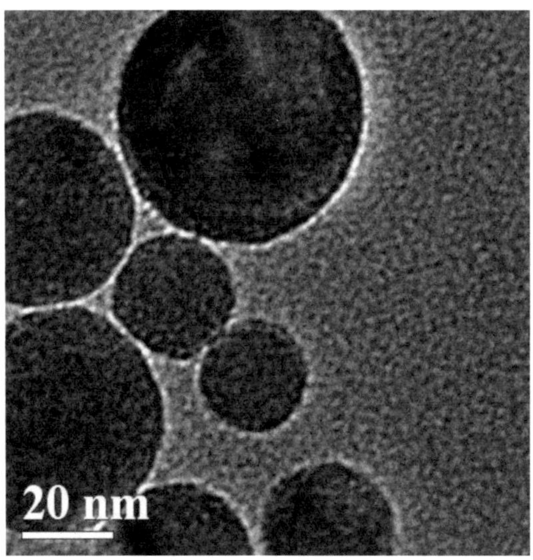

Fig. 5. HRTEM image of the precursor

1.2.4.4 DTA/TG analysis of the precursor

The carbon in the precursor can be removed through calcining process at proper temperature, which can be decided through DTA/TG analysis of the precursor in Fig.6. The

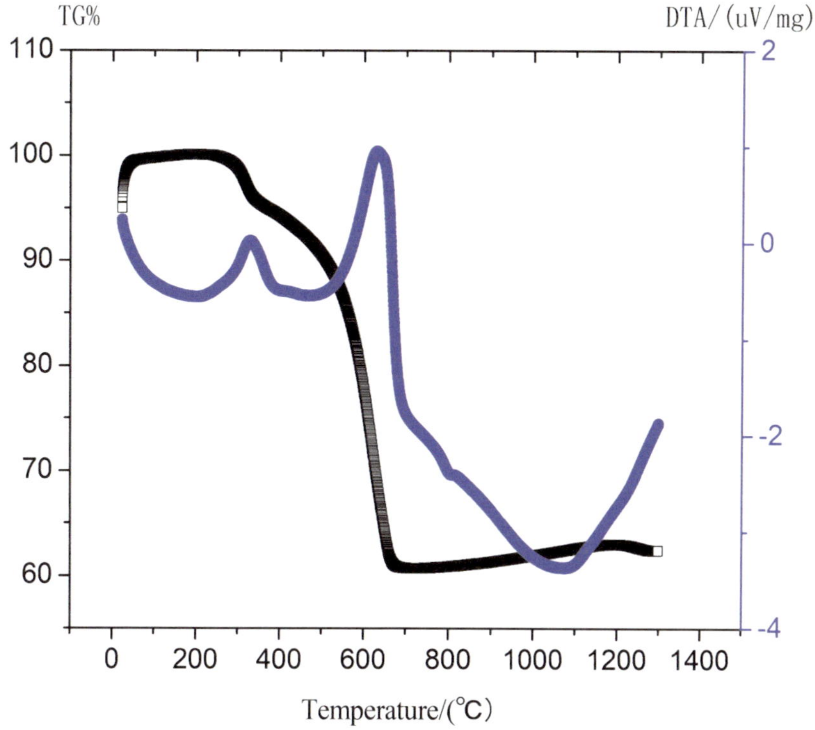

Fig. 6. TG-DTA curves of the precursor

thermal mass loss curve shows that the mass increases slightly at the beginning of elevating temperature, which is due to the flowing oxygen adsorbed by the oxygen-vacancy of the Al-O-N system. Subsequently the mass decreases from 327°C to 672°C . There are two obvious turning points at 327°C and 672°C , which are corresponding with the two obvious exothermic peaks of DTA curve, a weak exothermic peak at 327°C and a strong exothermic peak at 628°C . The mass loss of 3% at 327°C attributes to the slow oxidation of amorphous carbon and the mass loss of 34% at 672°C attributes to the rapid oxidation of nanographite, which shows that most of the carbon in the precursor is graphite. The above results are also consistent with the foregoing XRD analysis and Raman spectrum analysis of the precursor. The DTA/TG analysis shows the mass content of carbon is 38% and the Al_5O_6N is 62%. At the same time, the DTA/TG analysis result can also be considered as an evidence of calcining temperature schedule. According to Fig.6, carbon can be removed by calcining the precursor at the temperature about 600°C .

1.2.4.5 Influence of different calcining temperature

Fig. 7 shows the XRD patterns of the as-synthesized Al_5O_6N powders calcined at 600°C, 700°C and 800°C . Pattern (b) in Fig. 7 shows that the as-synthesized powder calcined at 600°C for 1 h is fully crystalline cubic structure and all of the reflections can be indexed to the crystal planes of a standard cubic Al_5O_6N crystal (JCPDS 48-0686). There exists no other crystalline peaks except Al_5O_6N in the pattern. According to Scherer's equation, the mean particle size of the Al_5O_6N crystal is estimated to be 28.21 nm, which is consistent completely with that of the precursor, seeing pattern in Fig.1. It indicates that the Al_5O_6N

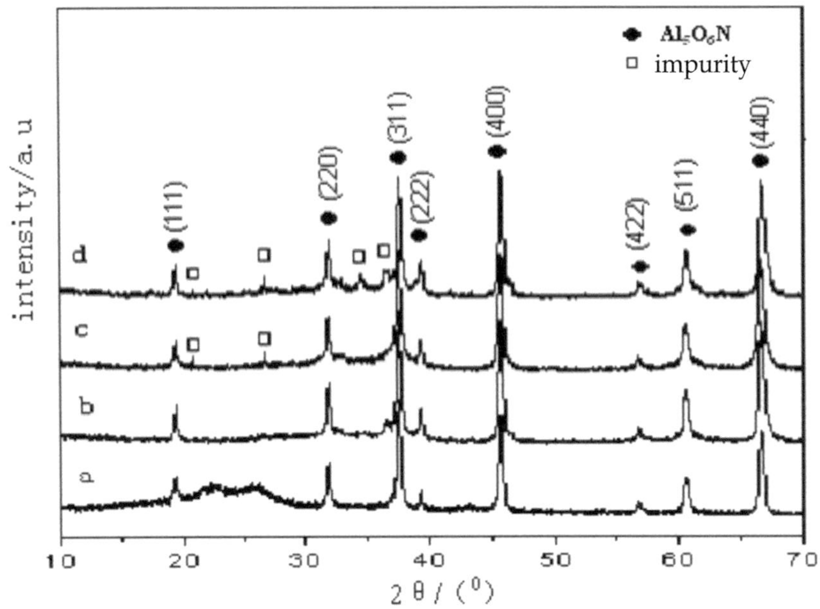

(a) the precursor (b) 600°C (c) 700°C (d) 800°C

Fig. 7. XRD patterns of products obtained by calcining the precursor at different temperatures

crystal has not grown up and the carbon indicates that the Al_5O_6N crystal has not grown up and the carbon is completely eliminated after calcined at 600°C for 1 h. Increasing calcining temperature has no obvious effect on the particle size, while some new crystalline diffraction peaks appear at 25.558, 34.878 and 37.568, seeing pattern (c) and (d) in Fig.7. These new peaks are identified to be a-Al_2O_3 crystal, which can be attributed to the oxidation of Al_5O_6N crystal at 700°C and 800°C for 1 h. Therefore, the calcining temperature should be controlled at about 600°C to assure that product has the pure Al_5O_6N phase.

1.2.4.6 TEM image of the sample

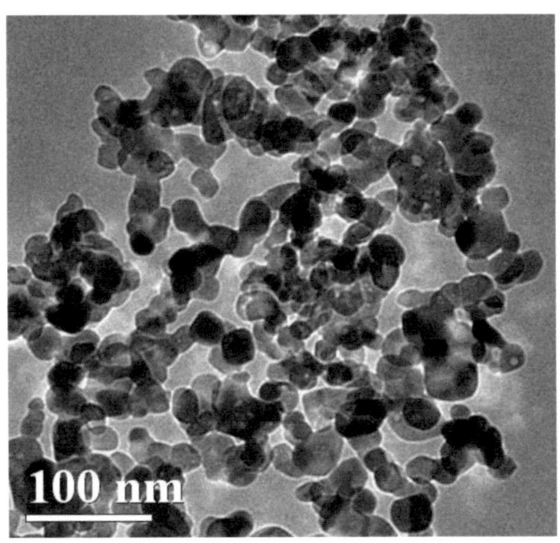

Fig. 8. TEM image of nano Al_5O_6N powder obtained by calcining the precursor at 600°C

Fig.8 shows the TEM image of the sample prepared in the optimum technology. It is concluded that the sample is particle-like with a uniform diameter from 30~40 nm and has no obvious accumulation.

1.2.4.7 XPS analysis of the sample

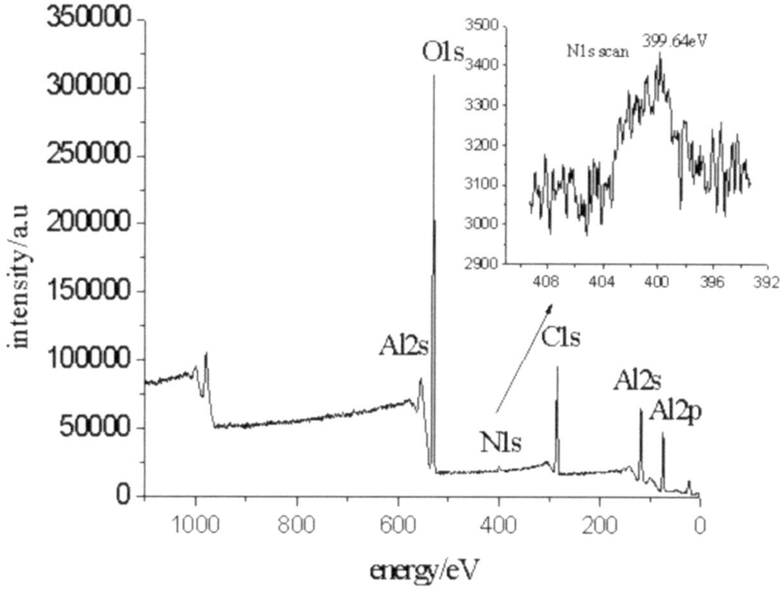

Fig. 9. XPS analysis of the nano Al_5O_6N powder

Fig.9 displays the XPS profile for the as-synthesized Al_5O_6N sample. The analysis results for the XPS spectrum areas show that the N1s region is near 400 eV, the Al2s region is near 119 eV and 582 eV, and the O1s region is near 530 eV. The binding energy peak for the N is broad, extending from 395.4 eV to 405.7 eV, which can be indexed to Al–N binding energy. Although the N1s peak in the XPS spectrum is weak, the peak position and its broaden are clear in the inset.

1.3 Discussion of mechanism for Al_5O_6N nano-powder

1.3.1 Mechanism for preparation of Al_5O_6N

1.3.1.1 Theoretical analysis

It is well known that the explosion is a very complex and rapid chemistry process. The strong hit and huge pressure produced in a split second become the largest barriers to analyze the reaction detailedly. An improved theory from the classic one about the detonation of the explosive compound contained Al insists that the first step of composite detonation of explosive is the detonate reaction of the energy-contained material:

$$C_7H_5N_3O_6 \longrightarrow 2.06CO + 1.03CO_2 + 1.88H_2O + 0.62\,H_2 + 1.5N_2 + 3.91C \quad (1\text{-}2)$$

Then the metal reacted with the product of the detonation at the high temperature and high pressure:

$$1.03CO_2+0.6\ Al \longrightarrow 0.34Al_2O_3+1.03CO \qquad (1\text{-}3)$$

$$1.88\ H_2O+1.26Al \longrightarrow 0.63\ Al_2O_3+1.88H_2 \qquad (1\text{-}4)$$

The overall reaction is :

$$C_7H_5N_3O_6+1.94Al \longrightarrow 0.97Al_2O_3+3.09CO+2.5H_2+1.5N_2+3.91C \qquad (1\text{-}5)$$

However, our experiment indicates that the product gained from the detonation of TNT/Al explosive under the protection of water is Al_5O_6N but not Al_2O_3.The possible mechanism of the formation of Al_5O_6N may be as follows:

In process [1], the reaction goes by the classic theory. Firstly, Al_2O_3 forms, and then Al_5O_6N is obtained followed by the replace reaction of N_2 at the high temperature.

$$TNT / Al \longrightarrow Al_2O_3 \longrightarrow Al_5O_6N$$

In process [2], Al_5O_6N forms after the detonate reaction immediately.

$$TNT / Al \longrightarrow Al_5O_6N$$

And if the process [2] comes into existence, is there a conflict with the classic theory of the detonation of explosive that has been approved?

The preparation of Al_2O_3 by detonation of explosive has been proved by the classic experiment and that Al_5O_6N gained under the protection of water has been proved by this experiment. The results look like opposite, unless there is a process [3].

In process [3], Al_2O_3 is obtained by the detonate reaction of TNT/Al without water. At the beginning, Al_5O_6N may produce. Then the reactions of oxygenation and denitrify at the high temperature occurred. Finally, Al_2O_3 is obtained.

$$TNT / Al \longrightarrow Al_5O_6N \longrightarrow Al_2O_3$$

In order to prove the process [1], we should testify whether Al_2O_3 could react with C or N_2 to form Al_5O_6N at the high temperature. This step will be tested by the proving experiment (1).

If process [1] doesn't come into existence, which means the Al_5O_6N could be formed directly without forming the intermediate Al_2O_3, then process [2] worked.

For process [3], we only need to check whether the final product of the oxidation reaction of Al_5O_6N is Al_2O_3 or not,which react without any protections at the high temperature. This step will be tested by the proving experiment (2).

The proving experiment (1): the negative-oxygen explosive TNT and the common Al_2O_3 were mixed uniformly with a mass ratio of 4:1 and the mixture was loaded into a mould and pressed into a column shape with a density of $1.60g/cm^3$. Then the sample was produced for characterization by repeating the operation of the synthesis process of nano Al_5O_6N.

The proving experiment (2): The synthesized Al_5O_6N was calcined at 1300°C with oxygen existing. Then, the gained sample was characterized by XRD.

1.3.1.2 The comparison of XRD of two detonations' products

Fig.11(a) and Fig.11(b) show the XRD patterns of the detonated products of TNT/Al and TNT/Al_2O_3.

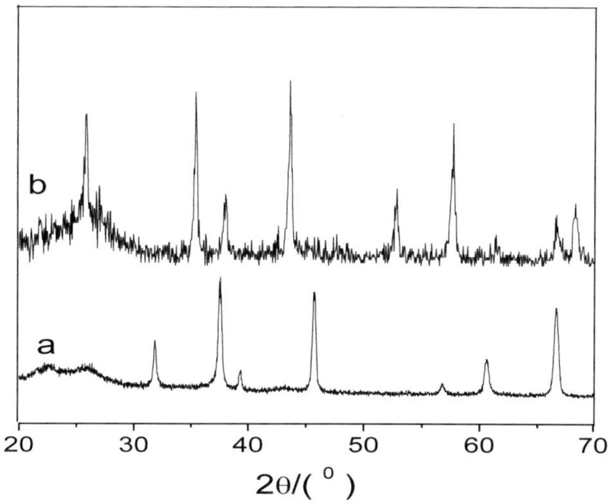

Fig. 11. XRD patterns of two different detonated products (a): TNT/Al (80/20); (b) TNT/ Al_2O_3 (80/20)

The position of the major diffraction peaks in Fig.4(a) has a strong inosculation with the standard JCPDS 48-0686, which confirms that the powder is mainly composed of Al_5O_6N. The main peak has an obvious broad band and the strongest diffraction appeared at 37.54^0 with the half band width of 0.440 can be indexed to (311) plane. According to Scherer equation, the average size of the particles is 28 nm, which matches the result of the electron microscope image. The whole spectrum has a higher diffraction mount and in $20\sim30^0$ exits some low and broad peaks, which indicates the product contains ultrafine imperfectly crystallized carbon in the detonate process, especially in the diffraction position at 25.5^0 where exists a low and broad graphite diffraction.

The position and intensity of the diffraction showed in fig.4 (b) match the standard JCPDS 46-1212 well, which confirms that the powder is α-Al_2O_3. The main peaks also have an obvious broad phenomena and the strongest diffraction for (311) plane located at 43.28^0, with the half band width is 0.440. According to Scherer equation, α-Al_2O_3 has a size of about 25 nm.

It is concluded that under the protection of water, the products are Al_5O_6N and Al_2O_3, which obtained from the detonation of explosive of TNT/Al and TNT/Al_2O_3 respectively.

1.3.1.3 The comparison of Raman spectra of two detonations' products

Fig.12 (a) and (b) show the laser of the detonate products from TNT/Al and TNT/Al_2O_3.

There is no obvious difference between the two Raman spectra of the products. Two wide peaks are located at 1344.81 cm^{-1} and 1575.94 cm^{-1}, and at the lower wave number side exists tailing. The whole spectra are unsymmetrical. The peak at 1344.81cm^{-1} is related to graphite's D-peak, formed from the A_{1g} vibration of aberrant carbon atoms induced by graphite micro-crystal border effect. The peak at 1575.94cm^{-1} is related to graphite's G-peak, formed from E_{2g} mold vibration of atom plane layer in the graphite's lattice, which suggested that the carbon in the mixture powder is graphite phase. According to the analysis of XRD, we can conclude that the micro-size carbon is not crystallized perfectly, but genus-graphite phase.

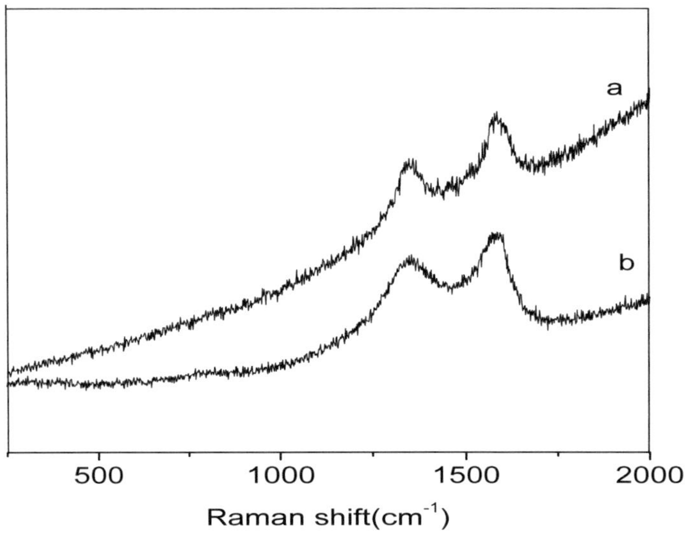

(a) TNT/Al (80/20) (b) TNT/ Al_2O_3 (80/20)

Fig. 12. Raman spectrum of two detonate products

Conclusion from the XRD and Raman analysis of the two detonate products is that in the explosion, Al_2O_3 could not react with N_2 to form Al_5O_6N at the high temperature with the coexistence of C.

1.3.1.4 The IR spectra of two detonations

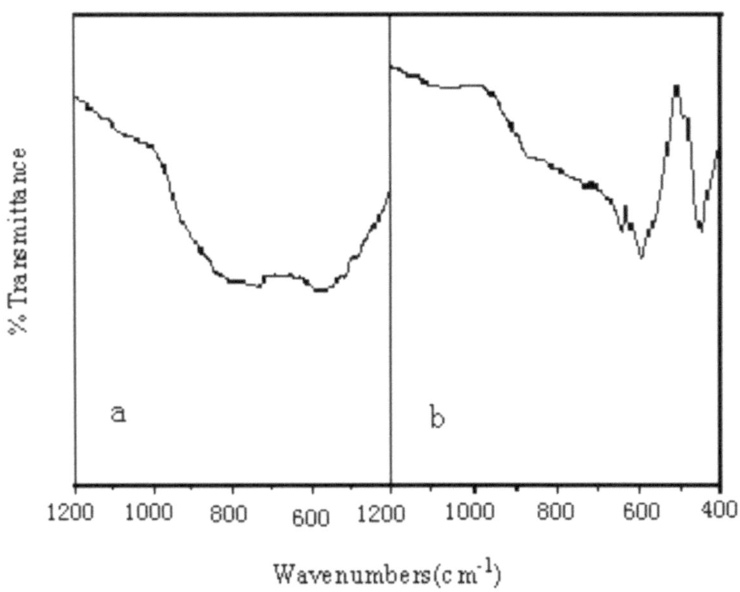

Fig. 13. The infrared spectra of two different detonations (a): TNT/Al(80/20) ; (b) TNT/ Al_2O_3(80/20)

In the Fig.13 (a), Al_5O_6N has a broad and messy absorption band in the area 400 cm^{-1} to 1200 cm^{-1}, which are the characteristic absorbing peaks of Al_5O_6N. Its board frequency band and strong absorption spectrum is induced by the nano-powder's huge specific surface area. Compared with the pure Al_2O_3, the specific surface area increases with the atom O replaced by N. Which induces the decrease of average coordination number of Al^{3+} and increase of unsaturated bonds. As a result, the nano Al_5O_6N would not have a single and merit bond vibration mode, but a rather broad distribution of vibration mode, and the vibration frequency also has a broad distribution under the infrared light field. Thence, the infrared spectrum of Al_5O_6N has a characteristic of peak broaden (Fig.13 (a)). The Fig.13 shows that Al_2O_3 exists obvious fine structures at 455.3 cm^{-1}, 598.7 cm^{-1} and 639.3 cm^{-1}, as Al^{3+} has a higher average coordination number and it is easier to generate a single and merit bond vibration mode which leads to the present of fine structure in the infrared spectrum. In the precondition of the similar size, the obvious differences between the two powders' capacity for infrared absorption makes a further proof of the different composition of the two powders.

Integrated on the comparative analysis of the XRD 、Raman 、IR, it is known that under the protection of water, the detonation product of TNT/Al（80/20）is made up of Al_5O_6N and type-graphite ultrafine carbon. It suggests that in the explosion, Al_2O_3 could not react with N_2 to form Al_5O_6N at the high temperature with the coexistence of C and N_2, and Al_5O_6N could only be produced by one step.

1.3.1.5 The discussion of the TNT/Al detonation reaction mechanism

According to the results of the two detonations, Al_2O_3 could not be replaced by N_2 at the high temperature and pressure. It is concluded that TNT/Al explosive detonation product, Al_5O_6N, is directly generated. The possible reaction mechanism is like this:

The explosive reaction forms a detonation field with ultrahigh temperature and pressure in a flash. All the materials that TNT contained exist in the form of fragment "atom". Aluminum is dissociated into gaseous ions in a moment and all of the fragments diffuse and mix under the disturbance of strong charge in the shock wave plasma space. Different atoms splash after being mixed in the effect of rapid cooling. The splash is selective and the splash ability is closely related to the electro-negativity of the element. The electro-negativity of different elements in the plasma space is showed in table 1[31, 32].

element	Al	H	C	N	O
X_P	1.61	2.18	2.55	3.04	3.44

Table 1. the electro-negativity(X_p) of different elements in the shock wave plasma space

Table 1 shows that the electro-negativity of O and N is larger than Al. When different Ions react in a splash, Al^{3+} with the smallest electro-negativity has priority to splash with O, N and generates Al_5O_6N. The excessive N atoms will form N_2 through a covalent bond in the splash process, in spite of its higher electro-negativity. Because the excessive N atoms are stable and could not attract electrons easily due to its half-full state P orbit. In addition, the diameter of N atom is small, which will lead to a large repulsion force among the electrons. Empathy, C and H elements formed C and H_2, and the total reaction equation of the mixed detonation of TNT and Al is:

$$C_7H_5N_3O_6+5Al \longrightarrow Al_5O_6N+2.5H_2+N_2+7C \qquad (1\text{-}6)$$

The excessive TNT reacts according to type (1-2).
The prepared Al_5O_6N powder was calcined at 1300°C for 2h with aerobic condition, and the Fig.14 shows the XRD spectrum of the generated powder.

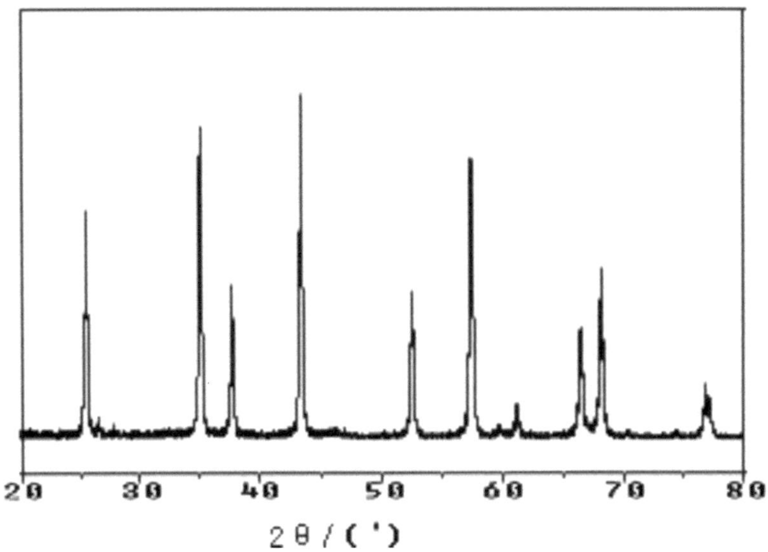

Fig. 14. XRD of the sample got from the calcined nano-Al_5O_6N

Test results from the samples show that Al_5O_6N powders calcined at high temperature can remove N and generate Alumina. Thence, it is entirely possible to rapidly oxidize and remove nitrogen at a higher temperature. When the rapid cooling occurs under the protection of water and C, the detonation stops by (1-6). However, in the absence of the protection of water and under aerobic condition, the solid-phase C is completely oxidized, and the Al_5O_6N is transformed to Al_2O_3 by oxidation and denitrification losing the protection of C.

$$4Al_5O_6N+3O_2 \longrightarrow 10Al_2O_3+2N_2 \qquad (1\text{-}7)$$

1.3.1.6 Verification of the mechanism

The crucial point to judge an analysis of a reaction mechanism is whether it can accurately predict the composition of the final product. According to the above mechanism analysis, 20g Al reacting with 33.6 g TNT due to the formula 1-6 would generate 36.3g Al_5O_6N and 12.4g C.
According to the explosion reaction formula 1-2, superfluous 46.6 g TNT would generate 9.6 g C, the explosion reaction of TNT with Al (80:20) could produce 58.3 g solid materials consist of 37.7% C and 62.3% Al_5O_6N.

Take advantage of the removal of carbon in the mixed-products in oxygen, we carry on TGA tests to the TNT/Al detonation product, and the results are showed in Fig.6.
The DTA-TG curves of the mixture show that there are two obvious exothermic peaks at 23~800°C . The mass loss of 38% at 672°C is attributed to the oxidation of nano-graphite. After the heat released, the curve goes to a platform, which indicates the nano-graphite is completely released, and Al_5O_6N is 62%. The contrast of the theoretical calculation and the actual detection is showed in table 2.

Composition	Carbon content (Wt%)	Al_5O_6N content (Wt%)
Theoretical calculation	37.7	62.3
Actual detection	38	62

Table 2. The contrast of theoretical prediction and actual detection to the TNT/Al detonation product's component

The theoretical analysis is basically consistent with the actual detection, and the analysis of the process of preparation of Al_5O_6N under water protection is further confirmed.
In a word, TNT/Al could directly generate Al_5O_6N by detonation reaction. In the classical theory, Al_5O_6N is transformed to Al_2O_3 by oxidation and denitrification at the high temperature without water protection as follows:

$$TNT / Al \longrightarrow Al_5O_6N \longrightarrow Al_2O_3$$

In this sense, the result does not contradict with the classical theory, but a further complement to it.

1.3.2 Discussion of mechanism of controlling agglomeration during the preparation of Al_5O_6N nanoparticles by shock wave plasma technics

The key point of preparation of nano-materials is to control the agglomeration, which is an integral process of the micro-particles' growth speed with time. As the speed or the time tends to 0, the agglomeration would be over. So, there are two ways to control the agglomeration, one is slow down the particles' growth speed; and the other is to reduce the growth time.
From the traditional compose techniques, high temperature is a necessary condition for preparation of Al_5O_6N, which is also the main driving force of agglomeration. Therefore, it is very difficult to control the growth speed of micro-particles under high temperature, which is the reason why Al_5O_6N nano-powder could not be prepared by conventional method. So, control the growth time is the only path to reduce agglomeration. Shock wave plasmas technique is chosen to prepare Al_5O_6N nano-powder because it could control the growth time during the composition at high temperature and further control of the agglomeration of the micro-particles.
Shock wave plasmas are obtained by instant compress of the gaseity substance by strong shock wave engendered from the explosion of materials with high energy. With the speed emission of the explosion products, shock wave plasmas field will orderly evolve as shown in Fig.15:

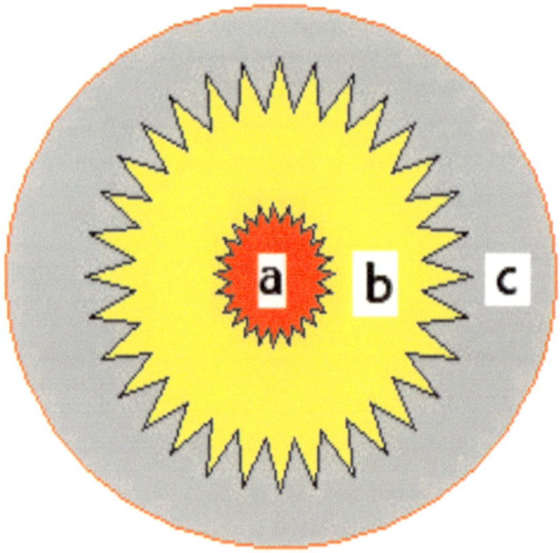

(a) plasmas field (b) splash field (c) crystal field

Fig. 15. Sketch map of orderly evolvement of shock wave plasmas field

The main body of TNT/Al detonator column explodes instantly under the shock wave of starting detonator, with the early pressure of 18~25GPa and temperature of 3000~3500K . The explosive detonation reaction forms a mass of gaseity matters and powerful shock waves.

In the exiguity area of the explosion, under the cooperation of super high pressure and strong shock wave, gaseity molecules absorb energy and bring on high temperature, which makes the atoms vibrate more acutely. Then the bonds among atoms break and form various lively free radicals in plasma state called plasma as showed in Fig.15 (a). In this area, fragments of C、H、O、N and Al irons radiate forth. Along with the expander of bulk, attenuation of shock wave energy and rapid decline of plasma temperature, products of explosion come into spatter area which is showed in Fig.15(b).

In spatter area, fragments of C, H, O, N and Al irons spatter and react selectively. O, N and Al form Al_6O_5N. H combines with O to compose H_2O. C combines with O to compose CO and CO_2. Some dissociate carbon atoms or radicals in the reaction area diffuse, collide and finally agglomerate. The products continue to radiate forth and come into crystal area with the further decline of temperature as showed in Fig.15(c).

Liquid matters diffused into crystal area crystal rapidly to form millions of nucleus due to sharp decline of temperature. The crystal nucleus is highly dispersed because of strong shock waves. At the same time, the forced soon cooling of water protection results in very short crystal time of goal production, which controls the growth time of crystal nucleus effectively. The powder keeps on diffusing and come into normal temperature area, where growth will stop immediately without heat drive of agglomeration.

From the rapid and orderly evolvement of shock wave plasma field, products of explosion existed as gaseity or liquid state in plasma and spatter area could not reunite under strong shock waves, and the agglomeration only happens in crystal area. However, the forced soon cooling of water protection make the process of temperature decline from crystal temperature to normal temperature finish in a second. This is a typical rapid quencher process, which makes the crystal area very narrow and crystal time very short. All these would control growth time of crystal nucleus effectively, and further control of the agglomeration. In addition, super-micro-carbon produced by explosion of detonator in balance of negative oxygen are highly dispersed under strong shock waves, which has some obstruct effects on coagulation of crystal nucleus during the crystal process and management later. It is proved by the phenomenon that the size of nano-particles has no obvious changes before and after calcining

In a word, strong concussion, ultra-high temperature and pressure and rapid forced cooling are the distinct virtues of shock wave plasmas technique. Products of explosion advance orderly along with plasma, spatter and crystal area instantly by rapid diffusion, which controls the growth time of micro-particles and further control their agglomeration under high temperature.

1.4 Thermal stability of Al_5O_6N nano-particles

1.4.1 TG/DTA curves of Al_5O_6N nanoparticles

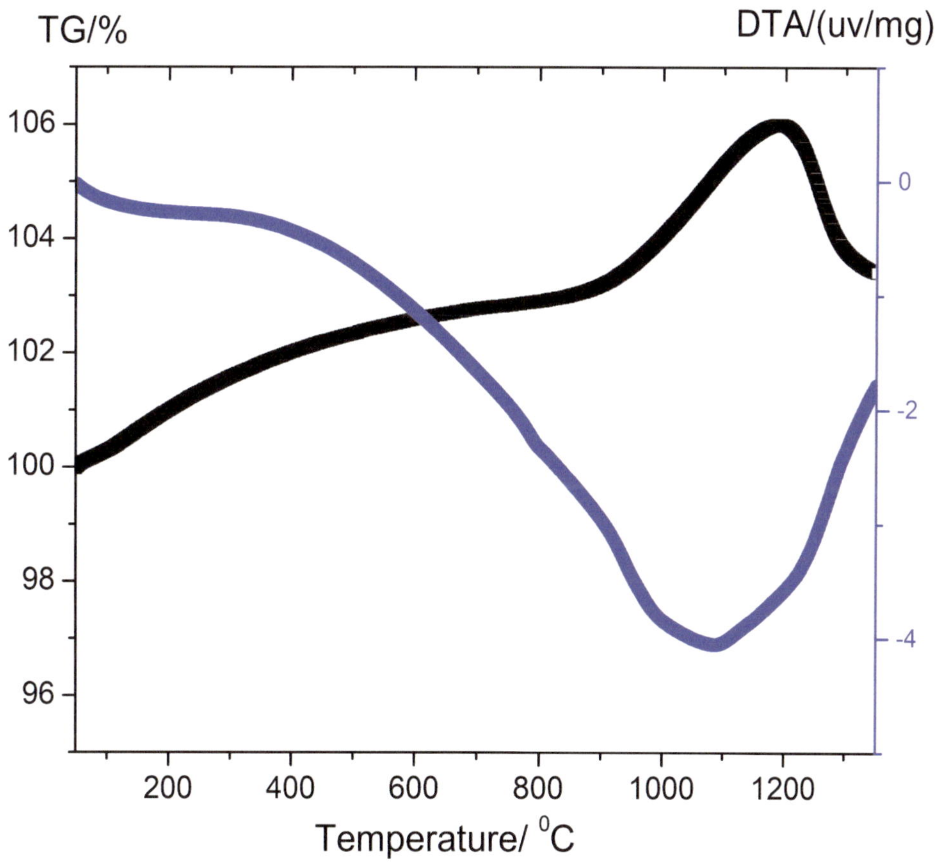

Fig. 16. DTA/TG curves of the nano- Al_5O_6N

DTA tests were performed to lucubrate the oxidation process of Al_5O_6N, and the results were showed in Fig.16.

The DTA-TG curves of the mixture show an obvious broad peak under the temperature from 23 to 1360°C .. There is only a little weight gain from room temperature to 950°C , which maybe have relation with N-doped effect. The effect results in many O holes in nano-powders and makes it possible for O to enter the inside of the powders. The nano-powder can absorb oxygen at low temperature, but atom O has a weak ability to enter the lattice. With the increase of the temperature, the powder's weight increases quickly until 1200°C . Then, the quality of the powder declines as the temperature rises, which may be due to the reaction of denitrify. The whole trend of the process is entirety similar to the DTA-TG curves of existing Al_5O_6N. The major difference is the denitrify temperature of Al_5O_6N bring forward 85°C than that of Al_5O_6N's 1285°C . Meanwhile, the DTA curves exist a visible endothermic peak in the heating process, which indicates the denitrify process of Al_5O_6N is an obvious endothermic process.

1.4.2 XRD patterns of the samples calcined at different temperature

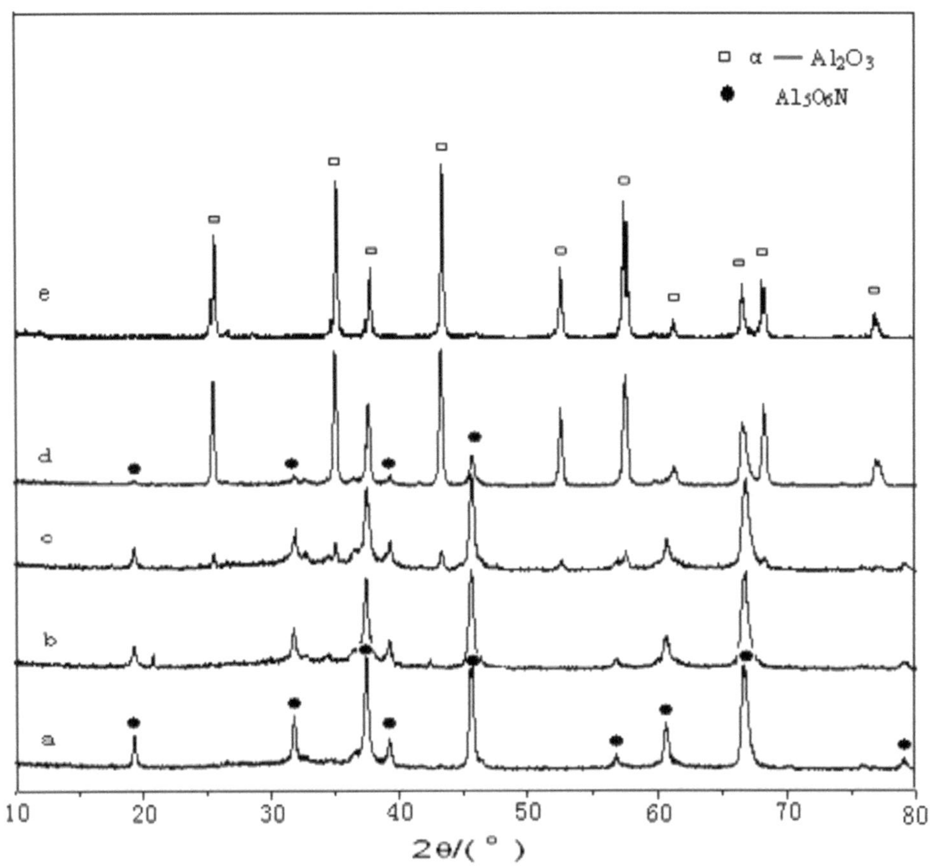

(a) Al_5O_6N (b) 1000°C 3h (c) 1100°C 3h (d) 1200°C 3h (e)1200°C 5h

Fig. 17. XRD patterns of different samples

The influence of different calcining temperature on crystal phase of the samples was studied by XRD, and the results are showed in Fig.17.
From Fig.17, we can see that the crystalline phase of nano- Al_5O_6N powder keep well in the condition of 1000°C for 3h,which indicates that in low temperature Al_5O_6N has certain antioxidant capacity. As the temperature rises to 1100°C , the peaks of Al_5O_6N still exist, but some low peaks of γ-Al_2O_3 appear at the same time. Integrate with TG, the reason could be that the enriched oxygen causes Al_2O_3 separate out and γ-Al_2O_3 is stable at low temperature. When the sample is calcined at 1200°C for 3h, the diffraction of γ-Al_2O_3 has a boost up and the peaks belongs to Al_5O_6N die out gradually, which suggests that most of the Al_5O_6N has been oxidated and trasformed into α-Al_2O_3. If the temperature keeps at 1200°C and calcination time prolongs to 5h, the peaks belong to Al_5O_6N disappear and all the powders change into α-Al_2O_3. So, we can conclude that when the temperature is below 1000°C , Al_5O_6N has certain stability, and the reactions of oxygenation and denitrify will occur when it comes to 1200 °C . The denitrify needs a few time, because the denitrify reaction in crystal is a diffuse process. According to the existing researches [33,34] to the dynamics of nitrogen oxides, the oxidation of nitrogen oxides can be divided into three stages; the control of reaction speed, mixed speed and diffusion speed. The advance of α-Al_2O_3 interface in the process of thermal oxidation above 1200°C is the main restrict step to the continuance of oxidation. The final result of thermal oxidation is the genaration of α-Al_2O_3. The reaction equation is:

$$4Al_5O_6N + 3\,O_2 = 10\,Al_2O_3 + 2N_2 \tag{1-8}$$

1.4.3 IR spectra of the samples calcined at different temperature

Fig.18 shows the results of IR characterization of the samples calcined in different conditions.

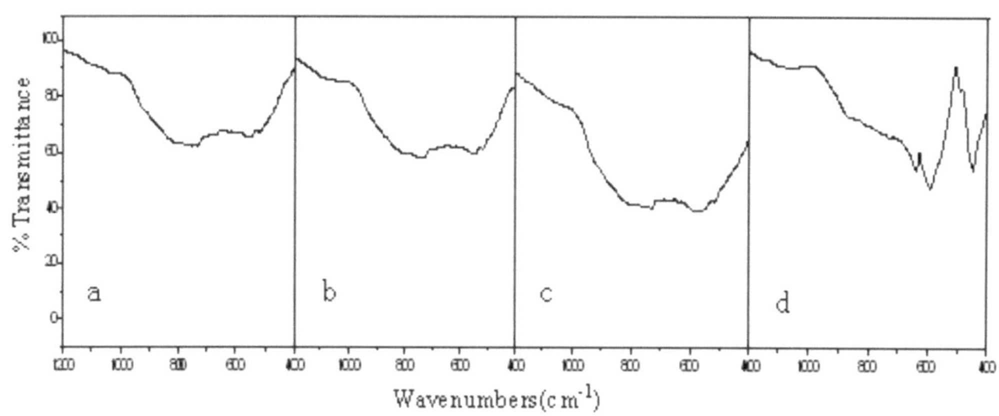

(a) nano-Al_5O_6N (b) 1000°C 3h (c) 1100 °C 3h (d) 1200°C 5h

Fig. 18. The infrared spectra of different samples

In Fig.18 (a), there is a broad and messy absorption band in the area 400 cm^{-1} to 1200 cm^{-1}, which belongs to Al_5O_6N. The nano-powder's huge BET surface area induces Al_5O_6N's broad frequency band and strong absorption spectra. Compared with the pure Al_2O_3, as the atom O is replaced by the atom N, the BET surface area increases, which induces the decrease of average coordination number of Al^{3+} and increase of its unsaturated bonds. As a result, the nano Al_5O_6N would not have a single and priority bond vibration mode, but a rather broad distribution of vibration mode, and so the vibration frequency also has a broad distribution under infrared irradiation. Thence, the infrared spectrum of Al_5O_6N has a width characteristic.

Related to XRD results, the main phase of the samples in Fig.18(b) and (c) is still Al_5O_6N with only a little γ- Al_2O_3, and there is no obvious change to the shape of the characteristic absorbing peaks of the mixture. Pure Al_2O_3 (Fig.18(d)) can be obtained after being calcined at 1200°C for 5h. The characteristic absorbing peaks completely disappeared, and the spectrum is obviously different from Al_5O_6N. Al_5O_6N has a broad and messy absorption band and doesn't have obvious fine structures. While Al_2O_3 in the similar size has obvious fine structures at 455.3 cm^{-1}, 598.7 cm^{-1} and 639.3 cm^{-1}, as Al^{3+} has a higher average coordination number and it is easier to generate a single and priority bond vibration mode which leads to the present of fine structure in the infrared spectrum. In the precondition of the similar size, the obvious difference between the two powders' composition makes a further difference in the capacity for infrared absorption.

1.5 The dynamics research of thermal oxidation to Al_5O_6N

According to the thermal analysis of the high-temperature oxidation process of Al_5O_6N, we can see that there is an obvious nitrogen and oxygen reaction during the process. To learn more about the nano- Al_5O_6N's activation energy of the nitrogen occurred in the thermal oxidation, we use Kissingrer [35] to solve the apparent activation energy (E) and pre-exponential factor (Z) of the Al_5O_6N's thermal oxidation:

$$\ln\frac{\beta}{T_r^2} = \ln\frac{R\cdot Z}{E} - \frac{E}{R}\cdot\frac{1}{T_r} \tag{1-9}$$

In the formula, β, Tr, E, R and Z represents the heating rate, the temperature of the peak, the activation energy, the gas constant and the pre-exponential factor, respectively. The Value of R equals to 8.314J/ (mol·K).

Tr can be learned from the thermal decompositon curve of different heating rate(β), mapping of $\frac{1}{T_r}$ to $\ln\frac{\beta}{T_r^2}$,we can obtain the activation energy(E) from the curve's slope and the pre-exponential factor(Z) from the intercept of the line.

We can also calculate the reactive rate constant(K) in the thermal oxidation process by Arrenius equation

$$K = Z\exp(-\frac{E}{T\cdot R}) \tag{1-10}$$

Use DTA analysis to research the effect of different heating rate to the decompound process of Al_5O_6N's denitrify. The calefactive rates are 5°C /min, 10°C /min, 20°C /min, 30°C /min and Fig.19 shows the results.

The curves indecate that as the increase of the heating rate the decomposition peaks of nano-Al_5O_6N offset to the pyrorigion. According to Kissingrer method equation to solve the apparent activation energy(E) and pre-exponential factor(Z) of the Al_5O_6N's thermal oxidation are showed in table 2.

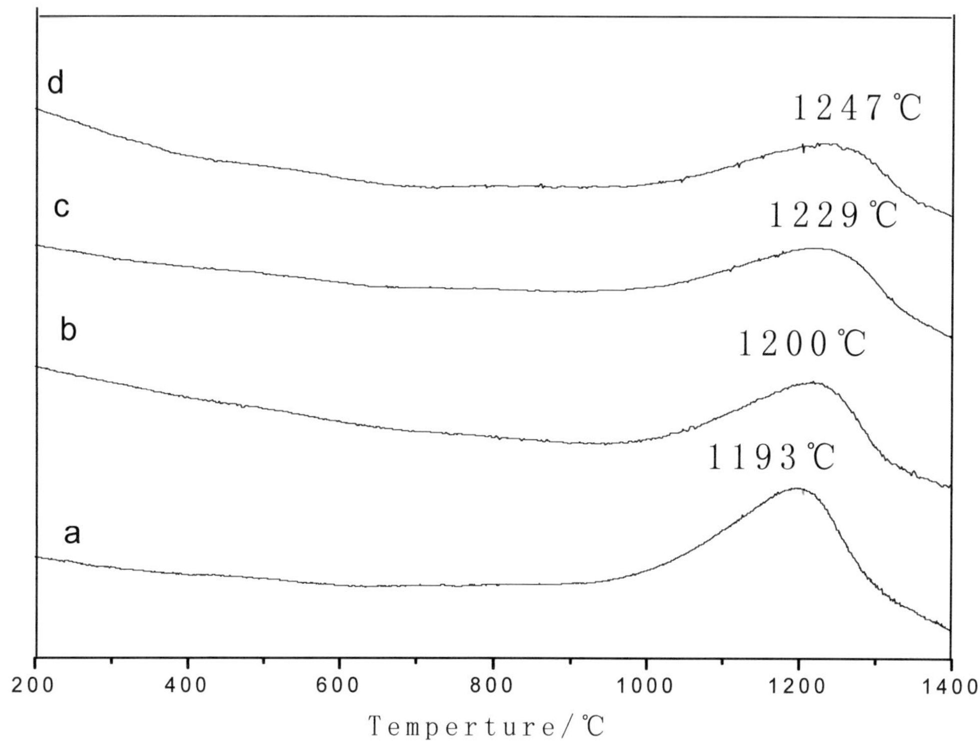

Fig. 19. The TG curves of different heating rate (a):5°C /min; (b):10°C /min; (c):20°C /min; (d): 30°C /min

Sample	t (°C)	E (KJ/mol)	ln Z_H
5	1193	60.25	38.49
10	1200		
20	1229		
30	1247		

Table 2. The data of DSC and dynamics constants of nano- Al_5O_6N's thermal oxidation

2. Acknowledgements

The authors are grateful for the financial support of Explosion Science and Technology Key Laboratory Foundation of China (KFJJ09-7).

3. References

[1] Li Fengsheng. Ultrafine Powder Technology [M]. Beijin: National Defence Industry Press. 2000.

[2] Wang Yi, Cui Zuolin, Zhang Zhikun. Preparation of single crystalline nano-tantalum powders by arc plasma method [J]. Rare metal materials and engineering. 2005,34(2): 309~311.

[3] Liu Wei, Zhang Xianping, Cui Zuolin and Zhang Zhikun. Nano-Au particles prepared by H_2 + Ar arc plasma[J]. Chinese journal of rare metals. 2004,28(6):1082-1084.

[4] Wei Zhiqiang, Ma Jun, Feng Wangjun, Qiao Hongxia and Yan Pengxun. The study on preparation of Ag nanopowders by plasma[J]. Precious Metals. 2004, 25(3): 29-33.

[5] Wei Zhiqiang, Wen Xianlun, Wu Xiancheng and Yan Pengxun. Ni nanopowder prepared by D.C. arc plasma method[J] · Journal of Lanzhou University(Natural Sciences). 2003, 39(5): 38-40.[6]Han Jinyi, Zhu Hongjie, Zhu Yihua, Yang Gwnbing and Hu Liming. Synthesis of Si3N4 ultrafine powder by radio frequency plasma chemical vapor deposition[J]. Chemical Industry Engineering Progress. 1995, 5: 29-33.

[7] Wang Yi, Cui Zuolin and Zhang Zhikun. Synthesis of spherical Titanium dioxide nanocrystals by O_2 + Ar arc Plasma method[J]. Journal of Inorganic Materials. 2004, 19(4): 912-916.

[8] Shang Shuyong, Mei Li, Li Lanying, Yin Yongxiang and Dai Xiaoyan. Research on synthesis of Aluminum Nitride ultra-fine powder by DC arc plasma[J]. Henan Chemical Industry. 2004, 6; 10-12.

[9] Zhao Huaqiao. Plasma Chemistry and Its Application [J]. University Chemistry, 1994, 9(4):1-8.

[10] Yamaguchi G,Yanagida H. Study on the reductive spinelanew spinel formula AlN-Al_2O_3 instead of the previous one Al_2O_3.[J].Chem Soc of Japan Bull,1959,32(11):1264.

[11] James W. McCauley, Parimal Patel, Mingwei Chen, Gary Gilde, Elmar Strassburger, Bhasker Paliwal, K.T. Ramesh, Dattatraya P. Dandekar. AlON: A brief history of its emergence and evolution.[J]. Journal of the European Ceramic Society,2009,29(2):223-236.

[12] Qiu C,Metselaar R.Phase rleations in aluminum carbide-aluminum nitride-aluminum oxide system. [J].J.Am.Chem.Soc,1997,80(8):2013.

[13] F.Y.C.Boey,X.L.Song,Z.Y.Gu,A.Tok. AlON phase formation in a tape-cast Al_2O_3/AlN composite.[J]. Journal of Materials Processing Technology,1999,89-90:478-480.

[14] Sang Sik Byeon, Kai Wang, Yeon Gil Jungand, Bon Heun Koo. Characteristic of AlON-Al2O3 coatings on Al6061 alloy by electrolytic plasma processing in aluminate and nitride electrolytes.[J]. Surface and Coatings Technology, 2010,204(20):3196-3199.

[15] Dariusz Zientara, Mirosław M. Bućko, Jerzy Lis. Dielectric properties of aluminium nitride–γ-alon materials.[J]. Journal of the European Ceramic Society. 2007, 27(13-15): 4051-4054.
[16] Paul W. Wang, Jin-Cherng Hsu, Yung-Hsin Lin, Huang-Lu Chen.Nitrogen bonding in aluminum oxynitride films.[J].Applied Surface Science. 2010, 256(13): 4211-4214.
[17] T. M. Hartnett, S. D. Bernstein, E. A. Maguire, R. W. Tustison. Optical properties of ALON (aluminum oxynitride).[J]. Infrared Physics & Technology. 1998, 39(4): 203-211.
[18] Huang Cun-xin, Li Jian-bao, Lei Mu-yun, Du Hong-bing. Research Progress and Current Trends in 3-5 Micron Optical Materials.[J].Journal of Synthetic Crystals. 2003,32(3): 276-281. (in Chinese)
[19] Staver A M et al · Ultrafine powders manufactured with the use of explosive energy[J] · Fizika Goreniya Ivzryza. 1984,20(5): 100.
[20] Liander H. ASEA Journal[J]. 1995(97): 28.
[21] Greiner N R, Phillips D S, Johnson J D. Diamonds in detonation soot[J]. Nature,1988, 333(2):440-442.
[22] Qinuma S, Tanaka K, Lida M, et al. Diamonds recovered from detonationproducts [J]. Shok Compression of Condensed Matter-1989, Schmidt SC(editonrs). Elsevier Science Publishers BU. 1990: 705.
[23] Jin Zengshou and Xu Kang. Nanoscale diamond synthesized by explosive detonation[J]. Energetic Materials, 1999,7(1): 38~44.
[24] Yun Shourong, Huang Fenglei, Ma Feng, Tong Yi, Chen Pengwan, Chen Quan, Zhou Gang and Zhao Sheng. Ultrafine diamond-21st century's new material [J]. World Sci-tech R & D. 22(1):39-46.
[25] Samara G A. Physica. 1985, (139-140), 4-5.
[26] Ree F H, Thiel M V. Statistical mechanical theory of condensed explosives[A]. Proceedings of the International Symposium on Pyrotechnology and Explosives. 1987: 12-15.
[27] Titov V M, Anisichkin V F, et al. Synthesis of ulrafine diamonds in detonation waves[A]. The Ninth Symposium(internation) on Detonation. Oregan. 1989: 407-416.
[28] Yamada K , Swaaoka A B · Very small spherical crystals of distorted diamond found in a detonation product of explosive/graphite mixtures and theirformation mechanism[J]. Carbon,1994,34(4): 665-673.
[29] Sun yebing, Hui Junming and CaoXinmao. Military explosive mixture[M]. Weapon Industry Press. Beijin: 1995.
[30] Kenjiro Yamada, J. Am. Soc. 79 (4) (1996) 1113–1118.
[31] Zhang Zuo-tai,SAIYIN Bater,LI Wen-chao.New Advances And Prospects of Research on AlON And Composites.[J].Materials Review.2003,17(12):40-43. (in Chinese)
[32] Sun Ye-bing,Hui Jun-ming,Cao Xin-mao. Military Mixed Explosive[M]. Enginery Industry Publishing House. Beijing, 1995:364-384.(in Chinese)
[33] A. Maghsoudipour, F. Moztarzadeh, M. Saremi, J. G. Heinrich. Oxidation behavior of AlN–Al_2O_3 composites.[J].Ceramics International, 2004,30(5): 773-783.

[34] D. Suryanarayana. Oxidation kinetics of aluminum nitride [J].J. Am.Ceram. Soc.1990.73 (4) :1108–1110.

[35] YUN Zhu-hui, ZHOU Zheng-mao. Data analysis for thermal analysis kinetics [J]. Chin.J.Explos.Propell.1983, 2: 24. (in Chinese)

6

Stopped-Flow Studies of the Formation of Organic Nanocrystals in the Reprecipitation Method

Daniel Oliveira[1], Koichi Baba[2], Winfried Teizer[1,3], Hitoshi Kasai[1,4], Hidetoshi Oikawa[1] and Hachiro Nakanishi[1]

[1]*Tohoku University,*
[2]*Osaka University,*
[3]*Texas A&M University,*
[4]*PRESTO, Japan Science and Technology*
[1,2,4]*Japan*
[3]*United States*

1. Introduction

Interest in organic nanoparticles is rapidly growing in the scientific community as it is now clear their potential impact in several attractive economic fields, in special their applications as pharmaceutically active organic compounds (Chen & Zhang, 2006), printing inks (Magdassi & Moshe, 2003) and color filters (Miyashita et al., 2008). Accordingly, a recent report published in BBC Research estimated that by 2015 the global demand for nano-related goods could be higher than US$ 2.4 trillion (McWilliams, 2010). Moreover, the author predicted that the economic power of the nanotechnology industry could be greater than those attained almost a decade ago by both the telecommunication and information technology industries. It is thus hardly surprising that chemical industries need new and reliable methods directed to the production of organic nanoparticles. Furthermore, sizes of prepared organic nanocrystals must be carefully controlled as the properties of materials are strongly dependent on the size of the particles of the materials (Auweter et al., 1999; Burda et al., 2005; Horn & Rieger, 2001).

Methods used for the preparation of organic nanoparticles are primarily based on either precipitation, milling, or through chemical reaction techniques. Unfortunately, the former two techniques proved to be less attractive in manufacturing organic nanosized materials, as milling techniques rely on applying an extremely high mechanical energy on the system leading to structural changes of the crystal (Bilgili et al., 2006; Peters, 1996); and synthetic approaches are based on considerable exhaustive, and therefore time consuming synthetic routes (Spatz et al., 1999; Sugimoto et al., 2002). As a consequence, methods which rely on solubility changes to induce nanoparticle formation are lately becoming more attractive. Among these methods (Frendler, 1987; Hayashi et al., 2007; Ibanez et al., 1998; Kasai et al., 1992; Kasai et al., 1996), a simple and convenient technique proved to be particularly effective for the preparation of organic nanoparticle dispersion, the so-called "reprecipitation method" (Kasai et al., 1992; Kasai et al., 1996).

In the reprecipitation method, a dilute solution of a target compound prepared on a water-soluble organic solvent is injected into vigorously stirred water as a poor solubility medium. As the organic solvent disperses, the sudden change in the solubility of the target compound causes it to precipitate in the form of dispersed nanoparticles in an aqueous medium. The reprecipitation method has the advantage of producing nanodispersed systems under mild conditions, a great benefit when considering that the vast majority of organic compounds are thermally unstable.

The physical properties of nanoparticles differ significantly from those of their related bulk crystals due to the changes in lattice state caused by an increase in the surface area of nanoparticles, and consequently, to lattice softening (Masuhara et al., 2002). It is believed that the lattice softening results in weaker interaction between adjacent molecules, ultimately resulting in wider band gaps (Masuhara et al., 2002). For that reason it is no surprise that a technique employed for organic nanoparticles formation must be able to precisely control the size of prepared crystals.

Even though there is vast literature available on the size-controlled formation, characterization, and application of a wide variety of nanoparticles, little attention has been paid to the kinetics of organic nanosized crystal growth; in particular, little is known about nucleation, as it is experimentally difficult to investigate such a phenomenon given the time frame of the process. Tiemann and coworkers systematically studied the early stages of nanoparticle growth (Tiemann et al., 2005; Tiemann et al., 2006; Tiemann et al., 2008) unfortunately, their investigations were restricted to inorganic nanoparticles, which as previously stated, indicates the limited understanding of organic nanoparticles when compared with their inorganic counterpart.

Following the initial work by Van Keuren et al. on the kinetics of the formation of organic nanocrystals (Van Keuren et al., 2001), Mori and coworkers were able to directly observe the formation of organic nanoparticles by the reprecipitation method using stopped-flow UV-Visible absorption spectroscopy (Mori et al., 2009). Authors quantitatively demonstrated that the kinetics of the reprecipitation process is governed by the classical nucleation theory (Chaikin & Lubensky, 1995; Debenedetti et al., 1996; Kashchiev & van Rosmalens, 2003; Laaksonen et al., 1995). Essentially, the classical nucleation theory states that nucleation succeeds through a supersaturated solution, where nuclei are formed by monomers clustering one molecule at a time to a single cluster.

This seminal work conducted by Mori and colleagues is of utmost importance as it correlated for the first time the mechanism of the reprecipitation process and a known nucleation model. However, it is not yet clear whether such a model can be used to explain the reprecipitation mechanism for a wider class of systems undergoing nanocrystallization. To better comprehend the mechanism of organic nanoparticle formation (and hence, control such formation), the classical nucleation theory must be tested to determine whether it can explain changes in nanocrystallization kinetics while varying several parameters of the reprecipitation method. In this chapter, the stopped-flow UV-Visible absorption spectroscopy technique will be used to systematically investigate the growth of several organic nanoparticles in the millisecond time regime and assert if the kinetics of the nanoparticle formation process obeys the classical nucleation theory.

Of utmost importance is not only to control organic nanocrystal formation, but also to precisely manipulate sizes of produced nanocrystals. Unfortunately no literature report shows the use of organic molecules as additives to induce size-controllable nanocrystallization of a desired organic compound. It will also be addressed in this chapter a novel technique to fabricate organic nanocrystals (here, perylene nanocrystals) with different sizes by employing N,N'-bis(2,6-dimethylphenyl)-3,4,9,10-perylenedicarboxyimide, or simply DMPBI, as an additive in the reprecipitation method. The choice of DMPBI as the additive to study the changes induced in the reprecipitation process of perylene was due to the structural similarity between both organic compounds, i.e., perylene and DMPBI. By monitoring the formation of perylene nanoparticles with the aid of stopped-flow UV-Visible absorption spectroscopy, results suggest that DMPBI molecules act as seeds in the nucleation process, allowing for seed-mediated perylene nanosized crystal growth, increasing nanocrystallization rates, and thus, altering the size of the prepared nanoparticles.

2. Experimental procedure

The sizes of perylene nanoparticles were obtained using the JEOL JSM-6700F scanning electron microscope (SEM) instrument, and the Malvern ZetaSizer Nano-ZS dynamic light scattering (DLS) technique. UV–visible absorption spectroscopy was performed using a Unisoku RSP-1000 stopped-flow spectrometer designed with a photodiode array detector. In order to simulate the reprecipitation method, the injected solution should be fixed at different ratios (Kasai et al., 1992; Kasai et al., 1996); therefore, a mixing ratio of 1:9 (water soluble solvent: water) was achieved by employing two syringes of different sizes. For each run, 41 μL (target compound in organic solvent) and 342 μL (water) were mixed in a spherical mixer and then used to fill a rectangular observation chamber (1 mm optical path length cell). Changes in absorbance were monitored every millisecond for a total of 500 ms. The dead time of the instrument was calculated to be about 4 ms (Tonomura et al., 1978). All collected data were referenced against those of purified water.

3. Results and discussion

Fig. 1 shows the UV-visible absorption spectra acquired with the stopped-flow instrument, where 0.10 mmol L^{-1} perylene dissolved in acetone was mixed with water and its absorption spectra were measured for 500 ms at 1 ms intervals (for clarity, only spectra obtained during the first 200 ms were plotted). Observation of the figure shows a clear pattern concerning the time-resolved absorption spectra of perylene, that is, the disappearance of the absorption band at 405 and 432 nm is occurring as a new absorption band centered at 448 nm is being formed. Earlier studies focused on the preparation of organic nanoparticles showed there is a red-shift in the perylene absorption band when nanocrystals were formed though the reprecipitation method (Kasai et al., 1992; Kasai et al., 1996). Authors concluded that the observed re-shift was due to the formation of aggregated species (An et al., 2002; Fu & Yao, 2001; Kang et al., 2007; Van Keuren et al., 2008; Wang et al., 2005). One can thus fairly assume that the increase in the intensity of the 448 nm band is due to the generation of perylene nanoparticles, while the decrease in the intensity of both the 405 and 432 nm band corresponds to the disappearance of perylene monomers. Also, it is important to point out the existence of an isosbestic point at 444 nm, indicating that only two species with distinct absorption properties are present in the mixture, that is, perylene monomers and perylene nanocrystals.

Using the data shown in the figure, it is possible to calculate the rate constant of nanoparticle formation by either the rate of increase in the area of the nanoparticle band (at

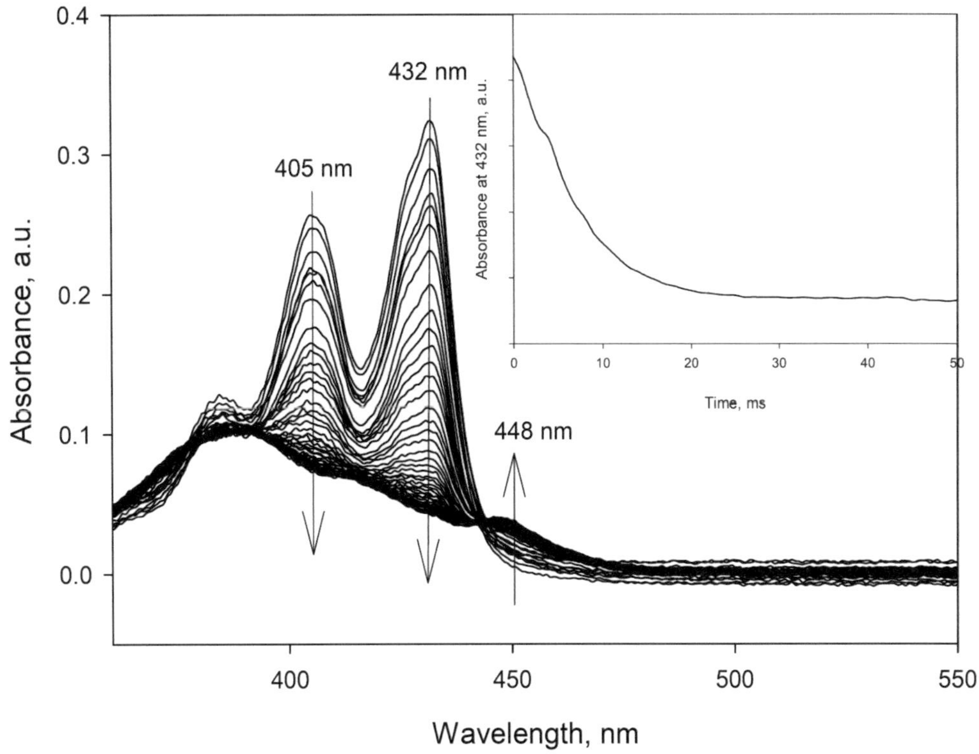

Fig. 1. Time-resolved absorption spectra of perylene in acetone (0.1 mmol L^{-1}) mixed with water. Inset shows the absorbance intensity at 432 nm as a function of time

448 nm) or by the rate of decrease in the area of the monomer band (either at 405, or at 432 nm). The latter band at 432 nm will be used throughout this chapter to calculate nanocrystallization rates because, as seen in the figure, the intensity of such band is much higher than the intensity of the nanoparticle band (and of the 405 nm band), and therefore, more accurate data is expected to be obtained. The Fig. 1 inset shows the intensity absorbance decay at 432 nm as a function of time during the first 50 ms after reprecipitation, corresponding to perylene monomers consumption. Calculation of rate of perylene nanoparticle formation is defined as the inverse of the time for the monomer absorbance to decrease to $1/e$ from its initial absorbance (Mori et al., 2009). It is important to point out that the formation of perylene nanocrystals by the reprecipitation method is completed within tens of milliseconds.

A key parameter in understanding the nanocrystallization mechanism of organic compounds is the supersaturation ratio, defined as C/C_e, where C is the initial target compound concentration and C_e is the solubility of that target compound in the water/organic solvent mixture. Therefore, one can change the type of target compound, the type of solvent, temperature, and concentration in order to gain better insight into the reprecipitation process. With that in mind, perylene solutions were prepared at three different concentrations (0.05, 0.1, and 0.2 mmol L^{-1}) and their time-resolved absorbance spectra measured every 5 degree Celsius from 5 to 50 ^{o}C. After spectral acquisition, the rate of nanoparticle formation at each concentration and temperature was calculated on the basis of the decrease in the intensity of the 432 nm monomer peak (as described above). Fig. 2 shows the rates of nanoparticle formation plotted against temperature for the three distinct perylene concentrations. In the figure, each data point is the average of three individual rate constant determinations.

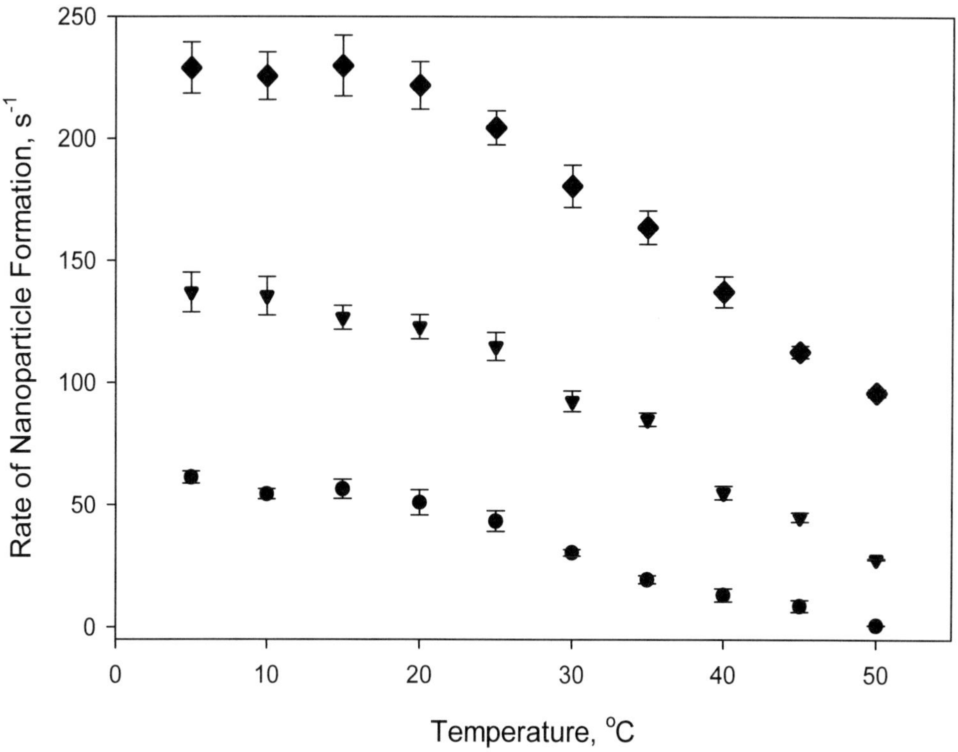

Fig. 2. Rates of nanoparticle formation as a function of temperature for perylene 0.05 mmol L^{-1} (circles), perylene 0.1 mmol L^{-1} (triangles) and perylene 0.2 mmol L^{-1} (diamonds). Solutions were prepared in acetone and mixed with water

Classical nucleation theory predicts that at high supersaturation ratios, molecules will rapidly aggregate and crystal growth will occur at a higher rate through the aggregation of the nucleated particles (Kashchiev & van Rosmalens, 2003). Moreover, the nanoparticle growth process is associated with a decrease in monomer concentration; as a consequence, additional amount of the organic compound must be added or temperature must be reduced to sustain supersaturation. Consequently, as temperature increases, nanoparticle formation rate decreases (at higher temperatures the solubility of the target compound in water increases), which leads to a decrease in supersaturation ratio and ultimately to a decrease in the rate.

Classical nucleation theory also predicts increasing critical nucleus size with increasing temperature, so that the formation of stable nuclei becomes increasingly harder at elevated temperatures. Such a phenomenon, accompanied by the higher cluster diffusivities at elevated temperatures, will result in lower rates of nanoparticle formation, as seen in Fig. 2. The distinctive pattern observed in Fig. 2 (lower nanoparticle formation rates at higher concentrations) is thus in agreement with the classical nucleation theory.

After the dependence of perylene nanoparticle formation as a function of concentration and temperature was successfully obtained and correlated with the classical nucleation theory, studies with other types of low-molecular-weight aromatic compounds shall provide additional information on the nanocrystallization mechanism of organic materials. Hence, the nanoparticle formation processes for anthracene and benzo[a]pyrene were also evaluated using the stopped-flow apparatus. Fig. 3 shows the temporal evolution absorption spectra obtained with the stopped-flow system for anthracene [Fig. 3(A)], and benzo[a]pyrene [Fig. 3(B)]. Both compounds were dissolved in acetone (0.1 mmol L^{-1}) and mixed with purified water in a 1:9 volume ratio. For simplicity, only the spectra acquired during the first 200 ms at 1 ms intervals were plotted.

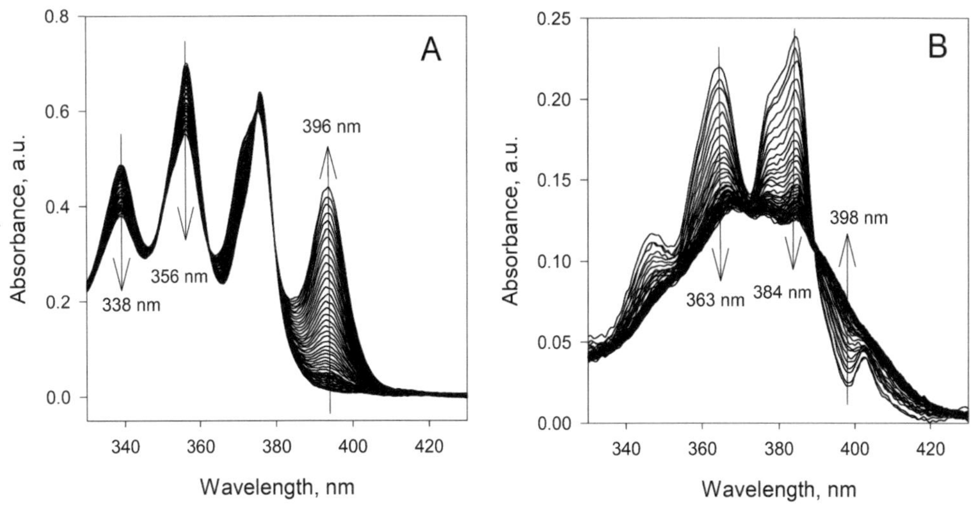

Fig. 3. Time-resolved absorption spectra of (A) anthracene in acetone (0.1 mmol L^{-1}) mixed with water; and (B) benzo[a]pyrene in acetone (0.1 mmol L^{-1}) mixed with water

Similarly as observed with perylene (Fig. 1), anthracene and benzo[a]pyrene shows strong time-dependent absorption spectra. The spectra of anthracene [(Fig. 3(A)] are marked by a decrease in the intensities of the absorption peaks at 338 and 356 nm (anthracene monomers), and by the formation of an absorption peak at 396 nm (anthracene nanoparticles). In addition, the absorption monomer peak at 376 nm gradually vanishes and blue shifts to a new anthracene nanoparticle peak at 371 nm. The time-dependent absorption spectra of benzo[a]pyrene [Fig. 3(B)] follow similar behavior, i.e., monomer bands at 363 and 384 nm decrease in intensity, accompanied by an increase in the intensity of the absorption peak at approximately 398 nm, the benzo[a]pyrene nanoparticle band.

Even though nanoparticle formation of anthracene and benzo[a]pyrene through the reprecipitation method was previously studied (Chung et al., 2006), no studies accounted the kinetics of the process. After confirmation that such nanocrystallization indeed occurs (illustrated on Fig. 3), further kinetic insight can be achieved if, as previously performed, the rate of nanoparticle formation is studied as a function of temperature. Fig. 4 shows such data for anthracene and benzo[a]pyrene, as well as for perylene (replotted from Fig. 2). Nanocrystallization rate was calculated on the basis of the 384 and 356 nm peak decays, for benzo[a]pyrene and anthracene, respectively. Each displayed data point is an average of three separate rate determinations. As observed in the figure, the rate of nanoparticle formation is clearly dependent on the type of target organic compound and the temperature of the process.

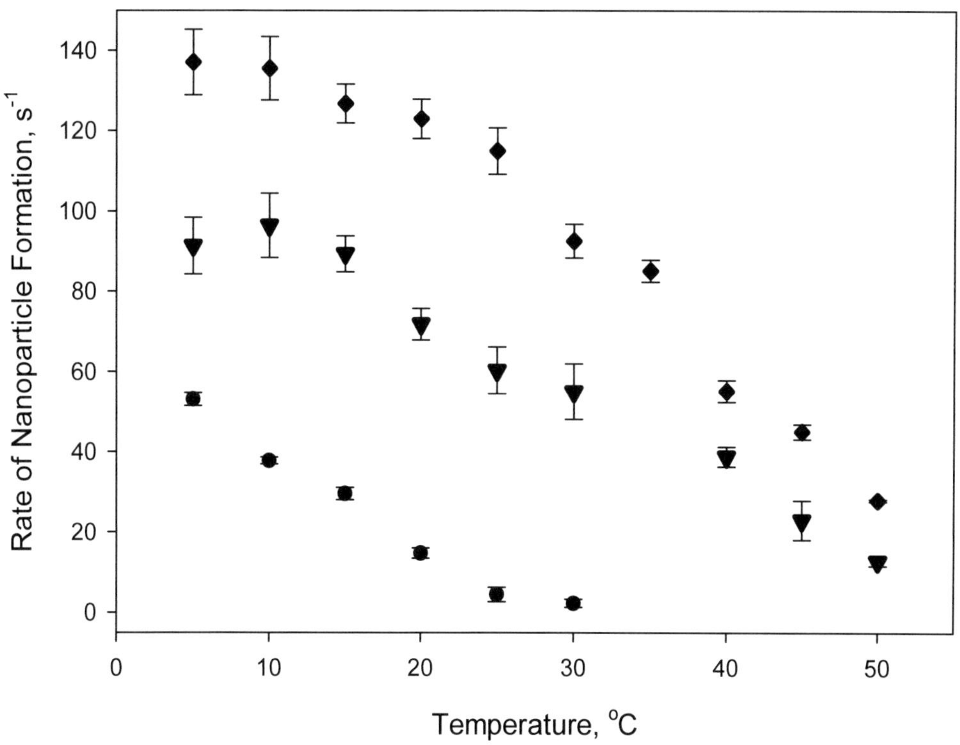

Fig. 4. Rates of nanoparticle formation as a function of temperature for perylene 0.1 mmol L^{-1} (diamonds), benzo[a]pyrene 0.1 mmol L^{-1} (triangles), and anthracene 0.1 mmol L^{-1} (circles). Solutions were prepared in acetone and mixed with water

It was previously stated that supersaturation ratio is a key parameter for understanding rates of nanocrystal formation. As supersaturation ratio is a function of organic compound solubility, the trend observed in Fig. 4 can be explained by correlating the solubility in water of the three organic compounds with their corresponding formation rates. Being perylene the least compound soluble in water [4.2×10^{-10} moles/L (Eisenbrand et al., 1970)] it will have the highest supersaturation ratio, and consequently the highest nanoparticle formation rate. In contrast, anthracene, the most soluble in water [2.4×10^{-7} moles/L (Eisenbrand et al., 1970)], will have the lowest supersaturation ratio and thus the lowest nanocrystallization rate. Possessing solubility between those of the two compounds, benzo[a]pyrene [(2.0×10^{-9} moles/L (Eisenbrand et al., 1970)] is expected to have an intermediate supersaturation ratio and hence an intermediate formation rate, as consistently observed in Fig. 4.

Examining the temperature influence on nanocrystallization rates, the solubility of benzo[a]pyrene increases nearly fivefold when the system is heated from 20 to 35 °C, namely from 2.0×10^{-9} moles/L (Eisenbrand et al., 1970) to 1.1×10^{-8} moles/L (Blyshak et al., 1989). Interestingly, for organic molecules with a higher water solubility (i.e., anthracene), the effect of temperature is even higher. At 35 °C, the solubility of anthracene is 3.5×10^{-7}

moles/L (Blyshak et al., 1989), which is one order of magnitude higher than that of benzo[a]pyrene at an equivalent temperature, and thus nanoparticle formation is observed at a considerable lower rate, as shown in Fig. 4.

Since the effects of concentration and solubility on nanocrystallization rates of organic materials turned out to be in accordance with the classical nucleation theory, the solvent on which organic compounds are dissolved can be varied to test its affect on nanocrystallization rates. Accordingly, perylene was dissolved in two other water miscible solvents, namely, ethanol and tetrahydrofuran (THF) and compared to results previously obtained with acetone as the solvent. Perylene solution of 0.1 mmol L^{-1} was prepared in both solvents, and by using the stopped-flow instrument, perylene nanoparticle formation rate was calculated and plotted in Fig. 5. Perylene 0.1 mmol L^{-1} in acetone was replotted from Fig. 2, where each data point is the average of three individual rate constant determinations.

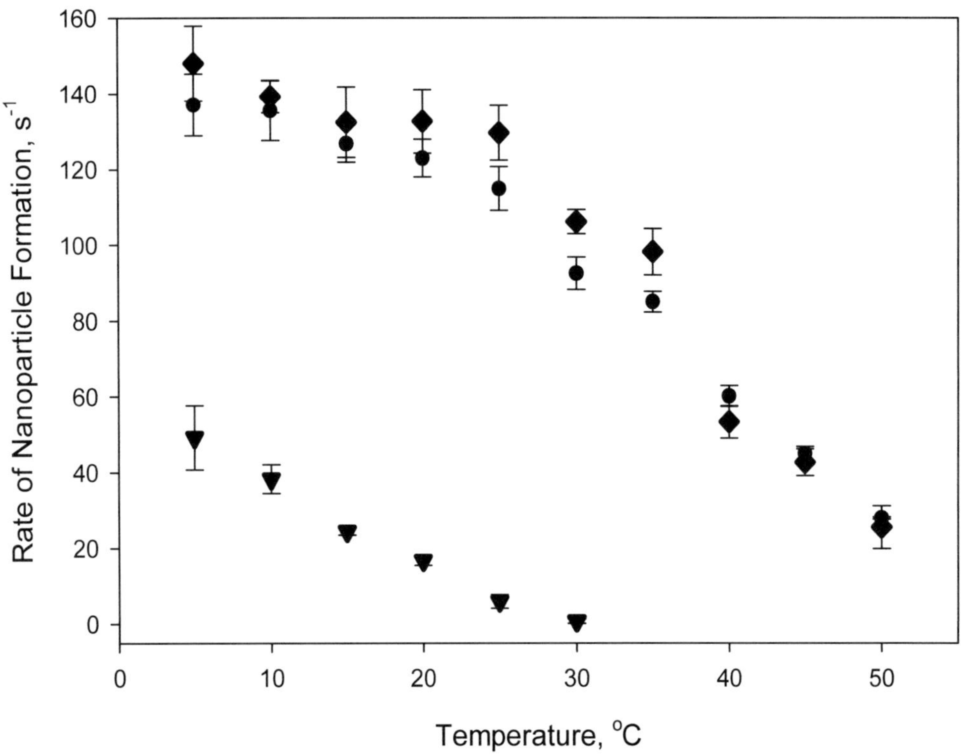

Fig. 5. Rates of nanoparticle formation as a function of temperature for perylene 0.1 mmol L^{-1} dissolved in ethanol (diamonds), acetone (circles), and tetrahydrofuran (triangles) mixed with water

Results visualized in Fig. 5 are strikingly similar with those obtained while varying the type of organic molecule (Fig. 4). Analysis of Fig. 5 shows, as previously observed in Fig. 4, that the dependence of rate of perylene nanoparticle formation on a given parameter (here the type of water-soluble solvent) is considerable strong. Consequently, on the basis of the classical nucleation theory, it is presumed that the same parameter can be used in understanding both results, i.e., solubility.

The solubilities of perylene in ethanol, acetone, and THF were experimentally determined by UV-visible spectroscopic analysis to be 3.7 x 10^{-4}, 4.1 x 10^{-3}, and 2.8 x 10^{-2} moles/L, respectively. As shown in Fig. 5, the rate of perylene nanocrystal formation is in accordance with the solubility of perylene in an organic solvent, that is, the rate increases in the order from the solvent where perylene is most soluble (THF) to the solvent where it is least soluble (ethanol). Bearing in mind that a fixed volume ratio of 1:9 (organic solvent: water) is used when injecting the perylene solution into the aqueous system, it is expected that the solubility of perylene in water will increase as the solubility of perylene in the organic solvent increases, as a consequence of the organic solvent acting as a solubilizing agent (Nyssen et al., 1987). This in turn leads to low supersaturation rates and thus low nanocrystallization rates, as observed in the THF system.

Indeed results presented in Figs. 2, 4 and 5 indicate that indeed organic nanocrystal formation follows the classical nucleation theory. Although it is tempting to conclude that the classical nucleation theory can be used to comprehensively describe organic material nanocrystallization, such proposal is considered to be over simplistic, as it does not take into account several other factors. As an example, a study conducted by Chung and coworkers shows the effect of organic solvent dielectric constant on the reprecipitation method (Chung et al., 2006). Authors demonstrated that cluster concentrations are higher for solvents with larger dielectric constants, because of better organic solvent dispersibilities of such solvents in water. Therefore, the solubility of organic solvent in water may have a prominent effect on the rate of nanocrystallization, as a result of the fast mixing between the two phases. The dielectric constants of ethanol, acetone, and THF are 24.6, 20.7, and 7.6 (Vogel et al., 1996), respectively (at room temperature). Fig. 5 shows the rates to be in accordance with the dielectric constants; ethanol, having the largest dielectric constant, offers a faster transport for perylene molecules to get in contact with the water phase (i.e., highest cluster concentrations) and thus the highest rates. Moreover, the similarity between the nanoparticle formation rates observed when perylene is dissolved in ethanol and acetone is better explained when taking dielectric constant into account. As the solubility of perylene in ethanol is one order of magnitude lower than that in acetone, one would expect the formation rate of perylene to be considerably higher than that shown in Fig. 5, which could be associated with the similar dielectric constants of the two solvents.

Despite the unfortunate lack of studies concerning the mechanism of nanosized organic crystal formation, it could be quantitatively suggested here that the kinetics of the nanoparticle formation process for several types of aromatic organic compounds under several different experimental gross conditions obeys the classical nucleation theory (Oliveira et al., 2009). However, for practical purposes, as important as understanding the mechanism of nanocrystal formation it is the capacity to manipulate its size.

The reprecipitation method has already been subject to several studies focused on the fabrication of size-controllable organic nanoparticles. Among the main parameters which can be controlled in order to manipulate the size of nanoparticles, concentration of the target compound (Katagi et al., 1996), medium temperature (Kasai et al., 1998), and also applied microwave irradiation (Baba et al., 2007) were found to be the most effectives. Now, a novel method to fabricate organic nanocrystals (exemplified by perylene) with different sizes by employing N,N'-bis (2,6-dimethylphenyl) - 3,4,9,10 – perylenedicarboxyimide (DMPBI) as an additive in the reprecipitation method is addressed in the remaining of this chapter.

It was found that adding DMPBI to a solution of perylene led to changes in its nanoparticle formation rate when undergoing reprecipitation in water. Fig. 6 shows the time resolved absorption spectra of 0.20 mmol L^{-1} perylene in acetone with the addition of 10% (0.02 mmol L^{-1}) DMPBI. It is seen that the basic features of perylene monomer disappearance and nanocrystal appearance are consistent with those illustrated in Fig. 1; also, no shift in either absorption bands is noticed. However, it is observed that monomer molecules are consumed faster, which indicates that the nanocrystallization of perylene is occurring at higher rates with the presence of 10% added DMPBI. Such statement can be readily reached by comparing the insets in Figs. 1 and 6, where a steeper decrease in the intensity of the 432 nm peak shown in Fig. 6 indicates faster rates of perylene monomer consumption in the presence of DMPBI.

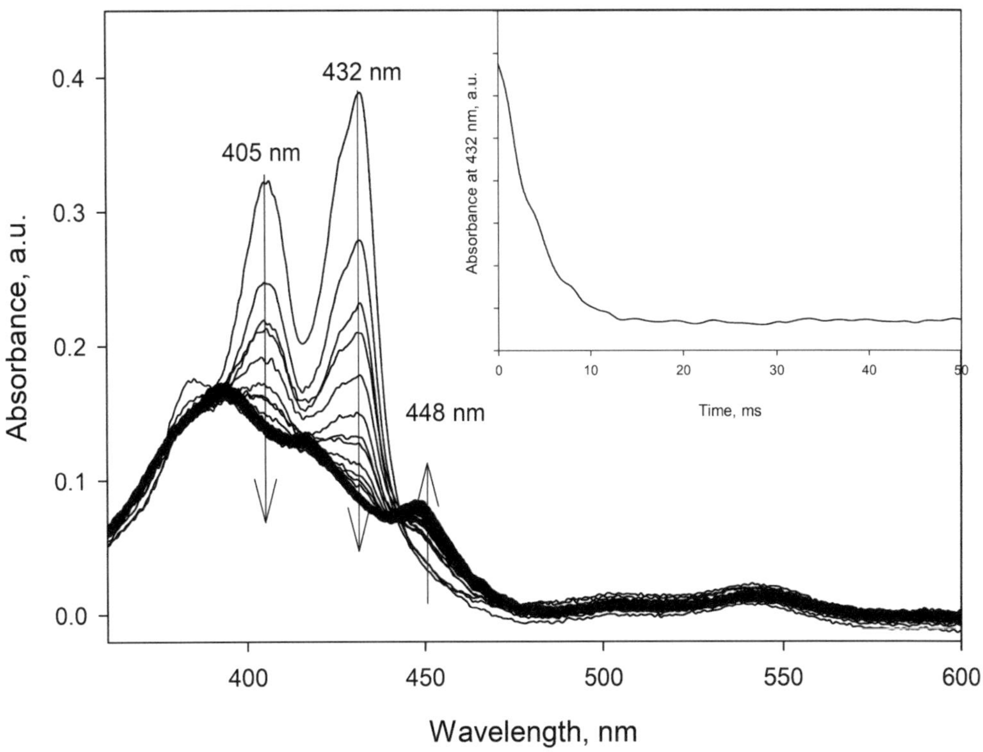

Fig. 6. Time-resolved absorption spectra of perylene 0.2 mmol L^{-1} in acetone with 10 % added DMPBI mixed with water. Inset shows the absorbance intensity at 432 nm as a function of time

Since the decrease in the intensity of the 432 nm band is used for nanoparticle formation rate calculation, the time resolved absorption spectra of DMPBI cannot be overlapping, and hence interfering, with that spectral region. To determine whether that is the case, Fig. 7 was plotted to visualize the temporal evolution absorption spectra obtained with the stopped-flow system for DMPBI (0.10 mmol L^{-1}) after reprecipitation in water. For reasons of simplicity, it is only plotted the spectra acquired during the first 200 ms at 1 ms intervals.

The absorption spectra for DMPBI, as seen in Fig. 7, is also strongly time dependent. Although a strong variation is noticed during the first tens of milliseconds after the mixing with water, a decrease in the area of the monomer band followed by an increase in the area

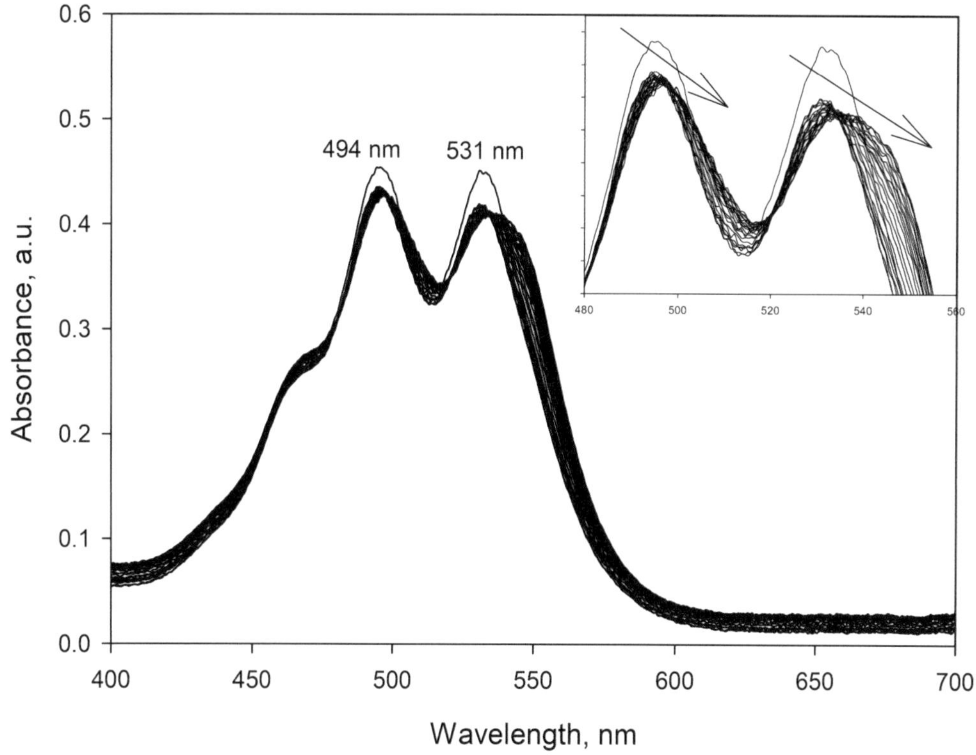

Fig. 7. Time-resolved absorption spectra of 0.1 mmol L^{-1} N,N'-bis (2,6-dimethylphenyl)-3,4,9,10 - perylenedicarboxyimide (DMPBI) mixed with water. Inset shows magnification of the absorbance profile from 480 to 560 nm

of the nanoparticle band is not as perceptible as in the case of perylene (Fig. 1), anthracene or benzo[a]pyrene (Fig. 3). Both absorption bands at 494 and 531 nm display similar behavior, that is, absorption monomer peaks are gradually disappearing and red shifting to a new DMPBI nanoparticle band. Calculation of DMPBI nanoparticle formation rate is an arduous task to accomplish because the decay in the intensity of absorbance is particularly low in magnitude. More importantly, the nanoparticle absorption band is created while overlapping with the monomer band, being those absorption maxima only a few nanometers apart. Consequently, kinetics decay curves necessary for the calculation of nanoparticle formation rates cannot be precisely isolated, and thus, obtained time decay curves are not suitable for accurate and reliable rate determination.

Observation of Fig. 7 clearly indicates that the DMPBI absorbance pattern does not interfere with perylene molecules absorbance (Fig. 1), especially at the 432 nm spectral range which is used for calculation of perylene nanocrystallization rates. As a result, rate constant calculation at 432 nm can be assigned solely to perylene molecules, and thus, the presence of DMPBI molecules in the reaction environment seems to accelerate rates of perylene nanocrystallization. In order to confirm if DMPBI is indeed responsible for the faster rate of perylene nanocrystal formation, a set consisting of seven solution of 0.20 mmol L^{-1} perylene in acetone with various amount of added DMPBI was prepared. Exact compositions of such solutions are displayed in Table 1. As listed, each solution has the same perylene concentration (0.20 mmol L^{-1}) but different DMPBI concentrations, ranging from 0.00 to 0.06 mmol L^{-1}.

	Concentration (mM)	
DMPBI %	Perylene	DMPBI
0%	0.200	0.000
1%	0.200	0.002
2%	0.200	0.004
5%	0.200	0.010
10%	0.200	0.020
20%	0.200	0.040
30%	0.200	0.060

Table 1. Composition of measured perylene solutions with added DMPBI percentage

Each solution displayed in Table 1 had its corresponding UV-visible absorption spectrum obtained with the stopped flow system after reprecipitation with water. Fig. 8 shows absorption spectra for the 7 solutions measured during the first 500 ms after mixing with water at 1 ms intervals; spectra A-F in the figure corresponds to solutions with 1 to 30% added DMPBI, respectively. The first solution shown on the table, i.e., 0% DMPBI, was not plotted on Fig. 8 as it is displayed in Fig. 1. It is readily observed from the decrease in the intensity of the 432 monomer peak that perylene undergoes nanocrystallization regardless of the DMPBI percentage in solution. Also, the intensity and shape of the 432 nm peak is clearly varying with addition of DMPBI. More importantly, inset in each graph indicates that perylene monomers are faster consumed in solution as the DMPBI concentration increases, acknowledged by steeper curve decays of the 432 monomeric peaks.

Intriguingly, when comparing not each individual absorbance spectrum for each solution as demonstrated on Fig. 8, but the initial and final absorbance of the perylene monomer peak, the 432 nm band, an interesting trend is immediately recognized. In Fig. 9, the 432 nm absorbance values at t = 0 ms (first measured absorbance value) and t = 500 ms (final acquired absorbance value) were plotted for all seven solutions of perylene with added DMPBI. It is clear that the intensity of the 432 nm peak decreases as the addition of DMPBI

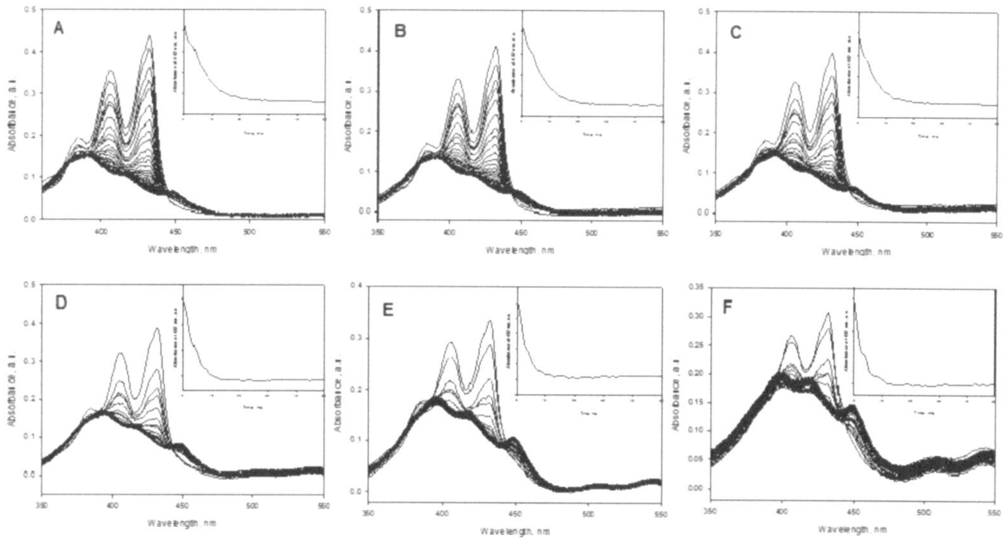

Fig. 8. Time-resolved absorption spectra of perylene in acetone after reprecipitation in water with 1%, 2%, 5%, 10%, 20%, and 30% added DMPBI (A - F, respectively). Refer to Table 1 for detailed information on the composition of solutions. Insets show absorbance decays for each added percentage of DMPBI at 432 nm (monomer consumption)

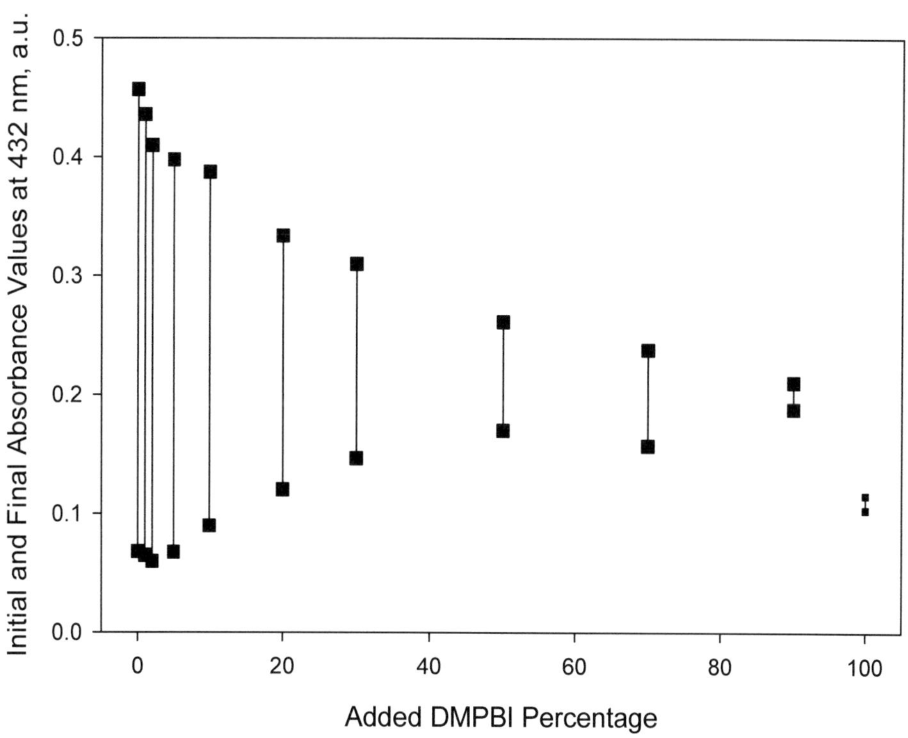

Fig. 9. Initial (t = 0 ms) and final (t = 500 ms) absorbance values at 432 nm for 0.2 mmol L^{-1} perylene in acetone mixed in water with 0%, 1%, 2%, 5%, 10%, 20%, and 30% added DMPBI

increases, more importantly, the initial value at 0 ms is progressive reaching lower absolute values as the DMPBI percentage raises from 0 to 30 percent. Because the concentration of perylene is fixed at 0.20 mmol L^{-1} for every solution, the initial measured absorbance value is expected to be constant throughout the addition of DMPBI, since as previously discussed in Fig. 7, there is no contribution in the 432 nm spectra range due to the presence of DMPBI. The fact that the initial absorbance value is constantly decreasing with rising additive concentration is a straightforward indication that DMPBI is causing the nanocrystallization process to occur at faster rates. As stated, the dead time of the stopped flow instrument is 4 ms, which means that the initial absorbance value is actually collected 4 ms after mixing the perylene/DMPBI solution with water. For that reason, the decline in initial absorbance values are believed to be due to reactions manifesting at faster nanocrystallization rates, for perylene monomer molecules are progressively consumed at accelerated rates during the 4 ms necessary for the stopped flow initial measurement.
As previously performed on Figs. 2, 4 and 5, rates of perylene nanocrystal formation were calculated using the 432 nm peak decay. Fig. 10 shows perylene nanocrystallization rates (A), as well as the mean crystal size for perylene nanoparticles (B), as a function of added DMPBI percentage. Crystal sizes plotted on the figure were determined with the aid of the dynamic light scattering (DLS) technique. Each displayed data point, on each graph, is the average of three individual determinations.

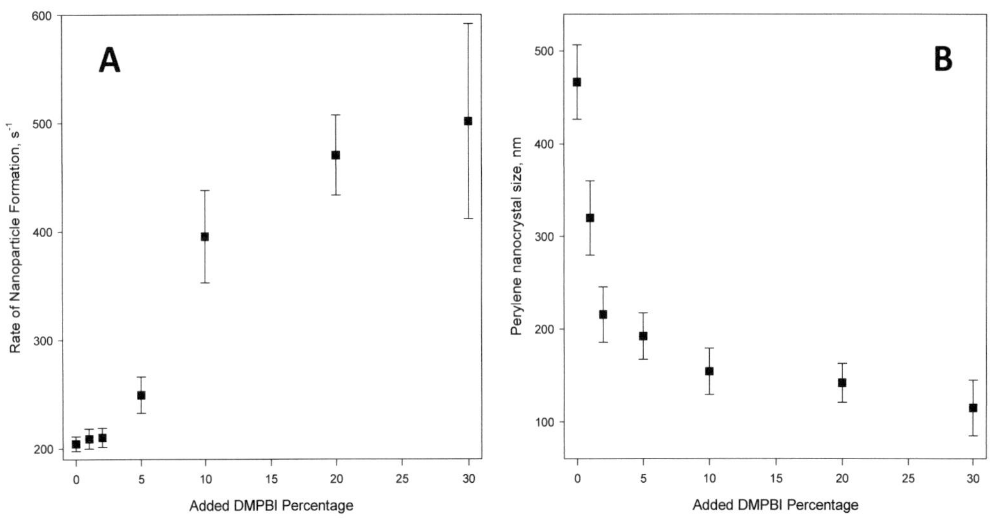

Fig. 10. (A) Rates of nanoparticle formation as a function of added DMPBI percentage for perylene 0.2 mmol L^{-1} , and (B) Perylene nanocrystals size measured with Dynamic Light Scattering (DLS) as a function of the percentage of added DMPBI.

Undoubtedly, the formation of perylene nanoparticles by the reprecipitation method is occurring at higher rates as the concentration of DMPBI in solution increases. It should be pointed out that in Fig. 8-F (30% DMPBI addition), the 432 nm perylene monomeric peak decay has a relative low intensity, leading to larger calculated errors on its formation rate, as illustrated in Fig. 10-A. In fact, standard deviations for calculated rates of perylene nanoparticle formation were not higher than 10% when the concentration of DMPBI was kept under 20 percent. However, at 30 percent DMPBI addition, standard deviation from three individual measurements was nearly 20%. The high errors associated with nanocrystallization rates at higher DMPBI percentages prompted experiments to be maintained at DMPBI concentrations not higher than 0.06 mmol L^{-1}, i.e., 30 percent.

The direct correlation between nanoparticle formation rates and nanoparticle sizes, makes it clear that addition of DMPBI to perylene solutions not only alters nanocrystallization rates, but is also an elegant way to manipulate sizes of perylene nanoparticles produced by the reprecipitation method, easily observed on Fig. 10-B. Comparing graphs A and B in Fig. 10, one notices that sizes of perylene nanoparticles are inversely proportional to added DMPBI concentration. When no DMPBI molecules are added to the system, perylene nanocrystals are reprecipitated with a mean size of 466.5 nm; in contrast, the addition of 30 percent DMPBI (refer to Table 1 for solution information) leads to a decrease in perylene mean particle size to about 115.1 nm.

It was suggested here that the kinetics of organic compounds'nanoparticle formation is in accordance with the classical nucleation theory. Nanoparticle formation rates of perylene, anthracene and benzo[a]pyrene (Figs. 2, 4, and 5) were calculated in solutions constituted of solvent and solute molecules only, going through what was previously named "homogeneous nucleation" (Kashchiev & van Rosmalens, 2003). In contrast, "heterogeneous nucleation"(Kashchiev & van Rosmalens, 2003) takes place in solutions having impurity molecules and/or foreign substrates that provide centers for nucleation to occur. In accordance, the introduction of DMPBI molecules, a foreign particle different from the nucleating crystalline phase, is sought to offer active centers for the heterogeneous nucleation of perylene nanoparticle to occur during the reprecipitation method.

It was formerly showed that at a given supersaturation level, homogeneous nucleation take place at lower rates than heterogeneous nucleation (Sear, 2007) since the presence of foreign substrates can act as nucleation-active centers which in turn could facilitate and thus accelerate the nucleation process. At higher nucleation rates, larger number of cluster (nuclei) are formed in solution and consequently there are less dissolved monomers available in solution for particle growth, ultimately leading to smaller sized particle.

Indeed Fig. 10 seems to indicates that DMPBI molecules indeed play the role of nucleation-active centers; indicated by the increase in nanocrystallization rates (Fig. 10-A), and consequently to the decrease in nanoparticle sizes (Fig. 10-B). Thus, it is clear that the addition of DMPBI to an organic compound solution in order to alter nanocrystallization rates is an effective method to manipulate sizes of organic nanocrystals produced by the reprecipitation method (Oliveira et al., 2010). An appropriate analogy for the role of DMPBI in organic nanoparticle formation is the extensive use of seeds in the fabrication of metallic nanoparticles, in which small metal particles are initially prepared to be later employed as "seeds" for the fabrication of larger size particle. Such process is commonly referred to "seeding growth method" (Kan et al., 2003; Yong et al., 2006), where the desired nanoparticle size can be successfully controlled by varying the ratio of seed to metal salt. In comparison, the size of prepared organic nanoparticles by the reprecipitation method seems to be controlled by simply altering the organic nanoparticle to foreign substrate (DMPBI) ratio.

To confirm nanoparticle sizes determined by DLS on Fig. 10-B, and to investigate the morphological difference in perylene nanocrystals with the addition of DMPBI, scanning electron microscopy (SEM) images were obtained for nanocrystals prepared using the stopped flow apparatus in both the presence and absence of DMPBI molecules. Fig. 11

displays SEM images of perylene nanocrystals without the addition of the organic additive, as well as with the addition of 1, 5, and 30 percent DMPBI (A-D respectively). Crystal sizes observed by SEM are similar to the average sizes obtained by DLS measurements, suggesting no association between dispersed perylene nanoparticles. Furthermore, the increasing addition of DMPBI to perylene solutions not only altered its nanocrystal size, but also modified its morphology. Spherical and cylindrical shaped nanocrystals could be seen even at 5 % DMPBI addition (Fig. 11-C), ultimately becoming the prominent shape at 30 % organic additive addition (Fig. 11-D).

Fig. 11. SEM images of perylene nanocrystals fabricated by the reprecipitation method using the stopped-flow apparatus. (A) No addition of DMPBI, (B) 1% added DMPBI, (C) 5% added DMPBI, and (D) 30 % added DMPBI . Scale bars represent 200 nm.

3. Conclusion

Collectively, it has been demonstrated that the stopped-flow UV-Visible absorption spectroscopy technique is a promising technique for studying the *in situ* formation of organic nanocrystals by the reprecipitation method within a time frame of few milliseconds. As the vast majority of academic and industry research on the mechanism of nanosized crystal formation is focused on inorganic compounds, it was valuable to address that the stopped-flow method can be used as a tool for studying the mechanism and kinetics of organic nanoparticle formation.

The results obtained suggested that the kinetics of the organic nanomaterial formation process by the reprecipitation method qualitatively obeys the classical nucleation theory. By correlating nanocrystals growth with a well established theory, the work presented here can have a meaningful influence on future developments regarding organic nanoparticle production.

Moreover, it was clearly elucidated that it is possible to manipulate sizes of organic nanomaterials produced by the reprecipitation method by the addition of a foreign organic substrate, namely, N,N'-bis(2,6-dimethylphenyl)-3,4,9,10–perylenedicarboxyimide (DMPBI). While providing active centers for nucleation, foreign particles facilitate the nucleation process, leading to considerable acceleration in the nanocrystallization process, and ultimately to smaller sized particles.

Having specific knowledge on the mechanism of organic nanosized material production while being able to precisely and accurately control its size, material scientists can directly contribute to advancements on society by providing novel methods directed to the production of high demanded nanomaterials. For instance, high performance organic pigments and dyes can be successfully produced with controllable sizes through methods described here. Such materials are extremely industry-attractive for its use on textiles, printing inks, paint and coatings, and plastic.

4. Acknowledgment

Authors acknowledge the New Energy and Industrial Technology Development Organization (NEDO) of the Japanese government for partially supporting this work.

5. References

An, B. K.; Kwon, S. K.; Jung, S. D. & Park, S. Y. (2002). Enhanced Emission and Its Switching in Fluorescent Organic Nanoparticles. *J. Am. Chem. Soc.*, Vol. 124, No. 48, (December 2002), pp. 14410-14415, ISSN 0002-7863

Auweter, H.; Haberkorn, H.; Heckmann, W.; Horn, D.; Luddecke, E.; Rieger, J. & Weiss, H. (1999) Supramolecular Structure of Precipitated Nanosize β-Carotene Particles. *Angew. Chem. Int. Ed.*, Vol. 38, No.15, (August 2009), pp. 2188-2191, ISSN 1433-7851

Baba, K.; Kasai, H.; Masuhara, A.; Okada, S.; Oikawa, H. & Nakanishi, H. (2007). Diacetylene Nanowire Crystals Prepared by Reprecipitation/Microwave-Irradiation Method. *Japanese Journal of Applied Physics*, Vol. 46, No. 11, (November 2007), pp. 7558-7561, ISSN 0021-4922

Bilgili, E.; Yepes, J. & Scarlett, B. (2006). Nano-milling of pigment agglomerates using a wet stirred media mill: Elucidation of the kinetics and breakage mechanisms. *Chemical Engineering Science*, Vol. 61, No. 1, (January 2006), pp. 149-157, ISSN 0009-2509

Blyshak, L. A.; Dodson, K.Y.; Patonay, G.; Warner, I.M. & May, W.E. (1989). Determination of cyclodextrin formation constants using dynamic coupled-column liquid chromatography. *Anal. Chem.*, Vol. 61, No. 9, (May 1989), pp. 955-960 , ISSN 0003-2700

Burda, C.; Chen, X.; Narayanan, R. & El-Sayed, M.A. (2005). Chemistry and Properties of Nanocrystals of Different Shapes. *Chemical Reviews*, Vol. 105, No. 4, (March 2005), pp. 1025-1102, ISSN 0009-2665

Chaikin, P. M. & Lubensky, T.C. (1995). *Principles of Condensed Matter Physics* (1st Ed.), Cambridge University Press, ISBN 0 521 43224 3, Cambridge, UK

Chen, W. & Zhang, J. (2006). Using Nanoparticles to Enable Simultaneous Radiation and Photodynamic Therapies for Cancer Treatment. *Journal of Nanoscience and Nanotechnology*, Vol. 6, No. 4, (April 2006), pp. 1159-1166, ISSN 1550-7033

Chung, H.R.; Kwon, E.; Oikawa, H.; Kasai, H. & Nakanishi, H. (2006). Effect of solvent on organic nanocrystal growth using the reprecipitation method. *Journal of Crystal Growth*, Vol. 294, No. 2, (September 2006), pp. 459-463, ISSN 0022-0248

Debenedetti, P. G. (1996). *Metastable Liquids* (1st Ed.), Princeton University Press, ISBN 0691085951, Princeton, NJ, USA

Eisenbrand, J. & Baumann, K. (1970). Über die bestimmung der wasserlöslichkeit von coronen, fluoranthen, perylen, picen, tetracen und triphenylen und über die bildung wasserlöslicher komplexe dieser kohlenwasserstoffe mit coffein. *European Food Research and Technology*, Vol. 144, No. 5, (April 1970), pp. 312-317, ISSN 1438-2377

Frendler, J.H. (1987). Atomic and molecular clusters in membrane mimetic chemistry. *Chemical Reviews*, Vol. 87, No. 5, (October 1987), pp. 877-899, ISSN 0009-2665

Fu, H. B. & Yao, J.N. (2001). Size Effects on the Optical Properties of Organic Nanoparticles. *J. Am. Chem. Soc.*, Vol. 123, No. 7, (February 2001), pp. 1434-1439, ISSN 0009-2665

Hayashi, K.; Morii, H.; Iwasaki, K.; Horie, S.; Horiishi, N. & Ichimura, K. J. (2007). Uniformed nano-downsizing of organic pigments through core–shell structuring. *Journal of Materials Chemistry*, Vol. 17, No. 6, (February 2007), pp. 527-530, ISSN 0959-9428

Horn, D. & Rieger, J. (2001). Organic Nanoparticles in the Aqueous Phase-Theory, Experiment, and Use. *Angew. Chem. Int. Ed.*, Vol. 40, No. 23, (December 2001), pp. 4330-4361, ISSN 1433-7851

Ibanez,A.; Maximov, S.; Guiu, A.; Chaillout, C. & Baldeck, P.L. (1998). Controlled Nanocrystallization of Organic Molecules in Sol-Gel Glasses. *Advanced Materials*, Vol. 10, No. 18, (December 1998), pp. 1540-1543, ISSN 0935-9648

Kan, S. H; Mokari, T.; Rothenberg, E. & Banin, U. (2003). Synthesis and size-dependent properties of zinc-blende semiconductor quantum rods. *Nature Materials*, Vol. 2, No. 3, (March 2003), pp. 155-158, ISSN 1476-1122

Kang, L.; Wang, Z.; Cao, Z.; Ma, Y.; Fu, H. & Yao, J. (2007). Colloid Chemical Reaction Route to the Preparation of Nearly Monodispersed Perylene Nanoparticles: Size-Tunable Synthesis and Three-Dimensional Self-Organization. *J. Am. Chem. Soc.*, Vol. 129, No. 23, (June 2007), pp. 7305-7312, ISSN 0009-2665

Kasai, H.; Nalwa, H.S.; Oikawa, H.; Okada, S.; Matsuda, H.; Minami, N.; Kakuta, A.; Ono, K.; Mukoh, A. & Nakanishi, H. (1992). A Novel Preparation Method of Organic Microcrystals. *Japanese Journal of Applied Physics*, Vol. 31, No. 8, (August 1992), pp. L1132-L1134, ISSN 0021-4922

Kasai, H.; Kamatani, H.; Okada, S.; Oikawa, H.; Matsuda, H. & Nakanishi, H. (1996). Size-Dependent Colors and Luminescences of Organic Microcrystals. *Japanese Journal of Applied Physics*, Vol. 35, No. 2, (February 1996), pp. L221-L223, ISSN 0021-4922

Kasai, H.; Oikawa, H.; Okada, S. & Nakanishi, H. (1998). Crystal Growth of Perylene Microcrystals in the Reprecipitation Method. *Bulletin of the Chemical Society of Japan*, Vol. 71, No. 11, (June 1998) 2597-2601, ISSN 0009-2673

Kashchiev, D. & van Rosmalens, G. M. Review: Nucleation in solutions revisited. *Crystal Research and Technology*, Vol. 38, No. 7-8, (July 2003), pp. 555-574, ISSN 0323-1300

Katagi, H.; Kasai, H.; Okada, S.; Oikawa, H.; Komatsu, K.; Matsuda, H.; Liu, Z. & Nakanishi, H. (1996). Size Control of Polydiacetylene Microcrystals. *Japanese Journal of Applied Physics*, Vol. 35, No. 10B, (October 1996), pp. L1364-L1366, ISSN 0021-4922

Laaksonen, A.; Talenquer, V. & Oxtoby, D. W. (1995). Nucleation: Measurements, Theory, and Atmospheric Applications. *Annual Review of Physical Chemistry*, Vol. 46, (October 1995), pp. 489-524, ISSN 0066-426

Magdassi, S. & Moshe, M. B. (2003). Patterning of Organic Nanoparticles by Ink-jet Printing of Microemulsions. *Langmuir*, Vol. 19, No. 3, (February 2003), pp. 939-942, ISSN 0743-7033

Masuhara, H.; Nakanishi, H. & Sasaki, K. (2003). *Single Organic Nanoparticles* (1st Ed.), Springer, ISBN 3-540-00187-5, New York, USA

McWilliams, A. (July 2010). Nanotechnology: A Realistic Market Assessment. *BBC Research*, 01.07.2010, Available from http://www.bccresearch.com/report/NAN031D.html

Miyashita, Y.; Baba, K.; Kasai, H.; Nakanishi, H. & Miyashita, T. (2008). A New Production Process of Organic Pigment Nanocrystals. *Molecular Crystals & Liquid Crystals*, Vol. 492, (October 2008), pp. 268-274, ISSN 1542-1406

Mori, J.; Miyashita, Y.; Oliveira, D.; Kasai, H.; Oikawa, H. & Nakanishi, H. (2009). Stopped-flow analysis on the mechanism of perylene nanoparticle formation by the reprecipitation method. *Journal of Crystal Growth*, Vol. 311, No. 3, (February 2009), pp. 553-555, ISSN 0022-0248

Nyssen, G. A.; Miller, E. T.; Glass, T. F. & Quinn II, C. R. (1987). Solubilities of hydrophobic compounds in aqueous-organic solvent mixtures. *Environmental Monitoring and Assessmet*, Vol. 9, No.1 , (July 1987), pp. 1-11, ISSN 0167-6369

Oliveira, D.; Baba, K.; Mori, J.; Miyashita, Y.; Kasai, H.; Oikawa, H. & Nakanishi, H. (2009). Nanocrystallization Mechanism of Organic Compounds in the Reprecipitation Method by Stopped-Flow Analysis. *Japanese Journal of Applied Physics*, Vol. 48, No. 10, (October 2009), pp. 105003-1-105003-5, ISSN 0021-4922

Oliveira, D.; Baba, K.; Mori, J.; Miyashita, Y.; Kasai, H.; Oikawa, H. & Nakanishi, H. (2010). Using an organic additive to manipulate sizes of perylene nanoparticles. *Journal of Crystal Growth*, Vol. 312, No. 3, (January 2010), pp. 431-436, ISSN 0022-0248

Peters, D. (1996). Ultrasound in materials chemistry. *Journal of Materials Chemistry*, Vol. 6, No. 10, (October 1996), pp. 1605-1618, ISSN 0959-9428

Sear, R. P. (2007). Nucleation: theory and applications to protein solutions and colloidal suspensions. *Journal of Physics: Condensed Matter*, Vol. 19, No. 3, (January 2007), pp. 033101-033129, ISSN 0953-8984

Spatz, J. P.; Herzog, T.; Mossmer, S.; Ziemann, P. & Moller, M. (1999). Micellar Inorganic–Polymer Hybrid Systems - A Tool for Nanolithography. *Advanced Materials*, Vol. 11, No. 2, (March 1999), pp. 149-153, ISSN 0935-9648

Sugimoto, T. (2000). *Fine Particles: Synthesis, Characterization, and Mechanisms of Growth* (1st Ed.), Mercel Dekker, ISBN 978-0824700010, New York, USA

Tiemann M.; Weiss, O.; Hartikainen, J.; Marlow, F. & Linden, M. (2005). Early Stages of ZnS Nanoparticle Growth Studied by In-Situ Stopped-Flow UV Absorption Spectroscopy. *ChemPhysChem*, Vol. 6, No. 10, (February 2005), pp. 2113-2119, ISSN 1439-4235

Tiemann, M.; Marlow, F.; Brieler, F. & Linden, M. (2006). Early Stages of ZnS Growth Studied by Stopped-Flow UV Absorption Spectroscopy: Effects of Educt Concentrations on the Nanoparticle Formation. *J. Phys. Chem. B*, Vol. 110, No. 46, (November 2006), pp. 23142-23147, ISSN 1089-5647

Tiemann, M.; Marlow, F.; Hartikainen, J.; Weiss, O. & Linden, M. (2008). Ripening Effects in ZnS Nanoparticle Growth. *J. Phys. Chem. C*, Vol. 112, No. 5, (February 2008), pp. 1463-1467 , ISSN 1932-7447

Tonomura, B. I.; Nakatani, H.; Ohnishi, M.; Yamaguchi-Ito, J. & Hiromi, K. (1978). Test reactions for a stopped-flow apparatus : Reduction of 2,6-dichlorophenolindophenol and potassium ferricyanide by L-ascorbic acid, *Analytical Biochemistry*, Vol. 84, No. 2, (February 1978), pp. 370-383, ISSN 0003-2697

Van Keuren, E.; Georgieva, E. & Adrian, J. (2001). Kinetics of the Formation of Organic Molecular Nanocrystals. *Nano Letters*, Vol. 1, No. 3, (March 2001), pp. 141-144, ISSN 1530-6984

Van Keuren, E.; Bone, A. & Ma, C. (2008). Phthalocyanine Nanoparticle Formation in Supersaturated Solutions, Langmuir, Vol. 24, No. 12, (June 2008), pp. 6079-6084, ISSN 0743-7463

Vogel, A. I.; Tatchell, A. R.; Furnis, B. S.; Hannaford, A. J. & Smith, P. W. G. (1996). *Vogel's Textbook of Practical Organic Chemistry* (5th Ed.), Prentice Hall, ISBN 0582462363, London, UK

Wang, F.; Han, M.Y.; Mya, K. Y.; Wang, Y. & Lai, Y.H. (2005). Aggregation-Driven Growth of Size-Tunable Organic Nanoparticles Using Electronically Altered Conjugated Polymers. *J. Am. Chem. Soc.*, Vol. 127, No. 29, (July 2005), pp. 10350-10355, ISSN 0002-7863

Yong, K.; Sahoo, Y.; Swihart, M. T. & Prasad, P. N. (2006). Growth of CdSe Quantum Rods and Multipods Seeded by Noble-Metal Nanoparticles. *Adv. Mater.*, Vol. 18, No. 15, (August 2006), pp. 1978-1982, ISSN 0935-9648

7

Self-assembly and Patterning of Nanocrystals

Yoshitake Masuda
National Institute of Advanced Industrial Science and Technology (AIST)
Anagahora, Shimoshidami, Moriyama-ku, Nagoya
Japan

1. Introduction

Nano/micro periodic structures have attracted much attention as next-generation devices[1,2] such as photonic crystals[3-5] in which the refractive index changes periodically to show a photonic band gap. Various scientific and engineering applications, such as control of spontaneous emission, zero-threshold lasing, sharp bending of light, and so on, are expected to become possible by using the photonic band gap and the artificially introduced defect states and/or light-emitters. The structures were prepared by semiconductor nanofabrication techniques such as lithography and etching processes[6,7], advanced wafer-fusion technique[8], lithographic layer-by-layer approach[9], holographic lithography[10], advanced silicon microelectromechanical systems[11], glancing angle deposition[12] or auto cloning technique[13], and theoretical studies were performed to estimate the properties of the structures. These studies confirmed the high potential of nano/micro periodic structures as future devices.

However, a simple process which requires a short time for fabrication, low energy and less amount of material needs to be developed to enable mass production. Additionally, the processes of patterning the structures need to integrate various elements for application to commercial devices. The regularity and feature edge acuity of periodic structures should also be improved in order to enhance the performance.

Nano/micro periodic structures can be prepared with short fabrication time and low energy by self-assembly of mono-dispersed particles in which particles and air (wall and air for inverse opal) are arranged periodically[14,15]. Self-assembly and patterning of nano/micro particles have attracted much attention recently[1,2,16,17]. Micropatterns of close-packed particle assemblies with high arrangement accuracy have been realized by using templates such as micromolds[18], grooves[19], cylindrical holes[20] or trenches[20]. However, substrates having micromolds or grooves are necessary in these processes, and feature edge acuity and regularity need to be improved further in order to fabricate various complicated structures for photonic devices.

In this section, nano/micropatterns of colloidal crystals were fabricated self-assembly. They were realized with several patterning methods in static solution systems[21-23] or in drying processes[24,25] without using molds or grooves. Microstructures constructed from particles such as micropatterns of particle layers, narrow particle wires, arrays of particle wires and so on were prepared under moderate conditions using self-assembled monolayers (SAMs). Additionally, 2D patterning of colloidal crystals were realized with two solution method.

The method has the advantage of both the static solution system and drying process. It offers excellent self-assembling performance for the fabrication of nano/micro periodic structures to be applied to next-generation devices.

2. Patterning of colloidal crystals in liquids[21-23]

1.1 Patterning of colloidal crystals in liquids

Novel processes to realize low-dimensional arrangement of SiO_2 particles were proposed[21-23]. Particle wires and a pattern of the close-packed particle monolayer were fabricated in the solution at room temperature. SAMs (self-assembled monolayers) were formed on Si substrates and modified to be suitable for templates of precise arrangement. Particles were arranged precisely in the desired positions in the solution using well controlled electrostatic interactions and chemical bond formation between particles and substrates.

1.2 SAM Preparation for patterning of coloidal crystals

Octadecyltrichlorosilane (OTS)-SAM was prepared by immersing the Si substrate in an anhydrous toluene solution containing 1 vol% OTS for 5 min under a N_2 atmosphere. SAMs were exposed for 2 h to UV light (184.9 nm) through a photomask. The UV-irradiated regions became hydrophilic due to Si-OH group formation, while the non-irradiated part remained unchanged, i.e., it was composed of hydrophobic octadecyl groups, which gave rise to patterned OTS-SAM. To check successful film formation and functional group change, water drop contact angles were measured for irradiated and non-irradiated surfaces. Initially deposited OTS-SAM had a water contact angle of 96°, while the UV-irradiated SAM surface was saturated (contact angle $< 5°$).

The patterned OTS-SAM was immersed in a toluene solution containing 1 vol% APTS (aminopropyltriethoxysilane) in air for 1 h. APTS molecules combined to silanol groups of SAM and hence, octadecyl / amino groups pattered SAM was fabricated. OTS-SAM exhibited a water contact angle of 96°, while that of the amino surface was 28°. These observations indicated successful fabrication of octadecyl / amino-groups patterned SAM.

1.3 Surface modification of SiO_2 particles

Silica particles (1μmφ, HIPRESICA UF, UNK, Ltd.) were immersed in a dicyclohexyl and sonicated for 10 min under a N_2 atmosphere for good dispersion. 1 vol% of trichlorocyanoethylsilane (TCES) was added to the dicyclohexyl solution under a N_2 atmosphere, and the solution was stirred gently for 30 min in order to chemisorb TCES onto the SiO_2 particle surfaces. SiO_2 particles with TCES were centrifuged several times to remove unreacted TCES using dicyclohexyl.

The SiO_2 particles with TCES were further dispersed in a tetrahydrofuran solution containing potassium tert-butoxide (t-BuOK) and 18-crown 6-ether for 48 h under an ambient atmosphere to oxidize the CN-groups to carboxyl groups. The solution was centrifuged several times using distilled water to remove t-BuOK, 18-crown 6-ether, and a tetrahydrofuran. SiO_2 particles modified with carboxyl groups were thus obtained.

SiO_2 particles covered by silanol groups or carboxyl groups were arranged selectively in silanol regions or amino regions of SAM using interactions between particles and SAMs. Zeta potentials of SiO_2 particles that have silanol groups and SiO_2 particles modified by carboxyl groups were measured (Zetasizer 3000HSA, Malvern Instruments Ltd.) as shown in Fig. 1. Zeta potentials measured in aqueous solutions (pH=7.0) for the surface of silicon substrate covered with silanol groups, phenyl groups (PTCS) and amino groups (APTS) are -38.23 mV, +0.63 mV and +22.0 mV, respectively.

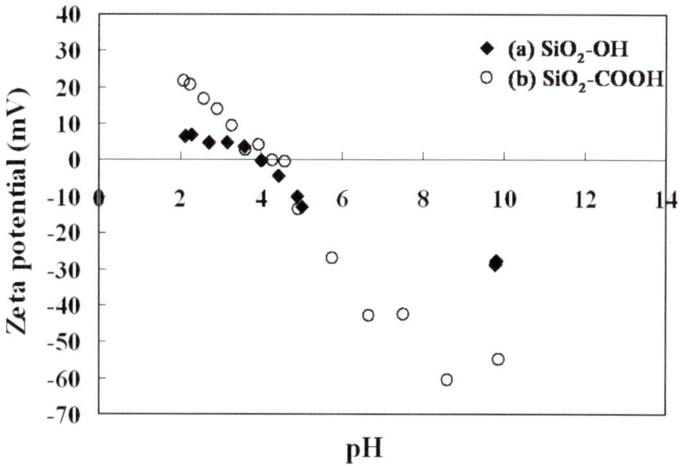

Fig. 1. Zeta potential of (a) SiO_2 particles and (b) SiO_2 particles modified with carboxyl groups.
Reprinted with permission from Ref. [23], Masuda, Y., Itoh, M., Yonezawa, T. and Koumoto, K., 2002, *Langmuir*, 18, 4155. Copyright @ American Chemical Society

1.4. Fabrication of particle wires employing selective arrangement process

OTS-SAM was modified by a diamond tip to form a line of silanol groups of approximately 0.5 µm width (Fig. 2). The diamond tip was contacted to OTS-SAM surface lightly and traced with low contact pressure in order to modify the SAM surface. The surface modified by a diamond tip, i.e., the white area in Figure 2 corresponds to silanol groups, showed low contact angle (<5°). This modified region was shown to be white compared with OTS-SAM region in a scanning electron micrograph (SEM; S-3000N, Hitachi, Ltd.). Octadecyl groups were broken mechanically by contact pressure with the diamond tip, and they possibly changed into silanol groups. The diamond tip was used to avoid contamination from a metal tip and the influence of a chemical reaction between the tip and the SAM.

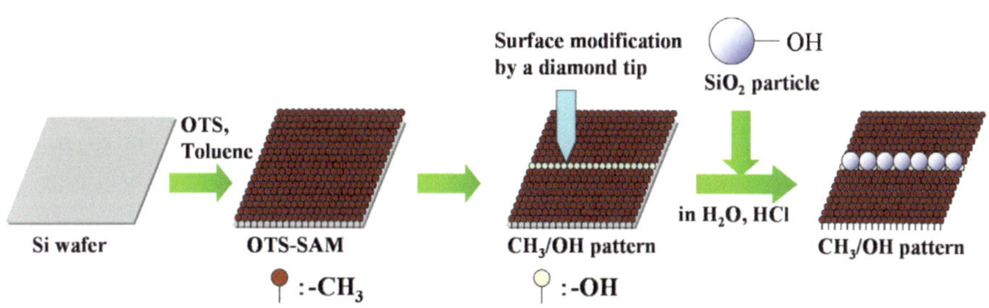

Fig. 2. Conceptual process for fabrication of a particle wire on a patterned SAM modified by a diamond tip.Reprinted with permission from Ref. [23], Masuda, Y., Itoh, M., Yonezawa, T. and Koumoto, K., 2002, *Langmuir*, 18, 4155. Copyright @ American Chemical Society

Patterned SAMs were immersed in the aqueous solution containing SiO_2 particles and a hydrochloric acid as a condenser, rinsed in water, and were observed by a SEM. SiO_2 particles were observed on lines of silanol groups selectively indicating particles were successfully arranged well (Fig. 3(a)). Because particles were not easily removed by sonication, it was judged that iloxane bonds had been formed by condensation of silanol groups between particles and a SAM.

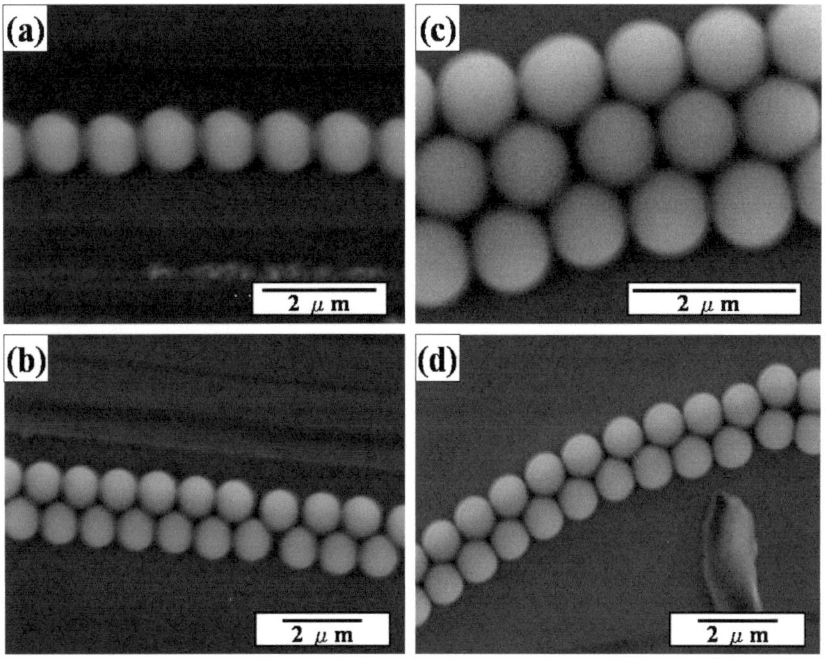

Fig. 3. SEM micrographs of (a) a single particle wire, (b) a double particle wire, (c) a triple particle wire and (d) a curved double particle wire
Reprinted with permission from Ref. [23], Masuda, Y., Itoh, M., Yonezawa, T. and Koumoto, K., 2002, *Langmuir*, 18, 4155. Copyright @ American Chemical Society

It is clearly seen that the accuracy of particle arrangement has been improved compared to our former experiments[21,22]. A double particle wire and a triple particle wire were likewise fabricated on wide silanol groups regions with about 1.4 μm and 2.2 μm in width, respectively (Fig. 3(b), (c)). The double particle wire that has a triangular lattice also demonstrates a high arrangement accuracy, though there is a defect in arrangement between seventh particle from left and eighth particle. Additionally, a curved double particle wire was fabricated on curved region of silanol groups. Curved double particle wires have not been reported previously, and they may have useful applications for an optical waveguide.
Accuracy of particle arrangement was evaluated from Fig. 3(a). Center position ((x_i, y_i) μm) of each particle was plotted to estimate the standard deviation (Fig. 4). The bottom left corner of Fig. 3(a) was set to be the origin of the x-y coordinate. The approximated straight line($f(x)$) and its slope (θ) are represented as follows.

$$f(x) = 0.0061x + 2.9609\,, \tag{1}$$

$$\cos\theta = 0.9940 \, , \tag{2}$$

Standard deviation from the approximated straight line is described by the expression,

$$\text{S (standard deviation)} = \frac{\left[\sum_i \left\{\cos\theta \cdot \left(f(x_i) - y_i\right)\right\}^2\right]^{1/2}}{n-1} \, , \tag{3}$$

where n is the number of particles ($n = 6$). S = 0.0126 was obtained. The accuracy of particle arrangement in Fig. 3(b) and (c) was estimated by the same manner. Standard deviation of seven particles from left in an upper particle line and an bottom particle line in Fig. 3(b) were estimated to be S = 5.66×10^{-3} and S = 3.84×10^{-3}. And standard deviation of an upper particle line, a middle particle line and a bottom particle line in Fig. 3(c) were estimated to be S = 8.11×10^{-4}, 8.27×10^{-3} and 2.30×10^{-2}, respectively.

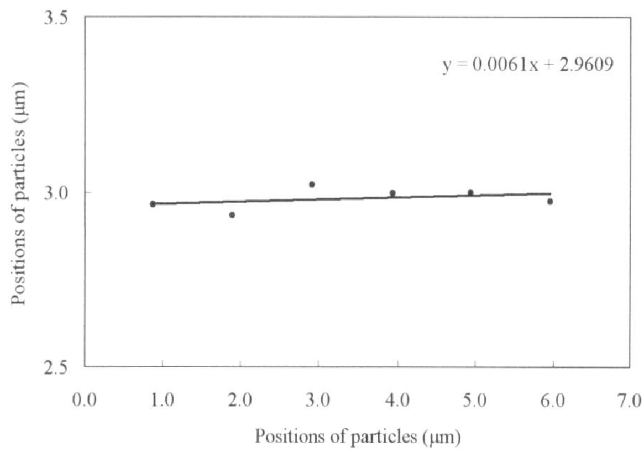

Fig. 4. Positions of particles in Fig. 3 (a) showing the accuracy of particle arrangement. Reprinted with permission from Ref. [23], Masuda, Y., Itoh, M., Yonezawa, T. and Koumoto, K., 2002, *Langmuir*, 18, 4155. Copyright @ American Chemical Society

1.5 Precise arrangement of particles on small-area silanol sites modified by AFM lithograph

OTS-SAM was modified to silanol groups by an AFM (atomic force microscope, Nanoscope E, Digital Instruments) to control position of arrangement accurately (Fig. 5). A source measure unit (SMU Model 236, Keithley) was installed in the AFM in order to control the electric current passing through the probe and a SAM. The SAM was biased positively, and the AFM probe was scanned with constant current mode (50 nA), and the scanned area was used as a template for arrangement. Scanning area (100 nmx100 nm) was set smaller than the diameter of the particles (500 nmφ) to facilitate precise arrangement of particles.

SiO_2 particles (500 nmφ, powder, Admatechs Co., Ltd., SO-E2) modified with carboxyl groups were sonicated for 10 min in tetrahydrofuran or dichloromethane, and this solution was refrigerated to -20 ℃ for 1 h. N,N'-dicyclohexylcarbodiimide was added to this solution as a condenser to form ester bonds between carboxyl groups of SiO_2 particles and silanol groups of a SAM. Modified OTS-SAM was then immersed in this solution for 2 h. The temperature of the solution was increased slowly to 25 ℃ and kept for 2 h. After having

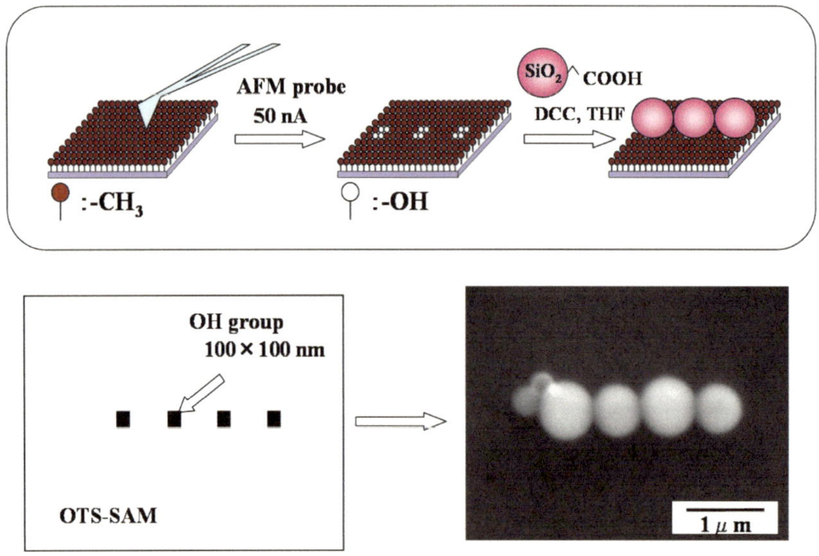

Fig. 5. Conceptual process and SEM micrograph of particle arrangement on a patterned SAM modified by AFM lithography.
Reprinted with permission from Ref. [23], Masuda, Y., Itoh, M., Yonezawa, T. and Koumoto, K., 2002, *Langmuir*, 18, 4155. Copyright @ American Chemical Society

been rinsed in water, a SAM was observed by a SEM. Particles were arranged in silanol regions and line of particle was fabricated (Fig. 5). Two-dimensional arrangement with required features can easily be realized with this technique, though it takes a long time to modify a SAM with an AFM probe. Particles weren't removed easily from a SAM by sonication, indicating that ester bonds were formed by condensation. The accuracy of particle arrangement in Fig. 5 was estimated to be $S = 1.17 \times 10^{-2}$. This might be decreased by decreasing the dimension of each silanol region.

In order to verify the formation of ester bonds between carboxyl groups and silanol groups, bromopropionic acid, whose molecule has a carboxyl group at one end and a bromo group at the other, was reacted with silanol groups of a Si substrate using the same reaction scheme as used to attach SiO_2 particles to silanol groups. After having been sonicated in acetone for 5 min, the substrate surface was analyzed by X-ray photoelectron spectroscopy (XPS; ESCALAB 210, VG Scientific Ltd., $1\text{-}3 \times 10^{-7}$ Pa, measurement area; 3 mmx4 mm). The X-ray source (MgK*a*, 1253.6 eV) was operated at 15 kV and 18 mA. The spectrum corresponding to Br 3d binding energy centering at 74.35 eV was observed. Although the observed binding energy is higher than that of KBr, this chemical shift must have been caused by carbon atoms neighboring bromine atoms. Since bromo groups can't react directly with silanol groups under the present conditions, the XPS result firmly indicates that carboxyl groups of bromopropionic acid reacted with silanol groups to form possible ester bonds.

1.6 Patterning of close-packed particle monolayers

SiO_2 particles modified with carboxyl groups were dispersed in water. Octadecyl / amino-groups patterned SAM was then immersed in the solution for several minutes (Fig. 6). The substrate was rinsed with water and observed with a SEM. SiO_2 particles were observed in silanol regions selectively forming a close-packed mono-particle layer (Fig. 7 (a)). Boundaries between the mono-particle layer and otcadecyl region is clearly observable, and a few particles are observed in octadecyl region. SiO_2 particles modified with carboxyl groups are charged negative, and amino groups of SAM are charged positive in water. Accordingly, particles are attracted to amino groups and form a mono-particle layer. Particles in the solution did not adhere to the mono-particle layer, since both the particles and the mono-particle layer have negative charges and repel each other. Particles were also deposited randomly in some areas (Fig. 7 (b)), and this suggests that it is difficult to obtain the pattern of the close-packed particle monolayer in a large area.

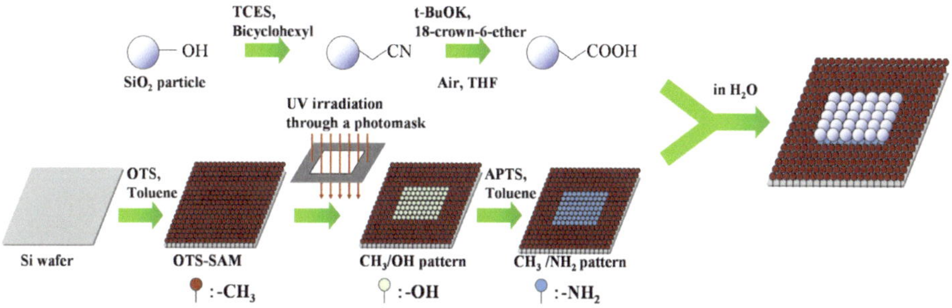

Fig. 6. Conceptual process for patterning of the close-packed particle monolayer.
Reprinted with permission from Ref. [23], Masuda, Y., Itoh, M., Yonezawa, T. and Koumoto, K., 2002, *Langmuir*, 18, 4155. Copyright @ American Chemical Society

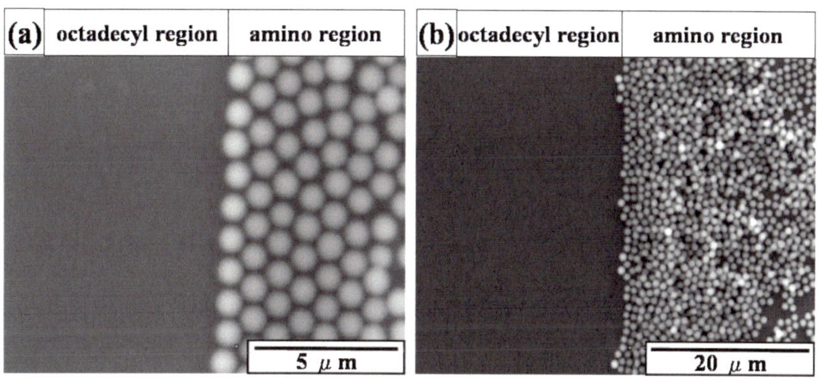

Fig. 7. SEM micrographs of (a) a close-packed particle monolayer of SiO_2 particles modified with carboxyl groups formed in the amino region and (b) randomly deposited particles in the amino region.
Reprinted with permission from Ref. [23], Masuda, Y., Itoh, M., Yonezawa, T. and Koumoto, K., 2002, *Langmuir*, 18, 4155. Copyright @ American Chemical Society

Additionally, octadecyl / amino-groups patterned SAM was immersed in the solution containing non-modified SiO_2 particles for several minutes (Fig. 8). While SiO_2 particles were observed in silanol region predominantly, the feature edge acuity of the pattern was lower than that of the pattern in which SiO_2 particles modified with carboxyl groups was used. This demonstrates applicability of surface modification of SiO_2 particles with carboxyl groups. Furthermore, octadecyl / silanol-groups patterned SAMs were immersed in solutions containing SiO_2 particles modified with carboxyl groups or non-modified SiO_2 particles for several minutes, respectively. Particles were not adhered to either octadecyl groups or silanol groups. This means that the difference in surface potential between SiO_2 particles modified with carboxyl groups and amino groups of SAM accelerate the adhesion of particles to the amino groups. Particles were attracted and adhered to amino groups predominantly by electrostatic interactions between particles and SAMs and chemical bonds weren't formed because pure water was used as a solution with no condensation agent. Surface of SiO_2 particles modified with carboxyl groups and amino SAM must have changed into $-COO^-$ and $-NH_3^+$, respectively, in water to attract each other.

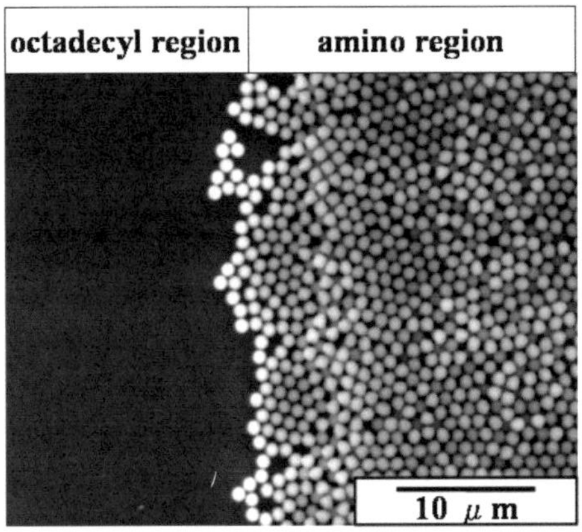

Fig. 8. SEM micrograph of randomly deposited SiO_2 particles in the amino region. Reprinted with permission from Ref. [23], Masuda, Y., Itoh, M., Yonezawa, T. and Koumoto, K., 2002, *Langmuir*, 18, 4155. Copyright @ American Chemical Society

2. Drying patterning of colloidal crystals and 2D arrays

2.1 Self-assembly patterning of colloidal crystals by drying patterning[24]

Particle wires were fabricated through self-assembly on hydrophilic regions of SAMs (self-assembled monolayer)[24]. An SAM of octadecyltrichlorosilane was formed on a silicon substrate and modified by UV irradiation to create a pattern of hydrophobic octadecyl and hydrophilic silanol groups. Ethanol or water containing particles (550 nmφ or 800 nmφ) was dropped onto a patterned SAM. The solution was separated into two droplets with a liquid bridge between the droplets along the hydrophilic regions of a patterned SAM. The droplets and the liquid bridge were used as a mold for fabrication of a two-dimensional pattern of colloid crystals. Particle wire was formed between two droplets and colloid crystals such as an opal structure were formed at both ends of the particle wire after drying the solution. The particle wires constructed from a close-packed structure or non-close-packed structure, i.e. square lattice, were fabricated through self-assembly at room temperature using this method.

The UV-irradiated regions became hydrophilic due to the formation of Si-OH groups, while the non-irradiated part remained unchanged, i.e. it was composed of hydrophobic octadecyl groups, which gave rise to patterned OTS-SAM. To confirm successful film formation and functional group change, water drop contact angles were measured for irradiated and non-irradiated surfaces. Initially deposited OTS-SAM had a water contact angle of 96°, while the UV-irradiated SAM surface was saturated (contact angle <5°). This observation indicated successful fabrication of SAM patterned with octadecyl/silanol groups (Fig. 9). Polystyrene particles in water (150 μl) (550 nmϕ particle or 820 nmϕ carboxylated particle, 10% wt, dispersed in water, Seradyn Co., Ltd.) were further dispersed in ethanol (3 ml) or water (3 ml), and poured onto a patterned OTS-SAM. The contact angles of the ethanol solution or water solution measured 10-20° or 96° on the OTS-SAM, respectively, while they were saturated (contact angle <5°) on silanol groups. The droplets were observed to separate into two drops and a bridge of solution was formed on a silanol line. The droplets and the liquid bridge were used as a mold for fabrication of a two-dimensional pattern of colloid crystals. After evaporation of the solution, substrates were observed by a scanning electron microscope (SEM; S-3000N, Hitachi, Ltd.).

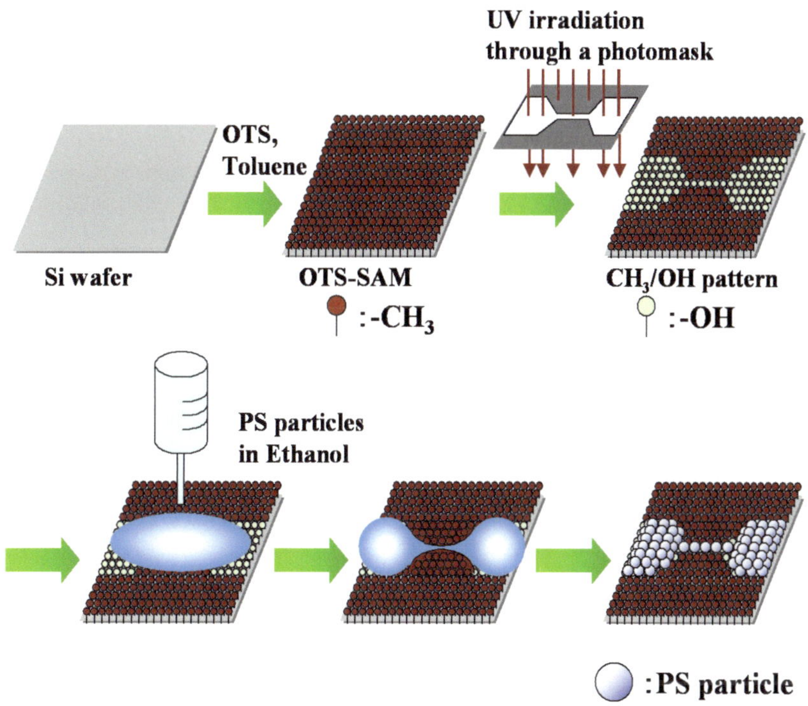

Fig. 9. Conceptual process for fabrication of a particle wire using a patterned SAM and a liquid bridge.
Reprinted with permission from Ref. [24], Masuda, Y., Tomimoto, K. and Koumoto, K., 2003, *Langmuir*, 19, 5179. Copyright @ American Chemical Society

The dispersibility of particles in the solution is very important for particle assembly and high dispersibility is necessary to form a close-packed structure. The zeta potentials of particles dispersed in the solutions were examined by direct measurement of electrophoretic mobility using an electrophoretic light scattering spectrometer (Zetasizer 3000HS, Malvern Instruments Co., Ltd.). The zeta potentials of polystyrene particles (550 nmϕ) in water, carboxylated particles (820 nmϕ) in water, polystyrene particles in ethanol and carboxylated particles in ethanol were determined to be -38.3 mV, -50.2 mV, -53.9 mV and -44.0 mV, respectively. Surface modification by carboxyl groups decreased the negative zeta potential in both solutions. Furthermore, particles in the ethanol had slightly low negative zeta potentials compared to those in water, i.e. the particles were slightly well-dispersed compared to those in water.

Polystyrene particles in water were poured onto a patterned OTS-SAM (Fig. 9), and observed after evaporation of the water. The water at the liquid bridge evaporated in about 24 h to form a particle wire, and droplets at the two ends completely evaporated in about 48 h. In this manner, particle wires constructed from a close-packed structure, i.e. triangular lattice, were produced from the water solution (Fig. 10a). The middle of a particle wire was narrower than its end (Fig. 10b, c). The width of the particle wire does not depend on the width of the silanol line, but rather on the interfacial tensions between solution and substrate, solution and atmosphere, and atmosphere and substrate. The silanol line was not used to decide the width of the particle wire, but rather the position of the liquid bridge and particle wire. Close-packed structures were also formed on large silanol regions (Fig. 10d, e). The right-hand area of Fig. 10d can be regarded as the {100} plane of the fcc structure and the left-hand area can be regarded as the {111} plane of the fcc structure or the {0001} plane of the hcp structure. The close-packed structure was thus considered to be an fcc structure.

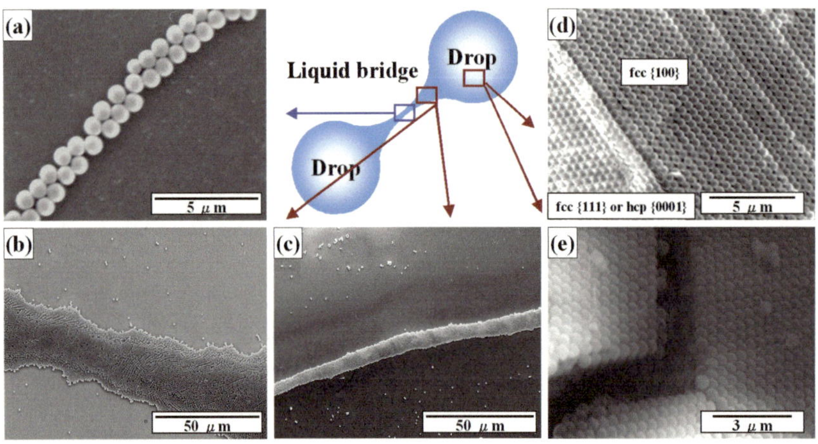

Fig. 10. SEM micrographs of particle structures fabricated from aqueous solution containing micro particles using a liquid bridge. (a) - (c) particle wires constructed from triangular lattice (close-packed structure) and (d) - (e) a close-packed 3D structure.
Reprinted with permission from Ref. [24], Masuda, Y., Tomimoto, K. and Koumoto, K., 2003, *Langmuir*, 19, 5179. Copyright @ American Chemical Society

Figure 11 shows particle wires and 3D structures fabricated from ethanol solution containing polystyrene particles. The ethanol at the liquid bridge evaporated in about 1 min to form a particle wire, and droplets at the two ends evaporated in about 20 min. The liquid bridge of ethanol evaporated faster than that of water for several reasons. The saturated vapor pressure of ethanol (59 mmHg (0.078 atm) at 25℃) is higher than that of water (24 mmHg (0.031 atm) at 25℃), explaining the difference in the evaporation rate of the two droplets. The ratio in evaporation rate of the ethanol liquid bridge to the ethanol droplets is higher than that of the water liquid bridge to the water droplets. This can be explained as follows: Water has high surface tension (71.8×10^{-3} N/m at 25℃) compared with ethanol

(22.0×10^{-3} N/m at 25℃). Ethanol existed along patterned hydrophilic regions with small meniscus at the angle between droplets and a liquid bridge. However, water formed large meniscus at the angle between droplets and a liquid bridge, causing a wide line width of water liquid bridge compared with ethanol on our patterned surfaces. Additionally, the water liquid bridge was higher than that of ethanol due to high surface tension. These made the cross-section area of water larger than that of ethanol. The thick liquid bridge evaporated slowly because of its large volume and low vapor pressure calculated from the Kelvin equation in which the smaller convex liquid surface gives rise to higher internal pressure and faster evaporation rate. Furthermore, the solution at the droplets flowed into a liquid bridge and this further complicated the evaporation mechanism.

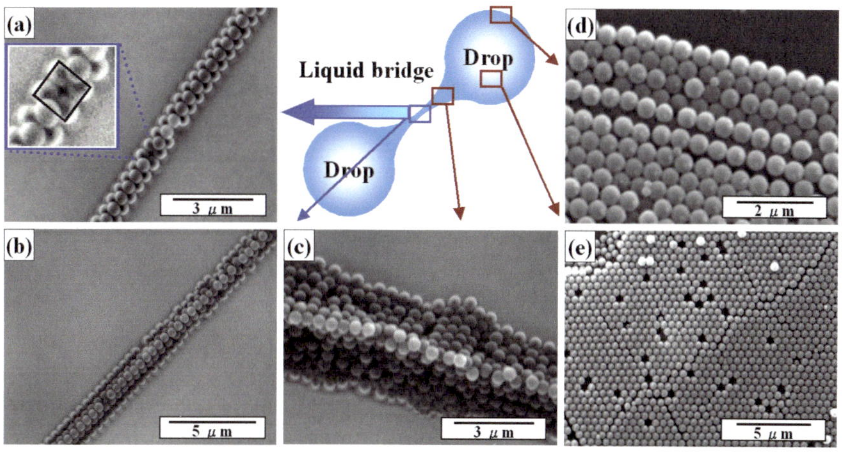

Fig. 11. SEM micrographs of particle structures fabricated from ethanol solution containing micro particles using a liquid bridge. (a), (b) particle wires constructed from square lattice, (c) particles deposited on edge of silanol line and (d), (e) close-packed 3D structures. Reprinted with permission from Ref. [24], Masuda, Y., Tomimoto, K. and Koumoto, K., 2003, *Langmuir*, 19, 5179. Copyright @ American Chemical Society

A narrow particle wire was formed at the center of the liquid bridge (Fig. 11a, b), and wide wires were formed at the edges of the liquid bridge (Fig.11c) along the silanol line. The particle wire in Fig. 11a is not a close-packed structure and is constructed from a square lattice, which is a metastable phase compared with a close-packed structure. High dispersibility of particles is necessary to form a close-packed structure in the solution. However, the particles dispersed well in ethanol and the dispersibility of particles in ethanol is similar to that in water. This shows that the non-close-packed structure was caused not only by the influence of dispersibility but also by many other factors. Movement and rearrangement of deposited particles is necessary to construct a close-packed structure. However, the ethanol evaporated quickly and suppressed the movement of particles by liquid bridge force. Additionally, adhesion between particles and a substrate, and cohesion between particles probably caused moderate suppression of the rearrangement of particles. Factors such as evaporation rate, interaction force between particles, and interaction force between particles and a substrate were important in the packing process. Close-packed 3D structures were also formed on large silanol regions (Fig. 11d, e), and they contained many defects (Fig. 11e). The ethanol evaporated so quickly that the particles did not rearrange well to form a close-packed structure during evaporation of ethanol. This is one of the factors of forming a loosely packed structure. To directly evaluate the effect of the evaporation rate, a similar experiment using ethanol was conducted in a small airtight container with small pinholes to allow the ethanol to evaporate slowly. The size and number of pin holes were adjusted for ethanol at the liquid bridge to evaporate in about 24 h. The particle wire constructed from a close-packed structure was formed after about 24 h, and droplets at both ends were dried after about 48 h. The close-packed 3D structures were formed in large silanol regions. The number of defects was smaller than that formed from the ethanol solution with a shorter time and was similar to that formed from water. The humidity in the container was close to 100% and the saturated vapor pressure of ethanol was 59 mmHg (0.078 atm) at 25℃. This showed that particles dispersed well in ethanol and the interaction between particles and a substrate was sufficiently weak to produce a close-packed structure in the drying process of 24 h. Additionally, the results showed that not only the interaction force between particles and that between particles and a substrate, but also the evaporation rate needs to be controlled to fabricate particle wires.

Accuracy of particle arrangement was evaluated from Fig. 11(a) as calculated in recent work[23]. The center position ((x_i, y_i) μm) of each particle in an upper layer was plotted to estimate the standard deviation. The bottom left corner of Fig. 11(a) was set to be the origin of the x-y coordinate.

The approximated straight line ($f(x)$) and its slope (θ) are represented as follows.

$$f(x) = 1.3965x - 5.3344\,, \tag{1}$$

$$\cos\theta = 0.5822\,, \tag{2}$$

Standard deviation from the approximated straight line is described by the expression,

$$\text{S (standard deviation)} = \frac{\left[\sum_i \left\{\cos\theta \cdot \left(f(x_i) - y_i\right)\right\}^2\right]^{1/2}}{n-1}\,, \tag{3}$$

where n is the number of particles ($n = 19$). The accuracy of the particle arrangement in Fig. 11(a) was estimated to be S = 1.63×10^{-3}. This is lower than that of the particle arrangement obtained in our previous work[23].

Particle wires were fabricated on hydrophilic regions of a patterned SAM. Ethanol or water containing particles was separated into two droplets with a liquid bridge between the droplets along hydrophilic regions of a patterned SAM. Particle wires constructed from a close-packed structure or non-close-packed structure were then formed through self-assembly between two droplets after drying of the solution.

2.2 Self-assembly 2D array of colloidal crystal wires by drying patterning[25,26]

An orderly array of particle wires constructed from a close-packed colloidal crystal were fabricated without preparation of patterned templates[25,26]. A substrate was immersed vertically into a SiO_2 colloidal solution, and the liquid surface moved downward upon evaporation of solution. Particles formed a mono/multi-particle layer, which was cut by the periodic drop-off of solution. The orderly array of particle wires was successfully fabricated, showing the suitability of the self-assembly process for the fabrication of nano/micro structures constructed from nano/micro particles or blocks. The mechanism of the assembly process and control of thickness, width and interval of particle wires were further discussed. Moreover, an array of particle wires constructed not from close-packed fcc (or hcp) structure but from two kinds of particles was realized to fabricate an array of particle wires with NaCl structure by this self-assembly process.

Octadecyltrichlorosilane (OTS)-SAM was prepared by immersing Si substrate in an anhydrous toluene solution containing 1 %vol OTS for 5 min under an N_2 atmosphere. The contact angles of the ethanol solution or water solution measured 10-20° or 96° on the OTS-SAM, respectively. The water solution showed a contact angle of 10-20° on a silicon wafer that was kept in air.

The OTS-SAM was immersed in ethanol solution (80 ml) containing SiO_2 particles (1000 nmφ, 10 mg). The bottom of the solution was heated at 70°C and a condenser tube was kept at the top of the solution to cool it. The temperature difference between the top and bottom of the solution was controlled so as to stir and move particles by convection. The surface of the solution moved on the OTS-SAM surface upon evaporation of the ethanol. Particles began to assemble at the surface of the solution (Fig. 12(a)) and the particle layer was fabricated by movement of the solution surface (Fig. 12(b)). Further evaporation of the solution caused separation of the particle layer and solution surface (Fig. 12(c)) because particles were not sufficiently supplied from the solution. The liquid surface then dropped off and the particle layer separated from the solution surface (Fig. 12(d)). The next particle layer was formed by the same procedure (Fig. 12(e)). Consequently, separated particle wires, i.e. an array of particle wires, were successfully fabricated by our newly developed method (Fig. 12(f), 12(a-d)).

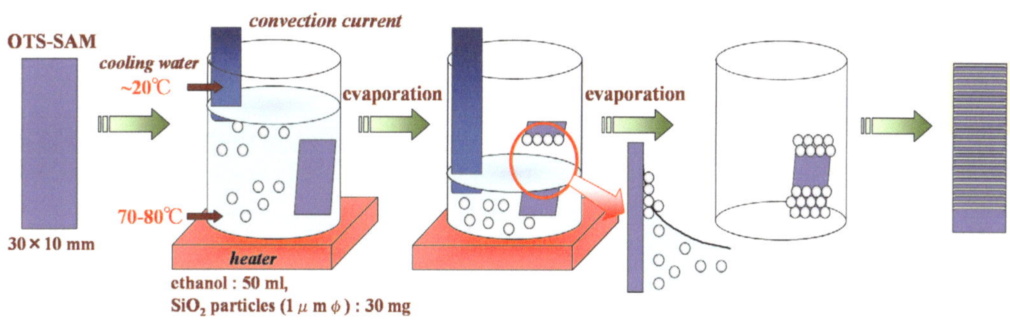

Fig. 12. Schematic for self-assembly process to fabricate an orderly array of particle wires constructed from a close-packed structure.
Reprinted with permission from Ref.[26], Masuda, Y., Itoh, T., Itoh, M. and Koumoto, K., 2004, *Langmuir*, 20, 5588. Copyright @ American Chemical Society

After having been immersed in the solution, which evaporates quickly, the substrates were observed using a scanning electron microscope (SEM; S-3000N, Hitachi, Ltd.), an optical microscope (BX51WI Microscope, Olympus Optical Co., Ltd.) with a digital camera (DP50, 5.8 megapixels, Olympus Optical Co., Ltd.) and a computer for capturing data, and a digital video camera recorder (DCR-TRV 50, Sony Corporation) with optical magnifying glass.

The width and interval of particle wires fabricated from the ethanol solution (80 ml) containing SiO_2 particles (10 mg) at 70°C were shown to be about 150 μm and 200 μm, respectively (Fig. 13(a-d)). Particle wires were constructed from a close-packed particle structure and their upper side showed high feature edge acuity (Fig. 13(c)). The array of particles finished suddenly as shown on the bottom side of the particle wires (Fig. 13(c-d)). These observations suggest that particle wires were formed from the upper side and were cut by the drop-off of solution, and are consistent with the procedure shown in Fig. 12.

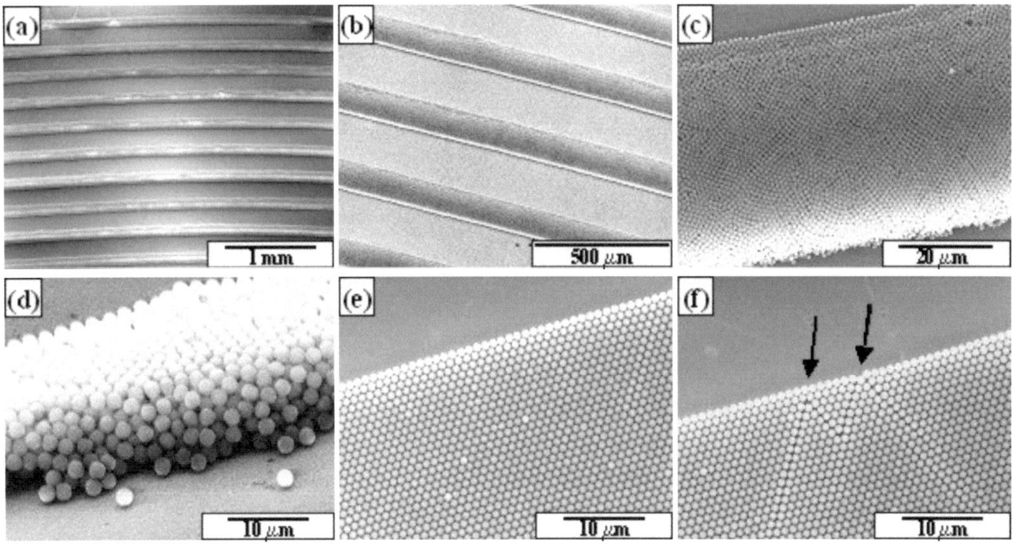

Fig. 13. SEM micrographs of array of particle wires constructed from (a-d) multiparticle layer or (e-f) monoparticle layer, (d) bottom side of a particle wire, (f) a monoparticle wire having two defects on the upper side.
Reprinted with permission from Ref.[26], Masuda, Y., Itoh, T., Itoh, M. and Koumoto, K., 2004, *Langmuir*, 20, 5588. Copyright @ American Chemical Society

Particle wires were formed on the OTS-SAM from the ethanol solution (80 ml) containing a small amount of SiO_2 particles (1000 nmφ, 1 mg) to fabricate thin wires constructed from a mono-particle layer (Fig. 13(e-f)). Particles were slowly supplied to particle wires from the solution. The number of particle layers was shown to be controlled by the change of particle number in the solution. The upper side of the particle wire in Fig. 13(f) has two defects, shown by arrows. The influence on the arrangement below the defects shows that particles were constructed from the upper side to form a close-packed structure. In addition, the disorder disappeared at the middle of the particle wire (Fig. 13(f)), which shows that the formation process of particle wire has self-recovery ability.

Figure 14 shows the formation process of particle wire. The upper particle line was formed (Fig. 14(a)) first and particles were supplied gradually from the solution to form a close-packed structure (Fig. 14(b)) to prepare a wide mono-particle wire (Fig. 14(c)). High dispersibility of particles in the solution and the effective meniscus force allowed us to prepare a highly-ordered close-packed structure.

Particle wires showed iridescent diffraction (Fig. 15) caused by the high regularity of the particle array shown in Fig. 13. Diffracted wave number was changed drastically by the diffraction angle.

A close-packed structure was easily obtained by the use of meniscus force compared to the site-selective deposition in the solution[21-23]. A two-dimensional ordered array can be fabricated without the preparation of a template although templates are required for the liquid mold method[24].

Control of the interval of particle wires was achieved by change in solution temperature (Fig. 16, 17). The interval of particle wire fabricated from the ethanol solution (80 ml) containing SiO_2 particles (20 mg) at 70°C (Fig. 16(a)) was about three times that of particle

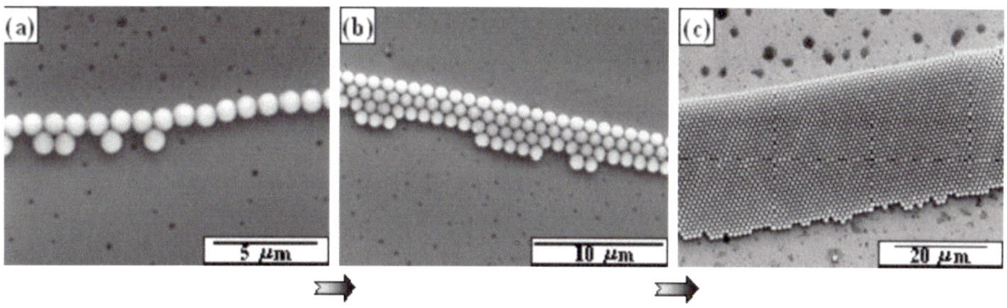

Fig. 14. SEM micrographs of the formation process of a particle wire.
Reprinted with permission from Ref.[26], Masuda, Y., Itoh, T., Itoh, M. and Koumoto, K., 2004, *Langmuir*, 20, 5588. Copyright @ American Chemical Society

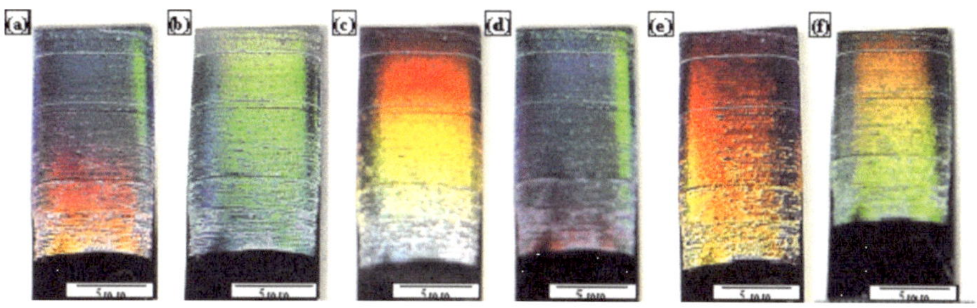

Fig. 15. Photographs of particle wires taken from different directions. (a) front view, (b-f) slightly cross shot of the same sample.
Reprinted with permission from Ref.[26], Masuda, Y., Itoh, T., Itoh, M. and Koumoto, K., 2004, *Langmuir*, 20, 5588. Copyright @ American Chemical Society

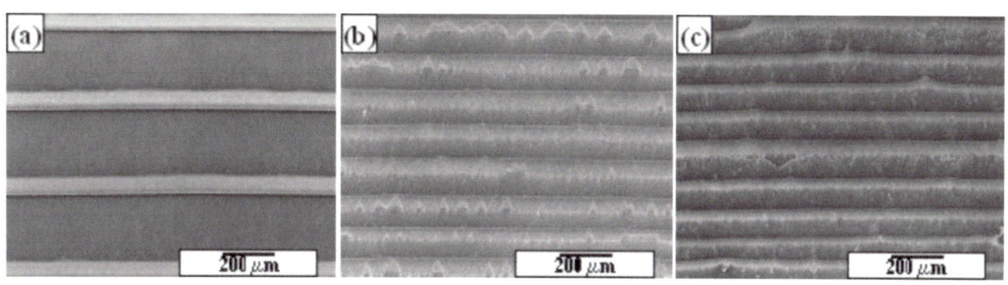

Fig. 16. SEM micrographs of array of particle wires constructed at (a) 70 °C, (b) 60 °C or (c) 50 °C.
Reprinted with permission from Ref.[26], Masuda, Y., Itoh, T., Itoh, M. and Koumoto, K., 2004, *Langmuir*, 20, 5588. Copyright @ American Chemical Society

wires fabricated at 60°C (Fig. 16(b)) or 50°C (Fig. 16(c)) (Fig. 17). The same tendency was confirmed in the dilute solution system (80 ml) containing SiO_2 particles (10 mg) (Fig. 17). Wide interval was caused from high descent speed of solution surface at high temperature. The regularity of particle wires at 70°C was much higher than that at 60°C or 50°C in which the movement of particles caused by convection and the descent speed of solution surface was low compared to that at 70°C.

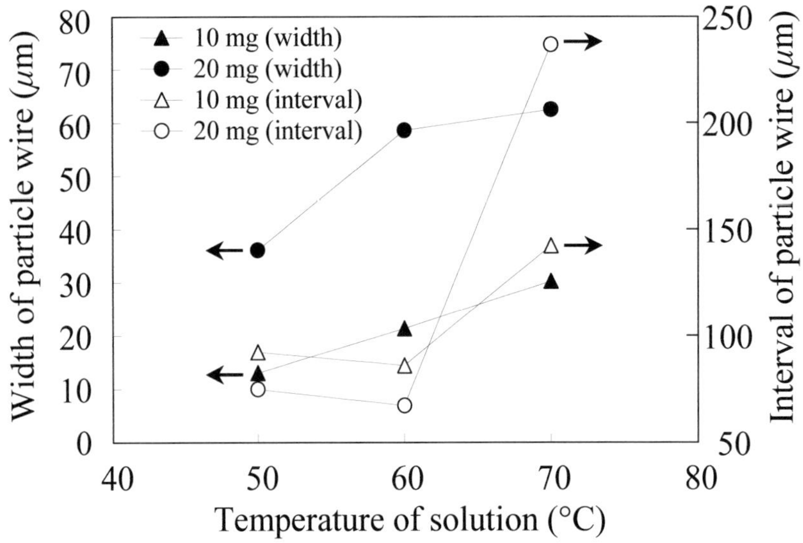

Fig. 17. Width and interval of particle wires as functions of temperature of solution. Reprinted with permission from Ref.[26], Masuda, Y., Itoh, T., Itoh, M. and Koumoto, K., 2004, *Langmuir*, 20, 5588. Copyright @ American Chemical Society

The width of particle wires was also controlled by change in solution temperature or particle concentration in the solution (Fig. 17). The width of particle wires was increased by the increase of solution temperature in both particle concentrations (10 mg or 20 mg), and the width at high particle concentration (20 mg) was about two times that at low particle concentration (10 mg). High descent speed of solution surface and high movement of particles caused by convection were brought about at high temperature. These factors probably cause the change in width; however, further precise control and clarification of mechanism are necessary. Descent speed of solution surface would be controlled by control of the lift speed of a substrate instead of control of the evaporation speed. Further improvement of the process is required for application to future devices.

Self-assembly of particle wires was realized using water instead of ethanol without an SAM (Fig. 18). The particle wires also showed iridescent diffraction and the diffracted wave number was changed drastically by the diffraction angle. A particle layer was reported to form on the whole area of OTS-SAM by the use of water, which evaporates slowly compared with ethanol. Fast movement of the solution surface caused by quick evaporation or lift of substrate is probably necessary for drop-off of the solution surface, which allows fabrication of separated particle wires. The condenser tube, which is used to cool the solution surface, was removed to evaporate water more quickly. This makes the descent speed of the water surface fast enough to separate the particle wires. Additionally, silicon wafer, which was kept in air to show a contact angle of 10-20° to water, was used instead of OTS-SAM. Because the contact angle of the water on the OTS-SAM (96°) was higher than that of ethanol (10-20°) the shape of the water surface was not suitable for making an array of particle wires. These improvements allowed us to fabricate an array of particle wires using water without an SAM. This process is environmentally friendly compared with the process using ethanol with OTS-SAM; however, the regularity and the feature edge acuity of particle wires were low compared to the process shown in Fig. 13 and thus further improvement is required in this system.

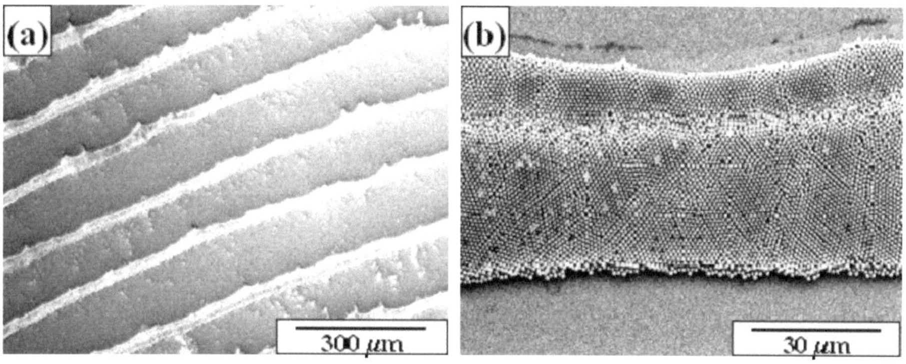

Fig. 18. SEM micrographs of array of particle wires fabricated from water solution. Reprinted with permission from Ref.[26], Masuda, Y., Itoh, T., Itoh, M. and Koumoto, K., 2004, *Langmuir*, 20, 5588. Copyright @ American Chemical Society

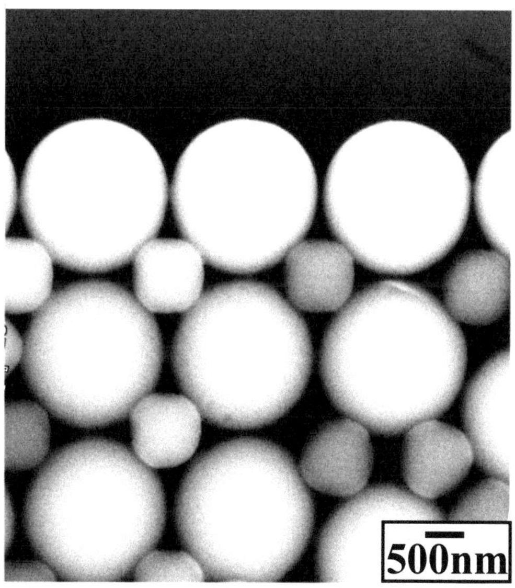

Fig. 19. SEM micrograph of array of particle wires constructed from NaCl structure. Reprinted with permission from Ref.[26], Masuda, Y., Itoh, T., Itoh, M. and Koumoto, K., 2004, *Langmuir*, 20, 5588. Copyright @ American Chemical Society

Array of particle wires constructed not from fcc (or hcp), but from NaCl structure (rock salt structure) in which each ion is 6-coordinate and has a local octahedral geometry, was realized by the use of two kinds of particles (SiO_2 modified with carboxyl groups: 2000 nmφ, SiO_2 modified with amino groups: 1000 nmφ; ratio of particle radii is 0.5). SiO_2 particles 2000 nmφ in diameter and SiO_2 particles 1000 nmφ in diameter were modified to have carboxyl groups or amino groups on their surfaces, respectively. The zeta potential of SiO_2 particles modified with carboxyl groups was measured to be -40 mV in an aqueous solution at pH 7.0[8] by direct measurement of electrophoretic mobility using an electrophoretic light-scattering spectrometer (Zetasizer 3000HS, Malvern Instruments Co., Ltd.). On the other hand, APTS (3-Aminopropyltriethoxysilane)-SAM was prepared by the immersion of silicon wafer in an anhydrous toluene solution containing 1 vol% APTS for 1 h in air to measure the zeta potential instead of SiO_2 particles modified with amino groups. Amino groups of APTS-SAM were measured to be +22.0 mV in aqueous solutions (pH = 7.0) by an electrophoretic light-scattering spectrophotometer (ELS-8000, Otsuka Electronics Co., Ltd.). Electrostatic interaction between SiO_2 particles modified with carboxyl groups and SiO_2 particles modified with amino groups was utilized for self-assembly of these particles to form ionic crystal such as NaCl structure. Although the ratio of ion radii is 0.611 (= $r^+(Na^+)/r^-(Cl^-)$) in NaCl, cations can contact to anions in the range of 0.414 - 0.732 in ratio of ion radii to form NaCl structure. Figure 19 shows the upper side of the particle wire and is the same arrangement as (100) face of NaCl structure. Each particle can be 6-coordinate when the same particle layer stacks on this layer with a slide of half lattice constant (sum of each particle radii). Small spherical particles (cations in NaCl structure) are shown as square or triangular in SEM micrographs because electrons from the SEM electron gun flowed from large particles into small particles to show the contacted area as white. Two kinds of dispersed particles would be adhered on the substrate and rearranged well to form 2-D layer. Particles were also deposited randomly in some areas, and this suggests that it is difficult to obtain NaCl structure in a large area. In addition, NaCl structure was prepared from two kinds of SiO_2 particles without surface modification (SiO_2: 2000 nmφ, SiO_2: 1000 nmφ; ratio of particle radii is 0.5) though the regularity of particle array was slightly lower than that prepared from SiO_2 particles with surface modification. NaCl-type arrangement of bimodal SiO_2 particles can be formed thermodynamically without electrostatic interaction because (100) face of NaCl structure constructed from large particles (x nmφ) and small particles ($(\sqrt{2}-1)x$ nmφ) was close-packed structure. This process should be improved to prepare NaCl structure without the use of surface modification of particles to apply for various kinds of particles. Additionally, the interaction between particles, substrates and liquid and the behavior of particles should be controlled well to form particle assembly having high regularity. Particles should be dispersed well in the solution to avoid aggregation, and attractive interaction between particles and a substrate should be decreased enough to accelerate rearrangement of particles on the substrate. Appropriate movement of particles accelerate formation of particles assembly, however, excess convection suppresses assembly of particles. This process has many factors should be improved to fabricate particle assembly having desired structure and high regularity. The fabrication of an array of particle wire constructed from NaCl structure shows high ability of self-assembly process to realize future devices such as photonic crystal in which various structures are required to effectively utilize PBG.

Particle wires constructed from a close-packed multi-particle layer or mono-particle layer were fabricated without patterned templates. The array shows iridescent diffraction caused by high regularity of the particle array. The mechanism of the process was discussed and the control of the thickness, width and interval of particle wires was realized by change of solution temperature and concentration of colloidal solution. Furthermore, an array of particle wires constructed from an ionic crystal such as NaCl structure was fabricated using electrostatic interaction between particles showing positive zeta potential and particles showing negative zeta potential. This shows the suitability of the self-assembly process for creating future devices such as photonic crystals.

3. Patterning of colloidal crystals and spherical assemblies by two-solution method

3.1 Self-assembly patterning of colloidal crystals by two-solution method[27]

Desired patterns of colloidal crystals having high feature edge acuity and high regularity were fabricated by two-solution method[27]. A micropattern of colloidal methanol prepared on a self-assembled monolayer in hexane was used as a mold for particle patterning, and slow dissolution of methanol into hexane caused shrinkage of molds to form micropatterns of close-packed SiO_2 particle assemblies. This result is a step toward the realization of nano/micro periodic structures for next-generation photonic devices by a self-assembly process.

Silicon substrate was immersed into toluene solution containing 1 vol % octadecyltrichlorosilane (OTS) molecules under nitrogen atmosphere for 5 min to prepare a hydrophobic OTS-SAM. OTS-SAM was irradiated by ultraviolet light (PL21-200, SEN Lights Co., 18 mW/cm^2, distance from a lamp 30 mm, 24°C, humidity 73 %, air flow 0.52 m^3/min, 100 V, 320 W) through a photomask for 10 min. UV irradiation modified hydrophobic octadecyl groups to hydrophilic silanol groups forming a pattern of octadecyl regions and silanol regions. Patterned OTS-SAM having hydrophobic octadecyl regions and hydrophilic silanol regions was used as a template for patterning of colloidal solution.

SiO_2 particles (1 μm in diameter) (0.002 – 0.2 mg) were thoroughly dispersed in methanol (20 μl) and dropped on a patterned OTS-SAM (Fig. 20). The solution was lightly repelled by hydrophobic regions and mainly exists on hydrophilic silanol regions. The substrate was then immersed into hexane and carefully swung to remove the residual solution. The solution was repelled well by octadecyl regions in hexane. The contact angle of the methanol solution on OTS-SAM was confirmed to increase from 51.6° in air to 129.5° in hexane (Fig. 20), indicating that the methanol solution tends to exist on silanol regions selectively.

Methanol was then gradually dissolved into hexane to shrink the colloidal solution mold containing particles. The shrinkage of the mold increased the concentration of particles in the solution. The particles then attracted each other by meniscus force to form a close-packed fcc (face-centered cubic) or hcp (hexagonal closest packing) structure during the drying process of methanol. Close-packed structures were thus formed on hydrophilic silanol regions selectively (Fig. 21, A to H).

2D micropatterns of multi particle-layers (Fig.21, A and B), partially double particle-layers (Fig. 21, C and D), or single particle-layers (Fig. 21, E to H) were successfully fabricated by changing the particle concentration in the solution and solution volume per unit of hydrophilic area (SiO_2/methanol 0.2 mg/20 μl for Fig. 21, A and B, 0.02 mg/20 μl for Fig. 21, C and D, 0.002 mg/20 μl for Fig. 21, E to H). 2D patterns of colloidal crystals with high regularity in particle assembly have not been prepared by our processes previously reported[21-25]. The feature edge acuity of patterns and regularity in particle assembly[21-25] presented here are clearly higher than those previously reported[21-25]. Figure 4-2B shows a magnified area of the patterns constructed from thick particle-layers (Fig. 21A). The edge of patterns (Fig. 21, A to H) shows high feature edge acuity due to close-packing induced by meniscus force in the drying process. The core area of the particle circle (Fig. 21D, Fig. 22)

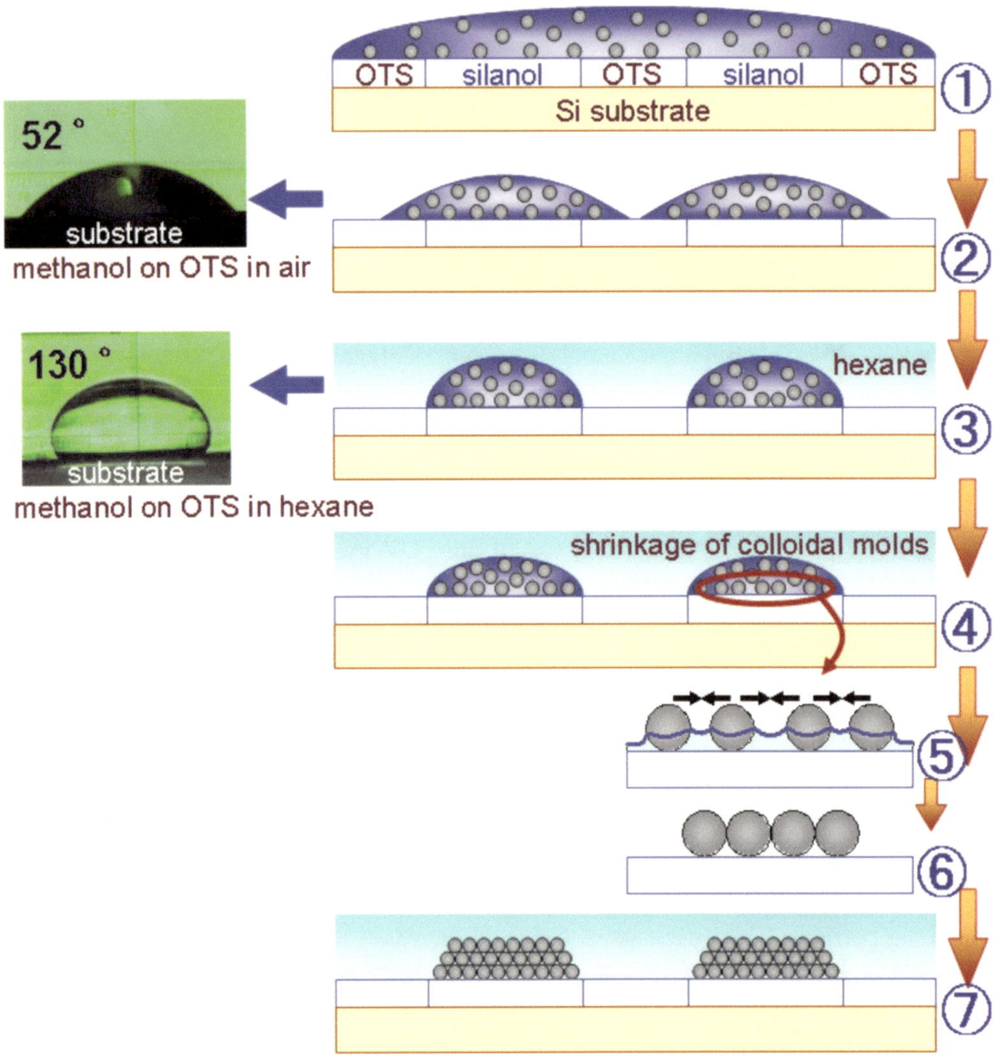

Fig. 20. Conceptual process for two-solution self-assembly method to fabricate patterned colloidal photonic crystals.
Reprinted with permission from Ref. [27], Masuda, Y., Itoh, T. and Koumoto, K., 2005, *Langmuir*, 21, 4478. Copyright @ American Chemical Society

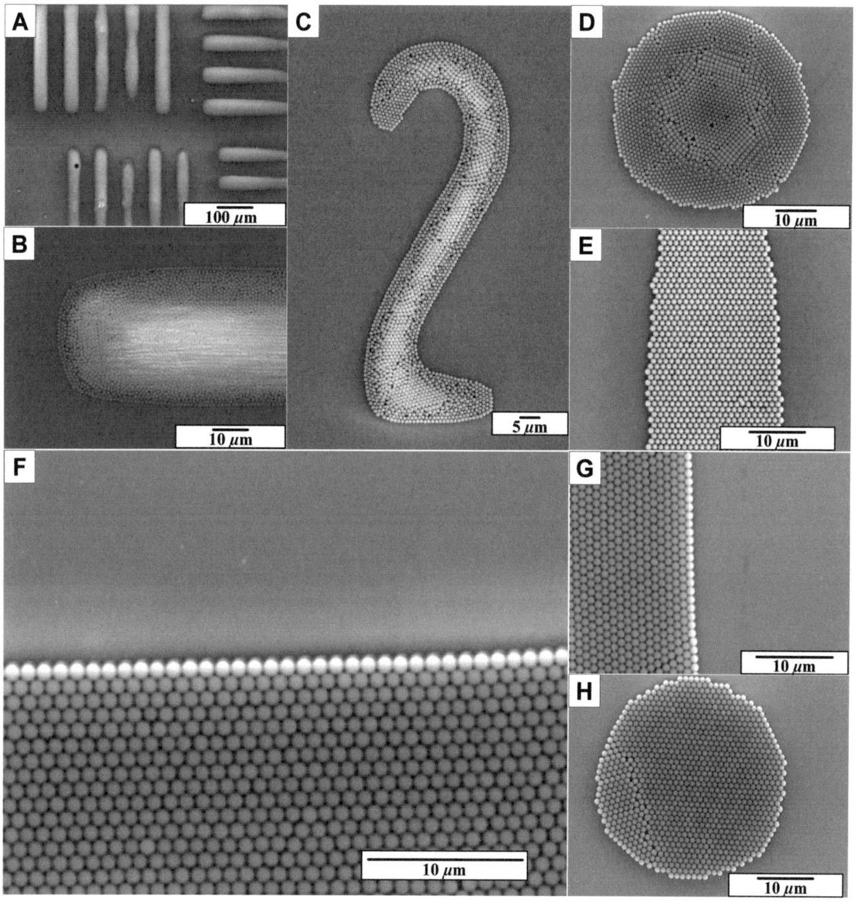

Fig. 21. SEM micrographs of patterned colloidal photonic crystals constructed from (A-B) multi particle-layers, (C-D) partially double particle-layers or (E-H) single particle-layers. Image (B) is a magnified area of (A).
Reprinted with permission from Ref. [27], Masuda, Y., Itoh, T. and Koumoto, K., 2005, *Langmuir*, 21, 4478. Copyright @ American Chemical Society

was a double particle-layer of close-packed hexagonal lattice, i.e., the arrangement of fcc{111}, and the outer shell of the circle was a single particle-layer of hexagonal lattice (fcc{111}). The boundary area of these two flat terraces, i.e., inner shell, was constructed from a nested structure of square lattice, i.e., the arrangement of fcc{100}, to form a gentle slope between the core double-layer and outer shell single-layer. The lattice constant of the square lattice at the inner shell increased gradually with distance from the core of the circle to form a gentle slope. The difference in height was caused by the assembling process and the shape of the liquid mold in which the center is higher than the outside. The particle arrangement in the patterns constructed from a single particle-layer (Fig. 21, E to H) were assigned to the arrangement in fcc{111} which is a close-packed structure. The border line of particle layer in Fig. 21E showed different shape with that in Fig. 21F, G because the border line of particle layer in Fig. 21F, G was fcc <1, 1, 0> and that in Fig. 21E was fcc <1/2, 1/2, 1> which is orthogonal to fcc <1, 1, 0>. There were far fewer defects in the particle patterns than

in our former processes[21-25] because of the effective meniscus force. The standard deviation for the edge of the pattern constructed from a single particle-layer (Fig. 21F) was calculated in the same manner reported previously[24]. The center position ((x_i, y_i) μm) of each particle at the edge was plotted to estimate the standard deviation. The particle at the far left in the edge line of Fig. 21F was set to be the origin of the x-y coordinate.
The approximated straight line ($f(x)$) and its slope (θ) are represented as follows.

$$f(x) = 0.0273x - 0.09\,, \tag{1}$$

$$\cos\theta = 0.9996\,, \tag{2}$$

The standard deviation from the approximated straight line is given by the expression,

$$\text{S (standard deviation)} = \sqrt{U} = \frac{\left[\sum_i \left\{\cos\theta \cdot \left(f(x_i) - y_i\right)\right\}^2\right]^{1/2}}{n-1}\,, \tag{3}$$

where n is the number of particles (n = 35). Unbiased variance (U) was used because the number of particles (n) is smaller than universe. The accuracy of the particle arrangement in Fig. 21F was estimated to be S = 8.75 × 10^{-3}. This is lower than that of the pattern constructed from a single particle-layer prepared in the solution using chemical reactions (S = 3.89 × 10^{-2}, Fig. 7 (a) in previous report[23]). The standard deviation for the edge of the pattern constructed from a single particle-layer (Fig. 21H) was calculated in the same manner. Distance (r_i) from center of a circle ((x_o, y_o) μm) to each particle ((x_i, y_i) μm) at the edge and its average ($\bar{r}$) are presented as follows.

$$r_i = \left((x_i - x_o)^2 + (y_i - y_o)^2\right)^{1/2}\,, \tag{4}$$

$$\bar{r} = \frac{\sum_{i=1}^{n}\left((x_i - x_o)^2 + (y_i - y_o)^2\right)^{1/2}}{n}\,, \tag{5}$$

x_o= 21.9 μm, y_o=17.7 μm, $\bar{r} = 16.53$ μm
The standard deviation from the approximated circle is given by the expression,

$$\text{S (standard deviation)} = \sqrt{V} = \left[\frac{\sum_{i=1}^{n}\left\{\left((x_i - x_o)^2 + (y_i - y_o)^2\right)^{1/2} - \bar{r}\right\}^2}{n}\right]^{1/2}\,, \tag{6}$$

where n is the number of particles (n = 112). Variance (V) was used instead of unbiased variance because the number of particles (n) is the same as universe, i.e., all particles at the edge. The accuracy of the particle arrangement in Fig. 21H was estimated to be S = 3.89 × 10^{-1}. This is higher than that in Fig. 21F because a perfect circle can't be constructed from a small number of particles which were packed in hexagonal arrangement.
The assembly process can be assumed from the details of structures and defects. Particle circles (Fig. 21, D and H) showed no defects at their core, implying that particles were

probably assembled from the core of the particle circle and not from the outer shell. The particle circle (Fig. 21D, Fig. 22) would be formed not layer-by-layer, and the upper layer at the core was also formed before the outer shell was assembled, since the first layer at the boundary area was not a close-packed assembly. A close-packed hexagonal lattice would be formed in the lower layer as shown in the single particle circle (Fig. 21H) in the case particles were assembled layer-by-layer. Other particle patterns (Fig. 21, A to C) would be assembled from the core area in the same manner.

Methanol solution containing particles was dropped onto a patterned OTS-SAM and dried in air without immersion into hexane for comparison. However, a particle pattern only with a low feature edge acuity was obtained. This indicates that the contact angle of methanol solution on OTS-SAM, i.e., the shape of the methanol solution mold, is important for particle patterning. Additionally, SiO_2 particles were dispersed in water instead of methanol, and a patterned OTS-SAM covered with the water colloidal solution was immersed into hexane, but the particle pattern was not obtained because the water solution cannot form a micropattern of the mold on hydrophilic regions selectively due to its high contact angle. These results suggest that the combination of methanol, hexane, OTS and silanol allowed us to fabricate micropatterns of particle assembly having high feature edge acuity and high regularity.

This two-solution self-assembly method successfully produced precise 2D patterns of colloidal photonic crystals, i.e., close-packed fcc or hcp particle assemblies having a photonic band gap, with short fabrication period and low energy. Various complicated patterns of colloidal crystals having high feature edge acuity and high regularity in particle assembly were fabricated, and the number of particle stacking layers, i.e., thickness of colloidal crystals, was demonstrated to be controlled. This process has the advantage of the drying process in which meniscus force can be utilized for close-packed particle assembly, and also the advantage of the static solution process in which particle assembling processes can be controlled well. The low dissolution rate of methanol into hexane and the appropriate contact angle of methanol on a substrate achieved by the combination of methanol, hexane, OTS and silanol allowed us to utilize the solution mold for particle patterning. Additionally, high dispersibility of particles and high repulsion force between particles and substrate suppressed the aggregation of particles, and the meniscus force between particles was effectively utilized to form a close-packed structure. The quality of SAM, shape and shrinkage rate of solution molds also greatly influence the feature edge acuity and the regularity of the particle assembly. The newly developed method achieved much higher regularity in particle assembly and feature edge acuity of the pattern than those previously reported[21-25]. The result is a step toward the realization of nano/micro periodic structures for next-generation photonic devices by the self-assembly process.

3.2 Self-assembly patterning of spherical colloidal crystals by two-solution method[28]

Micropatterns of spherical particle assemblies were fabricated by two-solution method[28]. Hydrophilic regions of a patterned self-assembled monolayer were covered with methanol solution containing SiO_2 particles and immersed in decalin to control the shape of droplets and gradually dissolve the methanol into decalin. Interfacing of methanol/decalin and shrinkage of methanol droplets were utilized to obtain meniscus force to form spherical particle assemblies; additionally, its static solution system allowed precise control of the conditions. Particles were assembled to form spherical shapes on hydrophilic regions of an SAM and

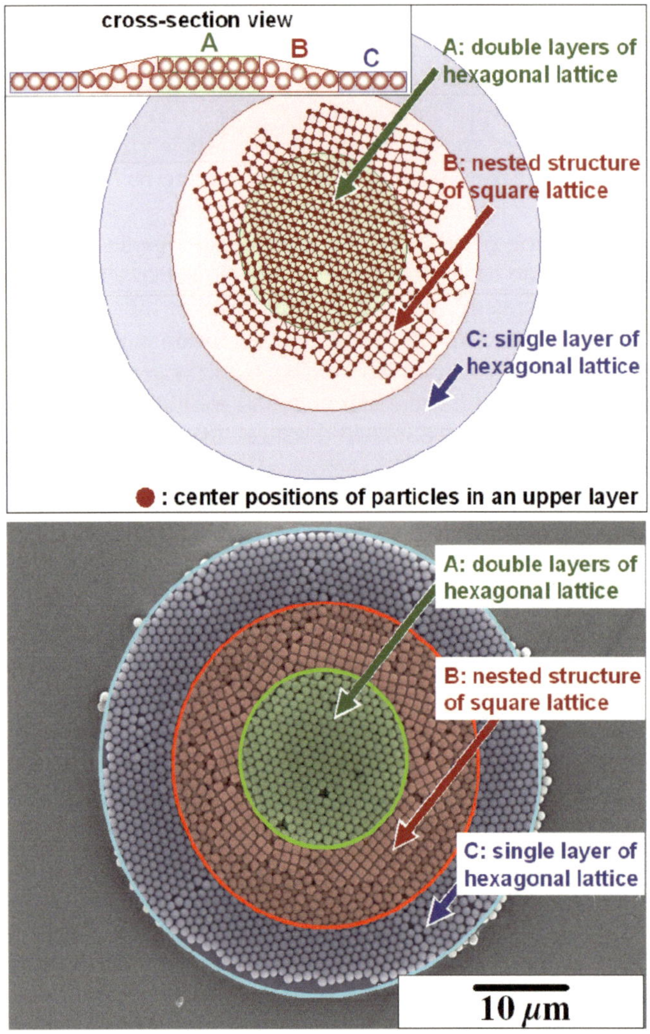

Fig. 22. Particle arrangement view (top) prepared from SEM micrograph (bottom) of patterned colloidal photonic crystals constructed from partially double particle-layers. Center positions of particles in an upper layer show hexagonal or square lattice arrangement regions. The cross-sectional view shows the difference in height caused from the difference in particle arrangement modes. The three colored areas show regions of (A) double layers of hexagonal lattice, (B) nested structure of square lattice and (C) single layer of hexagonal lattice.
Reprinted with permission from Ref. [27], Masuda, Y., Itoh, T. and Koumoto, K., 2005, *Langmuir*, 21, 4478. Copyright @ American Chemical Society

consequently, micropatterns of spherical particle assemblies were successfully fabricated through self-assembly. This patterned two-solution process has the advantages of both a drying process having meniscus force and a static solution process having high controllability. An Si wafer (p-type Si [100], NK Platz Co., Ltd.) was sonicated in water, ethanol or acetone for 10 min, respectively, and exposed for 2 h to UV light (184.9 nm) (low-pressure mercury lamp, NL-UV253, Nippon Laser & Electronics Lab.) to clean the surface. The OTS-SAM or HFDTS-SAM[29] were prepared by immersing the Si substrate in an anhydrous toluene(Aldrich Chemical Co., Inc.) solution containing 1 vol% OTS(Acros Organics) or HFDTS(Lancaster synthesis Ltd.) for 5 min under an N_2 atmosphere (Fig. 23). The substrate with the SAM was baked at 120°C for 5 min to remove residual solvent and promote chemisorption of the SAM.

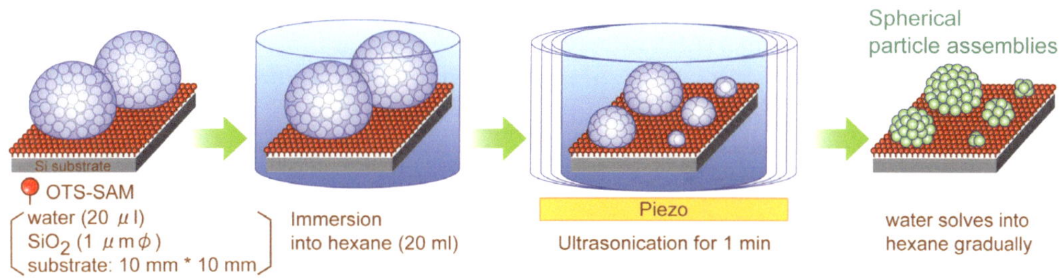

Fig. 23. Conceptual process for fabrication of spherical particle assemblies.
Reprinted with permission from Ref. [28], Masuda, Y., Itoh, T. and Koumoto, K., 2005, *Adv. Mater.*, 17, 841. Copyright @ Wiley-VCH Verlag GmbH & Co.

The HFDTS-SAM on the silicon substrate was exposed for 2 h to UV light through a photomask such as a mesh for transmission electron microscopy (Okenshoji Co., Ltd.) to be used as a template for micropatterning of spherical particle assemblies. UV-irradiated regions became hydrophilic due to silanol group formation, while the non-irradiated part remained unchanged. Formation of the SAMs and the modification to silanol groups by UV irradiation were verified using the water drop contact angle (θw). The initially deposited OTS-SAM or HFDTS-SAM showed a water contact angle of 105° or 112°, but the UV-irradiated surface of SAM was wetted completely (contact angle < 5°).

SiO_2 particles (1.13 $\mu m\varphi$, Hipresica UF N3N, CV: 3.57 %, specific gravity: 1.8 ± 0.1 g/cm^3, Ube-Nitto Kasei Co., Ltd.) (0.2 mg, 1.5 × 10^{27} particles) were thoroughly dispersed in water (20 μl) and dropped on a hydrophobic OTS(octadecyltrichlorosilane)-SAM (Fig. 23). The OTS-SAM with droplets was then immersed in hexane (20 ml, solubility of hexane in water at 20°C: 0.0013 g / 100 ml, specific gravity: 0.7) and ultrasonicated for 1 min. Large water droplets containing SiO_2 particles were separated into many small emulsions that kept them spherical on hydrophobic OTS-SAM. Water in the emulsions was gradually extracted to hexane to reduce the size of emulsions forming spherical particle assemblies[30]. After having been immersed for 12 h, spherical particle assemblies with different diameters were observed on OTS-SAM; it was also observed that the assemblies were constructed from various numbers of particles such as 3, 5, 6, 8 or many particles (Fig. 24, 25, 26). Quantities of 3, 5, 6 or 8 particles were assembled into triangular, pyramidal, octahedral or decahedral particle clusters, respectively. The number of particles in spheres can be controlled by the change of emulsion size or particle concentration in water[30]. The spherical shape of particle assemblies was caused by the high contact angle of water emulsion on hydrophobic OTS-SAM in hexane. Consequently, various sizes of spherical particle assemblies can be prepared using this emulsion process.

The particles would be strongly bound to the droplet interfaces by surface tension in emulsions[31]. In the formation process of small clusters of particles (Fig. 24), the water in emulsions dissolved into hexane to reduce droplet size and this restricted the area in which

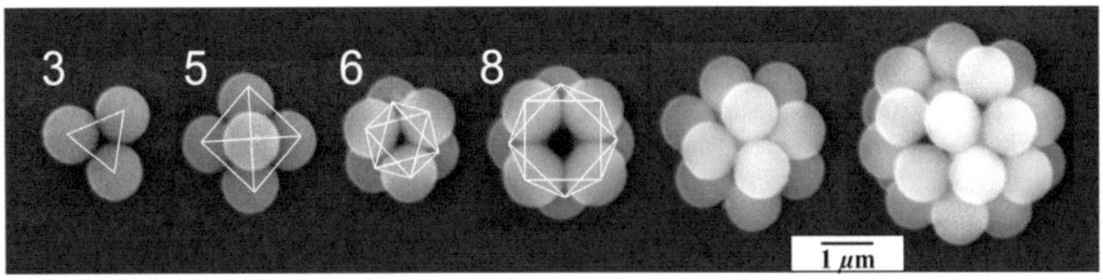

Fig. 24. SEM micrographs of spherical particle assemblies with different diameters. Particle assemblies constructed from a small number of particles such as 3, 5, 6, 8 and so on. Reprinted with permission from Ref. [28], Masuda, Y., Itoh, T. and Koumoto, K., 2005, *Adv. Mater.*, 17, 841. Copyright @ Wiley-VCH Verlag GmbH & Co.

Fig. 25. Spherical particle assembly and tilted images.
Reprinted with permission from Ref. [28], Masuda, Y., Itoh, T. and Koumoto, K., 2005, *Adv. Mater.*, 17, 841. Copyright @ Wiley-VCH Verlag GmbH & Co.

particles could exist[30]. Particles touched together by the reduction of emulsion size and formed a spherically packed assembly. Deformation of the interface then led to the rearrangement that formed close-packed particle assemblies. The clusters were formed using emulsions not dispersed in the solution but adsorbed on the flat substrate, and thus some of the clusters showed imperfect symmetry such as pyramidal, octahedral or decahedral, which have a large flat face touching the substrate.

Additionally, large spherical particle assemblies (for instance ~57 μmφ in Fig. 26) were prepared from large methanol emulsions (~100–300 μmφ) in decalin (decahydronaphthalene) without the use of ultrasonication, which makes emulsions smaller. Many linear disclinations, i.e., grain boundaries, were formed on the surface to reduce elastic strain energy because a close-packed triangular particle lattice composed of a particle surrounded by 6 particles in plane cannot cover a spherical surface[32]. Linear disclinations were composed of an alternative arrangement of the particle surrounded by 5 particles in plane (red, this particle can be assumed to have a charge of +1) and the particle surrounded by 7 particles in plane (yellow, charge –1). Both ends of a linear disclination were particles surrounded by 5 particles in plane to make the total charge of each linear disclination +1 as observed in previous report[32]. The minimum number of linear disclinations can be calculated for the large sphere (~57 μmφ) (Fig. 26) to be 22 (N : minimum number of linear disclinations $N = 12 + 0.41 r/d = 22$, r : radius of sphere ~28.5 μm, d : mean particle spacing 1.13 μm) from geometric calculations[32,33]. The surface area in a purple circle can be also calculated to be about 10.5 % of the total surface area of the sphere from the formula (surface area ratio: S_2/S_1 , surface area of the sphere: $S_1 = 4\pi r^2$, surface area in a purple circle: $S_2 = 2\pi r(r - r_1)$, radius of the purple circle shown in Fig. 26 $\sqrt{r^2 - r_1^2} = 17.5\mu\text{m}$). The minimum number of linear disclinations on the surface area in a circle can therefore be estimated to about 2.3(n : minimum number of linear disclinations in a purple circle

$n = N \cdot S_2/S_1$). However, many linear disclinations were observed in an SEM micrograph (Fig. 26 (Right)). It is suggested that our assembly method can be further optimized to reduce linear disclinations.

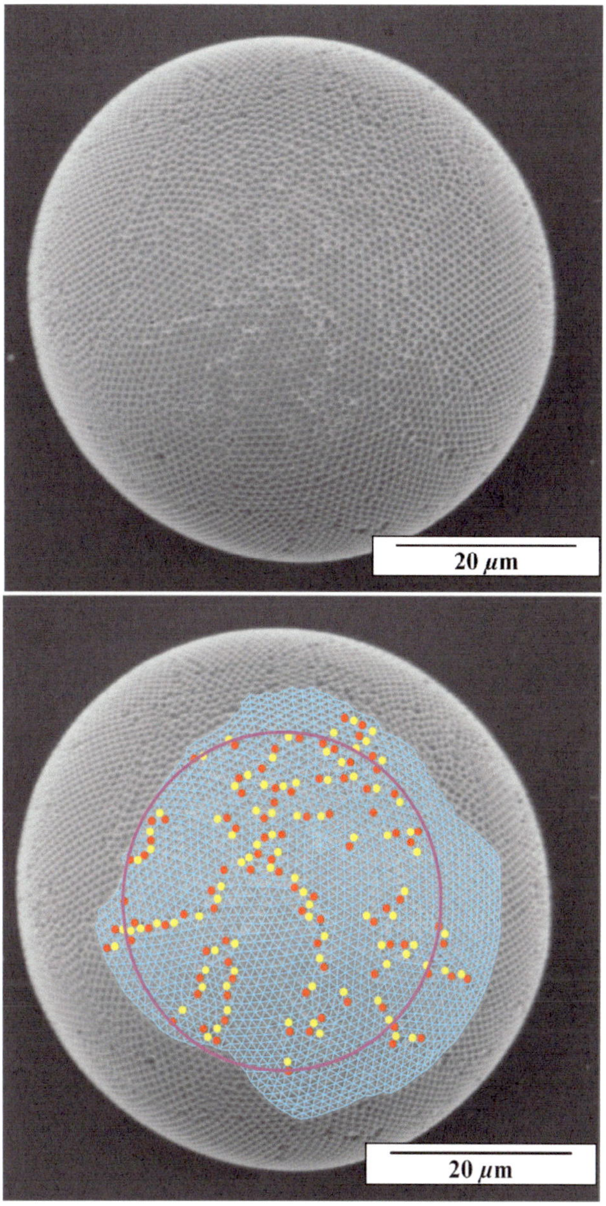

Fig. 26. Top: A large spherical particle assembly. Bottom: Linear disclinations on a large spherical particle assembly.
Reprinted with permission from Ref. [28], Masuda, Y., Itoh, T. and Koumoto, K., 2005, *Adv. Mater.*, 17, 841. Copyright @ Wiley-VCH Verlag GmbH & Co.

Furthermore, the upper sides of spherical particle assemblies were removed using a manipulator installed in an optical microscope (BX51WI Microscope, Olympus Optical Co., Ltd.) to evaluate the packing structure. Particle assemblies have a densely packed structure (Fig. 27). Particles would be completely rearranged to form a densely packed structure during the reduction of emulsion size due to the high dispersibility of particles in emulsions.

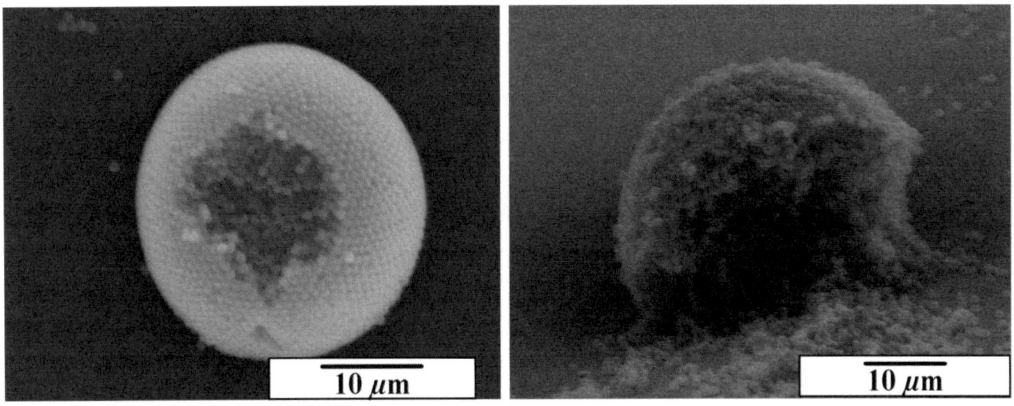

Fig. 27. SEM micrographs showing the inside of spherical particle assemblies formed from methanol emulsions in decalin.
Reprinted with permission from Ref. [28], Masuda, Y., Itoh, T. and Koumoto, K., 2005, *Adv. Mater.*, 17, 841. Copyright @ Wiley-VCH Verlag GmbH & Co.

SiO_2 particles (1.13 μmφ, 1 g/l) were thoroughly dispersed in methanol solution (10 μl) and dropped onto a patterned HFDTS(heptadecafluoro-1,1,2,2-tetrahydrodecyltrichlorosilane)-SAM[29] having hydrophobic HFDTS-SAM regions and hydrophilic silanol regions, photopatterned using a mesh for transmission electron microscopy as a photomask (Fig. 28). The patterned SAM covered with the solution was then carefully immersed in decalin so as not to remove the solution because the density of methanol (0.79) is lower than that of decalin (0.88) causing methanol to float on decalin. The patterned SAM was then gently vibrated to remove additional methanol solution and assist the movement of droplets to silanol regions. The methanol solution was selectively contacted on hydrophilic regions to form a micropattern of the solution, which became clearer after immersion for a few hours. The methanol solution containing particles formed a spherical shape because of the surface interaction between methanol, decalin and surface of a SAM and the buoyant force of methanol in decalin. Methanol in emulsions was gradually dissolved into exterior decalin phase to form particle assemblies. After having been immersed for 12 h, particle assemblies having a dome shape were formed at the center of each silanol region (Fig. 29). The distance between the centers of each spherical particle assembly was same to the distance between holes of a mesh. The diameter of spherical particle assemblies was about 18 μmφ. It was smaller than that of hydrophilic regions (about 100μm) due to low particle concentration in methanol and the shape of methanol droplets on a substrate. This indicates the arrangement regularity of spherical particle assemblies can be improved more by the use of the photomask having small holes to decide positions of each droplet precisely. Some extra particles, i.e., noise particles, were also deposited on hydrophobic regions. The process should be further optimized to control many factors such as volume of methanol solution on a substrate, quality of a SAM or aggregation of particles in the solution to avoid noise particles. Particle assemblies were shown to have a densely packed structure by destructive inspection using a manipulator. Consequently, the dot array of spherical particle assemblies was successfully fabricated by this self-assembly process.

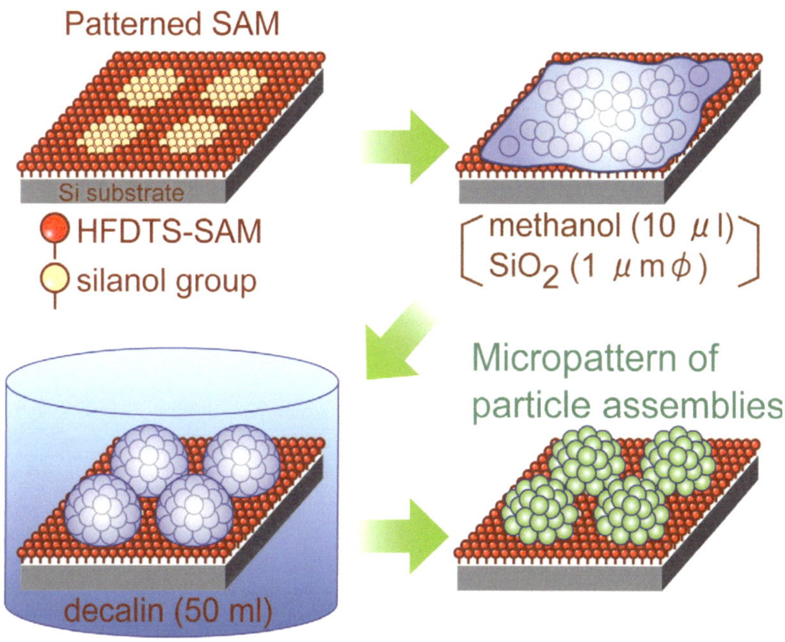

Fig. 28. Conceptual process of micropattern of spherical particle assemblies.
Reprinted with permission from Ref. [28], Masuda, Y., Itoh, T. and Koumoto, K., 2005, *Adv. Mater.*, 17, 841. Copyright @ Wiley-VCH Verlag GmbH & Co.

Micropatterns of particle assemblies were fabricated without the use of a template, having microstructures such as molds or grooves. A two-dimensional array of spherical particle assemblies was fabricated by self-assembly with this method. Interfacing between two solutions and shrinkage of droplets were utilized to obtain meniscus force to form spherical particle assemblies, and additionally, its static solution system allowed precise control of the conditions. These showed the high ability of self-assembly processes to prepare microstructures constructed from colloidal crystals. Further investigations of the solution-solution, solution-SAMs and solution-particles interfaces and the behavior of particles and solutions would allow us to develop this two-solution system to prepare desirable particle assembly structures.

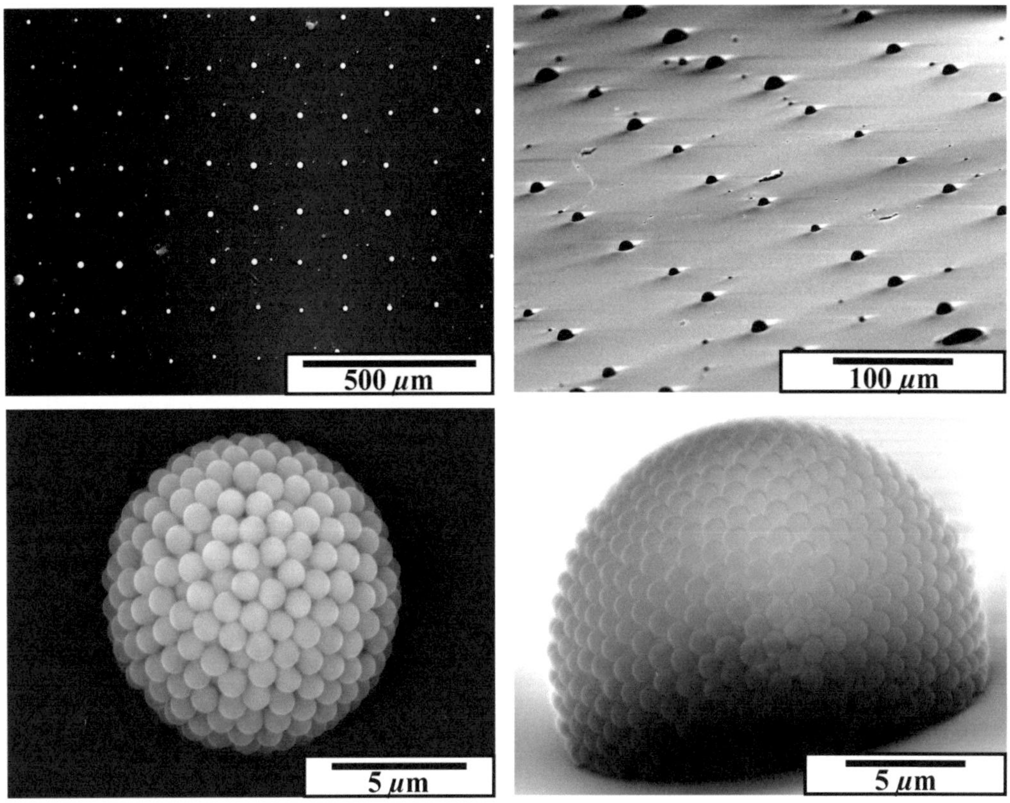

Fig. 29. SEM micrographs of micropattern of spherical particle assemblies. Top left: Micropattern of spherical particle assemblies. Bottom left: Magnified area of top left. Top right: Tilted micropattern of spherical particle assemblies. Bottom right: Magnified area of top right.
Reprinted with permission from Ref. [28], Masuda, Y., Itoh, T. and Koumoto, K., 2005, *Adv. Mater.*, 17, 841. Copyright @ Wiley-VCH Verlag GmbH & Co.

4. Summary

Self-assembly patterning of colloidal crystals was realized by liquid phase patternig, drying patterning and two-solution method. Micropatterns of colloidal crystals were fabricated under environmentally friendly conditions using self-assembled monolayers. They can be applied to photonic crystals, dye-sensitized solar cell, molecular sensors, gas sensors, nano-particle devices, etc. These showed high ability of self-assembled monolayers and self-assembly processes.

5. References

[1] Masuda, Y.; Koumoto, K. *J. Soc. Inor. Mater. Jpn.* 2000, *7*, 4-12.

[2] Masuda, Y.; Tachibana, K.; Itoh, M.; Koumoto, K. *Materials Integration* 2001, *14*, 37-44.

[3] Yablonovitch, E. *Phys. Rev. Lett.* 1987, *58*, 2059-2062.

[4] John, S. *Phys. Rev. Lett.* 1987, *58*, 2486-2489.

[5] Joannopoulos, J. D.; Meade, R. D.; Winn, J. N. *Princeton Press, Princeton, New Jersey* 1995.

[6] Chow, E.; Lin, S. Y.; Johnson, S. G.; Villeneuve, P. R.; Joannopoulos, J. D.; Wendt, J. R.; Vawter, G. A.; Zubrzycki, W.; Hou, H.; Alleman, A. *Nature* 2000, *407*, 983-986.

[7] Colombelli, R.; Srinivasan, K.; Troccoli, M.; Painter, O.; F. Gmachl, C.; Tennant, D. M.; Sergent, A. M.; Sivco, D. L.; Cho, A. Y.; Capasso, F. *Science* 2003, *302*, 1374-1377.

[8] Noda, S.; Tomoda, K.; Yamamoto, N.; Chutinan, A. *Science* 2000, *289*, 604-606.

[9] M. Qi; E. Lidorikis; P. T. Rakich; S. G. Johnson; J. D. Joannopoulos; E. P. Ippen; H. I. Smith *Nature* 2004, *429*, 538-542.

[10] M. Campbell; D. N. Sharp; M. T. Harrison; R. G. Denning; A. J. TurberReld *Nature* 2000, *404*.

[11] J. G. Fleming; S. Y. Lin *Optics Letters* 1999, *24*, 49-51.

[12] S. R. Kennedy; M. J. Brett; O. Toader; S. John *Nano Letters* 2002, *2*, 59-62.

[13] Kuramochi, E.; Notomi, M.; Kawashima, T.; Takahashi, J.; Takahashi, C.; Tamamura, T.; Kawakami, S. *Optical and Quantum Electronics* 2002, *34*, 53-61.

[14] Vlasov, Y. A.; Bo, X. Z.; Sturm, J. C.; Norris, D. J. *Nature* 2001, *414*, 289-293.

[15] A. Blanco; E. Chomski; S. Grabtchak; M. Ibisate; S. John; S. W. Leonard; C. Lopez; F. Meseguer; H. Miguez; J. P. Mondia; G. A. Ozin; O. Toade; H. M. van Driel *Nature* 2000, *405*, 437-440.

[16] Xia, N.; Gates, B.; Yin, Y.; Lu, Y. *Adv. Mater.* 2000, *12*, 693-713.

[17] Lopez, C. *Adv. Mater.* 2003, *15*, 1679-1704.

[18] Kim, E.; Xia, Y.; Whitesides, G. M. *Adv. Mater.* 1996, *8*, 245-247.

[19] Ozin, G. A.; Yang, S. M. *Adv. Func. Mater.* 2001, *11*, 95-104.

[20] Xia, Y. N.; Yin, Y. D.; Lu, Y.; McLellan, J. *Adv. Func. Mater.* 2003, *13*, 907-918.

[21] Masuda, Y.; Seo, W. S.; Koumoto, K. *Thin Solid Films* 2001, *382*, 183-189.

[22] Masuda, Y.; Seo, W. S.; Koumoto, K. *Jpn. J. Appl. Phys.* 2000, *39*, 4596-4600.

[23] Masuda, Y.; Itoh, M.; Yonezawa, T.; Koumoto, K. *Langmuir* 2002, *18*, 4155-4159.

[24] Masuda, Y.; Tomimoto, K.; Koumoto, K. *Langmuir* 2003, *19*, 5179-5183.

[25] Masuda, Y.; Itoh, M.; Koumoto, K. *Chem. Lett.* 2003, *32*, 1016-1017.

[26] Masuda, Y.; Itoh, T.; Itoh, M.; Koumoto, K. *Langmuir* 2004, *20*, 5588-5592.

[27] Masuda, Y.; Itoh, T.; Koumoto, K. *Langmuir* 2005, *21*, 4478-4481.

[28] Masuda, Y.; Itoh, T.; Koumoto, K. *Adv. Mater.* 2005, *17*, 841-845.

[29] Masuda, Y.; Sugiyama, T.; Koumoto, K. *J. Mater. Chem.* 2002, *12*, 2643-2647.

[30]Manoharan, V. N.; Elsesser, M. T.; Pine, D. J. *Science* 2003, *301*, 483-487.

[31] Levine, S.; Bowen, B. D.; Partridge, S. J. *Colloids and Surfaces* 1989, *38*, 325-343.

[32] Bausch, A. R.; Bowick, M. J.; Cacciouto, A.; Dinsmore, A. D.; Hsu, M. F.; Nelson, D. R.; Nikolaides, M. G.; Travesset, A.; Weitz, D. A. *Science* 2003, *299*, 1716-1718.
[33] Bowick, M. J.; Nelson, D. R.; Travesset, A. *Phys. Rev. B* 2000, *62*, 8738-8751.

8

Synthesis and Characterization of Dispersion Reinforced Sintered System Based on Ultra Fine and Nanocomposite Cu-Al_2O_3 Powders

Zoran Anđić[1], Aleksandar Vujović[1], Miloš Tasić[1], Marija Korać[2] and Željko Kamberović[2]
Scientific Research Center, Užice
Faculty of Metallurgy and Technology, Belgrade
Serbia

1. Introduction

The research in nanocrystal materials has intensified in the last years, primarily due to their attractive potential, i.e. properties which are significantly improved compared to large grain materials [Karch, et al., 1987; Milošević, 1999; Ristić, 2003]. Nanostructure materials rank in the group of ultrafine metastable structures which contain a high defect concentration (spotted defects, dislocations, grain boundaries, interphase boundaries, etc). These materials are structurally different from crystal and amorphous forms, due to the fact that grain boundaries and the interphase represent a specific condition of the solid matter, since the atoms on the boundary are subject to a periodical potential field of crystal from both sides of the boundary [Milošević, 1999]. Nanostructural materials can be synthesized in controlled processes by the following methods: highly energetic reactive milling [Ahn, 1996], solution precipitation (sol-gel [Ruys & Mai, 1999], hydrothermal synthesis [Byrappaa & Adschirib, 2007], electrochemical synthesis [Yuana et al., 2007]), internal oxidation [Afshar & Simchi, 2008] etc. The synthesis of nanoparticles is generally performed by the methods of "from the bottom - to the top approach" ("bottom-up approach"), such as the deposition from the gas or liquid phase, thermal decomposition (evaporation-condensation), and the methods based on "from the top - to the bottom approach" ("top-down approach"), wherein mechanical grinding processes are distinctive. When using the bottom-up method, nanoparticles are formed atom by atom or molecule by molecule, whereas when using the top-down approach, large volumes are gradually reduced in proportion to reach the nanometer dimensions [Hosokawa, 2007].
The synthesis of metal and alloy powders represents the starting stage in the production of sintered metal materials. For obtaining sintered products with the required properties, the starting material, powders of metals or alloys, are of decisive importance. Since the starting structure undergoes certain changes in further processing, but, basically, remains preserved in the structure of the final product [Anđić, 2007; Ristić, 2003], there is an increased necessity for a larger number of methods for producing powders. Although obtaining powders by a thermochemical method, where the input materials are in liquid phase, is not a new

procedure, the interest in this method has intensified recently, due to the development of contemporary materials with pre-set properties, especially in terms of nano-powder production [Jena et al., 2001; Lee et al., 2001; Wu et al., 2001].

This paper presents the synthesis and characterization of Cu-Al_2O_3 nanocomposite systems using a combination of procedures suitable for obtaining sintered materials with nanocrystal structure and homogenous distribution of dispersoids in the basic material matrix, characterized by a good combination of electro–mechanical properties. The obtained sintered materials with nanocrystal structure based on Cu-Al_2O_3 nanocomposite powders have a wide application in the field of electronics and electrical engineering. They are used as electrodes for spot welding, different contact materials, various switches, as well as thermal and electric conductors, microwave tubes, commutators for starting helicopter engines, relays, etc. In addition to this, there is a significant possibility of using these systems as coatings with low porosity and high adhesion degree [Anđić et al., 2006, 2007; Korać et al., 2007].

Disperse reinforcement has recently raised great theoretical and practical interest. It is known that the introduction of finely dispersed particles into a metal matrix could have significant reinforcing effects, which can be maintained at elevated temperatures. Consequently, dispersion reinforced materials of this type show good mechanical properties during exploitation at elevated temperatures and during preservation at room temperature after high temperature exposure. The improvement of mechanical properties is achieved without any significant loss in electric and heat conductivity [Naser et al., 1997; Trian et al., 2006; Trojanová et al., 1999]. For such reinforcement, ultra fine particles and nano particles of oxides are suitable, which, due to their hardness, stability and insolubility in a metal matrix present exceptionally good obstacles to dislocation motion at elevated temperatures. Dispersed phases present in metals before deformation, affect the structural design and behaviour of metals when heated. The deformed material structure created in the presence of secondary phases depends on the size and distance of particles. Small finely dispersed particles which are at a smaller distance cause the high density of evenly distributed dislocations. The borders of subgrains and individual subgrains are vaguely expressed, and the curve of the lattice, i.e. the discrepancy in the orientation of adjacent subgrains is too small for them to become the germs of new recrystallised grains, which consequently leads to the reduction in the speed of grain production, i.e. in recrystallisation speed, which may be totally obstructed. Finely dispersed particles at a smaller distance obstruct, i.e. reduce subgrain moveability so that the subgrains which might transform into germs are activated with great difficulty and they are turned into new recrystallised grains. Finely dispersed particles and their homogenous distribution in the base metal matrix cause stabilization of dislocation substructure formed during deformation and create significant reinforcing effects by a complex action of different mechanisms [Lianga et al., 2004; Plascencia & Utigard, 2005]. The third, $Cu_xAl_yO_z$, phase has a considerable influence since it appears in the structure due to the eutectic reaction on Cu-Al contact surfaces [Jena et al., 2001, 2004; Korać et al., 2010]. Accordingly, it has been concluded that binding mechanisms on Cu/$CuAlO_2$, Cu_2O/$CuAlO_2$ and Al_2O_3/$CuAlO_2$ inter-surfaces are considerably stronger compared to those on Cu/Al_2O_3 and Cu_2O/Al_2O_3 contact surfaces, and indicated that the formed third phase influences the stabilization of dislocation substructure and thereupon the improvement of mechanical properties and the achievement of a good combination of electro - mechanical properties of the sintered systems [Entezarian & Drew, 1996; Lee & Kim, 2004; Yi et al., 1999; Yoshino & Shibata, 1992].

The characteristics of starting powders significantly determine the later-stage processing and sintering properties and eventually determine the microstructure of the composite. Nano-powders give better performance in sintering due to their high surface area and, therefore, can tremendously improve the sintering process. The parts produced from nano-powders will have high density, hardness and fracture toughness [http://www.nanopowders.com/].

Sintering is an extremely complex process with a complex action of multiple mechanisms of material transfer and the distribution and change of particle shapes, which are all processes that are closely connected with the off-balance condition of the crystal lattice and that have a considerable influence on the properties of the obtained metal materials. The driving force of sintering is the free energy of the system, which is off-balance from the thermodynamic point of view. This results from the highly developed free surface of powder particles and the presence of balance (point defects) and off-balance defects (dislocations) [Anđić, 2007; Ristić, 2003]. Sintering of ultradispersed powders occurs due to particles sliding along their borders and a dislocation mechanism that is responsible for the creation of surplus vacancies. Surplus vacancy concentrations can reach a value which corresponds to the vacancy concentration in the area of temperatures close to the temperatures of material melting. On the basis of this, it can be concluded that diffusion activity during sintering of ultra dispersed particles in the area of really low temperatures (0.1-0.3Tm) is conditioned by the presence of unbalancing 'recrystallization' vacancies. High recrystallization speeds of ultradispersed particles are a subsequence of the process of recrystallization self-activation [Lapovok & Novikov, 1983; Mohorov et al., 1977; Ristić, 2003]. A considerable number of studies have examined fundamental issues related to the mechanism and kinetics of sintering [Lapovok & Novikov, 1983; Mohorov et al., 1977; Ristić, 1993, 2003], by using the existing phenomenological equations which have considerably contributed to the theory and technology of sintering and also enabled numerous technological processes in the production of various ceramic and metal-ceramic materials. On the basis of the kinetic analysis of the process of sintering powders from the Cu-Al_2O_3 system, it was concluded that sintering process occurs in two phases. Given that the volume diffusion requires higher activation energy, the results of the kinetic analysis indicate that the surface diffusion and grain- boundary diffusion occur in the first phase, whereas the mass transfer occurs in the second phase during the process of volume diffusion [Ristić, 1993, 2003]. The kinetic analysis results indicate higher values of activation energy due to the exceptionally small sizes of particles of nano-composite Cu-Al_2O_3 powder, i.e. their large surface and an extremely large surface between individual particles created at the very beginning of the process.

As noted, disperse phases present in the material prior to deformation influence structure designing, whereas uniform dispersion of nanoparticles of Al_2O_3 in a copper matrix provides completely new, improved, even unexpected properties. However, these unique properties essentially depend on the microstructure, which is a direct consequence of the synthesis procedure of Cu-Al_2O_3 nanocomposites. By hydrometallurgy and powder metallurgy along with prognosis of physical-chemical properties, a synthesis of new improved materials can be successfully performed with pre-set properties which are conditioned by the quality of starting powders i.e. by the improvement of their structure. In addition to the conventional methods of obtaining composites based on Cu-Al_2O_3, the paper shows the synthesis of these composites in a chemical way by liquid phase deposition, with a comparative analysis of the properties of the obtained nano-composite sintered samples characterized by a good ratio of electro - mechanical properties, suitable for operations at high temperatures.

2. The synthesis and characterization of ultrafine and nanocomposite Cu-Al_2O_3 powder

2.1 The synthesis of ultrafine and nanocomposite Cu-Al_2O_3 powder

Soluble nitrates of copper and aluminium were used, $Cu(NO_3)_2 \times 3H_2O$ and $Al(NO_3)_3 \times 9H_2O$ as a transient component, were used for the synthesis of nanocomposite Cu-Al_2O_3 powder by a thermochemical method. The synthesis process developed in five stages (Fig. 1), as follows:

- obtaining 50% of water solution in which $Cu(NO_3)_2 \times 3H_2O$ and $Al(NO_3)_3 \times 9H_2O$ are dissolved up to achieving the requested composition of Cu-Al_2O_3 nanocomposite system with 3 and 5wt.% of alumina,
- drying by spraying, using a modified house sprayer at a temperature of 180°C with the aim of obtaining composite particles of nitrate salts,
- annealing of the obtained loose mixture in air atmosphere at the temperature of 900°C for one hour with the aim of forming copper oxide and phase transformation of Al_2O_3 up to a thermodynamically stable α-Al_2O_3 phase,
- reduction of thermally treated powders in hydrogen atmosphere at the temperature of 400°C for one hour, whereas copper oxide is transformed into elementary copper and α -Al_2O_3 remains unchanged,
- homogenisation in a jar mill, type HM1, milling chamber dimensions with an internal diameter of 180mm, height of 160mm and volume of 4L.

2.2 The characterization of ultrafine and nanocomposite Cu-Al_2O_3 powder

The characterization of the produced powders consisted of determining the particle specific area, the pouring density and fluidness by the differential thermal and thermogravimetric analysis (DT-TGA), X-ray diffraction (XRD) and analytical electron microscopy (AEM) coupled with energy dispersive spectroscopy (EDS).

The fluidness and pouring density were determined by the standard method, using a Hall apparatus, and in accordance with the appropriate standards (ASTM B13 and ASTM B212). The specific area of the particles was determined by the gas absorption method (BET).

DTA-TGA was performed using a NETZCH STA model 409EP up to 1100°C. A Pt-Pt-Rd alloy S-type thermocouple was utilized. α-Al_2O_3 was used as the reference material.

The quantitative X-ray diffraction analysis of the powder was performed using a Siemens D500 PC difractometer, CuK_α radiation, in the range $2\theta = 0\text{-}100^\circ$ with a step of $(2\theta)\ 0.02^\circ$.

The analytical electron microscopy (AEM) coupled with energy dispersive spectroscopy (EDS) was carried out on a JEOL 200CX microscope on powders spread on conductive carbon tape.

The results of the determination of fluidness, pouring density and specific area of the nanocomposite Cu-Al_2O_3 powder particles, with different dispersoid contents, obtained by

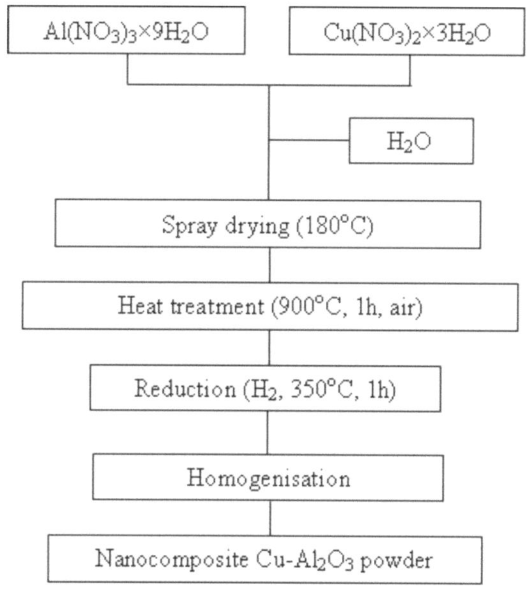

Fig. 1. Schematic presentation of the synthesis of Cu-Al_2O_3 nanocomposite powder by the thermochemical procedure

the thermochemical procedure are presented in Table 1. The results of the determination of fluidness, pour¬ing density and specific area of the obtained nano-structured composites with different amounts of Al_2O_3 dispersed in a copper matrix show that all investigated powders are not fluid and that mean values of pouring density and specific area are the same for different contents of Al_2O_3 (1.04 g/cm3 and 0.75 m^2/g, respectively).

Al_2O_3 (wt.%)	Fluidness	Pouring density (g/cm^3)	Specific area (m^2/g)
3	not fluid	1.04	0.75
5	not fluid	1.04	0.75

Table 1. Fluidness, pouring density and specific area of particles of the nanocomposite Cu-Al_2O_3 powder obtained by the thermochemical procedure (mean values)

The DTA-TGA curves of the nanocomposite Cu-Al_2O_3 powder with 5wt.% Al_2O_3, obtained by the thermochemical procedure are shown in Fig. 2. Two endothermic peaks may be observed on the DTA curve at approximately 150°C and 250°C, and they correspond to the evaporation and dehydratation of residual moisture, respectively. The exothermic peak at 324°C was accompanied by a mass increment of 5.88% and represents the beginning of fine copper powder oxidation. An intensive mass increase on the TG curve was recorded up to approximately 550°C, after which the TG curve levelled off, showing an insignificant mass increase of only a few percent. The overall mass increment during heating was 28.43%. Several exothermic peaks were observed at 684, 820, 885 and 938°C due to the phase transformations of Al_2O_3.

The XRD analysis of the spray-dried Cu-Al_2O_3 powder with 3wt.% of dispersoid is shown in Fig. 3. In accordance with the experimental set-up, only the peaks which correspond to copper and aluminium nitrates were registered in the structure.

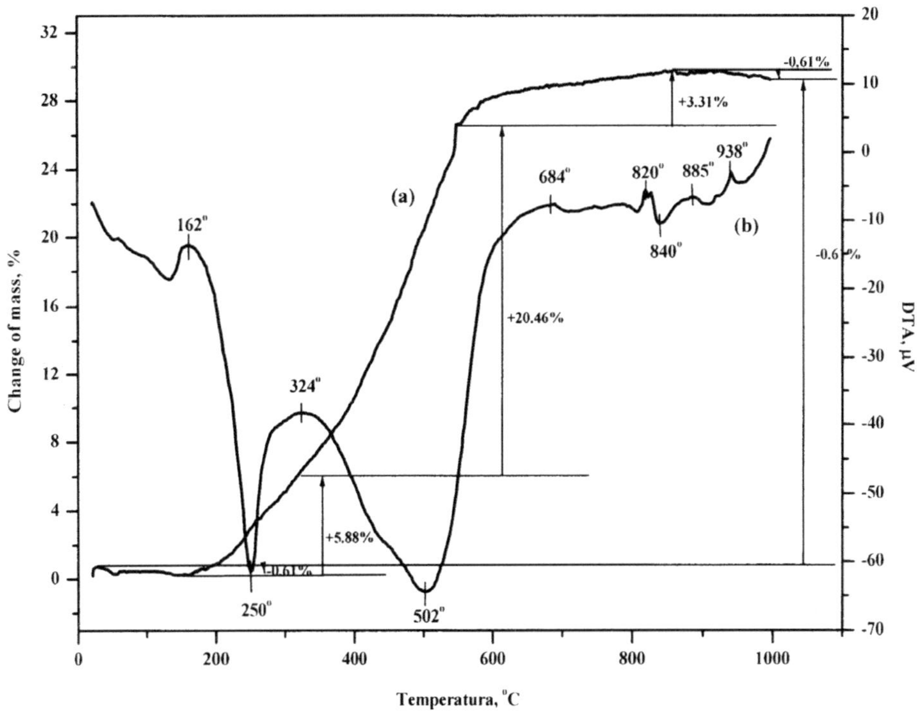

Fig. 2. DT-TGA curves of the Cu-5wt.%Al_2O_3 powder obtained by the thermochemical procedure

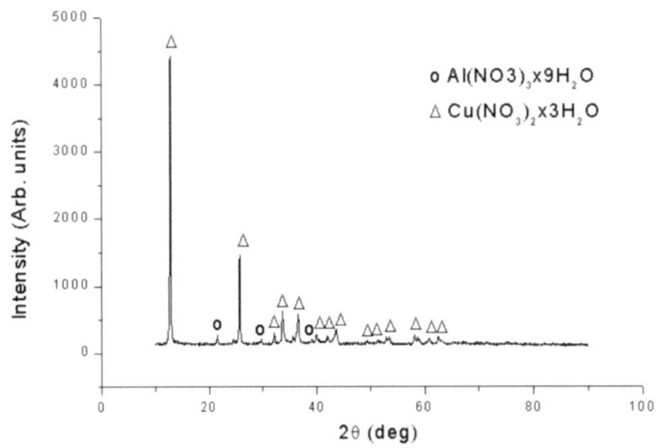

Fig. 3. XRD of the Cu-3wt.%Al_2O_3 powder after spray-drying

The X-ray diffraction analysis after annealing the dried powder is shown in Fig 4. The detected peaks correspond to CuO and Al_2O_3. An unidentified peak was also detected. According to previous studies, [Entezarian & Drew, 1996; Jena et al., 2001, 2004], this peak corresponds to the third phase, $Cu_xAl_yO_z$, which appears in the structure due to the eutectic reaction of (Cu+Cu_2O) with Al_2O_3. The formation of this phase is thermodynamically posible on Cu-Al contact surfaces. During the eutectic joining of copper and Al_2O_3, the eutecticum formed by heating up to the eutectic temperature expands and reacts with Al_2O_3 also forming $Cu_xAl_yO_z$, which is compatible with both phases on the intersurface. The formed third phase influences the nature of the dislocation structure and also the improvement of the mechanical properties and the creation of a good combination of mechanical and electrical properties of the sintered systems.

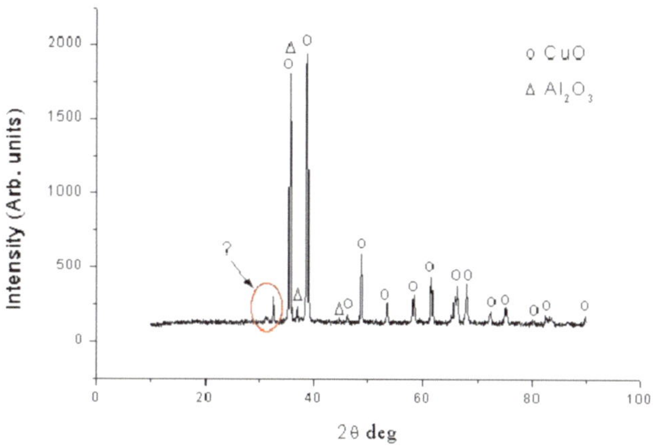

Fig. 4. XRD of the dried Cu-3wt.%Al_2O_3 powder after thermal treatment

XRD of powder after the reduction (Fig. 5) shows the presence of peaks which correspond to the elementary copper and Al_2O_3.

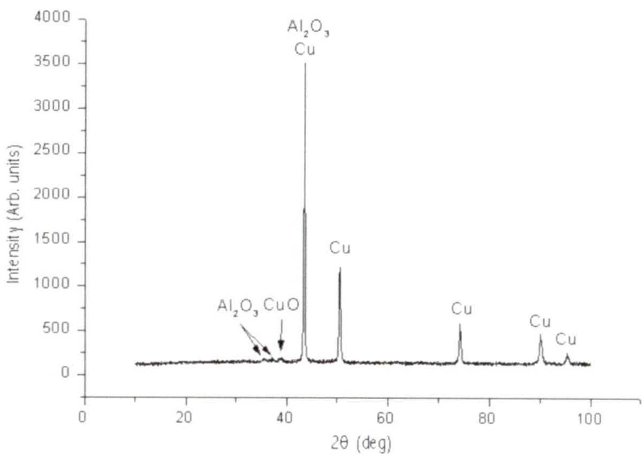

Fig. 5. XRD pattern of the Cu-3wt.%Al_2O_3 powder after reduction

The obtained powders were analyzed at the National Center for Electron Microscopy, University of California, Berkeley, by Analytical Electron Microscopy (AEM) with a suitable EDX (at the marked spot), as presented in Figure 6.

Particles of 20-50nm in size are clearly noticeable. The shape of particles is irregular, with the presence of individual particles of nodular shape. The surface morphology is rough. The agglomerates of the size >100nm, formed by the individual particles of the stated size are also observed. The agglomerates are of a spongy shape. Since the powders with exceptionally fine particles are obtained by the previously described procedure, the basic "condition" for the appearance of agglomerates is fulfilled. Namely, agglomeration of finer particles is a consequence of their large surface and high surface energy, respectively, and of the effect of attracting forces acting between them. The creation of attracting strains, the magnitude of which depends directly on the surface energy of the particles which are in contact, occurs on contact surfaces due to the atomic connections in the interface.
With the EDS composition of nanocomposite Cu-5wt.% Al_2O_3 powder is determined as shown in Figure 6. The EDS analysis of the spot marked in Figure 6. shows that the identified peaks correspond to Cu, Al and O. The intensity of peaks corresponds to the required composition of the examined systems. Therefore, the peak corresponding to copper is considerably higher than the peaks corresponding to aluminum and oxygen. The obtained results of the examinations are statistically processed and shown in Table 2.
The characterization of the obtained powder indicates a possibility of the synthesis of a nanocomposite Cu-Al_2O_3 system by a thermochemical method, starting from water solutions $Cu(NO_3)_2$ and $Al(NO_3)_3$.

Element	Weight%	Atom. %	Measurement discrepancy	Detection correction	k - factor	Absorption correction
O	6,087	20,191	0.189	0.376	2,702	1,000
Al	1,224	2,408	0.047	0.952	1,044	1,000
Cu	92,687	77,399	0.502	0.996	1,743	1,000

Table 2. Statistical data for the EDS analysis

3. Dispersion reinforced sintered system based on ultra fine and nanocomposite Cu-Al_2O_3 powders

After the characterization of powders, cold pressing by pressing force was performed from both sides with an apropriate tool with the dimensions of 8×32×2mm under compaction pressure of 500MPa. The hydraulic lab power press "ZIM", Russia, was used for pressing.
Sintering of the obtained samples was carried out in hydrogen atmosphere in isothermal conditions at two different temperatures, 800 and 900°C, for 30, 60, 90 and 120 minutes. Sintering was performed in a laboratory electric resistance tube furnace, with the power of 3kW and thermoregulation of ±1°C. The internal diameter of the furnace was 45mm and length 100cm. The maximum temperature in the working area (55cm) was 1,300±1°C.
The characterization of the Cu-Al_2O_3 sintered system included examinations of density, relative volume change, electrical and mechanical properties, examination of microstructure by the scanning electronic microscopy (SEM), the energetic dispersion spectroscopy (EDS), the transmission electron microscopy (TEM) and the high-resolution transmission electron microscopy (HRTEM), examinations by the focused ion beam (FIB) method and examinations of the selected area diffraction pattern (SADP).
The values of density, relative volume change, hardness and specific electrical resistence were determined by the standard methods in the above-mentioned examinations. The

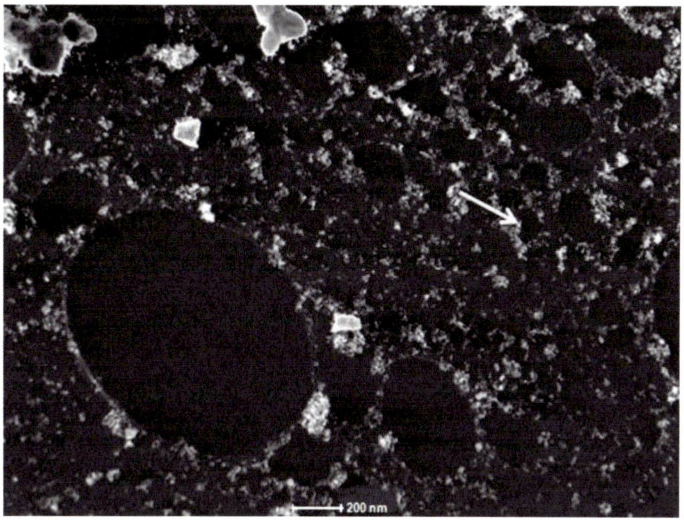

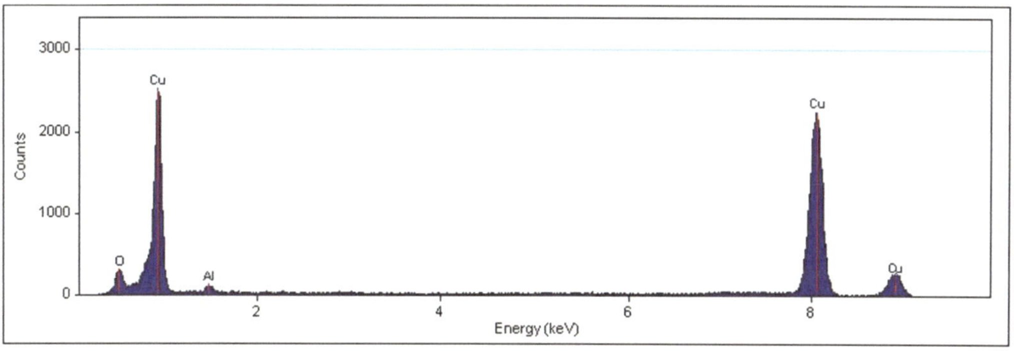

Fig. 6. AEM analysis of Cu-Al_2O_3 powder with 5wt.% of Al_2O_3

scanning electron microscopes, types JEOL JSM-T20 and JEOL 5300, as well as the analytical electron microscope (AEM), type JEOL 200CX with the ability to work in TEM and STEM modes between 80 and 200kV, were used for the examination of the microstructure of sintered systems. The contents of the examined systems were determined qualitatively by the energy dispersive spectroscopy (EDS - JEOL Superprobe 733, 20kV). HRTEM analysis was performed using Philips CM200/FEG.

The results of density examination, relative volume change, specific electric resistance and hardness are shown in Table 3..

The results of the research show that the density of the sintered samples at certain temperatures and times decrease with the increasing Al_2O_3 content. However, the density of the sintered samples increases with the increase in temperature. In the area of higher temperatures, when diffusion mobility of atoms is high enough, a complex action of all diffusion mechanisms of mass transfer occurs and they are responsible for the sintering process. With the increase in sintering temperature, the action of these mechanisms is more intensive, which directly affects the formation of contacts between certain particles, growth of contact surfaces, formation of closed pores and grain growth. What follows from this is that, after some time, the sintered density increases with temperature increase.

Temperature, °C	Time, min	Density $\bar{d}_s$, g/cm^3	$\Delta V/V_o$ (average)	Specific electric resistance ρ, $10^{-6}\Omega m$	Hardness HRB 10/40 (average)
Cu-3wt.%Al_2O_3					
800	15	5,58	0,1042	0,07413	88,2
	30	5,62	0,1194	0,07128	94,1
	60	5,70	0,1442	0,06581	102,1
	120	5,68	0,1448	0,06232	107,1
900	15	5,84	0,1821	0,06127	96,2
	30	6,14	0,1932	0,04027	101,9
	60	6,42	0,1933	0,03971	102,7
	120	6,44	0,1929	0,03927	102,3
Cu-5 wt.%Al_2O_3					
800	15	5,28	0,0612	0,08941	89,1
	30	5,34	0,0982	0,08827	101,2
	60	5,52	0,1191	0,08146	107,5
	120	5,58	0,1332	0,08007	109,1
900	15	5,94	0,1763	0,07413	99,1
	30	5,98	0,1824	0,06981	108,4
	60	6,14	0,1894	0,06218	118,5
	120	6,20	0,1888	0,06127	124,7

Table 3. Average density, $\Delta V/V_0$, specific electric resistance and hardness for sintered samples of Cu-Al_2O_3 with different alumina content

The relative volume change, as a measure of the system activity increases with the growth of the sintering temperature. The values of the relative volume change, at a certain temperature and sintering time, decrease with the increase in the Al_2O_3 content. Starting from general kinetic equations, and with the aim of an analysis of the sintering kinetics, the obtained results are in accordance with other research results and certainly confirm earlier investigations of the possibility of the use of existing phenomenological equations of sintering [Ristić, 2003].

The dependence of hardness and specific electric resistance on the temperature, time of sintering and alumina content are presented in Figs. 7 and 8, respectively.

The results of the examination of hardness of the sintered samples, as a measure of reinforcement of the highly conducting copper matrix, show that the growth of hardness value is a function of the decrease of specific electric resistance, i.e. of structural stabilization of the system. The results also point to the growth of hardness with Al_2O_3 content increasing, at a certain temperature and sintering time. The obtained results of hardness examination are the consequence of a relatively even distribution of Al_2O_3 dispersoids in the copper matrix. The relatively even distribution of alumina in the nanocomposite system, achieved during synthesis of powder by depositing from a liquid phase, causes stabilization of the dislocation structure and achievement of significant reinforcing effects by a complex action of several mechanisms. Thereby, reinforcement of the material with a small-grain structure can be caused by the reinforcement of the grain boundaries, dissolving reinforcement and by Orowan´s mechanism. Additionally, dislocations can disappear in the grain boundaries or cease to multiply, since the Frenk-Read source of dislocations cannot be activated in small-grained multi-phase materials, which represents an additional mechanism of reinforcement [Morris, D.G. & Morris M.A., 1992].

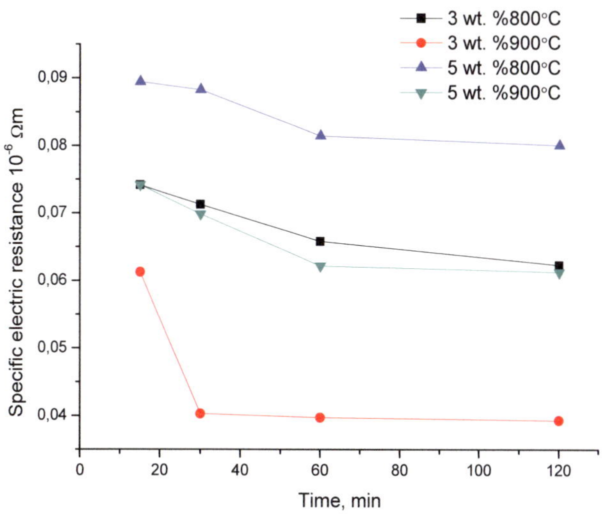

Fig. 7. Dependence of hardness on the sintering time at different temperatures and different alumina contents

With Al_2O_3 content increasing, the duration of the sintering process is increased. However, with increasing of the sintering temperature for a period of time, the value of specific electric resistance after sintering is decreased. In accordance with the stated and having in mind that the change of specific electric resistance represents a measure of structural stabilization of the system, it can be ascertained that at certain temperatures the structural stabilization of the system has not occurred, i.e. the structural stabilization process is not completed. Also, with the sintering temperature increasing, the duration of the sintering process is shortened (Fig. 8). Based on the value of specific electric resistance, as a measure of structural stabilization, for the system with 3wt.% Al_2O_3 during sintering at 800°C, sintering lasts 120 minutes, whereas for the same system during sintering at 900°C the sintering process lasts 30 minutes.

The analysis of the mechanical and electrical properties of Cu-Al_2O_3 sintered systems showed that for the system with 3wt.%Al_2O_3 structural stabilization had finished after 30 min of sintering at 900°C, with significant reinforcment effects. For the other systems, structural stabilization was not completed even after 120 min. The optimal solution for the production of dispersed reinforced Cu-Al_2O_3 is the process conducted at 900°C for 30 min with 3wt.%Al_2O_3, which was confirmed by microstructural analysis, presented in Figs. 9-11.

Considering the fact that the structure takes a central place in the complex of the materials science as a parameter through which the connection between the properties and synthesis of new materials is expressed i.e. since it directly determines mechanical as well as electrical

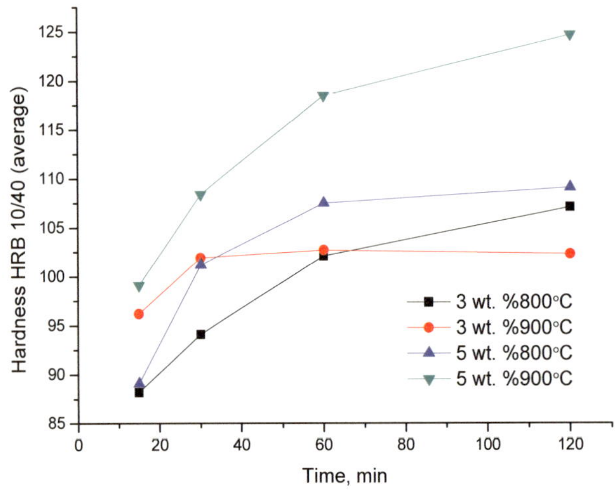

Fig. 8. Dependence of specific electric resistance on sintering time at different temperatures and different alumina contents

properties, the following development of microstructural constituents through all individual stages of the technological process is of an exceptional importance. The analysis of the microstructure of the corresponding sintered samples confirms the stated stipulations. In Fig. 9 a microstructure survey of the sample sintered at 800°C for 30 minutes is given, and it is clearly seen that the structural stabilization process is not completed. The microstructure is characterized by the formation of closed pores, which is typical of a medium stage of sintering, and also, in certain areas, by achieving contacts between certain particles, which is typical of the starting sintering stage. In Figs. 10 and 11. a survey of the microstructure of samples sintered at 900o°C for 15 and 30 minutes is given. The shown microstructures are characteristic of the medium (Fig. 10), i.e. final stages (Fig. 11) of sintering, which is confirmed by the analysis of the structural stabilization of the system based on the value of specific electric resistance of the sintered samples. Apart from that, a relatively even distribution of pores can be seen in the examined samples, which, among other things, significantly contributes to the reinforcement of the highly conductive copper matrix.

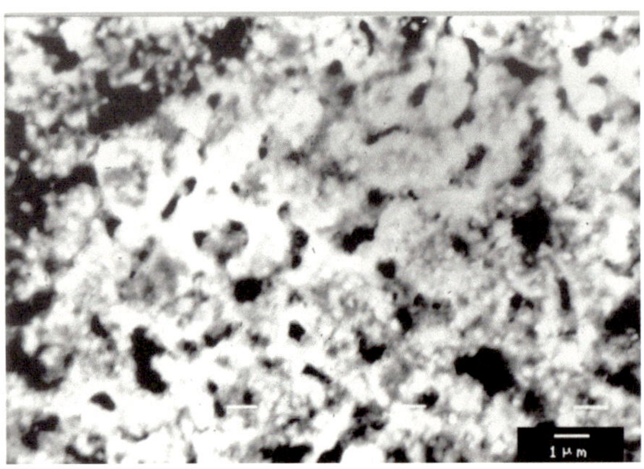

Fig. 9. SEM of the sintered Cu-3wt.%Al_2O_3 system (800°C, 30 min)

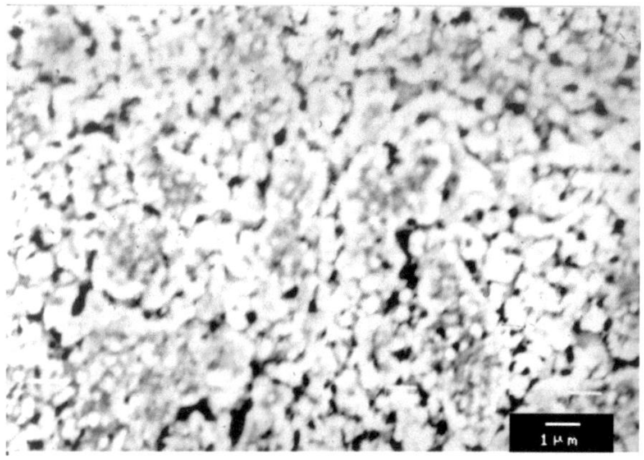

Fig. 10. SEM of the sintered Cu-3wt.%Al_2O_3 system (900ºC, 15 min)

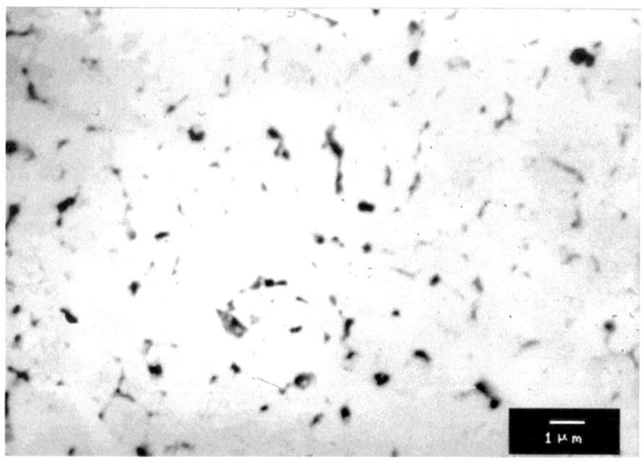

Fig. 11. SEM of the sintered Cu-3wt.%Al_2O_3 system (900ºC, 30 min)

In order to determine the distribution of elements in the structure, the surface analysis of the sample was performed by EDS. The SEM microphotograph of the examined sample with the surface on which the surface scanning was performed is shown in Fig. 12, and the results of the examination of the sample of the sintered Cu-3wt.%Al_2O_3 system by EDS are given in Fig. 13.

The results of surface scanning show a homogeneous distribution of elements in the structure. In Fig. 13.a it can be seen that copper covers almost the entire surface of the sample. The results of surface scanning of aluminium (Fig. 13.b) and oxygen (Fig. 13.c) show that these two elements are present less in the structure of the sintered sample and that the surfaces they occupy are inter-lapping, which corresponds to the existence of an Al_2O_3 dispersoid in the structure. Beside aluminium and oxygen, an inter-lapping of all three elements is also noticeable, which leads to the assumption of the presence of a $Cu_xAl_yO_z$ phase [Jena, et al., 2004]. Finally, an apparatus with an exceptionally high resolution is necessary for the detailed characterization of the third. $Cu_xAl_yO_z$ phase.

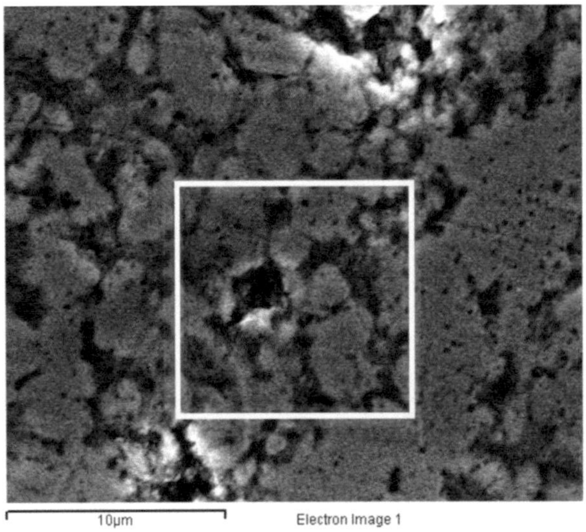

Fig. 12. SEM of the sintered Cu-3wt.%Al_2O_3 sample with the surface scanning area marked

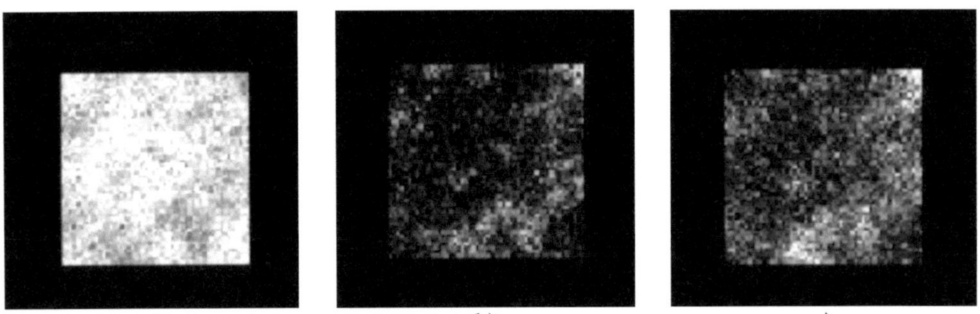

Fig. 13. Surface scanning of the sintered Cu-3wt.%Al_2O_3 sample by EDS for: a) copper, b) aluminium, c) oxygen

The characterization of the Cu-Al_2O_3 sintered systems based on the powders obtained by the thermochemical procedure, apart from the stated examination, has included examinations by the Focused Ion Beam (FIB), examinations with Transmission Electronic Microscopy (TEM), examinations of selected diffractive fields (SADP) and High Resalution Transmission Electronic Microscopy Examinations (HRTEM), which were carried out at the National Center for Electronic Microscopy, University of California, Berkeley (USA).

The FIB analysis of the sintered Cu-Al_2O_3 system based on the powders obtained by the thermochemical procedure at different magnifications (Fig. 14) did not indicate, even at considerably higher magnifications, the existence of a phase rich in alumina. In accordance with [Jena et al., 2001], bright fields are identified, i.e. a phase rich in copper, as well as gray fields, which lead us to the possibility of the existence of the third, $Cu_xAl_yO_z$ phase. In addition to this, the stated examinations by Scanning Electronic Microscopy [Jena, et al., 2001] are in accordance with the results of the surface EDS analysis showing that the peaks are identified and correspond to copper, aluminum and oxygen, and finally, having in mind that their intensity corresponds to the demanded composition, this additionally points to the

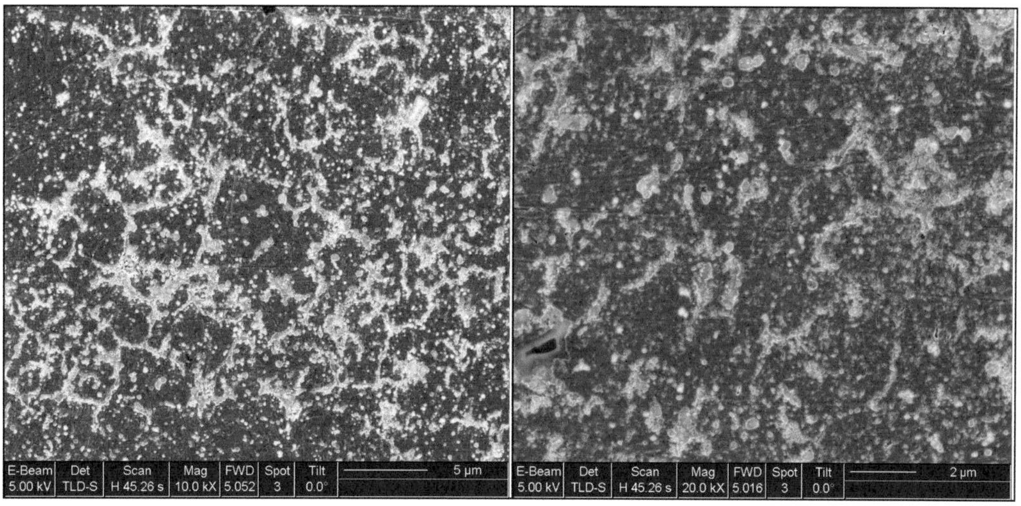

Fig. 14. FIB of the sintered Cu-5wt.% Al_2O_3 system (900°C, 120 nm)

possibility of the existence of the third, $Cu_xAl_yO_z$ phase. The formation of this phase is thermodynamically possible on Cu-Al contact surfaces. During the eutectic joining of copper and Al_2O_3, the eutecticum formed by heating up to the eutectic temperature expands and reacts with Al_2O_3 creating $Cu_xAl_yO_z$, which is compatible with both phases on the inter-surface. In accordance with [Entezarian & Drew, 1996; Jena et al., 2001], the process of formation of the third phase is developed through the following reactions: $CuO + H_2O \rightarrow Cu_2O + H_2O$, $Cu_2O + Al_2O_3 \rightarrow 2CuAlO_2$ and/or $CuO + Al_2O_3 \rightarrow CuAl_2O_4$, whereas Misra and Chaklader [Misra & Chaklader, 1963] have published that CuAlO2 is stable in the air with the temperature range from 800°C to 1000°C, whereas $CuAl_2O_4$ is transformed into $CuAlO_2$ at the temperature of app. 1,000 °C.

Additionally, from microphotographs of the examined samples, homogenous distribution of the present phase is clearly noticeable as well as the size of microstructural constituents in the range of 50 - 250nm (Fig. 14).

As noted, finely dispersed particles and their homogenous distribution in the base metal matrix cause stabilization of dislocation substructure formed during deformation and the achievement of significant reinforcing effects by A complex action of different mechanisms [Lianga et al., 2004; Plascencia & Utigard, 2005]. The third, $Cu_xAl_yO_z$ phase has a considerable influence since it appears in the structure due to the eutectic reaction on Cu-Al contact surfaces [Jena et al., 2001, 2004; Korać et al., 2010]. Consequently, it has been concluded that the binding mechanisms on $Cu/CuAlO_2$, $Cu_2O/CuAlO_2$ and $Al_2O_3/CuAlO_2$ inter-surfaces are considerably stronger compared to Cu/Al_2O_3 and Cu_2O/Al_2O_3 contact surfaces, indicating that the formed third phase has an influence on the stabilization of dislocation substructure and thereupon the improvement of mechanical properties and achievement of a good combination of electro - mechanical properties of the sintered systems [Entezarian & Drew, 1996; Lee & Kim, 2004; Yi et al., 1999; Yoshino & Shibata, 1992].

A typical TEM pair bright field (BF) - the centered dark field (CDF) of nanocomposite Cu-5wt.%Al_2O_3 sintered system based on powders synthesized by the thermochemical procedure are shown in Figure 15 (a and b).

The TEM characterization of nanocomposite Cu-5wt.%Al_2O_3 sintered system based on powders synthesized by the thermochemical procedure reveals a microstructure with several interesting characteristics. In Fig. 15a and 15b, a typical TEM pair bright field (BF) - centered dark field (CDF) of nanocomposite Cu-5wt.% Al_2O_3 sintered system is shown, with the well developed crystals of copper of 100 nm size exposed to twinning despite their small size. The conditions for twinning are achieved when a great number of obstacles are created in the crystal, which hamper the moving of dislocations, dislocation plaits or already present twins. Since dislocations are piled up at the obstacles, internal tension is increased in the local regions, and this, along with external tension, causes the creation of twins. Having in mind that the decrease in dislocations mobility represents a condition for creating a twin embryo, the clearly noticeable presence of twins in Fig. 15a and 15b, indicates a decreased mobility of dislocations, i.e. stabilization of dislocation substructure, which is an elementary precondition for improving mechanical properties. i.e. for the reinforcement of metal materials. Additionally, when the deformed metal material is heated up to the sufficiently high temperature, recrystallization and grain growth occur, whereas in the stated processes and especially for the grain growth, grain boundaries movement is a basic mechanism which causes a relieved formation of twins.

Typical models of selected diffractive fields (SADP) of the examined nanocomposite Cu-5wt.% Al_2O_3 sintered systems based on the powders synthesized by the thermochemical procedure are shown in Figure 15c - at the point (marked by arrow in Fig. 15a) and Fig. 15d - from the overall area.

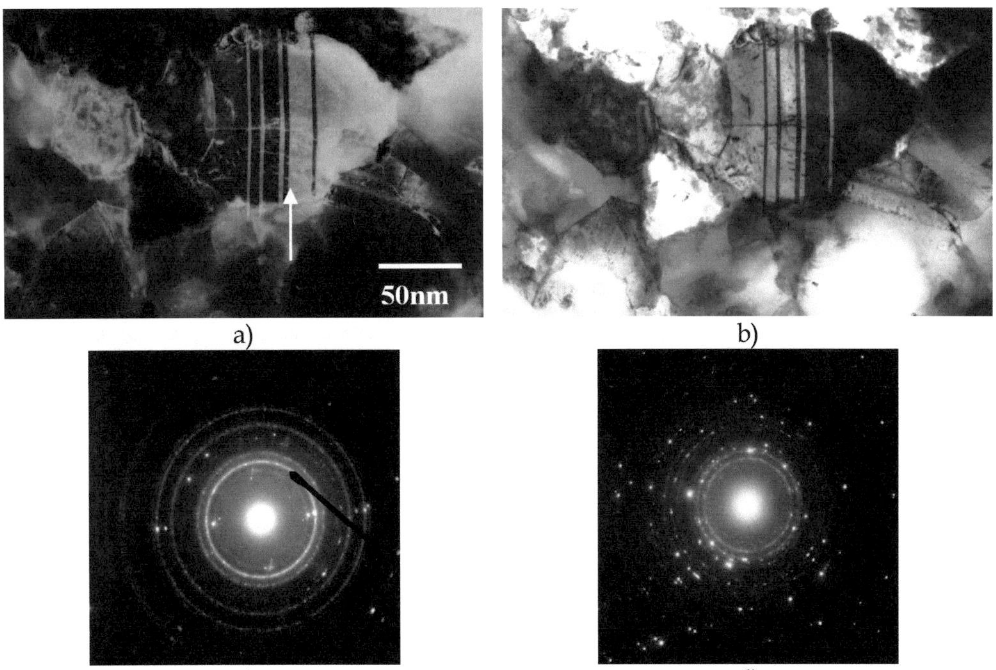

Fig. 15. (a and b) TEM BF - CDF pair of the nanocomposite Cu-5wt.%Al_2O_3 sintered system, (c) SADP centered at the point (marked by arrow in Fig. 15a), (d) SADP taken from the overall area

SADP clearly shows spots and rings, where spots refer to the individual crystals of copper, and sharp circles (rings) originate from nanocrystals of alumina dispersed in the copper matrix. Except for copper and alumina, diffractive fields are not identified, which is typical of some other phenomena (e.g. appearance of the third phase). The reason for that could be the common limitations of the technique, i.e. the determined critical value below which the phases cannot be identified. The presence of nano twins within the crystals of copper is partially confirmed by the selected diffractive fields taken from the overall surface (Figure 15d). Finally, apart from the stated, the obtained results of the examinations confirm the crystal nature of copper and dispersed alumina.

The high resolution transmission electron microscopy (HRTEM) of nano composite Cu-5wt.% Al_2O_3 sintered system based on powders synthesised by the thermochemical method is shown in Figure 16.

The HRTEM photos of sintered samples indictae a change in the lattice parameter on the grain boundary, where'raster' and 'fingerprint' variations can be detected. This leads to the conclusion that, due to the presence of alumina, the eutectic reaction (Cu+Cu_2O) with Al_2O_3 occurred and the third, $Cu_xAl_yO_z$ phase was formed, which is possible on Cu-Al contact surfaces from the thermodynamic point of view.

Bearing in mind the results of all of the above-mentioned examinations, the reinforcement of the copper matric occurs due to the to mechanisms: dispersion and reinforcement due to the dispersion of the fine particles of Al_2O_3 in the matrix, and the reinforcement of grain boundaries due to the appearance of the third phase.

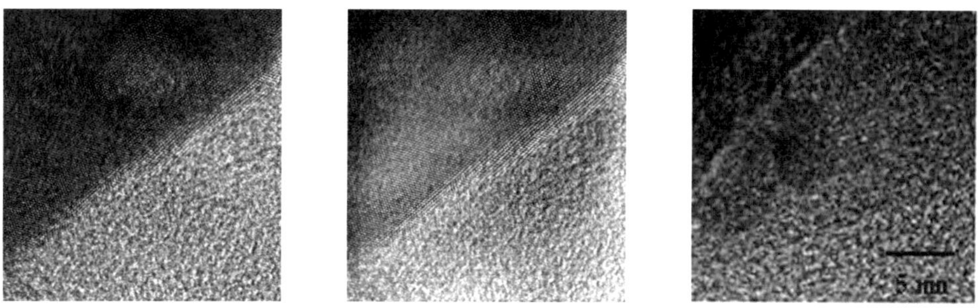

Fig. 16. HRTEM of nanocomposite Cu-5wt.% Al_2O_3 sintered system

4. Conclusion

Characterization of the obtained powder indicates a possibility of the synthesis of nanocomposite Cu-Al_2O_3 system by the thermochemical procedure, starting from water solutions $Cu(NO_3)_2$ and $Al(NO_3)_3$. AEM analysis indicates the presence of individual particles of 20-50 nm size. The shape of particles is irregular, with the presence of individual particles of nodular shape. The surface morphology is rough. Apart from that, the presence of an agglomerate with the magnitude of >100nm and of a sponge shape is noticeable. The obtained nanocomposite powders, with the structure basically preserved together with the final product, provided a significant reinforcement effect in the obtained sintered system. This is a consequence of the homogenous distribution of the elements in the structure, accomplished during the synthesis of powder and presence of the third phase which causes stabilization of dislocation substructure and achieves the relevant reinforcing effect.

The analysis of the mechanical and electric properties of the sintered Cu-Al_2O_3 systems based on powders obtained by the thermochemical method shows that in the system with 3wt.%Al_2O_3, sintered at 900°C, structural stabilisation occurs only after 30 minutes with considerable reinforcement effects. Since in other systems, the structural stabilisation process was not completed even after 120 minutes, the system with 3wt.% of dispersoids sintered at 900°C for 30 minutes seems to be the optimum solution for the production of dispersively reinforced Cu-Al_2O_3 systems. This statement is confirmed by the corresponding analysis of the microstructure.
In accordance with the previous statements, the EDS analysis of the sintered systems surface as well as FIB analysis show a homogenous distribution of elements, i.e. present phases. FIB analysis also indicates the size of microstructural constituents within the range of 50-250nm. TEM analysis of the sintered systems reveals the presence of copper crystals of 100 nm in size exposed to twinning, thus pointing to stabilization of the dislocation substructure. SADP of the examined nanocomposite Cu–Al_2O_3 sintered system shows spots and rings, where spots refer to the individual crystals of copper, and sharp circles (rings) originate from nanocrystals of alumina dispersed in the copper matrix. HRTEM analysis indicates the changes in the lattice parameter, which leads to the conclusion that the eutectic reaction occurred and the third, $Cu_xAl_yO_z$ phase was formed on the grain boundary, which will be the subject of future studies based on a more precise approach to physical chemistry of system surfaces and of thermodynamic examinations of the influences of finely dispersed Al_2O_3 on the formation of the third phase and the increase in the system surface energy.
In the end, all the above-mentioned examinations show that the reinforcement of Cu-Al_2O_3 system occurs via two mechanisms, which are: dispersion and reinforcement mechanism due to the homogenous dispersion of fine particles of Al_2O_3 in the matrix, and the mechanism of grain boundary reinforcement due to the appearance of the third phase.

5. Acknowledgement

The authors wish to thank Ministry of Science and Technological development of Republic of Serbia for financial support, projects TR19032 "Synthesis of nanostructured powders designed for production of the new dispersion strengthened sintered materials in system Cu-Al_2O_3" and TR34033 "Innovative synergy of by-products, waste minimization and cleaner technologies in metallurgy" and Prof. dr Velimir Radmilović, National Center for Electron Microscopy, Lawrence Berkley National Laboratory, Berkley for help with instrumental techniques and interpretation of results.

6. References

Afshar, A. & Simchi, A., (2008). *Scripta Materialia*, 58, 966.

Ahn, J., Song, I. & Hahn, Y., (1996). *Mater. Trans. JIM*, *37*, 733

Anđić, Z., Korać, M., Tasić, M., Raić, K. & Kamberović, Ž. (2006). The synthesis of ultra fine and nanocomposite powders based on copper, silver and alumina, *Kovove Mater.*, 44, pp. 145-150

Anđić, Z., Korać, M., Kamberović, Ž., Vujović A.& Tasić, M. (2007). Analysis of the Properties of a Cu-Al_2O_3 Sintered System based on Ultra Fine and Nanocomposite Powders, *Science of Sintering*, 39, pp. 145-152

Anđić, Z., (2007). D. Sc. Thesis, Faculty of Technology and Metallurgy, University of Belgrade, Serbia

Byrappaa, K. & Adschirib, T., (2007). Hydrothermal technology for nanotechnology, *Prog. Cryst. Growth Charact. Mater.*, 53, pp. 117-166

Entezarian, M. & Drew, R., (1996). Direct bonding of copper to aluminum nitride, *Materials Science and Engineering, A: Structural Materials: Properties, Microstructure and Processing*, A212(2), pp. 206-212

Hosokawa, M., Nogi, K., Naito, M. & Yokoyama, T., (2007). *Nanoparticle Technology Handbook*, Elsevier, ISBN 978-0-444-53122-3, Oxford, UK

Jena, P., Brocchi, E. & Motta, M., (2001). In-situ formation of Cu-Al_2O_3 nano-scale composites by chemical routes and studies on their microstructures, *Materials Science and Engineering*, A313m, pp. 180-186

Jena, P., Brocchi, E. & Motta, M., (2001). Characterization of Cu–Al_2O_3 nano-scale composites synthesized by in situ reduction, *Materials Science and Engineering*, C15, pp. 175-177

Jena, P., Brocchi, E., Solórzano, I. & Motta, M., (2004). Identification of a third phase in Cu-Al_2O_3 nanocomposites prepared by chemical routes, *Materials Science and Engineering*, A317, pp. 72-78

Karch, J., Birringer, R. & Gleiter, H., (1987). Ceramics ductile at low temperature, *Nature*, 330, pp. 556-558

Korać, M., Anđić, Z., Tasić, M., & Kamberović, Ž. (2007). Sintering of Cu-Al_2O_3 nano-composite powders produced by a thermochemical route, *J. Serb. Chem. Soc.*, 72(11), pp. 1115-1125

Korać, M., Kamberović, Ž., Anđić, Z., Filipović, M. & Tasić, M., (2010). Sintered Materials Based on Copper and Alumina Powders Synthesized by a Novel Method, *Science of Sintering*, 42(1), pp. 81-90

Lapovok, V., & Novikov, V., (1983). *Fizika tverdogo tela*, 25, 1848, Russia

Lee, D. & Kim, B., (2004). Nanostructured Cu–Al_2O_3 composite produced by thermochemical process for electrode application Materials Letters, *Materials Letters*, 58, pp. 378-383

Lee, D., Ha, G. & Kim, B., (2001). Synthesis of Cu-Al_2O_3 nano composite powder, *Scripta mater.*, 44, pp. 2137-2140

Lianga, S., Fana, Z., Xua, L. & Fangb, L., (2004). Kinetic analysis on AlO/Cu composite prepared by mechanical activation and internal oxidation, *Composites Part A Applied Science and Manufacturing*, 35(12), pp. 1441-1446

Milošević, O., (1999). Nanostructured materials and their development in light of the triad "synthesis - structure - properties", *Sintering and sintered materials in the ligt of the triad "synthesis – structure - properties"*, Monograph of material science, SANU, 38, pp. 55-65

Misra, S. & Chaklader A., (1963). System Copper Oxide-Alumina, *J. Am. Ceram. Soc.*, 46, 509

Mohorov, I., Trusov L. & Chiszhnik, S., (1977). *Ultrdispersnoe metallicheskie sredoe*, Atomizdat, Moskva, Russia

Morris, D.G. & Morris M.A., (1992). In Structural Applications of Mechanical Alloying, F.H. Froes, J.J. DeBarbadillo, (Ed.), ASM, Metals Park, Ohio

Naser, J., Ferkel, H. & Riehemann, W., (1997). Grain stabilisation of copper with nanoscaled Al_2O_3-powder, *Materials Science and Engineering*, A 234-236, pp. 470-473

Plascencia, G. & Utigard, T., (2005). High temperature oxidation mechanism of dilute copper aluminium alloys, *Corros. Sci.*, 47, pp. 1149-1163

Ristić, M., (1993). Principles of materials science, P. Miljanić, Eds., Serbian academy of science and arts, Department of technical science, Monofraphs, Vol. DCXVII, No. 36, Belgrade

Ristić, M., (2003). Fundamental problems of material science, TFČ and SANU, Čačak

Ruys, A. & Mai, Y., (1999). The nanoparticle-coating process: a potential sol–gel route to homogeneous nanocomposites, *Mater. Sci. Eng.*, A 265, pp. 202-207

Trian, B., Liua, P., Songa, K., Lia, Y., Liua, Y., Rena, F. & Sua, J., (2006). Microstructure and properties at elevated temperature of a nano-Al_2O_3 particles dispersion-strengthened copper base composite, *Mater. Sci. Eng.*, A 435-436, pp. 705-710

Trojanová, Z., Ferkel, H., Luká, P., Naser, J. & Riehemann, W., (1999). Thermal stability of copper reinforced by nanoscaled and microscaled alumina particles investigated by internal friction, *Scripta Materialia*, 40(9), pp. 1063-1069

Wu, Y., Zhang, Y., Huang, X. & Guo, J., (2001). Preparation of platelike nano alpha alumina particles, *Ceramics International*, 27, pp. 265-268

Yi, S., Trumble, K. & Gaskell, D., (1999). Thermodynamic analysis of aluminate stability in the eutectic bonding of copper with alumina, *Acta. Mater.*, Vol. 47, No. 11, pp. 3221-3226

Yoshino, Y. & Shibata, T., (1992). Structure and Bond Strength of a Copper-Alumina Interface, *J. Amer. Ceramic Soc.*, 75 (10), pp. 2756-2760

Yuana, G., Jiang, H., Lina, C. & Liaoa, S., (2007). Shape and size controlled electrochemical synthesis of cupric oxide nanocrystals, *J. Cryst. Growth*, 303(2), pp. 400-406

Nano Powder Industries, http://nanopowders.com/applications.htm

9

Alkaline-Earth Metal Carbonate, Hydroxide and Oxide Nano-Crystals Synthesis Methods, Size and Morphologies Consideration

Mohammad Amin Alavi and Ali Morsali
Department of Chemistry, Faculty of Sciences, Tarbiat Modares University
Islamic Republic of Iran

1. Introduction

Inorganic materials are of fundamental interest and technological importance due to their broad application in materials chemistry. It is well documented that the properties of inorganic nano-materials depend strongly on their morphologies, thus, the design and controlled synthesis of nanostructures with different morphological configurations and size distribution on a large scale is very important from the viewpoint of both basic science and technology [1–4]. Barium carbonate ($BaCO_3$) is an important mineral as a more thermodynamically stable crystal modification among the heavy metal carbonates and an important material in industry to produce barium salts, pigment, optical glass, ceramic, electric condensers and barium ferrite [5]. Strontium carbonate is an important raw material in modern electronic industry [6] and the glass industry [7]. Furthermore, strontium carbonate has only one crystal-phase, so it has been widely studied as a model system for bio-crystallization [8–13]. CaO is a material having a wide range of applications, being of continuous interest in the field of materials research. Pure CaO is an oxide with cubic lattice structure [14] with anisotropic catalytic properties and is often investigated as a component in catalytic powder materials or cements [15]. CaO is also known as dopant able to stabilize cubic zirconia [16] or hafnia [17], and fluently modify the refractive index of silicate glasses [18]. Due to its wide band gap (7.1 eV) [19], high dielectric constant (11.8) [20] and ability to form solid solutions and ternary crystalline phases CaO and their ternary alloys can be considered as interesting dielectric gate materials, exhibiting high mechanical and radiation resistance [21]. The magnesium oxide (MgO) is *a* very suitable material for insulation applications due to their low heat capacity and high melting point [22]. Nano-MgO is a functional material that has been widely used in various areas [23, 24]. Recently, it has been reported [25, 26] that MgO has *a* good bactericidal performance in aqueous environments due to the formation of superoxide (O_2^-) anions on its surface. Klabunde and co-workers [27, 28] demonstrated that nano- MgO exhibits high activity against bacteria, spores and viruses after adsorption of halogen gases because of its large surface area, abundance in crystal defects and positively-charged particles which can result in strong interactions with negatively-charged bacteria and spores [29].
There are different methods for synthesis of these nano-materials and some of them are common. Here we tried to describe a summary of these methods and in the separated section their explanation are expanded one by one.

2. $BaCO_3$ nanostructures

Xu et al. [30] synthesized $BaCO_3$ nanowires in the Triton X-100/cyclohexane/water reverse micelles. Reverse micelle or microemulsion (soft template) is increasingly used to prepare nanowires and nanorods, and works as a microreactor that can compartmentalize reactants at the nanometer level in discrete water. The shape, size and size distribution of nanowires and nanorods prepared by the reverse micelle could be controlled by using different reaction temperature, surfactant, additives, surfactant concentration and mole ratio of water to surfactant (ω_0) [31]. Triton X-100 (Tx-100) is a nonionic surfactant that forms a nonspherical micelle in cyclohexane. Fig. 1 shows the TEM micrographs of $BaCO_3$ nanowires with different magnifications after aging for 48 h. It can be clearly seen that many wires are disordered pack with length up to several tens of micrometers[30].

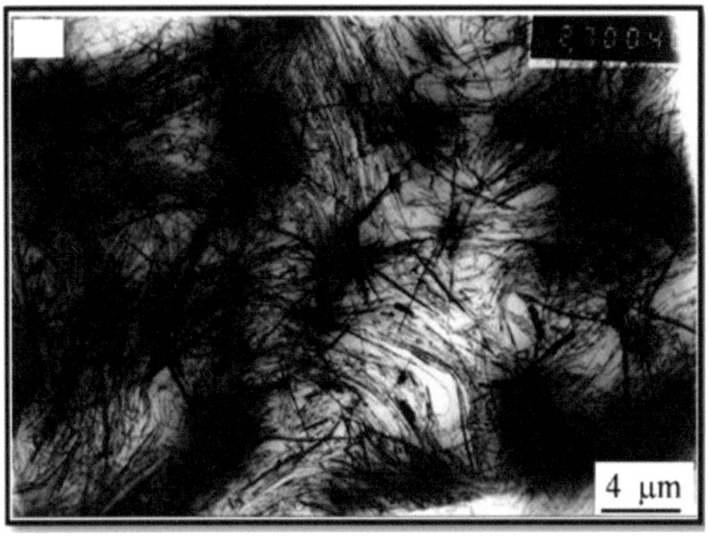

Fig. 1. TEM images of $BaCO_3$ samples formed in Tx-100 reverse micelles [30]. Reproduced with permission of Elsevier.

Microemulsion method was used by Momchilova et al. [32] for preparation nano-sized particles of $BaCO_3$ and microemulsion is applied as a special microreactor to limit the nano-sized particles growth. The shape of the microreactor depends on reaction conditions [33]. This method increases the homogeneity of the chemical composition at nano-level and facilitates the preparation of nano-particles with comparatively equal sizes [34]. The specific properties of the nano-particles make them suitable for microelectronics, ceramics, catalysis, medicine, cosmetics, as piezoelectric materials, conductors and etc. Photographs of nano-sized particles of $BaCO_3$ are presented in Fig. 2. The nano-particles synthesized were with spherical shape and almost equal sizes varying from 20 to 30 nm. The electron-microscopy analysis revealed that processes of particles agglomeration took place. Since the opalescence observed with the organic phase from the experiments was not very good, the agglomeration was supposed to occur during the electron microscopy analysis itself.

Chu et al. [35] synthesized $BaCO_3$ nanobelts and nanorods by Microemulsion-based method. The introduction of microemulsions has provided a relatively simple and powerful method for controlling the size, shape and surface texture of nanoparticles. In such a microenvironment, the nanomaterials are encapsulated into the closed shells of surfactant molecules. The size and shape of nanomaterials could be well controlled by the size of capsules through varying with the different ratios of water and oil. $BaCO_3$ one-dimensional structures were all successfully synthesized in CTAB/cyclohexane/H_2O and NP_{10}/cyclohexane/H_2O microemulsion, respectively. Belt-like $BaCO_3$ nanoparticles were synthesized with solvent hydrothermal treatment. When the 0.18 ml of 0.1 M Ba^{2+} existed in CTAB microemulsion system, only nanowhiskers of $BaCO_3$ were obtained with 200–400 nm in length and 10–20 nm in diameter. Concentration of surfactant in the microemulsion is another important parameter that affects the morphology of as-obtained samples. When $[NP_{10}]$=0.0639 M, all the samples were composed of nanorods. Increasing value of concentration resulted in the formation of short nanorod of $BaCO_3$ samples. With the increasing of CTAB concentration, the water pool becomes smaller, and the shape changes from rod to ellipsoidal. Solvent hydrothermal treatment of a resulting microemulsion at 140 °C resulted in the formation of $BaCO_3$ nanobelts with orderly edges, diameter of 200 nm and length of 1.8–3.4 μm (Fig. 2). The stripes in the nanobelts confirm that the belts are very thin [35].

Fig. 2. TEM images of $BaCO_3$ obtained by hydrothermal treating for 12 h. $[NP_{10}]$=0.127 M and T=140 °C [35]. Reproduced with permission of Elsevier.

The barium carbonate nanoparticles were synthesized nanoparticles by flame spray pyrolysis (FSP) [36]. FSP is usually applied for the preparation of metal and metal-oxide nanoparticles [37, 38]. Recently it has been shown that also nanosized salts such as carbonates, phosphates and halogenides can readily be made by flame synthesis [39–41]. The as-prepared material consisted of non-agglomerated crystalline $BaCO_3$ nanoparticles (Fig. 3). TEM analysis further revealed the formation of "bean-like" shaped particles with length of about 100 nm and width of approximately 50 nm. The corresponding BET-particle diameter for a spherical particle was 70 nm (specific surface area (SSA): 20.5m^2/g) in good agreement with the TEM observations [36].

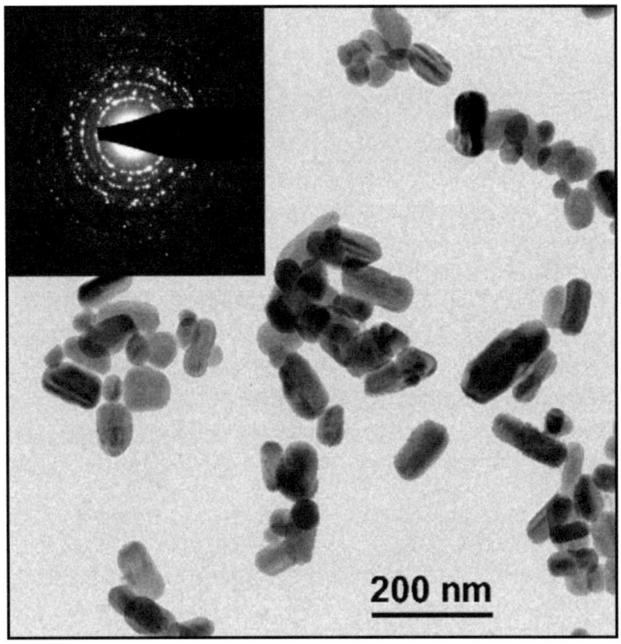

Fig. 3. TEM image and ED pattern (inset) of flame-made $BaCO_3$ nanoparticles [36]. Reproduced with permission of Elsevier.

Sun et al. [5] synthesized single crystalline $BaCO_3$ with different morphologies and sizes by a facile and practicable approach in the presence of polyvinylpyrrolidone (PVP) as a guide reagent at room temperature. Some polymer-assisted syntheses, such as polyacrylic acid (PAA), polyvinyl alcohol (PVA), and polyvinylpyrrolidone (PVP), have been used to prepare a lot of nanomaterials. It is implied that these polymers either changed the surface chemistry of the crystals, or as a guide reagent for crystal growth. Especially, it was found that PVP had excellent effects on the anisotropic growth of Ag, Au nanowires, etc. [42]. The morphologies and microstructure of the as-synthesized products were further investigated by TEM. The bundles of rods, dendritic structures and nanorod morphologies are indeed obtained. The morphology of the bundles of rods (Fig. 4) is similar to the CeO_2 prepared by Bai et al., which was described as comet-like [43]. In the process of the reaction, the quantity of CO_3^{2-} increased gradually with the pH value of the reaction system increasing. Therefore the number of Ba^{2+} is large relative to CO_3^{2-} under low pH conditions, and certain facets with relatively higher free energies will preferentially form the active sites and show higher growth rate, thus, the bundles of rods and dendritic structures are obtained, which indicate that the preferential orientation is strongly dependent on the pH value of the aqueous solution [5].

Zhu et al. [44] synthesized $BaCO_3$ nanostructures in water/ethylene glycol mixed solvents by microwave-assisted method. The EG is a polar solvent which has a reducing ability. It has been widely used in materials synthesis [45–49]. EG is miscible with water at any ratio, and the addition of EG to water can easily change the physicochemical properties. The morphologies of the samples were investigated with SEM and TEM. Fig.5 shows SEM image of the prepared $BaCO_3$ by oil bath heating in water (10 mL)/EG (10 mL) mixed solvents at 80 °C for 30 min., from which one can see $BaCO_3$ nanorods with diameters of

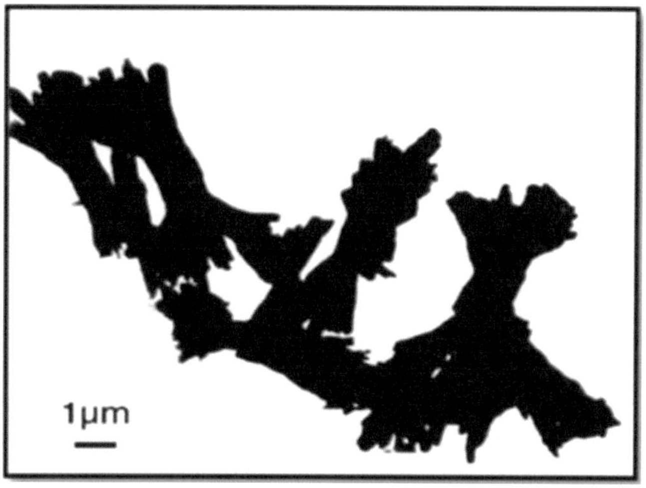

Fig. 4. TEM images of synthesized $BaCO_3$ by PVP-assisted method at pH = 12 [5]. Reproduced with permission of Elsevier.

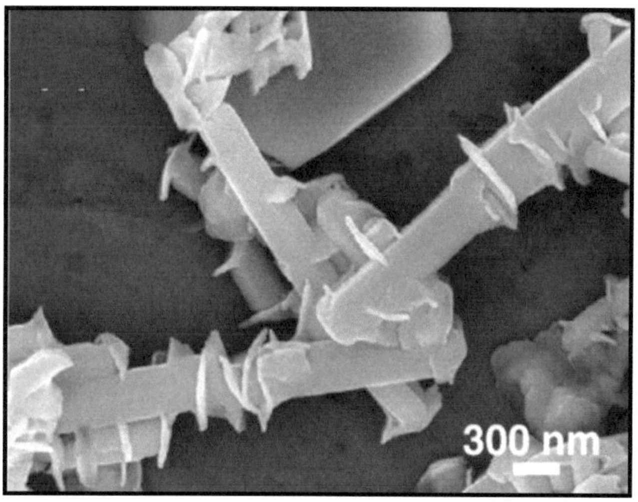

Fig. 5. SEM image of as prepared $BaCO_3$ powder by oil bath heating in water (10 mL)/EG (10 mL) mixed solvents at 80 °C for 30 min [44]. Reproduced with permission of Elsevier.

about 250 nm and lengths of about 1μm. It is interesting that the nanosheets grew perpendicularly on the nanorods. The sizes of $BaCO_3$ nanorods were relatively uniform. [44]. $BaCO_3$ nanostructures was also fabricated by liquid phase precipitation method [50]. The liquid phase precipitation method is one of the earliest methods for the synthesis of inorganic particles. V. K. Lamer et al. [51] reported the synthesis of monodisperse colloidal sulfur in ethanol/water mixed solvents in 1950. After that, the application of this method in the synthesis of inorganic materials has been fast growing [52]. The precipitation reactions involve the nucleation, growth, ripening, or agglomeration processes. The separation of nucleation and growth is the key step for the preparation of high quality crystals. The growth mechanism, such as Ostwald ripening [53–55] and aggregation especially oriented attachment [56, 57], will dramatically affect the size, morphology, and properties of the products. In the Ostwald ripening process [53–55], the larger particles will grow at the expense of the smaller ones. The "oriented attachment" mechanism was reported by Penn and Banfield [56, 57]. Fig. 6a shows TEM image of the sample prepared using $NaHCO_3$ as the CO_3^{2-} source, from which one can see $BaCO_3$ rods with diameters of several hundred nanometers and lengths of several micrometers. When Na_2CO_3 was used as the CO_3^{2-} source the rods and nanoparticles co-existed. When using $(NH_4)_2CO_3$ as the CO_3^{2-} source, a completely different shape (bundle and flower) of $BaCO_3$ was observed. These $BaCO_3$ bundles and flowers were assembled from nanosheets. These results indicate that the type of CO_3^{2-} source has an influence not only on the crystallinity of $BaCO_3$, but also on the morphology of $BaCO_3$. In the aqueous solution, Na_2CO_3 is a relatively strong base, and $(NH_4)_2CO_3$ is a weak base. The pH of the solution has an influence on the nucleation and growth of $BaCO_3$. This may explain the different morphologies of $BaCO_3$ obtained using different CO_3^{2-} sources. They also investigated the effect of the surfactant on the morphology of $BaCO_3$. When P123 was used, more dense flowers were observed. When SDBS was used the major morphology was bundle-like, and some fragments were also observed as a minor morphology. When CTAB was used, the loosely assembled flowers were obtained (Fig. 6b). Therefore, the type of the surfactant has an effect on the morphology of $BaCO_3$.

Fig. 6. a) TEM images of as prepared $BaCO_3$ nanostructures by liquid phase precipitation method using a) $NaHCO_3$ as the CO_3^{2-} source and b) SEM images of the sample prepared using 20 mL CTAB (9.100 g/L) [50]. Reproduced with permission of Elsevier.

In recent our work [58] ultrasound irradiation was used for synthesis of $BaCO_3$ nanostructures. Sonochemical synthesis has been used in the preparation of many materials [59–86]. Ultrasound induces chemical changes due to cavitation phenomena involving the formation, growth, and instantaneously implosive collapse of bubbles in liquid, which can generate local hot spots having a temperature of roughly 5000 °C, pressures of about 500 atm, and a lifetime of a few microseconds [87]. These extreme conditions can drive chemical reactions such as oxidation, reduction, dissolution, and decomposition, which have been developed to fabricate a variety of metal, oxide, sulfide, and carbide nanoparticles [88–92]. The morphology, structure and size of the samples are investigated by Scanning Electron Microscopy (SEM). Fig. 7a shows SEM image of as prepared $BaCO_3$ 1D nanostructures that concentration of Ba^{2+} ion was 0.1M, the power of ultrasound device was 30-60 W and the sonication time was 1h. As we described [58] there are three parameters that we changed and the best morphology, size and good distribution was obtained for the above sample (Fig. 7a). As illustrated in Fig. 7b we tried to synthesize better nanorods of $BaCO_3$ so we used 2g PVA (Polyvinyl Alcohol) as stabilizer in the same condition as Fig.7a and so we can gain the nanorods with approximately 95 nm width and about 1μm length.

Fig. 7. SEM images of as prepared $BaCO_3$ 1D nanostructure by sonochemical method a) without PVA and b) in the presence of 2g PVA [58]. Reproduced with permission of Elsevier.

Chen et al. [93] synthesized thorny $BaCO_3$ dendritic structures via a simple PEG-assisted method. They reported that low molecular weight polyethylene glycol (PEG), M_w 1000 g mol^{-1}, could be used for the self-assembly of complex $BaCO_3$ superstructures through a facile mineralization process under ambient conditions. SEM images of synthesized $BaCO_3$ nanostructures illustrated in Fig. 8 shows the influence of the amount of PEG on morphologies of $BaCO_3$ particles at 40 °C after 72 h under the synthesis condition. Bundles of rods structures $BaCO_3$ were observed in the control experiments without adding polymer additives (Fig. 8a). However, when the amount of PEG is increased to 0.2, 1, and 1.5 g L^{-1}, the bole becomes longer and the branches become shorter, and there are many small thorns on the surface. When the amount of PEG becomes larger (from 0.2 to 1.5gL^{-1}), the thorns become more (Fig. 8b), typical for the interaction between the hydroxy group of PEG and the crystallizing $BaCO_3$, which effectively suppresses the branch crystal growth [93].
The effects of reaction time and citric acid contents on the morphologies of $BaCO_3$ via PVP-assisted method was studied by Sun et al. [94]. Usually, organic additives and/or templates with complex functionalization patterns can control crystal growth via face selective adsorption. Therefore, to extend the application of $BaCO_3$ and to deepen the comprehension of its crystal growth behaviors, it is necessary to have a suitable choice of "capping reagent" possessing both stability and simplicity for the preparation of $BaCO_3$ with distinguished shapes and especially good uniformity. It is generally accepted that the coordination reagent kinetically controls the growth rates of various faces of nanocrystals through selective adsorption and desorption on these surfaces [95]. It implied that these polymers either changed the surface chemistry of the crystals, or as a guide reagent for crystal growth. Moreover, PVP in their experimental procedures may also serve as a stabilizer, which inhibits $BaCO_3$ particles from aggregating. The morphologies and microstructure of the as-synthesized products were further investigated by TEM. The products with dumbbell and pillar morphologies are obtained with pH 13, citric acid (CA)/Ba^{2+} = 4 and a reaction time of

Fig. 8. SEM images of as prepared $BaCO_3$ dendritic structures via a PEG-assisted method at 40 °C after 72 h a) without PEG and b) presence of 0.2 g L^{-1} PEG [93]. Reproduced with permission of Elsevier.

15 min, 2 h, 8 h. Reaction time seems an important factor that can affect the morphology of the products. As shown in Fig. 11a, the products with dumbbell morphology, smooth surface and uniform size distribution were obtained at 15 min, just like being placed in an orderly arrangement. When the reaction time is prolonged, the products with pillar morphology appear with a length of several hundred nanometers. The size of the particles was decreased gradually with reaction time, the phenomenon may be attributed to the partial dissolution of the particles. The products with different morphologies can be synthesized through controlling the molar ratio of citric acid to the concentration of the Ba^{2+} ions. Increasing CA/Ba^{2+} results in more carboxylic groups, when the molar ratio of CA to Ba^{2+} rises up to 5 and the other conditions are kept the same, the products with dumbbell-, double-pillar- and pillar-like are obtained at different times. The products with homogeneous dumbbell-like morphology were obtained at 15 min (Fig. 9b). It is clearly shown that the middle bending of dumbbell is small relative to those in Fig. 9a. When the reaction time is prolonged to 2 h, the products with double-pillar-like morphology appear with a length of 280–420 nm [94].

Thongtem et al. [96] used sonochemical method for synthesis of $BaCO_3$ nanoparticles. TEM images and SAED patterns of $BaCO_3$ are shown in Fig. 10. TEM shows that the products were composed of dispersed round nanoparticles with the sizes of 40–100 nm for $BaCO_3$ (as illustrated in fig 10). Their SAED patterns appear as diffuse and hollow concentric rings of bright spots, caused by the diffraction of transmitted electrons through the nanocrystals with different orientations. Interplanar spaces were calculated using diameters of the diffraction rings [97], and compared with those of the JCPDS database [98]. They correspond to the (111), (002), (112), (221), (132) and (113) planes for $BaCO_3$. Their sizes were measured from 150 particles on TEM images. The products were synthesized in all sizes, ranging from the smallest to the largest with the average of 55.20±9.60 nm, 65.00±10.04 nm and 89.56±16.10 nm for $BaCO_3$ by the 1 h, 3 h and 5 h ultrasonic irradiation, respectively. To form $BaCO_3$ nanoparticles, $Ba(NO_3)_2$ reacted with Na_2CO_3 in ethylene glycol (EG) under ultrasonic irradiation. Once the $BaCO_3$ nuclei formed in ethylene glycol by the assistance of ultrasonic irradiation, they did not fully develop. They grew into a number of nanoparticles via the diffusion process in EG. These nanoparticles became larger when the lengths of times were longer. They were still retaining their nanosize, although the reaction time was lengthened to 5 h [96].

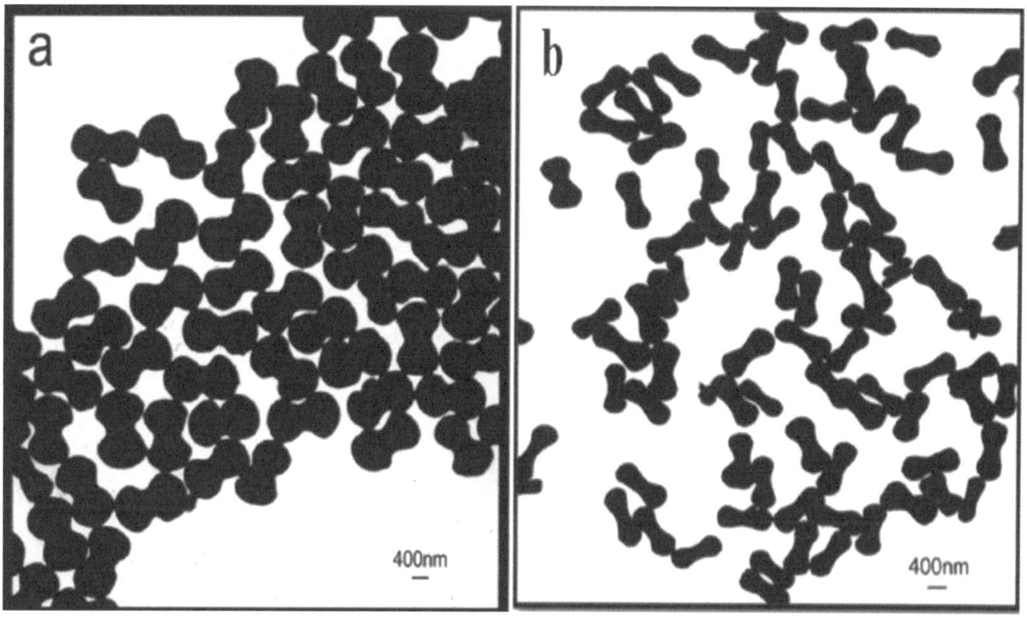

Fig. 9. TEM images of as prepared $BaCO_3$ nanostructures via PVP-assisted method with pH 13, at reaction times = 15 min (a) CA/Ba^{2+} = 4 and b) CA/Ba^{2+} = 5 [94]. Reproduced with permission of Elsevier.

3. $SrCO_3$ and $Sr(OH)_2$ nanostructures

Hu et al. [99] synthesized $SrCO_3$ nanostructures by Microemulsion-Mediated Solvothermal method. A reverse micelle or microemulsion is a transparentand isotropic liquid medium with nanosized water pools dispersed in a continuous phase and stabilized by surfactant, for example, cationic surfactant CTAB (cetyltrimethylammonium bromide), anionic surfactant AOT (sodium bis(2-ethylhexyl) sulfosuccinate) or nonionic surfactant Triton- 100 (octylphenol-poly(ethylene glycol) ether), and cosurfactant (such as 1-pentanol or 1-butanol) molecules at the water/oil interface. Accordingly, reverse micelles or microemulsions that are thermodynamically stable systems and isotropic on a molecular scale have the ability to solubilize proper solution. As the nanosized water pools, they have been widely used as spatially constrained microreactors for controlled synthesis of nanoparticles with a desired narrow size distribution. They reported the synthesis of $SrCO_3$ nanostructures with various morphologies in a simple cationic surfactant-CTAB-microemulsion system under solvothermal conditions. By carefully controlling fundamental experimental parameters such as the molar ratio H_2O to CTAB (defined w) and the concentration of reactants, $SrCO_3$ nanostructures with morphologies of rodlike, whiskerlike, ellipsoidlike, and spherelike can be efficiently achieved. Fig.11 shows a typical TEM image of obtained $SrCO_3$ nanorods and also the growth mechanism of SrCO3 nanostructures [99].

Mesoporous $SrCO_3$ spheres in room-temperature ionic liquid was reported [100]. Room temperature ionic liquids (ILs), which are liquid organic salts at room temperature, have been very attractive in both academia and industry because of their special characteristics, such as a negligible vapor pressure, wide liquid temperature range, thermal stability, high conductivity, etc., and they are regarded as environmentally benign solvents [101]. TEM image in Fig. 12a includes a number of $SrCO_3$ spheres, and the high-magnification TEM image (Fig. 12b) indicates that the spheres are composed of nanoparticles, which form the special loose

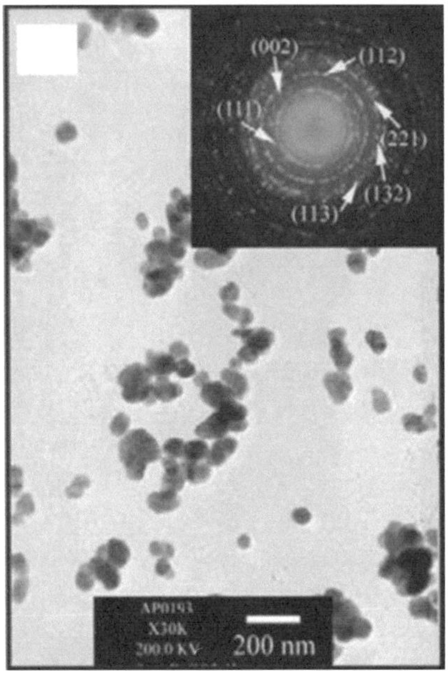

Fig. 10. TEM images and SAED patterns of as synthesized $BaCO_3$ by sonochemical method at 80 °C for 1 h [96]. Reproduced with permission of Elsevier.

Yang et al. [102] synthesized flower-like $SrCO_3$ nanostructures by hydrothermal method. The morphology of the products characterized by field emission scanning electron microscopy (FESEM) is shown that most of the flower-like nanostructures consist of $SrCO_3$ nanorod bundles. The TEM characterization of the above-mentioned sample was shown in Fig. 13. The morphology of flowerlike $SrCO_3$ nanostructures consisting of nanorods with diameter of 100–300 nm and length of about several micrometres is observed from Fig. 13.

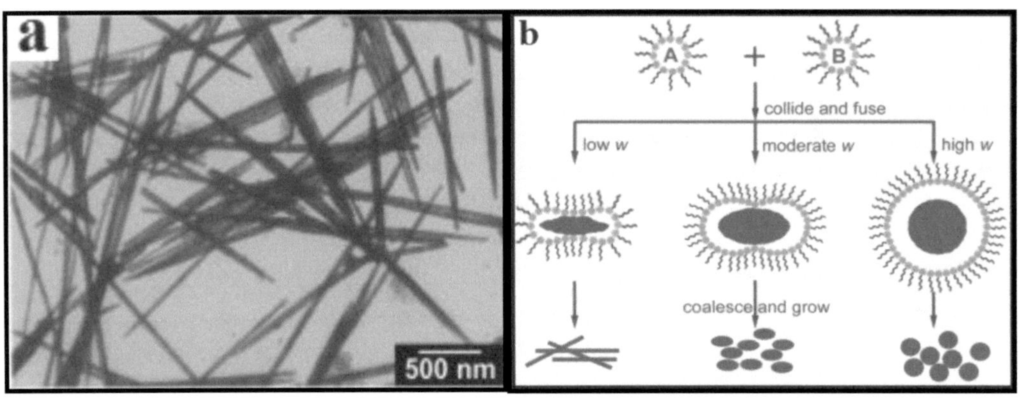

Fig. 11. a) TEM image of obtained $SrCO_3$ nanorods at $w= 5$ and $Sr(NO_3)_2$ concentration of 0.5 M and b) the growth mechanism of $SrCO_3$ nanostructures [99]. Reproduced with permission of American Chemical Society.

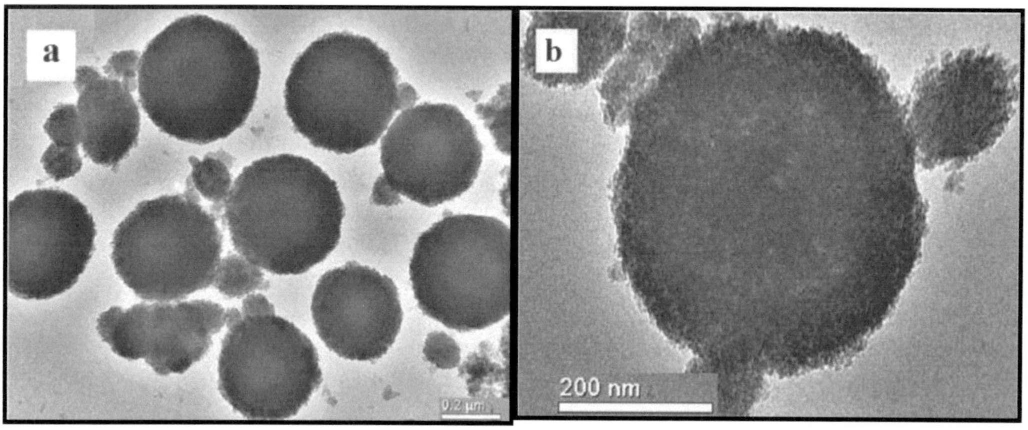

Fig. 12. a) TEM image of the obtained $SrCO_3$, b) typical TEM image of mesoporous $SrCO_3$ spheres with high magnification [100]. Reproduced with permission of Elsevier.

Fig. 13. TEM image of the flower-like $SrCO_3$ nanostructures synthesized by the hydrothermal process at 200 °C [102]. Reproduced with permission of Elsevier.

Yu et al. [103] investigated on the shape evolution of $SrCO_3$ particles in the presence of poly-(styrene-alt-maleic acid). PSMA was used as a crystal modifier to adjust the morphology of $SrCO_3$ particles by varying the concentration of PSMA. It was found that different shapes of $SrCO_3$ particles could be successfully obtained. Fig. 14 shows SEM micrographs of $SrCO_3$ particles obtained from aqueous solution in the absence and presence of PSMA at room temperature. It could be seen from Fig. 14a that, in the absence of PSMA, the as-obtained particles appeared bundle-like aggregates consisting of many small $SrCO_3$ needles aligned radially towards both ends. Further observation showed that there existed many fragments ruptured at the middle parts of the bundles. Therefore, it could be inferred that the middle parts of the bundles were more fragile than their radial branches. When a small amount of PSMA (0.01 gL^{-1}) was added into the reaction system, the morphology of $SrCO_3$ particles obviously changed (as shown in Fig. 14b).

Fig. 14. SEM images of $SrCO_3$ particles obtained in the presence of PSMA at room temperature, $[SrCO_3]$ = 2mM, pH = 10 a) without PSMA and b) 0.01 gL^{-1} PSMA [103]. Reproduced with permission of Elsevier.

One-dimensional $SrCO_3$ nanostructures were prepared and characterized by Microwave method [104]. In this work ethylenediamine (EDA) as a coordination molecular template was used for synthesis of $SrCO_3$ nanostructures. The main advantages of microwave heating are rapid volumetric heating, fast heating rate, and short reaction time, leading to new routes to realize materials synthesis in a short time. Fig. 15a shows SEM images of the typical branch-like morphology of $SrCO_3$ and fig. 15b shows the top view of the hexagonal cone, from which one can clearly see the six crystal planes [104].

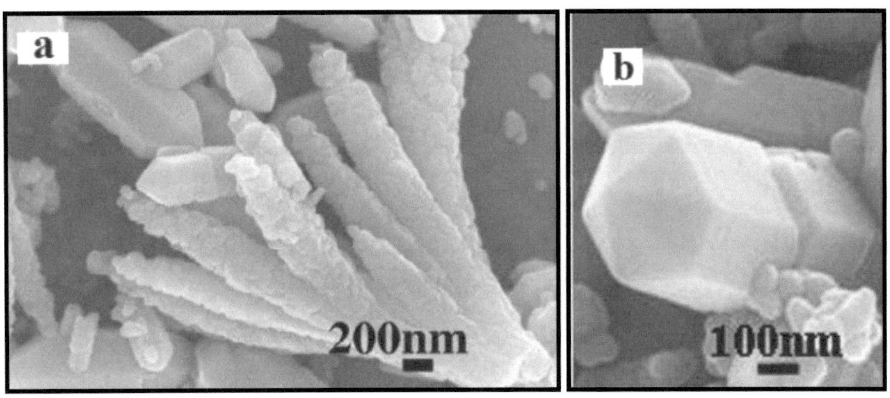

Fig. 15. SEM images of $SrCO_3$ prepared by microwave heating an aqueous solution of $(NH_4)_2CO_3$, $Sr(NO_3)_2$ and EDA at 90 °C for a) 40 min and b) 5 min [104]. Reproduced with permission of Elsevier.

Huang et al. [105] investigated on crystallization of strontium carbonate in alcohol or water solution containing mixed nonionic/anionic surfactants. They use Pluronic F127 ($EO_{97}PO_{68}EO_{97}$) and sodium dodecyl sulfate (SDS). Fig. 16 shows SEM images of $SrCO_3$

nanostructures in the presence of surfactant [105]. Comparing with SEM images of obtained product in the present and absence of surfactant F127, it has little effect on morphology of product (as shown in fig. 16a) but by adding the 0.02 M SDS to the reaction the morphology changes as illustrated in fig 16b.

Fig. 16. SEM images of samples produced in aqueous solution a) in the presence of 2.0 g/L F127 and b) in the presence of 2.0 g/L F127/0.02 M SDS [105]. Reproduced with permission of Elsevier.

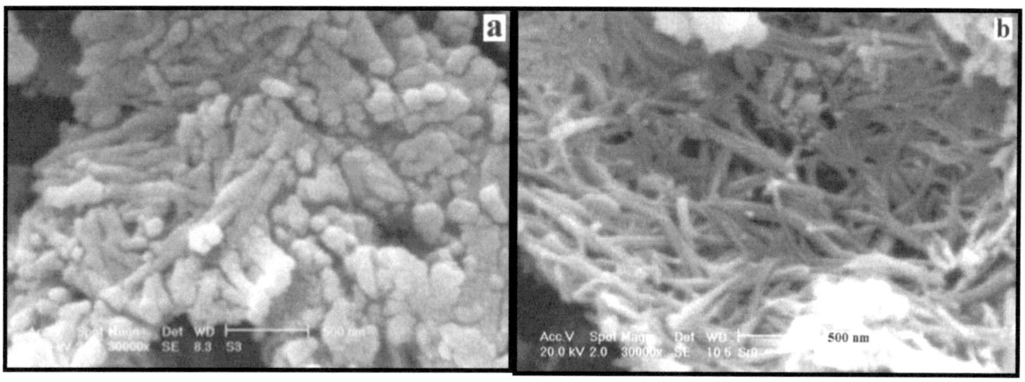

Fig. 17. SEM images of the mixture of $Sr(OH)_2$ and $SrCO_3$ nanoparticles $[Sr(CH_3COO)_2]$= 0.1 M, Sonication time= 2 h a) Power of ultrasound device=30-60 W and b) presence of 1g $NaNO_3$ and Power of ultrasound device=120-150 W [106]. Reproduced with permission of Elsevier.

Nanostructures of the mixture of $Sr(OH)_2$ and $SrCO_3$ were prepared by sonochemical method [106]. Our search shows that there is no report for synthesis of $Sr(OH)_2$ nanostructures and this is the first report. Fig. 17 shows SEM images of obtained mixture of $Sr(OH)_2$ and $SrCO_3$. When the sonication time increases to 2 hr nanoparticles aggregate and grow to prepare 1D nanostructures (fig. 17a). For synthesis of 1D nanostructures we used $NaNO_3$ as an alkali salt and as it has been shown in fig. 17b the nanorods of $Sr(OH)_2$ and $SrCO_3$ with suitable length are produced.

Thongtem et al. [96] used sonochemical method for synthesis of $SrCO_3$ nanoparticles. TEM shows that the products were composed of dispersed round 128 nanoparticles with the sizes of 20-50 nm for $SrCO_3$ (fig. 18). Their SAED patterns appear as diffuse and hollow concentric rings of bright spots, caused by the diffraction of transmitted electrons through the nanocrystals with different orientations. Interplanar spaces were calculated using diameters of the diffraction rings [97], and compared with those of the JCPDS database [98]. They correspond to the (111), (002), (112), (221), (132) and (113) planes for $SrCO_3$. Their sizes were measured from 150 particles on TEM images. The products were synthesized in all sizes, ranging from the smallest to the largest with the average of 29.83±4.26 nm, 34.02±5.26 nm and 37.20±5.86 nm for $SrCO_3$ by the 1 h, 3 h and 5 h ultrasonic irradiation, respectively. To form $SrCO_3$ nanoparticles, $Sr(NO_3)_2$ reacted with Na_2CO_3 in ethylene glycol (EG) under ultrasonic irradiation. Once the $SrCO_3$ nuclei formed in ethylene glycol by the assistance of ultrasonic irradiation, they did not fully develop. They grew into a number of nanoparticles via the diffusion process in EG. These nanoparticles became larger when the lengths of times were longer. They were still retaining their nanosize, although the reaction time was lengthened to 5 h [96].

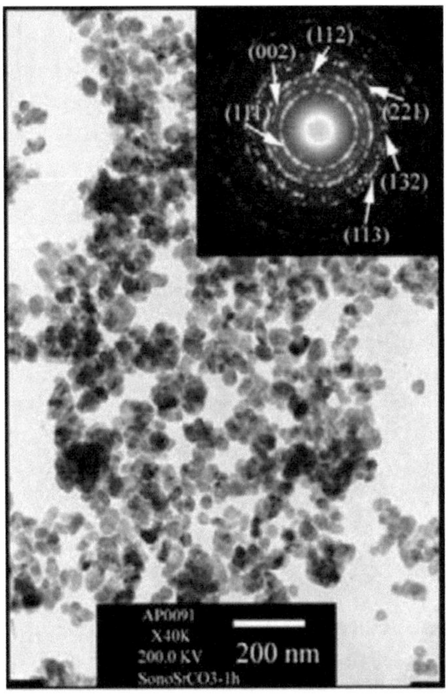

Fig. 18. TEM images and SAED patterns of as prepared $SrCO_3$ [96]. Reproduced with permission of Elsevier.

4. $Ca(OH)_2$ and CaO nanostructures

Tang et al. [107] prepared nano-CaO by thermal-decomposition method. They used $Ca(NO_3)_2.4H_2O$ as precursor, NaOH aqueous solution as precipitant, and ethylene glycol as medium. The $Ca(OH)_2$ that was obtained through the above mixing calcinated at 500°C and nano-CaO was prepared.

CaO Nanopods were synthesized [108] and also applied for high temperature CO_2 capture. $CaCO_3$ was precipitated from a saturated aqueous solution of $Ca(OH)_2$ by bubbling through

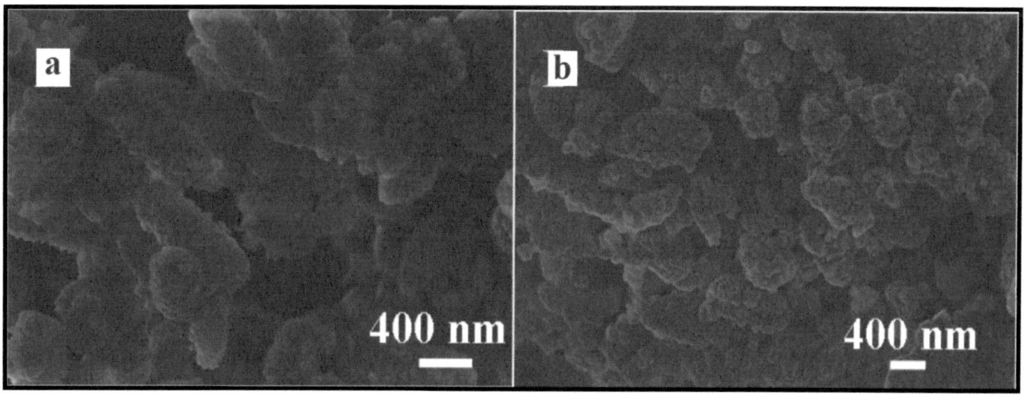

Fig. 19. FESEM images a) $CaCO_3$ nanopods prepared in the presence of P123 surfactant and b) CaO derived from $CaCO_3$ nanopods [108]. Reproduced with permission of American Chemical Society.

Fig. 20. SEM images of as prepared nanoparticles of a) $Ca(OH)_2$ and b) CaO [109]. Reproduced with permission of Taylor & Francis

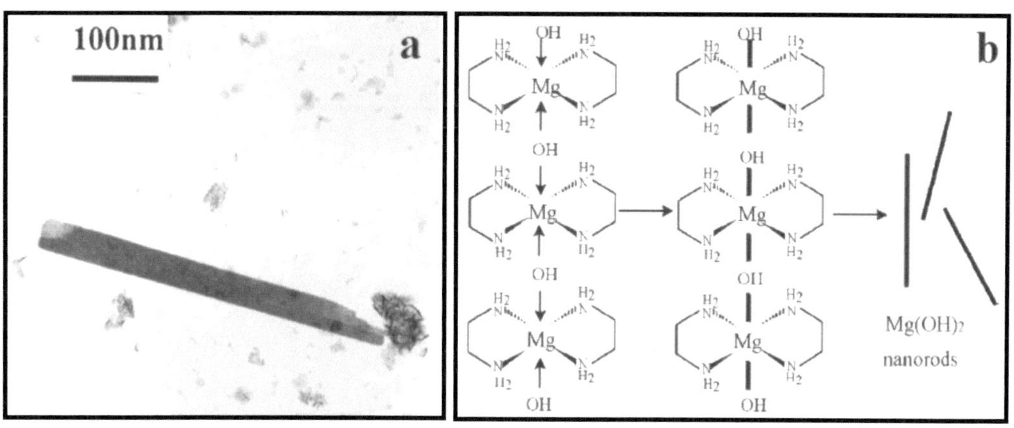

Fig. 21. a) TEM image and b) Schematic diagram of the mechanism of formation of $Mg(OH)_2$ nanorods from Mg^{2+}-ethylenediamine complexes [110]. Reproduced with permission of Wiley-VCH Verlag GmbH & Co. KGaA.

CO_2 gas in the presence of the triblock copolymer, P123 ($PEO_{20}PPO_{70}PEO_{20}$), an amphiphilic and neutral surfactant. The standard recipe involved dissolving 1.0 g of P123 in 100 mL of distilled water. This was done at room temperature with vigorous stirring. And then CaO were prepared for analysis by calcining the $CaCO_3$ in a tube furnace at 700 °C for 3 h under N_2 (1 L/min). FESEM images of precursor ($CaCO_3$) and CaO nanopods are shown in fig. 19.
In the recent our study [109] we synthesized $Ca(OH)_2$ and CaO nanostructures by simple sonochemical method. By reaction of calcium acetate and sodium hydroxide under ultrasound irradiation and also by using PVA and sodium nitrate as additives different morphologies of $Ca(OH)_2$ are obtained. Then the prepared precursor is calcinated at 600 °C to synthesize of CaO nanoparticles. Fig. 20 shows a typical SEM image of both $Ca(OH)_2$ and CaO nanoparticles.

5. $Mg(OH)_2$ and MgO nanostructures

Li et al. [110] synthesized $Mg(OH)_2$ nanorods from Mg powder and distilled water (5 mL) with ethylenediamine as the solvent at 180°C under pressure in an autoclave. Fig. 21 shows TEM image and also the preparation mechanism of $Mg(OH)_2$ nanorods.
The Li group also prepared rod-like $Mg(OH)_2$ nanocrystallites under hydrothermal conditions and converted it to MgO nanorods by thermal dehydration. Fig. 22 shows TEM images of as prepared $Mg(OH)_2$ nanorods by hydrothermal method and also MgO nanorods that is obtained after thermal dehydration of $Mg(OH)_2$ precursor.

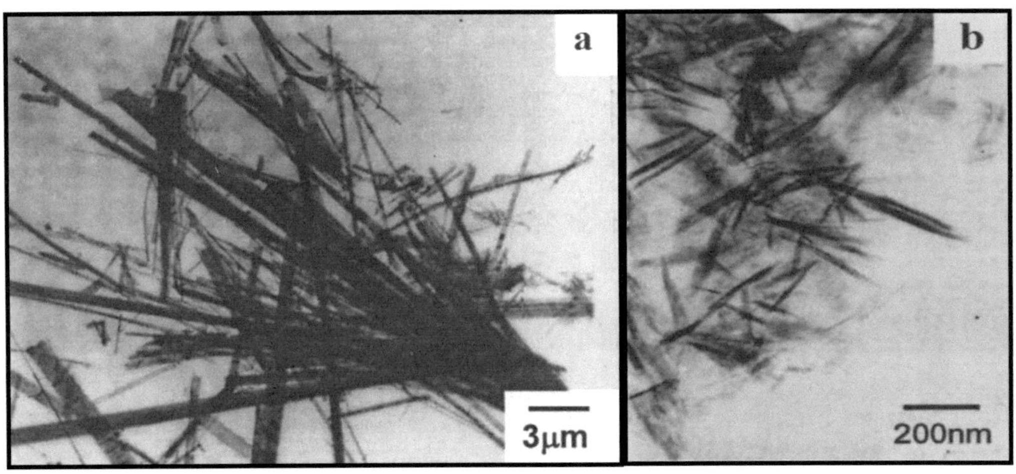

Fig. 22. TEM images of as prepared, a) $Mg(OH)_2$ and b) MgO nanorods [111]. Reproduced with permission of Elsevier.

$Mg(OH)_2$ nanotubes using $Mg_{10}(OH)_{18}C_{12}.5H_2O$ nanowires as precursors by solvothermal method were synthesized [112]. Fig. 23 shows TEM image of $Mg(OH)_2$ nanotubes that are obtained after solvothermal treatment of precursor in the presence of ethylenediamine as a solvent.
Ma et al. [113] investigated on formation of uniform MgO nanobelts from in situ Mg_3N_2 precursor. Mg ribbons were put in an alumina boat inserted into a quartz tube reactor heated by a resistivity-heating furnace. The furnace temperature was raised to 650 °C at a ramp rate of 10 °C/ min and was held for 2 h under a constant nitrogen (N_2) flow in a

flowing rate of 500 standard cubic centimeters per minute [113]. Fig. 24 shows SEM image of MgO nanobelts that are obtained from calcination of Mg_3N_2 precursor under nitrogen/oxygen atmosphere (N_2=O_2, 10:1 in volumes).

Fig. 23. TEM image of $Mg(OH)_2$ nanotubes [112]. . Reproduced with permission of The Royal Society of Chemistry.

Fig. 24. SEM image of MgO nanobelts [113]. Reproduced with permission of Elsevier.

The size and morphologies of magnesium hydroxide nanoparticles precipitated in dilute aqueous solution was studied [114]. The samples were obtained by precipitation, at a controlled temperature, of a magnesium salt solution of concentration 0.75 mol/l by addition of an alkaline solution (NaOH or NH_4OH) of concentration 1.5 mol/l. Both

Fig. 25. SEM image of sample No. 4 that $Mg(NO_3)_2$ and NH_4OH are used as Mg and base source, respectively [114]. Reproduced with permission of Elsevier.

reactants were simultaneously added into an ultrasonicated bath following the controlled-double jet precipitation technique, with the help of peristaltic pumps working at a discharge of 3 ml/min. A 450 ml volume of alkaline water [pH 10, either NH_4OH or NaOH] initially present in the reactor, vigorous stirring was applied during the addition of the reactants, as well as during the ageing of the precipitate in the mother liquor. The suspension was allowed to age at the synthesis temperature for one day, then subsequently at room temperature for 2 days [114]. Fig. 25 shows SEM image of one of the samples that the platelet shape $Mg(OH)_2$ is prepared by this method.

The growth of dumbbell-shaped MgO nanowhiskers was done by depositing Mg vapor generated *via* carbothermal reduction onto boron powder [115]. The whiskers and balls of these dumbbells have diameters of 50–150 nm and 1.4–2 mm respectively. The ball surface and the interface between ball and whisker are very smooth but the whisker surface is rough, which indicates that whiskers formed in advance and balls formed later by depositing silicon oxide and other vapor species onto the grown whisker surface. Fig.26 shows TEM image and SAED pattern of MgO dumbbell-shape whiskers.

Qu et al. [116] investigated on controlled growth of three morphological structures of magnesium hydroxide nanoparticles by wet precipitation method. It has been shown that the alkali solution concentration is of prime importance besides the complex dispersant: the use of lower concentration aqueous ammonia (5 wt%) promotes the formation of needle or rod morphology, while the synthesis driven with higher concentration aqueous ammonia (25 wt%) promotes the formation of platelet-shaped particles. Those behaviors are attributed to the mechanism of nuclei growth [116].

MgO nanostructured microspheres by an interfacial reaction in a solid-stabilized emulsion were reported and SEM image of obtained magnesium oxide is illustrated in Figure 27. The average diameter of the MgO microspheres was 5.3 μm, a little larger than that of their precursors. The diameter of the MgO microspheres was in a narrow range of 5.1 to 5.4 μm. The surfaces of the MgO microspheres also had leaf-like nanostructures with a leaf thickness of about 65 nm (measured by SEM images).

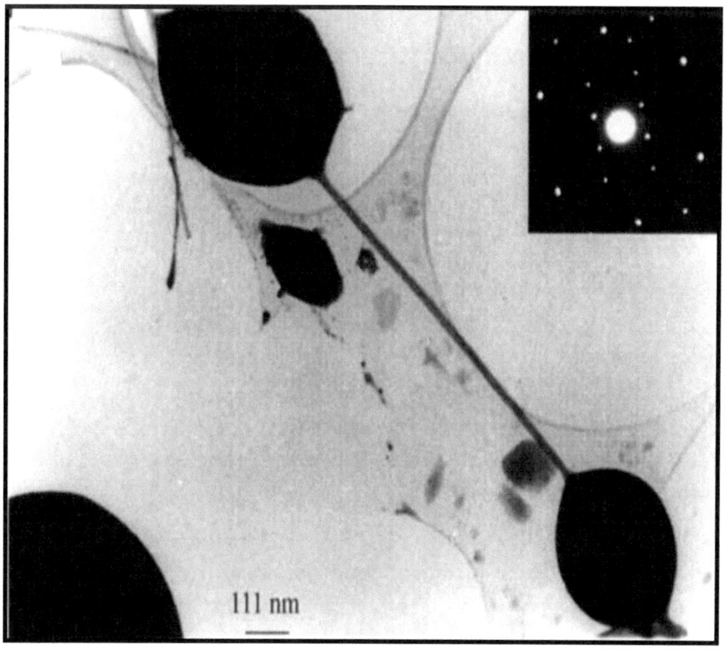

Fig. 26. TEM image and SAED pattern (inset) of dumbbell-shape whiskers of MgO [115]. Reproduced with permission of Elsevier.

Fig. 27. SEM image of MgO microspheres [117]. Reproduced with permission of Elsevier.

Kim et al. [118] investigated on the growth of MgO nanowires by annealing treatment of Au-coated substrates. They applied thermal annealing treatment to the Au-coated substrates before they grow the MgO structures by a thermal evaporation of MgB_2 powders. They obtained MgO nanowires by controlling the predeposition annealing temperature. The produced nanowires were of cubic MgO structures with diameters in the range of 40–200 nm (as illustrated in fig.28).

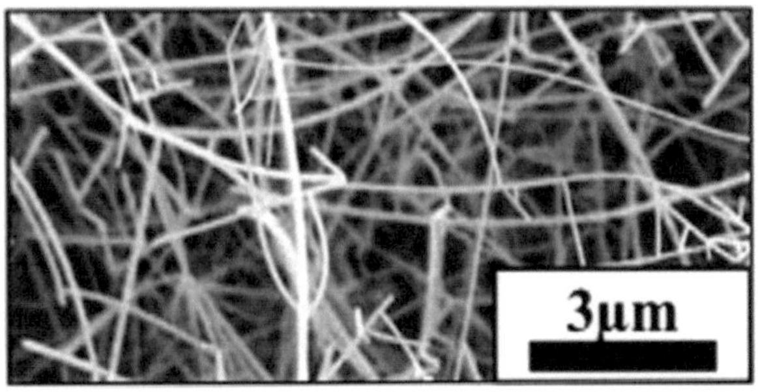

Fig. 28. SEM image of MgO nanowires with the predeposition annealing at 700 °C [118]. Reproduced with permission of Elsevier.

Nano-cube MgO was prepared by means of a domestic microwave oven [119]. The purpose of this preparation is to investigate a simple and rapid synthesis method using magnesium (Mg) chip and steel-wool as starting materials for the growth of the MgO with nano-cube shape. Fig. 29 shows the HR-TEM image of a nano-cube MgO. The resolved spacing of 0.21 nm corresponds to the (200) facets of MgO. The surface of the nano-cube MgO is clean and atomically sharp without any sheathed amorphous phase [119].

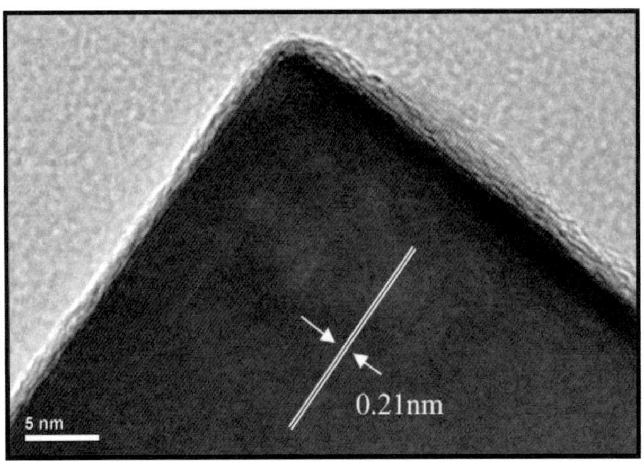

Fig. 29. HR-TEM image of the obtained MgO nano-cube [119]. Reproduced with permission of Elsevier.

Wu et al. [120] investigated on self-assembled growth of MgO nanosheet arrays via a micro-arc oxidation technique. They present a relatively simple method of fabricating aligned MgO sheet-like nanostructures on the surface of magnesium alloys via a promising micro-arc oxidation (MAO) technology. SEM observations reveal that many sheet-like structures form on the substrate surface. The nanosheets are slightly curved and approximately 300–400 nm in width and tens of nanometers in thickness. Most of them stick together to form nanosheet arrays. A large quantity of flower-like branched structures was also observed on the substrate. Fig. 30 shows these unique structures.

Preparation of $Mg(OH)_2$ nanoparticles in water-in-oil microemulsions [121] displays a platelet shape and the thickness is *ca.* 20 nm. The particle size is larger than that of crystallites size because of the coalescing of crystallites. The particle size of different samples indicated that small particles were obtained in small micelles (at low ω_0). The particle sizes from SEM images were seemingly larger than the mean sizes of micelles as a result of the coalescence of nanoparticles in different micelles caused by the collision of micelles [121]. Rezaei et al. [122] synthesized nanocrystalline magnesium oxide with high surface area. The pore size distribution, as calculated by the *BJH* method from the desorption branch of the nitrogen isotherm, reveals that the MgO calcined at 600 and 700 °C contain small mesopores with pore size between 3.7–7.7 nm and large mesopores with a pore size of about 32.2 nm. The former mesopores exist among small primary nanoparticles and the latter textural mesopores are formed by the aggregation of the small nanoparticles. The increase in calcination temperature influences the pore size distribution. It is seen that with increasing calcination temperature, the pore size distributions were shifted to larger sizes. The sample calcined at 800 °C showed a broad pore size distribution. In this sample the small mesopores were transformed to larger ones due to aggregation of the small nanoparticles at 800 °C [122]. TEM analysis of above samples shows that in this condition plate like structure of nanocrystalline MgO is prepared.

$Mg(OH)_2$ nanostructures and MgO nanoparticles were prepared by sonochemical method [123]. Fig. 31a shows SEM image of MgO nanoparticles that is obtained by calcination of as prepared $Mg(OH)_2$ nanostructures from reaction of $Mg(OAc)_2$ and NaOH under ultrasound irradiation and also fig. 32b. shows SEM image of as prepared plate like structure of $Mg(OH)_2$ by increasing ultrasound device power up to 150 W. Particle size histogram was prepared for MgO particles after heating of $Mg(OH)_2$ nanostructures at 400 °C for investigation of the size distribution of the nanoparticles. Most of the particles possess sizes in range from 55 to 75 nm.

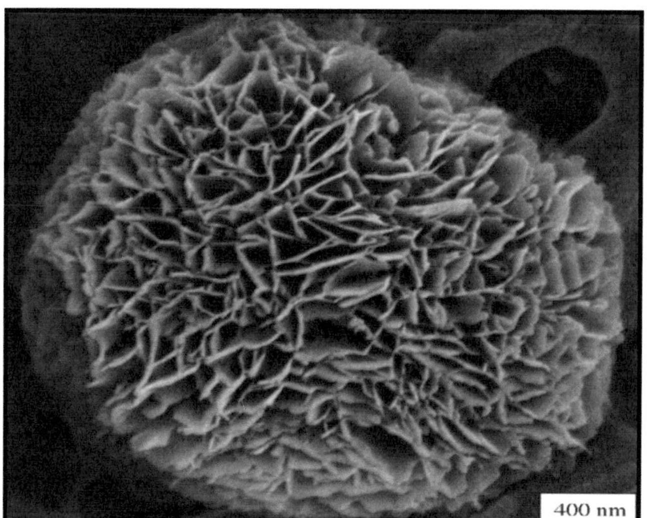

Fig. 30. SEM images of micrometer scale flower-like structures of MgO [120]. Reproduced with permission of Elsevier.

Fig. 31. SEM images of a) MgO nanoparticles and b) plate like structure of $Mg(OH)_2$ [123]. Reproduced with permission of Elsevier.

6. Conclusion

Because of importance of some of alkaline-earth metal compounds in industry and medicine, in this chapter we tried to summarize different methods that are used to synthesis nanostructures of $BaCO_3$, $Sr(OH)_2$, $SrCO_3$, $Ca(OH)_2$, CaO, $Mg(OH)_2$ and MgO. As morphology of these compounds affect on application, so preparation methods have an important role on applicability of above compounds. Considering nanostructures discussed in this chapter, to obtain 1D nanostructure it can be used many compounds as template or stabilizer such as ethylenediamine, polyvinyl alcohol or some methods such as hydrothermal, reverse micelle, but in some cases as a nature of alkaline-earth metal just nanoparticles are obtained.

7. Acknowledgments

Supporting of this investigation by Tarbiat Modares University is gratefully acknowledged.

8. References

[1] W. Feng, L.-D. Sun, Y.-W. Zhang, C.-H. Yan, *Coord. Chem. Rev.* 254 (2010) 1038.
[2] L. Qi. *Coord. Chem. Rev.* 254 (2010) 1054.
[3] D.B. Kuang, A.W. Xu, Y.P. Fang, H.Q. Liu, C. Frommen, D. Fenske, *Adv. Mater.* 15 (2003) 1747.
[4] F. Kim, S. Connor, H. Song, T. Kuykendall, P.D. Yang, *Angew. Chem. Int.* Ed. 43 (2004) 3673.
[5] Sa Lv, Ping Li, Jie Sheng, Wendong Sun, *Mater. Lett.* 61 (2007) 4250.
[6] K. Omata, N. Nukui, T. Hottai, Y. Showa, M. Yamada, *Catal. Commun.* 5 (2004) 755.
[7] US Patent No. 5, 338, 706.
[8] J. Küther, G. Nells, R. Seshadri, M. Schaub, H.J. Butt, W. Tremel, *Chem. Eur. J.* 4 (1998)1834.
[9] J. Küther, R. Seshadri, G. Nells, W. Assenmacher, H.J. Butt, W. Mader, W. Tremel, *Chem. Mater.* 11 (1999) 1317.

[10] J. Küther, M. Bartz, R. Seshadri, G.B.M. Vaughan, W.J. Tremel, *Mater. Chem.* 11 (2001) 503.
[11] M. Sastry, A. Kumar, C. Damle, S.R. Sainkar, M. Bhagwat, V. Ramaswamy, *Cryst. Eng. Commun.* 21 (2001) 1.
[12] D. Rautaray, S.R. Sainkar, M. Sastry, *Langmuir* 19 (2003) 888.
[13] L. Qi, K. Xi, J. Ma, *Acta Chim. Sinica* 1 (2003) 126.
[14] N.H. de Leeuw, J.A. Burton, *Phys. Rev., B.* 63 (2001) 195417.
[15] D. Kulkarni, I.E. Wachs, *Appl. Catal. A.* 237 (2002) 121.
[16] M. Boulouz, L. Martin, A. Boulouz, A. Boyer, Mater. Sci. Eng., B, *Solid-State Mater. Adv. Technol.* 67 (1999) 122.
[17] C.E. Curtis, L.M. Doney, J.R. Johnson, *J. Am. Ceram. Soc.* 37 (1954) 358.
[18] H. Doweidar, *J. Non-Cryst. Solids.* 277 (2000) 98.
[19] R. C. Whited, C. J Flaten and W. C. Walker. *Solid State Commun.* 13 (1973)1903.
[20] A. Yamasaki and T. Fujiwara, *Phys. Rev. B.* 66 (2002) 245108.
[21] A. M. Stoneham, *J. Non-Cryst. Solids.* 303 (2002) 114.
[22] M. C. Wu, J. S. Corneille, C. A. Estrada, J. W. He, D. W. Goodman, *Chem. Phys. Lett.* 182 (1991) 472.
[23] S. K. Shukla, G. K. Parashar, A. P. Mishra, P. Misra, B. C. Yadav, R. K. Shukla, L. M. Bali, G. C. Dubey, *Sens. Actuators B.* 98 (2004) 5.
[24] L.D. Zhang, J.M. Mo, Nanometer Materials, Science Press, Liaoning, 1994, pp. 303 (in Chinese).
[25] J. Sawai, H. Kojima, H. Igarashi, A. Hashimoto, S. Shoji, T. Sawaki, A. Hakoda, E. Kawada, T. Kokugan, M. Shimizu, *World J. Microb. Biot.* 16 (2000) 187.
[26] O. Yamamoto, J. Sawai, T. Sasamoto, Int. *J. Inorg. Mater.* 2 (2000) 451.
[27] O. B. Koper, J. S. Klabunde, G. L. Marchin, K. J. Klabunde, P. Stoimenov, L. Bohra, *Curr. Microbiol.* 44 (2002) 49.
[28] P. K. Stoimenov, R. L. Klinger, G. L. Marchin, K. J. Klabunde, *Langmuir.* 18 (2002) 6679.
[29] L. Huang, D. Li, Y. Lin, M. Wei, D. G. Evans, X. Duan, *J. Inorg. Biochem.* 99 (2005) 986.
[30] D. Kuang, A. Xu, Y. Fang, H. Ou, H. Liu, *J. Cryst. Growth* 244 (2002) 379.
[31] G. D. Rees, R. Evans-Gowing, S. J. Hammond, B. H. Robinson, *Langmuir* 15 (1999) 1993.
[32] C. Karagiozov, D. Momchilova, *Chem. Eng. Process.* 44 (2005) 115.
[33] K. Song, J. Kim, *Powder Technol.* 103 (2000) 268.
[34] J. Shmidt, *TU Berlin,* 2000.
[35] L. Li, Y. Chu, Y. Liu, L. Dong, L. Huo, F. Yang, *Mater. Lett.* 60 (2006) 2138.
[36] R. Strobel, M. Maciejewski, S. E. Pratsinis, A. Baiker, *Thermochim. Acta* 445 (2006) 23.
[37] L. Mädler, *KONA* 22 (2004) 107.
[38] S.E. Pratsinis, *Prog. Energy Combust. Sci.* 24 (1998) 197.
[39] R.N. Grass, W.J. Stark, *Chem. Commun.* (2005) 1767.
[40] M. Huber, W.J. Stark, S. Loher, M. Maciejewski, F. Krumeich, A. Baiker, *Chem. Commun.* (2005) 648.
[41] S. Loher, W.J. Stark, M. Maciejewski, A. Baiker, S.E. Pratsinis, D. Reichardt, F. Maspero, F. Krumeich, D. Günther, *Chem. Mater.* 17 (2005) 36.
[42] Y.G. Sun, Y.D. Yin, Brian T. Mayers, Thurston Herricks, Y.N Xia, *Chem. Mater.* 14 (2002) 4736.

[43] J.G. Bai, Z.D. Xu, Y.F. Zheng, H.Y. Yin, *Mater. Lett.* 60 (2006) 1287.
[44] M-G. Ma, Y-J. Zhu, J-F. Zhu, Z-L. Xu, *Mater. Lett.* 61 (2007) 5133.
[45] G. Viau, F. Fiévet-Vincent, F. Fiévet, J. Mater. Chem. 6 (1996) 1047.
[46] P. Toneguzzo, G. Viau, O. Acher, F. Fiévet-Vincent, F. Fiévet, *Adv. Mater.* 10 (1998) 1032.
[47] G. Viau, R. Brayner, L. Poul, N. Chakroune, E. Lacaze, F. Fiévet-Vincent, F. Fiévet, *Chem. Mater.* 15 (2003) 486.
[48] C. Feldmann, H.O. Jungk, *Angew. Chem. Int. Ed.* 40 (2001) 359.
[49] J.F. Zhu, Y.J. Zhu, *J. Phys. Chem., B* 110 (2006) 8593.
[50] M-G. Ma, Y-J. Zhu, G-F. Cheng, Y-H. Huang, *Mater. Lett.* 62 (2008) 3110.
[51] V.K. Lamer, R.H. Dinegar, *J. Am. Chem. Soc.* 72 (1950) 4847.
[52] B.L. Cushing, V.L. Kolesnichenko, C.J. O'Connor, *Chem. Rev.* 104 (2004) 3893.
[53] J.A. Marqusee, J. Ross, *J. Chem. Phys.* 79 (1983) 373.
[54] S. Sugimoto, *J. Colloid Interface Sci.* 63 (1978) 16.
[55] D.B. Dadyburjor, E. Ruckenstein, *J. Cryst. Growth* 40 (1977) 279.
[56] R.L. Penn, J.F. Banfield, *Geochim. Cosmochim. Acta* 63 (1999) 1549.
[57] J.F. Banfield, S.A. Welch, H.Z. Zhang, T.T. Ebert, R.L. Penn, *Science* 289 (2000) 751.
[58] M. A. Alavi, A. Morsali, *Ultrason. Sonochem.* 15 (2008) 833.
[59] K.S. Suslick, *Science* 247 (1990) 1439.
[60] D.N. Srivastava, N. Perkas, G.A. Seisenbaeva, Y. Koltypin, V.G. Kessler, A. Gedanken, *Ultrason. Sonochem.* 10 (2003) 1.
[61] X.L. Zhang, Y.H. Kim, Y.S. Kang, *Curr. Appl. Phys.* 6 (2006) 796.
[62] A. Askarinejad, A. Morsali, *Mater. Lett.* 62 (2008) 478.
[63] A. Askarinejad, A. Morsali, *Ultrason. Sonochem.* 16 (2009) 124.
[64] N. Soltanzadeh, A. Morsali, *Polyhedron* 281 (2009) 1343.
[65] A. Aslani, A. Morsali, V. T. Yilmaz, C. Kazak, *J. Mol. Struct.* 929 (2009) 187.
[66] S. Khanjani, A. Morsali, *J. Mol. Struct.* 935 (2009) 27.
[67] A. Morsali, H. H. Monfared, A. Morsali, *Inorg. Chim. Acta,* 362 (2009) 3427.
[68] A. Askarinejad, A. Morsali, *Chem. Eng. J.* 153 (2009) 183.
[69] A. Aslani, A. Morsali *Inorg. Chim. Acta,* 362 (2009) 5012.
[70] Z. R. Ranjbar, A. Morsali, *J. Mol. Struc.* 936 (2009) 206.
[71] N. Soltanzadeh, A. Morsali, *Ultrason. Sonochem.* 17(2010) 139.
[72] M. J. Soltanian Fard-Jahromi, A. Morsali, *Ultrason. Sonochem.* 17(2010) 435.
[73] H. Sadeghzadeh, A. Morsali, P. Retailleau, *Polyhedron,* 29 (2010) 925.
[74] H. Sadeghzadeh, A. Morsali, *Ultrason. Sonochem. In Press,* Available online 24 February 2010.
[75] H. Sadeghzadeh, A. Morsali, V. T. Yilmaz, O. Büyükgüngör *Ultrason. Sonochem,* 17 2010) 592.
[76] A. R. Abbasi, A. Morsali, *Ultrason. Sonochem.* 17 (2010) 572.
[77] H. Sadeghzadeh, A. Morsali, V. T. Yilmaz, O. Büyükgüngör, *Inorg. Chim. Acta,* 363 (2010) 841.
[78] M. Khanpour, A. Morsali, P. Retailleau, *Polyhedron,* 29 (2010) 1520.
[79] A. R. Abbasi, A. Morsali *Ultrason. Sonochem.* 17 (2010) 704.
[80] S. Khanjani, A. Morsali, *J. Mol. Liq.* 153 (2010) 129.
[81] Z. Darvishi, A. Morsali, *Ultrason. Sonochem. In Press,* Available online 4 June 2010.

[82] A. R. Abbasi, A. Morsali, *Ultrason. Sonochem. In Press,* Available online 30 June 2010.
[83] M. S. Yazdan Parast, A. Morsali, *Ultrason. Sonochem. In Press,* Available online 6 July 2010.
[84] L. Aboutorabi, A. Morsali, *Ultrason. Sonochem. In Press,* Available online 27 July 2010.
[85] L. Aboutorabi, A. Morsali, *Inorg. Chim. Acta,* 363 (2010) 2506.
[86] H. Sadeghzadeh, A. Morsali, *CrystEngComm.* 12 (2010) 370.
[87] A. Gedanken, *Ultrason. Sonochem.* 11 (2004) 47.
[88] V.G. Pol, A. Gedanken, *J. Chem. Mater.* 15 (2003) 1111.
[89] T. Gao, Q.H. Li, T.H.Wang, *Chem. Mater.* 17 (2005) 887.
[90] N.A. Dhas, A. Zaban, A. Gedanken, *Chem. Mater.* 11 (1999) 806.
[91] T. Gao, T.H.Wang, *Chem. Commun.* 22 (2004) 2558.
[92] N.A. Dhas, *Appl. Phys. Lett.* 72 (1998) 2514.
[93] Q. Zhang, C. Chen, L. Fang, *Mater. Chem. Phys.* 111 (2008) 191.
[94] S. Lv, J. Sheng, S. Zhang, W. Sun, *Mater. Res. Bull.* 43 (2008) 1099.
[95] X.L. Hu, Y.J. Jie, *Langmuir* 20 (2004) 1521.
[96] T. Thongtem, N. Tipcompor, A. Phuruangrat, S. Thongtem, *Mater. Lett.* 64 (2010) 510.
[97] K.W. Andrews, D.J. Dyson, S.R. Keown. Interpretation of electron diffraction patterns, 183 New York: Plenum Press; 1971.
[98] Powder Diffract. File, JCPDS-ICDD, 12 Campus Boulevard, Newtown Square, PA 179 19073–3273, U.S.A. (2001).
[99] M. Cao, X. Wu, X. He, C. Hu, *Langmuir* 21 (2005) 6093.
[100] J. Du, Z. Liu, Z. Li, B. Han, Y. Huang, J. Zhang, *Microporous Mesoporous Mater.* 83 (2005) 145.
[101] (a) T. Welton, *Chem. Rev.* 99 (1999) 2071;
(b) K.R. Seddon, *Nat. Mater.* 2 (2003) 363.
[102] S. Li, H. Zhang, J. Xu, D. Yang, *Mater. Lett.* 59 (2005) 420.
[103] J. Yu, H. Guo, B. Cheng, *J. Solid State Chem.* 179 (2006) 800.
[104] M-G. Ma, Y-J. Zhu, *Mater. Lett.* 62 (2008) 2512.
[105] G. Guo, G. Yan, L. Wang, J. Huang, *Mater. Lett.* 62 (2008) 4018.
[106] M. A. Alavi, A. Morsali, *Ultrason. Sonochem.* 17 (2010) 132.
[107] Z-X. Tang, D. Claveau, R. Corcuff, K. Belkacemi, J. Arul, *Mater. Lett.* 62 (2008) 2096.
[108] Z.Yang, M. Zhao, N. H. Florin, A. T. Harris, *Ind. Eng. Chem. Res.* 48 (2009) 10765.
[109] M. A. Alavi, A. Morsali, *J. Exp. Nanosci.* 5 (2010) 93.
[110] Y. Li, M. Sui, Y. Ding, G. Zhang, J. Zhuang, and C. Wang, *Adv. Mater.* 12 (2000) 818.
[111] L. Yan, J. Zhuang, X. Sun, Z. Deng, Y. Li, *Mater. Chem. Phys.* 76 (2002) 119.
[112] W. Fan, S. Sun, L. You, G. Cao, X. Song, W. Zhang and H. Yu, *J. Mater. Chem.* 13 (2003) 3062.
[113] R. Ma, Y. Bando, *Chem. Phys. Lett.* 370 (2003) 770.
[114] C. Henrist, J.-P. Mathieu, C. Vogels, A. Rulmont, R. Cloots, *J. Cryst. Growth* 249 (2003) 321.
[115] Y. Chen, J. Li, Y. Han, X. Yang, J. Dai, *Ceram. Int.* 29 (2003) 663.
[116] J. Lv, L. Qiu, B. Qu, *J. Cryst. Growth* 267 (2004) 676.
[117] Y. He, *Mater. Lett.* 60 (2006) 3511.
[118] H. W. Kim, S. H. Shim, *Chem. Phys. Lett.* 422 (2006) 165.

[119] N. Takahashi, *Solid State Sci.* 9 (2007) 722.
[120] T. Qiu, X.L. Wu, F.Y. Jin, A.P. Huang, P.K. Chu, *Appl. Surf. Sci.* 253 (2007) 3987.
[121] J. Wu, H. Yan, X. Zhang, L. Wei, X. Liu, B. Xu, *J. Colloid Interface Sci.* 324 (2008) 167.
[122] F. Meshkani, M. Rezaei, *Powder Technol.* 196 (2009) 85.
[123] M.A. Alavi, A. Morsali, *Ultrason. Sonochem.* 17 (2010) 441.

Part 2

Nanostructuring and Function of Nanocrystals

10

Nano/Micro-Patterning of Metal Oxide Nanocrystals

Yoshitake Masuda
National Institute of Advanced Industrial Science and Technology (AIST)
Anagahora, Shimoshidami, Moriyama-ku, Nagoya
Japan

1. Introduction

Metal oxides have recently been fabricated in solutions without high-temperature sintering in order to reduce energy consumption and allow application for various substrates having low heat resistance. The fabrication of metal oxide thin films from solutions has been encouraged by the development of environment-friendly chemistry such as Green & sustainable chemistry[1-6], Bioinspired materials chemistry[7], Biomimetic materials chemistry[7], Soft-solution processing[8-10], Soft chemistry ("Chimie douce" in French)[11], Liquid phase deposition[7,12], Chemical bath deposition (CBD)[12,13], Electroless deposition (ED) with catalyst[12,13], Successive ion layer adsorption and reaction (SILAR) [12,13], Sol-gel process[14,15], Hydrothermal reaction[16], Electrodeposition[17,18] and so on. Solution processing of metal oxides allows us to prepare metal oxide thin films on the surface of solids such as substrates, particles, and fibers. Metal oxide nano/microstructures can also be fabricated by applying these solution systems to electronic or photonic devices.

Many kinds of lithography or patterning techniques have been developed to prepare patterns of thin films, for instance, X-ray/electronbeam lithography and photolithography,[19] microcontact printing,[20,21] wet etching,[22] ink-jet printing,[23] embossing,[24,25] slip-pressing,[26] charge-based printing,[27] micromolding,[28] and cold welding[29]. However, etching or lift-off processes are required in many of these methods, which causes degradation of performance, increases waste and energy consumption, and makes the process complicated. Additionally, etching or lift-off processes cannot be applied to corrosion-resistant metal oxides. The deposition of metal oxides only on desired areas of a substrate is thus required for the pattering of metal oxide thin films.

In this section, Liquid Phase Patterning (LPP) of metal oxides was reported [30]. Self-assembled monolayer (SAM), which can modify the surface of solids with various functional groups, was used as the template to enable molecular recognition for LPP. Solution systems were developed and applied to LPP of ceramic thin films on patterned SAMs by the proposed novel LPP processes.

2. Liquid phase patterning of metal oxides

2.1 SAM prearation for patterning of metal oxides

Self-assembled monolayer (SAM) can modify the surface of solids such as a substrate, particles, or fibers with various functional groups, and the molecular recognition of functional groups of SAM was the key technique for LPP. Patterned SAMs were used as templates in our LPP processes to deposit metal oxide thin films on desired areas of substrates.

An Si wafer (p-type Si [100], NK Platz Co., Ltd.) was sonicated in water, ethanol or acetone for 10 min, respectively, and exposed for 15 min to UV light (184.9 nm and 253.7 nm) (low-pressure mercury lamp 200 W, PL21-200, 15 mW/cm^2 for 254 nm, SEN Lights Co.) to clean the surface. UV light (PL21-200) has stronger power than that used in former studies (NL-UV253, Nippon Laser & Electronics Lab.). The OTS(octadecyltrichlorosilane, $C_{18}H_{37}SiCl_3$)-SAM or APTS(aminopropyltrimethoxysilane, $H_2NC_3H_5Si(OCH_3)_3$) were prepared by immersing the Si substrate in an anhydrous toluene (Aldrich Chemical Co., Inc.) solution containing 1 vol% OTS (Acros Organics) for 15 min or APTS (TCI) for 2 h under an N_2 atmosphere [31,32](Fig. 1). The substrate with the SAM was baked at 120°C for 5 min to remove residual solvent and promote chemisorption of the SAM. The control of preparation conditions such as humidity is very important to fabricate organically modified surfaces which realize site-selective deposition.

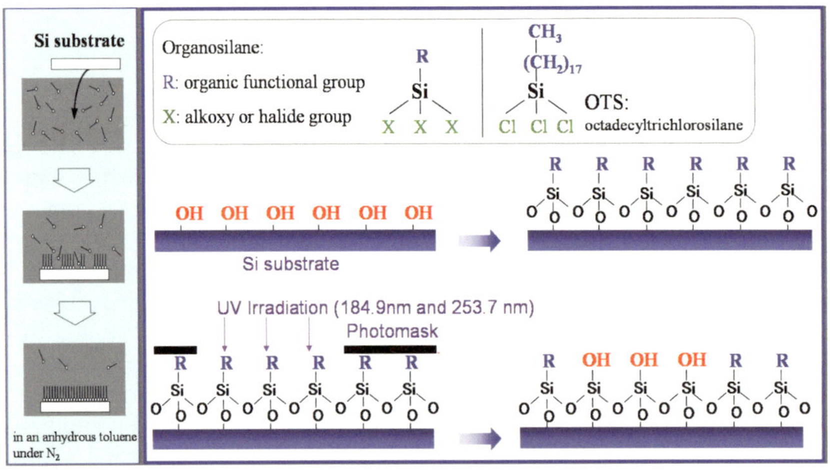

Fig. 1. Conceptual process for fabrication of a self-assembled monolayer.
Reprinted with permission from Ref.[31], Masuda, Y., Gao, Y. F., Zhu, P. X., Shirahata, N., Saito, N. and Koumoto, K., 2004, *J. Ceram. Soc. Japan*, 112, 1495. Copyright @ The Ceramic Society of Japan

The SAMs on the silicon substrates were exposed for 15 min to UV light through a photomask to be used as a template for micropatterning of ZnO crystals. UV-irradiated regions became hydrophilic due to silanol group formation, while the non-irradiated part remained unchanged. Formation of the SAMs and the modification to silanol groups by UV irradiation were verified using the static water drop contact angle (θw) (a contact angle meter CA-D, Kyowa Interface Science Co., Ltd.) and X-ray photoelectron spectroscopy (XPS) (ESCALAB 210, VG Scientific Ltd.). The X-ray source (MgKα, 1253.6 eV) was operated at 15 kV and 18 mA, and the analysis chamber pressure was 1-3×10-7 Pa. The initially deposited OTS-SAM or APTS-SAM showed a static water contact angle of 105° or 63°, but the UV-irradiated surface of SAM was wetted completely (contact angle < 5°) [33] (Fig. 2). The spectrum peak corresponding to the N 1s binding energy centered at 399.5 eV was observed for the surface of the APTS-treated Si substrate on which APTS-SAM was formed, however, it wasn't detected from the surface after UV irradiation. These experiments show the decomposition and removal of SAMs from the surface of substrates.

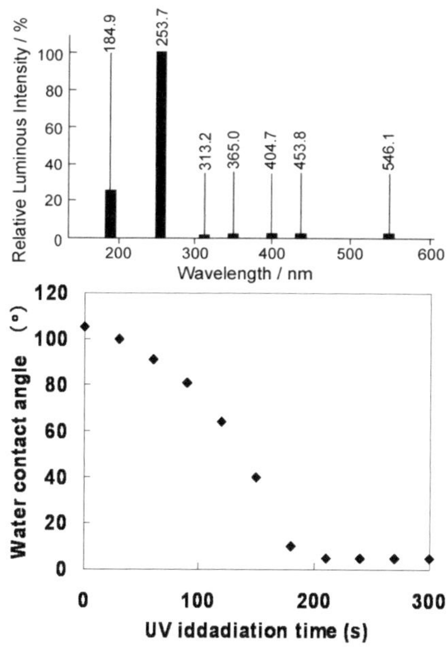

Fig. 2. Relative luminous intensity of UV lamp and water drop contact angle of OTS-SAMs as a function of UV-irradiation time.
Reprinted with permission from Ref. [33], Masuda, Y., Kinoshita, N., Sato, F. and Koumoto, K., 2006, *Cryst. Growth Des.*, 6, 75. Copyright @ American Chemical Society

2.2 Liquid phase patterning of amorphous TiO_2 thin films[34-37]

Patterned OTS-SAM was immersed in an anhydrous toluene (99.8%, water < 0.002%, Aldrich) solution containing 0.1 M TDD(titanium dichloride diethoxide) for 30 min under an N_2 atmosphere using a glove box (Fig. 3) [34-37]. All glassware was dried in a dry box at 50°C before use. The estimated partial pressure of H_2O in an N_2 atmosphere is below 0.1 hPa. Chlorine atoms of TDD react with H_2O and change into OH, which further react with silanol groups of SAM resulting in the formation of Ti-O-Si bonds[38]. The ethoxy group, OC_2H_5, of TDD is hydrolyzed into hydroxyl groups which are further condensed to form Ti-O-Ti bonds[38]. The thickness of films can be easily controlled by varying the soaking time. After SAM substrates had been rinsed with toluene and preserved in air, thin films appeared on the silanol surfaces of OTS-SAM but were not observed on octadecyl surfaces[34] (Fig. 4). A micropattern of amorphous TiO_2 thin films was thus fabricated on a patterned OTS-SAM. Line width measurements at 15 equally spaced points on each line indicated an average printed line width of 23.3 μm. Line edge roughness, as measured by the standard deviation of the line width, was ~0.5 μm, representing a ~2.1% variation (i.e., 0.5/23.2) in the nominal line width[34]. X-ray diffraction measurements (XRD) (Rigaku RU-200) with CuK*α* radiation (40 kV, 30 mA) for as-deposited thin films showed that they were composed of amorphous phases. The ratio of oxygen to titanium was evaluated after 20 min of Ar^+ ion sputtering to avoid the influence of the contaminated layer on the surface. The 1s peak of O can be deconvoluted into two curves (ratio of 529.7 eV (films) and 531.3 eV (silicon oxide) is 1 :

0.22). The ratio of oxygen to titanium was estimated to be 2.2 : 1. Small amounts of chlorine and carbon were also detected (Ti : O : Cl : C = 1 : 2.2 : 0.17 : 0.37)[34].

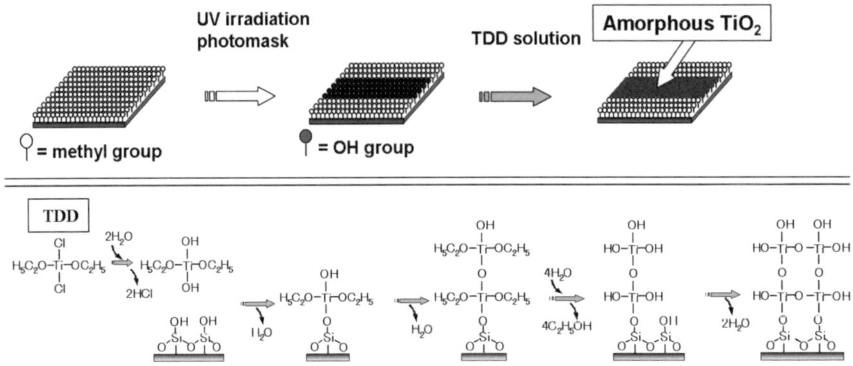

Fig. 3. Conceptual process for selective deposition of amorphous TiO_2 thin film using a self-assembled monolayer.
Reprinted with permission from Ref. [34], Masuda, Y., Sugiyama, T., Lin, H., Seo, W. S. and Koumoto, K., 2001, *Thin Solid Films*, 382, 153. Copyright @ Elsevier B.V.

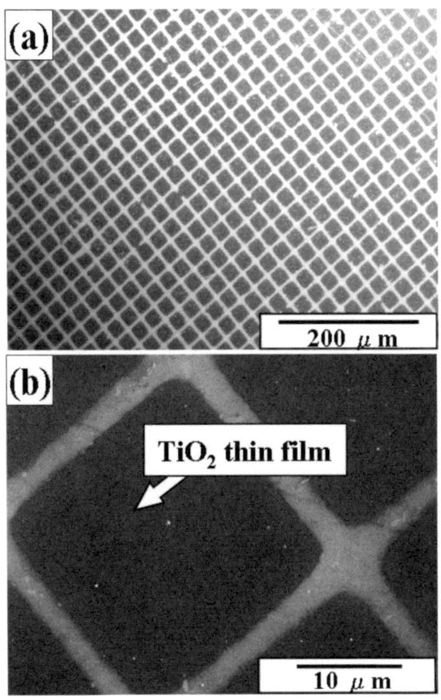

Fig. 4. SEM micrographs of (a) a micropattern of amorphous TiO_2 thin films and (b) magnified area of (a).
Reprinted with permission from Ref. [34], Masuda, Y., Sugiyama, T., Lin, H., Seo, W. S. and Koumoto, K., 2001, *Thin Solid Films*, 382, 153. Copyright @ Elsevier B.V.

2.3 Liquid phase patterning of crystalline anatase TiO_2 thin films using a seed layer[39]

The concept of LPP using a seed layer which accelerates the deposition of thin films was proposed[39]. The deposition process of anatase TiO_2 from an aqueous solution was evaluated in detail using a quartz crystal microbalance, and it was found that the nucleation and initial growth of anatase TiO_2 were accelerated on amorphous TiO_2 thin films compared with silanol, amino, phenyl, or octadecyl groups. Amorphous TiO_2 thin films were deposited on silanol regions of a patterned OTS-SAM (Fig. 5(1-a,b)) from a TDD solution (Fig. 5(2-a,b)). This substrate was immersed in an aqueous solution containing a Ti precursor at pH 1.5 for 1 h to be used as a template for LPP. Anatase TiO_2 was selectively deposited on amorphous TiO_2 regions to form thin films. Consequently, a micropattern of anatase TiO_2 thin film which had high feature edge acuity was successfully fabricated in an aqueous solution (Fig. 5(3-a,b)). The center of the anatase TiO_2 thin film region was 61 nm higher than the octadecyl regions, and the thickness of the anatase TiO_2 thin film was estimated to be 36 nm

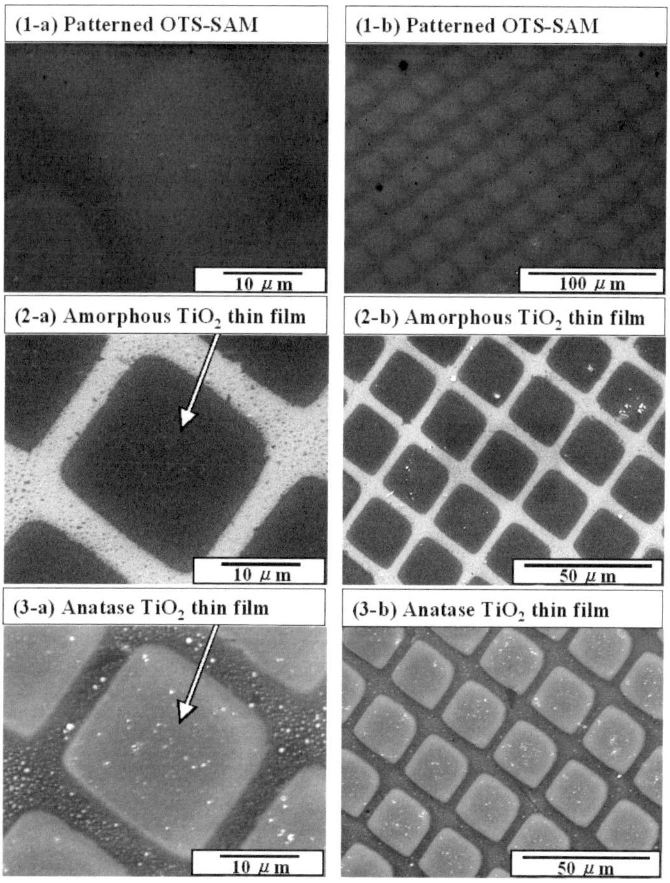

Fig. 5. SEM micrographs of [(1-a), (1-b)] a patterned OTS-SAM, [(2-a), (2-b)] a micropattern of amorphous TiO_2 thin films, and [(3-a), (3-b)] a micropattern of anatase TiO_2 thin films deposited at pH 1.5.

Reprinted with permission from Ref. [39], Masuda, Y., Ieda, S. and Koumoto, K., 2003, *Langmuir*, 19, 4415. Copyright @ American Chemical Society considering the thickness of amorphous TiO_2 thin film (27 nm)[34] and OTS molecules (2.4 nm)[39]. This result is similar to that estimated by QCM measurement (36 nm). The AFM image showed the film roughness to be 3.7 nm (horizontal distance between measurement points: 6.0 μm), which is less than that of amorphous TiO_2 thin film (RMS 9.7 nm, 27 nm thick, horizontal distance between measurement points: 6.0 μm))[34]. Additionally, the roughness of the octadecyl group regions was shown to be 0.63 nm (horizontal distance between measurement points: 1.8 μm)[39]. This study showed the good performance of the LPP process using a seed layer and the importance of quantitative analysis of the deposition process.

2.4 Liquid phase patterning of crystalline anatase TiO_2 thin films using site-selective elimination[40]

Ammonium hexafluorotitanate ($[NH_4]_2TiF_6$) (purity 96%, 1.031 g) and boric acid (H_3BO_3) (purity 99.5%, 0.932 g) were dissolved separately in deionized water (50°C, 50 ml)[40]. An appropriate amount of HCl was added to the boric acid solution to control pH, and ammonium hexafluorotitanate solution was added. Solutions (100 ml) with 0, 0.1 or 0.6 ml of HCl showed pH 3.8, 2.8 or 1.5, respectively. Supersaturation of solution can be changed by pH value as discussed in ref. 55. TiO_2 thin films can be formed fast by the deposition of homogeneously nucleated particles at high pH condition such as pH 3.88, and uniform films can be obtained slowly by heterogeneous nucleation at low pH condition. SAMs were immersed in the solution (100 ml) containing 0.05 M $(NH_4)_2TiF_6$ and 0.15 M (H_3BO_3) at pH 1.5, 2.8 or 3.8 and kept at 50°C for 4 h. to deposit anatase TiO_2. Ultrasonication was done during immersion period. Deposition of TiO_2 proceeded by the following mechanism:

$$TiF_6^{2-} + 2H_2O \rightleftarrows TiO_2 + 4H^+ + 6F^- \ldots \quad (a)$$

$$BO_3^{3-} + 4F^- + 6H^+ \longrightarrow BF_4^- + 3H_2O \ldots \quad (b)$$

Equation (a) is described in detail by the following two equations:

$$TiF_6^{2-} \xrightarrow{nOH^-} TiF_{6-n}(OH)_n^{2-} + nF^- \xrightarrow{(6-n)OH^-} Ti(OH)_6^{2-} + 6F^- \ldots \quad (c)$$

$$Ti(OH)_6^{2-} \longrightarrow TiO_2 + 2H_2O + 2OH^- \ldots \quad (d)$$

Fluorinated titanium complex ions gradually change into titanium hydroxide complex ions in an aqueous solution as shown in Eq. (c). Increase of F^- concentration displaces the Eq. (a) and (c) to the left, however, produced F^- can be scavenged by H_3BO_3 (BO_3^{3-}) as shown in Eq. (b) to displace the Eq. (a) and (c) to the right. Anatase TiO_2 was formed from titanium hydroxide complex ions ($Ti(OH)_6^{2-}$) in Eq. (d), and thus the supersaturation degree and the deposition rate of TiO_2 depend on the concentration of titanium hydroxide complex ions. The high concentration of H^+ displaces the equilibrium to the left in Eq. (a), and the low concentration of OH^-, which is replaced with F^- ions, suppresses ligand exchange in Eq. (c) and decreases the concentration of titanium hydroxide complex ions at low pH such as pH 1.5. The solution actually remained clear at pH 1.5, showing its low degree of supersaturation. On the other hand, the solution at high pH such as pH 2.8 or 3.8 became turbid because of homogeneously-nucleated anatase TiO_2 particles caused by a high degree of supersaturation. Anatase TiO_2 thin film was formed by heterogeneous nucleation in the solution at pH 1.5, while the film was formed by heterogeneous nucleation and deposition of homogeneously nucleated particles at pH 2.8 or 3.8.

After having been immersed in the solution with ultrasonic treatment, the substrates were rinsed with distilled water. Thin films were observed on the silanol group regions to form nano/micro-scaled patterns at pH 3.88, 2.8 or 1.5. The films were deposited for 4 h from the solution without the addition of HCl. Thin films were observed as being dark in an optical micrograph. Separated parallel lines 200–400 nm in width at 100–200 nm intervals were successfully fabricated with this method. The length of the separated parallel lines reached more than 100 μm. A cross section of the lines was shown as a semicircle, and the thickness of the center of the lines was estimated to be about 100 nm by AFM observation (Fig. 6). Feature edge acuity of the pattern was higher than that of the pattern fabricated by our lift-off process or by the site-selective immersion method.[41] Site-selective deposition was realized at any pH conditions such as pH 3.88, 2.8 or 1.5. Patterns which have higher feature edge acuity can be obtained at low pH conditions because films were formed slowly without the deposition of homogeneously nucleated particles.

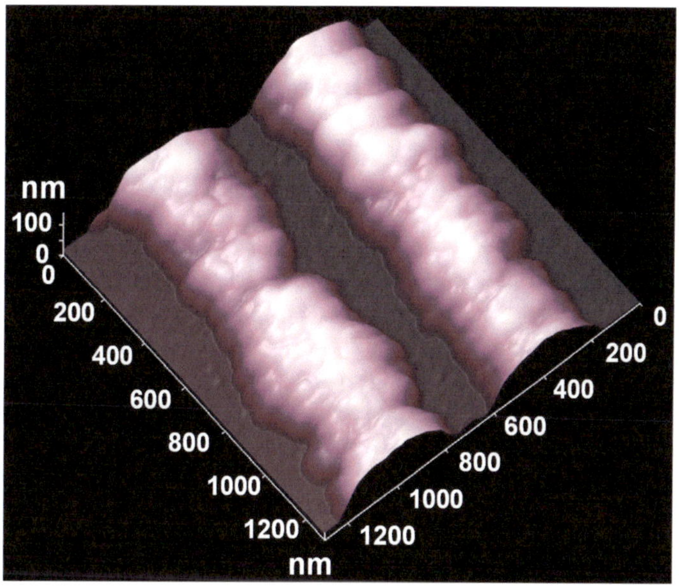

Fig. 6. AFM image of a nanopattern of anatase TiO_2 fabricated by site-selective elimination. Reprinted with permission from Ref. [40], Masuda, Y., Saito, N., Hoffmann, R., De Guire, M. R. and Koumoto, K., 2003, *Sci. Tech. Adv. Mater.*, 4, 461. Copyright @ American Chemical Society

For the adhesion of TiO_2 films to silanol groups, the pH of the deposition solution is critical. Pizem *et al.*[13] reported that adherent TiO_2 films formed from solutions similar to those used here at pH=3.9, but that the films were less adherent at pH=2.9. They related this difference in adherence to an increased electrostatic attraction of TiO_2 to the oxidized surface of silicon at the higher pH. In our experiment, there also were less adherence of TiO_2 at pH 2.8 and 1.5 than that at pH 3.5 due to low supersaturaiton shown in Eq. (a) and low electrostatic attraction.[13]

However, some depositions were observed on octadecyl group regions in SEM micrographs. One probable cause is that pinholes and other defects in the films provide at least some degree of access of water to underlying unreacted OH groups in the OTS films. Once exposed to the solution, these sites can act as nucleation points for TiO_2 growth. Because the depositions are performed at elevated temperatures, it is likely that pinholes and defects will continually open and close on the OTS film surfaces due to thermal motions of alkyl chains in the films. The TiO_2 precursors formed in these defects would act as points for eventual growth of TiO_2 over the entire SAM-covered region. This would provide a weakly bound TiO_2 film on the OTS film regions due to the limited number of connections to the underlying silanol sites in the film regions. In fact, Sagiv and others have shown that macroscopic defects induced in alkylsiloxane films can readily be accessed by solution species.[42-44] More recently, Dressick and coworkers demonstrated that solvent accessibility to underlying substrates in aromatic siloxane films is also important[45-47] and may be an even greater factor in controlling the properties of those films, which may account for our previous selectivity observation using phenylsiloxane films, as shown in the lift-off process.
In the lift-off process[48], thin films were formed on the entire area of patterned SAM that has silanol group regions and phenyl (or octadecyl) group regions. After being dried, the substrate was sonicated in water to lift off thin films on phenyl (or octadecyl) group regions selectively. Thin films on phenyl (or octadecyl) group regions were peeled off along the cracks that formed during the drying process. Thin films on phenyl (or octadecyl) group regions without cracks were not peeled off because depositions strongly connected to each other to form solid timber (monolith). The lift-off along cracks decreased the feature edge acuity of the pattern in this method. Thin films were formed on silanol group regions selectively and site-selective deposition was realized with our newly developed method. This resulted in high feature edge acuity of the patterns compared to our previous works.[48]
Additionally, the micropattern of thin films was also fabricated by the site-selective immersion method[41]. A solution containing a Ti precursor contacted the hydrophilic regions during the experiment and briefly came into contact with the hydrophobic regions. The solution on the hydrophilic surface was replaced with a fresh solution by continuous movement of bubbles. Thus TiO_2 was deposited and a thin film was grown on the hydrophilic regions selectively. This technique can be applied for the formation of many kinds of films from any solution and to fabricate micropatterns for many kinds of thin film because the technique creates the difference in contact time of the solution between hydrophilic regions and hydrophobic regions. However, it is difficult to form a solution layer on nano-scaled hydrophilic regions selectively and replace it with a fresh solution by continuous movement of bubbles while avoiding contact of the solution on hydrophobic regions. This prevents fabrication of nano-scaled pattern with this method. On the other hand, site-selective deposition was realized in the solution with our newly developed method using the difference of adhesive strength of depositions to substrates. Heterogeneously nucleated deposition and homogeneously nucleated particles and/or clusters can be removed from octadecyl group regions even if these regions are designed in nano-scale order in which depositions are smaller. This allowed us to realize high feature edge acuity of the patterns compared to site-selective immersion.[39,41]
The distribution of elements on the surface of the substrates was evaluated by energy dispersive X-ray analysis (EDX; EDAX Falcon, EDAX Co. Ltd.), which is built into SEM. Titanium was detected from thin films selectively and oxygen was detected mainly from silanol group regions by EDX. Other elements, except for silicon from the substrate, were

not observed from the thin film and substrate by EDX. Oxygen was detected from not only the deposited thin film but also from the natural oxide layer (amorphous SiO_2 layer) formed on all surface areas of a silicon substrate. These observations showed predominant deposition of titanium oxide on silanol group regions.

The deposited thin films were also investigated using an X-ray diffractometer (XRD; RAD-C, Rigaku) with CuKα radiation (40 kV, 30 mA) and Ni filter plus a graphite monochromator. Thin films deposited at pH 3.8 for 4 h showed an XRD pattern of anatase-type TiO_2 having orientation similar to that of films deposited in the solution at pH 1.5 or 2.8[41]. The diffraction from parallel to c-plane such as (004) was observed as being strong compared to that of the randomly oriented powder diffraction pattern. Pizem *et al.* postulated that the commonly observed [001] orientation of anatase films could be due to the slight polarity of the planes parallel to the [001] axis, unlike other low-index planes of this structure such as {100}, (110), and (210). The orientation and crystal growth mechanism are further discussed in a separate article.[49]

Thin films were further evaluated by X-ray photoelectron spectroscopy (XPS; ESCA-3200, Shimazu Corporation, 1×10^{-5} Pa). The X-ray source (MgKα, 1253.6 eV) was operated at 8 kV and 30 mA. The spectral peaks corresponding to Ti 2p (458.7 eV) were observed from thin films deposited on the silanol region. This binding energy is higher than that of Ti metal (454.0 eV), TiC (454.6 eV), TiO (455.0 eV), TiN (455.7 eV) and Ti_2O_3 (456.7 eV), and similar to that of TiO_2 (458.4 - 458.7 eV).[50-52]. This suggests that the titanium atoms in thin films are positively charged relative to that of titanium metal by formation of direct bonds with oxygen. On the other hand, this spectrum was not observed from octadecyl group regions. The O 1s spectrum was observed from the silanol regions and divided into O 1s (530.2 eV) and O 1s (532.3 eV). O 1s (532.3 eV) can be assigned to the silicon oxide layer on the surface of the silicon wafer (532.0 eV[51]), whereas the binding energy of O 1s (530.2 eV) is similar to that of TiO_2 (529.9 eV[52], 530.1 eV[51]) as observed by Shin *et al.*[53]. This shows that oxygen is negatively charged compared with neutral oxygen molecules (531.0 eV), possibly through the formation of chemical bonds with Ti. The ratio of titanium to oxygen was estimated from the Ti $2p_{3/2}$ (458.7 eV) spectrum and O 1s (530.2 eV) spectrum to be Ti:O = 1:2.0.

2.5 Liquid phase patterning of magnetite particulate thin films using Pd catalyst[54,55]

Catalyst solution containing Na_2PdCl_4 (0.38 mM) and NaCl (0.01 M) in a 0.01 M 2-morpholinoethane sulfonate pH 5 aqueous buffer was prepared[54,55]. The details of preparation of this solution are described in the reference[56]. Hydrolyzed Pd colloids were formed in this solution.[57] The patterned APTS-SAM was immersed into the colloidal dispersion of catalyst at 25 °C for 30 min and catalyzed APTS-SAM was rinsed with water.

Catalyzed SAM was immersed in an aqueous solution containing iron(III) nitrate (0.0025 M) and dimethylamine-borane (DMAB) (0.03 M) and kept at 80 °C using a water bath for 30 min to deposit magnetite particulate thin film.[58] DMAB was used to reduce nitrate ions, giving rise to OH- ions and hence raising the solution pH to precipitate Fe_3O_4.

A black colored iron oxide film selectively deposited onto regions of the APTS-SAM that had not been exposed to UV radiation following application of the Pd catalyst dispersion. Figures 7 show an optical microscope image and a SEM image of as-deposited films, respectively. Black contrast represents a deposited film in an optical microscope image, whereas the white contrast shows deposited films in SEM images. The EDX mapping images shown in Figure 7(c) indicate the films deposited on the amino-surface regions, showing mapping images consisting mainly of iron and oxygen. Thickness of the films was easily controlled in the range from several ten nano meter to several micro meter by change of immersion period. The XRD pattern of the thin film deposited on the whole surface of the APTS-SAM clearly indicates that it is a magnetite (Fe_3O_4) film composed of randomly oriented crystallites of about 20 nm in diameter, which was evaluated using the Scherer equation. These evaluations show the successful fabrication of a micropattern of crystalline magnetite films in an aqueous solution using a patterned APTS-SAM and Pd colloid catalysts adsorbed on amino-group ($-NH_2$) regions of a SAM.

Fig. 7. (a) optical microscope image, (b) SEM image and (c) characteristic X-ray images [Fe] of a micropattern of crystalline Fe_3O_4.
Reprinted with permission from Ref. [54], Nakanishi, T., Masuda, Y. and Koumoto, K., 2004, *Chem. Mater.*, 16, 3484. Copyright @ American Chemical Society

2.6 Liquid phase patterning and morphology control of crystalline ZnO[33]

Zinc acetate ($Zn(CH_3COO)_2$, Kishida Chemical Co., Ltd.) was dissolved into water to be 15 mM at 50°C, and ammonia (28% solution, Kishida) was then added to be 30, 60 or 90 mM ($[NH_3]$ / $[Zn]$ = 2.0, 4.0 or 6.0) with stirring as complexing agent[33]. These solutions showed pH = 7.04, 7.50 or 8.93, respectively. Zinc ions reacted with ammonium ions (NH_4^+) formed from ammonia to form tetra amine zinc(II)[59] $[Zn(NH_3)_4]^{2+}$. ZnO was crystallized from the reaction between $[Zn(NH_3)_4]^{2+}$ and OH^-. The solution became clouded shortly after adding ammonia due to homogeneous nucleation of ZnO crystals. Morphology of ZnO crystals was controlled by the ratio of ammonia to zinc acetate, i.e., super-saturation degree for crystallization. Patterned OTS-SAMs were immersed downward into the solution containing zinc acetate (15 mM) and ammonia (30 mM) as complexing agent ($[NH_3]$ / [Zn] = 2.0) at 50°C for 3 h (Fig. 8).

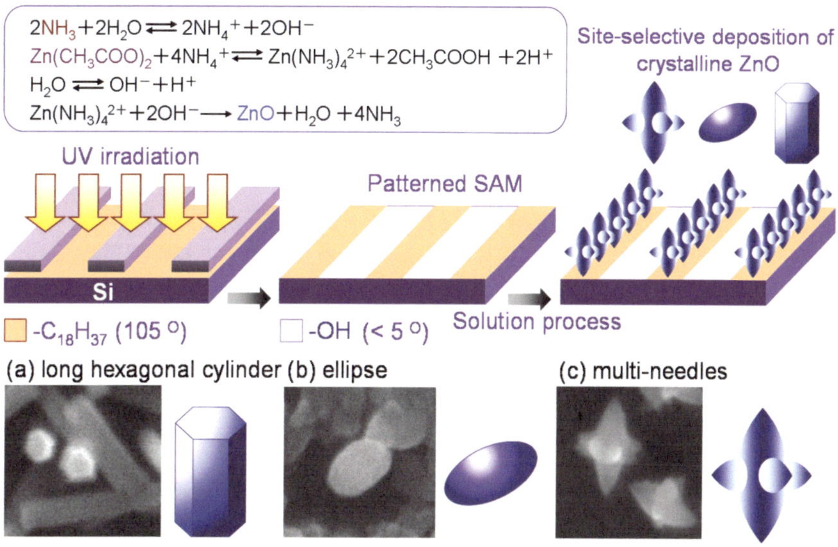

Fig. 8. Conceptual process for self-assembly patterning of light-emitting crystalline ZnO nanoparticles in an aqueous solution.
Reprinted with permission from Ref. [33], Masuda, Y., Kinoshita, N., Sato, F. and Koumoto, K., 2006, *Cryst. Growth Des.*, 6, 75. Copyright @ American Chemical Society

ZnO crystals having long hexagonal cylinder shape were homogeneously nucleated to make the solution turbid shortly after adding ammonia. Crystals showed sharp hexagonal facets of about 100 nm in diameter and larger than 500 nm in length. The morphology indicated high crystallinity of ZnO nanoparticles. The nanoparticles were deposited and further grown on hydrophobic octadecyl group regions of a patterned SAM selectively. Consequently, a micropattern of light-emitting ZnO crystals was successfully fabricated in an aqueous solution without Pd catalyst. ZnO crystals were also deposited on hydrophobic regions of patterned SAMs such as DTS-SAM, HTS-SAM, PTS-SAM, MTS-SAM, PTCS-SAM or APTS-SAM. This showed that the method is highly versatile and offers good potential for the fabrication of devices.

ZnO crystals were deposited on hydrophobic SAM regions such as OTS-, APTS- or other SAMs rather than hydrophilic silanol regions. Zeta potential of ZnO crystals deposited on a silicon substrate was measured to be 10 mV at pH 8.1 and ZnO crystals should thus have positive zeta potential not less than 10 mV in the solution at pH 7.04. SAM of OTS, silanol and APTS showed zeta potential of -3 mV, -38.2 mV or +22.0 mV, respectively. ZnO having positive zeta potential should be deposited on silanol regions having negative zeta potential rather than other SAMs, if the site-selective deposition was caused only by electrostatic interactions. The site-selective deposition of ZnO crystals would be caused by not only electrostatic interactions as shown by the relation of zeta potentials. ZnO crystals having long hexagonal cylinder shape were deposited on a hydrophilic silicon substrate to evaluate the surface of crystals. The substrate covered with many deposited ZnO crystals exhibited high water contact angle (WCA 140°). The deposited ZnO crystals were found from the experiment to have hydrophobic surfaces. Surfaces of naked ZnO crystals would be hydrophilic because of surface hydroxyl groups and they would become hydrophobic by being covered with organic molecules having hydrophobic functional groups. CH_3COO^- ions coming from $Zn(CH_3COO)_2$ might be adsorbed to ZnO crystal surfaces by the interaction between Zn and $-COO^-$ to cover the surface with hydrophobic $-CH_3$ groups and some of $Zn(CH_3COO)_2$ would exist in the surface layer of ZnO crystals. Additionally, deposited ZnO crystals having long hexagonal cylinder shape became hydrophilic (< 10°) and their zeta potential shifted positively by UV irradiation in air. ZnO crystals deposited on a silicon substrate showed zeta potential of 10 mV at pH 8.1, 0 mV at pH 8.8 and –15 mV at pH 9.2, while they shifted to 20 mV at pH 8.1, 10 mV at pH 8.8 and 7 mV at pH 9.2 by UV irradiation. The decomposition of CH_3COO^- ions and the breakage of the bond between CH_3COO and Zn would be caused by light excitation, ozone and active oxygen by UV irradiation in air. This finding suggests that organic molecules, such as CH_3COO^- ions, which show negative zeta potential and can be removed by UV irradiation, would be absorbed onto the surfaces of ZnO crystals. Furthermore, ZnO crystals were confirmed to deposit on a hydrophobic polyethylene terephthalate surface rather than on a hydrophilic polyethylene terephthalate surface modified by UV irradiation in the same solution. Additionally, organic molecule was reported to adsorb to growing ZnO crystals, in which poly (ethylene oxide)-blockpoly (methylacrylic acid) (PEO-b-PMAA) was adsorbed preferentially to {0001} face of ZnO to retard crystal growth perpendicular to this face[60]. Consequently, site-selective deposition was achieved by the effective molecular recognition caused by combination of the forces composed mainly of hydrophobic interactions between functional groups of SAMs and ZnO crystal surfaces.

Patterned SAMs were also immersed into the solution containing zinc acetate (15 mM) and ammonia (60 mM or 90 mM) as complexing agent ($[NH_3]$ / [Zn] = 4.0 or 6.0) for 3 h. ZnO crystals having ellipse or multi-needle shape (two large needles and four small needles) were homogeneously nucleated to make the solution turbid shortly after adding ammonia. Nucleation and deposition of ZnO crystals were accelerated by addition of ammonia. Each ZnO crystal was about 500 nm in size. The crystals were deposited and further grown on hydrophobic regions of patterned SAMs selectively. Micropatterns of light-emitting ZnO crystals having ellipse or multi-needle shape were fabricated on patterned SAMs such as OTS-SAM, DTS-SAM, HTS-SAM, PTS-SAM, MTS-SAM, PTCS-SAM or APTS-SAM in aqueous solutions.

XRD spectra of ZnO crystals having ellipse or multi-needle shape showed dominant peaks corresponding to ZnO (0002) planes revealing that ZnO crystals were deposited with a high degree of orientation of their c-axes perpendicular to the substrate. Enhanced (0002) and (10-10) peaks from ZnO crystals having long hexagonal cylinder shape showed that crystals were deposited to make (0002) or (10-10) planes parallel to the substrate. Crystals having high crystallinity and high purity with no additional phase were shown to be prepared in an aqueous solution with precise control of their morphologies without the use of Pd catalyst. The aqueous solution system showed high ability for fabricating nano/micro devices composed of crystalline materials. ZnO crystals are well known to grow along the c-axis. The orientations evaluated from XRD patterns were consistent with SEM observations and were shown to be controlled precisely by the solution conditions.

Photoluminescence properties of ZnO crystal patterns were further evaluated. Micropatterns of ZnO crystals were observed by an optical microscope (Fig. 9) and strong visible-luminescence from ZnO crystals excited by 330–385 nm light was observed by a photoluminescence microscope. ZnO crystals showed strong UV luminescence (around 390 nm) attributed to band-edge luminescence and visible-light luminescence caused from oxygen vacancy (450–600 nm)[61,62]. All of the crystals showed photoluminescencedue to high purity and high crystallinity with optimal oxygen vacancy, and this caused the bright visible-photoluminescence image. Luminescence properties can be controlled by changing the crystalline morphologies. ZnO crystals deposited from an aqueous solution were shown to have high visible-light-emitting properties.

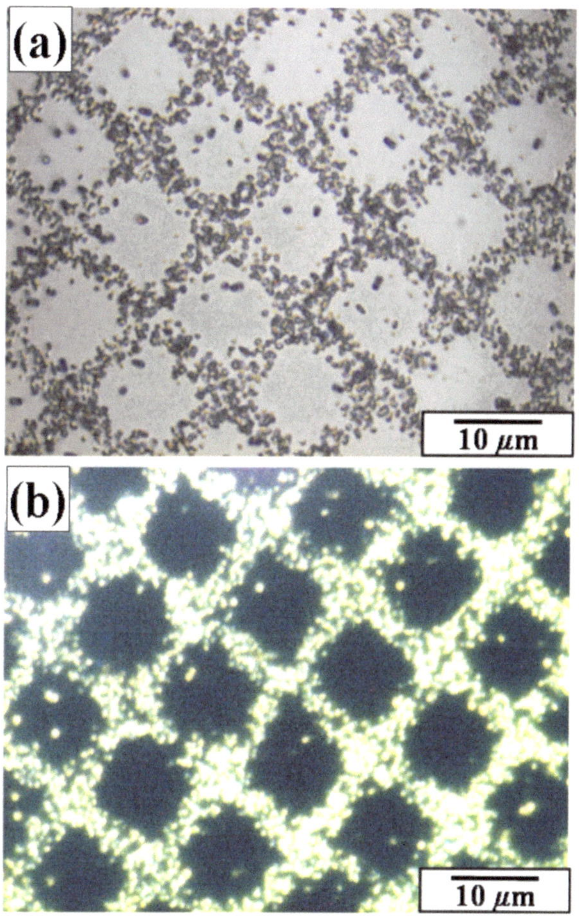

Fig. 9. (a) Optical microscope image and (b) photoluminescence image of patterned ZnO particles under white light or UV light (330–385 nm).

Reprinted with permission from Ref. [33], Masuda, Y., Kinoshita, N., Sato, F. and Koumoto, K., 2006, *Cryst. Growth Des.*, 6, 75. Copyright @ American Chemical Society

2.7 Liquid phase patterning of Y_2O_3: Eu thin films[63]

The patterned APTS-SAM was immersed in an aqueous solution containing $Y(NO_3)_3 \cdot 6H_2O$ (4 mM), $Eu(NO_3)_3 \cdot 6H_2O$ (0.4 mM) and NH_2CONH_2 (50 mM) at 25 °C[63]. The solution was heated to 77 °C gradually as shown in Fig. 10 since urea (NH_2CONH_2) decomposes to form ammonium ions (NH_4^+) above 70 °C (Eq. (a)). The decomposition of urea at elevated temperature plays an essential role in the deposition of yttrium oxide. The aqueous solution of urea yields ammonium ions and cyanate ions (OCN^-) at temperatures above 70 °C[64] (Eq. (a)). Cyanate ions react rapidly according to Eq. (b). Yttrium ions are weakly hydrolyzed[65,66] in water to $YOH(H_2O)_n^{2+}$ (Eq. (c)). The resulting release of protons (H^+) and/or hydronium ions (H_3O^+) accelerates urea decomposition (Eq. (b)). The precipitation of the amorphous basic yttrium carbonate ($Y(OH)CO_3 \cdot xH_2O$, x=1) can take place through the reaction in Eq. (d)[67,68]. The controlled release of cyanate ions by urea decomposition causes deposition of basic yttrium carbonate once the critical supersaturation in terms of reacting component is achieved. Since the decomposition of urea is quite slow, the amount needed to reach supersaturation within a given period of time must be considerably higher than the stoichiometric amount of yttrium ions, as revealed by previous studies of lanthanide compounds[69].

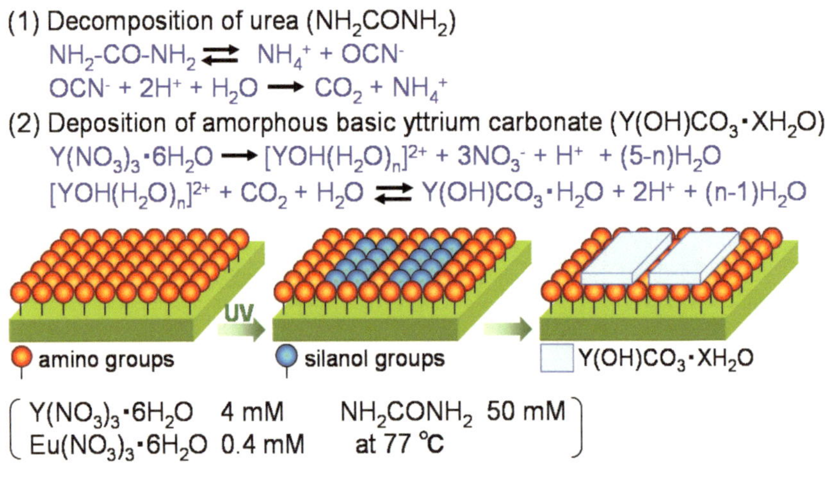

Fig. 10. Conceptual process for site-selective deposition of visible-light emitting Y_2O_3:Eu thin films using a self-assembled monolayer.
Reprinted with permission from Ref. [63], Masuda, Y., Yamagishi, M. and Koumoto, K., 2007, *Chem. Mater.*, 19, 1002. Copyright @ American Chemical Society

The temperature of the solution increased gradually and reached 77 °C in about 80 min. The solution was kept at ~ 77 °C during deposition. The pH of the solution increased from 5.2 to 5.8 in about 90 min and then gradually decreased to 5.6. Temperature and pH increased for the initial 90 min and became stable after 90 min. The average size of particles homogeneously nucleated in the solution at 100 min was about 227 nm and increased to 262 nm at 150 min, 282 nm at 180 min, 310 nm at 210 min, and 323 nm at 240 min. Particles nucleated and grew after the solution temperature exceeded 70 °C because urea decomposes above 70 °C to form carbonate ions[64] which causes deposition of basic yttrium carbonate[65-68]. The particles grew rapidly at the beginning of the growth period and then their growth rate decreased exponentially. The decrease in growth rate was caused by the decrease of supersaturation degree influenced by a decrease in solution concentration.

Yttrium carbonate films were observed to deposit on amino regions of a patterned SAM after the immersion in an aqueous solution (Fig. 11). Deposits showed white contrast, while silanol regions without deposition showed black contrast in SEM observation. Narrow lines of depositions having 10–50 μm width were successfully fabricated in an aqueous solution. Patterned APTS-SAM showed high ability for site-selective deposition of yttrium carbonate in solution systems.

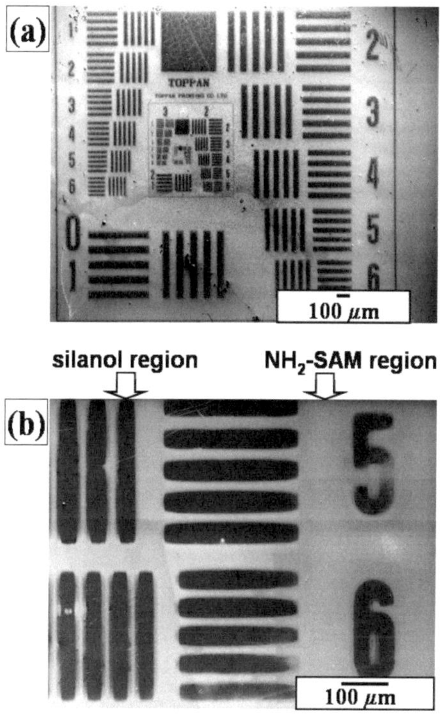

Fig. 11. (a) SEM micrograph of patterned Y_2O_3:Eu thin films and (b) magnified area of (a). Reprinted with permission from Ref. [63], Masuda, Y., Yamagishi, M. and Koumoto, K., 2007, *Chem. Mater.*, 19, 1002. Copyright @ American Chemical Society

Yttrium carbonate films were also deposited on the hydrophobic octadecyl surface of OTS(octadecyltrichlorosilane)-SAM having water contact angle (WCA) of 116 ° and as-purchased silicon wafer having WCA of about 20-50 ° which was kept in a plastic case in air. On the other hand, the films were not deposited on UV irradiated silicon wafer having WCA < 5 °. The super hydrophilic surface of WCA < 5 ° suppressed film deposition, whereas the hydrophobic surface and medium surface of WCA > 20-30 ° accelerated film deposition possibly because of hydrophobic interaction between deposition and substrate surface. This is consistent with a former study[32]. Yttrium carbonate was deposited both on bare single crystal Si wafers, and on Si wafers coated with sulfonate-functionalized organic self-assembled monolayers.

Yttrium, europium, oxygen and carbon were observed from as-deposited thin films on amino regions, while silicon and oxygen were detected from non-covered silanol regions by

EDX. The molecular ratio of yttrium to europium was determined to be 100 : 8. It was close to that of $Y(NO_3)_3 \cdot 6H_2O$ to $Eu(NO_3)_3 \cdot 6H_2O$, i.e., 100 : 10, in the solution because the chemistry of $Eu(NO_3)_3$ is similar to that of $Y(NO_3)_3$ to incorporate europium in the precipitation. The content of europium was in the range we had expected. Y_2O_3:Eu with atomic ratio Y : Eu = 100 : ~ 8 was reported to have strong photoluminescence[70,71]. Carbon was detected from yttrium carbonate. Silicon and oxygen were detected from silicon wafer covered with a natural oxide layer (amorphous SiO_2).

Amino regions were covered with thin films composed of many large particles (about 100–300 nm in diameter) and very high roughness (RMS 25.6 nm). Silanol regions, on the other hand, showed only nano-sized small particles (about 10–50 nm in diameter) and very low roughness (RMS 1.7 nm). The high site-selectivity of deposition and the big difference in surface morphology and roughness were clearly shown by AFM observation. The thickness of the films was estimated from AFM scans across deposited and undeposited regions of the substrate. It increased with immersion time after 45 min (0 nm at 45 min, 60 nm at 70 min and 100 nm at 90 min). The average growth rate (70 nm/h = 100 / 90 min) was higher than that previously reported (2 nm/h = 35 nm / 15 h)[68]. An amorphous yttrium basic carbonate film was deposited at 80 °C from aqueous solutions of $YNO_3 \bullet 5H_2O$ and urea on Si wafers coated with sulfonate-functionalized organic self-assembled monolayers in previous studies. The thickness was then evaluated by TEM after the treatment with ultrasonication for half an hour in distilled water. The difference of growth rate was caused mainly by the difference of the substrate treatment by ultrasonication. Additionally, the thickness of our film was smaller than the particle size in the solution (227 nm at 100 min). Heterogeneous nucleation and attachment of initial particles of yttrium carbonate occurred without the attachment of aggregated large particles. The yttrium carbonate was then grown on the substrate to form a film of 100 nm thickness after immersion for 90 min. The particles of about 100 nm in height were removed by ultrasonication for 30 min and the film of several nm in height remained as reported[68].

Yttrium was not detected by XPS from the substrate immersed for 45 min, however, it was clearly observed from that immersed for 90 min. This indicates that the deposition began between 45 and 90 min after immersion. The solution temperature reached 70 °C in ~ 45 min and then the solution began to decompose and release carbonate ions, causing the deposition of basic yttrium carbonate. The deposition mechanism evaluated by XPS is consistent with the change of solution temperature, decomposition temperature of urea and chemical reaction of this system. The binding energy of Y 3d5/2 spectrum from the deposition (158.2 eV) was higher than that of metal yttrium (155.8 eV)[72]. The spectrum shifted to lower binding energy (156.7 eV) after annealing at 800 °C in air for 1 h and is similar to that of Y_2O_3 (157.0 eV)[73]. The binding energies of Y 3d5/2 spectra in as-deposited films and annealed films were higher than that of metal yttrium possibly due to the chemical bonds formed between yttrium ions and oxygen ions. The chemical shift of Y 3d5/2 binding energy by annealing is consistent with crystallization of as-deposited films to crystalline Y_2O_3. C 1s spectra were detected at 289.7 eV and 284.6 eV from as-deposited films. The C 1s spectrum at 289.7 eV then disappeared by the annealing. C 1s at 284.6 eV was assigned to surface contamination and C 1s at 289.7 eV was detected from as-deposited yttrium carbonate. The disappearance of C 1s at 289.7 eV is consistent with the phase transition from yttrium carbonate to Y_2O_3.

As-deposited film was shown to be an amorphous phase by XRD measurement. The film showed no diffraction peak after annealing at 400 °C for 1 h, however, it showed 222, 400 and 440 diffraction peaks of crystalline cubic Y_2O_3[74] without any additional phase after annealing at 600 °C for 1 h and the intensities of diffraction peaks increased further by annealing at 800 °C for 1 h. The film was shown to be a polycrystalline Y_2O_3 film constructed from randomly deposited Y_2O_3 particles without crystal-axis orientation. The crystal structure model and diffraction pattern of Y_2O_3 were calculated from the crystal structure data of ICSD #23811. The crystallization by annealing confirmed from XRD measurement is consistent with XPS evaluation.

Y_2O_3 films were attemped to remove from the silicon substrate by debonding with scotch tape or by ultrasonication for 5 min in water. However, the films maintained their bonds with the substrate, indicating that strong adhesion had formed between films and substrate. The thin film annealed at 800 °C for 1 h, i.e., crystalline Y_2O_3:Eu thin film, was shown to be excited by 230–250 nm (center: 243 nm) and emit red light photoluminescence centered at 611 nm in the fluorescence excitation spectrum (Fig. 12a). Neither the as-deposited film nor the film annealed at 400 °C for 1 h showed photoluminescence, on the other hand, the films annealed at 600 °C or 800 °C for 1 h emitted light centered at 617 nm by 250 nm in fluorescence emission spectra (Fig. 12b). The fluorescence intensity of the film annealed at 800 °C was stronger than that of the film annealed at 600 °C. Fluorescence intensity increased by the phase transformation from amorphous yttrium carbonate to yttrium oxide and crystal growth by the heat treatments, and is consistent with the crystallization observed by XRD. The spectra are described by the well-known $^5D_0$–^7F_J line emissions (J = 0, 1, 2, ...) of the Eu^{3+} ion with the strongest emission for J = 2 at 612 nm. The thin film annealed at 800 °C produced visible red light photoluminescence by excitation from Nd: YAG laser (266 nm) (Fig. 12, inset). The white square shows the edges of the Y_2O_3:Eu thin film and the red color shows visible red emission from the irradiated area on the substrate.

2.8 Liquid phase patterning of In_2O_3 thin films[75]

Synthesis and patterning of Pd nanoparticles: A catalyst dispersion[57,76] containing Na_2PdCl_4 (0.38 mM) and NaCl (0.01 M) in a 0.01 M 2-morpholinoethane sulfonate pH 5 aqueous buffer was prepared (Fig. 13) as described in references[56,57,76]. Hydrolyzed Pd colloids were formed in this solution[57]. Light-scattering measurements indicated that the catalyst dispersion contained colloid particles of about 30 nm in diameter. Pd nanoparticles showed negative zeta potential (-30.5 eV) at pH 5. APTS-SAM showed positive zeta potential[77] at pH 5 because of protonation of the amino group ($-NH_2$) to $-NH_3^+$. Silanol groups of UV irradiated APTS-SAM, on the other hand, showed negative zeta potential at pH 5 caused by deprotonation of the silanol group (-Si-OH) to $-Si-O^-$. The patterned APTS-SAM was immersed in the colloidal dispersion of catalyst at 25 °C for 30 min and the catalyzed APTS-SAM was rinsed with water. Pd colloids adsorbed on amine groups of APTS-SAM by electrostatic interactions between the negative surface charge of Pd colloids and positive surface charge of APTS-SAM[54] and formed covalent bonds[57], while electrostatic repulsion force kept Pd catalyst particles away from silanol group regions having negative zeta potential.

Pd nanoparticles deposited on amino group regions of a patterned SAM had a diameter of about 30 nm and surface roughness (RMS) of about 1 nm as shown by AFM observation

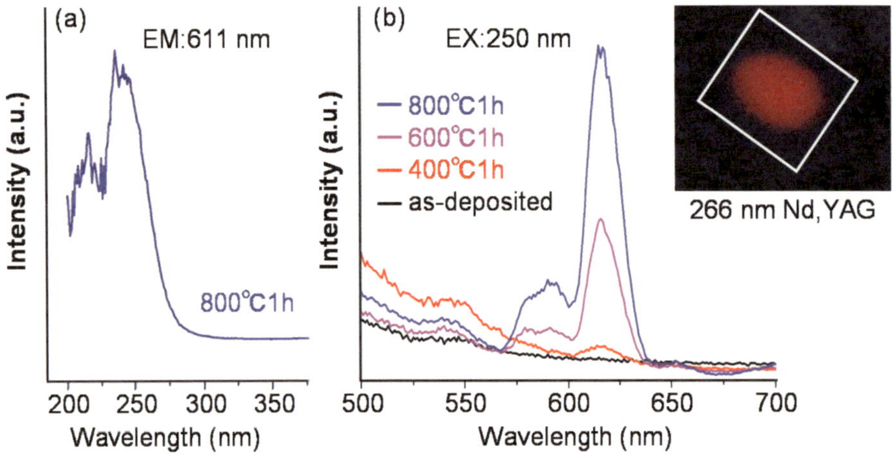

Fig. 12. (a) Fluorescence excitation spectrum (emission: 611 nm) for Y_2O_3:Eu thin film after annealing at 800 °C for 1 h. (b) Fluorescence emission spectra (excitation: 250 nm) for Y_2O_3:Eu thin films before and after annealing at 400, 600 or 800 °C for 1 h. Inset: Photoluminescence image for Y_2O_3:Eu thin film annealed at 800 °C for 1 h (excitation: 266 nm).
Reprinted with permission from Ref. [63], Masuda, Y., Yamagishi, M. and Koumoto, K., 2007, *Chem. Mater.*, 19, 1002. Copyright @ American Chemical Society

(Fig. 13). Pd was found to be adsorbed on amino group regions uniformly to form a thin catalytic layer with small surface roughness. Even application, small thickness and small surface roughness of the Pd layer are significant for deposition of a uniform transparent indium oxide layer.

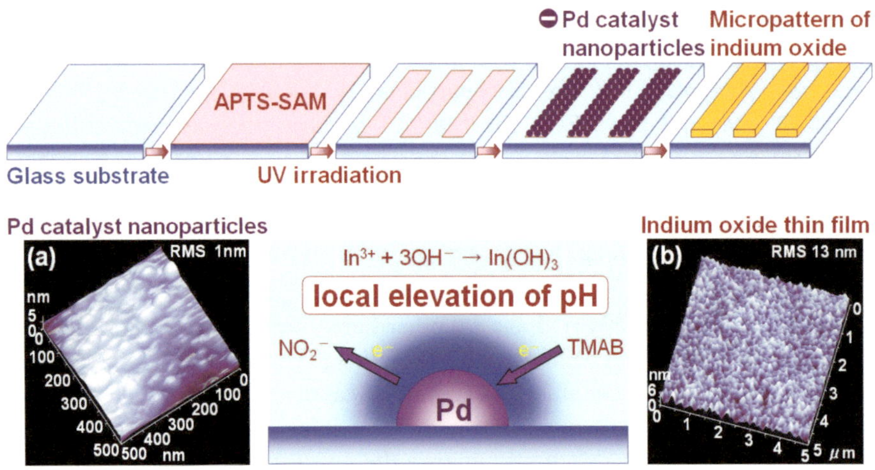

Fig. 13. Conceptual process for micropatterning of indium oxide thin films on a glass substrate. AFM images of (a) Pd catalyst nanoparticles and (b) indium oxide thin films (lower stand).
Reprinted with permission from Ref.[75], Masuda, Y.; Kondo, M.; Koumoto, K., 2009, *Cryst. Growth Des.*, 9, 555. Copyright @ American Chemical Society

Furthermore, site-selective adsorption of Pd was clearly observed by TOF-SIMS mapping[54]. Bright regions due to Pd (m/z = 104, 105, 106, 108, and 110) were observed on the APTS-SAM surface, while no Pd signal was seen on the silanol group surface. This result clearly indicates that site-selective adsorption of Pd catalyst occurred on the APTS-SAM surface.

Deposition control of $In(OH)_3$ thin films by Pd catalyst: The patterned SAM having Pd catalytic nanoparticles on amino group regions was immersed in an aqueous solution containing $In(NO_3)_3$ and TMAB at 65 °C for 1 h (Fig. 13).

Nitrate ions were generated by dissolution of indium nitrate in water according to:

$$In(NO_3)_3 \rightarrow In^{3+} + 3NO_3^- \cdots \quad \text{(a)}$$

Pd catalyst is indispensable for Eq. (b) and (c). Oxidation of reducing agent, TMAB, is promoted by Pd catalyst to generate electrons in Eq. (b). Nitrate ions are reduced to nitrite ions by receiving electrons according to Eq. (c). Hydroxide ions are generated by oxidation-reduction reactions near Pd catalyst according to Eq. (c). Local elevation of pH thus occurs near Pd catalysts.

$$(CH_3)_3NBH_3 + 2H_2O \rightarrow BO_2^- + (CH_3)_3N + 7H^+ + 6e^- \ldots \quad \text{(b)}$$

$$NO_3^- + H_2O + 2e^- \rightarrow NO_2^- + 2OH^- \ldots \quad \text{(c)}$$

$In(OH)_3$ nucleates and grows at high pH according to Eq. (d).

$In(OH)_3$ is thus deposited in Pd-adsorbed regions of patterned SAM (Fig. 13).

$$In^{3+} + 3OH^- \rightarrow In(OH)_3 \ldots \quad \text{(d)}$$

Transparent conducting In_2O_3 thin film is fabricated by annealing at 300 °C in a reduced atmosphere (3% - H_2 / N_2) for 1 h according to Eq. (e).

$$2In(OH)_3 \rightarrow In_2O_3 + 3H_2O \ldots \quad \text{(e)}$$

Liquid phase patterning of $In(OH)_3$ thin films: After having been immersed in the solution containing $In(NO_3)_3$ and TMAB, the patterned SAM having Pd catalytic nanoparticles on amino regions was rinsed with distilled water and dried in air. The thin film was clearly shown by SEM observation to deposit on amino group regions selectively (Fig. 14). Silanol group regions and amino group regions of SAM were shown to be black or white, respectively. Magnified SEM micrograph (b) shows the surface morphology of deposited thin films. The thin films were continuous films without micrometer-scale cracks.

The distribution of elements on the surface of the substrates was evaluated by EDX. Indium was detected from thin films on amino group regions selectively and appeared black in EDX mapping images (Fig. 14). On the other hand, silicon was detected mainly from silanol group regions which were not covered with depositions and exposed bare silicon substrate (Fig. 14). These observations showed the site-selective deposition of thin films containing indium on amino group regions.

Surface morphology was further evaluated by AFM conducted at room temperature under ambient air. Thin films were observed on the amino group regions selectively to form micro-scale patterns. The surface of the thin films showed a uniform morphology and low surface roughness RMS = 13 nm (Fig. 13). The thin films deposited on a glass substrate were found to be transparent, which would be caused by the low surface roughness which reduces diffuse reflection. In-plane particle size was estimated to about 10 - 25 nm in diameter. Film thickness at the edge of the thin film was estimated to be 84 nm.

As-deposited thin film was shown to be crystalline $In(OH)_3$ (JCPDS No. 16-0161) with no additional phase by XRD evaluation. 002 and 004 diffraction peaks of $In(OH)_3$ only were detected. $In(OH)_3$ thin film was shown to have high c-axis orientation. Crystalline size was estimated to be 17.4 nm by using the 002 diffraction peak. This was consistent with the in-plane particle size estimated by AFM observation. Thus, each particle comprising the thin film would be a single crystal.

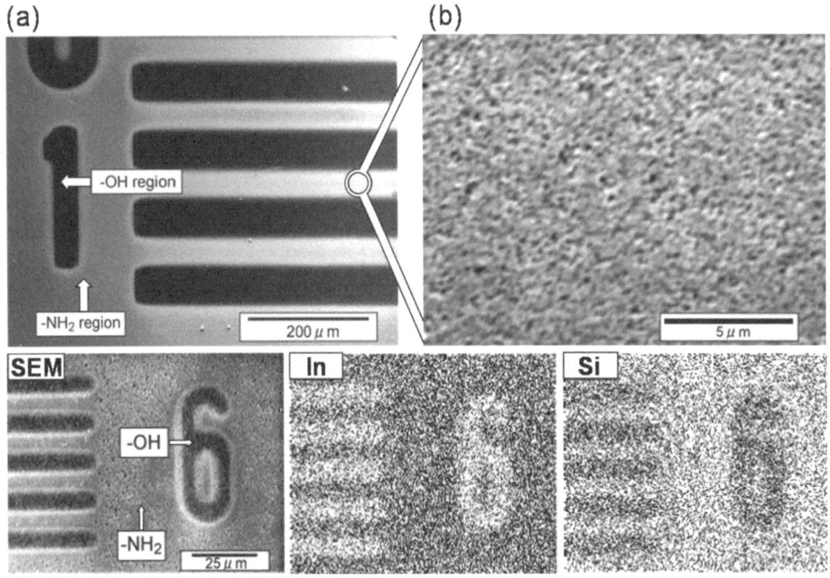

Fig. 14. SEM micrographs of (a) micropattern of indium oxide thin films and (b) magnified area of (a) (upper stand). SEM micrograph and EDX images for In and Si (lower stand). Reprinted with permission from Ref.[75], Masuda, Y.; Kondo, M.; Koumoto, K., 2009, *Cryst. Growth Des.*, 9, 555. Copyright @ American Chemical Society

Micropatterning of In_2O_3 thin films and its optical properties: $In(OH)_3$ thin film was annealed at 200, 250 and 300 °C in air. $In(OH)_3$ thin films transformed into single-phase crystalline In_2O_3 above 250 °C. The film annealed at 300 °C showed 222, 400, 332, 431 and 440 diffraction peaks of In_2O_3 (JCPDS No. 44-1087) with no additional phase. Crystalline size was estimated to be 6.8 nm using the 222 diffraction peak, which was about 0.4 times smaller than that of $In(OH)_3$.

Thin films maintained their uniform surface morphology in the annealing at 250 °C in air and showed a low surface roughness RMS = 11 nm. In-plane particle size was estimated to be about 10 - 25 nm in diameter. Film thickness on the edge of the thin film was estimated to be 45 nm. Shrinkage of film thickness would be caused by volume decrease during crystallization to In_2O_3.

The thin films deposited on a glass substrate showed transparency of 60 - 70% in the visible light region and this would be caused by the low surface roughness which reduced diffuse reflection to 5 - 15%.

The optical band gap energy for direct transition in In_2O_3 thin films was estimated to be 3.7 eV assuming that all of the interband transition was direct transition.
Electrical property of In_2O_3 thin films: $In(OH)_3$ thin film was annealed at 300 °C in a reduced atmosphere (3% - H_2 / N_2) for 1 h instead of atmospheric heating to induce oxygen vacancies to increase the carrier concentration. 222, 400, 332, 431 and 440 diffraction peaks of In_2O_3 (JCPDS No. 44-1087) were observed from the thin film after annealing with no additional phase. Crystalline size was estimated to be 9.8 nm after annealing at 300 °C in a reduced atmosphere using the 222 diffraction peak, which was about 0.6 times smaller than that of $In(OH)_3$ and slightly larger than that annealed at 300 °C in air.
Carrier concentration and Hall moqbility were evaluated to be 2.1×10^{19} cm^{-3} and 5.2 cm^2 V^{-1} s^{-1}, respectively by Hall effect measurement. Specific resistance was evaluated to be 5.8×10^{-2} Ωcm by the Van der Pauw method. These electrical properties are similar to those of In_2O_3 thin films prepared by the sol-gel method[78] (carrier concentration: 1.7×10^{19} cm^{-3}, Hall mobility: 5.9 cm^2 V^{-1} s^{-1}, specific resistance: 6.1×10^{-2} Ωcm). Electrons would be scattered by grain boundaries in thin films and this would decrease Hall mobility and increase specific resistance. Increase of carrier concentration and decrease of grain boundaries by optimization of the reduction conditions and film formation process would allow us to obtain higher Hall mobility and lower specific resistance, thus improving the electrical properties.

XPS analysis:

(a) XPS analysis for patterning of Pd nanoparticles: Pd was detected from amino group regions of a patterned SAM by XPS. Detection of Pd from amino group regions is consistent with AFM observation and the deposit observed in the AFM image was shown to be Pd colloids particles. On the other hand, Pd was not observed in silanol regions, i.e., UV-irradiated regions of APTS-SAM, even with XPS which is a highly surface sensitive analysis method. Pd adsorption on silanol regions would be less than the detection limit of XPS. High site-selectivity of Pd deposition only on amino group regions of a patterned SAM was shown in XPS analysis.
Pd 3d $_{5/2}$ and 3d $_{3/2}$ were observed for APTS-SAM at 337.2 eV, 342.5 eV, respectively. The binding energy of Pd 3d $_{5/2}$ observed is higher than that of Pd metal (334.6 eV[79], 335.1 eV[80-82]). Referring to an earlier report[57], the spectrum of Pd 3d $_{5/2}$ can be deconvoluted into 3 peaks with peak positions corresponding to Pd-N (338.7 eV)[83], Pd-Cl (337.8 eV)[84-86] and Pd-O (336.9 eV)[87] to be Pd-N : Pd-Cl : Pd-O = 0.01 : 0.11 : 0.88[54] (peak ratio). Pd on the APTS-SAM was mainly combined with O as Pd-O (336.9 eV). This result indicates that the surface of the Pd colloid layer on APTS-SAM consists of Pd-O or Pd-OH as well as a small amount of Pd-Cl. Although the Calvert group reported that the hydrolyzed Pd particles form covalent bonds with other SAMs which have amine groups[57], the Pd-N bond was not observed in our XPS experiment because of the relatively low depth analyzed by the method (the escape depth of the photoelectrons at the binding energy corresponding to Pd 3d 5/2 is few nanometers), but does not exclude that Pd-N bonds might have been present also in our case.
N 1s spectra of amino group surfaces were detected before and after the immersion into Pd nanoparticle solution, though the spectrum intensities were very low. N1s binding energy of the amino group surface covered with Pd nanoparticles (400.4 eV) was higher than that before immersion (399.6 eV). The positive shift of N1s was about 0.8 eV and is similar to that observed between Pd nanoparticles and the amino group of 3-aminopropyltriethoxysilane-

SAM (about 0.8 eV)[88]. The shift of N 1s would be caused by the decrease of electron cloud density around nitrogen atoms and suggests the formation of chemical bonds between nitrogen atoms and Pd ions. The amino group is a strong electron donor and can coordinate to transition metal ions due to the lone pair electrons of nitrogen atoms. On the other hand, the outermost electron of soft metal ion Pd (II) is constructed from $4d^8 5s^0 5p^0$ and has an empty lower energy orbit that can accept electrons. Thus, Pd would form strong bonds with nitrogen rather than oxygen and chloride[88].

(b) XPS analysis for patterning of $In(OH)_3$ thin films: XPS spectral peaks corresponding to In $3d_{5/2}$ (445.1 eV), In $3d_{3/2}$ (452.8 eV) and O 1s were observed from $In(OH)_3$ thin films deposited on the amino group regions.

The binding energy of In $3d_{5/2}$ is higher than that of In metal (443.1[89], 443.6[90,91], 443.8[92], 444.3 eV[93]) and In_2O_3 (444.5[94,95], 444.6[96], 444.7[95,97], 444.8[91], 444.9[98] eV), and similar to that of $In(OH)_3$ (445.0[80], 445.2 eV[99]). This suggests that the indium atoms in thin films are positively charged relative to that of indium metal by formation of direct bonds with oxygen. The binding energy of In $3d_{5/2}$, which is similar to that of $In(OH)_3$ rather than In_2O_3, is consistent with XRD evaluation. On the other hand, this spectrum was not observed from silanol group regions, revealing site-selective deposition of $In(OH)_3$ thin film.

(c) XPS analysis for patterning of In_2O_3 thin films: The spectral peak corresponding to In $3d_{5/2}$ was shifted to a lower binding energy, 444.9 eV, by annealing at 250 °C in air. This was within the range of that of In_2O_3 and consistent with crystallization into In_2O_3 revealed by XRD evaluation.

Additionally, $In(OH)_3$ thin film was annealed at 300 °C in a reduced atmosphere (3% - H_2 / N_2) for 1 h instead of atmospheric heating. In $3d_{5/2}$ was shifted to a lower binding energy, 445.0 eV. This was similar to biding energy of $3d_{5/2}$ in In_2O_3 and indicated crystallization of $In(OH)_3$ into In_2O_3.

2.9 Liquid phase patterning of crystalline anatase TiO_2 using a superhydrophilic surface[100]

A glass substrate coated with an F doped SnO_2 transparent conductive film (FTO, SnO_2: F, Asahi Glass Co., Ltd., 9.3-9.7 Ω/ϖ, 26 × 50 × 1.1 mm) showed a water contact angle of 96°. The UV-irradiated surface was, however, wetted completely (contact angle 0–1°). The contact angle decreased with irradiation time (96°, 70°, 54°, 35°, 14°, 5° and 0° for 0 min, 0.5 min, 1 min, 2 min, 3 min, 4 min and 5 min, respectively). This suggests that a small amount of adsorbed molecules on the SnO_2: F substrate was removed completely by UV irradiation. The surface of the SnO_2: F substrate would be covered by hydrophilic OH groups after irradiation. Consequently, the SnO_2: F substrate was modified to have a patterned surface with hydrophobic regions and super-hydrophilic regions.

Aqueous solutions containg ammonium hexafluorotitanate ($[NH_4]_2TiF_6$) and boric acid (H_3BO_3) were kept at 50°C for 25h. The substrates were immersed into the solutions at 50°C for 2 h to form a micropattern of TiO_2.

After having been immersed in the solution, the substrate was rinsed with distilled water and dried in air. The initial FTO surface appeared to be blue-green under white light due to light diffracted from the FTO layer. On the other hand, TiO_2 films deposited on the super-hydrophilic surface appeared to be yellow-green. The color change would be caused by deposition of transparent TiO_2 film which influenced the wavelength of the diffracted light.

The micropattern of TiO_2 was shown by SEM evaluation to be successfully fabricated (Fig. 13). TiO_2 deposited on super-hydrophilic regions showed black contrast, while the initial

FTO regions without deposition showed white contrast. The average line width is 55 μm. Line edge roughness[34], as measured by the standard deviation of the line width, is ~2.8 μm. This represents a ~5% variation (i.e., 2.8/55) in the nominal line width, similar to the usual 5% variation afforded by current electronics design rules. The minimum line width of the pattern depends on the resolution of the photomask and wavelength of irradiated light (184.9 nm). It would be improved to ~ 1 μm by using a high-resolution photomask.

The FTO layer was a particulate film having a rough surface (Fig. 15-b1, b2). Edged particles of 100 - 500 nm in diameter were observed on the surface. The micropattern of TiO_2 thin film was covered by an assembly of nano crystals of 10 - 30 nm in diameter (Fig. 15-a1, a2). The nano crystals would be anatase TiO_2 which grew anisotropically. The TiO_2 film also had large structural relief of 100 - 500 nm in diameter. As the thin TiO_2 film was deposited on the edged particulate surface of the FTO layer, the surface of TiO_2 had large structural relief.

The morphology of the TiO_2 layer and FTO layer was further observed by fracture cross section profiles (Fig. 16). The polycrystalline FTO layer prepared on a flat glass substrate was shown to have a thickness of ~ 900 nm, and a high roughness of 100 - 200 nm on the surface (Fig. 16a). Nano TiO_2 crystals were deposited on the super-hydrophilic FTO surface (Fig. 16a), whereas no deposition was observed on the initial FTO surface. The super-hydrophilic FTO surface was covered with an array of nano TiO_2 crystals (Fig. 16b, c), which had a long shape of ~ 150 nm in length and ~ 20 nm in diameter. These observations were consistent with TEM and XRD evaluations[99]. Nano TiO_2 crystals would grow along the c-axis and thus enhance the 004 X-ray diffraction peak and 004 electron diffraction peak. They formed a long shape having a high aspect ratio of 7.5 (150 nm in length / 20 nm in diameter) as shown in the SEM fracture cross section profile (Fig. 16b, c) and TEM micrograph[99]. The orientation of nano TiO_2 crystals with their long axis perpendicular to the FTO layer (Fig. 16b, c) would also enhance the 004 diffraction peak.

The film deposited on the substrate was evaluated by XRD analysis. Strong X-ray diffractions were observed for films deposited on FTO substrates and assigned to SnO_2 of FTO films. The 004 diffraction peak of anatase TiO_2 was not observed clearly for TiO_2 film on FTO substrates because both of the weak 004 diffraction peak of TiO_2 and the strong diffraction peak of FTO were observed at the same angle. Glass substrates with no FTO coating were immersed in the solution. Weak X-ray diffraction peaks were observed at 2θ = 25.3, 37.7, 48.0, 53.9, 55.1 and 62.7° for the films deposited on glass substrates. They were assigned to 101, 004, 200, 105, 211 and 204 diffraction peaks of anatase TiO_2 (ICSD No. 9852). A broad diffraction peak from the glass substrate was also observed at about 2θ = 25°.

The intensity of the 004 diffraction peak was stronger than that of the 101 diffraction peak for the film obtained by the liquid phase crystal deposition method, though the intensity of 101 was stronger than that of 004 for anatase TiO_2 powders with no orientation (ICSD No. 9852). The integral intensity or peak height of 004 was 2.6 times or 2.2 times that of 101, respectively, suggesting high c-axis orientation of anatase TiO_2 crystals. Crystallite size perpendicular to the 101 or 004 planes was estimated from the full-width half-maximum of the 101 or 004 peak to be 9 nm or 17 nm, respectively. Elongation of crystals in the c-axis direction was also suggested by the difference of crystallite size. These evaluations were consistent with high c-axis orientation observed by TEM and electron diffraction[99]. Crystallite size estimated by XRD was similar to that in TiO_2 under layer rather than that of acicular crystals observed by TEM. TiO_2 thin film prepared on a glass would be constructed of not acicular crystals but polycrystals in under layer.

Fig. 15. SEM micrograph of a micropattern of anatase TiO_2 films on SnO_2: F substrates. (a1) Surface of anatase TiO_2 films deposited on super-hydrophilic region. TiO_2 was formed on super-hydrophilic region which was cleaned by UV irradiation before the immersion. (a2) Magnified area of (a1) showing surface morphology of anatase TiO_2 film. (b1) Surface of SnO_2: F substrate without TiO_2 deposition. TiO_2 was not formed on non-cleaned region. (b2) Magnified area of (b1) showing surface morphology of SnO_2: F substrate.
Reprinted with permission from Ref.[100], Masuda, Y. and Kato, K., 2008, *Chem. Mater.*, 20, 1057. Copyright @ American Chemical Society

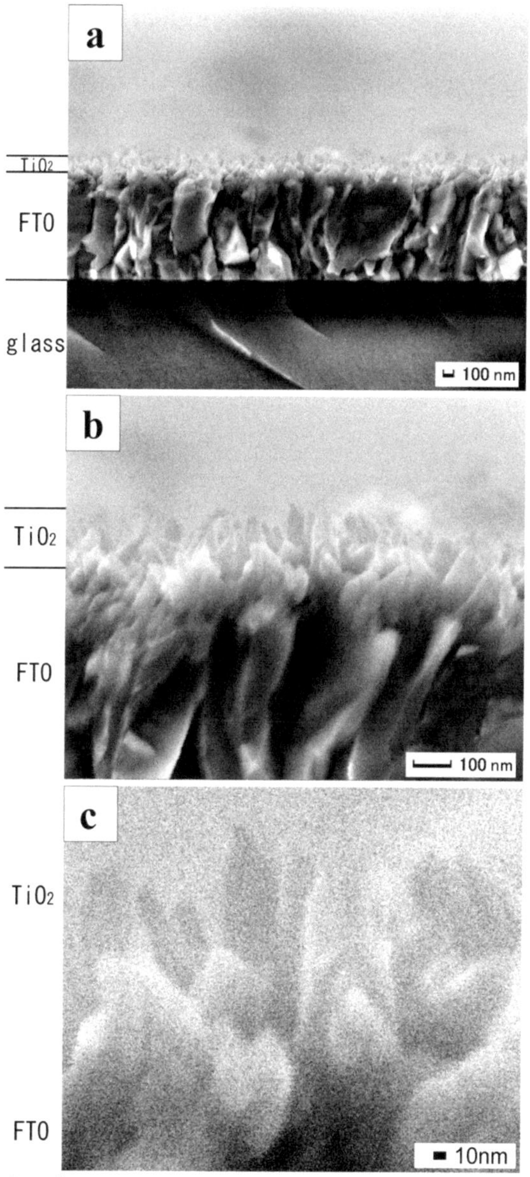

Fig. 16. SEM micrographs of anatase TiO_2 films on SnO_2: F substrates. (a) Fracture cross section of TiO_2 films. (b, c) Magnified area of (a) showing morphology of nano TiO_2 crystals. Reprinted with permission from Ref.[100], Masuda, Y. and Kato, K., 2008, *Chem. Mater.*, 20, 1057. Copyright @ American Chemical Society

3. Summary

A novel concept "Liquid Phase Patterning" was proposed based on scientific knowledge obtained from investigations of interactions and chemical reactions between functional groups of SAMs and ions, clusters and homogeneously nucleated particles in solutions. Nano/micropatterns of metal oxides such as TiO_2, Fe_3O_4, ZnO, etc. were successfully fabricated on a patterned SAM. Mechanisms were further discussed. These studies showed high performance and high potential of solution chemistry for inorganic materials. The novel concepts and technologies in these studies would open new doors for next generation ceramic science.

4. References

[1] http://www.epa.gov/greenchemistry/whats_gc.html.

[2] Anastas, P. T.; Warner, J. C. *Green Chemistry:Theory and Practice*; Oxford University Press: New York, 1998.

[3] http://www.meti.go.jp/policy/chemistry/.

[4] *Green chemistry workshop, Ministry of Economy, Trade and Industry* Japan, 1999.

[5] http://www.rsc.org/is/journals/current/green/greenpub.htm.

[6] http://www.gdch.de/.

[7] Mann, S. *Biomimetic Materials Chemistry*; VCH Publishers: Weinheim, 1996.

[8] Yoshimura, M.; Livage, J. *MRS Bulletin* 2000, *25*, 12-13.

[9] Yoshimura, M.; Suchanek, W. L.; Byrappa, K. *MRS Bull.* 2000, *25*, 17-25.

[10] Yoshimura, M. *J. Mater. Res.* 1998, *13*, 796-802.

[11] Figlarz, M. *Chem. Scr.*, 1988, *28*, 3-7.

[12] Niesen, T. P.; DeGuire, M. R. *J. Electroceram.* 2001, *6*, 169-207.

[13] Pizem, H.; Sukenik, C. N.; Sampathkumaran, U.; McIlwain, A. K.; De Guire, M. R. *Chem. Mater.* 2002, *14*, 2476-2485.

[14] Lev, O.; Wu, Z.; Bharathi, S.; Glezer, V.; Modestov, A.; Gun, J.; Rabinovich, L.; Sampath, S. *Chem. Mater.* 1997, *9*, 2354-2375.

[15] Klein, L. *Sol-Gel Optics: Processing and Applications*; Kluwer Acadmic Publishers: Boston, 1994.

[16] Byrappa, K.; Yoshimura, M. *Handbook of Hydrothermal Technology*; LLC/Noyes Publications: Park Ridge, NJ, 2000.

[17] Yoshimura, M.; Suchanek, W. *Solid State Ionics* 1997, *98*, 197-208.

[18] Yoshimura, M.; Suchanek, W.; Han, K. S. *J. Mater. Chem.* 1999, *9*, 77-82.

[19] Cho, Y. R.; Lee, J. H.; Song, Y. H.; Kang, S. Y.; Hwang, C. S.; Jung, M. Y.; Kim, D. H.; Lee, S. K.; Uhm, H. S.; Cho, K. I. *Mater. Sci. Eng. B* 2001, *79*, 128-132.

[20] Jeon, N. L.; Clem, P. G.; Nuzzo, R. G.; Payne, D. A. *Journal of Materials Research* 1995, *10*, 2996-2999.

[21] Aizenberg, J.; Braun, P. V.; Wiltzius, P. *Phys. Rev. Lett.* 2000, *84*, 2997-3000.

[22] Nashimoto, K.; Haga, K.; Watanabe, M.; Nakamura, S.; Osakabe, E. *Appl. Phys. Lett.* 1999, *75*, 1054-1056.

[23] Mott, M.; Song, J. H.; Evans, J. R. G. *J. Am. Ceram. Soc.* 1999, *82*, 1653-1658.

[24] Stutzmann, N.; Tervoort, T. A.; Bastiaansen, C. W. M.; Feldman, K.; Smith, P. *Adv. Mater.* 2000, *12*, 557-562.

[25] Matsuda, A.; Matsuno, Y.; Tatsumisago, M.; Minami, T. *J. Am. Ceram. Soc.* 1998, *81*, 2849-2852.

[26] Bauer, W.; Ritzhaupt-Kleissl, H. J.; Hausselt, J. *Ceram. Inter.* 1999, *25*, 201-205.

[27] Jacobs, H. O.; Whitesides, G. M. *Science* 2001, *291*, 1763-1766.

[28] Kim, E.; Xia, Y. N.; Whitesides, G. M. *Nature* 1995, *376*, 581-584.

[29] Kim, C.; Burrows, P. E.; Forrest, S. R. *Science* 2000, *288*, 831-833.

[30] Masuda, Y. *http://staff.aist.go.jp/masuda-y/index.html.*

[31] Masuda, Y.; Gao, Y. F.; Zhu, P. X.; Shirahata, N.; Saito, N.; Koumoto, K. *J. Ceram. Soc. Japan* 2004, *112*, 1495-1505.
[32] Masuda, Y. *J. Jpn. Soc. Powder Powder Metallurgy* 2007, *54*, 854-862.
[33] Masuda, Y.; Kinoshita, N.; Sato, F.; Koumoto, K. *Cryst. Growth Des.* 2006, *6*, 75-78.
[34] Masuda, Y.; Sugiyama, T.; Lin, H.; Seo, W. S.; Koumoto, K. *Thin Solid Films* 2001, *382*, 153-157.
[35] Masuda, Y.; Jinbo, Y.; Yonezawa, T.; Koumoto, K. *Chem. Mater.* 2002, *14*, 1236-1241.
[36] Masuda, Y.; Wang, D. J.; Yonezawa, T.; Koumoto, K. *Key Eng. Mater.* 2002, *228-229*, 125-130.
[37] Wang, D. J.; Masuda, Y.; Seo, W. S.; Koumoto, K. *Key Eng. Mater.* 2002, *214*, 163-168.
[38] Lee, L. H. *Journal of Colloid and Interface Science* 1968, *27*, 751-760.
[39] Masuda, Y.; Ieda, S.; Koumoto, K. *Langmuir* 2003, *19*, 4415-4419.
[40] Masuda, Y.; Saito, N.; Hoffmann, R.; De Guire, M. R.; Koumoto, K. *Sci. Tech. Adv. Mater.* 2003, *4*, 461-467.
[41] Masuda, Y.; Sugiyama, T.; Koumoto, K. *J. Mater. Chem.* 2002, *12*, 2643-2647.
[42] Gun, J.; Sagiv, J. *J. Colloid Interface Science* 1986, *112*, 457-472.
[43] McGovern, M. E.; Kallury, K. M. R.; Thompson, M. *Langmuir 1994* 1994, *10*, 3607-3614.
[44] Duchet, J.; Chabert, B.; Chapel, J. P.; Gerard, J. F.; Chovelon, J. M.; Jaffrezic-Renault, N. *Langmuir* 1997, *13*, 2271-2278.
[45] Dressick, W. J.; Chen, M.-S.; Brandow, S. L. *J. Amer. Chem. Soc.* 2000, *122*, 982-983.
[46] Dressick, W. J.; Chen, M.-S.; Brandow, S. L.; Rhee, K. W.; Shirey, L. M.; Perkins, F. K. *Appl. Phys. Lett.* 2001, *78*, 676-678.
[47] Dressick, W. J.; Nealey, P. F.; Brandow, S. L. *Proc. SPIE 2001* 2001, *4343*, 294-305.
[48] Koumoto, K.; Seo, S.; Sugiyama, T.; Seo, W. S.; Dressick, W. J. *Chem. Mater.* 1999, *11*, 2305.
[49] Masuda, Y.; Sugiyama, T.; Seo, W. S.; Koumoto, K. *Chem. Mater.* 2003, *15*, 2469-2476.
[50] Collins, R. J.; Shin, H.; De Guire, M. R.; Heuer, A. H.; Sukenik, C. N. *Appl. Phys. Lett.* 1996, *69*, 860-862.
[51] Huang, D.; Xiao, Z. D.; Gu, J. H.; Huang, N. P.; Tuan, C.-W. *Thin Solid Films* 1997, *305*, 110-115.
[52] Zhang, F.; Mao, Y.; Zheng, Z.; Chen, Y.; Liu, X.; Jin, S. *Thin Solid Films* 1997, *310*, 29-33.
[53] Shin, H.; Collins, R. J.; DeGuire, M. R.; Heuer, A. H.; Sukenik, C. N. *J. Mater. Res.* 1995, *10*, 699-703.
[54] Nakanishi, T.; Masuda, Y.; Koumoto, K. *Chem. Mater.* 2004, *16*, 3484-3488.
[55] Nakanishi, T.; Masuda, Y.; Koumoto, K. *J. Cryst. Growth* 2005, *284*, 176-183.
[56] Brandow, S. L.; Dressick, W. J.; Marrian, C. R. K.; Chow, G. M.; Calvert, J. M. *J. Electrochem. Soc.* 1995, *142*, 2233-2243.
[57] Dressick, W. J.; Dulcey, C. S.; Georger, J. H.; Calabrese, G. S.; Calvert, J. M. *J. Electrochem. Soc.* 1994, *141*, 210-220.
[58] Izaki, M.; Shinoura, O. *Adv. Mater.* 2001, *13*, 142-145.
[59] Call, R. L.; Jaber, N. K.; Seshan, K.; Whyte, J. R. *Solar Energy Mater.* 1980, *2*, 373-380.
[60] Oner, M.; Norwig, J.; Meyer, W. H.; Wegner, G. *Chem. Mater.* 1998, *10*, 460-463.
[61] Wu, X. L.; Siu, G. G.; Fu, C. L.; Ong, H. C. *Appl. Phys. Lett.* 2001, *78*, 2285-2287.
[62] Kang, J. S.; Kang, H. S.; Pang, S. S.; Shim, E. S.; Lee, S. Y. *Thin Solid Films* 2003, *443*, 5-8.
[63] Masuda, Y.; Yamagishi, M.; Koumoto, K. *Chem. Mater.* 2007, *19*, 1002-1008.
[64] Shaw, W. H. R.; Bordeaux, J. J. *J. Am. Chem. Soc.* 1955, *77*, 4729-4733.
[65] Ryabchikov, D. E.; Ryabukhin, V. A. *Analytical Chemistry of Yttrium and the Lanthanide Elements*; Humphrey Science: Ann Arbor, MI, 1970.

[66] Baes, C. F.; Mesmer, R. E. *The Hydrolysis of Captions*; Wiley: New York, 1976.
[67] Aiken, B.; Hsu, W. P.; Matijevic, E. *J. Am. Ceram. Soc.* 1988, *71*, 845-853.
[68] Agarwal, M.; DeGuire, M. R.; Heuer, A. H. *Appl. Phys Lett.* 1997, *71*, 891-893.
[69] Matijevic, E.; Hsu, W. P. *J. Colloid Interface Sci.* 1987, *118*, 506-523.
[70] Sharma, P. K.; Jilavi, M. H.; Nass, R.; Schmidt, H. *J. Lumin.* 1999, *82*, 187-193.
[71] Kwaka, M. G.; Parkb, J. H.; Shon, S. H. *Solid State Commun.* 2004, *130*, 199-201.
[72] Fuggle, J. C.; Martensson, N. *J. Electron Spectrosc. Relat. Phenom.* 1980, *21*, 275.
[73] Wagner, C. D. *Practical Surface Analysis*; 2 ed.; John Wiley, 1990; Vol. 1.
[74] Paton, M. G.; Maslen, E. N. *Acta Crystallographica* 1967, *1*, 1948-23.
[75] Masuda, Y.; Kondo, M.; Koumoto, K. *Crystal Growth & Design* 2009, *9*, 555-561.
[76] Dressick, W. J.; Kondracki, L. M.; Chen, M. S.; Brandow, S. L.; Matijevic, E.; Calvert, J. M. *Colloids and Surfaces A-Physicochemical and Engineering Aspects* 1996, *108*, 101-111.
[77] Masuda, Y.; Koumura, T.; Okawa, T.; Koumoto, K. *J. Colloid Interface Sci.* 2003, 263, 190-195.
[78] Tahar, R. B. H.; Ban, T.; Ohya, Y.; Takahashi, Y. *J. Appl. Phys.* 1997, *82*, 865-870.
[79] Hilaire, L.; Legare, P.; Holl, Y.; Maire, G. *Solid State Commun.* 1979, *32*, 157-160.
[80] Wagner, C. D.; Riggs, W. M.; Davis, L. E.; Moulder, J. F.; Muilenberg, G. E. *Handbook of X-ray Photoelectron Spectroscopy*; Perkin-Elmer Corp., Physical Electronics Div., Eden Prairie: Minnesota, 1979.
[81] Weightman, P.; Andrews, P. T. *J. Phys. C-Solid State Phys.* 1980, *13*, L815-L819.
[82] Weightman, P.; Andrews, P. T. *J. Phys. C-Solid State Phys.* 1980, *13*, L821-L825.
[83] Nefedov, V. I.; Zakharova, I. A.; Moiseev, I. I.; Porai-koshits, M. A.; Vargoftik, M. N.; Belov, A. P. *Zh. Neorg. Khim.* 1973, *18*, 3264-3268.
[84] Nefedov, V. I.; Kokunov, Y. V.; Buslaev, Y. A.; Poraikos.Ma; Gustyako.Mp; Ilin, E. G. *Zhurnal Neorganicheskoi Khimii* 1973, *18*, 931-934.
[85] Choudary, B. M.; Kumar, K. R.; Jamil, Z.; Thyagarajan, G. *Journal of the Chemical Society-Chemical Communications* 1985, 931-932.
[86] Sakurada, O.; Takahashi, H.; Taga, M. *Bunseki Kagaku* 1989, *38*, 407-412.
[87] Datye, A. K.; Bravo, J.; Nelson, T. R.; Atanasova, P.; Lyubovsky, M.; Pfefferle, L. *Applied Catalysis a-General* 2000, *198*, 179-196.
[88] Bazzicalupi, C.; Bencini, A.; Bianchi, A.; Giorgi, C.; Valtancoli, B. *Coordination Chemistry Reviews* 1999, *184*, 243-270.
[89] Ouchene M., S. C., Belin E., Gheorghiu A., Theye M. *J. Non-Cryst. Solids* 1983, *59&60*, 625-628.
[90] Bertrand, P. A. *J. Vac. Sci. Technol.* 1981, *18*, 28-33.
[91] Kazmerski, L. L.; Jamjoum, O.; Ireland, P. J.; Deb, S. K.; Mickelsen, R. A.; Chen, W. *J. Vac. Sci. Technol.* 1981, *19*, 467-471.
[92] Sen, P.; Hegde, M. S.; Rao, C. N. *Appl. Surf. Sci.* 1982, *10*, 63.
[93] Wagner, C. D. *Discuss. Faraday Soc.* 1975, *60*,, 291.
[94] Fan, J. C. C.; Goodenough, J. B. *J. Appl. Phys.* 1977, *48*, 3524-3531.
[95] Lin, A. W. C.; Armstrong, N. R.; Kuwana, T. *Anal. Chem.* 1977, *49*, 1228-1235.
[96] Clark, D. T.; Fok, T.; Roberts, G. G.; Sykes, R. W. *Thin Solid Films* 1980, *70*, 261-283.
[97] Cahen, D.; Ireland, P. J.; Kazmerski, L. L.; Thiel, F. A. *J. Appl. Phys.* 1985, *57*, 4761-4771.
[98] Hewitt, R. W.; Winograd, N. *J. Appl. Phys.* 1980, *51*, 2620-2624.
[99] Faur, M.; Faur, M.; Jayne, D. T.; Goradia, M.; Goradia, C. *Surf. Interface Anal.* 1990, *15*, 641-650.
[100] Masuda, Y.; Kato, K. *Chem. Mater.* 2008, *20*, 1057-1063.

11

Designing Nanocrystal Electrodes by Supercritical Fluid Process and Their Electrochemical Properties

Dinesh Rangappa and Itaru Honma
Institute of Multidisciplinary Research for Advanced Materials, Tohoku University
2-1-1, Katahira, Aoba-ku, Sendai 980-8577
Japan

1. Introduction

Modern society's dependence on the electrical appliances and portable electronics has made rigorours development in battery technology. After three decades of development in battery technology, the Li-ion battery technology has emerged as one of the most popular battery technology [1-3]. They are widely used in varieties of electronic devices due to their good cycle life, high energy density and high capacity over any battery technology. The Li-ion battery technology that now dominates much of the portable battery business has matured enough over the last five years to be considered for the short-term implementation in Hybrid Electric Vehicles (HEV) and Electric Vehicles (EV) applications [1]. However, high cost, safety hazards, and chemical instability of cathode materials have been major concern in the battery industries, which prohibits wider application in the automotive industry [4]. In addition, the volatility of the Li-ion battery technology, has made a massive product recall from the battery industries in recent years. Therefore, increased public awareness about these battery issues solemnly demanding for Li ion batteries with reliability and safety materials.

Among the battery components, the cathode materials are the one which are crucial in determining the high power, safety, longer life and cost of the battery that satisfy the requirement of the larger battery system that can be applicable to electric vehicles, power tools, energy storage equipment and so on [5, 6]. In this context, the olivine-type materials based on lithium transition metal phosphates ($LiMPO_4$ with M=Fe, Mn, Co, Ni) has been emerging as a potential cathode candidate for high power batteries [7]. When compared to the well known layered structure (e.g. $LiCoO_2$) and spinel structure (e.g. $LiMn_2O_4$) based cathode materials, the olivine structured $LiMPO_4$ cathode materials exhibit a flat voltage profile at 3.45-5.1 V vs Li+/Li. Among these, the $LiFePO_4$ and $LiMnPO_4$ cathodes shows the theoretical discharge capacity of about 170 mAhg^{-1}, which make this material a safe cathode material with good cycle life and high power density [5, 6]. However, $LiMPO_4$ materials are basically insulating in nature, hence, has a very poor lithium ionic and electronic conductivity (10^{-9} to 10^{-12} S cm^{-1}). This is because of separation of MO_6 octahedron by the PO_4 tetrahedra, and the one-dimensional chains formed by the edge-sharing LiO_6 octahedra along the b-axis of the orthorhombic structure. Since the first report by Padhi et al., a lot of effort has been devoted to optimize the electronic conductivity of the material and successfully solved the problem by adding different conductive additives during or after the synthesis. One way to overcome this difficulty was to add carbon, either by use of carbon additive to synthesize the material or adding the carbon source to the precursors during the synthesis process and convert it to thin carbon coating on the surface of cathode particles by carbonization [7]. Other efforts to improve the electronic conductivity includes the addition of dispersed metal powders, doping with several elements and coating with conductive polymers [8-11].

In addition to above conductive coating development, in recent years the efforts to overcome the inherent deficiency of $LiMPO_4$ materials have been focused on reducing the particle size to nanometer scale [12-15]. Some reports indicate that Li^+ diffusion capability may become a controlling step, when the electronic conductivity of the olivine phosphate such as $LiFePO_4$ is optimized [4]. The rate capability of the electrode increases with the decrease of particle size due to the short length of the electronic/ionic transportation and the large specific surface available for electrochemical reaction [16-18]. For example, the lithium-ion diffusion coefficient, DLi, in the cathode $LiFePO_4$ was estimated at 10-14 cm^2s^{-1} from electrochemical impedance spectroscopy. Based on this value, a discharge process of one minute requires a lithium-ion diffusion length of below 7.5 nm, i.e., a particle size below 15 nm [19]. Therefore, unlike in bulk $LiMPO_4$ the Li+ diffusion will be very fast and full capacity can be achieved with nano $LiMPO_4$. Tremendous efforts have been made in recent years to reduce the particle size by employing conventional as well as solution-based synthetic method [6-18]. Different process engineering of $LiMPO_4$ electrode materials had been reported in the literature is shown in the Figure 1, with processing conditions such as reaction temperature and time. Increase in the reaction temperature and reaction time makes the process commercial not feasible due to high cost. Most of the high temperature or energy consuming process, including conventional solid state reaction, usually resulted in the large particle size distribution with particle size ranging from 200-1000 nm due to the high temperature and high energy treatments (Figure 1). Among the various synthetic strategies, the solution based route have witnessed great progress in size control and electrochemical performances [20-26]. For example, solvothermal method has been used to synthesize the $LiFePO_4$ nanoparticles and nanoplates by using organic solvents such as tetraethyleneglycol, benzyl alcohol and ethylene glycol [27-28]. This method shows disadvantages in terms of the expensive solvents and high process temperatures. Murugan et al reported microwave-solvothermal synthesis for phospho-olvine nanorods. This is also one of the suitable processes for rapid one-pot synthesis of electrode materials. However, the particle size was in the range of 50-100 nm. Since 2001, the hydrothermal method has been widely used to prepare the size and morphology controlled $LiMPO_4$ materials [20]. This method offered many advantages such as quick, easy to perform, low cost, energy-saving and easily scalable method to prepare the fine $LiMPO_4$ particles. Conversely, this method requires additional heat treatment step in order to improve the cyrstallinity as well as conductive coating.

Recently, we have reported a rapid, one-pot supercritical fluid (SCF) process for the preparation of size, and morphology controlled $LiMPO_4$ nanocrystals [29-30]. The SCFs provides a number of benefits for preparing size and morphology controlled nanocrystals. These days, the SCFs have been widely studied as a new kind of reaction media for nanomaterials synthesis owing to their unique properties such as gas like diffusivity, low

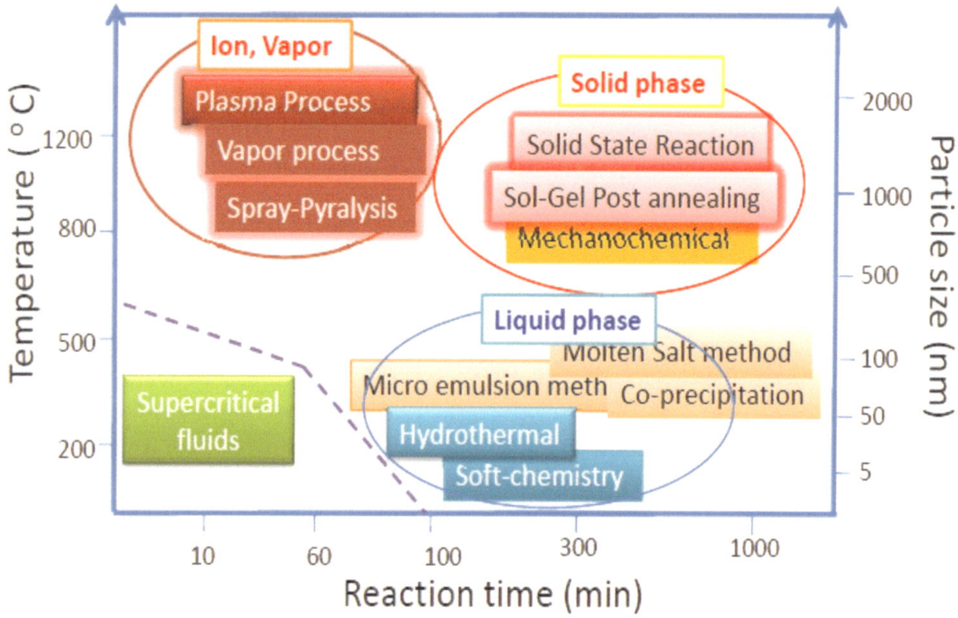

Fig. 1. Process engineering of electrode nanomaterials, the process involving high temperature and high energy species are not feasible for commercialization point of view.

viscosity and the density closer to that of liquid [29-32]. The surface tension completely vanishes above the critical point of the fluid; this is of particular interest in controlling the surface and interface chemistries of the nanostructures. Further, the SCFs forms homogeneous phase with gas or organic compounds. This makes the SCFs a unique reaction medium for the preparation of organic-inorganic hybrid nanostructures.

Progress in the size and morphology control of $LiMPO_4$ nanocrystals have opened up various $LiFePO_4$ nano structures such as irregular nanoparticles, nano rods, nano plates, nanowires and porous nanostructures and so on [33-35]. Organization of such nano scale building blocks into complex hierarchical architectures via self-assembly is of great interest, because of their size and shape dependent physical and chemical properties. However, there are only few reports on the preparation of $LiFePO_4$ hierarchical architectures with well defined size and morphologies.

In this chapter, we present the systematic study of designing $LiFePO_4$ electrode nanocrystals with controlled size, morphology, hierarchical nanostructures and properties by supercritical ethanol (SCE) process. We present the effect of different reaction conditions such as reaction time, temperature, precursor concentration and precursor to the surfactant ratio on the nano structure formation and their surface chemistry. We demonstrate that the supercritical fluid process not only suitable for controlling size, morphology and crystal structures in one pot process, but also facilitates the formation of self assembled hierarchical architectures such as rhombus and dumbbell like structures. We as well demonstrate the decoration of carbon nanotube with $LiFePO_4$ nanocrystals under one-pot supercritical fluid process. The formation mechanism, hierarchical structure, surface chemistry and electrochemical properties of nanostructure electrode will be presented taking $LiFePO_4$ as model nanocrystals electrode material.

2. Experimental section

2.1 Synthesis and self assembly of $LiFePO_4$ nanostructures

Considering the cost-effective process development, we have used common and cheap solvent ethanol as SCF medium. All the starting material and chemicals were purchased from Wako, chemicals. Firstly, precursor solution was prepared by dissolving $FeCl_2 .4H_2O$ (0.4 mmol) in 20 ml oleylamine (20-30 mmol) and 16 ml ethanol at a constant temperature of 60 °C for one hour. Solutions of o-H_3PO_4 (0.4 mmol) and Li acetyl acetonate (0.4 mmol) were prepared by dissolving each compound separately in 2 ml of ethanol. These solutions were slowly added to the above mixture under constant stirring. In a typical synthesis $LiFePO_4$ nanostructures, about 5 ml of precursor solution was charged into 10 cc^3 volume stainless steel reactor and heated up to 300-400 °C temperatures and 38 MPa pressure for 4-20 min. Then, the reaction was terminated by quenching the reactor with cold water bath. The resultant $LiFePO_4$ nanostructures were collected by repeated washing and centrifugation with ethanol, followed by drying in a vacuum dry oven at 120 °C for 12 hours. In order to prepare conductive $LiFePO_4$/C, the required amount of as prepared $LiMPO_4$ nanocrystals were dispersed in ethanol. To this 5 weight percent of acetylene black and 10 weight percent of conductive commercial carbon nanotube called VGCF was added as a carbon source under constant magnetic stirring for 6 hours. Later, solution was dried off by evaporating the ethanol at 60 °C. The dried powder sample was collected and subjected to heat treatment at 600 °C for 4 hours under Ar and H_2 gas atmosphere.

Similarly, we prepared the $LiFePO_4$ decorated multi wall carbon nano tube (MCNT) using above precursor but, in the absence of oleylamine. The starting precursor solution was mixed with suitable MCNT (20 weight % of the $LiFePO_4$) and mixed well under magnetic stirrer for 1 hour. Then, the solution was treated 400 °C under SCE conditions and product was obtained as described in the above paragraph.

2.2 Materials characterization

The crystal structure was examined by X-ray diffraction (XRD) analysis with a Bruker AXS D8 Advance instrument using Cu Kα radiation. The morphology was observed by high-resolution transmission electron microscopy (HRTEM; JEOL JEM-2010F). Infrared (IR) spectra of the as prepared materials that were recorded by an FT/IR-6200 IR spectrophotometer (JASCO Corp., Tokyo, Japan).

2.3 Electrochemical cell preparation and characterization

The electrochemical properties of $LiFePO_4$/C nanostructure were studied by assembling a beaker type three electrode cell. The samples were dried overnight at 100 °C in a vacuum before assembling the cell. The dried $LiFePO_4$/C sample was mixed and ground with acetylene black and Teflon (poly(tetrafluoroethylene)) binder in the weight ratio of 90:5:5. The prepared paste was spread uniformly on a 0.1 cm^2 stainless steel SUS sheet (100 mesh) by manually pressing on to the substrate. The cathode loading was 4-5 mg/cm2. Li metal on stainless steel SUS mesh was used as a counter and reference electrodes. The electrolyte consists of the solution of 1 M $LiClO_4$ in ethylene carbonate (EC)/diethyl carbonate (DEC) (1/1 by volume). The cell assembly was carried out in a glove box filled with high purity argon gas. The charge-discharge tests were performed with a Solartron Instrument Model 1287 controlled by a computer in the potential range of 2.0–4.5V versus Li under different current densities.

3. Results and discussion

3.1 Crystal structural analysis $LiFePO_4$ nanostructures

The powder X-ray diffraction analysis was carried out to confirm the crystal structure and $LiFePO_4$ formation. The crystal structure of the triphylite $LiFePO_4$ has been well studied and documented in the literature [7-15]. The pure $LiFePO_4$ belongs to the olivine of lithium ortho-phosphates with an orthorhombic lattice structure in the space group Pnmb. The structure consists of corner-shared FeO_6 octahedra and edge-shared LiO_6 octahedra running parallel to the b-axis, which are linked together by the PO_4 tetrahedra [36-38]. In our previous study, we have reported the preparation of size, and morphology controlled $LiMPO_4$ nanocrystals by using oleylamine as both reducing as well as capping agent [29]. In this study, we further developed the reaction conditions and obtained $LiFePO_4$ hierarchical structure, by optimizing the oleylamine concentration and reaction time.

Figure 2 shows the XRD patterns of the $LiFePO_4$ nanostructure prepared at 400 °C temperature, for 10 and 20 minute reaction time under supercritical ethanol conditions. All the peaks in the XRD pattern were indexed to an orthorhombic structure with Pnmb space group (JCPDS 40-1499). The peak intensity of the $LiFePO_4$ hierarchical nanostructure formed

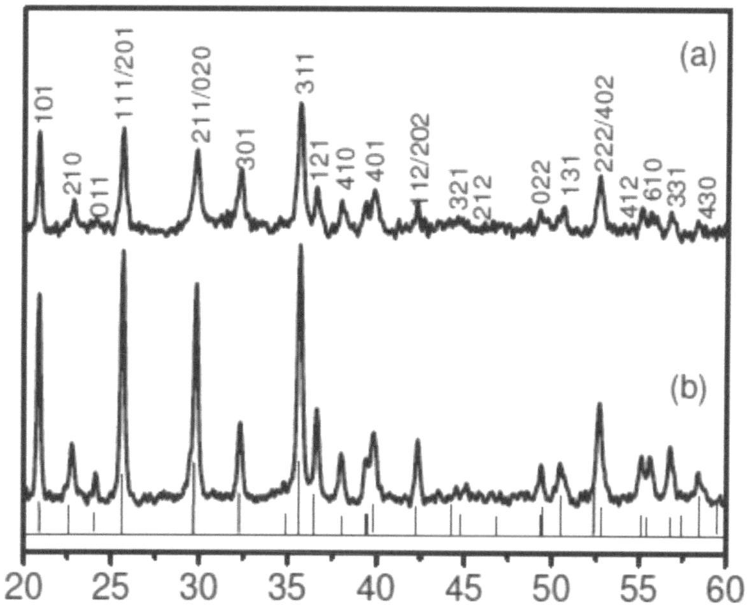

Fig. 2. XRD patterns of the $LiFePO_4$, (a) nano-rods obtained by 8 min reaction; and (b) hierarchical nanostructure obtained by 20 min reaction time at 400 °C.

at 15-20 minutes reaction time is higher than the $LiFePO_4$ nanocrystals formed at 8-minute reaction time as shown in the Figure 2 a and b. This difference in the peak intensity can be attributed to the growth of the $LiFePO_4$ nanstructures with increased reaction time. Effect of reaction time on the hierarchical nanostructural formation is discussed in later section. The XRD patterns clearly show formation of single phase $LiFePO_4$ under supercritical ethanol conditions. The Figure 3 shows the XRD patterns of $LiFePO_4$ nanocrystals prepared at different reaction temperature. The olivine structure $LiFePO_4$ nanocrystals with good crystallinity were obtained at 250 °C temperature. Generally, a high reaction temperature (>

700 °C) or postheat treatment are necessary to prepare the pure and crystalline $LiFePO_4$ particles [13, 20]. It is well known that the $LiFePO_4$ particles prepared by the solution processes without heat treatment possess amorphous $LiFePO_4$ or impurities such as Li_3PO_4 or $Fe_2(PO_4)OH$ or α-Fe_2O_3 along with the $LiFePO_4$ [39]. However, recent progress in solution based techniques demonstrated that using the suitable solution process it is possible to obtain crystalline $LiFePO_4$ nanocrystals around 300-400 °C. By employing SCF process for nanocrystals electrode preparation, the crystalline $LiFePO_4$ nanocrystals were obtained at 250 °C temperature in 10 min reaction time without any post-heat treatment. This is one of the lowest possible temperature reported to obtain the crystalline single phase $LiFePO_4$ nanocrystals. As we can see in the Figure 3, the increased reaction temperature (250-400 °C) had the significant influence on the crystal phase and crystallinity of the prepared samples. The nanocrystals synthesized below 250 °C were not crystalline and included some intermediate phases such as Li_3PO_4. However, as the reaction temperature was increased above 250 °C, the crystallinity was improved and the impurity phases disappeared showing only olivine $LiFePO_4$. Therefore, by taking the advantage of SCE conditions, we could obtain the single phase $LiFePO_4$ nanocrystals at the lowest temperature of 250 °C. The reaction temperature and time played an important role in the formation of well crystalline $LiFePO_4$ nanoparticles under SCE conditions. Because, the temperature above critical point favors the crystallinity of particles, when compared to the subcritical temperature. In addition, the growth of the nanocrystals was controlled by controlling the reaction time between 10-20 minutes. This also had influenced by the rapid heating and homogeneous reaction atmosphere achieved during the SCE process.

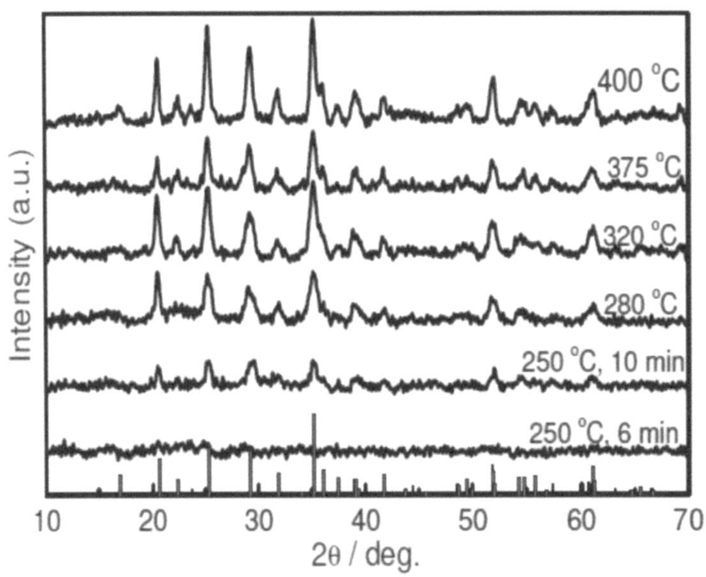

Fig. 3. XRD patterns of the $LiFePO_4$ nanostructure prepared under different reaction temperature and 10 minutes reaction time.

3.2 Surface chemistry of $LiFePO_4$

The surface nature of the $LiFePO_4$ nanostructure has been studied using FTIR spectroscopy and X-ray photoelectron spectroscopy analysis. FTIR spectra provides the information on the organic modification of $LiFePO_4$ nanocrystal surface, which is important in understanding the formation of self assembled hierarchical nanostructure of $LiFePO_4$. The XPS spectra provide information about the valence state of iron in the nanostructure. This is the direct evidence to understand the effect of oleylamine as reducing agent in controlling the oxidation of Fe^{2+} to Fe^{3+}.

3.2.1 FTIR vibrational spectra

The FTIR absorption spectra of $LiFePO_4$ is shown in Figure 4. There are two classes of vibrational modes reported for $LiFePO_4$ [40]. Internal mode, originate in the intra-molecular

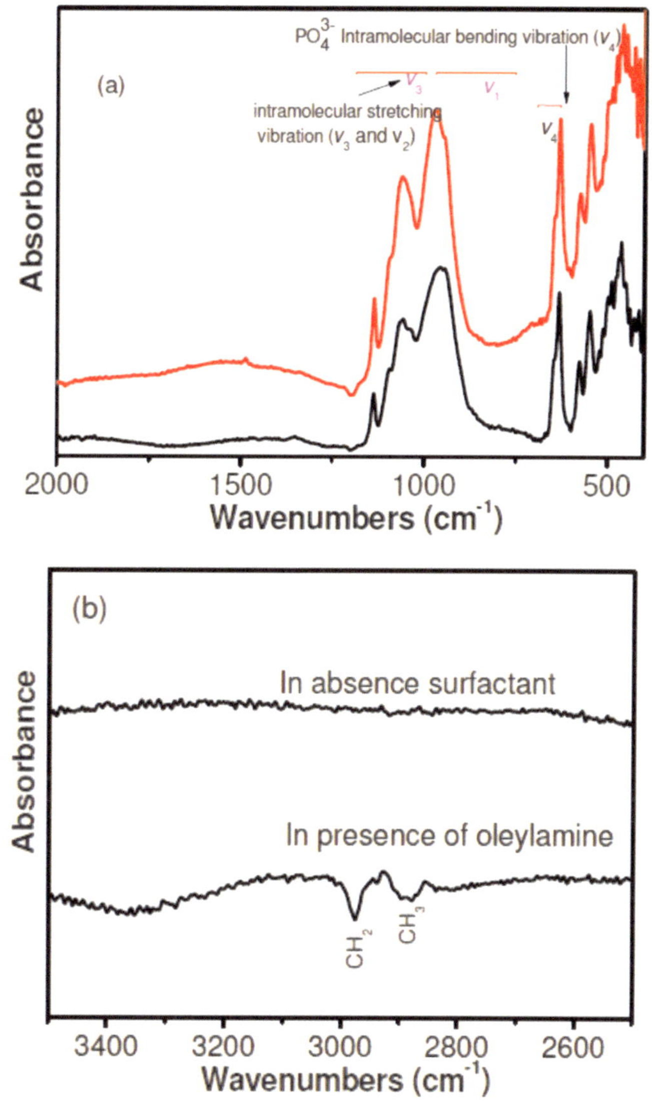

Fig. 4. FTIR spectra of the $LiFePO_4$ nanostructure prepared (a) without oleylamine (top spectrum) and with oleylamine (bottom spectrum) as the surfactant; (b) shows the absence and presence of the C-H stretching mode of methyl and methylene groups, respectively.

vibrations of PO_4^{3-} anion. The vibration of each PO_4^{3-} anion is correlated to those of the other PO_4^{3-} ions in the unit cell, producing a rich vibrational multiplet structure. The bands observed at 943 cm^{-1} and 1042 cm^{-1}, 1068 cm^{-1} and 1140 cm^{-1} are attributed to the intra-molecular stretching (V_1 *and* V_3) motions of the phosphate anions, respectively. The bands in the region of 643-633 cm^{-1} can be assigned to the bending modes (V_4) of phosphate anions. The bands in the region of 500-665 cm^{-1} can be assigned to the lithium ion motion. In addition to the confirmation of phosphate anion, we can also study the surface chemistry of nanocrystals using FTIR spectra, especially, when nanocrystals are prepared by surfactant assited synthesis strategy. The Figure 4b shows evidence of capping the $LiFePO_4$ nanocrystals surface with oleylamine ligand molecules by covalent bond formation. As we can see in the FTIR spectra, the organic modified $LiFePO_4$ shows bands in the region 2800-2960 cm^{-1} which are attributed to the C-H stretching mode of methyl and methylene groups. This confirms that $LiFePO_4$ nanostructure surface was coated with organic ligand, which plays a key role in self assembly and hierarchical nanostructure formation of $LiFePO_4$.

3.2.2 XPS spectra of $LiFePO_4$ nanostructure

XPS spectra of $LiFePO_4$ sample is shown in Figure 5. This data confirms the presence of the Fe as Fe^{2+} in $LiFePO_4$. The components of the Fe 2p doublet (Fe $2p_{3/2}$ and Fe $2p_{1/2}$), because of spin-orbit splitting, are observed at 712.6 and 726.3eV, respectively, with an energy separation (¢*E*Fe) of 13.5 eV (Figure 5a). The data are in excellent agreement with values of Fe^{2+} reported in the literature [41-42]. These results confirm, within the limit of detection, that no significant oxidation of Fe^{2+} to Fe^{3+} on the surface of the nanostructures occurred during the reaction and the subsequent purification. This shows the importance of oleylamine as reducing as well as capping agent in order to keep the divalent Fe^{2+} in the $LiFePO_4$ nanostructure. The Figure 5b indicates the presence of PO^{3-}_4 as the P 2p core level with a binding energy of 133.5 eV. This typical value of binding energy represents P bonded to O [41-44]. By comparing the reported binding energy values of different compounds (132.2-132.9 eV), the P 2p core level at 133.5 eV in this work is attributed to P^{5+} state resulting from the PO^{3-}_4 group. Our XPS data are in consistent with what is expected for $LiFePO_4$ on the basis of the structure and chemistry.

3.4 Size and morphology

The influence of oleylamine concentration on the formation of smaller $LiMPO_4$ nanocrystals has been reported in our previous report [29]. We could control the particle size and morphology by controlling the oleylamine and ethanol ratio in the starting precursor solution. For example, $LiFePO_4$ nanocrystals less than 20 nm and 15 nm were obtained, when the oleylamine to the ethanol volume ratio was 20:20 and 20:10 respectively. In the present study, we further extended this work and optimized the oleylamine ratio and reaction time to obtain the self assembled hierarchical nanostructure of $LiFePO_4$. Figure 6 shows the transmission electron microscopy images of the $LiFePO_4$ nanostructure formed under different reaction time and increased amount of oleylamine. When the concentration of oleylamine was increased above 20 mmol, the nano-rods (10-20 nm width) formation was achieved by adjusting reaction time to 8 min at temperature 400 °C (Figure 6a). These nano-rods started to self assemble by oriented attachment as the reaction time was increased to 10 minutes (Figure 6b). A further increase in the reaction time to 15-20 minutes led to the formation of hierarchical nanostructures via self assembly and shows the different morphology such as dumbbell and rhombus shape, as seen in Figure 8b-f.

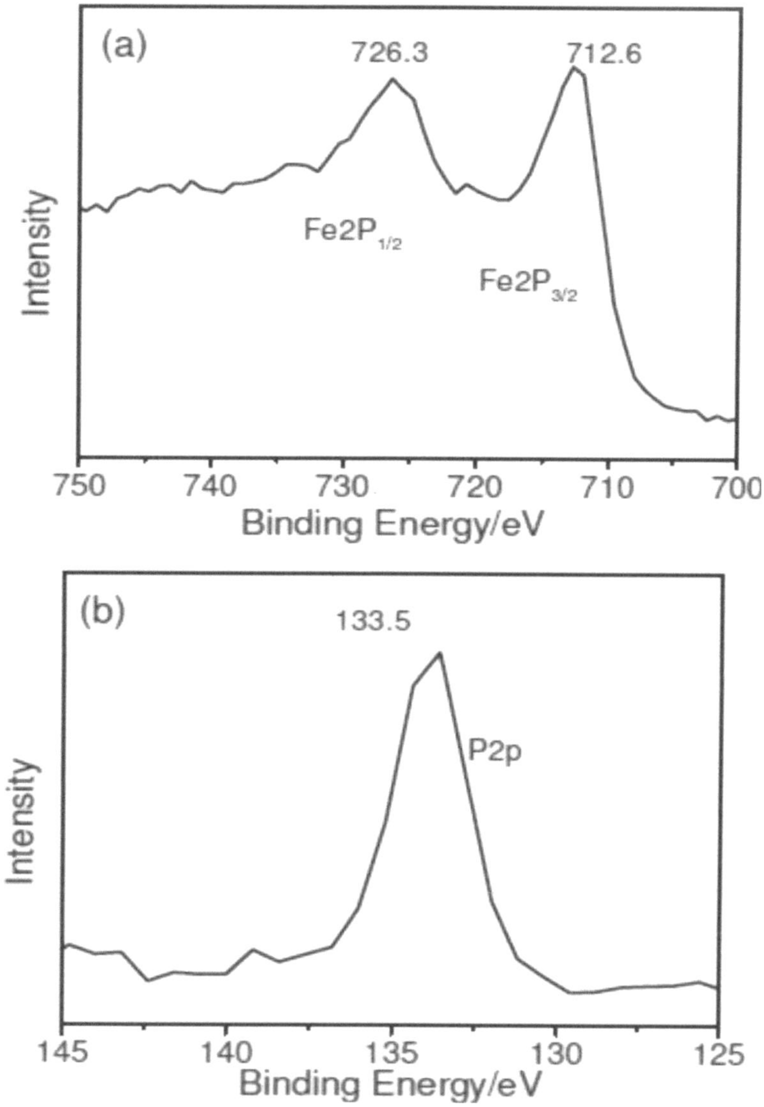

Fig. 5. XPS spectra of the Fe 2p and P 2p core level of the $LiFePO_4$ hierarchical nanostructure synthesized by SCE method.

3.5 Formation of hierarchical nanostructure of $LiFePO_4$

The formation of $LiFePO_4$ hierarchical nanostructure and their formation mechanism was studied in detail by carrying out the time dependent experiments under SCE conditions. The samples were prepared at the different reaction time such as 8 min, 10 min, 15 min and 20 min at 400°C, keeping all the other parameters same. Figure 7 shows the schematic illustrations of formation mechannism and growth of the $LiFePO_4$ nano-rods followed rhombus and dumbbell shape hierarchical nanostructure by the nano rods self assembly. We already had confirmed the formation of the sphere like $LiFePO_4$ nanoparticles with in the 4 min reaction time at 400 °C temperature in our previous study [29]. Now, by increasing

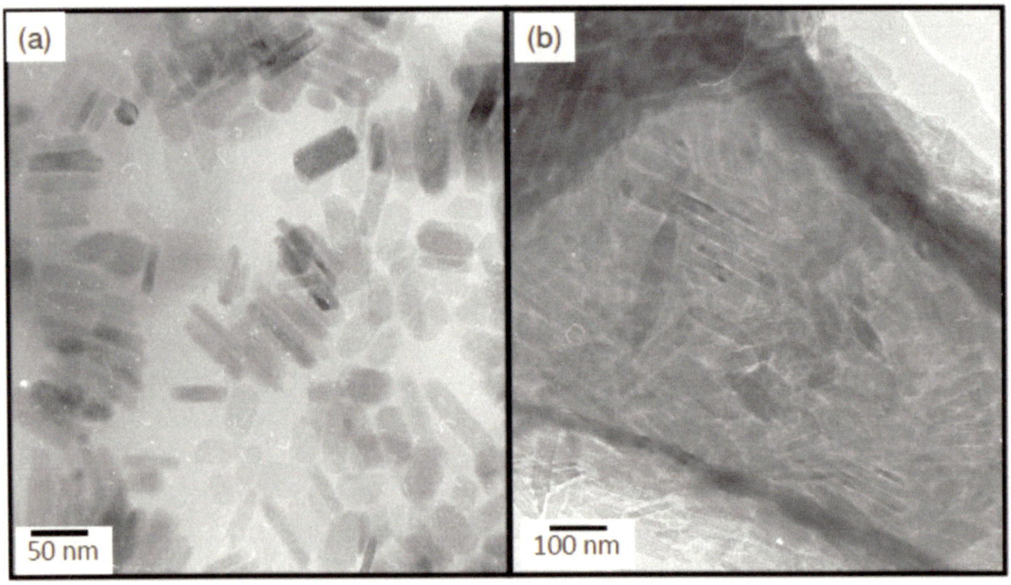

Fig. 6. TEM images of nano rods formed at 8 min reaction time (a), and oriented attachments of nano-rods at 10 minutes reaction time and 400 °C temperature under SCE process.

the reaction time up to 8 min, the nanoparticles continued to grow via elongation into 1D rod like structure as shown in Figure 8a. There are two types rods coexisting at this crystal formation stage, namely, small short nano rods and long rods like structure. As reaction proceeds to 10 minutes, these nanorods aligned by oriented attachement. Figure 8b shows the 4-5 nanorods are aligned by oriented attachment. Finally, further increase with reaction time 15-20 minutes leads to the formation of selfassembled hierarchical nanostructures as shown in Figure 8c-f. We observed the formation of rhombus shape nanostructure at 15 min and with a further increase in the reaction time about 20 minutes led to the dumbbell shape nanostructures formation. The formation of hierarchical nanostructure by self assembly and oriented attachment growth mechanism was also observed by different researchers [45-46]. We have reported formation of $LiFePO_4$ flower like hierarchical microstructure by

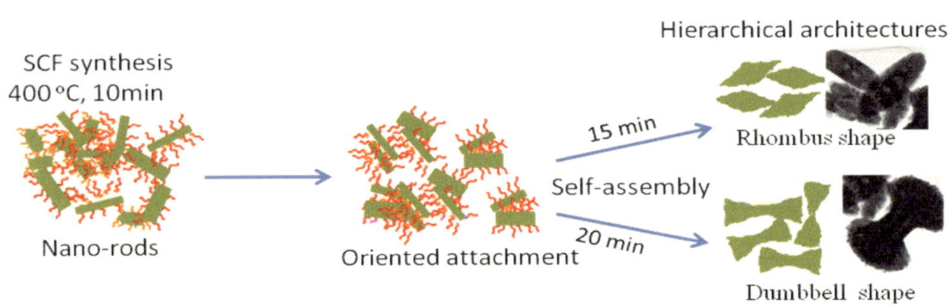

Fig. 7. Schematic illustrations of formation mechannism and growth of the $LiFePO_4$ nanorods and rhombus and dumbbell shape hierarchical nanostructure by nano rods self assembly.

solvothermal reaction using ethylene glycol and hexane as mixed solvent [46]. Zhou et al have successfully prepared highly hierarchical plate like $FeWO_4$ micro-crystals by a simple solvothermal route using ethylene glycol as a capping agent [45]. It was suggested that solvent or surfactant such as ethylene glycol or oleyleamine plays an important role in directing the growth and self-assembly of such unique structures along certain growth direction. It seems that similar reaction mechanism occurs in the present study where oleylamine acts as a soft template in directing the growth of nanoparticles to nano-rods at the early stages through forming hydrogen bonds, and these nano rods further aligned by orientation to form dumbbell and rhombus shape hierarchical structure when the reaction time was prolonged to 15-20 min in the presence of high concentration of oleylamine. As it is well known, oleylamine acts as a surfactant and capping agent, which may greatly affect the size, morphology and microstructure of the products. Thus, the oleylamine molecules may adsorb onto the surface of these nano-rods. It is the interaction of these adsorbed oleylamine molecules that led to the decrease in the rods length and finally formation of hierarchical nanostructure with prolonged reaction time.

3.7 Electrochemical property of $LiFePO_4$

The electrochemical performance of the $LiFePO_4$ hierarchical nanostructures were studied by charge-discharge measurement. Before the electrochemical measurements, the $LiFePO_4$ sample was heated up to 600 °C in Ar and H_2 atmosphere for 4 hours. As we know from FTIR spectra, the surfactant remains more or less on the surface of the final product, even after being washed several times with ethanol. Considering this, the product was annealed to remove the left surfactant from the surface. In addition, annealing can transform the surfactant into carbon to enhance the electronic conductivity. Therefore, oleylamine not only act as a surfactant to control the size and morphology, but also as a carbon source. Figure 9 displays the charge and discharge curves of the sample at the current density of 0.1 C in the potential range from 2-4.5V. The $LiFePO_4$ hierarchical nanostructure showed a flat discharge voltage at approximately 3.4 V, which represents the typical electrochemical action of Li^+ insertion into $FePO_4$. The charge discharge plateaus of $LiFePO_4$ hierarchical nanostructures is shorter than the $LiFePO_4$ bulk spherical particles that is reported in the literature [21]. The $LiFePO_4$ charge discharge plateau represents the two phase region, where the $FePO_4$ and $LiFePO_4$ coexist during the Li insertion. There are different explanations about the slope of charge-discharge capacity curves. One is the formation of solid solution of Li_xFePO_4 below the certain critical size, for example 40 nm [47]. The second explanation is that pseudo-capacitive effect, which means the charge storage of Li ions from faradaic processes occurring at the surface of the materials. The $LiFePO_4$ hierarchical flower like microstructures formed by the self-assembly of 200-400 nm nano-rods under solvothermal reaction showed the pseudo-capacitive behavior [46]. The pseudo-capacitive effect has been observed for other $LiFePO_4$ hierarchical microstructures such as the dumbbell like structure [48]. However, in the present study, the sample shows very small slope region when compared to those results. The capacity of hierarchical nanostructure offer capacity as high as 154 mAh/g at 0.1-0.5 C with good power capability. This material retains about 70% capacity at 2C with a very good cyclic ability and no noticeable fade as seen in Figure 9b. Therefore, the electrochemical performance of the $LiFePO_4$ hierarchical nanostructure shows promising results that were observed for the $LiFePO_4$ hierarchical microstructures prepared by different synthetic routes.

Fig. 8. TEM images of (a) nanorods, (b-c) aligned nanorods, (d, f) rhombus, and (e) dumbbell like hierarchical microstructures of the $LiFePO_4$ obtained at the different reaction time.

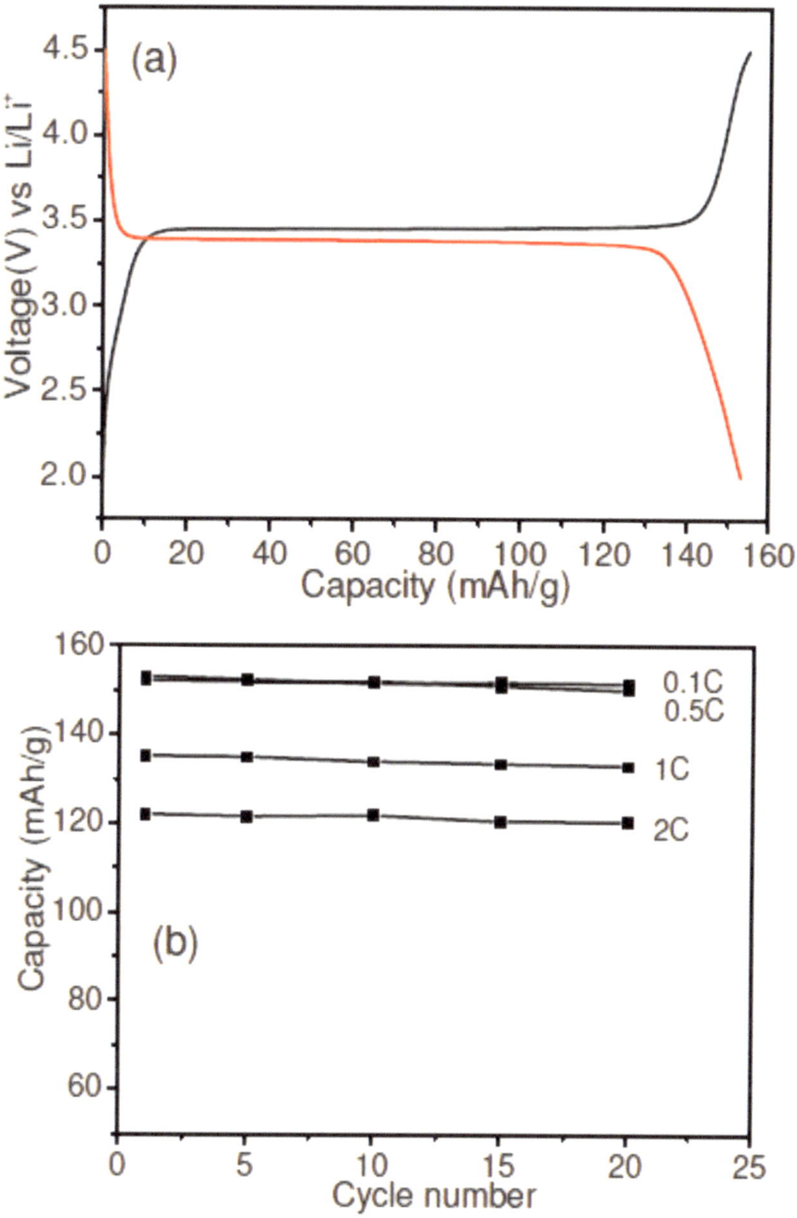

Fig. 9. Charge-discharge profile of the $LiFePO_4$ hierarchical nanostructure measured at 0.1 C (a), along with the rate capability and cyclic performance (b).

3.6 The decoration of carbon nanotube with $LiFePO_4$ nanocrystals

Preparation of carbon nano-tube and electrode materials hybrid nanostructure is very helpful in improving the electro activity of electrode materials such as Si, Sn, TiO_2 and so on, especially in improving the cycle life and discharge/charge rate capability of electrode materials [49-52]. Designing the $LiFePO_4$ electrode nanostructure with conductive carbon shows the improved performance of cathode material. Furthermore, addition of conductive metal oxide to olivine nanostructure cathode material can create the electronic conducting networks. The combination of both the nano size and conducting network is effective as the diffusion length for both electrons and ions is reduced to only several nanometers. Recently, the synthesis of C-$LiFePO_4$:RuO_2 nano-composite was reported and this composite showed a high rate capability when used as cathode materials for lithium ion batteries [53]. However, the RuO_2 is an expensive material; a cost-effective alternate is desired for such nanostructure. Carbon is one of the best choices because of its high electronic conductivity, good lithium permeation, and electrochemical stability. The carbon-coating technique is widely applied for olivine cathode materials. However, the synthesis of such nano composite is complicated and the thickness of carbon shell needs to be controlled to a few nanometers and high temperature in such process is involved.

In recent years, the SCF has been the most popular choice of the reaction media for the preparation of single metal/metal oxide decorated carbon nano-tube composites [50, 54]. The unique features of SCF such as low viscosity, high diffusivity, near zero surface tension and strong solvent power facilitates the formation of hybrid materials with special structures and functions under SCFs conditions. For example, Wai et al. decorated metal nanoparticles (Pd, Ru, Rh) on CNTs and also fabricated Pd, Ni, and Cu nano-wires and nano-rods from the hydrogen reduction of organo-metallic precursors in supercritical CO_2 using CNTs as templates [49-50] for catalytic application. Preparation of Ru metal nanocrystals decorated CNTs under supercritical water condition was reported by Sun et al. [54]. In addition, several other metal oxides decorated CNT were prepared by the decomposition of metal salts under supercritical carbon dioxide and ethanol solution at a low temperatures 100-150°C. However, one-pot preparation of CNT with electrode nanocrystals such as $LiFePO_4$ has not been reported. In this section, we have focused our studies on the decoration of MCNT with the $LiFePO_4$ nanocrystals in one-pot SCE synthesis method.

The $LiFePO_4$ decorated MCNT were synthesized by one-pot reaction of starting precursor solution with commercial MCNT under SCE conditions as shown in the schematic representation in the Figure 10. At first, the starting precursor solution was mixed with suitable MCNT (20 weight % of the $LiFePO_4$) and mixed well under magnetic stirrer for 1 hour. During the mixture, the precursor solution was adsorbed on to the surface of MCNT. The $LiFePO_4$ nanocrystals were deposited on the surface of MCNT when the solution was heated up to 400 °C temperatures and 38 MPa pressure for 10 min. During the SCE reaction, the precursor solution decomposes and the $LiFePO_4$ nanocrystals were deposited on the surface of the MCNT as shown in the Figure 10.

The $LiFePO_4$ decorated MCNT hybrid composite material produced herein have been characterized by different analytical methods such as X-RD and HRTEM to confirm the formation of $LiFePO_4$ decorated MCNT hybrid structure under supercritical ethanol process. The Figure 11, shows the XRD pattern of the $LiFePO_4$ decorated MCNT hybrid composite material. The XRD pattern indicates the formation of the single phase $LiFePO_4$ on the MCNT support. All the peaks observed can be indexed to the olivine structured $LiFePO_4$ with a *Pnmb space group* (JCPDS 40-1499). The peaks for MCNT was not observed as surface of the MCNT is completely covered with nanocrystals. The particle size of the single nanocrystals calculated from the XRD data using Scherrers equation is about 50 nm. This data is in good agreement with the size observed in the TEM images.

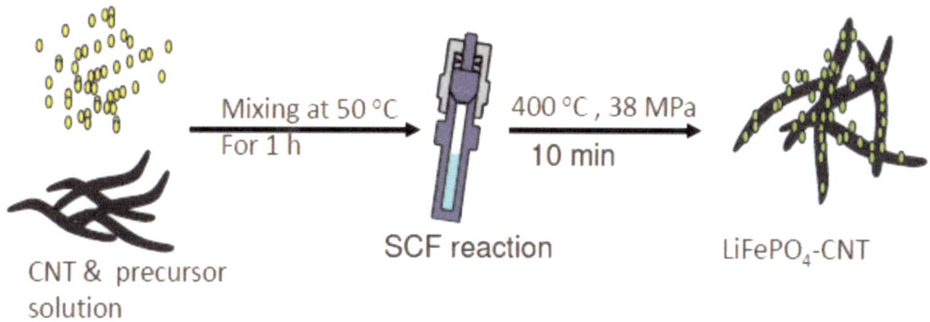

Fig. 10. Schematic illustration of preparation of the $LiFePO_4$ decorated MCNT by using SCE process.

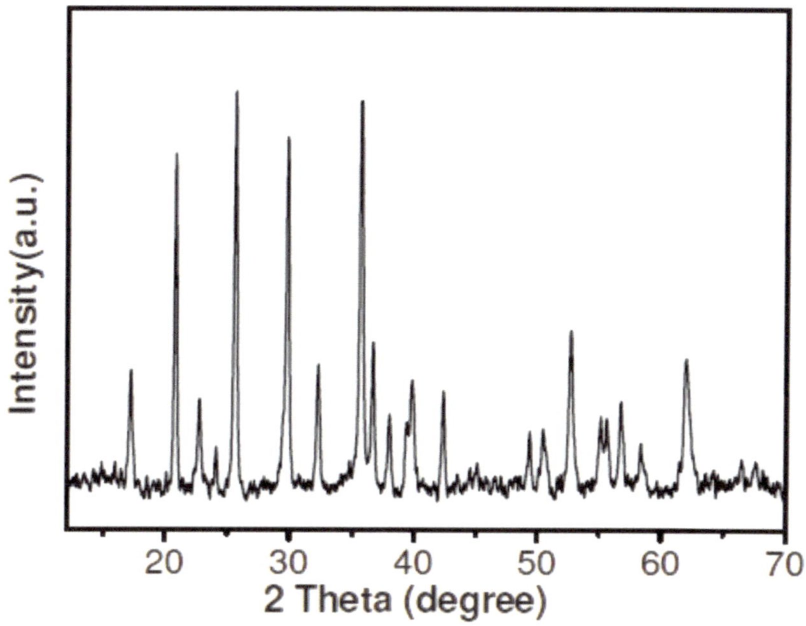

Fig. 11. XRD pattern of the $LiFePO_4$ decorated MCNT nano composite synthesized under SCE conditions at 400 °C for 10 min reaction.

The TEM images of $LiFePO_4$ nanocrystals deposited on the MCNT are shown in Figure 12. The TEM images show that the $LiFePO_4$ nanocrystals are deposited on the surface of MCNT having a plate like morphology with average particle size of 50-100 nm. The size were bigger than the $LiFePO_4$ nanocrystals prepared in the presence of oleylamine without MCNT support as described in the previous section. However, we can further control the particle size by optimizing the reaction conditions and/or using the oleylamine as capping agent and this study is under progress. It was reported that carbon nanotubes surface could be wetted by fluid whose surface tension does not exceed about 200 m N m_1 [50]. It is well known that, the surface tension of the SCF is near zero. Therefore, carbon nanotubes can be wetted during the process to prepare $LiFePO_4$/MCNT composites under SCE conditions. We have confirmed the stability of $LiFePO_4$ nanocrystals coated on the MCNT surface by repeated washing and prolonged ultrasonic treatment. It was found that the nanoparticles could not be separated from MCNTs even after ultra-sonication treatment; this indicates that attachment of the nanocrystals on MCNTs is very strong.

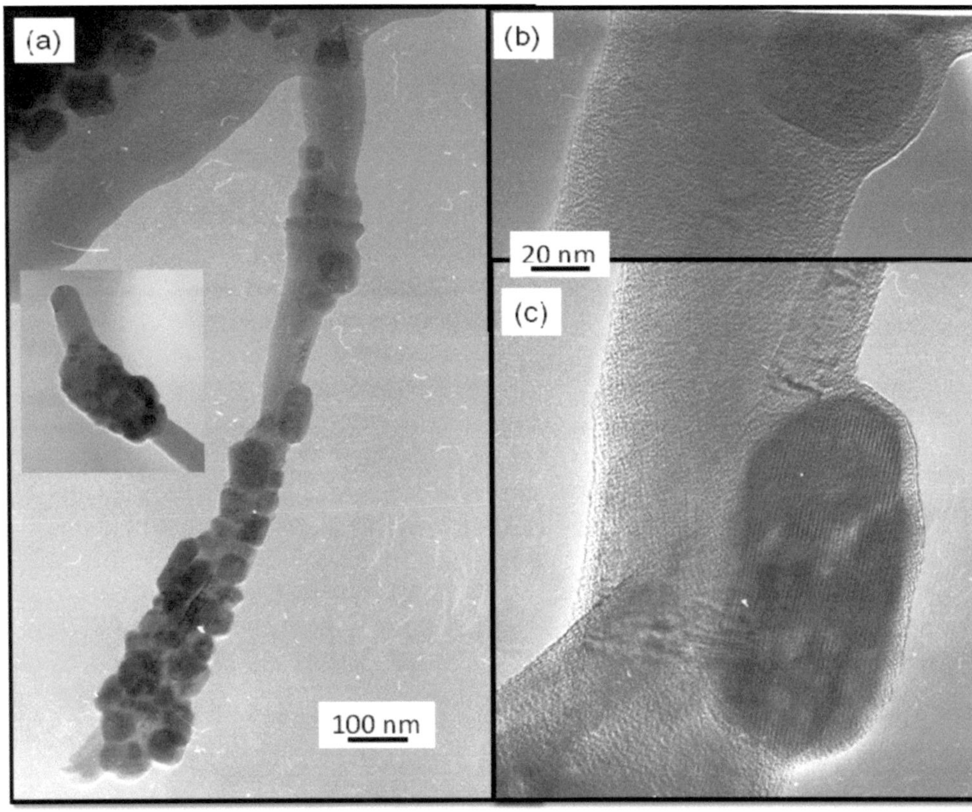

Fig. 12. TEM images of the $LiFePO_4$ nanocrystals deposited on the MCNT (a); and the HRTEM images of the single $LiFePO_4$ nanocrystals deposited on the MCNT (b-c) under SCE conditions at 400 °C for 10 min reaction.

The electrochemical properties of the $LiFePO_4$ nanocrystals deposited MCNT samples were studied by measuring the charge-discharge performance at 0.1 C. The as prepared sample was just mixed with 10 weight % acetylene black and 5 weight % conductive PEDOT polymer before preparing the electrode. Figure 13 shows the charge-discharge profile of the $LiFePO_4$ nanocrystals deposited on MCNT samples. The initial charge-discharge capacity of about 120 mAh/g at 0.1C rate was observed for this material. This result are interesting considering that the nano composite was prepared in one-pot synthesis by the SCE process at 400 °C, without any high temperature heat treatment and carbon coating methods. This capacity is comparable to some of the high temperature processed $LiFePO_4$ results [10,11]. The measured specific capacity was about 70 % of theoretical capacity. This sample also showed good cyclic performance with a very little change in capacity after 30 cycles (data is not shown here). Considering the low processing temperature, the $LiFePO_4$ nanocrystals deposited on the MCNT sample shows a promising discharge capacity. A study to optimize the synthesis conditions and electrochemical performance of this material is under progress and will be reported in near future.

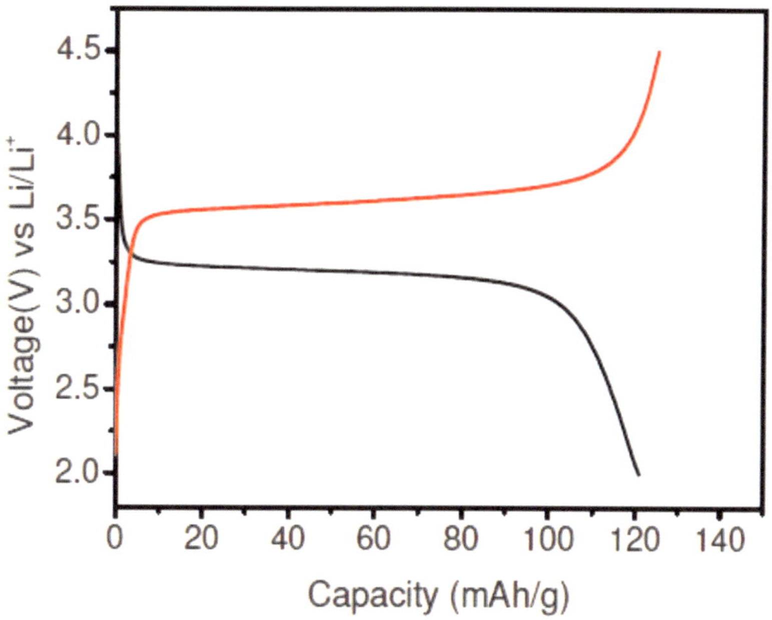

Fig. 13. Charge-discharge profile of the $LiFePO_4$ decorated CNT nano composite measured at 0.1C.

3. Conclusion

In summary, we have successfully demonstrated the designing the $LiFePO_4$ electrode hierarchical nanostructures using SCE process. The self assembly of hierarchical nanostructures such as rhombus and dumbbell like structures were obtained by controlling different reaction conditions such as reaction time and precursor to the surfactant ratio. These parameters had significant effect on the nano structure formation and their surface chemistry. It was observed that the SCE process is not only suitable for the controlling size, morphology and crystal structures in one pot process, but also facilitates the formation of self assembled hierarchical architectures. The formation mechanism studies shows that oriented attachment and growth were responsible for the hierarchical nanostructures formation under supercritical conditions, in the presence of oleylamine as reducing and capping agent. The charge-discharge capacity of hierarchical nanostructure sample was about 154 mAh/g at 0.1-0.5 C, which is about 90 percent of theoretical capacity of the material. This material also showed a good rate capability and retains about 70% capacity at 2C with a very good cyclic stability. The electrochemical performance of the $LiFePO_4$ hierarchical nanostructure shows promising results that were observed for the $LiFePO_4$ hierarchical microstructures prepared by different synthetic routes.

We also demonstrated the decoration of MCNT with $LiFePO_4$ nanocrystals under one-pot SCE process. The formation $LiFePO_4$ decorated MCNT bybrid nanocmposite and their morphology was confirmed by XRD and TEM analysis. About 50-100 nm size nanocrytals were deposited on the surface of the MCNT was observed. The electrochemical properties of this composite show charge-discharge capacity about 120mAh/g at 0.1C. Further studies should be carried out in order to improve the electrochemical performance of this hybrid structure.

4. Acknowledgment

The authors would like to thank Japanese Society of Promotion of Science (JSPS) and NEDO, Minister of Economy, Trade and Industry, Japan for financial support for this study.

5. References

[1] Tarascon, J.-M. Armand, M. *Nature,* (2001) 414, 359–367.
[2] Arico, A.S, Bruce, P. Scrosati, B. Tarascon J.M. Schalkwijk W.V. (2005), 4, 366-377.
[3] Nazar, L. F. & Crosnier, O. In *Lithium Batteries Science and Technology* (eds Nazri, G.-A. & Pistoia, G.) 112–143 (Kluwer Academic/Plenum, Boston, 2004).
[4] Tarascon, J.M. Recham, N. Armand, M. Chotard, J.N. Barpanda, P. Walker, W. Dupont, L. Chem. Mater. (2010) 22, 724-739.
[5] Ellis, B.L. Lee, KZ.T. Nazar, L.F. *Chem. Mater.* (2010) 22, 691-714.
[6] Padhi, A.K. Nanjundaswamy, K.S. Goodenough J.B. *J. Electrochem. Soc.* (1997), 144, 1188.
[7] Ravet, N. Goodenough, J. B. Besner, S, Simoneau, M. Hovington, P. Armand, M. Procceings of the 196th ECS Meeting, Honolulu, october 1999.
[8] Huang, H. Yin, S.-C. Nazar, L. F. *Electrochem. Solid State Lett.* (2001) 4, 10, A170.
[9] Yamada, A. Chung, S. C. Hinokuma K. *J. Electrochm. Soc.* (2001) 148, A224.
[10] Wang, Y. Wang, J. Yang, J. Nuli, Y. *Adv. Funct. Mater.* (2006) 16, 2135.
[11] Franger, S. Cras, F. L. Bourbon, C. Rouault, H. *Electrochem. Solid-State Lett.* (2002) *5*, A231.
[12] Yang, S. Zavalij, P. Y. Whittingham, M. S. *Electrochem. Commun.*, (2001) *3*, 505.
[13] Shiraishi, K. Dokko, K. Kanamura, K. *J. Power Sources*, (2005) *146*, 555.
[14] Wang, G. X. Bewlay, S. Needham, S. A. Liu, H. K. Liu, R. S. Drozd, V. A. Lee, J.-F. Chen, J. M. *J. Electrochem. Soc.* (2006), *153*, A25.
[15] Park, K. S. Kang, K. T. Lee, S. B. Kim, G. Y. Park, Y. J. Kim, H. G. *Mater. Res. Bull.* (2004), *39*, 1803.
[16] *Poizot, P.* Laruelle, S. Grugeon, S. Dupont, L. Tarascon, J-M. *Nature,* (2000), 407, 496-499.
[17] *D. Larcher*, C. Masquelier, *D.* Bonnin, Y. Charbre, V. Masson,. J.B. Leriche *J. Electrochem. Soc. (2003) 150*, A133.
[18] *Armstrong, R.; Armstrong,* G.; Canales, J.; Garcia, *R.*; Bruce, P. *Adv. Mater. (2005), 17, 862.*
[19] Wang, Y. Li, H. He, P. Hosono, E. Zhou, H., Nanoscale, (2010), 2, 1294-1305.
[20] Yang, S. Zavalij, P. Y. Whittingham, M. S. *Electrochem. Commun.*(2001), 3, 505- 508
[21] Xiao, J. Xu, W. Choi, D. Zang, J-G. *J. ElectroChem.Soc.* (2010), 157, A141-A147,
[22] Drezen, T.; Kwon, N.; Bowen, P.; Teerlinck, I.; Isono, M.; Exnar, I. *J. Power Sources* (2007), 174, 949- 953,
[23] Delacourt, C.; Poizot, P.; Morcrette, M.; Tarascon, J.-M.; Masquelier, C. *Chem. Mater.* (2004), 16, 93- 99.

[24] Lim, J.; Kim, D.; Mathew, V.; Kim, J; Phys. Scr. (2010), T139, 014060,
[25] Murugan, A.V. Muraliganath, T. Manthiram, A. *Electrochem. Comm.* (2008), 10, 903
[26] Wang, D. Buqa, H. Crouzet, M. Deghenghi, G. Drezen, T. Exnar, I. Kwon, N-H. Miners, J. H. Poletto, L. Gratzel, M.*J. Power Sources,* (2009), 189, 624.
[27] Kim D. H. and Kim, J. *Electrochem. Solid-State Lett.*, (2006), 9, A439
[28] Murugan, A. V. Muraliganth T. Manthiram, A. *Electrochem. Commun.*, (2008), 10, 903
[29] Rangappa, D.; Sone, K.; Kudo, T.; Honma, I.; *Chem. Commun.* (2010) 46, 7548.
[30] Rangappa, D.; Ichihara, M.; Kudo, T.; Honma, I.; *Journal of Power Sources,* (2009), 194, 1036–1042.
[31] Veriansya, B.; Kim, J-D.; Min, B. K.; Shin, Y. H.; Lee, Y-W.; Kim. J.; *The J. Supercritical Fluids,* (2010), 52, 76-83,
[32] Rangappa, D.; Ohara. S.; Naka, T.; Kondo, A.; Ishii, M.; Adschiri, T. *J. Am.Chem.Soc.* (2007), 129, 36, 11064
[32] Adschiri, T.; Hakuta, Y.; Arai, K.; *Ind. Eng. Chem. Res.* (2000), *39,* 4901.
[33] Ellis, B. Kan, W. H. Makahnouk, W. R. M. Nazar, L. F. *J. Mater. Chem.* (2007), 17, 3248.
[34] Dominko, R. Bele, M. Goupil, J. M. Gaberscek, M. Hanzel, D. Arcon, I. Jamnik, J. *J. Chem. Mater.* (2007), 19, 2960.
[35] Saravanan, K. Reddy, M. V. Balaya, P. Gong, H. Chowdari, B.V. R. Vittal, J. J. *J. Mater. Chem.* (2009), 19, 605.
[36] Franger, S. Cras, F. L. Bourbon, C. Rouault, H. *Electrochem. Solid-State Lett.* (2002), *5,* A231.
[37] Islam, M. S. Driscoll, D. J. Fisher C. A. J. and Slater, P. R. *Chem. Mater.* (2005), 17, 5085.
[38] Zaghib, K. J. B. Goodenough, A. Mauger, Gendron F. Julien, C. M. *Chem. Mater.*, (2007), 19, 3740
[39] Dokko, K. Shiraishi, K. Kanamura, K.*J. Electrochem. Soc.* 152 (2005), 11, A2199.
[40] Christopher, M. B, Roger, F. *Spectrochimica Acta Part A* (2006), *65,* 44.
[41] Herstedt, M.; Stjerndahl, M.; Nyten, A.; Gustafsson, T.; Rensmo, H.; Siegbahn, H.; Ravet, N.; Armand, M.; Thomas, J. O.; Edstrom, K. *Electrochem. Solid-State Lett.* (2003), *6,* A202.
[42] Dedryvere, R. Maccario, M. Croguennec, L. Cras, F.L. Delmas, C. Gonbeau, Chem. Mater. (2008), 20 7, 164-7170.
[43] Briggs, D.; Seah, M. P. *Practical Surface Analysis by Auger and XPS;* Wiley: New York, (1985), p 37.
[44] Morgan, W. E.; Van Wazer, J. R.; Stec, W. J. *J. Am. Chem. Soc.*(1973), *95,* 751.
[45] Zhou, Y.-X. Yao, H.-B. Zhang, Q. Gong, J.-Y. Liu, S.-J. Yu, S.-H. *Inorg. Chem.*, (2009), 48, 1082.
[46] Rangappa, D.; Sone, K.; Kudo, T.; Honma, I.; *Journal of Power Sources,* (2010) 195, 6167–6171.
[47] Gibot, P. Cabanas, M. C. Laffont, L. Levasseur, S. Carlach, P. Hamelet, S. Tarascon, J. M. Masquelier, C. *Nat. Mater.* (2008), 7, 741.
[48] H. Yang, X-L. Wu, M-H. Cao and Y-G. Guo, *J. Phys. Chem. C,* (2009), 113, 8 3345.

[49] Ye XR, Lin YH, Wang CM, Wai CM. Adv Mater (2003) 15, 4, 316

[50] Ye XR, Lin YH, Wang CM, Engelhard MH, Wang Y, Wai CM. J Mater Chem, (2004), 14, 5, 908-13.
[51] Fu, L. Liu, Z. M. Liu, Y. Q. Han, B. X. Wang, J. Q. and Cao, L. C. Adv. Mater. (2004) 16, 350
[52] Sakamoto J.S. Dunn, B. J. Electrochem. Soc. (2002), 149, A26
[53] Hu, Y.-S. Guo, Y.-G. Dominko, R. Gaberscek, M. Jamnik, J. Adv.Mater. (2007), 19, 1963.
[54] Sun ZY, Liu ZM, Han BX, Wang Y, Du JM, Xie ZL, Adv Mater (2005) 17, 7, 928-32

12

Photoelectron and Photosensitization Properties of Silicon Nanocrystal Ensembles

Elizaveta A. Konstantinova[1,2], Vyacheslav A. Demin[2] and Pavel K. Kashkarov[1,2]
[1]*Moscow State University, Moscow*
[2]*Russian Research Center Kurchatov Institute, Moscow*
Russia

1. Introduction

Due to their unique properties, silicon nanocrystals get wide applications in various fields of science and engineering (Bisi et al., 2000). Recently, it has been revealed that photoexcitation of nanocrystals in microporous silicon (micro-PS) layers leads to the generation of singlet oxygen on the surface of the samples (Kovalev et al., 2002). It is known that the ground state of the oxygen molecule is represented by the triplet state ($^3O_2$, where the superscript indicates the spin multiplicity; the total spin of the triplet molecule is S_{TO}= 1). During energy absorption, the oxygen molecule transforms into an excited singlet state ($^1O_2$, S_{SO}= 0) (Halliwell & Gutteridge, 1999). In this state, the oxygen molecule exhibits the highest reactivity and enters into oxidation reactions with many substances. This property of singlet oxygen $^1O_2$ is widely used in biomedicine and, in particular, in photodynamic therapy of cancer (Halliwell & Gutteridge, 1999). It should be noted that direct excitation of molecular oxygen from the triplet state to the singlet one is forbidden by the selection rules for the orbital and spin quantum numbers. In order to transform molecular oxygen into the singlet state, it is common practice to use organic dyes that serve as photosensitizers (Kumar et al., 2009). An important property of micro-PS lies in the fact that upon photoexcitation of silicon nancrystals excitons are generated at rather high concentrations (the quantum yield of exciton photoluminescence reaches a few percent (Bisi et al., 2000)). It has been demonstrated that the energy can be effectively transferred from excitons to $^3O_2$ molecules adsorbed on the surface of silicon nanocrystals with the subsequent transformation of these molecules into an excited state (Gross et al., 2003; Kovalev et al., 2002). Therein lies the essence of the mechanism of photosensitization of molecular oxygen. The energy exchange between excitons and $^3O_2$ molecules occurs through the exchange of electrons (Gross et al., 2003) (the Dexter mechanism (Dexter, 1953)). Compared to organic dyes, the use of micro-PS as a photosensitizer of molecular oxygen offers a number of advantages, for example, the relatively simple and available technique for synthesizing this material and its nontoxicity. Evidently, it is important to determine the concentration of the generated singlet oxygen for the practical use of this effect. In this work, we used electron paramagnetic resonance (EPR) spectroscopy for this purpose. EPR spectroscopy allows us to study the interaction of spin centers on the surface of silicon nanocrystals (silicon dangling bonds) with the paramagnetic molecules of triplet oxygen and thus to measure a decrease of their relative amount. The EPR-diagnostics of the $^1O_2$ oxygen generation, suggested in the present work, is based on the change in the spin–lattice (T_1) and spin–spin (T_2) relaxation times of spin centers, i.e., silicon dangling bonds (the so-called P_b centers (Bisi et al., 2000)). In this respect, in the present study, these times were measured by the pulsed EPR technique. The results obtained were confirmed by directly measuring the change in the amount of $^3O_2$ oxygen upon photoexcitation of micro-PS with the use of EPR technique in the Q band. It should be noted that the photosensitization of O_2 molecules on the micro-PS surface is one of the possible channels for nonradiative recombination of excitons in silicon nanostructures. Therefore, to achieve the maximum efficiency of $^1O_2$ generation in microporous silicon layers, it is important to comprehensively study the transfer and scattering of the excited silicon nanocrystals energy and to determine the characteristic times of these processes. To solve this problem, we analyze the decay kinetics of the photoluminescence (PL) of silicon nanocrystals (nc-Si) in micro-PS layers. The PL intensity of such samples is known to be well described by the so-called stretched exponent (Maly et al., 1996)

$$I_{\rm PL} = I_{\rm PL}^{(0)} \exp\left(-(t/\tau)^{\beta}\right), \qquad (1)$$

where τ is the average PL lifetime and β is the nonexponentiality parameter. The values of τ and β depend on the excitation wavelength λ, temperature (Pavesi & Ceschini, 1993), and intensity and some other parameters (Chen et al., 1992) and are τ = 5–50 μs and β = 0.5–0.7 for PS samples with porosity P=70%. Note that Eq. (1) describes the PL relaxation kinetics in both ensembles of interconnected nc-Si in porous silicon layers (Mihalcescu et al., 1996; Pavesi & Ceschini, 1993) and other solid-state nanostructures, such as closely spaced (1–2 nm) nc-Si in a SiO_2 matrix (nc-Si/SiO_2) (Germanenko et al., 2001; Kanemitsu, 1996), CdSe-ZnSe superlattices (Chen et al., 1992), InGaN layered structures (Pophristic et al., 1998), and nanoporous SiGe solid solutions (Lebib et al., 1999). Numerous attempts were made to explain the causes of the stretched exponent of PL relaxation in the systems given above. For example, the authors of (Chen et al., 1992; Kanemitsu, 1996) relate this character of luminescence drop to the presence of disordered regions in these objects, which can be caused by a shape distribution of nanocrystals or a variation in their spatial arrangement. These factors induce a scatter of the recombination energies and probabilities of electron-hole pairs, which is likely to result in luminescence kinetics of type (1). Other researchers (Germanenko et al., 2001; Linnros et al., 1999) assume that the distortion of a luminescence monoexponential decrease is mainly caused by energy transfer between neighboring nanocrystals. This energy transfer can occur because of exciton migration from smaller to larger nanocrystals owing to a smaller energy gap in the latter nanocrystals. Thus, the exciton lifetime depends on the transfer rate; as a result, the kinetics of PL intensity drop "stretches". This assumption is supported by the study of the PL time characteristics of nc-Si ensembles separated by sufficiently thick (about 5 nm) spatial SiO_2 barriers (Vinciguerra, et al., 2000). In this case, nc-Si crystals are supposed to be well isolated; therefore, no energy exchange occurs between them and the lifetime is constant for nc-Si of a given size. Indeed, the PL kinetics detected in such systems turns out to be monoexponential with a characteristic time of 0.2–0.8 ms (Vinciguerra, et al., 2000). However, recent theoretical investigation (Chen, 2003) predicts a stretched exponent for PL kinetics even for individual (isolated) nanocrystals in the case where the rate of trapping of photoexcited charge carriers is higher than the rate of their recombination. Note that Chen (Chen, 2003) assumed that nonradiative carrier capture centers should occur in every nc-Si. However, this assumption was not supported experimentally for luminescent PS samples, since EPR spectroscopy data demonstrate that the number of defects in 1 cm^3 (about 10^{17}) is much smaller than the number of nc-Si crystals in this volume (about 10^{19}–10^{20} or 10^{21} in the case of a fractal PS structure (Moretti et al., 2007; Nychyporuk et al., 2005)). Thus, most nc-Si do not have trapping sites, which requires an alternative explanation for the nonexponential kinetic curves of PL decay. Therefore the purpose of this work is (i) to experimentally investigate the process of singlet oxygen generation in nc-Si ensembles using electron paramagnetic resonance and photoluminescence spectroscopy and (ii) to theoretically study the PL decrease kinetics and the steady-state PL intensity of ensembles of nc-Si with allowance for exciton migration–assisted energy transfer in them, and (iii) to compare the results obtained in terms of the developed theory with the experimental data on the luminescence and photosensitization properties of nc-Si ensembles.

2. Sample preparation and experimental technique

Silicon nanocrystals in the PS layers were formed by a standard method of the electrochemical etching of silicon single crystals in a solution based on hydrofluoric acid: HF(48%) : C_2H_5OH=1 : 1 (e.g., see the review (Bisi et al., 2000)). The *p*-type crystalline silicon with the (100) surface orientation and the resistivity ρ=10…20 Ω·cm and 10…20 mΩ·cm was used as a substrate for micro-PS and meso-PS. The current density was 50 mA/cm^2 (micro-PS1 and meso-PS) and 70 mA/cm^2 (micro-PS2), the etching time was 60 min. After the completion of the pore formation process, PS layer was lifted from the substrate by a short-time increase in the current density to 700 mA/cm^2. The sample thickness measured with an optical microscope was 50 μm. Porosity of the samples was determined gravimetrically and was approximately 70 % for micro-PS1 and meso-PS samples, and 85 % for micro-PS2 samples. As determined by transmission electron microscopy, the average size of the nanocrystals was about 2-4 nm (micro-PS) and ~10 nm (meso-PS). A part of micro-PS samples were dried in air (PS1 and PS2 samples). Upon drying in air at room temperature, the PS films were failed by the surface tension force of water and transformed into granules with a lateral size of about 0.5 mm (coarsegrained (CG) PS). Fine-grained (FG) PS was prepared by mechanical milling in air of a CG PS powder placed in an ampule with quartz balls on a vibratory mill at a vibration frequency of 17 Hz and an amplitude of 7 mm for 2 min.

The EPR measurements were performed on a Bruker ELEXSYS-580 EPR spectrometer (operating frequency is 9.5 and 35 GHz, X- and Q-band, recpectively). Pulsed EPR measurements were made at the regime of "spin-echo" (time resolution ~5 ns). The admission of oxygen and the subsequent evacuation to the residual pressure $p \sim 10^{-5}$ mbar were performed "in situ" using oil-free vacuum equipment. For Q-band measurements soldered ampoules with PS in oxygen ambient were used because of technical features. The samples were illuminated with light from a halogen lamp at the maximum intensity I_{exc} ~2 W/cm^2 immediately in the resonator of the EPR spectrometer.

The PL of the PS samples were excited by nitrogen laser radiation with a wavelength of 337 nm at pulse duration of 10 ns and pulse energy of 10 μJ. PL spectra were measured on a Solar TII spectrometer equipped with a Hamamatsu MS-101H CCD camera. PL kinetics

were recorded with an MDR-12 monochromator, Hamamatsu R237 photomultiplier tube, and an Agilent 54642A digital oscilloscope.
The experiments were carried out at room temperature.

3. Results and discussion

3.1 Investigation of the EPR spectra of silicon nanocrystalls in the PS1 layers in vacuum, nitrogen and oxygen

Figure 1a shows the EPR spectra of micro-PS samples in oxygen in the dark and under illumination, which were measured at the low power P_{mw} = 0.64 mW of incident microwave radiation. The parameters of the observed EPR signal are characteristic of P_b-centers, which are the dangling bonds of silicon at the Si/SiO_2 interface (Bisi et al., 2000; Cantin et al., 1995).

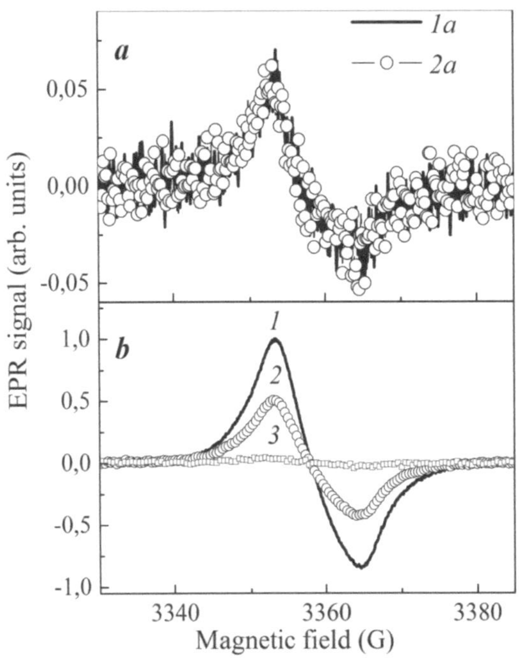

Fig. 1. (a) EPR spectra of PS measured at P_{mw} =0.64 mW in an oxygen atmosphere (*1a*) in the dark and (*2a*) under illumination. (b) EPR spectra of PS in an oxygen atmosphere (*1*) in the dark and (2) under illumination with the intensity I_{exc} = 650 mW/cm^2 or (*3*) in a vacuum

As can be seen in Fig. 1a, a noticeable change in the EPR signal amplitude of micro-PS in oxygen did not occur under illumination (curve *2a*), as compared with that in the dark (curve *1a*). At the same time, a considerable variation in the EPR signal amplitude of micro-PS under analogous conditions was detected at P_{mw} =200 mW (Fig. 1b; curves *1* and *2*). It is well known that, at a sufficiently high intensity of microwave radiation its absorption by P_b centers was saturated; this manifested itself in a decrease in the EPR signal amplitude (Cantin et al., 1995). This effect was observed in our measurements performed with micro-PS in a vacuum (Fig. 1b, curve *3*). Note that the relaxation of P_b-centers from an excited state to the ground state occurred by energy transfer to crystal lattice phonons in the case of micro-PS samples in a vacuum. At the same time for micro-PS in an oxygen atmosphere, the magnetic dipoles of $^3O_2$ molecules adsorbed on the surface of silicon nanocrystals interacted with the magnetic moments of excited P_b-centers to induce their rapid relaxation to the ground state. Consequently, the relaxation time of P_b-centers for micro-PS in an oxygen atmosphere decreased; the saturation effect of microwave absorption weakened; and, as a result, the EPR signal amplitude considerably increased (Fig. 1b, curve *1*). The photosensitization of oxygen molecules occurred under the illumination of micro-PS layers placed in an oxygen atmosphere; consequently, the concentration of $^3O_2$ molecules decreased. In turn, this caused an increase in the relaxation time of P_b-centers; as a result of this, the EPR signal amplitude decreased (Fig. 1b, curve *2*). It should be noted that the change in the EPR signal amplitude upon photoexcitation of PS is almost completely reversible in switching on–switching off cycles if the defect formation under illumination is reduced to a minimum. The samples under investigation satisfy this condition, because they were preliminarily oxidized upon illumination in air for several minutes, which resulted in a considerable decrease in the defect generation rate.

In order to confirm the decisive role of oxygen in the decrease of the EPR signal amplitude upon photoexcitation of nc-Si, we measured the EPR spectra of micro-PS in an atmosphere of nitrogen because of N_2 molecules are diamagnetic (Fig. 2). As can be seen in Fig.2, a noticeable change in the EPR signal amplitude of micro-PS in nitrogen ambient did not occur under illumination (curve *2*), as compared with that in the dark (curve 1). The amplitude of both spectra is small because of saturation effect. Therefore the relaxation of spins takes place only throw spin-lattice relaxation canal, the spin-spin relaxation is suppressed because of diamagnetic nature of N_2 molecules. Therefore namely triplet oxygen molecules are responsible for the effective spin relaxation process and an elimination of saturation effect.

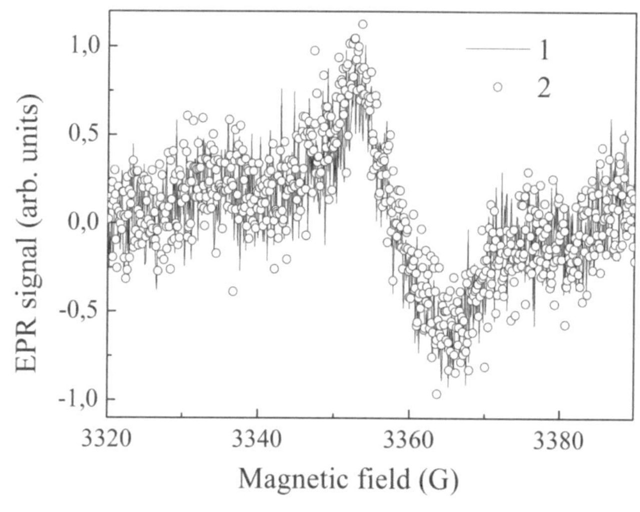

Fig. 2. EPR spectra of micro-PS in an nitrogen atmosphere: (1) in the dark and (2) under illumination. Experimental conditions: P_{mw} =200 mW, I_{exc} =650 mW/cm^2, and p_{O2} =1 bar.

Control experiments for meso-PS show that there is no change in EPR spectra amplitude of meso-PS under illumination of the samples in oxygen ambient (Fig.3). This fact points out an absence of singlet oxygen generation effect and confirms the decisive role of excitons in energy transfer process from photoexcited excitons in nc-Si to oxygen molecules adsorbed on their surface. Indeed, the average size of nanocrystals in meso-PS layers is approximately 10 nm. In such systems the exciton binding energy is small in comparison with thermal energy at room temperature (26 meV).

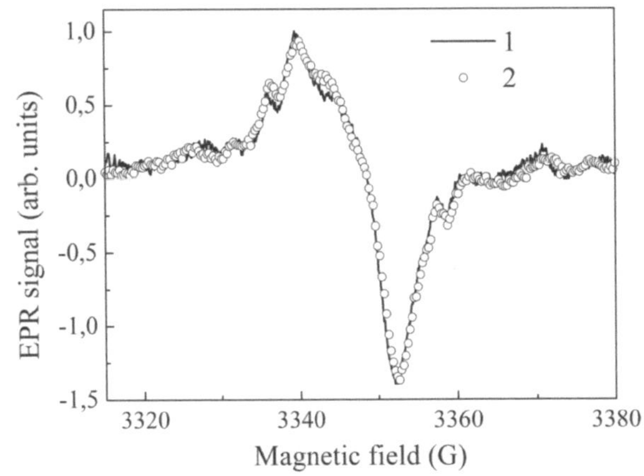

Fig. 3. EPR spectra of meso-PS in oxygen: (1) in the dark and (2) under illumination. Experimental conditions: P_{mw} =200 mW, I_{exc} = 650 mW/cm^2, and p_{O2} = 1 bar.

The above data were obtained at high microwave powers P_{mw} . In order to evaluate the limits of applicability of the EPR technique in the study of the generation of singlet oxygen upon photoexcitation of silicon nanocrystals, let us analyze the influence of the microwave power on the EPR signal amplitude. The dependences of the EPR signal amplitude on the square root of the microwave radiation power $I_{EPR}(\sqrt{P_{mw}})$ are plotted in Fig. 4. It can be seen from this figure that the dependences $I_{EPR}(\sqrt{P_{mw}})$ obtained in vacuum and oxygen under illumination and in the dark coincide at low microwave powers ($P_{mw} \leq 0.5$ mW) and differ substantially at high microwave powers due to the saturation effect. Actually, at low microwave powers P_{mw}, the probability W of induced resonance transitions between the Zeeman energy levels per unit time is so low that, even in the case of the electron–phonon mechanism of relaxation of the P_b center (dominant in vacuum), its characteristic lifetime in the excited state is shorter than the characteristic time 1/W of absorption of a microwave photon. Consequently, at low microwave powers P_{mw}, the saturation effect is absent and, correspondingly, the EPR signal amplitudes are identical for micro-PS in vacuum and the oxygen atmosphere irrespective of the illumination (Fig. 4).

Consequently, at low microwave powers P_{mw}, the saturation effect is absent and, correspondingly, the EPR signal amplitudes are identical for micro-PS in vacuum and the oxygen atmosphere irrespective of the illumination (Fig. 4). The last circumstance along with the reversibility of the amplitude of the EPR spectrum after illumination is switched off is an additional argument indicating that the above decrease in the amplitude at high

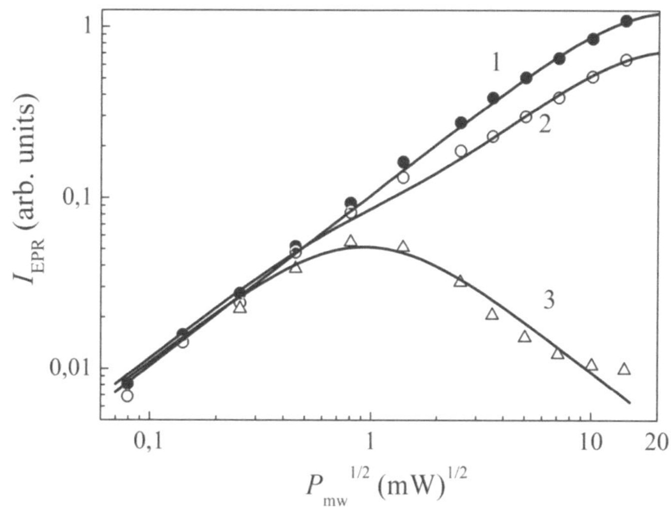

Fig. 4. Saturation curves for micro-PS in oxygen (p_{O2} =1 bar): (1) in the dark, (2) under illumination, and (3) in vacuum (p_{O2} =10^{-6} bar). The approximation dependences were obtained in the framework of the Bloch theory with due regard for the specific features of the detection system. The errors coincide with the sizes of experimental points.

microwave powers P_{mw} (Fig. 1) is not associated with the decrease in the number of spin centers in the sample. However, as was noted above, the decrease in the concentration of triplet oxygen either upon evacuation or in the course of photosensitization of oxygen molecules leads to an increase in the characteristic relaxation times of P_b centers and, as a consequence, to a decrease in the absorption of the microwave power. Indeed, the saturation curve for micro-PS in oxygen under illumination (Fig. 4, curve 2) lies lower than that obtained under dark conditions (Fig. 4, curve 1) (effect of generation of $^1O_2$ molecules), and the saturation curve for the samples in vacuum (Fig. 4, curve 3) is characterized by a lower amplitude as compared to curve 2, passes through a maximum, and falls off with an increase in the microwave power P_{mw} (Fig. 4). In order to explain this behavior of the dependences $I_{EPR}\left(\sqrt{P_{mw}}\right)$, we invoke the Bloch theory, according to which the shape of the absorption line g(ω) in the EPR spectrum is described by the following relationship (Carrington & McLachlan, 1979):

$$g(\omega) = \frac{H_1^2 T_2}{s + T_2^2(\omega - \omega_0)^2}, \qquad (2)$$

where H_1 is the amplitude of the magnetic field strength of the microwave wave, ω_0 is the frequency of the EPR transition, $s = 1 + \gamma^2 H_1^2 T_1 T_2$ is the so-called saturation factor, and γ is the electron gyromagnetic ratio. It should be noted that the dependences of the EPR signal amplitude on the microwave radiation power are shown in Fig. 4 in the form $I_{EPR}\left(\sqrt{P_{mw}}\right)$ universally accepted in EPR spectroscopy (see, for example, (Laiho et al., 1994; Poole & Horacio, 1987)). As can be seen from Fig. 4, the dependences $I_{EPR}\left(\sqrt{P_{mw}}\right)$ exhibit a linear behavior due to the linearity of the detection of the EPR signal with respect to the field in the microwave wave incident on the detector (Schottky diode) over the entire range of the microwave radiation powers used [11]. Then, the dependence of the diode current j on the microwave power that is incident on the diode and equal to the difference between the quantities P_{mw} and ΔP_{mw} (the power absorbed by the sample in the cavity) can be written in the form

$$j \propto \sqrt{P_{\text{mw}} - \Delta P_{\text{mw}}} \approx \sqrt{P_{\text{mw}}} - \frac{\Delta P_{\text{mw}}}{2\sqrt{P_{\text{mw}}}} \tag{3}$$

because $\Delta P_{mw} << P_{mw}$ (Poole & Horacio, 1987). By using expressions (2) and (3), the obvious relationship $\sqrt{P_{\text{mw}}} \propto H_1$, and the fact that $\Delta P_{\text{mw}} \propto g(\omega)$ (Carrington & McLachlan, 1979), we obtain the expression for the recorded EPR signal $I(\omega; \omega_0)$ in the following form:

$$I(\omega;\omega_0) \propto \frac{\Delta P_{\text{mw}}}{\sqrt{P_{\text{mw}}}} \propto \frac{H_1 T_2}{s + T_2^2(\omega - \omega_0)^2} \tag{4}$$

Taking into account that the EPR spectrum, as a rule, is recorded in the form of the first derivative $I'(\omega; \omega_0)$ of expression (4) with respect to ω, we have

$$I'(\omega;\omega_0) \propto -\frac{2H_1T_2^3(\omega - \omega_0)}{\left[s + T_2^2(\omega - \omega_0)^2\right]^2}. \tag{5}$$

The width and the amplitude of the EPR spectrum determined by formula (5) are represented in the form

$$\Delta\omega_{pp} = \frac{2}{T_2}\sqrt{\frac{s}{3}}, \tag{6a}$$

$$I_{EPR} = \frac{9}{4\sqrt{3}} \frac{H_1 T_2^2}{\left(1 + \gamma^2 H_1^2 T_1 T_2\right)^{3/2}} = \frac{a\sqrt{P_{\text{mw}}}}{\left(1 + bP_{\text{mw}}\right)^{3/2}}, \tag{6b}$$

where the parameters a and b determine the position of the maximum in the dependence $I_{EPR}\left(\sqrt{P_{mw}}\right)$; that is,

$$P_{\text{mw}}^{\max} = \frac{1}{2b}, \quad I_{EPR}^{\max} = \frac{2a}{\sqrt{27b}}. \tag{7}$$

Relationships (2) and (4)–(6) are valid in the case of a homogeneous broadening of the EPR line due to a finite lifetime of spin centers in the excited state. However, the EPR line of porous silicon is inhomogeneously broadened and can be described by the Gaussian distribution (Carrington & McLachlan, 1979)

$$N(\Delta\omega) = \frac{T_2^*}{\sqrt{2\pi}} \exp\left[-\left(\frac{\Delta\omega T_2^*}{2}\right)^2\right],$$

where $\frac{1}{T_2^*} = \sqrt{\langle(\Delta\omega)^2\rangle}$ is the root-mean-square deviation of the Larmor precession frequency in the volume of the sample. Then, the experimentally observed EPR spectrum can be written in the following from (Carrington & McLachlan, 1979):

$$I^{real}(\omega;\omega_0) = \int_{-\infty}^{+\infty} I(\omega;\omega_0 + \Delta\omega)N(\Delta\omega)d\Delta\omega, \tag{8}$$

where $I(\omega;\omega_0 + \Delta\omega)$ is a homogeneously broadened spectrum defined by expression (4). An analysis of relationship (8) demonstrates that changes in the characteristics (for example, the amplitude) of the spectrum $I(\omega;\omega_0 + \Delta\omega)$ result in the corresponding changes in the characteristics of the spectrum $I^{real}(\omega;\omega_0)$; i.e., there is a one-to-one correlation between these spectra. Therefore, in order to avoid complications of calculations, the dependences $I_{EPR}\left(\sqrt{P_{mw}}\right)$ for micro-PS in vacuum and oxygen in the dark were approximated by expression (6b) that involves the fitting parameters a and b and holds true for the homogeneously broadened spectrum. In this case, the calculated curves and the experimental data are in satisfactory agreement (Fig. 4). The fitting parameters were determined to be a_v=0.09 relative units and b_v=0.4 mW^{-1} for the sample in vacuum and a_d=0.10 relative units and b_d = 1.1 × 10^{-3} mW^{-1} for the sample in the oxygen atmosphere under the dark conditions. According to formulas (7), the saturation curve $I_{EPR}\left(\sqrt{P_{mw}}\right)$ reaches a maximum at $P_{mw}^{\max}$ = 1.25 mW for micro-PS in vacuum and $P_{mw}^{\max}$ = 450 mW for porous silicon in oxygen. The dependences $I_{EPR}\left(\sqrt{P_{mw}}\right)$ for the samples of PS in oxygen under illumination were approximated by the sum of the saturation curves for PS in the oxygen atmosphere in the dark and in vacuum (Fig. 4): $I_{EPR}^{light} = \alpha \cdot I_{EPR}^{vac} + \beta \cdot I_{EPR}^{dark}$. In this expression, the quantity α determines the fraction of nanocrystals involved in the photosensitization of oxygen and the quantity β determines the fraction of nanocrystals that do not participate in this process ($\alpha + \beta = 1$). Actually, nc-Si with sizes that do not exceed 2–4 nm are electron donors for triplet oxygen molecules, because these nanocrystals contain excitons due to the quantum size effect. Therefore, their surface under illumination is predominantly covered by $^1O_2$ molecules that do not contribute to the paramagnetic relaxation. The other part of nanocrystals (with larger sizes), as in the case of meso-PS (see above), do not make a contribution to the photosensitization of oxygen. This is equivalent to the relaxation of P_b centers in the atmosphere of $^3O_2$ molecules under the dark conditions. It follows from the aforesaid that the quantity α also determines the percentage of oxygen molecules transforming from the triplet state to the singlet state. The best approximation is achieved at α =0.41 and β =0.59. This means that, in our case, approximately 41% of the total number of oxygen molecules transform into the singlet state. The quantity α is conveniently expressed through the experimental data. Indeed, since $\alpha + \beta$ =1, we have

$$I_{EPR}^{light} = \alpha \cdot I_{EPR}^{vac} + (1-\alpha) \cdot I_{EPR}^{dark}$$

and, hence,

$$\alpha = \frac{I_{EPR}^{dark} - I_{EPR}^{light}}{I_{EPR}^{dark} - I_{EPR}^{vac}}. \quad (9)$$

The quantity α calculated from formula (9) at high microwave powers P_{mw} corresponds to the fraction of oxygen molecules that transform into the singlet state upon illumination of micro-PS layers. For example, by using the data presented in Fig. 4, we obtain α =0.42 at P_{mw}^{max} = 200 mW, which almost completely coincides with the value α = 0.41 determined by approximating the experimental data from Fig. 4 with use of the theoretical curves calculated from relationship (6b). Therefore, a considerable fraction (approximately 40%) of the molecules of triplet oxygen can transform into the singlet state upon photoexcitation of nc-Si. Evidently, the efficiency of generation of $^1O_2$ molecules depends on the amount of $^3O_2$ molecules that surround a silicon nanocrystal. Figure 5a shows the dependence of EPR signal amplitudes on the pressure of oxygen (p) for PS layers in the dark and under illumination. The curve for the sample in oxygen under illumination is also described by $I_{EPR}^{light} = \alpha \cdot I_{EPR}^{vac} + \beta \cdot I_{EPR}^{dark}$; however, in this case, the quantities α and β are also functions of p. Indeed, the time of energy transfer from an exciton to the $^3O_2$ molecule shortened as the concentration of oxygen molecules surrounding a nanocrystal was increased (Gross et al., 2003). In this case, the relaxation time of $^1O_2$ remained unchanged (direct singlet–triplet energy exchange between oxygen molecules is spin forbidden). Consequently, the fraction of $^1O_2$ molecules adsorbed on the surface of silicon nanocrystals increased. At the same time, this fraction depends on the quantity α (see above), which is responsible for the dependence of α on p. Using Eq. (9), we can express the value of α in terms of experimental data shown in Fig. 5a. The amplitude of an EPR signal at p = 10^{-5} mbar was chosen as I_{EPR}^{dark}. The dependence of the fraction of photosensitized $^1O_2$ molecules on p thus obtained can be directly converted into the concentration of $^1O_2$ molecules (N_{SO}) taking into account the initial triplet oxygen concentration in silicon pores, which is equal to $2.7 \cdot 10^{19}$ cm^{-3} at p_{O2} =1 bar (the Avogadro number divided by the molar volume). Figure 5b shows this result.

An increase in the intensity of illumination of porous silicon nanocrystals caused an increase in amount of $^1O_2$ molecules photosensitized on the surface of nc-Si (Gross et al., 2003) As a result of this, the EPR signal amplitude decreased (Fig. 6). In this case, a sharp decrease in the value of I_{EPR} was observed up to I_{exc}=600 mW/cm^2; as I_{exc} was further increased, I_{EPR} reached an approximately constant value. The latter was likely due to the fact that, at the specified value, the predominant fraction of oxygen molecules that covered a nanocrystal occurred in a singlet state. Figure 6 also shows the dependence of N_{SO} on I_{exc}, as calculated using Eq. (9) with consideration for the initial concentration of triplet oxygen.

3.2 Investigation of the photosensitization of oxygen molecules by the pulsed EPR technique

The EPR diagnostics of the generation of $^1O_2$ oxygen molecules in ensembles of silicon nanocrystals (considered in the preceding section) is based on a change in the relaxation times of spin centers. In this respect, the relaxation times T_1 and T_2 of spin centers in the samples under investigation were measured by the pulsed EPR technique based on the spin echo phenomenon (Hahn, 1950). It should be noted that, since the relaxation time T_1 characterizes the return of the net magnetization component parallel to the constant

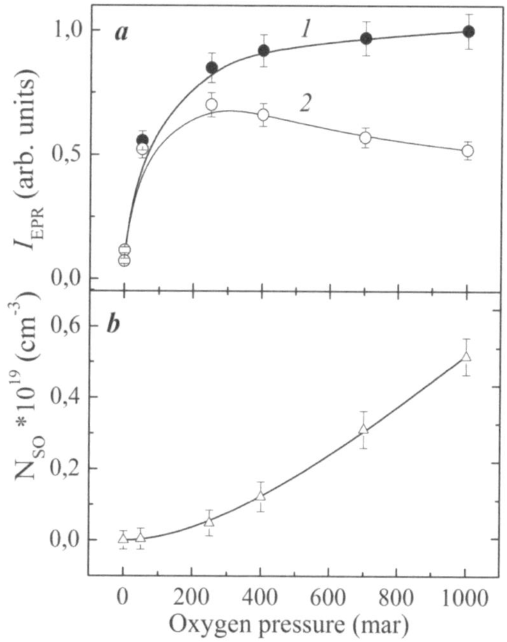

Fig. 5. (a) The amplitude of the EPR spectra of micro-PS samples vs. p_{O2} (*1*) in the dark and (*2*) under illumination. (b) The concentration of $^1O_2$ molecules photosensitized in the PS layers vs. p_{O2} . The parameters are I_{exc}=650 mW/cm^2 and P_{mw}=200 mW.

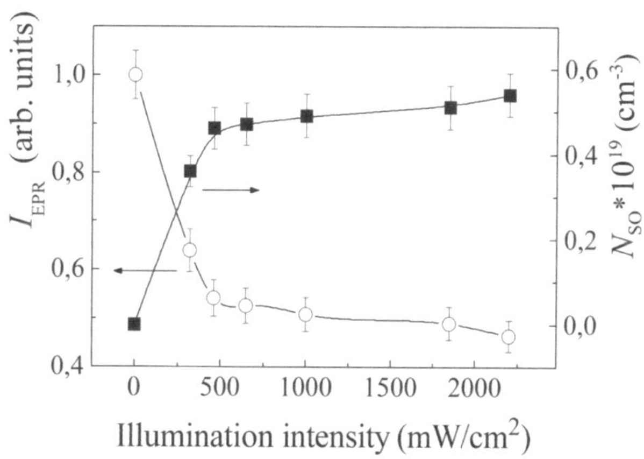

Fig. 6. Amplitude of the EPR spectra of micro-PS and the concentration of photosensitized $^1O_2$ molecules vs. I_{exc} measured at P_{mw} =200 mW and p_{O2} =1 bar.

magnetic field to its equilibrium thermal value M_0, this quantity is also referred to as the longitudinal relaxation time. The relaxation time T_2 characterizes the relaxation of the transverse magnetization component to zero, which does not affect the total Zeeman energy of spins but determines the width of the EPR line. The relaxation times T_1 and T_2 can be represented by the following relationships (Carrington & McLachlan, 1979):

$$\left.\begin{aligned} \frac{1}{T_1} &= \gamma^2\left[\overline{H_x^{*2}} + \overline{H_y^{*2}}\right]\frac{\tau_c}{1+\omega_0^2\tau_c^2}, \\ \frac{1}{T_2} &= \gamma^2\left[\tau_c\overline{H_z^{*2}} + \frac{1}{2}\left(\overline{H_x^{*2}} + \overline{H_y^{*2}}\right)\frac{\tau_c}{1+\omega_0^2\tau_c^2}\right], \end{aligned}\right\} \quad (10)$$

where H_x^*, H_y^*, H_z^* are the components of the fluctuating magnetic field $H^*(t)$. The field $H^*(t)$ disturbs the regular spin precession in the constant field and is responsible for the exponential decay of the components of the spin vector S with the times T_1 and T_2. It is assumed that the field $H^*(t)$ is characterized by zero mean value and fluctuates with the characteristic correlation time τ_c (Carrington & McLachlan, 1979). The analysis of the system of equations (10) demonstrates that the relaxation times are related by the expression

$$\frac{1}{T_2} = \frac{1}{T_2'} + \frac{1}{2T_1}. \quad (11)$$

In expression (11), $\frac{1}{T_2}$ is the EPR line width (in the absence of saturation by microwave radiation), which characterizes the homogeneous broadening of the Zeeman energy levels. This broadening is associated with the two factors: the lifetime T_1 of the spin in the excited state (the contribution $\frac{1}{2T_1}$) and the time characterizing the interactions that broaden the EPR line but do not lead to the spin flip, because they depend only on the fluctuating magnetic field component parallel to the constant magnetic field H_0. In our case, the fluctuating magnetic field $H^*(t)$ is governed by the anisotropy of the g factor of the spin center in PS and fluctuations of the dipole moment of the triplet oxygen molecules physically sorbed on the surface of nc-Si. In the former case, the interaction of the P_b centers with phonons through the spin-orbit interaction can be represented as the interaction of the spin center with a randomly varying magnetic field. This type of relaxation dominates in vacuum. In the presence of triplet oxygen molecules at the nanocrystal surface in the vicinity of the P_b center, the mechanism of magnetic dipole-dipole interaction between the spins of the P_b center and the $^3O_2$ molecule dominates. As a result, the P_b centers appear to be in the magnetic field induced by the oscillating magnetic moment of the physically sorbed $^3O_2$ molecules. It is this type of relaxation that dominates in the oxygen-containing atmosphere and is responsible for the increase in the amplitude of the EPR spectrum in the oxygen atmosphere as compared to vacuum (Fig. 1b). In this case, the lifetime T_1 of the P_b center in the excited state and, hence, the relaxation time T_2 should decrease in the oxygen atmosphere (see expression (11)). Actually, the analysis of the experimental data obtained by the pulsed EPR technique indicates the presence of this tendency. Figure 7 depicts the relaxation curves for the longitudinal (Fig. 7a) and transverse (Fig. 7b) components of the net magnetization of the micro-PS samples with the characteristic time T_1 and T_2, respectively.

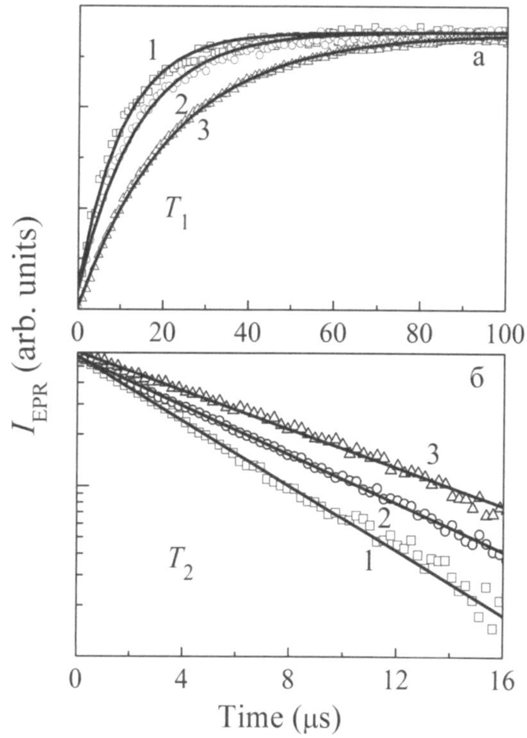

Fig. 7. Kinetics of relaxation of the (a) longitudinal and (b) transverse components of the net magnetization for the micro-PS sample in oxygen (p_{O2} = 1 bar): (1) in the dark, (2) under illumination, and (3) in vacuum (p_{O2} = 10^{-6} mbar). The experimental points are approximated by exponential functions.

It should be noted that, for PS, the net magnetization of the sample is the sum of the contributions from the spin moments of all P_b centers. With the aim of determining the relaxation times, the experimental curves were approximated by the dependences $M_0 \cdot \left(1 - 2\exp(-\frac{t}{T_1})\right)$ and $M_0 \cdot \exp(-\frac{t}{T_2})$, which made it possible to derive the relaxation times T_1 and T_2, respectively. The error in the measurements was equal to 5–7%. The paramagnetic relaxation times thus obtained for micro-PS are listed in Table 1. It can be seen from this table that the illumination of the micro-PS samples in oxygen leads to an increase in the relaxation time T_1 and, correspondingly, in the relaxation time T_2. The observed increase in the relaxation times of spin centers suggests that a number of oxygen molecules transform from the paramagnetic triplet state (participating in the dipole–dipole spin relaxation process) into the diamagnetic singlet state upon photosensitization of oxygen on the surface of nc-Si. Therefore, the pulsed EPR data confirm the generation of $^1O_2$ molecules upon photoexcitation of nanocrystals in micro-PS layers. Within the limits of experimental error, the illumination of the meso-PS samples in oxygen is not accompanied by a variation in the relaxation times of spin centers (Table 1). This indicates the absence of generation of singlet oxygen in meso-PS and agrees with the data obtained by the EPR technique under continuous microwave irradiation (see Section 3.1). Apart from the micro-PS and meso-PS samples, the Si samples were also investigated for comparison by the pulsed EPR technique. The relaxation times T_1 determined for these samples in measurements in vacuum are given in Table 1. It should be noted that the relaxation times T_1 for c-Si are in agreement with those available in the literature (Lepine, 1972). For the nc-Si, the relaxation times T_1 were determined for the first time. The analysis of the relaxation times T_1 for all the samples under investigation permitted us to reveal the following tendency: the spin-lattice relaxation process in micro-PS is retarded as compared to that in meso-PS and c-Si. In turn, the relaxation times T_1 for meso-PS are longer than those for c-Si (Table 1).

		In vacuum	In oxygen in the dark	In oxygen under illumination
Microporous silicon	T_1	22.4 ± 1.6	10.5 ± 0.7	13.8 ± 1.0
	T_2	7.7 ± 0.5	4.6 ± 0.3	6.1 ± 0.4
Mesoporous silicon	T_1	17.2 ± 1.2	11.1 ± 0.8	11.4 ± 0.8
	T_2	4.3 ± 0.3	3.4 ± 0.2	3.4 ± 0.2
c-Si	T_1	4.5 ± 0.3		

Table 1. Relaxation times of spin centers (in microseconds) at the surface of the studied samples in vacuum and in oxygen.

The retardation of the spin–lattice relaxation with a decrease in the size of the structure can be associated with the decrease in the efficiency of the electron–phonon interaction in silicon nanocrystals as compared to the bulk phase of the material under investigation. Indeed, as the size of the object decreases, the phonon spectrum undergoes substantial changes, including a partial degeneracy of phonon modes (Roodenko et al., 2010). In turn, this affects the character of the interaction between electrons and phonons that are confined in nanostructures and, in particular, leads to a decrease in the efficiency of energy transfer from excited spins to the lattice.

3.3 EPR spectroscopy of molecular oxygen

The EPR spectra of $^3O_2$ molecules in pores of micro-PS are shown in Fig. 8. It should be noted that the investigation of triplet oxygen by the EPR technique was performed in the millimeter (Q) band of microwave radiation in view of a large width (approximately 10 kG) of the EPR spectrum of $^3O_2$ molecules (due to the short lifetime in the excited state (Vahtras et al., 2002)). The presence of several lines in the EPR spectrum (Fig. 8) is determined by the strong interaction of the rotational angular momentum K and the spin angular momentum S of the oxygen molecule.

The total angular momentum J takes on values K and K ± 1 (because S = 1). As a rule, the recorded EPR spectrum contains six lines designated as C, E, F, G, K, and J, which have the

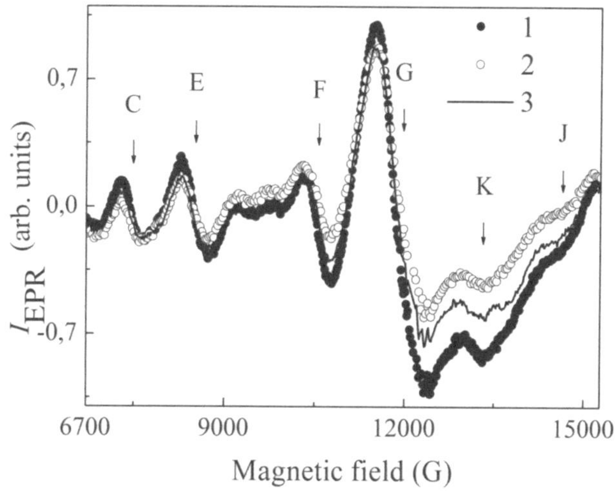

Fig. 8. EPR spectra of molecular oxygen in layers of micro-PS: (1) in the dark, (2) under illumination, and (3) within 5 min after the cessation of illumination. Experimental conditions: P_{mw} =200 mW, I_{exc}=650 mW/cm^2, and p_{O2} =0.5 bar.

highest intensities (Vahtras et al., 2002). They correspond to the following magnetic dipole transitions J, $M_J \rightarrow J'$, M_J': {1, -1 → 1, 0} (C); {2, 1 → 2, 2} (E); {2, 0 → 2, 1} (F); {6, -2 → 4, -1} (G); {2, -1 → 2, 0} (K) and {2, 1 → 2, 2} (J). It can be seen from Fig. 8 that the illumination of the samples results in a decrease in the amplitude of the EPR spectrum. This suggests a decrease in the concentration of triplet oxygen molecules. The data obtained can be explained by the transition of oxygen molecules to the singlet state and can be considered a direct proof of the generation of $^1O_2$ molecules in micro-PS layers. It should be noted that the EPR signal amplitude after the illumination is switched off only partially regains its initial value (before illumination) (Fig. 8). This can be explained by the fact that a considerable number of $^1O_2$ molecules pass from pores of the sample into the closed volume of the measuring cell, i.e., into the gaseous medium, in which the lifetime of the singlet state increases to approximately 50 min as compared to approximately 500 μs in pores of silicon (Gross et al., 2003). Since the area under the EPR line is proportional to the number of spin centers, it is easy to estimate the fraction of the newly formed singlet oxygen molecules from the relationship $\alpha = 1 - \frac{S_{light}}{S_{dark}}$, where S_{light} and S_{dark} are the areas under the EPR lines of $^3O_2$ molecules under illumination and in the dark, respectively. Therefore, approximately 30% of oxygen molecules upon photoexcitation of nanocrystals in micro-PS layers transform into the singlet state. It should be noted that similar experiments with meso-PS revealed no changes in the amplitude of the EPR spectrum of triplet oxygen molecules in pores of mesoPS. This indicates that $^1O_2$ molecules are not formed in this material (see Sections 3.1, 3.2).

3.4 Kinetic equations for the relaxation processes in nc-Si ensembles

We now develop kinetic equations for an ensemble of excited nanocrystals in PS samples in vacuum. To his end, we consider a model fragment of the PS microstructure, namely, a chain of spherical nc-Si crystals having a diameter changing in small limits and forming a quantum filament of a variable cross section (Fig. 9a). This model representation agrees with the real structure of micro-PS, whose layers consist of nc-Si crystals no more 5 nm in size. Figure 9 shows micrographs of micro-PS films produced by the anodic oxidation of p-type silicon wafers. They clearly exhibit nc-Si chains forming silicon filaments of a variable cross section (Cullis & Canham, 1991). We number neighboring nanocrystals in a chain beginning from a zeroth crystal with the maximum size as 0, …, m – 1, m, m + 1, …, M, where m = 0, 1, 2, …, M. In the case of a fractal PS model (Moretti et al., 2007; Nychyporuk et al., 2005), m has the meaning of an ordinal number that is repeated when the nanofragment (nanocrystal) scale decreases, $d_m = d_0k^{-m}$, where k > 1 is the linear scale reduction coefficient and d_0 is he diameter of the largest crystal. Let N_m be the number of nc-Si crystals with ordinal number m; then, N_m can be written as $N_m = N_{0m} + N_{1m} + N_{2m}$, where N_{0m}, N_{1m}, and N_{2m} are the numbers of unexcited nc-Si crystals and nc-Si crystals containing one and two excitons, respectively. The number of nc-Si crystals with more than two excitons is negligibly small as compared to N_{2m} because of the rapid Auger recombination processes that occur when the number of excitons in a nanocrystal is larger than one.

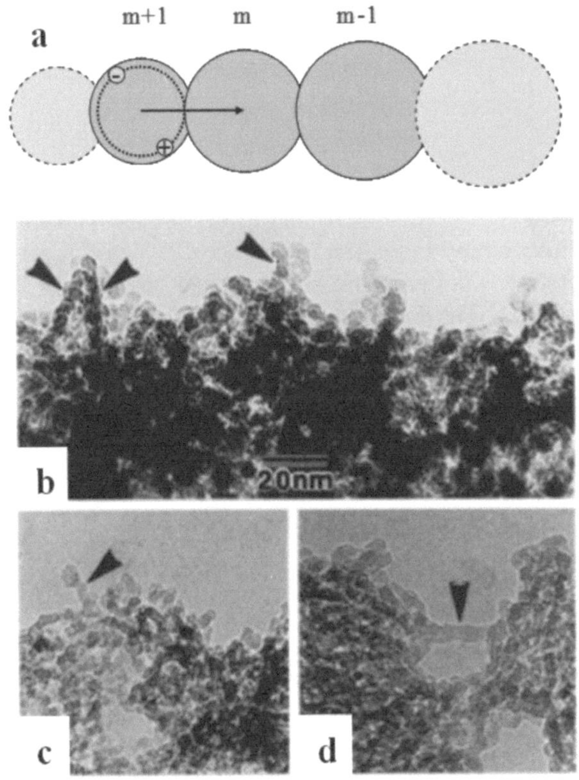

Fig. 9. (a) Model for a fragment of one of the quantum filaments of a variable cross section that form the PS microstructure and (b–d) transmission electron microscopy micrographs of PS layers (Cullis & Canham, 1991). The arrows indicate columnar structures.

Moreover, N'_m is the number of nc-Si crystals with point defects (no less than one per crystal; N'_m is part of the total number of nc-Si crystals N_m). In general, all primed quantities relate to the processes that take place in nc-Si crystals with defects. Then, a set of kinetic equations describing the energy relaxation for N_{0m}, N_{1m}, and N_{2m} after the end of optical excitation is (point above a quantity means differentiation with respect to time t)

$$\left.\begin{aligned} \dot{N}_{1m} &= -\alpha_{rm}N_{1m} - \alpha'_m N'_{1m} - D_m N_{1m} N_{0m-1} - D^*_m N_{1m} N_{1m-1} + D_{m+1} N_{1m+1} N_{0m} - D^*_{m+1} N_{1m+1} N_{1m}, \\ \dot{N}_{0m} &= \alpha_{rm} N_{1m} + \alpha'_m N'_{1m} + D_m N_{1m} N_{0m-1} + D^*_m N_{1m} N_{1m-1} - D_{m+1} N_{1m+1} N_{0m} + \alpha_A N_{2m}, \\ \dot{N}_{2m} &= -\alpha_{Am} N_{2m} + D^*_{m+1} N_{1m+1} N_{1m}. \end{aligned}\right\} \quad (12)$$

The meaning of the parameters entering into the set of nonlinear differential equations (12) is as follows: α_{rm} and α'_m are the rates of radiative and nonradiative recombination, respectively. Obviously, the latter takes place only in nc-Si with defects. The dimension of the α coefficients is μs^{-1}. D_m and D^*_m are the coefficients of exciton migration from nc-Si with ordinal number m to nc-Si with ordinal number m–1, i.e., toward larger nanocrystals (toward a decrease in the energy gap due to the quantum size effect; Fig. 9a). D^*_m describes the efficiency of exciton migration into a nanocrystal having an exciton before the transfer event. Since the dimension of coefficient D is cm^3 μs^{-1}, it can be formally interpreted as the product of the translational thermal exciton velocity into the transfer cross section. Finally, α_{Am} is the Auger recombination rate (μs^{-1}). In total, all three equations of set (12) yield zero, since

$$\frac{dN_{0m}}{dt} + \frac{dN_{1m}}{dt} + \frac{dN_{2m}}{dt} = \frac{dN_m}{dt} = 0$$

Let us explain the physical meaning of the terms of type $D_m N_{1m} N_{0m-1}$. This term describes the exciton migration from nc-Si with concentration N_{1m} (i.e., from nc-Si crystals having ordinal number m and containing one exciton) to nc-Si with concentration N_{0m-1} (neighboring unexcited larger nc-Si crystals). In the third equation of the set for nc-Si crystals with concentration N_{2m}, we neglected the terms responsible for the exciton migration from these nanocrystals ($-D_m N_{2m} N_{0m-1} - D^*_m N_{2m} N_{1m-1}$) and radiative recombination ($-\alpha_{rm} N_{2m}$) assuming that the Auger recombination is a much faster process. Indeed, α_{Am} for nc-Si is 10^{-8} –10^{-9} s (Proot et al., 1992; Delerue et al., 1993), whereas the characteristic migration and radiative recombination times fall in the microsecond range, as follows from an analysis of the experimental PL relaxation kinetics (see below). Moreover, in the third equation of set (12), we neglected the terms of type $-\alpha'_m N'_{2m}$, which describe nonradiative recombination at defects in nc-Si crystals with two excitons. This assumption is reasonable, since the probability of the presence of two excitons in one nc-Si crystal with a defect is negligibly low. Indeed, an exciton in an nc-Si crystal with a defect undergoes nonradiative annihilation before the transfer of another exciton in this nc-Si crystal at a high probability, since the time of carrier trapping by a neutral defect (10^{-9} –10^{-11} s) for the nanocrystals that luminesce in the red spectral region (which have an energy gap $1.3 < Eg < 2.2$ eV) is extremely short (Delerue et al., 1993). If we consider the case of nanocrystals that do not luminesce in this spectral region (for which α' is comparable with the radiative recombination rate (Delerue et al., 1993), $\alpha'_m N'_{2m}$ term and the terms describing radiative recombination and the exciton migration from these nc-Si crystals may be neglected due to their smallness compared to term $\alpha_{Am} N_{2m}$ (Delerue et al., 1993). We now consider the kinetic equation that describes the nonradiative recombination of excitons in nc-Si crystals with defects,

$$\dot{N}'_{1m} = -\alpha' N'_{1m} - \alpha_{rm} N'_{1m} - D_m N'_{1m} N_{0m-1} - D^*_m N'_{1m} N_{1m-1} - D^*_{m+1} N_{1m+1} N'_{1m} + D_{m+1} N_{1m+1} N'_{0m} \quad (13)$$

We substitute $(N'_m - N'_{1m})$ for N'_{0m}, group the terms, and obtain the expression

$$\dot{N}'_{1m} = -\left[\alpha'_m + \alpha_{rm} + D_m N_{0m-1} + D^*_m N_{1m-1} + (D^*_{m+1} + D_{m+1}) N_{1m+1}\right] N'_{1m} + D_{m+1} N_{1m+1} N'_m \quad (14)$$

We analyze this expression by analogy with the procedure performed for $\dot{N}_{2m}$ and conclude that all terms in the square brackets may be neglected compared to α'_m. Below, we will prove the legitimacy of these assumptions for luminescent nc-Si crystals. Allowing for these assumptions, we write kinetic equations for nc-Si crystals in which rapid relaxation processes take place:

$$\left.\begin{aligned} \dot{N}'_{1m} &= -\alpha'_m N'_{1m} + D_{m+1} N_{1m+1} N'_m, \\ \dot{N}_{2m} &= -\alpha_{Am} N_{2m} + D^*_{m+1} N_{1m+1} N_{1m}. \end{aligned}\right\} \quad (15)$$

Let us analyze the processes that occur in such nc-Si crystals immediately after removing an optical excitation in detail. If the initial values of N'_{1m} and N_{2m} are sufficiently high, they decrease rapidly in 10^{-6} –10^{-9} s via nonradiative and Auger recombination (Delerue et al., 1993). Beginning from the time when the recombination rates become equal to the rate of filling of these nanocrystals due to the exciton migration from nc-Si crystals with larger ordinal numbers, these processes manifest themselves as negative-feedback systems to some extent. For example, if the inequality $\alpha'_m N'_{1m} > D_{m+1} N_{1m+1} N'_m$ holds true at a certain time, concentration N'_{1m} decreases ($\dot{N}'_{1m} < 0$). As a result, the absolute value of the rate of loss of excitons N'_{1m} (which is proportional to $\alpha'_m N'_{1m}$) decreases and $D_{m+1} N_{1m+1} N'_m$ remains constant (since it is independent of N'_{1m}). Finally, the rates of loss and supply of excitons in nc-Si crystals become the same and the equality $\dot{N}'_{1m} = 0$ again takes place. Similarly, if $\dot{N}'_{1m} > 0$, rate $\alpha'_m N'_{1m}$ increases in absolute value and again becomes equal to the rate of exciton supply to the ensemble of nc-Si crystals due to migration from neighboring nanocrystals. The same is true for the number of nc-Si crystals with two excitons (N_{2m}). Thus, the approximate equalities

$$\left.\begin{aligned} \dot{N}'_{1m} &\approx 0, \\ \dot{N}_{2m} &\approx 0. \end{aligned}\right\} \quad (16)$$

hold true due to the substantial difference in the rates of rapid recombination processes and slow migration processes at any time beginning from approximately 10^{-6} s after removing an optical excitation. With the first equation from set (16), we can express N'_{1m} and substitute it into the first equation of set (12), and the second equation of set (16) demonstrates that $N_{2m} << N_{1m}$. Indeed, we have

$$N_{2m} = \left(\frac{D^*_{m+1} N_{1m+1}}{\alpha_{Am}} \right) N_{1m} << N_{1m}. \tag{17}$$

Based on inequality (17), we neglect N_{2m} compared to N_{1m} in an expression for the number of unexcited nc-Si crystals:

$$N_{0m} = N_m - N_{1m} - \cancel{N_{2m}}.$$

As a result of this simplification, the substitution of N'_{1m}, and the regrouping of the terms, we reduce the first equation of set (12) to the form

$$\begin{aligned} \dot{N}_{1m} &= -\alpha_{rm} N_{1m} - D_m N_{1m} N_{m-1} + (D_m - D^*_m) N_{1m} N_{1m-1} - \\ &-(D_{m+1} + D^*_{m+1}) N_{1m+1} N_{1m} + D_{m+1} N_{1m+1} (N_m - N'_m) \end{aligned} \tag{18}$$

Although set (12) is substantially simplified, it is rather difficult to directly solve Eq. (18) even using numerical methods. First, the total number of equations can be rather large, since m varies from zero to M. The value of parameter M can easily be estimated in a fractal model, since it represents the order of the minimum size nc-Si crystal with a diameter $d_M = d_0 k^{-M}$. For estimation, we assume d_M = 0.6 nm (which is the double Si–Si bond length) and d_0 = 5 nm (see Figs. 9b–1d) and obtain

$$M = \frac{\ln(d_0 / d_M)}{\ln k} \approx \frac{2.1}{\ln k}.$$

Thus, M depends on coefficient k of linear scale reduction, which can be only insignificantly higher than unity in the case of crystals having almost the same sizes; as a result, M can be arbitrarily large. For example, at k = 1.02 we have M = 107. Second, the character of the dependence of recombination rate α and, especially, migration coefficient D on ordinal number m is unknown. Thus, to solve the set of equations (18), we have to know the dependences of these parameters on the nc-Si crystal size over the entire size range (in our case, from 0.6 to 5 nm). Nevertheless, equations (18) become much simpler if we assume "local quasi-identity" of nc-Si crystals, i.e., if we assume that neighboring nc-Si crystals with numbers m–1, m, and m+1 and several farther neighbors have similar shapes and sizes. The larger the number of almost identical nc-Si crystals in a chain, the more adequate this assumption. In a fractal model of the structure of PS, this assumption means that k is close to unity. Then, the number of neighbors of any nc-Si crystal is $n = k^{DF}$, where D_F is the fractal dimension of nc-Si (which ranges from 2.2 to 2.6) (Nychyporuk et al., 2005). Taking D_F = 2.4 and k = 1.02, we obtain $n \approx 1.05$. The structure described by these values of parameters n and k is represented by long filaments made of almost the same nc-Si crystals (the difference in the neighboring crystal sizes is approximately 2%) intersecting with other filaments every 20 nc-Si crystals. As follows from Figs. 9b–9d, this structural model is close to reality. With the assumption of local quasi-identity of nc-Si crystals, we can omit all ordinal indices in Eq. (18) assuming that

$$\begin{aligned} N_{m-1} &\approx N_m \approx N_{m+1} \equiv N, \\ N_{1m-1} &\approx N_{1m} \approx N_{1m+1} \equiv N_1, \\ D_{m-1} &\approx D_m \approx D_{m+1} \equiv D, \\ \alpha_{rm-1} &\approx \alpha_{rm} \approx \alpha_{rm+1} \equiv \alpha_r, \end{aligned} \tag{19}$$

and so on. As a result, we can rewrite Eq. (18) in the form

$$\dot{N}_1 = -\alpha_r N_1 - DN_1N + (D - D^*)N_1^2 - (D + D^*)N_1^2 + DN_1(N - N') \tag{20}$$

and eventually obtain the following kinetic equation for the relaxation of an optical excitation in an ensemble of nc-Si crystals with allowance for exciton transfer in them:

$$\begin{aligned} \dot{N}_1 &= -\alpha N_1 - 2D^* N_1^2, \\ \alpha &\equiv \alpha_r + DN', \end{aligned} \tag{21}$$

where we introduced a designation for the sum $\alpha_r + DN'$. Note that term $-2D^*N_1^2$ in Eq. (21), which causes a nonexponential decrease in the PL intensity, depends only on the migration coefficient, which describes the exciton transfer to an excited neighboring crystal (which contains an exciton before the transfer event). Thus, it is ultrafast Auger recombination processes that determine the nonexponential relaxation of an ensemble of coupled nc-Si crystals. Note that the values of coefficients D and D^* should be close. The small difference between them can be caused by an additional interaction of excitons located in neighboring nc-Si crystals (in the case of D^*). In the dipole approximation, we have $D=D^*$, since the trace of the symmetric dipole–dipole interaction tensor is zero (Carrington & McLachlan, 1979). Therefore, the energy of this interaction is also zero upon averaging over a spherically symmetric exciton charge distribution. The difference in coefficients D and D^* can be related to the presence of a quadrupole (or higher moments of) interaction of excitons or an exchange interaction in the case of a substantial overlapping of their wavefunctions. Obtained ordinary differential equation (21) is a Bernoulli equation with constant coefficients and an exponent $\gamma=2$ of the highest degree of the unknown function. This equation is solved analytically using the substitution $y = N_1^{1-\gamma} = 1/N_1$. With allowance for the initial condition $N_1(t=0) = N_1^{(0)}$, the solution to Eq. (21) has the form

$$N_1(t) = \frac{\alpha N_1^{(0)}}{\alpha \exp(\alpha t) + 2D^* N_1^{(0)}(\exp(\alpha t) - 1)}. \tag{22}$$

Obviously, the product $\alpha_r N_1(t)$ describes the PL intensity decrease kinetics for an ensemble of nc-Si crystals of a given size. Let us note some advantages of Eq. (22) over stretched exponent (1). The absence of migration ($D = D^* = 0$) leads to the usual monoexponential character of a decrease in the PL intensity with increasing radiative lifetime, according to the experimental kinetics obtained for nc-Si crystals separated by a thick oxide SiO_2 layer (Vinciguerra et al., 2000). The form of the normalized kinetics $\frac{\alpha_r N_1(t)}{N_1^{(0)}}$ depends on $N_1^{(0)}$ excitation level, which was repeatedly noted by many researchers (see, e.g., (Chen et al., 1992)). At $t \to \infty$, the $N_1(t)$ curve has a monoexponential drop, which agrees well with the experimental data (see below). Defect concentration N' contributes substantially (in the product with D; see Eq. (21)) to optical excitation relaxation rate α, in contrast to the case without migration. If excitons did not move along an nc-Si network ($D = D^* = 0$), the normalized PL intensity decrease kinetics would not depend on the defect concentration, which is in conflict with the experimental data (see below). Thus, energy transfer via exciton migration does occur in an ensemble of coupled nc-Si crystals. Note that, if we do not neglect terms α_{rm} and $D_m N_{m-1}$ compared to α'_m in Eq. (14), we can obtain a more accurate expression $\alpha \equiv \alpha_{acc}$,

$$\alpha_{acc} = \alpha_r + \frac{\alpha'}{\alpha' + \alpha_r + DN} DN'. \tag{23}$$

However, a numerical analysis with the Mathcad software package demonstrates that the factor multiplying DN' in Eq. (23) differs from unity by several fractions of a percent for luminescent silicon nanocrystals. The experimental PL relaxation curves of micro-PS samples can be approximated by analytical dependence (22) using α and $\zeta = 2D^* N_1^{(0)}$ as fitting parameters. The results of this process will be presented below. It is interesting to find the value of diffusion coefficient D^* using the well-known value of approximation parameter ζ. To this end, we have to determine $N_1^{(0)}$ initial excitation level. As will be shown in the next section, such a calculation can be performed in terms of the proposed model.

3.5 Photoluminescence of porous silicon under steady-state photoexcitation conditions

We write the first and second equations from set (12) for a steady-state case ($\frac{d}{dt} = 0$) in the presence of an optical excitation in the system $g_m\left[\mu s^{-1}\right] = \sigma_m I_{ex}$ (where σ_m is the absorption cross section of nanocrystals with ordinal number m (i.e., nanocrystals of a certain size; see Section 3.4) and I_{ex} is the radiation intensity):

$$\left.\begin{aligned} \dot{N}_{1m} &= g_m N_{0m} - \alpha_{rm} N_{1m} - \alpha'_m N'_{1m} - D_m N_{1m} N_{0m-1} - D^*_m N_{1m} N_{1m-1} + \\ &\quad + D_{m+1} N_{1m+1} N_{0m} - D^*_{m+1} N_{1m+1} N_{1m} = 0, \\ \dot{N}_{2m} &= g_m N_{1m} + D^*_{m+1} N_{1m+1} N_{1m} - \alpha_{Am} N_{2m} = 0. \end{aligned}\right\} \tag{24}$$

Note that an analysis of the last equation in set (24) immediately yields the inequality $N_{2m} << N_{1m}$; whence, it also follows that $N'_{2m} << N'_{1m}$. Indeed, the equality

$$\frac{N'_{2m}}{N_{2m}} = \frac{N'_{1m}}{N_{1m}} = \frac{N'_m}{N_m}$$

is obvious in the steady-state case. It reflects the fact that these relations are determined by the fraction of defect-containing nc-Si crystals of the total number of nc-Si crystals in the same manner. Based on the inequalities, we may neglect N_{2m} and N'_{2m} in the following expressions:

$$N_{0m} = N_m - N_{1m} - \cancel{N_{2m}} \text{ and } N'_{0m} = N'_m - N'_{1m} - \cancel{N'_{2m}}$$

We substitute the latter expression into the equation for nc-Si crystals with defects, which is analogous to the first equation in set (24) in which N_{0m} and N_{1m} substitute for N'_{0m} and N'_{1m}, and have

$$N'_{1m} = \frac{(g_m + D_{m+1}N_{1m+1})N'_m}{g_m + \alpha_{rm} + \alpha'_m + D_m N_{m-1} - (D_m - D^*_m)N_{1m-1} + (D_{m+1} + D^*_{m+1})N_{1m+1}}. \tag{25}$$

Based on the causes described in the previous section, we neglect all quantities in the denominator as compared to α'_m in the last expression and substitute it in the first equation in set (24). We now group the terms, take into account the consideration about the smallness of N_{2m} given above, and write the first equation of set (24) in the form

$$\begin{aligned} 0 = g_m(N_m - N'_m) - N_{1m}(g_m + \alpha_{rm} + D_m N_{m-1}) + (D_m - D^*_m)N_{1m-1}N_{1m} + \\ + D_{m+1}N_{1m+1}(N_m - N'_m) - (D_{m+1} + D^*_{m+1})N_{1m+1}N_{1m} \end{aligned} \tag{26}$$

This is the general form of an equation for the number of photoexcited nc-Si crystals of the mth order in a fractal chain in the steady-state case. This equation can be solved by an iteration method beginning from $m = 0$. However, this process is related to the difficulties described in the previous section. At the same time, Eq. (26) can be solved in the local nc-Si quasi-identity approximation, which is represented by Eqs. (19). As a result, we obtain the following quadratic equation for N_1, in which ordinal indices are omitted:

$$\begin{aligned} & 2D^*N_1^2 + \alpha_{ex}N_1 - g(N - N') = 0, \\ & \alpha_{ex} \equiv g + \alpha = g + \alpha_r + DN', \end{aligned} \tag{27}$$

where we introduced a designation for $g + \alpha_r + DN'$ similarly to Eq. (21). The nonnegative solution to Eq. (27) has the form

$$N_1^{(0)} = \sqrt{\left(\frac{\alpha_{ex}}{4D^*}\right)^2 + \frac{g(N - N')}{2D^*}} - \frac{\alpha_{ex}}{4D^*}. \tag{28}$$

Note that a more accurate (by several fractions of a percent) solution can be obtained if we do not neglect $g_m + \alpha_{rm} + D_m N_{m-1}$ as compared to α'_m in Eq. (25). In this case, N' in Eq. (28) is everywhere replaced by the product

$$\varepsilon' N' \equiv \frac{\alpha'}{g + \alpha' + \alpha_r + DN} N',$$

in which correcting coefficient ε' is insignificantly smaller than unity for relatively low optical excitation levels as compared to the significant rate of nonradiative recombination at a defect α' for luminescent nc-Si crystals.
In the limiting case $D^*=0$, Eq. (28) is reduced to the form

$$N_1^{(0)}\Big|_{D^*=0} = \frac{g(N - N')}{g + \alpha_r}, \tag{29}$$

that is, it coincides with the solution to set (24) in the absence of the terms describing migration. According to Eq. (29), steady-state PL intensity of nc-Si ensembles $I_{PL} \propto N_1^{(0)}$ weakly depends on the defect concentration in them, since the fraction of defect-containing nc-Si crystals of the total number of nc-Si crystals $(\frac{N'}{N})$ is low (10^{-3} –10^{-2}). However, according to the experimental data, the PL spectrum amplitude decreases significantly during the photostimulated oxidation of PS samples (see below), whereas the defect concentration increases only by a factor of 1.5–2 (see below). Therefore, as in the case of the kinetic curves of PL decay (Section 3.4), the motion of excitons along a network of coupled nc-Si crystals also substantially affects the steady-state PL intensity. To express diffusion coefficient D^* through parameters α_{ex} and ζ, we multiply both sides of Eq. (28) by $2D^*$

$$\zeta \equiv 2D^* N_1^{(0)} = \sqrt{\left(\frac{\alpha_{ex}}{2}\right)^2 + 2D^* g(N - N')} - \frac{\alpha_{ex}}{2}, \tag{30}$$

as a result, we have

$$D^* = \frac{\zeta(\alpha_{ex} + \zeta)}{2g(N - N')} \approx \frac{\zeta(\alpha_{ex} + \zeta)}{2gN}, \tag{31}$$

where we neglected N' compared to N in the last equality (see above). Thus, the exciton migration coefficient averaged over the entire ensemble of nc-Si crystals of a given size can be calculated by Eq. (31) using approximation parameters α and ζ of the kinetic dependences of PL decay and determining photoexcitation level g and the nc-Si concentration in the ensemble. Although the latter is most difficult, it can be performed in terms of a fractal PS model. Indeed, the number of nc-Si crystals of a given size (in this case, about 3 nm, which corresponds to an energy gap of 1.6 eV (Delerue et al., 1993)) can readily be estimated from the following expression for the number of nc-Si crystals of order m: $N_m = N_0 k^{D_F m}$, where N_0 is the number of nc-Si crystals of the zeroth order, k=1.02 is the linear scale reduction coefficient (according to the local quasi-identity approximation; see Section 3.4), D_F=2.4 is the fractal dimension of PS layers, and m ranges from zero to M = 107 (see Section 3.4). The total number of nc-Si crystals of all orders is

$$N_{tot} = \sum_{m=0}^{M} N_m = \frac{k^{MDF} - 1}{k^{DF} - 1} N_0 \approx 3300 N_0.$$

We assume that $N_{tot}=10^{21}$ cm^{-3} and determine N_0. The order corresponding to an nc-Si diameter d_m=3 nm is expressed as

$$m = \frac{\ln(d_0 / d_m)}{\ln k} = \frac{\ln(5/3)}{\ln 1.02} \approx 26,$$

as a result, we obtain

$$N_{26} = \frac{N_{tot}}{3300} 1.02^{2.4 \cdot 26} \approx 10^{18} cm^{-3}.$$

Similarly, knowing the total number of nc-Si crystals with defects (about 10^{17} cm^{-3}) or oxygen molecules (at p_{O2} =1 bar, n=2.4·10^{19} cm^{-3}) and taking into account that their specific number per nanocrystal is proportional to the nanocrystal surface area, we can estimate the number of defects (about 10^{15} cm^{-3}) or O_2 molecules (about 2·10^{17} cm^{-3}) per ensemble of nc-Si crystals of a given size (with an ordinal number m=26). At a nitrogen laser radiation intensity of 300 mW/cm^2 and an quantum energy of 3.7 eV, I_{ex} = 5·10^{17} photons/(cm^2 s) are incident on a sample. Assuming that the absorption cross section of a PS sample is σ=10^{-14} cm^2 (Bisi et al., 2000), we find that the optical photoexcitation rate is $g = I_{ex}\sigma \approx 0.005$ μs^{-1}. Allowing for α ≈0.008 μs^{-1}, we obtain $\alpha_{ex} = g + \alpha \approx 0.013$ μs^{-1}. The characteristic value of α was taken from the results of approximation of the kinetic dependences for coarse-grained PS powders with a porosity *P*= 70% (table 2). The approximation parameter is ζ =0.038 μs^{-1} (table 2).

	PS 1 (*P* = 70%)		PS 2 (*P* = 85%)	
	CG	FG	CG	FG
α^{-1}, μs	126 ± 13	128 ± 13	117 ± 12	126 ± 13
ζ^{-1}, μs	26 ± 3	49 ± 5	104 ± 10	72 ± 7

Table 2. Parameters of the approximation of the kinetic PL decay curves shown in Fig. 15 by Eq. (22).

Substituting these values of parameters into Eq. (31), we finally obtain the migration coefficient for coarse-grained (*P*=70%) PS samples, D^* =3.9·10^{-13} cm^3/s. Using Eq. (28) and the values of all necessary parameters determined above, we can construct a relationship between the number of excited nc-Si crystals $N_1^{(0)}$ and, e.g., the level of optical excitation or the defect concentration (for nanocrystals with defects; see Fig.10). It is seen that, in the presence of nonradiative recombination centers, the PL of the entire ensemble ($I_{\rm PL} = \alpha_r N_1^{(0)}$) is quenched in all nc-Si crystals, which agrees with the conclusions in (Delerue et al., 1993) (Fig. 10a). The dependences of the PL intensity of a sample in vacuum on the level of optical excitation (Fig. 10b) agree qualitatively with the experimental curves (see, e.g., (Gross et al., 2003)). Based on the developed theory, we can estimate the radiative lifetimes of the PL of an ensemble of nc-Si crystals. Indeed, when analyzing experimental kinetics, we obtain approximation parameter α, which is expressed through radiative recombination rate α_r as $\alpha \equiv \alpha_r + DN'$ (see Eq. (21)). The concentration of point defects in as-deposited micro-PS films is low, so that $DN' << \alpha_r$; therefore, we may assume $\alpha \approx \alpha_r$ for estimation.

3.6 Theoretical analysis of the singlet oxygen generation efficiency

We now consider an ensemble of nc-Si crystals of a certain size under conditions of steady-state photoexcitation in an oxygen-containing medium and still assume that the local nc-Si quasi-identity approximation holds true (see Section 3.4). Then, the condition of dynamic equilibrium between the number of O_2 molecules in the singlet (n_{SO}) and triplet (n_{TO}) states is

$$\left.\begin{aligned} \dot{n}_{SO} &= \alpha_{ET} N_1 \frac{n_{TO}}{N} - \alpha_{SO} n_{SO} = 0, \\ n_{SO} + n_{TO} &= n \end{aligned}\right\}, \tag{32}$$

where α_{ET} is the probability of energy transfer per unit time from an excited Si nanocrystal to one oxygen molecule adsorbed on its surface. The first term takes into account the fact that the interaction efficiency is proportional to the number of $^3O_2$ molecules per nc-Si ($\frac{n_{TO}}{N}$).

Quantity α_{SO}^{-1} is the oxygen molecule lifetime in the singlet state in PS pores, which is about 500 μs (Gross et al., 2003), and n is the concentration of O_2 molecules. Here, for simplicity, the number of excited nc-Si crystals (N_1) is written without superscript (0), which means that a steady-state case is analyzed. Using Eq. (32), we can easily express the stationary concentration of $^3O_2$ molecules,

$$n_{TO} = \frac{\alpha_{SO} N n}{\alpha_{SO} N + \alpha_{ET} N_1}. \tag{33}$$

For an oxygen-containing medium, the term $\alpha_{ET} N_1 \frac{n_{TO}}{N}$ is added to the left-hand side of Eq. (27); this term is responsible for the energy transfer from annihilating excitons to triplet oxygen molecules on the nc-Si surface. After the substitution of Eq. (33) for n_{TO} and some manipulations, we write Eq. (27) in the form

$$\begin{aligned} &2D^* \alpha_{ET} N_1^3 + \left(\alpha_{ET}\alpha_{ex} + 2D^*\alpha_{SO} N\right) N_1^2 + \\ &+\left(\alpha_{ET}\alpha_{SO} n + \alpha_{ex}\alpha_{SO} N - \alpha_{ET} g(N-N')\right) N_1 - \alpha_{SO} g(N-N')N = 0. \end{aligned} \tag{34}$$

This is a cubic equation for N_1, which can be solved numerically. We used the Mathcad software package. With Eq. (33) and the solution to Eq. (34), we obtain the following expression for the concentration of photosensitized $^1O_2$:

$$n_{SO} = n - n_{TO} = \frac{\alpha_{ET} N_1 n}{\alpha_{SO} N + \alpha_{ET} N_1}. \tag{35}$$

Figure 11 shows the calculated dependence of the fraction of singlet oxygen on the nc-Si photoexcitation rate at various molecular oxygen concentrations.
Hereafter, by default we use the following parameters for numerical simulation, unless otherwise specified (see Section 3.4): $D \approx D^* = 3.9 \cdot 10^{13}$ cm^3/s, $N = 10^{18}$ cm^{-3}, $N' = 10^{15}$ cm^{-3}, $\alpha_r^{-1} = 125$ μs, $\alpha_{ET}^{-1} = 40$ μs (Gross et al., 2003), and $\alpha_{SO}^{-1} = 500$ μs (Gross et al., 2003). The curves in Fig. 11 are similar to the experimental dependence of the concentration of $^1O_2$ molecules on the photoexcitation intensity obtained by EPR (see Fig.6). When oxygen partial pressure p_{O2} increases, the concentration of forming $^1O_2$ increases (Fig. 12a); however, the fraction of O_2 molecules in the singlet state of the total number of O_2 molecules decreases beginning from a certain value of p_{O2} (Figs. 11 and 12b). Indeed, until the number of O_2 molecules per Si nanocrystal in the ensemble under study is smaller than a certain value, the fraction of $^1O_2$, which is only determined by the relation between the characteristic times of excitation, relaxation, and energy transfer, remains constant with increasing pressure. When the specific number of O_2 molecules reaches a certain threshold value (more than unity) per nc-Si crystal, this crystal cannot excite all oxygen molecules adsorbed on its surface at the same efficiency. As a result, the fraction of $^1O_2$ decreases when the concentration of molecular oxygen increases further (Fig. 12b). The difference between the curves shown in Fig. 12a and the curve obtained experimentally by EPR (see Fig.5b) can be caused by the integral character of EPR: a microwave power signal of absorption by all spin centers (Pb-centers) in a sample is measured in this technique, whereas many of them do not interact with O_2 molecules at a low oxygen concentration and, thus, are insensitive to a change in the $^1O_2$ concentration.

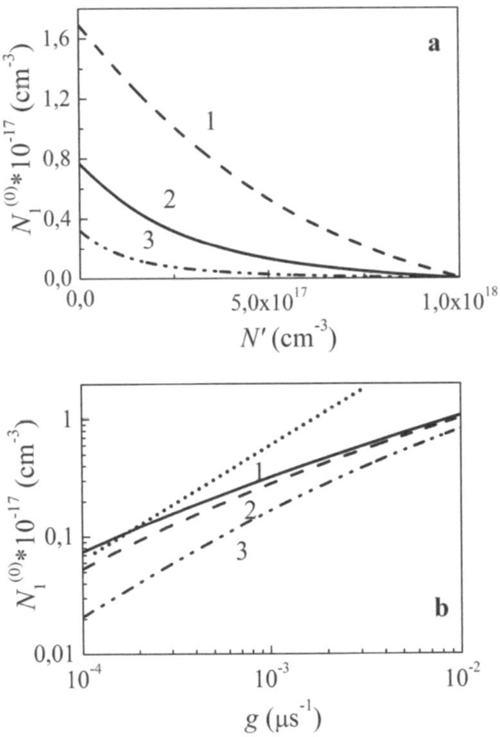

Fig. 10. Calculated dependences of the steady-state PS PL intensity in vacuum ($I_{PL} \propto N_1^{(0)}$) on (a) the concentration of nc-Si crystals with defects N' at g=(1) 0.025, (2) 0.005, and (3) 0.001 μs^{-1} and on (b) g at N'=(1) 10^{15} , (2) $2 \cdot 10^{16}$, and (3) 10^{17} cm^{-3}. The total number of nc-Si crystals in the ensemble is 10^{18} cm^{-3}. A linear dependence is shown as a dashed line for clarity.

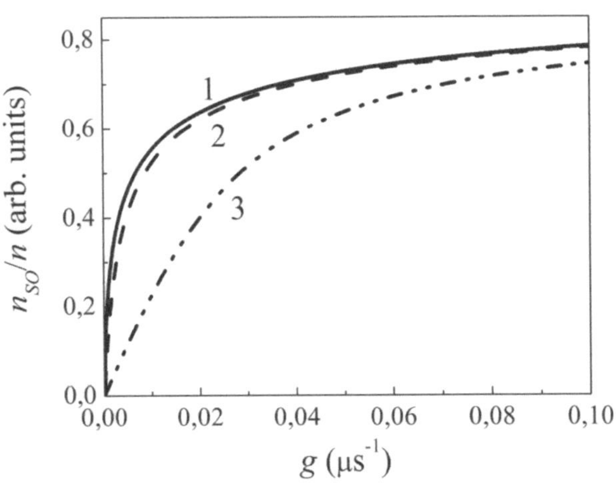

Fig. 11. Calculated dependence of the fraction of photosensitized $^1O_2$ in PS layers on the photoexcitation level at an O_2 molecule concentration n=(1) $2 \cdot 10^{17}$, (2) $2 \cdot 10^{18}$, (3) $2 \cdot 10^{19}$ (3) cm^{-3}.

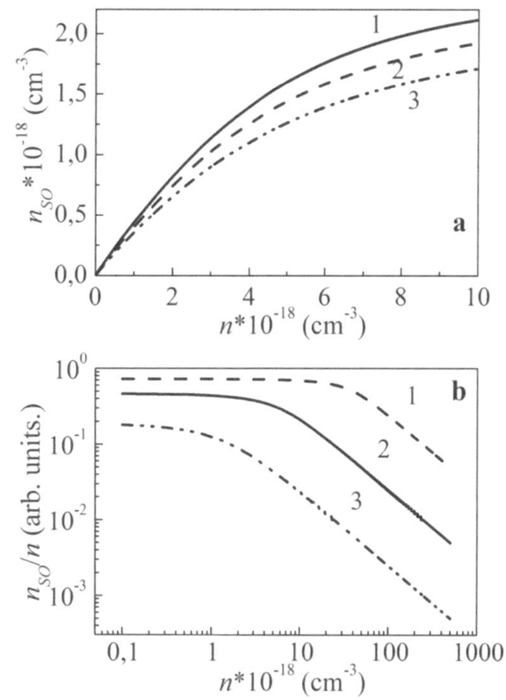

Fig. 12. Calculated dependences illustrating the effect of the molecular oxygen concentration on the $^1O_2$ generation efficiency in PS layers at (a) defect concentration N'=(1) 10^{16}, (2) $5 \cdot 10^{16}$, and (3) 10^{17} cm^{-3} and (b) g=(1) $5 \cdot 10^{-2}$, (2) $5 \cdot 10^{-3}$, and (3) $5 \cdot 10^{-4}$ µs^{-1}.

Hence, the EPR diagnostics of singlet oxygen generation is inapplicable in the case of a low concentration of O_2 molecules. As noted above, the presence of a defect in a luminescent nc-Si crystal fully suppresses its luminescence due to fast (10^{-9} –10^{-11}) nonradiative trapping of nonequilibrium charge carriers. Therefore, this nanocrystal cannot excite an oxygen molecule on its surface, since the energy transfer time is several tens of microseconds (Gross et al., 2003). This conclusion is reflected in a decrease in the fraction of photosensitized $^1O_2$ when the number of nc-Si crystals with defects increases (Fig. 13). When concentration N' is equal to the number of nc-Si crystals in the ensemble under study (10^{18} cm^{-3}), the $^1O_2$ generation efficiency vanishes (Fig. 13). In the next section, we will show that the calculated dependences have an experimental support.

3.7 Effect of the powder granule size of porous silicon on the luminescence properties

Figure 14 shows micrographs of the FG PS powders taken on a scanning electron microscope. Submicronscale granules are clearly visible apart from micronscale granules (Fig. 14a). Moreover, many microgranules consist of a large number of smaller aggregated particles (Fig. 14b), with the number of such agglomerates increasing with the time of sample storage in air. Indeed, nc-Si crystals in PS layers oxidize in time and their surfaces become hydrophilic. Such PS granules coalesce and form coarse particles due to the

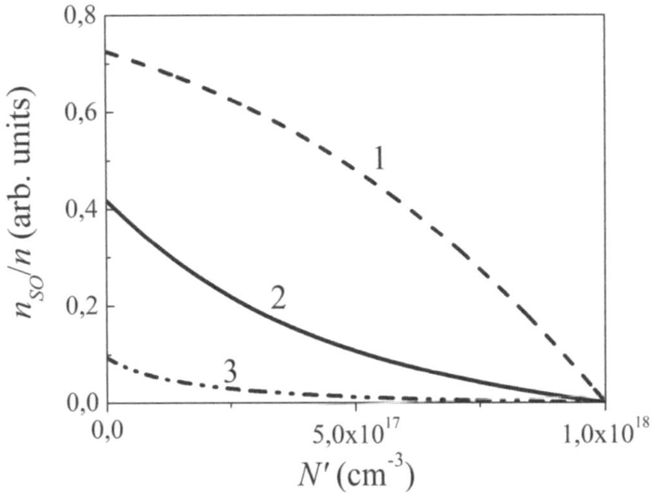

Fig. 13. Calculated dependence of the fraction of photosensitized $^1O_2$ in PS layers on the concentration of nc-Si crystals with defects at g=(1) $5\cdot10^{-2}$, (2) $5\cdot10^{-3}$, and (3) $5\cdot10^{-4}$ μs^{-1}. The O_2 molecule concentration is n=$2\cdot10^{17}$ cm^{-3}

hydrogen bonds appearing between them. The image contrast of particles with a diameter of several tens of nanometers in Fig. 14 is low because of their strong static charging. As follows from Fig. 14, simple mechanical milling in a vibratory mill decreases the PS granule size to submicron sizes. This is also indicated by a change in the powder color from dark red-brown (in the case of a CG powder) to light yellow (FG powder), which is likely to be caused by Rayleigh scattering by particles several tens of nanometers in size and structural relaxation (i.e., the relief of microstresses, along which granules predominantly fail). As follows from the EPR spectroscopy data, the concentration of Pb-centers increases almost twofold, from $1.4\cdot10^{17}$ to $2.8\cdot10^{17}$ cm^{-3}. We assume that this increase leads to a decrease in the PL decay time and, hence, the PL intensity because of an increase in the probability of nonradiative exciton recombination by defects (Delerue et al., 1993).

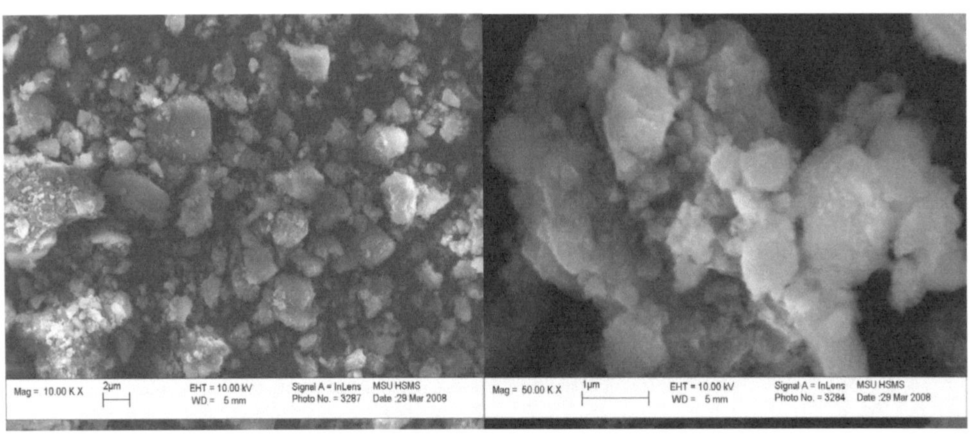

Fig. 14. Scanning electron microscopy micrographs of PS FG powder granules with different scale.

An analysis of the relaxation times demonstrates that, for PS1 samples, they increase in the powders subjected to FG milling (Fig. 15a), which is in conflict with the assumption made above.

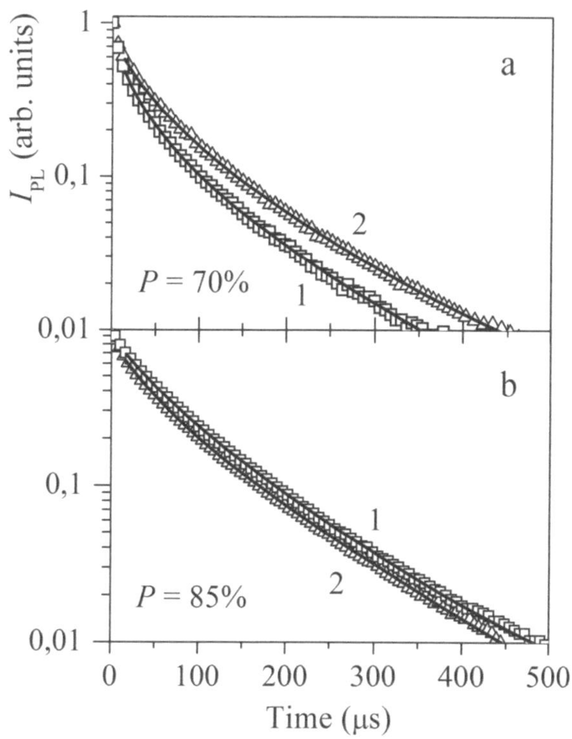

Fig. 15. PL kinetics of (1) CG and (2) FG (a) PS1 and (b) PS2 powders in a vacuum p_{O2} =1 mbar recorded at a wavelength λ=760 nm. Approximating dependences calculated using the proposed model (solid lines).

This experimental fact can be explained in terms of a model of exciton migration along a network of intersecting nc-Si crystals with a preferred motion direction from smaller to larger crystallites (against the gradient of the energy gap). Both radiative and nonradiative recombination is possible in each of the intermediate nc-Si crystals. The experimental fact of an increase in the PL relaxation time for the FG PS powders as compared to the CG powders can be explained by a decrease in the number of possible exciton migration ways when PS films are milled to granules with sizes varying from several tens to several hundreds of nanometers. For every fraction of nc-Si crystals having a certain size, a nonradiative relaxation channel is partly suppressed due to exciton motion from these nanocrystals; as a result, the exciton lifetimes at the PL wavelength corresponding to this nc-Si crystal size increase. It is important that the properties are modified upon crystal refinement only for nc-Si crystals on a fracture surface and that the internal porous structure of PS granules remains unchanged. Then, the detected change in the PL properties of the samples is related to the fact that the number of surface nc-Si crystals becomes comparable with their number in the volume of a PS granule when its size decreases from approximately 50 μm (etched layer thickness) to several tens of nanometers. Therefore, this effect can be detected experimentally. Indeed, the simplest model considerations yield the following estimate of the ratio of the number of surface nc-Si crystals having a spherical shape to their number in the volume of a spherical granule: $\varphi = \frac{6r}{R}$, where r is the nanocrystal radius and R is the granule radius. Assuming $2r$=3 nm and $2R$ = 90 nm, we find that 20% of all nc-Si crystals that form a PS granule of the given size are at its surface; at $2R$ = 1000 nm, the fraction of surface nc-Si crystals is only about 2%. To test the hypothesis of restricted exciton migration, we prepared samples (PS2) with a higher porosity (85%). At such a high porosity, PS2 nanocrystals are known to be almost completely isolated from each other (Bisi et al., 2000); therefore, exciton migration along an nc-Si network is substantially restricted as compared to the PS1 samples. Thus, milling of PS2 samples should not significantly change their luminescence properties. Indeed, the kinetic curves of a decrease in the PL intensity for the CG and FG PS2 powders are almost the same (Fig. 15b), and the nonexponential segments at short times that are mainly caused by exciton migration are almost absent (cf. Figs. 15a, 15b). The solid lines in Fig. 15 represent approximating dependences calculated by Eq. (22). The table 2 gives the approximation parameters for the kinetic curves that were discussed above (α, $\zeta = 2D^* N_1^{(0)}$). The relative error in determining these parameters is about 10% and reflects the variation in the values of α and ζ when the approximation range and the initial value of an approximating curve change. Parameter ζ, which describes the migration rate, is seen to decrease upon fine milling of PS1 powders, which indicates a decrease in the efficiency of exciton motion from nc-Si crystals in the ensemble under study. Using the parameters determined in Section 3.4 by Eq. (31), we can calculate migration coefficients D^* for CG and FG PS1 samples; they are found to be $3.9 \cdot 10^{13}$ and $1.3 \cdot 10^{13}$ cm^3/s, respectively. Thus, upon fine milling of CG PS samples with a porosity of 70%, diffusion coefficient D^* averaged over an ensemble of nc-Si crystals decreases approximately threefold. The weak change in parameter α upon fine milling of PS samples indicates a low value of DN' compared to α_r (see Eq. (21)) in spite of an approximately twofold increase in the defect concentration. Parameter $\zeta = 2D^* N_1^{(0)}$ turns out to be higher than α_r, which leads to the expected result $N_1^{(0)} >> N'$. Indeed, $N_1^{(0)}$ can reach several tens of percent of the number of all nc-Si crystals N for a sufficiently high level of optical excitation (see Fig. 10); then, we have $N_1^{(0)} \sim 0.1N >> N'$. This inequality reflects the following well-known empirical fact: the number of defects is substantially smaller than the number of nc-Si crystals. For example, a fractal model of the PS microstructure yields $N'/N \approx 10^{15}$ cm^{-3} $/10^{18}$ $cm^{-3} = 10^{-3}$ for the ensemble of nc-Si crystals under study (see Section 3.4). The developed theory can be used to determine the actual radiative PL lifetime of an ensemble of nc-Si crystals, in contrast to formal approximation parameter τ in Eq. (1) for an stretched exponent. Indeed, when analyzing the experimental kinetics, we obtain the value of approximation parameter α_r, which is expressed through radiative recombination rate α_r ($\alpha \equiv \alpha_r + DN'$). As follows from our earlier considerations, DN' is low compared to α_r for the as-prepared PS samples; therefore, we can assume $\alpha \approx \alpha_r$ for estimation. Another conclusion drawn from an analysis of the approximation parameters of the decrease in the PS PL intensity consists in the fact that all assumptions regarding the smallness of the terms describing exciton migration as compared to nonradiative recombination rate α' for luminescent nc-Si crystals are justified. Indeed, these kinetic curves were obtained at a wavelength λ = 760 nm, which corresponds to E_g≈1.6 eV. At this energy gap, the characteristic rate of nonradiative recombination at a defect is more than 10^4 μs^{-1} (Delerue et al., 1993), whereas the exciton migration rate $\zeta = 2D^*$ is no more than 0.04 μs^{-1}. Therefore, any products of type DN are substantially lower than α' even if the total number of nc-Si crystals is larger than the number of initially excited nanocrystals ($N > N_1^{(0)}$).

3.8 Effect of the degree of oxidation on the photosensitization activity of porous silicon

To study the efficiency of $^1O_2$ generation on the nc-Si surface, it is important to know its dependence on the photostimulated oxidation of PS samples during the photosensitization of molecular oxygen. Figure 16a shows the EPR spectra of coarse-grained PS1 powders that were recorded in an oxygen atmosphere as a function of the exposure time during irradiation by the light of a galogen lamp. The spectrum amplitude (and P_b-center concentration, respectively) is seen to monotonically increase with exposure time t_{exp} (Fig.16b). For PS powders to be applied in photodynamic therapeutic methods, it is

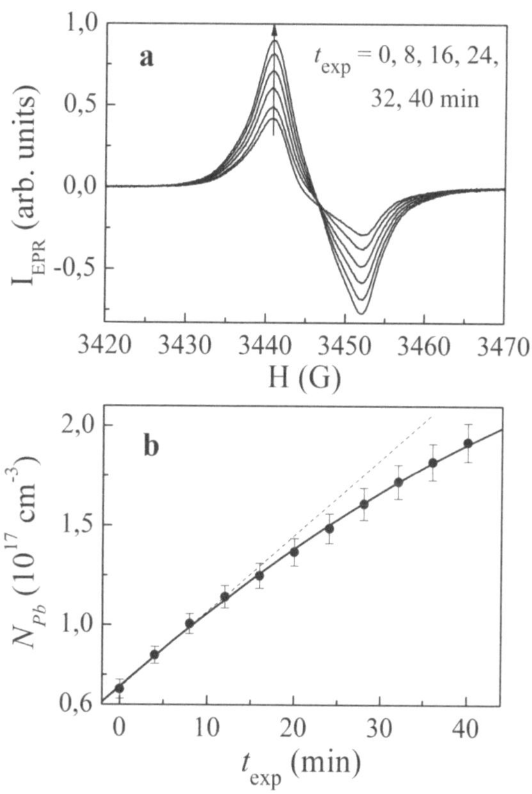

Fig. 16. (a) EPR spectra of PS1 powders and (b) the corresponding concentrations of P_b-centers as a function of exposure time t_{exp} in an O_2 atmosphere. (b): (solid line) for clarity and (dashed line) linear dependence at P_{mw}=2.0 mW.

important to study the effect of the number of defects on the efficiency of singlet oxygen generation. Figure 17a shows the PL spectra of CG PS1 powders recorded in vacuum at various times t_{exp} of preliminary holding under the light in an oxygen atmosphere. The PL intensity decreases monotonically according to an increase in the contribution of nonradiative recombination, and the spectral maximum shifts toward short wavelengths due to an increase in the quantum confinement of excitons in Si nanostructures upon an increase in the surface SiO_2 oxide thickness (Bisi et al., 2000).

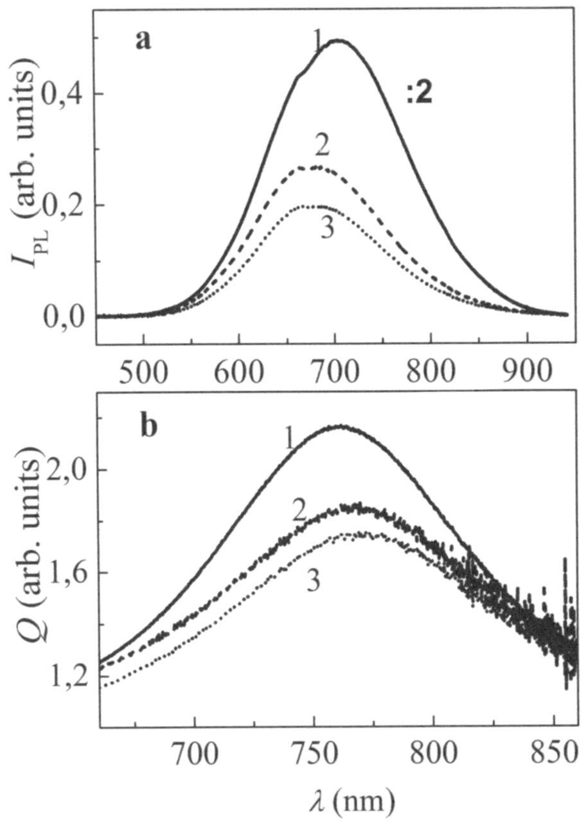

Fig. 17. (a) PS1 PL spectra and (b) quenching function $Q(\lambda)$ during illumination in an O_2 atmosphere at t_{exp}= (1) 0, (2) 6, and (3) 12 min.

Using the dependence of the concentration of P_b-centers in an ensemble of nc-Si crystals on the exposure time as a normalized curve (Fig. 16b), we can plot the dependence of the PL intensity at a certain wavelength on the defect concentration (Fig. 18a). The curve in Fig. 18a is seen to agree quantitatively with the calculated curves in Fig. 10a, which indicates that the initial principles of the developed exciton migration theory are consistent.

When the number of defects increases, the $^1O_2$ generation efficiency degrades, which is indicated by the monotonic decrease in the quenching function amplitude $Q(\lambda) = I_{PLvac}/I_{PLair}$ air with increasing time t_{exp} (Fig.17b). The quenching function is the ratio of the PS PL spectrum amplitudes in vacuum and an oxygen-containing medium and characterizes the intensity of exciton recombination followed by energy transfer to oxygen molecules (Gross et al., 2003). Note that the quantity

$$1 - Q^{-1} = \frac{I_{vac} - I_{air}}{I_{vac}} \qquad (36)$$

approximately determines the fraction of oxygen molecules in the singlet state on the assumption that the entire energy of PL quenching of PS samples is consumed for the excitation of molecular oxygen.

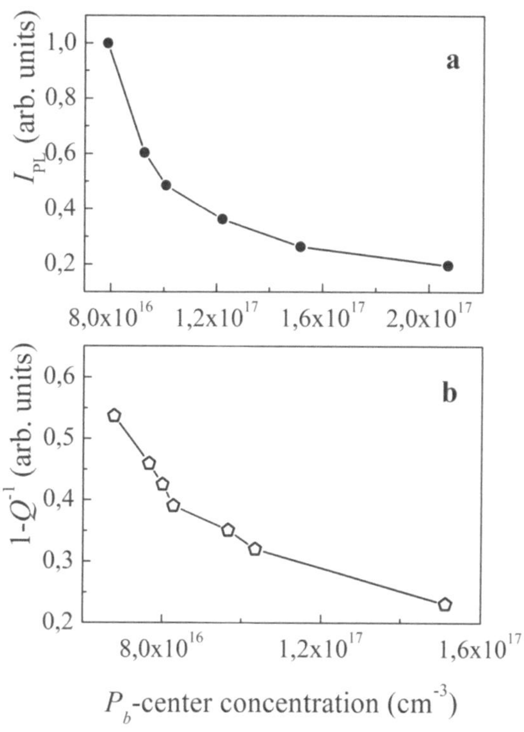

Fig. 18. Effect of the defect concentration on the PL intensity (λ=740 nm) and the fraction of photosensitized O_2 (λ=760 nm). The dependences were obtained upon processing the data some of which are shown in Fig. 17.

Figure 18b shows quantity (36) as a function of the concentration of the defects that form during photostimulated oxidation at a wavelength λ=760 nm. The photosensitized activity of PS layers is seen to monotonically decrease with increasing number of P_b-centers, which points to the suppression of the luminescent and, hence, photosensitized properties of nanocrystals with point defects. As in the case of the PL intensity, the calculated curves shown in Fig. 13 agree qualitatively with the experimental curve in Fig. 18b. Thus, the photostimulated oxidation of PS samples leads to a monotonic decrease in the $^1O_2$ generation efficiency, which is important for practical application. After photosensitization, ensembles of nc-Si crystals transform into harmless amorphous SiO_2, which does not exert a phototoxic effect on biological tissues.

4. Conclusions

The generation of singlet oxygen was investigated and its concentration upon photoexcitation of silicon nanocrystals in microporous silicon layers was determined using EPR spectroscopy. It was demonstrated that a considerable part (approximately 40%) of the molecules of triplet oxygen can transform into the singlet state in the course of photosensitization of oxygen molecules in ensembles of silicon nanocrystals. The longitudinal (T_1) and transverse (T_2) relaxation times of spin centers increase under illumination of the samples in an oxygen atmosphere owing to the decrease in the efficiency of the dipole-dipole interaction between silicon dangling bonds and molecules of triplet oxygen as a result of the transformation of a number of the latter molecules into an excited (diamagnetic) state. It was shown that the amplitude of the EPR spectrum of triplet oxygen molecules in pores of silicon decreases upon photoexcitation of the samples, which suggests the formation of singlet oxygen molecules.

Taking into account exciton migration, we developed a quantitative model for the relaxation processes that occur in ensembles of silicon nanocrystals. With this model, we theoretically described the photosensitization of singlet oxygen on the surface of silicon nanocrystals under various external conditions. We experimentally studied the effect of the granule size of a porous silicon powder on its luminescence properties. Upon mechanical fine milling of as-deposited PS films, both the PL intensity and the concentration of point defects increase. The effect of the photostimulated oxidation of PS layers on the photosensitization of molecular oxygen on the surface of the nanocrystals forming the PS was investigated. The photosensitized activity of silicon nanocrystals was found to degrade monotonically during their oxidation. With the developed model, we explained the experimental data by restricted energy transfer between coupled silicon nanocrystals because of breaks in exciton migration trajectories when the granule size of a PS powder decreases and by an increase in the efficiency of nonradiative recombination during the photostimulated oxidation of the nanocrystals, which leads to the suppression of singlet oxygen generation. Moreover, using this model, we can approximate the kinetic curves of PL decay at a high accuracy, determining the actual exciton radiative recombination time in the nanocrystals. Our data indicate that PS is promising for the methods of nontoxic photodynamic therapy of a number of diseases, including oncological ones.

5. Acknowledgment

This work was performed on the equipment of the Center of User's Facilities of the M.V. Lomonosov Moscow State University. We are grateful to prof. V.Yu. Timoshenko and K. Lips for helpful discussions.

6. References

Bisi, O.; Ossicini, S. & Pavesi, L. (2000). *Porous Silicon: a Quantum Sponge Structure for Silicon Based Optoelectronics. Surface Science Reports*, Vol.38, No.1-3, (April 2000), p.p. 1-126, ISSN 0167-5729.

Cantin, J.; Schoisswohl, M.; Bardeleben, H.; Hadj, N. & Vergnat, M. (1995). *Electron-Paramagnetic-Resonance Study of the Microscopic Structure of the Si (100)-SiO2 Interface.*

Physical Review B: Condensed Matter, Vol.52, No.16, (April 1995), p.p. R11599-R11602, ISSN 1098-0121.

Carrington, A. & McLachlan, A. (1979). *Introduction to Magnetic Resonance with Applications to Chemistry and Chemical Physics*, Chapman and Hall, ISBN 047026572, New York, United States of America.

Chen, R. (2003). *Apparent Stretched-Exponential Luminescence Decay in Crystalline Solids.* Journal of Luminescence, Vol.102–103, (May 2003), p.p. 510-518, ISSN 0022-2313.

Chen, X.; Henderson, B. & O'Donnell, K. (1992). *Luminescence Decay in Disordered Low-Dimensional Semiconductors.* Applied Physics Letters, Vol. 60, No.21, (March 1992), p.p. 2672-2674, ISSN 0003-6951.

Cullis, A. & Canham, L. (1991). *Visible Light Emission due to Quantum Size Effects in Highly Porous Crystalline Silicon.* Nature, Vol.353, No. 6342, (August 1991), p.p. 335-338, ISSN 0028-0836.

Delerue, C.; Allan, G. & Lannoo, M. (1993). *Theoretical Aspects of the Luminescence of Porous Silicon.* Physical Review B: Condensed Matter, Vol.48, No.15, (October 1993), p.p. 11 024-11036, ISSN 1098-0121.

Dexter, D. (1953). *A Theory of Sensitized Luminescence in Solids. Journal of Chemical Physics*, Vol.21, (September 1953), p.p. 836-850, ISSN 0021-9606.

Germanenko, I.; Li, S. & El-Shall, M. (2001). *Decay Dynamics and Quenching of Photoluminescence from Silicon Nanocrystals by Aromatic Nitro Compounds.* Journal of Physical Chemistry B, Vol.105, No.1, (January 2001), p.p. 59-66, ISSN 1520-6106.

Gross, E.; Kovalev, D.; Künzner, N.; Diener, J.; Koch, F.; Timoshenko, V. & Fujii, M. (2003). *Spectrally Resolved Electronic Energy Transfer from Silicon Nanocrystals to Molecular Oxygen Mediated by Direct Electron Exchange.* Physical Review B: Condensed Matter, Vol.68, No.11, (September 2003), p.p. 115405-1-115405-11, ISSN 1098-0121.

Hahn, E. (1950). *Spin Echoes.* Physical Review, Vol.80, No.4, (November 1950), p.p. 580-594, ISSN 1098-0121.

Halliwell, B. & Gutteridge, J. (1999). *Free Radicals in Biology and Medicine*, 3th ed., Oxford University Press, ISBN 0-19-850044-0, Oxford, United Kingdom.

Kanemitsu, (1996). *Photoluminescence Spectrum and Dynamics in Oxidized Silicon Nanocrystals: Nanoscopic Disorder System.* Physical Review B: Condensed Matter, Vol.53, No.11, (September 1996), p.p. 13513-13520, ISSN 1098-0121.

Kovalev, D.; Gross, E.; Künzner, N.; Koch, F.; Timoshenko, V. & Fujii, M. (2002). *Resonant Electronic Energy Transfer from Excitons Confined in Silicon Nanocrystals to Oxygen Molecules.* Phys. Rev. Lett. Vol.89, No.13, (September 2002), p.p. 137401-1-137401-4, ISSN 0031-9007.

Kumar, V.; Tripathi, M.; Kumar, M. & Shukla, G. (2009). *Photosensitization Studies in Selected Dyestuffs.* E-Journal of Chemistry, Vol. 6, No.3, (Juli, 1999), p.p. 659-664, ISSN 0973-4945.

Laiho, R.; Vlasenko, L.; Afanasiev M. & Vlasenko, M. (1994) *Electron Paramagnetic Resonance in Heat-Treated Porous Silicon.* Journal of Applied Physics, Vol.76, No.7, (June 1994), p.p. 4290-4293, ISSN 0021-8979.

Lebib, S.; Bardeleben, H.; Cernogora, J.; Fave, J. & Roussel, J. (1999). *Time-Resolved Photoluminescence Study of the Red Emission in Nanoporous SiGe Alloys.* Journal of Luminescence, Vol.80, No.1-4, (March 1999), p.p. 153-157, ISSN 0022-2313.

Lepine, D. (1972). *Spin-Dependent Recombination on Silicon Surface.* Physical Review B: Solid State Vol.6, No.2, (July 1972), p.p. 436-441, ISSN 1098-0121.

Linnros, J.; Lalic, N.; Galeckas, A. & Grivickas, V. (1999). *Analysis of the Stretched Exponential Photoluminescence Decay from Nanometer-Sized Silicon Crystals in SiO_2.* Journal of Applied Physics, Vol.86, No.11, (August 1999), p.p. 6128-6134, ISSN 0021-8979.

Maly, F.; Trojanek, P.; Kudma, J.; Hospodkova, A.; Banas, S.; Kohlova, V.; Valenta, J.; & Pelant, I. (1996). *Picosecond and millisecond dynamics of photoexcited carriers in porous silicon.* Physical Review B: Condensed Matter, Vol.54, No.11, (September 1996), 7929-7936, ISSN 1098-0121.

Mihalcescu, I.; Vial, J. & Romestain, R. (1996). *Carrier Localization in Porous Silicon Investigated by Time-Resolved Luminescence Analysis.* Journal of Applied Physics, Vol.80, No.4, (May 1996), p.p. 2404-2411, ISSN 0021-8979.

Moretti, L.; De Stefano, L. & Rendina, I. (2007). *Quantitative Analysis of Capillary Condensation in Fractal-Like Porous Silicon Nanostructures.* Journal of Applied Physics, Vol.101, No.2, (January 2007), p.p. 024 309-1-024 309-5, ISSN 0021-8979.

Nychyporuk, T.; Lysenko, V. & Barbier, D. (2005). *Fractal Nature of Porous Silicon Nanocrystallites.* Physical Review B: Condensed Matter, Vol.71, No.11, (March 2005), p.p. 115 402-1-115 402-5, ISSN 1098-0121.

Pavesi, L. & Ceschini, M. (1993). *Stretched-Exponential Decay of the Luminescence in Porous Silicon.* Physical Review B: Condensed Matter, Vol.48, No.23, (December 1993), p.p. 17625-17628, ISSN 1098-0121.

Poole, C.P. & Horacio, F. (1987). *Theory of Magnetic Resonance, 2nd Ed.*, Wiley, ISBN: 978-0-471-81530-3C, New York, United States of America.

Pophristic, M.; Lang, F.; Tran, C.; Ferguson, I. & Karlicek, R. (1998). *Time-Resolved Photoluminescence Measurements of InGaN Light-Emitting Diodes.* Applied Physics Letters, Vol.73, No.24, (October 1998), p.p. 3550-3552, ISSN 0003-6951.

Proot, J.; Delerue, C. & Allan, G. (1992). *Electronic Structure and Optical Properties of Silicon Crystallites: Application to Porous Silicon.* Applied Physics Letters, Vol.61, No.16 , (August 1992), p.p. 1948-1950, ISSN 0003-6951.

Roodenko, K.; Goldthorpe, I.; McIntyre, P. & Chabal, Y. (2010). *Modified Phonon Confinement Model for Raman Spectroscopy of Nanostructured Materials.* Physical Review B: Condensed Matter, Vol.82, No.11, (September 2010), p.p. 115210-1-115210-11, ISSN 1098-0121.

Smith, R. & Collins, S. (1992). *Porous Silicon Formation Mechanisms.* Journal of Applied Physics, Vol. 71, No.8, (January 1992), p.p. R1-R22, ISSN 0021-8979

Vahtras, O.; Loboda, O.; Minaev, B.; Ågren H. & Ruud K. (2002). *Ab Initio Calculations of Zero-Field Splitting Parameters.* Chemical Physics, Vol.279, No.2-3, (June 2002), p.p. 133-142, ISSN 0301-0104.

Vinciguerra, V.; Franzo, G.; Priolo, F.; Iacona, F. & Spinella, C. (2000). *Quantum Confinement and Recombination Dynamics in Silicon Nanocrystals Embedded in Si/SiO_2 Superlattices.* Journal of Applied Physics, Vol.87, No.11, (February 2000), p.p. 8165-8173, ISSN 0021-8979.

13

Magnetic Nanostructures for Biomedical Applications: An Iron Nitride Crystal/Cationic Lipid Nanocomposite for Enhanced Magnetically Guided RNA Interference in Cancer Cells

Yoshihisa Namiki[1], Satoshi Matsunuma[2], Tetsutaro Inoue[2], Shigeo Koido[3], Akihito Tsubota[1], Yoji Kuse[4] and Norio Tada[1]

[1]*Institute of Clinical Medicine and Research, The Jikei University School of Medicine 163-1 Kashiwa-shita, Kashiwa, Chiba, 277-8567*

[2]*Division of Research and Development, Hitachi Maxell 1-1-88 Ushitora, Ibaraki, Osaka, 567-8567*

[3]*Divisoin of Gastroenterology, The Jikei University School of Medicine 163-1 Kashiwa-shita, Kashiwa, Chiba, 277-8567*

[4]*Giken Parts CO. LTD. 290 Sugimoto-cho, Tenri, Nara, 632-0078*

Japan

1. Introduction

Cancer is a major cause of death throughout the world, especially in the industrially developed nations. Minimally invasive and effective treatments are thus urgently required. Small interfering RNA (siRNA) is a possible candidate because it selectively decomposes and silences the target genes (Fire et al., 1998) associated with disease progression. However, the obstacle to the development of siRNA as a therapeutic agent results from the absence of an effective system for delivering siRNA to the target region (Li et al., 2006; Toub et al., 2006).

In this context, we previously devised functional nanoparticles composed of iron oxide crystals as the siRNA-delivery system to magnetically augment the accumulation of siRNA at target regions (Namiki et al., 2009). Among the magnetic materials available for biomedical applications, including gene delivery, iron oxide forms, such as magnetite (Fe_3O_4) and maghemite (γ-Fe_2O_3), have been widely investigated (Xie et al., 2009; Lin et al., 2008; Laurent et al., 2008; Gupta et al., 2005) because of their comparatively low cytotoxicity and dependable degree of magnetism.

In the field of information technology, meanwhile, various kinds of magnetic storage devices, such as floppy disks, hard disks and magneto-optical disks have been developed. Among the ferromagnetic materials used in ultra-high density storage devices, iron nitride has several advantages, as follows (Sugita et al., 1994; Takahashi et al., 1999; Hattori et al., 2001). (1) Iron nitride displays a greater degree of magnetization compared with iron oxide, even when the nanocrystal-size is small. (2) Iron nitride is an iron-based material, and iron is cheap and abundant. (3) Iron nitride is a much cheaper ferromagnetic material than other ferromagnetic materials, such as FePt (Sun et al., 2000). (4) Iron nitride is made from non-toxic and decomposable chemical elements such as iron and nitride. The spherical iron nitride nanocrystal is therefore an attractive material, not only as an ultra-high density data storage device, but also as a gene therapy vector in biomedical technology fields.

The absorption force of magnetic nanoparticles to a magnet can be enhanced by the optimization of the magnetic materials. Furthermore, for biomedical applications, iron nitride is expected to be a good magnetic material candidate if it is able to display a similar low degree of cytotoxicity as iron oxide. However, iron nitride nanoparticles for biomedical applications have not been previously reported. We have devised an innovative $Fe_{16}N_2$/cationic lipid nanocomposite and tested it in the magnetically guided silencing of target genes in cancer cells. Through an interdisciplinary fusion of the advances in various scientific fields, we were able to develop a novel gene delivery tool. This tool is comprised of magnetic nanoparticles composed of an iron nitride nanocrystal-core and a cationic lipid-shell with high affinity genes.

In this chapter, we describe the enhanced magnetism and gene silencing effect of a magnetic nanocomposite made from iron nitride instead of the conventionally used iron oxide, and we report the potential of the iron nitride/cationic lipid nanocomposite as magnetically guided gene therapeutic vector which delivers siRNA into cells. Initially, we describe the preparation and purification procedures used for this nanocomposite. We prepared a magnetic fluid in which oleic acid-coated iron nitride nanocrystals were dispersed in chloroform. Through a further development of our previously devised procedure (Namiki et al., 2009), the oleic acid-coated iron nitride was assembled with the cationic lipid via hydrophobic interaction. Subsequently, we report the morphological and physiological properties of this composite, termed LipoFeN. In addition, siRNA designed to knockdown target genes and cationic lipid-coated iron nitride nanocrystals were mixed to form the complexes through electrostatic force. Finally, we show the degree of magnetic attraction, gene silencing effect and cytotoxic effect of this siRNA-loaded nanocomposite.

2. Preparation and morphology of the iron nitride/metallic oxide core shell nanostructure

Initially, magnetite nanocrystals with an approximate 20 nm diameter were synthesized as follows. $FeCl_3$ (0.04 mol; Wako, Tokyo, Japan) and $FeCl_2$ (0.02 mol; Wako) were dissolved in 25 ml of distilled water. When 25 ml of ammonium hydroxide (25%) were vigorously mixed with this solution, magnetite slurry was precipitated. Subsequently, on the surface of the dried magnetite nanocrystals, yttrium and aluminum were deposited in the solution-dispersed magnetite nanocrystals. These dried particles were deoxidized with hydrogen gas and then continuously treated with NH_3 gas. Finally, the iron nitride/metallic oxide core shell nanostructure was obtained (Sasaki et al., 2005). The shell layer, which is composed of yttrium-oxide and aluminum oxide, has the roles of preventing iron nitride oxidization and decomposition, and thus of maintaining the magnetism of the iron nitride. The nanostructure was examined using transmission electron microscopy (TEM), which revealed that this nanocomposite has an average diameter of 26 nm and approximate 3.5 nm shell thickness (Fig. 1).

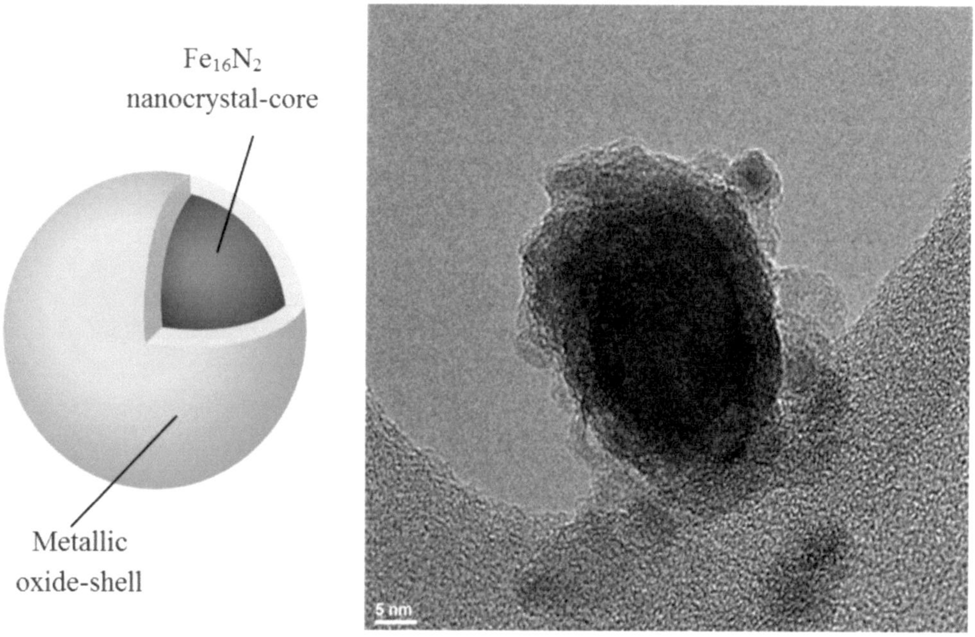

Fig. 1. Schematic representation and transmission electron micrograph of the $Fe_{16}N_2$/metal oxide core shell nanocomposite.

3. Physiological properties of the iron nitride/metallic oxide core shell nanostructure

The crystal structure of the particles was examined with an X-ray diffractometer (XRD). The surface condition was investigated using X-ray photoelectron spectroscopy (XPS). Magnetic measurements were carried out using a vibration sample magnetometer (VSM) by applying the maximum field of 1270 kA/m (16 kOe).

Data was acquired using a coupled Theta: 2-Theta scan on a Rigaku Ultima-III diffractomer equipped with a copper x-ray tube, parafocusing optics, computer-controlled variable slits, and a diffracted beam monochromator. Figure 2 shows the best matches after comparing the background-subtracted raw data with that of the ICDD/ICSD diffraction database. As expected, the primary phase is $Fe_{16}N_2$. There is also a small amount of maghemite. Note that the primary maghemite peak, near 36 degrees on 2-Theta, is somewhat shifted relative to the reference pattern. This indicates that the Fe:O ratio may be somewhat different than 2:3. Note also that no metallic iron was detected. This phase is strongly overlapped with $Fe_{16}N_2$, so trace amounts of Fe in $Fe_{16}N_2$ might not be detected.

XPS data is quantified using relative sensitivity factors and a model that assumes a homogeneous layer. The analysis volume is the product of the analysis area (spot size or aperture size) and the depth of information. Photoelectrons are generated within the X-ray penetration depth (typically some number of microns), but only the photoelectrons within the top three photoelectron escape depths are detected. Escape depths are on the order of 15-35 Å, which leads to an analysis depth of ~50-100 Å. Typically, 95% of the signal originates from within this depth.

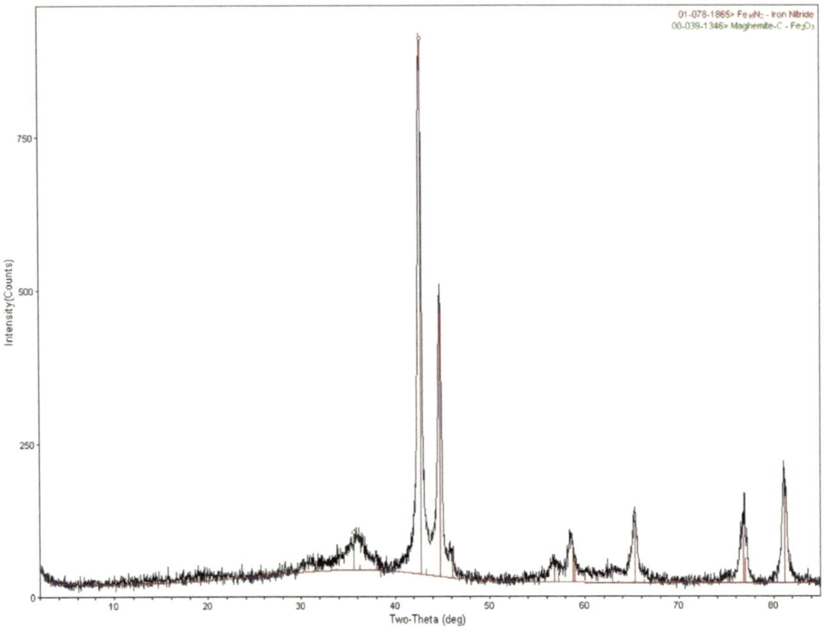

Fig. 2. X-ray diffraction peaks originating in the $Fe_{16}N_2$/metal oxide core shell nanocomposite were clearly observed.

Instrument	PHI Quantum 2000
X-ray source	Monochromated Alka 1486.6eV
Acceptance Angle	±23°
Take-off angle	45°
Analysis area	1400mm x 300mm
Charge Correction	C1s 284.8 eV

Table 1. Analytical parameters of XPS.

Table 2 provides the atomic concentrations of the elements detected. The values given are normalized to 100% using the elements detected. Detection limits are approximately 0.05 to 1.0 atomic %. The major factors affecting the detection limit are the element itself, interference and background. Chemical state assignments for a given element were made by consulting the reference data from the literature. Chemical state assignments must be considered as tentative in cases where the available reference data is limited or in cases where similar binding energies are observed for a number of different chemical states. Non-linear least squares (NLLS) curve fitting has been applied to selected high resolution spectra to assist in possible chemical state assignment. The results of the NLLS fits are shown on the individual spectra. It should be noted that the curve fit routines are mathematical devices, and for a given peak a number of mathematically valid results may exist.

1. The particle surfaces are predominantly composed of aluminum, iron, oxygen and carbon, with low levels of yttrium and trace levels of fluoride, magnesium and chlorine (Table 2).
2. Aluminum is found within the cap of the binding energies consistent with Al_2O_3 (Fig. 3b). While various reports on YxAlyOx have comparable aluminum binding energies, the yttrium data below suggests that such a species is not present in these samples.

3. Yttrium is found at a binding energy (157.8eV, Fig. 3c) most consistent with that of $Y(OH)_3$ (157.7eV). Its binding energy is higher than that for the Y_2O_3 reference (~156.3eV). The yttrium binding energy for an $Y_3Al_5O_{12}$ reference run was found to be 157.1eV. Thus, the yttrium in this sample is most likely present in the form $Y(OH)_3$.
4. Iron is found predominantly as iron oxide, likely Fe_2O_3, based in part on the XRD data. Additionally, low levels of Fe nitride are present (Fig. 3b, Table 2). It is noted that XRD found the opposite (high levels of iron nitride with trace levels of Fe_2O_3). This discrepancy is due to sampling depth differences (i.e. XRD samples the entire particle, whereas XPS samples only the top 60-80Å).
5. Trace levels of fluorine and chlorine are found as inorganic fluorides and chlorides, respectively. These may be associated with iron, magnesium, yttrium or aluminum. The levels are insufficient to allow confirmation using the respective spectra.
6. Carbon is found predominantly as a hydrocarbon (C-C, C-H), with lower levels of oxigen-containing species, potentially including CO_3 (data not shown). The low carbon levels observed coupled with its chemistry indicates that much, possibly all, of the carbon is present as adventitious carbon, which had adsorbed onto sample surfaces from the atmosphere.
7. Magnesium does not exhibit any significant shift in its binding energies in different chemical states.
8. Oxygen is present as a complex mixture of metal oxides/hydroxides and organic oxygen.
9. Note that the presence of low levels of nitrogen cannot be confirmed in this system due to interference from a secondary yttrium signal (data not shown). However, based on the low levels of iron nitride present (from the iron data) and the low nitrogen/iron ratio (based on XRD having determined the iron nitride to be $Fe_{16}N_2$), nitrogen would only be present within the XPS information depth at concentrations of 0.07%, i.e. below the XPS detection limit for this element.

The particle surfaces are composed predominantly of Al_2O_3, with lower levels of $Y(OH)_3$ on Fe_2O_3. Only low levels of iron nitride ($Fe_{16}N_2$ based on the XRD data) are observed, indicating that XPS does not reach deep into the particle core (Table 3). Additionally, trace levels of inorganic chlorides, fluorides and magnesium are present.

C	N	O	F	Mg	Al	Cl	Fe	Y
9.9	[b]?	62.1	0.7	0.6	16.0	0.2	8.3	2.3

Table 2. Atomic concentrations (in %)[a]

[a] Normalized to 100% of the elements detected. XPS does not detect H or He.
[b] The presence of N could not be confirmed due to the interference from a secondary Y signal.

FeN	Fe_2O_3
7	93

Table 3. Iron chemical state (in % of total Fe[a])

[a] The values in this table are percentages of the total atomic concentrations of the corresponding element shown in Table 2.

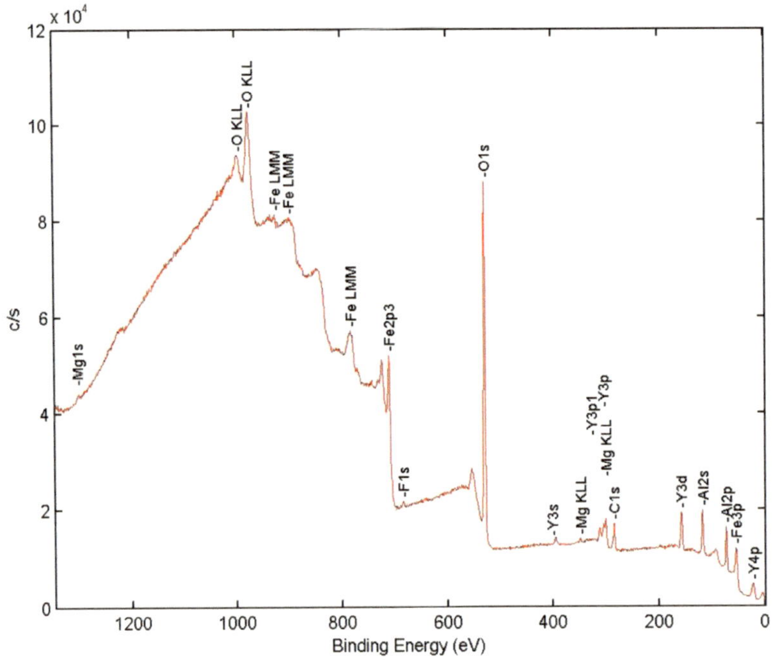
x 10^4
c/s
Binding Energy (eV)
-Mg1s
-O KLL
-O KLL
-Fe LMM
-Fe LMM
-Fe LMM
-Fe2p3
-F1s
-O1s
-Y3s
-Mg KLL
-Y3p1
-Mg KLL
-Y3p
-C1s
-Y3d
-Al2s
-Al2p
-Fe3p
-Y4p
(a)

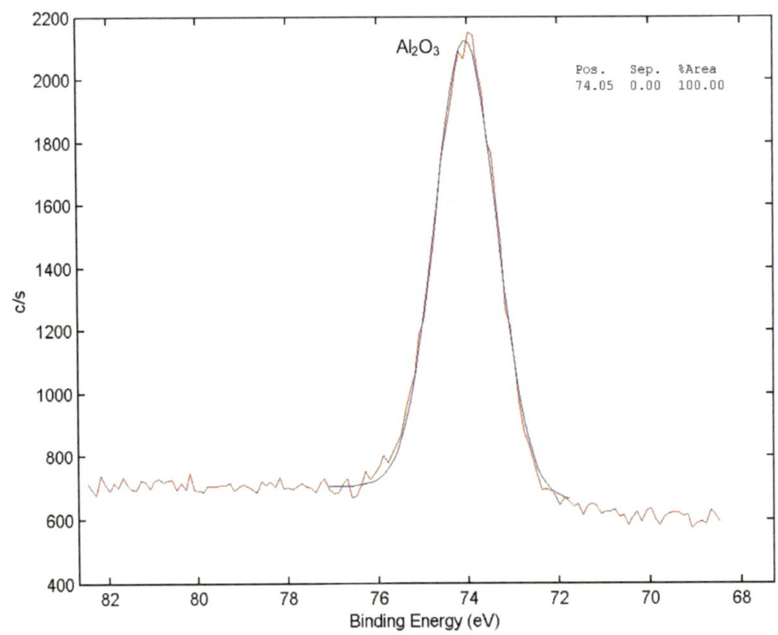
Al_2O_3
Pos. Sep. %Area
74.05 0.00 100.00
c/s
Binding Energy (eV)
(b)

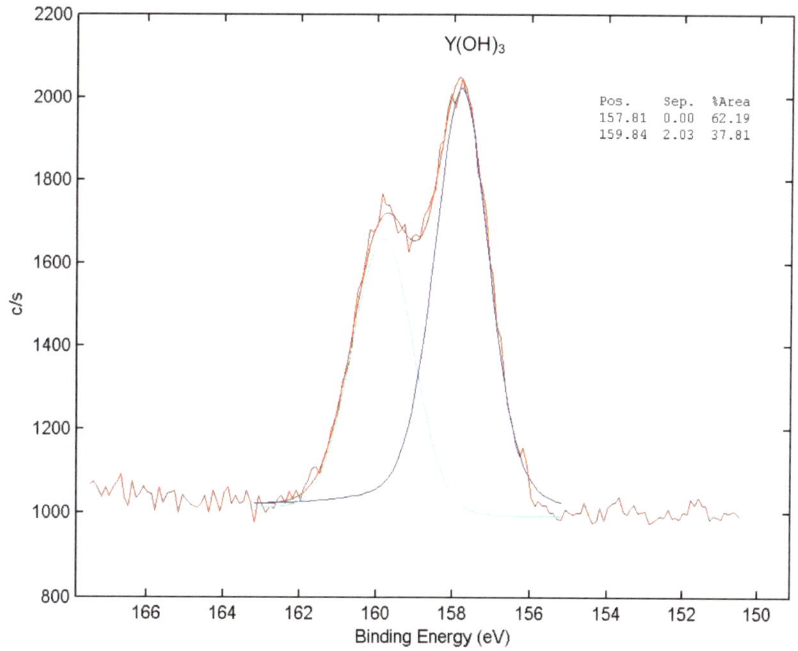

(c)

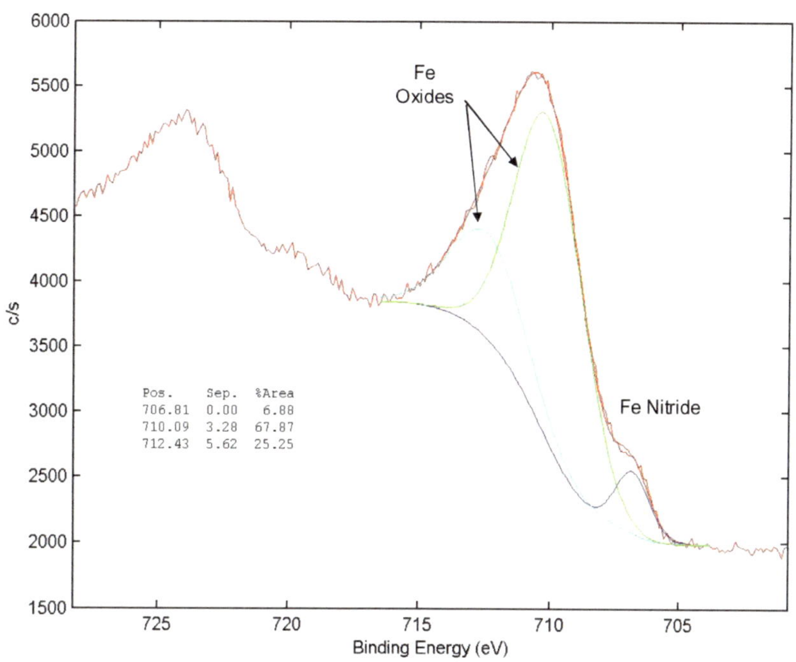

(d)

Fig. 3. The X-ray photoelectron spectroscopy profiles indicated that the metalic oxide shell was composed of aluminum (Al)-, yttrium (Y)- and iron (Fe)- oxides.

In terms of magnetism, $Fe_{16}N_2$/metalic oxide nanocomposite displayed saturation magnetization (103.1 emu/g) and coercive force (3055 Oe), while Fe_3O_4 displayed 50.8 emg/g and 0 Oe (Fig. 4).

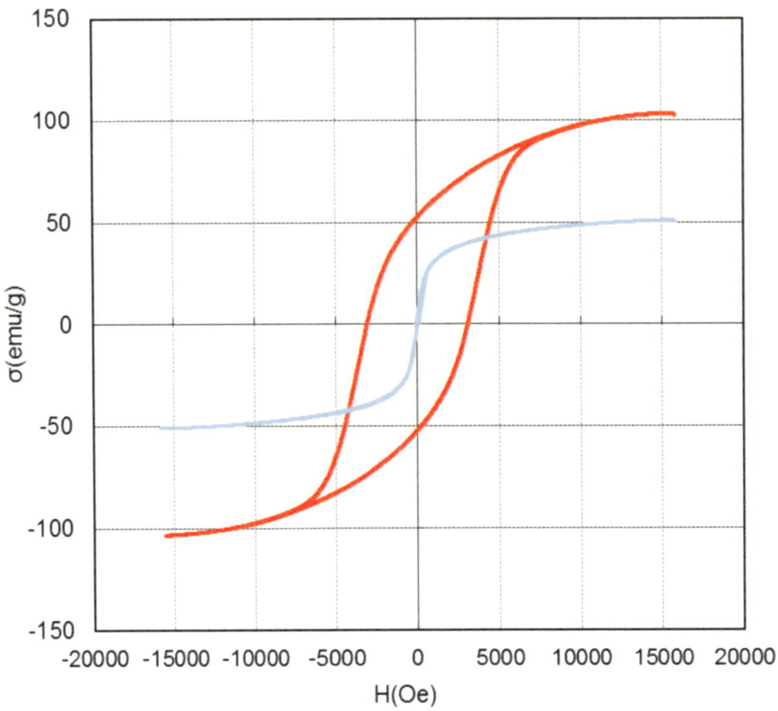

Fig. 4. The magnetization curve of the $Fe_{16}N_2$/metalic oxide nanocomposite and Fe_3O_4. $Fe_{16}N_2$/ metallic oxide, *red*; Fe_3O_4, *blue*.

4. Preparation of iron nitride/metallic oxide/oleic acid-based magnetic fluid

The air-dried iron nitride/metallic oxide core shell nanostructure was well dispersed in ammonium hydroxide (25%). This mixture was then heated to 95 °C and 55 ml of 10% oleic acid (Wako)/dodecan (Wako) were added. In this process, the FeN-nanocomposites were coated with hydrophilic ammonium oleate, and the ammonium oleate began to decompose at 78 °C, with the evaporation of ammonia gas thus changing the FeN-nanocomposites coated with hydrophobic oleic acid upon continuous heating. As a result, a distinct phase separation between the upper organic portion, in which the oleic acid-coated FeN-nanocomposites were stably dispersed, and the lower aqueous portion appeared. Most of the aqueous phase was removed using a pipette and the heating of the residue was continued until the remaining water had been completely evaporated. The oleic acid-coated FeN-nanocomposites were flocculated by the addition of 100 ml of acetone, and a permanent magnet was used to collect this magnetic flocculate while the supernatant liquid was poured off. Moreover, this flocculate was then washed with 80 ml of acetone to eliminate excess oleic acid and centrifuged at 800 G. Acetone was completely removed from the resulting precipitation under reduced pressure (10 mmHg) at room temperature.

Twenty micrograms of these dried oleic acid-coated FeN-nanocomposites were dispersed in 5 ml of chloroform, so a chloroform-based magnetic fluid was finally obtained (Fig. 5, 6).

Fig. 5. Dispersion of the oleic acid-coated $Fe_{16}N_2$/ metalic oxide nanocomposite in chloroform. The magnetic fluid (*right*) was composed of the oleic acid (*green*)-coated iron nitride/ metallic oxide nanocomposite (*dark gray/light gray*) dispersed in chloroform (*yellow*).

5. Preparation of the iron nitride/metallic oxide/oleic acid/cationic lipid nanocomposite

One microgram of a lipid mixture (DC-6-14 (Sogo Pharmacochemical, Tokyo, Japan): DOPE (Wako) = 1: 0.4; molar ratio) was dissolved in 0.5 ml of chloroform. The lipid mixture, along with the chloroform-based magnetic fluid containing 0.2 mg of oleic acid-coated $Fe_{16}N_2$/ metalic oxide nanocomposite, was introduced into a pear-shaped flask. Chloroform

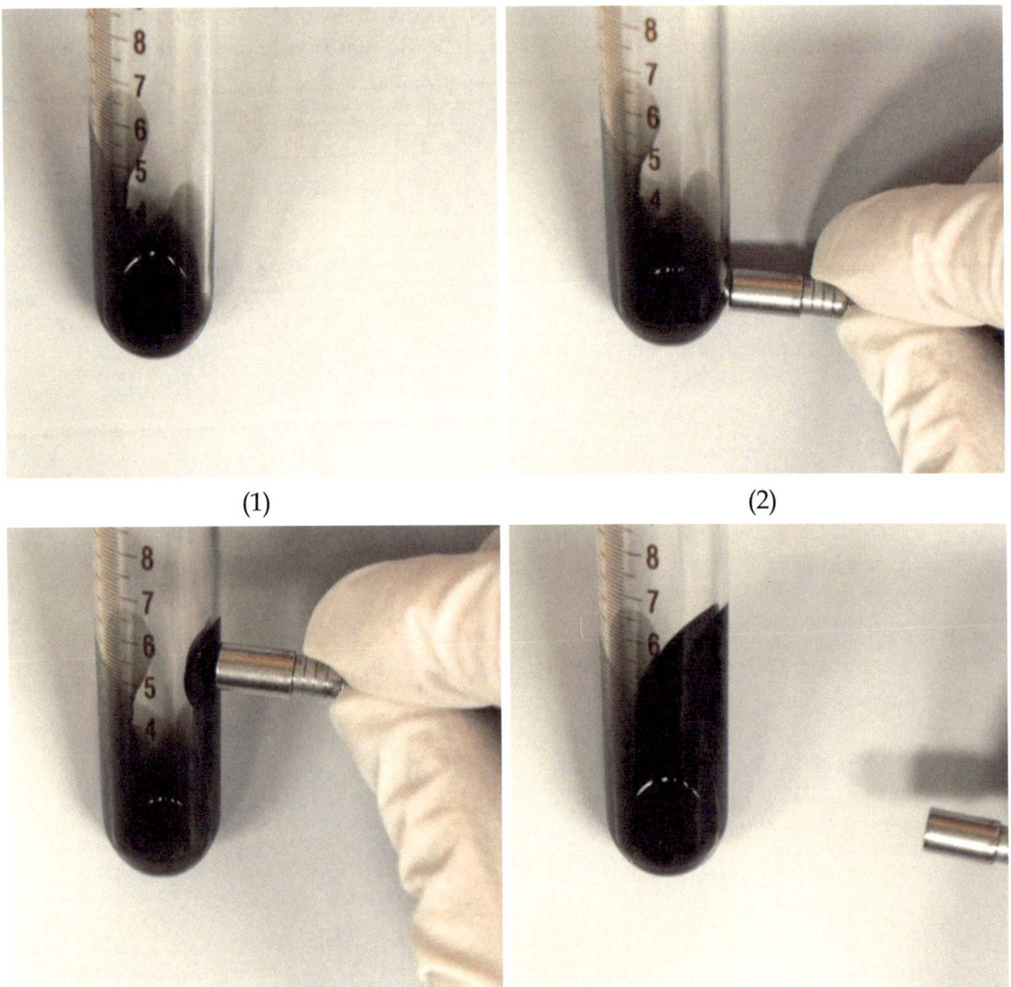

Fig. 6. Magnetic attraction of the chloroform-based magnetic fluid composed of the oleic acid-coated $Fe_{16}N_2$/metalic oxide nanocomposite.

was added to this flask until the total volume reached 1 ml and mixed well. After the addition of 1 ml of distilled water, the mixture was agitated on a vortex mixer for 1 min and sonicated by a cup-horn type sonicator (Sonifier 250D, Branson, CT) with a maximum output for 2 min at 37°C. The flask containing a gray homogeneous suspension was attached to a rotary evaporator and rotated at 2.5 G. The contents were evaporated at 50 mmHg under strong sonication at 37°C. In this process, chloroform was evaporated off to yield a gel and the gel immediately collapsed to give a clear gray-colored suspension. The remaining chloroform was removed from this suspension at 10 mmHg for 10 min (room temperature). This suspension was sonicated with a maximum output for 30 min at 37°C, and finally, iron nitride/metallic oxide/oleic acid/cationic lipid nanocomposite were obtained through hydrophobic interactions (Fig. 7).

6. Purification of the iron nitride/metallic oxide/oleic acid/cationic lipid nanocomposite from free cationic liposome.

The magnetally attractive stainless steel (SUS 430) granule (size 0.4 mm) was pressed and sintered, until finally the sintered magnetic metal filter was obtained (diameter x height: 9.0 x 10 mm). This filter was inserted in the cylinder. Both poles of twin neodymium magnets, which were placed horizontally and oppositely arranged, were magnetically attracted on the column at the region of the stainless filter (Fig. 8). Under a magnetic field, magnetic nanoparticles were captured by the stainless steel filter of the syringe column, and magnetic crystal-free lipid vesicles lacking magnetic nanocrystals were eliminated. In the absence of a magnetic field, magnetitc nanoparticles were eluted from the filter (Fig. 9).

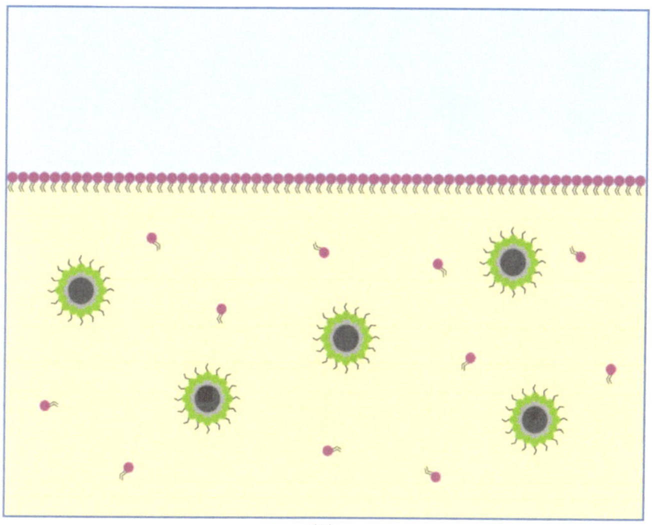

(1)

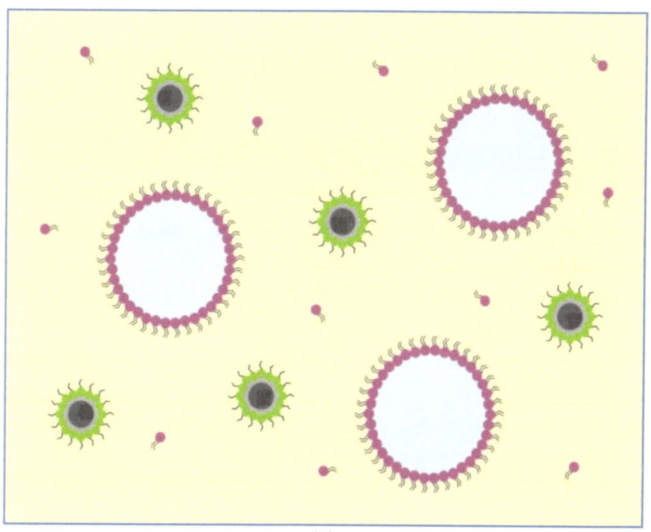

(2)

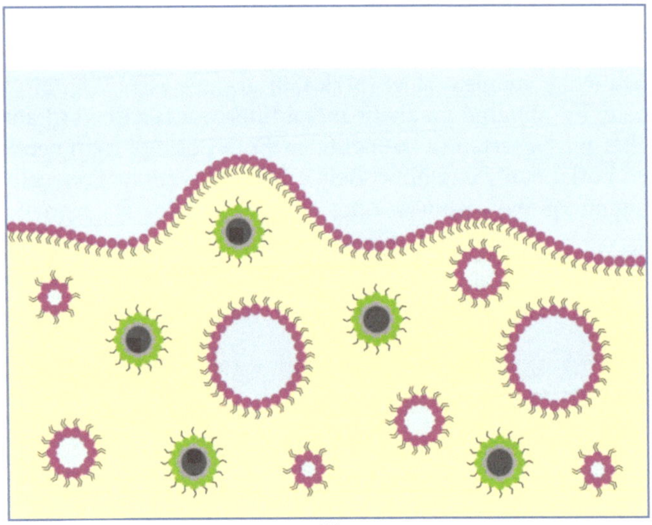

(3)

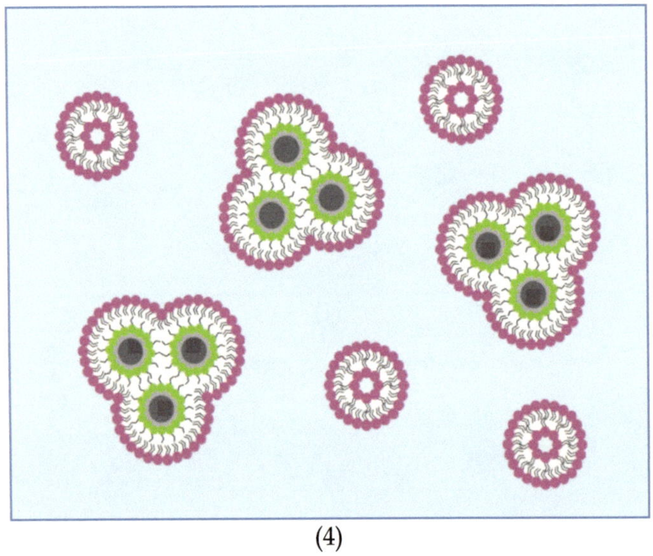

(4)

Fig. 7. Schematic representation of the preparation of the cationic lipid-coated ferromagnetic nanocomposite.
1) The magnetic fluid was composed of the oleic acid (*green*)-coated iron nitride/metallic oxide nanocomposite (*dark gray/light gray*) dispersed in chloroform (*yellow*). The hydrophobic chloroform-based magnetic fluid and phospholipids (*pink*) dissolved in chloroform were mixed. When this mixture and distilled water (*blue*) were poured in a pear-shaped flask, an upper layer of aqueous phase and a lower layer of organic phase containing oleic acid-coated magnetic nanocrystals appeared.
2) After vigorous agitation and sonication, micro droplets were stabilized by a phospholipid layer, and "inverted micelles" formed.
3) Removal of chloroform by strong vacuuming leads to a transformation of these inverted

micelles into a viscous gel state.
4) After the induction of gel-collapse through the continuous vacuum, the hydrophobic groups of amphiphilic phospholipids and oleic acid joined together to form the hydrophilic, cationic lipid-coated ferromagnetic nanocomposite.

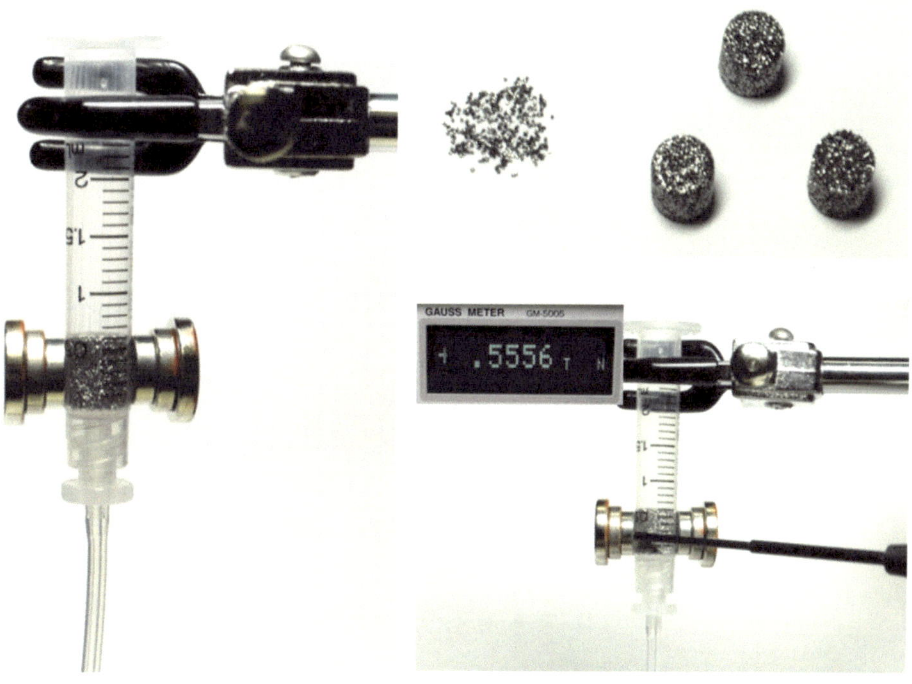

Fig. 8. Photograph of the stainless filter used for the purification of the cationic lipid-coated iron nitride/metallic oxide nanocomposite.

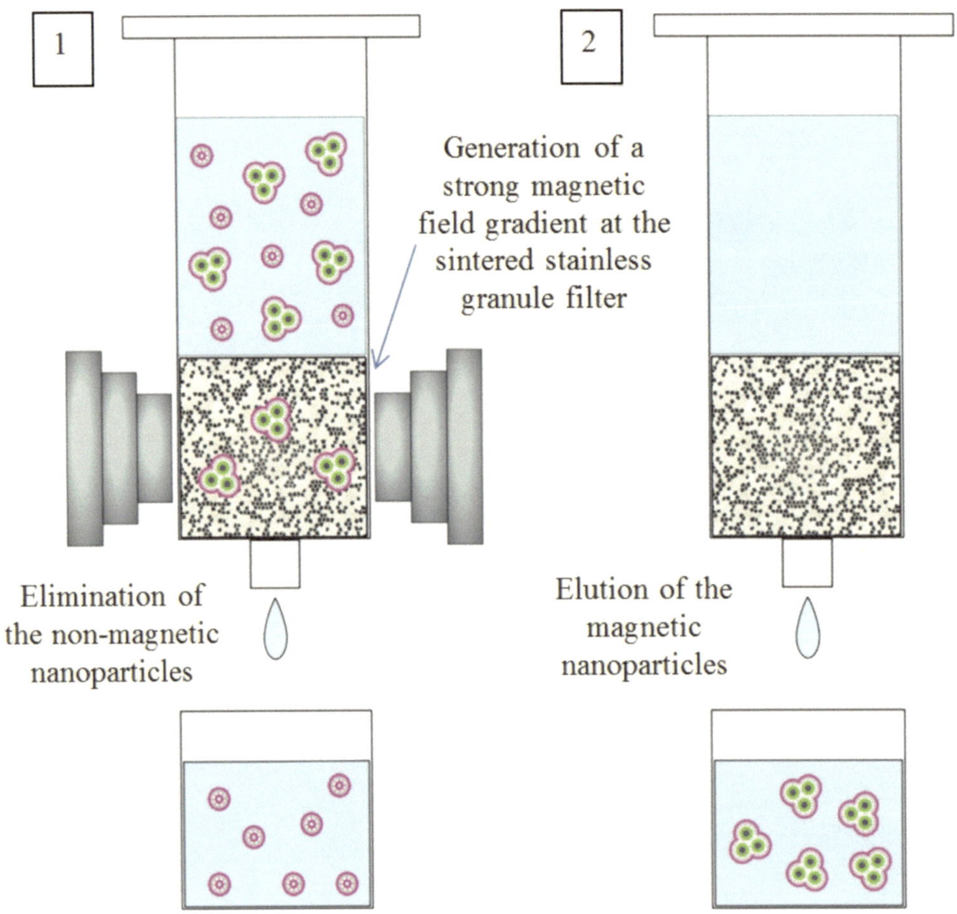

Fig. 9. Schematic representation of the stainless filter used for the purification of the cationic lipid-coated iron nitride/metallic oxide nanocomposite.

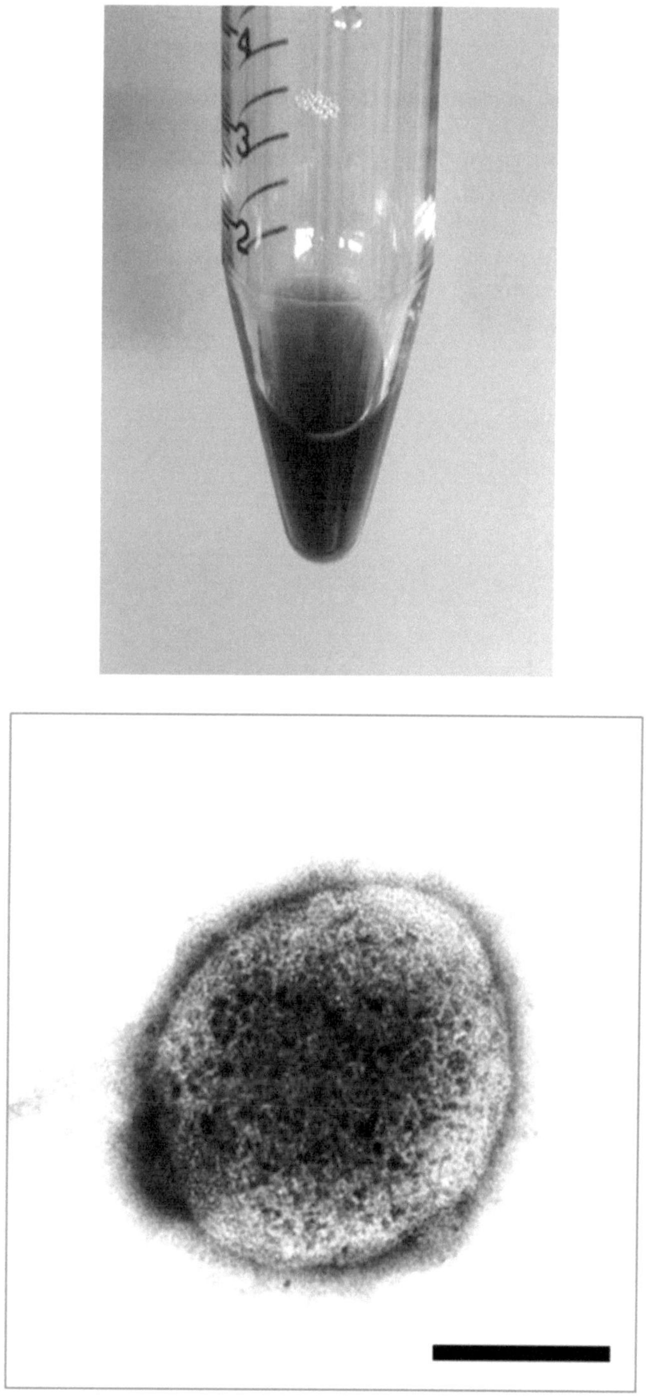

Fig. 10. Photograph and transmission electron micrograph (negative staining: Woodcock et al., 1991) of the cationic lipid-coated iron nitride/metallic oxide, bar: 200 nm.

7. Magnetic attraction of the iron nitride/metallic oxide/oleic acid/cationic lipid nanocomposite

The speed of magnetic attraction of the FeN-based nanocomposite and iron oxide based nanoparticles was compared (0 - 7.5 min). One hundred microliters of the aqueous solution containing 100 μg of each nanoparticle was placed in each of the wells of a 96 well microplate and a magnetic field (200 mT) was irradiated under the well (Fig. 11). The magnetic attraction of these nanoparticles was recorded using a digital camera (Fig. 12). We confirmed that LipoFeN (iron nitride crystal-lipid nanostructures) were magnetically guided 1.5 to 2 times more quickly than PolyMag™ (iron oxide crystal-polymer nanostructures: OZ Biosciences, Marseille, France) (Fig. 13).

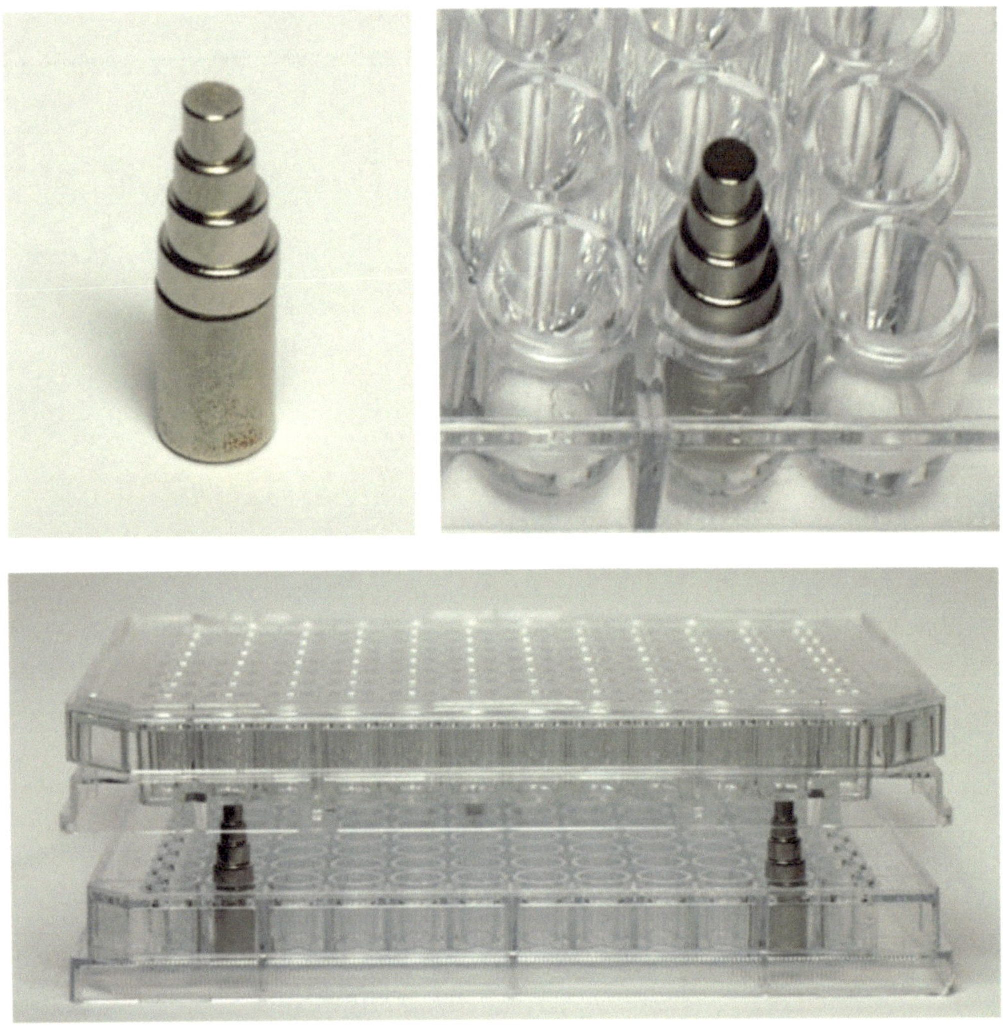

Fig. 11. The magnetic guidance of the magnetic nanoparticles using 96 well microplates.

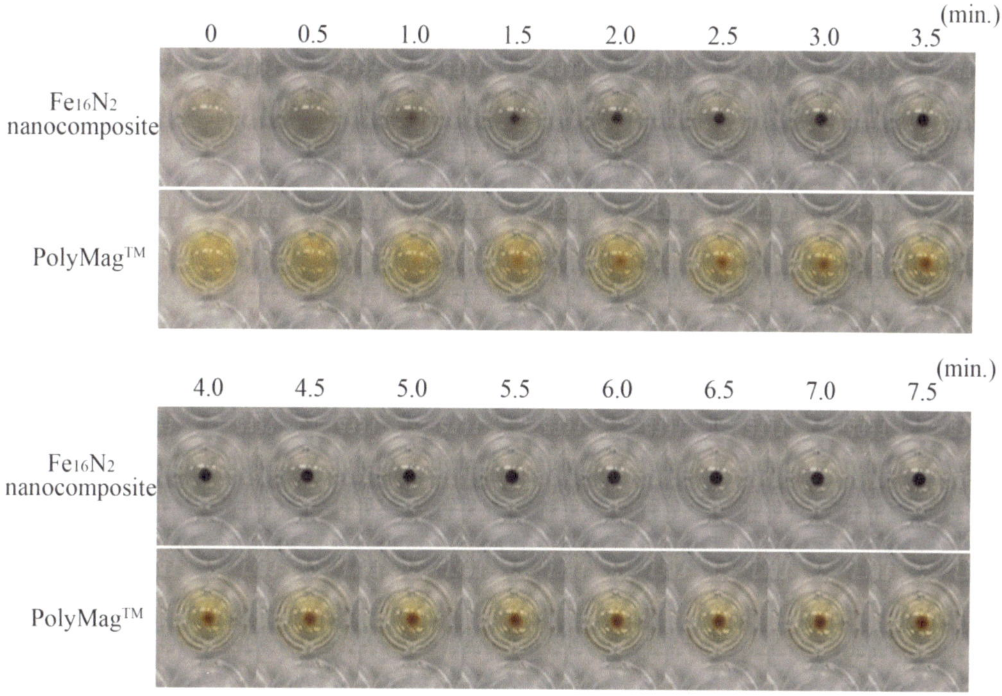

Fig. 12. The speed of the magnetic attraction of the cationic lipid-coated iron nitride/metallic oxide nanocomposite compared with iron oxide-based nanoparticles.

8. The gene silencing in cells induced by the magnetically guided siRNA delivered by the iron nitride/metallic oxide/oleic acid/cationic lipid nanocomposite

Under a magnetic field, the gene silencing effect of of LipoFeN, LipoMag (iron oxide crystal-lipid nanostructures) and PolyMag™ loaded with the luciferase-targeting siRNA were compared in a luciferase-expressing cancer cell line (Fig. 14). A human colon cancer cell line stably expressing luciferase (HT-29-luc2: Caliper, Alameda, CA) was cultured in McCoy's 5A (Invitrogen, Carlsbad, CA) complete medium containing 10% heat-inactivated fetal bovine serum (Sigma). Cells were incubated at 37 °C in a humidified atmosphere of 5% CO_2.

Cells (1 x 10^4/well) were cultured in 96 well plates for 12 hr. Either 25 μl of $siRNA^{Luc}$ (Table 4) or 25 μl of $siRNA^{NC}$ was mixed with 25 μl of each type of magnetic nanoparticles for 10 min at room temperature. Cell culture medium was removed from the 96 well plate and cells were washed twice with PBS. Cells were incubated with the obtained 50 μl of siRNA-magnetic nanoparticle complex at 37 °C using a magnetic plate (Fig. 15). At 5 min after the initiation of the incubation, these complexes were removed, cells were washed twice with PBS and culture medium was dispensed to each well. Finally, luciferase activity was quantified at 48 hr after the incubation with these complexes. Luciferase activity was quantified as described previously (Takahashi et al., 1997). Briefly, cells were homogenized

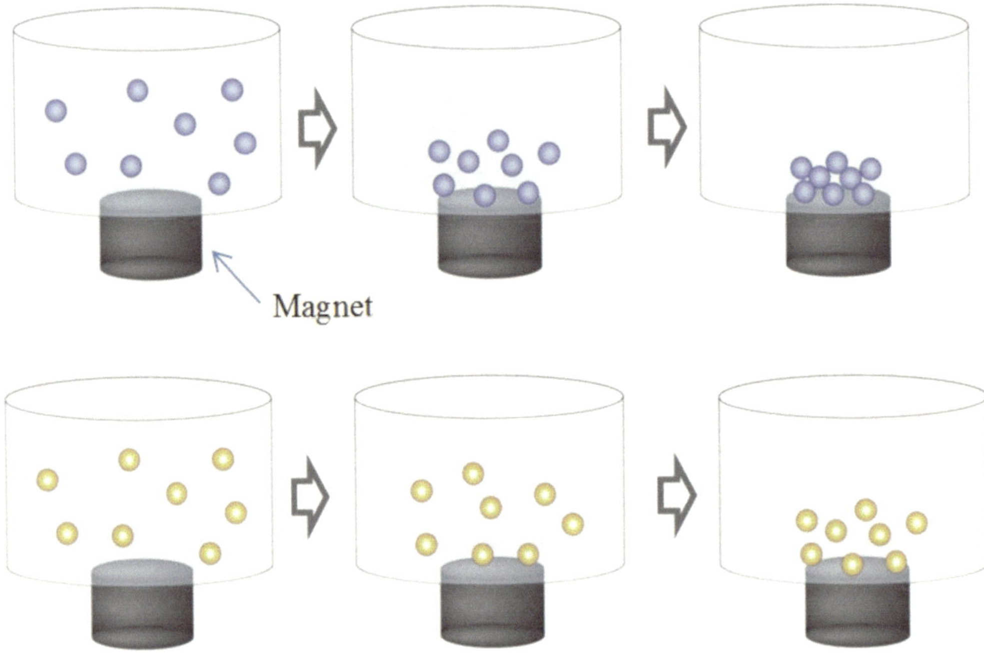

Fig. 13. Schematic representation of the magnetic attraction of the iron nitride-based LipoFeN (*blue*) compared with the iron oxide-based PolyMag (*brown*).

in cell lysis reagent (Toyo ink, Tokyo, Japan), and incubated for 15 min at room temperature. After centrifugation, 2.5 μl of the supernatant were mixed with 100 μl of luciferase assay reagent (Toyo ink), and the emission of light was measured with a luminometer (lumat LB9501; Berthold Designs, Berlin, Germany). For analysis of the total protein content, 10 μl of supernatant was assayed using the bicinchonic acid method, as described previously (Namiki et al., 1998).

Sense sequence: 5'-CGUACGCGGAAUACUUCGAdTdT-3' Antisense sequence: 5'- UCGAAGUAUUCCGCGUACGdTdT-3' Target DNA sequence: AACGTACGCGGAATACTTCGA

Table 4. The sequence of siRNA targeted to the luciferase gene and the target sequence of luciferase DNA (Elbashir et al., 2001). All Stars NC siRNA (siRNA^{NC}: Qiagen, Valencia, CA), which was used as the negative control nonsense siRNA, was designed to be a non-targeting sequence bearing no significant homology to the sequence of the human or murine transcripts. The negative control was used to eliminate non-specific off-target effects. These siRNAs were the most thoroughly tested and validated as a positive control or a negative control available from Qiagen.

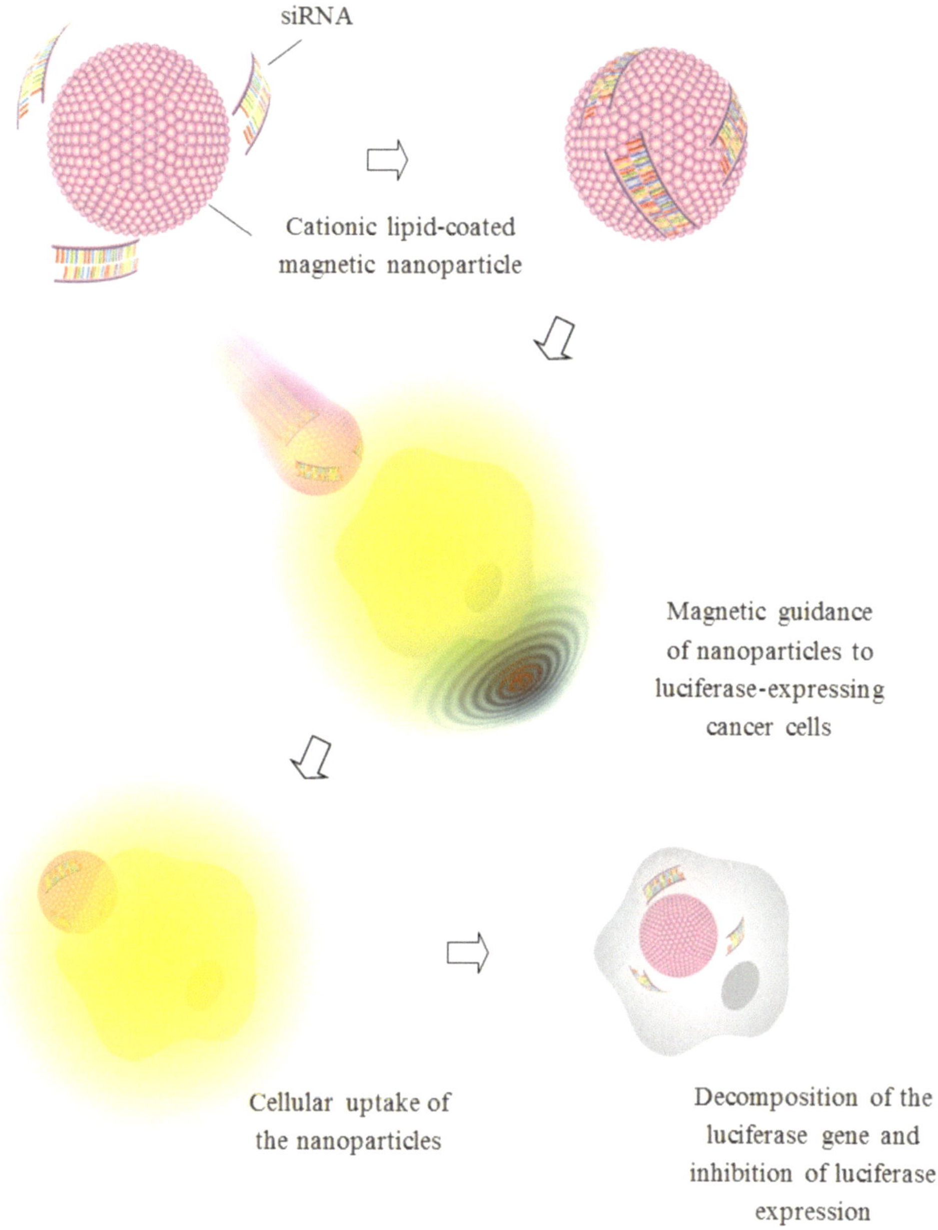

Fig. 14. Schematic representation of the magnetic guidance of siRNA loaded with the magnetic nanoparticles and of the silencing of the luciferase gene.

The gene silencing effect (the % knockdown of luciferase activity) of each type of magnetic nanoparticle was calculated (Fig. 16) using the following formula.
[the % knockdown of luciferase activity]= 100 - 100 x [the activity of luciferase for samples transfected with siRNA^{luc}] / [the activity of luciferase for samples transfected with siRNA^{NC}].

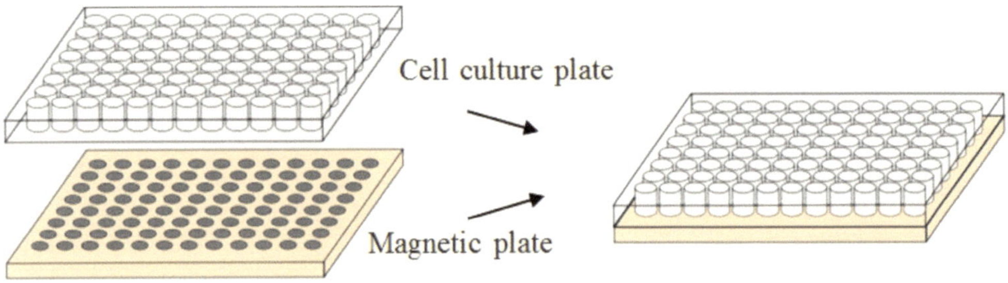

Fig. 15. Schematic representation of the magnetic attraction of the magnetic nanoparticles to the cultured cells using a magnetic plate (200 mT).

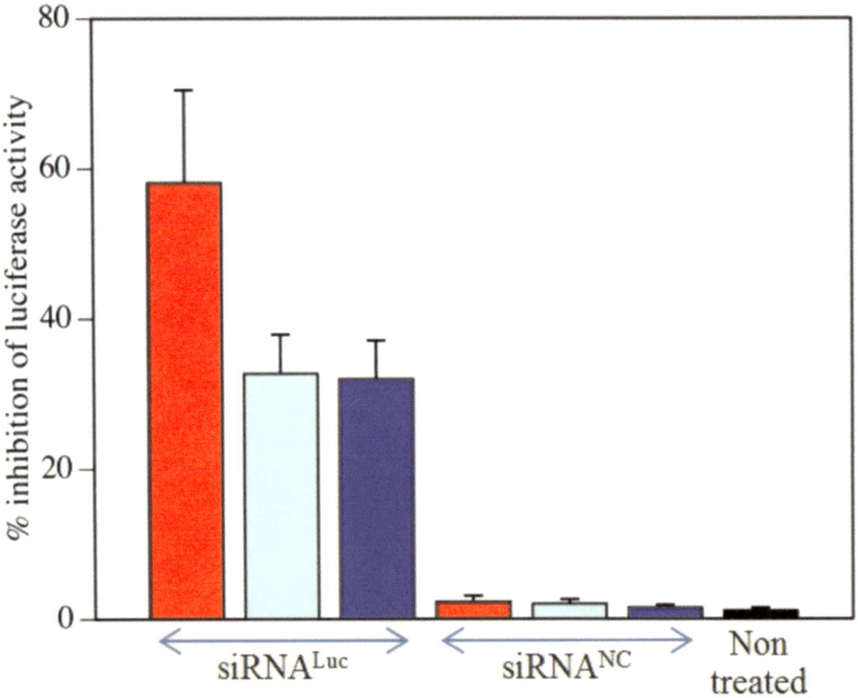

Fig. 16. The magnetically guided target gene silencing of the $Fe_{16}N_2$/cationic lipid (*red*), iron oxide/cationic lipid (*light blue*) and iron oxide-based PolyMag™ (*blue*) in cancer cells. Cells were incubated with LipoFeN/siRNA^{Luc}, LipoFeN/siRNA^{NC}, LipoMag/siRNA^{Luc}, LipoMag/siRNA^{NC}, (all of which included 90-720 ng of total lipids and 7.8-31.3 ng of siRNA), PolyMag™/siRNA^{Luc} or PolyMag™/siRNA^{NC} (both of which included 12.5-100 nl of PolyMag™ and 7.8-31.3 ng of siRNA). At 48 h after the incubation with the complexes for 5 min, the degree of the knockdown of luciferase-expression was evaluated. LipoFeN containing 360 ng of total lipid and 15.6 ng of siRNA^{Luc} per 1 x 10^4 cells displayed a greater gene silencing effect than LipoMag (containing 360 ng of total lipid) and PolyMag™ under each optimized condition. The data given are the averages of two different experiments, each performed in triplicate (n= 6); bars, SD.

9. Cytotoxicity of the iron nitride/metallic oxide/oleic acid/cationic lipid nanocomposite

We evaluated the cytotoxicity of magnetic nanoparticles by the LDH (lactate dehydrogenase) release assay. This assay is recommended instead of the MTT assay (Heeg et al., 1985) when the magnetic lipid nanoparticle-mediated cytotoxicity is measured (Soenen et al., 2007). In particular, the LDH release assay (Korzeniewski et al., 1983; Decker et al., 1988) is used for the quantitative analysis of cell death be measuring the LDH activity liberated from cells with cell membrane injuries, such the cell membrane lysed by drugs. This assay can be used for the determination of the toxicity of various agents, utilizing cultured cells. In the presence of NAD, LDH catalyzes the conversion of lactate to pyruvate. Lactate is oxidized to pyruvate and an equivalent amount of NAD is reduced to NADH. The NADH formed reduces nitrotetrazolium blue in the presence of diaphorase to produce blue colored Diformazan (Fig. 17), the absorbance wavelength of which is 560 nm.

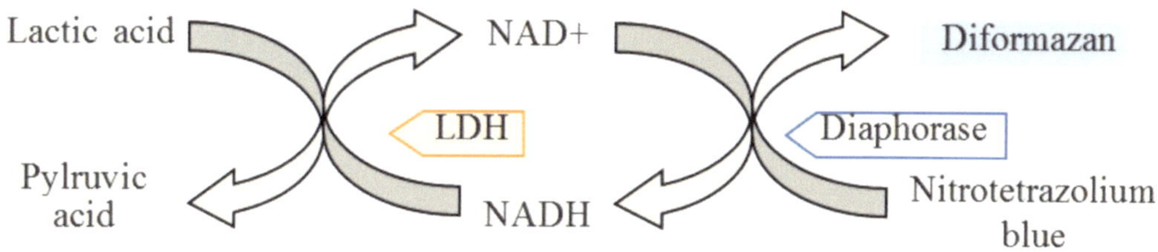

Fig. 17. Principle of the LDH release assay

HT-29-luc2 cells (1 x 10^4 cells/well of a 96 well microplate) were incubated with each type of magnetic nanoparticles or with culture medium alone for 5 min under the magnetic field. After the removal of each medium, cells were washed twice with PBS and cultured with culture medium for 48 hr in a 5% CO_2 incubator at 37 °C. After the removal of the culture medium, cells were washed twice with PBS and 75 µl of 0.2% Tween 20/PBS was added to the well. The plate was allowed to stand for 15 min at room temperature and moderately agitated. The plates were centrifuged for 3 min at 1000 rpm and 50 µl of supernatant was collected. Five milliliters of color reagent containing nitrotetrazolium blue (3.7 mg), diaphorase and NADH, and 5 ml of buffer solution containing DL-lithium lactate (25 mg) were mixed to obtain the coloring solution. Fifty microliters of this coloring solution and 50 µl of supernatant were mixed. At 45 min after the mixture (room temperature), 100 µl of hydrochloric acid (0.5 mol/l) were added to each well. Finally, the absorbance of the reaction mixture was measured with a microplate reader at 560 nm after termination of the coloring reaction. The percentage of cell viability was calculated (Fig. 18) using the following formula.

$$\% \text{ cell viability} = S/N \times 100$$

(S: Absorbance of the sample; N: Absorbance of non-treated negative control cells)

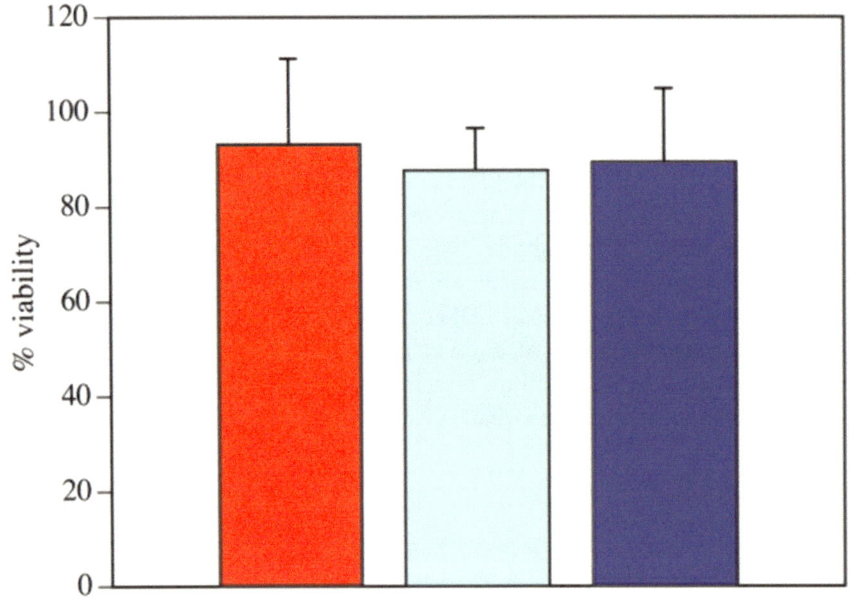

Fig. 18. The cytotoxicity of the iron nitride-based nanocomposite compared with the iron oxide-based nanocomposite. To eliminate non-specific cytopathic effects, we confirmed that cell viability was over 85% of the non-treated cell at 48 h after gene silencing under the optimized conditions of siRNA-transfection using an LDH release assay procedure. Cells were incubated with LipoFeN/siRNALuc (red), LipoMag/siRNALuc (light blue) or PolyMag™/siRNALuc (blue) in a similar manner as Figure 16. The data given are the averages of two different experiments, each performed in triplicate (n= 6); bars, SD.

10. Conclusion

We were able to successfully enhance the gene silencing effect using a ferromagnetic iron nitride-based nanocomposite instead of an iron oxide-based nanocomposite. In summary, the newly devised ferromagnetic crystal-lipid nanostructures comprise a highly efficient, magnet-guided delivery system of siRNA. Using the $Fe_{16}N_2$/cationic lipid nanocomposite achieved a better gene silencing effect than the major commercially available iron oxide-based magnetic vector, PolyMag™, even with a short incubation time, and the expression of the target gene was effectively inhibited. The formulations of iron nitride crystal-lipid nanostructures clearly have the potential for widespread use as high-performance gene delivery systems for cancer cell-targeted RNA interference.

11. Acknowledgments

We dedicate this work to the late T. Terada and the late K. Nariai. We would like to acknowledge Yukiko Ishii, Yoko Yumoto and Shuichi Nakagawa for their assistance. This work was supported by a Funding Program for Next Generation World-Leading Researchers (LS114), from JSPS and by an Industrial Technology Research Grant, Program (08C46049a), from the NEDO of Japan. Pacific Edit reviewed the manuscript prior to submission.

12. References

Fire, A. et al. Potent and specific genetic interference by double-stranded RNA in Caenorhabditis elegans. Nature 391, 806–811 (1998).

Li, C.X. et al. Delivery of RNA interference. Cell Cycle 5, 2103–2109 (2006).

Toub, N. et al. Innovative nanotechnologies for the delivery of oligonucleotides and siRNA. Biomed. Pharmacother. 60, 607–620 (2006).

Namiki, Y. et al. A novel magnetic crystal-lipid nanostructure for magnetically guided *in vivo* gene delivery. Nat. Nanotechnol. 4, 598-606 (2009).

Xie, J. et al. Iron oxide nanoparticle platform for biomedical applications. Curr. Med. Chem. 16, 1278-1294 (2009).

Lin, M.M. et al. Development of superparamagnetic iron oxide nanoparticles (SPIONS) for translation to clinical applications. IEEE Trans. Nanobioscience 7, 298-305 (2008).

Laurent, S. et al. Magnetic iron oxide nanoparticles: synthesis, stabilization, vectorization, physicochemical characterizations, and biological applications. Chem. Rev. 108, 2064-2110 (2008).

Gupta, A.K. et al. Synthesis and surface engineering of iron oxide nanoparticles for biomedical applications. Biomaterials 26, 3995-4021 (2005).

Sugita, Y. et al. Magnetic and Mössbauer studies of single-crystal $Fe_{16}N_2$ and Fe-N martensite films epitaxially grown by molecular beam epitaxy. J. Appl. Phys. 76, 6637–6641 (1994).

Takahashi, H. et al. Perpendicular uniaxial magnetic anisotropy of $Fe_{16}N_2$ (001) single crystal films grown by molecular beam epitaxy. IEEE Trans. Magn. 35, 2982–2984 (1999).

Hattori, T. et al. Magnetic particles of $Fe_{16}N_2$ fine particles. J. Magn. Soc. Jpn. 25, 927–930 (2001).

Sun, S. et al. Monodisperse FePt nanoparticles and ferromagnetic FePt nanocrystal superlattices. Science 287, 1989-1992 (2000).

Sasaki, Y. et al. Development of NanoCAP Technology for High-Density Recording. IEEE Trans. Magn. 41, 3241-3243 (2005).

Woodcock, C.L. et al. Ultrastructure of chromatin: negative staining of isolated fibers. J. Cell Sci. 99, 99-106 (1991).

Takahashi, T. et al. Induction of the suicide HSVtk gene by activation of the Egr-1 promoter with radioisotopes. Hum. Gene Ther. 8, 827–833 (1997).

Namiki, Y. et al. Gene transduction for disseminated intraperitoneal tumor using cationic liposomes containing non-histone chromatin proteins. Gene Ther. 5, 240–246 (1998).

Elbashir, S.M. et al. Duplexes of 21-nucleotide RNAs mediate RNA interference in cultured mammalian cells. Nature 411, 494-498 (2001).

Korzeniewski, C. et al. An enzyme-release assay for natural cytotoxicity. J. Immunol. Methods 64, 313-320 (1983).

Decker, T. et al. A quick and simple method for the quantitation of lactate dehydrogenase release in measurements of cellar cytotoxicity and tumor necrosis factor activity. J. Immunol. Methods 115, 61-69 (1988).

Heeg, K. et al. A rapid colorimetric assay for the determination of IL-2-producing helper T cell frequencies. J Immunol. Methods 77, 237-246 (1985).

Soenen, S.J. et al. Optimal conditions for labeling of 3T3 fibroblasts with magnetoliposomes without affecting cellular viability. Chembiochem. 8, 2067-2077 (2007).

14

Tailoring the Morphology and the Optical Properties of Semiconductor Nanocrystals by Alloying

Young-Kuk Kim and Chul-Jin Choi
Korea Institute of Materials Science
Republic of Korea

1. Introduction

Colloidal semiconductor nanocrystals (NCs) or quantum dots (QDs) based on semiconductors open possibilities for applications in new areas (Bawendi et al., 1990; Alivisatos, 1996; Markovich et al., 1999). Since hot injection synthesis of mono-dispersed spherical nanocrystals using organometallic precursors was introduced by an MIT group (Murray et al., 1993), extensive works have been devoted to make variation of their size and shape, compositions. In this chapter, we will focus on the effect of the alloying or impurity doping on the shape and the optical properties of the NCs.

In the first section, we will introduce and discuss the effect of alloying on the morphology of semiconductor NCs. The effects of Pb^{2+} addition on the morphological development of CdSe nanocrystals was discussed. The addition of Pb ions in the initial precursor solution for the synthesis of alloyed CdSe NCs changed the morphology of CdSe nanocrystals to branched rod with high aspect ratio. The branched nanocrystals are mainly composed of wurzite phase grown along [001] direction and the length of rods in each branched nanocrystal can be increased by increasing amount of Pb^{2+} addition to accelerate anisotropic growth of nanocrystals. The luminescence, however, mostly arises from trap-related recombination and significantly red-shifted by Pb^{2+} addition. Moreover, it will be shown that the zinc blend-wurzite polymorphism was popularly observed in these branched CdSe NCs and the twinned structure gave rise to highly asymmetric line shape of photoluminescence spectra. Based on the known optical properties, we will further discuss the application of the branched CdSe NCs to the photovoltaic devices.

In the second section, we discussed the effect of the alloying on the optical properties of Cd-free I-III-VI_2 group semiconductor nanocrystals. Although, many works have been focused on II-VI semiconductors such as CdSe, CdS, CdTe, etc. due to their high quantum yield of fluorescence with tunable emission wavelength, incorporation of toxic heavy metals such as Cd, Pb, Hg in the II-VI semiconductors have limited their public application. As alternatives to II-VI semiconductors, several materials system including III-V semiconductor NCs (e.g. InP) and impurity-doped ZnS NCs have been developed. However, each material system have some drawbacks such as requirement of expensive pyrophoric compounds as precursors, chemical stability, tunability of emission wavelength, etc. Ternary chalcopyrite-type I-III-VI_2 group semiconductor NCs were also synthesized as less-toxic new candidates to II-VI semiconductor NCs. I-III-VI_2 group semiconductors with appropriate band-gap energies (e.g. 1.5eV for $CuInS_2$ and 1.68eV for $CuGaSe_2$) are direct transition semiconductors favoring ionic bonding similar to II-VI semiconductors. However, in contrast to II-VI group semiconductors which make luminescence by direct exciton recombination, the dominant luminescence mechanism for the chalcopyrite-type I-III-VI_2 group semiconductor NCs in the visible spectrum is known to be "donor-acceptor" pair recombination (DAP) which requires defects to play an important role in the luminescence process. Recently, several results have been reported for synthesis of non-aggregated I-III-VI_2 group semiconductor NCs of a few nanometers with fluorescence quantum yield of a few percent. We discussed the luminescence mechanism of those nanocrystals and delineate the effect of alloying on their luminescence. The photoluminescence of ZnS alloyed NCs show large blue-shift with respect to pure I-III-VI_2 NCs. It was shown that the emission wavelength and intensities of I-III-VI_2 NCs can be tuned by alloying. In addition, we intended to report a one-pot synthesis route to highly luminescent I-III-VI_2-based core-shell NCs with high quantum yield through defect control and shell formation by refluxing with zinc acetate and palmitic acid. Zinc acetate played an important role in formation of inorganic shell and size regularization of I-III-VI_2 NCs. We showed that Cu-deficiency in precursor solution is essential to enhancing quantum yield and surface modification by refluxing with zinc acetate and palmitic acid enables a reproducible synthesis of monodispersed highly luminescent I-III-VI_2 NCs with less toxic and highly stable precursors.

2. Tailoring the morphology of CdSe nanocrystals by alloying with Pb

The size and shape of semiconductor nanocrystals play an important role to determine their electronic and optic properties. In particular, anisotropic nanocrystal exhibit unique performance such as enhanced solar cell efficiency, polarized luminescence, etc. Thus, anisotropic nanocrystals with variety of compositions such as CdS, CdTe, CdSe, etc. were synthesized with hot injection method (Manna et al., 2000; Zhang & Yu, 2006; Li & Wang, 2003). In particular, tetrapod-shaped CdTe nanocrystals were synthesized with high shape selectivity and high size uniformity (Zhang & Yu, 2006). In contrast to this, CdSe nanocrystals were relatively hard to synthesize with high shape selectivity, since the energy difference between the wurzite and the zinc blende is too small to facilitate tetrapod-shaped crystals (Zhang & Yu, 2006; Asokan et al., 2007). Recently, Asokan et al. reported synthesis of CdSe tetrapods with high shape selectivity and high size uniformity (Asokan et al., 2007). Most of syntheses of anisotropic CdSe nanocrystals involve special surfactants such as alkylphosphonic acid, quaternary ammonium salt compounds, etc. These surfactant ligands preferentially bind to certain particle facets to reduce growth rate of these facets.

In addition, doping in semiconductor nanocrystals can adjust their optical and electronic properties (Erwin, 2005). For example, doping foreign atoms can introduce characteristic luminescence peaks in photoluminescence spectra of nanocrystals and reduce the lasing threshold of semiconductor nanocrystals. Thus, works on doped nanocrystals are mainly devoted to tune the optical and electronic properties of nanocrystals. To our knowledge, reports on the morphological development of nanocrystal by doping are quite rare (Hsu & Lu, 2008). In this study, we report the effect of Pb^{2+} addition on the morphological development of CdSe nanocrystals and their structure-property relationships were analyzed.

2.1 Synthesis of (Cd,Pb)Se alloyed nanocrystals

CdSe nanocrystals were synthesized by injection of selenium precursor solution into hot Cd precursor solution based on the synthesis method for spherical CdSe nanocrystals (Asokan et al., 2005). The Cd precursor solution was synthesized by dissolving 4mmol CdO to the mixture of 10mmol oleic acid and 20ml 1-octadecene (ODE). Sometimes, lead (II) acetate hydrate was intentionally added instead of the same amount of CdO maintaining total concentration of metal salts. Subsequently, the mixtures were heated to 150°C under N_2 gas flow and metallic precursors were completely dissolved to form clear solution of metal oleate (Solution A). Separately, 8mmol selenium powders were dissolved in 7.71ml trioctylphosphine (TOP) under dry N_2 atmosphere in the glove box and the selenium solution was swiftly injected to Cd-based precursor solution A at 225°C. Aliquots were taken at different time interval after injection and rapidly cooled to room temperature. Before further characterization, acetone was added to precipitate CdSe-based nanocrystals and the precipitates were purified by repetition of centrifugation and re-dispersion to toluene. Overall reaction occurred in this synthesis reaction can be written as $(1 - x)[Cd(A)_{n1}] + x[Pb(A)_{n1}] + Se(B)_{n2} \Rightarrow Cd_{1-x}Pb_xSe + n_1A + n_2B$, where A and B corresponds to oleic acid and trioctylphosphine, respectively.

2.2 Effect of Pb addition on the morphology of CdSe nanocrystals

The crystal structure and phase formation of nanocrystals were characterized by powder X-ray diffraction (XRD) taken using RINT2000 (Rigaku, Japan) with Cu-$K\alpha$ line as the X-ray source. Fig. 2.1(a) shows XRD patterns of nanocrystals grown at 225°C for 300 seconds with various initial abundance of Pb^{2+} in the precursor solution. Irrespective of their initial abundance of Pb in the solution, XRD patterns of nanocrystals are consistent with that of hexagonal wurzite CdSe and displaying no trace of impurity peaks. The more Pb^{2+} ions were contained in the initial precursor solution, the narrower peaks in the XRD patterns became. Uniquely, the width of (002) peaks in the XRD pattern of CdSe nanocrystals was drastically decreased with increasing initial Pb^{2+} content in the reaction mixtures. The sharp (002) peak in the XRD pattern implies a preferential growth of CdSe nanocrystals along [001] direction. This preferential growth of CdSe nanocrystals resulted in highly anisotropic shape such as rod, tetrapod, branched crystals, etc. Furthermore, the narrower (002) peak with increasing Pb content in the reaction mixture indicates that longer nanocrystals were synthesized from the precursor solution containing more Pb^{2+}.

The high resolution TEM (HR-TEM) image of Pb-containing nanocrystals presented in Fig. 2.2(a) showed highly entangled and branched morphology. Analysis of the lattice image showed that the planar spacing along axis of nanorods is about 0.35nm, which corresponds to planar spacing of (002) planes in the wurzite CdSe. Again, it was confirmed that CdSe crystals were preferentially grown along [002] direction. The junction point of the two branches is composed of zinc blende phase as indicated by dotted circle in Fig. 2.2(a).

Fig. 2.2(b) showed the abundance of Pb in the grown nanocrystals, Pb/(Cd+Pb) measured by induction coupled plasma-mass spectroscopy (ICP-MS). Pb content increased linearly with the initial amount of Pb in the precursor solution. Although Pb^{2+} can be completely soluble in CdSe within 40% (Hankare et al., 2005), the solubility of Pb in CdSe nanocrystals is highly limited due to the fast kinetics of nanocrystal growth at 225°C. The Pb content in the CdSe nanocrystals is decreased with increasing reaction time at 225°C as shown in Fig. 2(b). That is, a relatively large amount of Pb is incorporated in nanocrystals just after Se injection at 225ºC and smaller amount of Pb is incorporated in the nanocrystals during the growth of each branch.

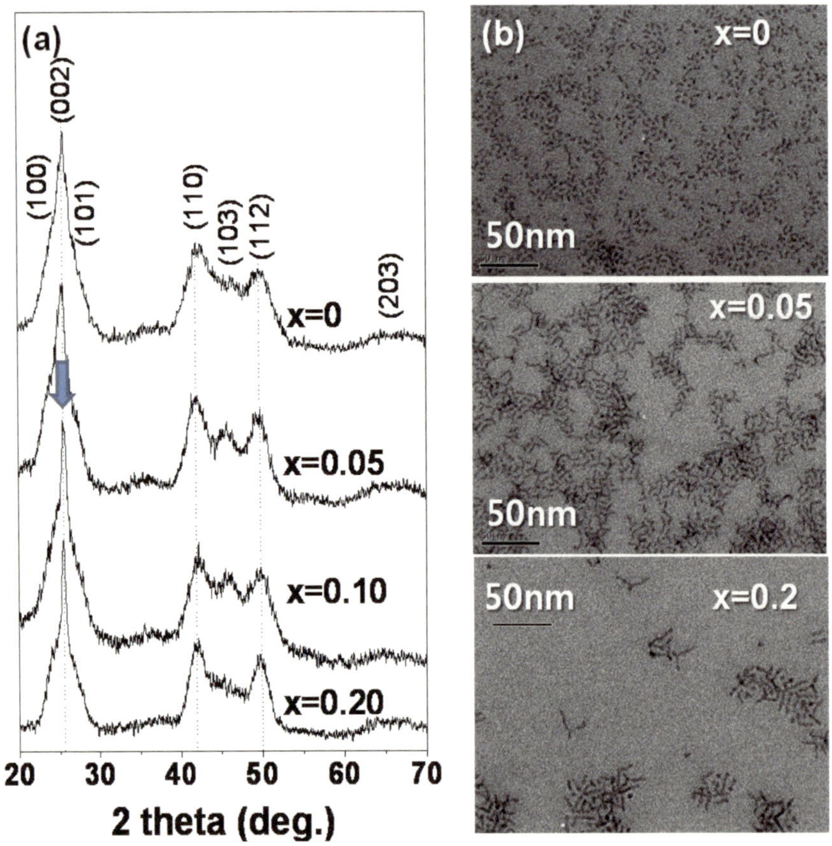

Fig. 2.1. (a) X-ray diffraction (XRD) patterns and TEM images of nanocrystals grown at 225ºC with various initial molar abundance of Pb, x=Pb/(Cd+Pb). Peak indices in the XRD pattern were assigned according to wurzite CdSe.

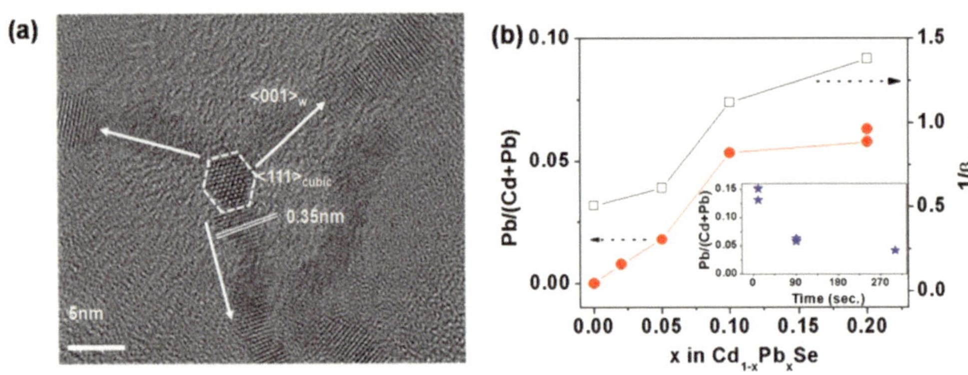

Fig. 2.2. (a) High resolution TEM images for nanocrystals grown from $Cd_{0.8}Pb_{0.2}Se$ precursor solution and (b) the abundance of Pb in the grown nanocrystals, Pb/(Cd+Pb).

Since the reciprocal value of FWHM (β) for (002) peaks in the XRD patterns is proportional to length (L) of each branch in the nanocrystals according to Scherrer's formula ($L=0.94\lambda/\beta\cos\theta$), the monotonous increase in $1/\beta_{(002)}$ indicates linear increase in the length of branches of the nanocrystals by Pb addition. Thus, the initial abundance of Pb controls the morphological development of nanocrystals. Raman spectroscopy has been used to study the structure of nanocrystals. Specimen for Raman spectra were prepared by dropping solution containing nanocrystals onto silicon wafer. After drying, Ar^+ laser (λ=514.5nm, power less than 10mW) was focused on a spot of 50μm diameter in each sample and Raman spectra were collected with a triple Raman spectrometer (SPEX-1430). Fig.2.4(a) showed a Raman spectrum of the nanocrystals grown at 225°C. The first order Raman band was identified at ω=206cm^{-1} for nanocrystals grown from Pb-free precursor solution for 90 seconds at 225°C and ω=207cm^{-1} for nanocrystals grown from Pb-containing precursor solution under the same growth condition. The observed Raman spectra are highly asymmetric and shifted from the value of bulk CdSe (ω=210cm^{-1}). The asymmetric line shape of the Raman band is commonly observed in low dimensional nanostructures and can be fitted well with two Lorentzian peak functions which is explained with two effects. The more pronounced peak at ω=206cm^{-1} can be assigned to be a CdSe longitudinal optical (LO) mode and the shoulder on the low frequency side of LO mode is related to surface optical phonon (SO) mode (Lange et al., 2009; Comas et al., 2002).

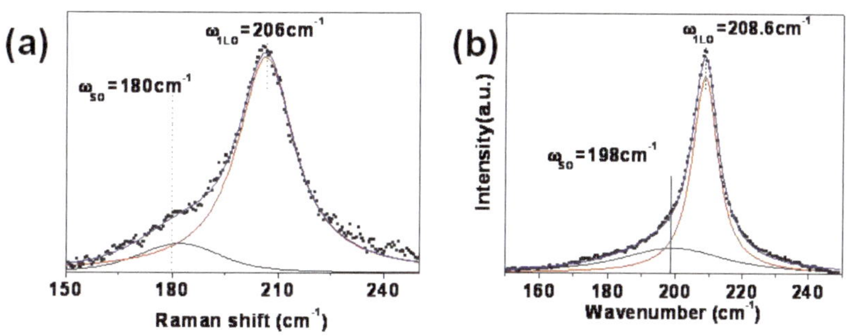

Fig. 2.3. Raman spectra of the nanocrystals grown at 225°C from (a) Pb-free precursor solution and (b) solution containing Pb with x=0.2.

The SO mode can be observed in the Raman scattering process due to non-spherical geometry of nanocrystals and it is highly dependent on the aspect ratio of nanocrystals. As shown in Fig. 2.3(b), the frequency of SO mode for nanocrystals grown from Pb-containing precursor solution shifted to high frequency. That is, the frequency of SO mode for nanocrystals grown from Pb-free precursor solution is much lower than that for nanocrystals grown from Pb-containing precursor solution at the same growth conditions. Lange *et al.* showed that Raman-active SO mode of CdSe nanorods is shifted to high frequency, as the aspect ratio (length/diameter) increases (Comas et al., 2002). Again, this indicated that the nanocrystals with high aspect ratio can be synthesized from Pb-containing precursor solution.

We noted that the emission spectra of nanocrystals were symmetric and sharp, which indicates narrow size distribution of the grown nanocrystals. Most of emission spectra from the branched nanocrystals detected at the room temperature were from band-edge emission as shown in Fig. 2.4(a). However, they showed drastic reduction in their intensity with increasing amount of Pb in the precursor solution. In general, emission from semiconductor nanocrystals is composed of band-edge emission and emission from surface trap states (Saunders et al., 2008; Jing et al., 2009). Temperature-dependent PL spectroscopy is often used to study the radiative and the non-radiative relaxation process in nanocrystals. Contrast to the room temperature PL spectrum, the low temperature PL spectrum of the nanocrystals grown from Pb-containing precursor solution exhibited two peaks: a narrow peak at 620 nm due to band-edge emission (BE) and a broad intense peak highly red-shifted from the band edge, centered at about 800 nm, which results from trap-related emission (TE). In the low temperature PL spectrum, majority of the emission spectrum emanated from TE. The large surface-to-volume ratio of branched nanocrystals with high aspect ratio increased the occurrence of surface trap states and reduced the band-edge emission.

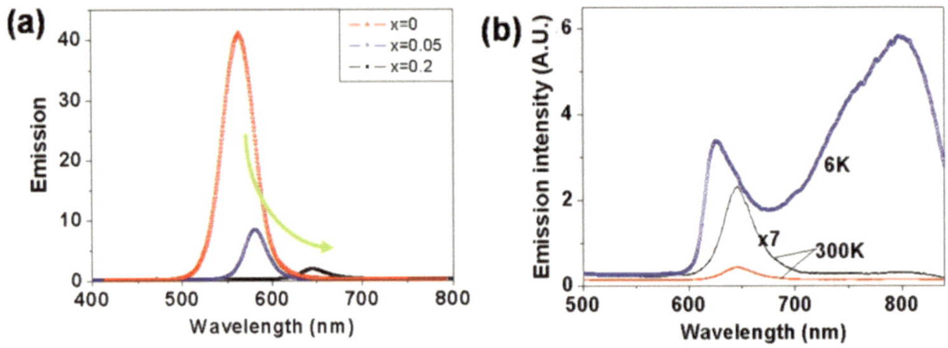

Fig. 2.4. (a) Room temperature photoluminescence of $Cd_{1-x}Pb_xSe$ with various Pb-content(x) and (b) temperature dependence of photoluminescence spectra for $Cd_{1-x}Pb_xSe$ (x=0.2) nanocrystals grown at 225°C.

Two mechanisms working on the formation of branched nanocrystals have been suggested (Kudera et al., 2008). The formation of branched nanocrystals was commonly based on the initial formation of octahedral crystallite and the subsequent growth of anisotropic crystallites from the facets of the octahedra. One model for the branched nanocrystal formation is the *octa-twin* model which was developed at the early 1990s for tetrapods of ZnS and ZnO. This model describes the initial octahedral crystallites as cores composed of eight wurzite domains with tetrahedral shape. In this model, branches are grown on {0001} facets of wurzite cores and form branched nanocrystals with multi-pods. The other mechanism suggested is the *polymorphism* model which describes the formation of branched nanocrystals as initial formation of octahedral cores with zinc blende structure and further growth of branches on the {111} facets of cubic cores. Since the formation energy difference between cubic zinc blende structure and hexagonal wurzite structure for CdSe is quite small, co-existence of both phases in a crystal is not rare. In particular, during the synthesis of CdSe, injection of Se precursors can drop the temperature and zinc blende nuclei readily form by the temperature effect. In addition to temperature effect, the change in crystalline phase can be justified by the affinity of ligands to certain facets (Manna et al., 2000). Special ligands such as alkylphosphonic acid can tightly bind the facets of wurzite and induce the formation of wurzite crystallites.

For $Cd_{1-x}Pb_xSe$ alloyed nanocrystals, those two mechanisms cannot explain the formation of branched nanocrystals without modification, since oleic acid was used as capping ligands

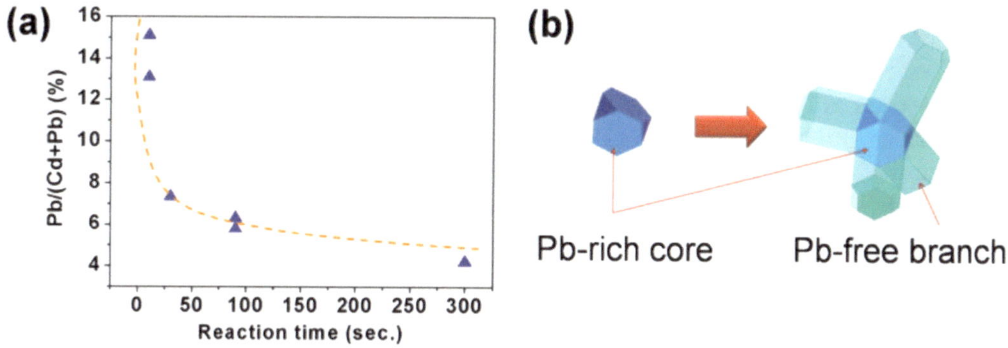

Fig. 2.5. (a) Variation of the incorporated Pb content with respect to reaction time and (b) a schematic illustration of core-branch growth in $Cd_{1-x}Pb_xSe$ (x=0.2) nanocrystals.

and nearly isotropic nanocrystals were synthesized in the absence of Pb. Fig. 2.5(a) shows the variation of Pb-content in the branched nanocrystals, $Cd_{1-x}Pb_xSe$ (x=0.2). At the initial nucleation-growth stage, the concentration of Pb was *ca.* 15% and rapidly decreased to be less than 8% after this stage. That is, a relatively large amount of Pb is incorporated in nanocrystals just after Se injection at 225 °C and the incorporation of Pb was highly suppressed during the growth of each branch. One possible explanation on the formation of branched CdSe nanocrystals by alloying with Pb is that nuclei with cubic zinc blende structure were formed by incorporation of Pb during the nucleation step and further growth wurzite branch on (111) facet of cubic crystallites as suggested in the *polytypism* model []. In particular, the low Pb content in the CdSe nanocrystals after prolonged reaction at 225°C can be explained by "self-purification effect" (Dalpian & Chelikowsky, 2006), which assess migration of impurities or lattice defect to the surface of nanocrystals (Erwin et al, 2005). However, further study is now in progress to elucidate the formation mechanism of branched nanocrystals in Cd-Pb-Se system.

3. Tailoring the optical properties of nanocrystals by alloying

The semiconductor nanocrystals were characterized with strong emission of the visible light and their continuous tuning by size due to quantum confinement of charge carriers. Synthesis of semiconductor nanocrystals (NC) has been intensely pursued due to their importance in optoelectronic, photovoltaic and biological applications. The applicability of the luminescent nanocrystals to the above application can be determined by their luminance quantum efficiency. Much effort has been endowed to enhance the quantum efficiency of semiconductor nanocrystals. The quantum efficiency of semiconductor nanocrystals is highly related with passivation of dangling bonds on the nanocrystal surface. Surface capping of nanocrystals with a wide band gap semiconductor can successfully improve the luminescence of nanocrystals and these core-shell NCs have shown to be generally more robust against chemical degradation or photo-oxidation (Xie et al., 2005). Such core-shell nanostructures reduce dangling bonds on nanocrystal surface by epitaxial growth of shell layers and are of practical interest for bio-imaging, LED, etc. Recently, semiconductor nanocrystals with quantum yield larger than 60% can be obtained with this core-shell approach. Basically, it needs that the core and shell materials should have similar crystal structure and lattice parameters to produce particles with high crystallinity without formation of interfacial defects. In order to improve quantum efficiency by improving crystallinity of nanocrystal through reducing interfacial lattice mismatch, multishell structure such as $CdS/Zn_{0.5}Cd_{0.5}S/ZnS$ was developed. Xie et al. synthesized multishell nanocrystals with high fluorescence quantum yield of 70-85% (Xie et al., 2005). However, this core-shell approach requires a careful shell growth process using highly toxic and stench precursor such as diethyl zinc, trimethylsilyl sulfide, etc.

Recently, alloy nanocrystals with composition gradient along the radial direction were synthesized by several research groups using the difference in reactivity of cationic or anionic component (Bae et al., 2008; Bailey & Nie, 2003). In particular, Bae *et al.* synthesized CdZnSeS nanocrystals by changing ratios of Cd to Zn and Se to S with the total concentrations of the Cd-Zn pair and Se-S pair fixed to be equal (Bae et al., 2008). The nanocrystals prepared in the single step synthesis are also structurally stable due to the chemical composition (or energy level) gradient in the radial direction, which effectively relieves the lattice mismatch and confines electrons and holes within the core.

As an alternative to II-VI semiconductors, various semiconductors with less-toxic elements have been proposed (Xu et al., 2008; Ryu et al., 2008; Zhong, 2008). Among these, $CuInS_2$ with bulk band gap of 1.5eV is selected as an important candidate for optical application. Recently, several works on the synthesis of $CuInS_2$ (CIS) nanocrystals and efficient route to highly luminescent $CuInS_2/ZnS$ core shell nanocrystals were reported.[5-7] Zhong *et al.* reported synthesis of $CuInS_2$ nanocrystals using copper acetate, indium acetate and 1-dodecanethiol by thermolysis of intermediate, CuIn(SR)x compound (Zhong, 2008). Moreover, Xie *et al.* reported synthesis of spherical $CuInS_2$ nanocrystals by modifying chemical reactivity of indium precursors and enhancement of their photoluminescence by formation of ZnS shell layers (Xie, 2008). However, inorganic shell capping of nanocrystals for high luminescence and photostability often requires repetitive formation of monolayers and precise process control. Recently, Ryu *et al.* reported a novel shell formation route for InP nanocrystals by refluxing with zinc acetate and 1-dodecanethiol (Ryu et al., 2009). The addition of zinc acetate played an important role in the formation of ZnS shell layers as well as in etching the defective surface of InP NCs by formation of acetic acid resulted from its reaction with long chain carboxylic acid. Moreover, Uehara *et al.* improve the photoluminescence of $CuInS_2$ NCs by introducing lattice defects through highly off-stoichiometric composition from $CuInS_2$ (Uehara et al., 2008. This improvement in PL was attributed to the large population of donor or acceptor defects required for "donor-acceptor pair recombination (DAP)" of excited charge carriers. Recently, Hamanaka *et al.* assessed that the fluorescence in $CuInS_2$ NCs was originated from the deep surface trap recombination rather than DAP by considering the large Stokes shift of $CuInS_2$ NCs (Hamanaka et al., 2008). Thus, the origin of photoluminescence in $CuInS_2$ NCs is not fully investigated. We studied the photoluminescence of $CuInS_2$-based nanocrystals with the time- and the temperature-dependent photoluminescence which provides the radiative and the non-radiative process in semiconductor nanocrystals. Herein, we developed a novel surface capping route for $CuInS_2$ NCs, which is quite effective to remove surface trap sites and enhances the luminescence quantum yield. Furthermore, we reported the change of the dominant photoluminescence mechanism in $CuInS_2$ NCs by both controlling copper stoichiometry and surface modification. In addition to this, we report the tuning of emission wavelength by alloying.

3.1 Synthesis of highly luminescent Cd(Se,S) alloy nanocrystals

3.1.1 Synthesis of nanocrystals

The Cd precursor solution was synthesized by dissolving 8mmol CdO to the mixture of 10mmol oleic acid and 20ml 1-octadecene (ODE). Subsequently, the mixtures were heated to 150°C under N_2 gas flow and metallic precursors were completely dissolved to form clear solution of metal oleate (solution A). Separately, selenium powders and sulfur powders with different ratio were dissolved in 2ml trioctylphosphine (TOP) and 10ml ODE under dry N_2 atmosphere in the glove box maintaining the total amount of [Se]+[S] fixed and the mixed solution was swiftly injected to Cd-based precursor solution A at 300°C. Aliquots were taken at different time interval after injection and rapidly cooled to room temperature. Before further characterization, acetone was added to precipitate CdSeS nanocrystals and the precipitates were separated and purified by repetition of centrifugation and re-dispersion to toluene.

3.1.2 Nanocrystals with composition-gradient

The synthesis of $CdSe_xS_y$ nanoparticles was carried out by a single injection of a premixed solution of Se/S in TOP and ODE solvents to the cadmium oleate solution heated at 300°C. In this study, excess total cadmium relative to chalcogen (Se,S) ratio of 1~8 times. Therefore Se and S precursors compete due to their difference in reactivity with Cd. Then, following reactions are expected to take place:

$$Cd(L)n + TOPSe \rightarrow (CdSe)(L)m(P) \quad (1)$$

$$Cd(L)n + TOPS \rightarrow (CdS)(L)m(P) \quad (2)$$

$$X(CdSe)(L)m(P)p + Y(CdS)(L)m(P)p \rightarrow CdSexSy(L)m(P) \quad (3)$$

where L= oleic acid and/or solvent, P = TOP, TOPS, or TOPSe, and k_1 and k_2 = rate constants of reactions 1 and 2. Reaction 1 and 2 is for formation of CdSe and CdS, respectively. The reaction 3 is for formation of alloy nanocrystals of CdSeS. It is known that Cd precursor reacted with Se faster than S to form CdSe preferably and leave surplus S after the reaction (Bailey et al., 2003).

Under Cd-rich conditions, the chalcogen precursors in with the limited supply would be completely consumed while the reactant in large excess would maintain a nearly constant concentration during the reaction and alloyed nanocrystals with a composition gradient structure can be produced (Bailey et al., 2003).

Fig. 3.1 shows photoluminescence (PL) spectra of CdSeS nanocrystals prepared with different cation to chalcogen concentration in the precursor solution. The peak intensity of PL spectrum grows to be intensified and the center of peak moves to low energy side. If the concentration of Cd and chalcogen precursors are similar or [Cd]/([Se]+[S]) ~ 1, the concentration of [Cd] is changed with reaction time. In this case, the composition of nanocrystal is determined only by initial composition ratio of [Se] to [S] even the concentration of Cd in precursor solution changed with time and nearly homogeneous alloy of CdSeS can be obtained. Contrary to this, in the case of the large ratio of Cd to chalcogen or a large excess amount of cadmium precursors, the concentration of cadmium would be maintained without significant change during the reaction. Since Se is more reactive to Cd compared with S, Cd-Se bondings are formed in the initial stage of reaction. As the reaction

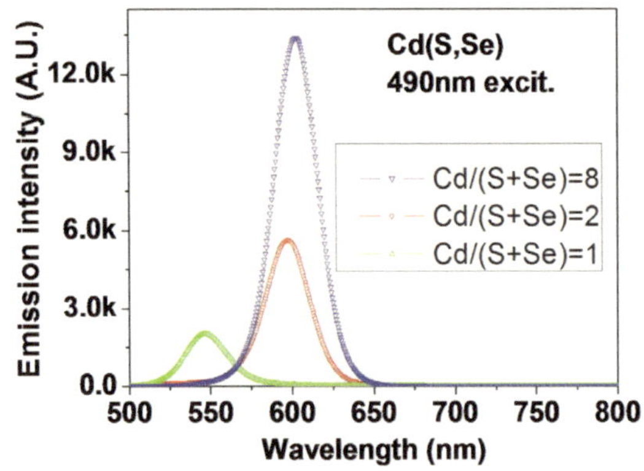

Fig. 3.1. Photoluminescence of CdSeS nanocrystal prepared with different ratio of [Cd] to [Se]+[S]

proceeds, the concentration of Se is depleted and number of Cd-S bondings is increases. After completion of reaction, cores of grown nanocrystals are Se-rich and the concentration of S is increases along radial direction of nanocrystals. That is alloy nanocrystals with composition gradient are formed with this reaction conditions. Since the band gap of CdSe (E_g~1.8eV) is smaller than that of CdS (E_g~2.3eV), the band gap of alloy nanocrystal grow larger along radial direction of nanocrystals. Then, a core-shell like nanostructures can be formed and luminescence of nanocrystal is expected to be enhanced under this condition. As shown in Fig. 3.1, the center of PL spectrum shifts to longer wavelength as the ratio of cation to chalcogen increases. In case of [Cd]/(Se]+[S]) ~ 1, homogeneous alloy nanocrystals are expected to be formed and the peak position of the PL spectrum is positioned in short wavelength due to larger band gap of CdSeS alloy compared with CdSe. In case of [Cd]/(Se]+[S]) >> 1, alloy nanocrystals with composition gradient along radial direction are expected to be formed and the peak position in the PL spectrum is determined by Se-rich composition of core of CdSeS alloy dots. In addition to this, since the composition gradient nanostructure in alloy nanocrystals provides effective confinement of electrons and holes, the intensity of the PL spectrum is highly enhanced. The maximum quantum yield of these alloy nanocrystals was larger than 90% by comparing photoluminescence of an organic dye, Rhodamine 6G.

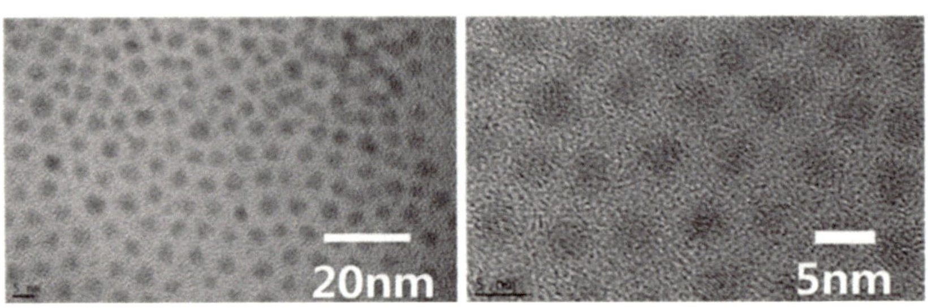

Fig. 3.2. TEM image of CdSeS nanocrystal grown for 2 minutes at 300°C under N_2 flow with [Cd]/(Se]+[S]) >> 1.

Fig. 3.2 shows TEM images of alloy nanocrystals grown with [Cd]/(Se]+[S]) >> 1. Highly uniform and spherical nanocrystals with average size 4nm were formed. The distance between surfaces of neighboring nanocrystals is about 2nm which is similar to the length of oleic acid. Figure 3.3(a) shows powder diffraction profiles of CdSeS nancrystal grown with [Cd]/(Se]+[S]) >> 1. The peak position of alloyed nanocrystals moved to higher angle which means smaller lattice parameters compared with CdSe. The calculated lattice parameters of grown nanocrystals are between that of CdSe and CdS, which confirms the formation of alloy nanocrystals and as the ratio of [Se] to [S] become smaller, the lattice parameter also become smaller. The maximum of photoluminescence of alloy nanocrystals can be changed by changing initial ratio of [Se] to [S] as shown in Fig. 3.3(b). That is, the luminescence wavelength can be tuned by changing initial ratio of [Se] to [S].

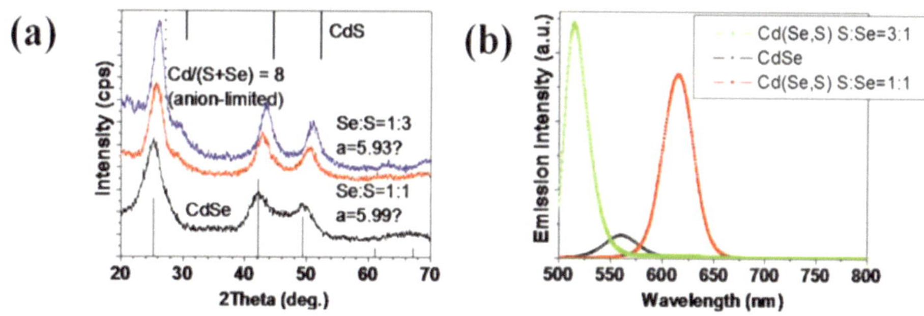

Fig. 3.3. (a) powder X-ray diffraction profiles (b) Photoluminescence spectra of CdSeS alloy nanocrystal synthesized with different ratio of [Se] to [S]

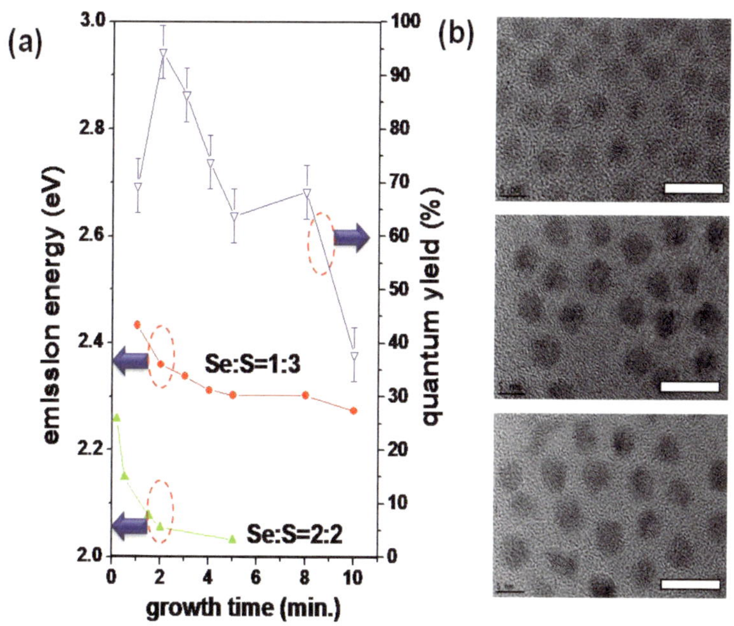

Fig. 3.4. (a) Temporal change of optical emission energy and quantum yield (b)TEM image of CdSeS alloy nanocrystals after 2 (upper), 4(middle), 8 minutes (lower).

Figure 3.4 shows temporal development of photoluminescence spectra of Cd(Se,S) alloy nanocrystals with initial Se-to-S ratio of 1:3. During the initial 2 minutes after injection of selenium precursor solution, the quantum yield of nanocrystals rapidly increased with time and photoluminescence spectra moves to lower energy side, which implies that "composition-gradient" structure is developed in the nanocrystals, which is structure composed of Se-rich core part and S-rich shell part. After 2 minutes at 300°C, the quantum yield decreased and emission energy showed red-shift, monotonously. The quantum yield of nanocrystals rapidly decreased to be 65% and emission energy showed red-shift to be 2.3eV after 5 minutes. During this stage of reaction, nanocrystal shows slight size growth as shown in the TEM image of Fig. 4(b). The change in optical spectra and size of nanocrystals during the growth stage with time from 2 minutes to 5 minutes indicates destruction of "composition-gradient" structure and homogenization of Se and S distribution in the nanocrystals. At 300°C, Se and S can diffuse into the other lattice site and exchange position. As a result of homogenization, high quantum yield obtained by CdS-rich surface layer was decreased. Further reaction at 300°C showed more rapid decrease in the quantum yield with small change in emission energy. TEM image showed some non-spherical and irregular nanocrystals. As shown in Fig. 3.5, photoluminescence spectra were rapidly broadened after 8 minutes at 300°C. Li *et al.* showed that Ostwald ripening of the core and core/shell nanocrystals is inevitable at above 240°C (Li et al, 2003). That is, broadening and photoluminescence spectra and irregular size distribution of nanocrystals indicates Ostwald ripening is dominant growth mechanism in this growth stage. The photoluminescence spectra of Cd(Se,S) nanocrystals grown at 300°C show decrease after reaction for 2 minutes without significant change in their width. As mentioned earlier, one possible origin of this change in PL intensity is redistribution of Se and S atoms inside nanocrystals. It is inferred that a composition-gradient of Se and S was formed in nanocrystals during the reaction at 300°C for 2 minutes and the radial distribution of Se and S atoms in nanocrystals was homogenized by prolonged reaction at the same temperature by diffusion. For CdSe-based nanocrystals, the temperature of 300°C is sufficient to further diffusion of atoms in the nanocrystal. A schematic illustration shown in Fig. 3.6 described the redistribution of anions and Ostwald ripening during the prolonged reaction at 300°C inferred from the PL spectra.

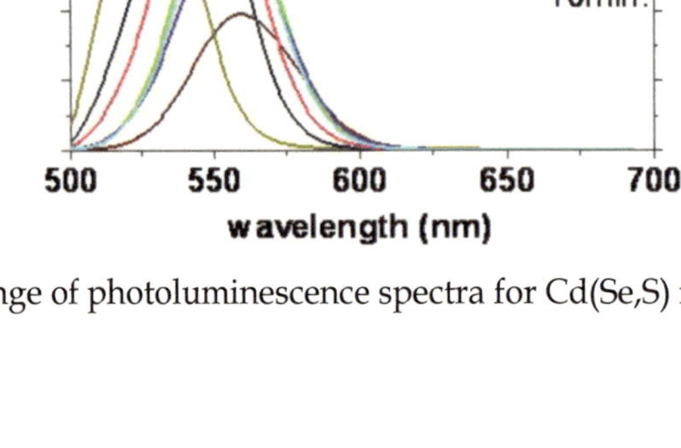

Fig. 3.5. Temporal change of photoluminescence spectra for Cd(Se,S) nanocrystals grown at 300°C.

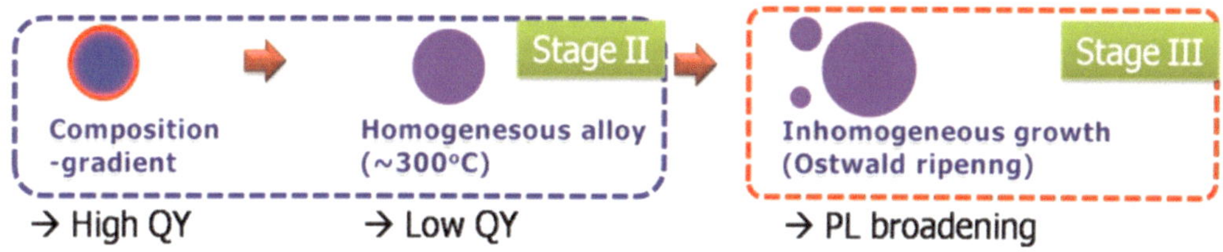

Fig. 3.6. A schematic illustration describing the origin of change in the PL spectrum for nanocrystals synthesized at 300°C

This suggest that fast cooling after injection can retard homogenization of atomic distribution and conserve the composition-gradient in nanocrystals. Figure 3.7(a) shows temporal change of quantum yield of nanocrystals which were fast cooled to various temperature after injection of Se and S mixed precursor solutions at 300°C. Without fast cooling to lower temperature, the quantum yield of nanocrystals initially reached to maximum value and rapidly diminished to be less than 40% after 8 minutes. In contrast to this, nanocrystals fast cooled to 225°C after precursor injection showed no significant decrease during 30 minutes. Moreover, the PL spectra for nanocrystals reacted at higher temperature showed clear red shift and broadening contrary to those reacted at lower temperature as displayed in Fig 3.7(b).

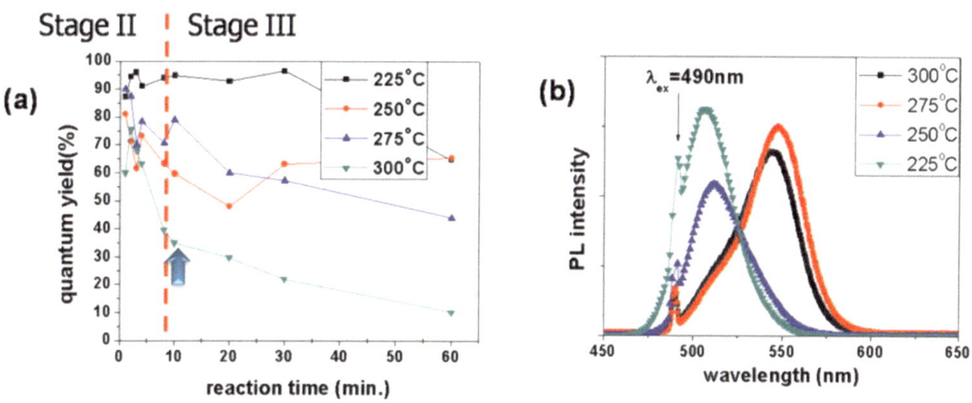

Fig. 3.7. Temporal change of (a) quantum yield and (b) photoluminescence spectra for Cd(Se,S) nanocrystals grown at various temperatures for 20 minutes.

3.2 Photoluminescence of $CuInS_2$ nanocrystals: Effect of alloying

3.2.1 Synthesis of $CuInS_2$ nanocrystals

The $CuInS_2$ NCs were synthesized using copper acetate, indium acetate and 1-dodecanethiol (DDT) according to the protocol reported by Zhong *et al* (Zhong et al., 2008). In detail, the amount of Cu(II) acetate was varied from 0.2mmol to 1mmol and indium acetate with quantity of 1mmol was used. Those precursors were dispersed in 30ml of 1-octadecene under inert atmosphere at room temperature and subsequently heated to 240°C. After reaction for 15 minutes, the solution was cooled to room temperature. To construct core-shell structure, 1mmol of zinc acetate was added to the $CuInS_2$ -based core solution after cooling to room temperature without purification and separation of synthesized NCs. In some experiment, we substitute cadmium acetate for small part of zinc acetate maintaining total amount of metal acetate constant to be 1mmol. For colloidal stability of NCs, 3mmol of palmitic acid mixed with 10ml of 1-octadecene was also added to the solution. Subsequently, the mixture was heated to 230°C under inert N_2 atmosphere and reacted without further addition of precursors. The mixed solution became uniform at above 120°C and further dwelling at 230°C to enhance the luminescence. After refluxing for 5 hours, DDT with quantity equivalent to zinc acetate was injected to the solution and maintained for an hour to form ZnS shell at the same temperature. Aliquots were taken at the end of each steps and rapidly cooled to room temperature. Before further characterization, polar solvent such as acetone, methyl alcohol, etc. was added to precipitate Cu-In-S_2-based NCs and the precipitates were purified by repetition of centrifugation and re-dispersion to n-hexane.

3.2.2 Effect of composition and surface modification

Reports have shown that the main fluorescence process in CIS NCs is closely related to lattice defects such as copper vacancy, indium substitute anti-site defect, etc.(Uehara et al., 2008). This implies that controlling crystal defect can enhance the fluorescence of $CuInS_2$ nanocrystals and intentional formation of some crystal defects was proved to be efficient to high fluorescence quantum yield. Fig. 3.8(a) shows various fluorescence spectra obtained from $Cu_{1-x}InS_2$ (x = 0 ~ 0.8) NCs with various different starting compositions.

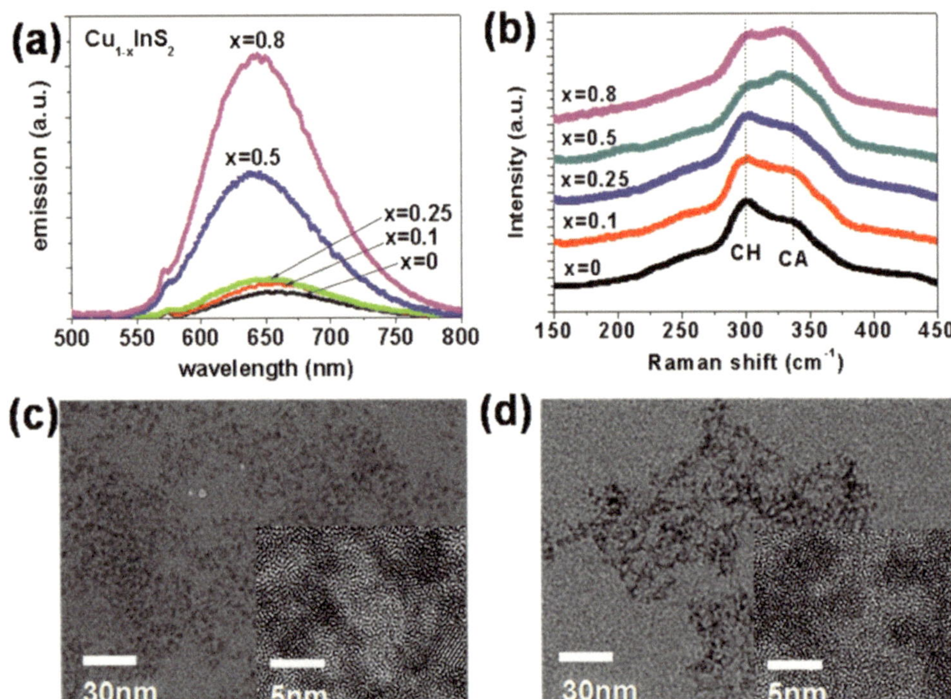

Fig. 3.8. Photoluminescence (PL) and (b) Raman spectra of $Cu_{1-x}InS_2$ nanocrystals prepared with different degree of copper deficiency in the starting solutions. (c-d) TEM images of $Cu_{1-x}InS_2$ nanocrystals with different starting composition of x=0 and x=0.8, respectively.

The luminescence intensity of CIS NC was highly enhanced and the wavelength of emission spectra showed blue-shift with minimizing copper content in the starting solution. As mentioned earlier, incorporation of crystal defect in CIS NCs by highly Cu-deficient composition in the starting solution can be ascribed to the source of improved luminescence. Uehara *et al.* explained the blue-shift of fluorescent spectra by the lowering of valence band due to Cu-deficiency in CIS NCs. The composition of CIS NCs prepared from precursor solution with Cu:In=0.2:1 was measured to be Cu:In=0.19:1 from induction-coupled plasma-mass spectrometry (ICP-MS). Thus, Cu-deficiency in starting solution resulted in large Cu deficiency in CIS NCs. In addition, the Raman spectra of Cu-deficient CIS NCs displayed in Fig. 3.8(b) show a characteristic peak of "CuAu"-type ordering (CA) at 340cm^{-1} as well as "chalcopyrite (CH)" peak at 300cm^{-1}. Since the CA peak indicates the clustering of defect pair at high concentration of defects (Uehara et al., 2008;Alvarez-Garcia et al., 2005), CA peak in the Raman spectrum of Cu-deficient Cu-In-S NCs also supports a large number of crystal defects induced by copper deficiency in the starting solutions. In particular, the more intense "CA" peaks compared with "CH" peaks in the Raman spectra implied that the highly Cu-deficient composition in the starting solution gave rise to higher concentration of lattice defects in the synthesized NCs. As observed in the TEM image in Fig. 3.8(c) & (d), the grown NCs was quite irregular in shapes and showed broad size distribution irrespective of Cu-deficiency in the starting solutions.

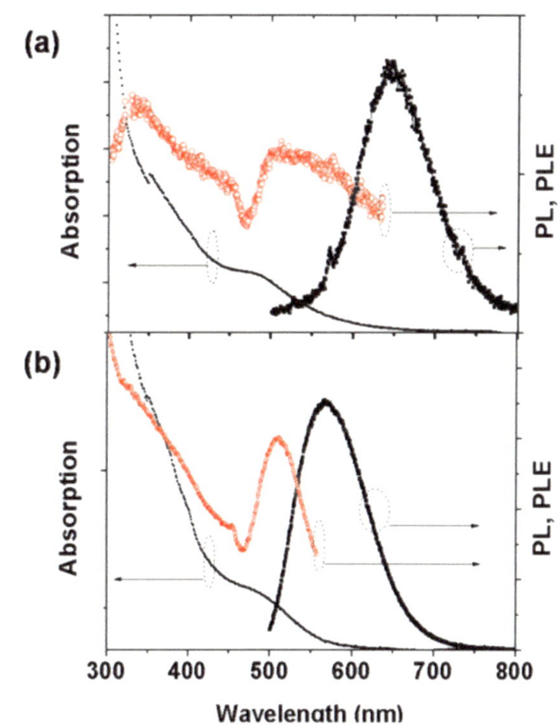

Fig. 3.9. Absorption, PL and PLE of (a) as-synthesized $Cu_{1-x}InS_2$ nanocrystals (x=0.2) and (b) $Cu_{1-x}InS_2$ nanocrystals (x=0.2) after refluxing with zinc acetate and palmitic acid for 5 hours at 230°C.

According to Zhong *et al.*'s work, $CuInS_2$ NCs were synthesized from the thermal decomposition of intermediate compound, $CuIn(SC_{12}H_{25})_x$ complex and prepared NCs were highly aggregated due to the oriented attachment of small particles (Zhong et al., 2008). As a result of irregular shape and poly-dispersed size distribution, the absorption spectrum of NCs showed broad shoulder with long wavelength trail and photoluminescence excitation spectrum (PLE) showed broad intensity distribution shown in Fig. 3.9(a).

The non-radiative recombination of electron-hole pairs by the surface states has been attributed to the major origin of the low quantum yield in semiconductor NCs. In addition, the passivation of surface defects by further shell growth on the semiconductor NCs was proved to be effective for high luminescence properties. Ryu *et al.* have suggested that zinc acetate acts as a source of ZnS shell on InP and acetic acid produced from dissociation of zinc acetate etches the defective surface of as-synthesized InP cores (Ryu et al., 2009). In this study, impressed by Ryu *et al.*'s work, we tried to improve the photoluminescence (PL) of CIS NCs by modifying their surface through refluxing with zinc acetate and palmitic acid. Fig. 3.9(b) shows optical spectra of CIS NCs after treatment with zinc acetate and palmitic acid for 5 hours at 230°C. The long wavelength trail in the absorption spectrum was efficiently suppressed and photoluminescence excitation spectrum (PLE) showed narrow intensity distribution. Since the long wavelength trail in the absorption spectra and broad PLE is mainly caused by inhomogeneous size distribution, more uniform size distribution of NCs was obtained after refluxing with zinc acetate and palmitic acid as inferred from the narrow absorption band in the optical spectra.

In addition, TEM images of NCs shown in Fig. 3.10(a) also support narrowing of the size distribution after refluxing with zinc acetate and palmitic acid. The low resolution image showed equally spaced dots, and oblate nanocrystals with crystalline lattice was observed from the high resolution image. Thus, the refluxing of CIS core solution with zinc acetate and palmitic acid resulted in a digestive ripening of CIS NCs. The aggregated CIS NCs were supposed to be broken into quasi-monodispersed NCs by etching with acetic acid produced from the reaction between zinc acetate and palmitic acid. Fig. 3.10(b) shows temporal evolution of the photoluminescence spectra of CIS NCs after refluxing with zinc acetate and palmitic acid at 230°C. The fluorescence quantum yield of NCs was highly enhanced to be above 40%, which was initially 3% for pristine CIS NCs. In particular, the quantum yield was improved up to 70% by further injection of DDT and subsequent refluxing for an hour at the same temperature. Since CIS core solution contained large amount of sulfur to suppress Cu_xS formation (Zhong et al., 2008) and as-synthesized CIS cores were refluxed without further purification or removing unreacted precursors, formation of ZnS shell is expected even without further injection of sulfur precursor such as DDT during the refluxing. In addition, the crystal structure of $CuInS_2$ and ZnS is quite similar and lattice mismatch between them is about 2%. Thus, as a result of passivation of surface defects and confinement of charge carriers by coherent shell formation, fluorescence of the CIS NCs was intensified by simple refluxing with zinc acetate and palmitic acid. Moreover, we supposed that etching of defect-rich surface with acetic acid also contributed to high quantum yield in CIS NCs refluxed with zinc acetate and palmitic acid. We have checked the effect of Cu-deficiency on the luminescence of CIS/ZnS NCs. Fig. 3.10(c) shows PL spectra of CIS/ZnS NCs prepared with different copper content in the starting solutions. The quantum yield of $Cu_{1-x}InS_2$/ZnS NCs was measured to be 3%, 16%, 70% for x=0, x=0.5, x=0.8, respectively. This implies that refluxing CIS NCs with zinc acetate and palmitic acid does not remove

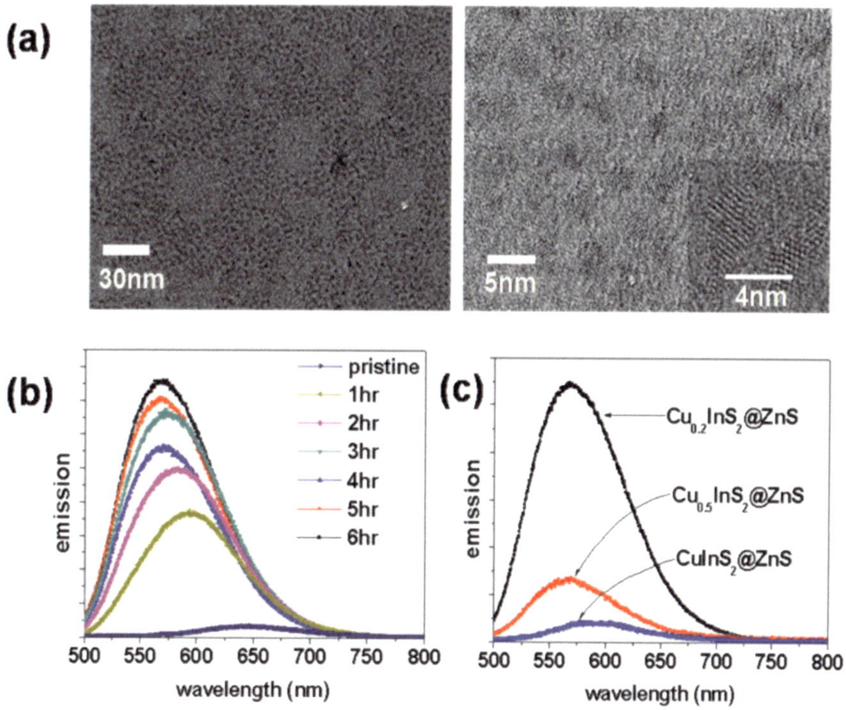

Fig. 3.10. (a) TEM images of $Cu_{1-x}InS_2$ nanocrystals (x=0.8) treated with zinc acetate and palmitic acid for 5 hours at 230°C. (b) Temporal evolution of PL of $Cu_{1-x}InS_2$ nanocrystals (x=0.2) (c) PL of $Cu_{1-x}InS_2$ NCs with various starting composition (x=0, 0.5, 0.8) after refluxing with zinc acetate and palmitic acid for 5 hours at 230°C.

copper defects located in the interior of CIS NCs, but passivate surface defects of the nanocrystals. Thus, it is concluded that Cu-deficiency is essential to high quantum yield of CIS NCs even after inorganic capping with ZnS.

As mentioned earlier, the origin of photoluminescence in copper indium sulfide nanocrystals was not unambiguously determined. In this work, we investigated the photoluminescence of nanocrystals with time-decay PL to obtain the detailed information on the phtoluminescence process. PL dynamics contains information about the recombination of photoinduced carriers in the nanocrystals. Fig. 3.11(a) demonstrates time-decay PL data of CIS NCs with various copper deficiency. The nanocrystals with small copper deficiency showed a much rapid decay contrary to those with large copper deficiency. In contrast to this, the surface modification of CIS NCs decreased their PL decay rate. In order to analyze the time-decay PL more quantitatively, we analyzed the multi-exponential PL decay by reconstructing the distribution of decay times with a minimum of *a priori* assumptions on the PL decay (Schlegel et al., 2002; Zenkevich et al., 2005). Since the recovery of decay time distribution from the decay data is an ill-posed problem, a regularization method should be used. We analyzed the multi-exponential PL decay by reconstructing the distribution of decay times with Tikhonov regularization method (Kapitonov et al., 1999; Hansen, 1994). Decay time distribution was computed using n=400 exponential terms with decay times uniformly spaced on the logarithmic scale. Fig. 3.11(b)

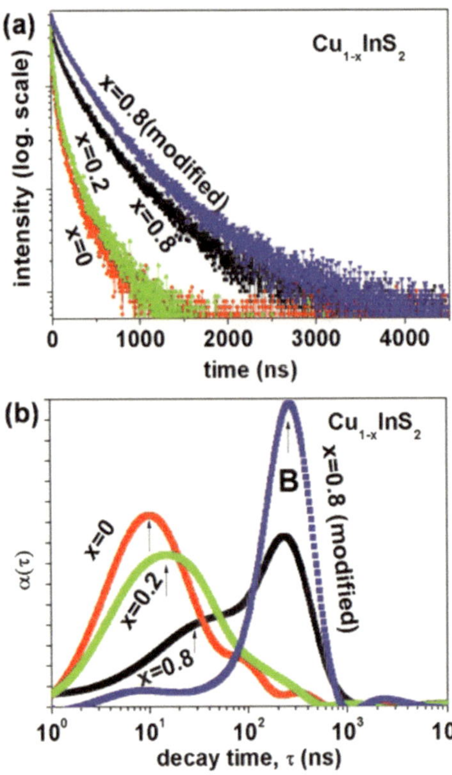

Fig. 3.11. (a) Time-dependent photoluminescence and (b) reconstructed decay time distribution of $Cu_{1-x}InS_2$ nanocrystals with various copper deficiency (x) and the surface modified $Cu_{1-x}InS_2$ (x=0.8) nanocrystals.

showed decay time distributions reconstructed from the time-resolved PL data. It is known that the decay time distribution in the photoluminescence of semiconductor nanocrystals can be modeled by log-normal statistics (van Driel et al., 2007). The decay time distribution for CIS NCs with small copper deficiency exhibited a log-normal distribution centered at τ_{max} = 3ns and for x=0 and τ_{max} = 15ns for x=0.2. In contrast to this, a bimodal decay time distribution was shown at τ_{max} =35ns and τ_{max} =237ns for x=0.8. That is, the contribution of the fast decay component was highly reduced and the slow decay component was dominant for the copper-deficient CIS NCs. Thus, the reconstruction of decay time distribution with a regularization method provides more detailed information on the excited state decay process compared with direct numerical fitting with stretched exponential function. In addition, the fast decay component in the decay time distribution was highly minimized by the surface modification of copper-deficient CIS NCs through formation of ZnS shell and etching of their surface.

It is noted that the decay time component of initially populated core-state recombination of exciton was known to be about 3ns (Dean, 1973). In addition, the large Stokes shift and the broad PL spectra eliminate the possibility of the direct exciton recombination in CIS NCs. The suppression of the fast decay component in the time-resolved PL by the surface modification of CIS NCs strongly indicated a contribution of trap-related emission. The decay time of PL in the semiconductor nanocrystals in the range of 15 through 25ns at room temperature was often ascribed to trap-related emission as reported for CdSe and CdSe/ZnS (Wang et al., 2003; Schlegel et al., 2002). The broad distribution of decay time for the fast decay component possibly resulted from the dispersion in trap energy levels. Thus, it is concluded that both trap-related emission and donor-acceptor recombination contribute to the luminescence of the CIS NCs and emission from the latter process is dominant for the CIS NCs with large Cu deficiency.

The role of trap states was further evidenced by the temperature-dependent photoluminescence spectra shown in Fig. 3.12(a). The PL spectra of CIS NCs with x=0.8 was quenched when the temperature was increased at above 100K. Contrary to this, PL spectra of the surface modified CIS NCs remain constant at least up to 300K. It is known that the quenching of the emission occurs at high temperature by ionization of the trapping centers if trapped electrons or holes play a role in the non-radiative recombination. Then, the thermal activation energy of quenching process can be closely related to the ionization energy of the trapping centers (Binsma et al., 1982). The thermal change in photoluminescence of semiconductor were often described as eq(1) which considers capture cross-section of carriers at the recombination centers (Krustok et al., 2000):

$$I(T) = I_0 / (1 + \phi_1 T^{3/2} + \phi_2 T^{3/2} \exp(-E_T / k_B T)) \qquad (1)$$

Here, I_0, ϕ_1, ϕ_2 are the three fitting parameters and E_T is the thermal quenching energy. By numerical fitting with eq(1), the thermal change of the integrated PL intensity of bare CIS NCs was well fitted and the thermal quenching energy of bare CIS NCs was estimated to be 84meV. Thus, the temperature-dependent PL study showed that the trap states incorporated in the uncapped CIS NCs were effectively removed by the surface modification through refluxing with zinc acetate and palmitic acid.

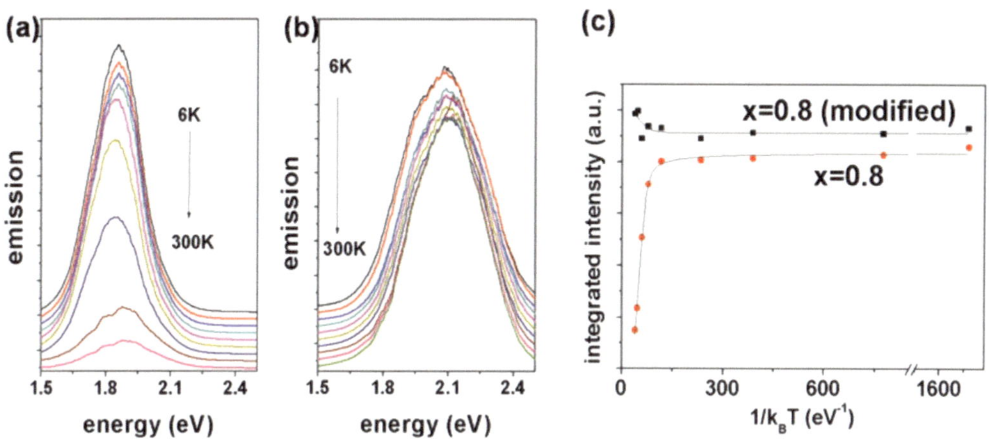

Fig. 3.12. Temperature-dependent photoluminescence of (a) $Cu_{1-x}InS_2$ nanocrystals (x=0.8) and (b) the surface modified $Cu_{1-x}InS_2$ (x=0.8) nanocrystals; (c) Thermal change of the integrated intensity of photoluminescence spectra of both the as-synthesized and the surface modified $Cu_{1-x}InS_2$ (x=0.8) nanocrystals.

3.2.3 Effect of alloying on the photoluminescence of $CuInS_2/ZnS$ nanocrystals

One of important features in copper deficient CIS/ZnS core-shell nanocrystals is large blue-shift of luminescence intensity after refluxing nanocrystals with zinc acetate and palmitic acid. In contrast to digestive ripening of CdSe quantum dots demonstrating red-shift in PL spectra, digestive ripening of CIS NCs by refluxing with zinc acetate and palmitic acid showed blue-shift in PL spectra. Fig. 3.10(b) shows temporal development of photoluminescence excitation spectra of CIS NCs during the surface treatment with zinc acetate and palmitc acid. The blue shift of luminescence peak implies that the band gap of CIS NCs increases with increasing reflux time. Several reports have suggested various origin of blue-shift in PL such as size reduction by etching with acetic acid (Ryu et al., 2009), alloying of CIS NCs with ZnS (Uehara et al., 2008) and strain field caused by lattice mismatch between core material and shell material in the "core/shell" NCs. In order to test the feasibility of these suggestions, we refluxed $Cu_{1-x}InS_2(x=0.8)$ core solution with the cadmium acetate as well as zinc acetate and palmitic acid. Fig. 3.13(a) showed the PL spectra and their temporal evolution during refluxing. Since the total amount of metal acetates added was maintained irrespective of the relative portion of cadmium acetate in the reaction mixture, the amount of free acetic acid generated from reaction between metal acetate and palmitic acid can be considered to be constant. Hence, if the contribution of size reduction by etching with acetic acid is significant, addition of cadmium acetate is hard to modify the blueshift in PL. However, as the amount of cadmium acetate was increased, the blue shift of emission wavelength was minimized. In particular, the blueshift in PL after refluxing at 230°C was effectively suppressed by applying equal amount of cadmium acetate and zinc acetate. Thus, the influence of etching with acetic acid on the shift in the fluorescence wavelength is negligible. In addition, possibility of the contribution of alloying with wider band gap semiconductor can be eliminated. Since both CdS (Bulk Eg=2.4eV) and ZnS (Bulk Eg=3.6eV) have much wide band gap compared with $CuInS_2$ (Bulk Eg=1.5eV), alloying or doping of Cd or Zn into the lattice of CIS NCs usually exhibit clear blue-shift in PL. Then, the contribution of alloying to the blue-shift in PL observed in this study is not plausible.

It is known that lattice strain can change band gap energy significantly by a coherent overgrowth of shell material on a small core (Smith & Mohs, 2009). The lattice mismatch between core and shell of nanocrystals was estimated to be about 2% for CIS/ZnS NCs. The lattice stress applied to a CIS core by a ZnS shell overgrowth is compressive as estimated from lattice parameters of each materials (CIS: a=0.5517nm, ZnS: a=0.5345nm). The values of band deformation potential (in eV/unit dilatation) for CIS are known to be -11.0 and +8.27 for conduction band and valence band, respectively (Quintero et al., 1992). Then, energy gap of CIS changed to -19.27eV by unit dilatation. The compressive stress applied to CIS NCs induces contraction of CIS lattice and increase the band gap. As a result of this, the PL spectra shifted to high energy or short wavelength. In addition, stress applied to a CIS core by a CdS shell is tensile considering lattice parameter of CdS (a=5.820nm). Then, the position of PL spectra moves to low energy or long wavelength by capping with CdS. In fact, CdS capping on $CuInS_2$ NCs showed clear redshift in PL spectra.Thus, the position of PL spectra for CIS NCs refluxed with mixture of cadmium acetate, zinc acetate and palmitc acid should move to long wavelength since sign of stress on CIS NCs by capping changes from compressive to tensile through increasing amount of cadmium acetate. Hence, it is concluded that the noticeable blue shift in PL spectra of CIS NCs after refluxing with zinc acetate and palmitic acid is a result of lattice strain applied to CIS cores by formation of inorganic shell layers.

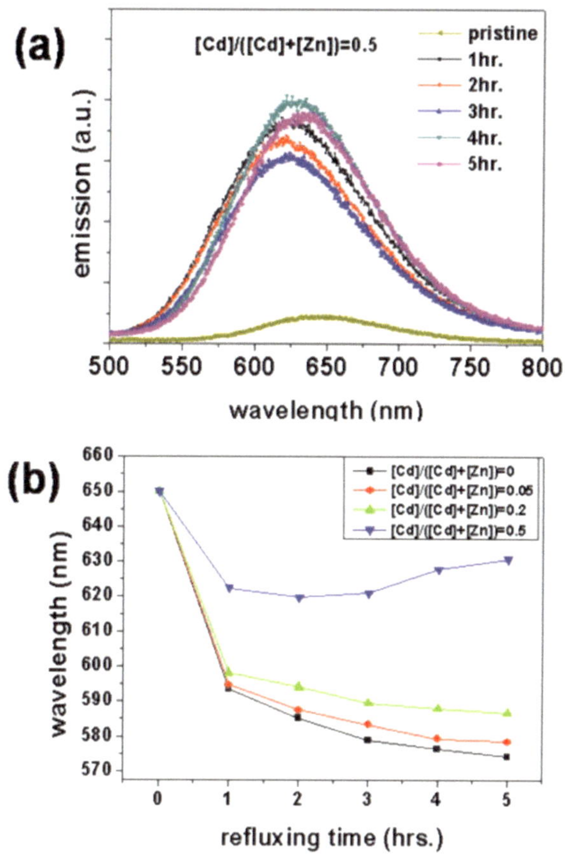

Fig. 3.13. Temporal change of (a) photoluminescence spectra and (b) emission wavelength of $Cu_{1-x}InS_2$ nanocrystals (x=0.8) the surface modified with alloyed capping layers.

4. References

Alivisatos, A. P. (1996). Semiconductor Clusters, Nanocrystals, and Quantum Dots, *Science* Vol. 271, pp.933-937

Alvarez-Garcia, J.; Barcones, B.; Perez-Rodrigues, A.; Romano-Rodrigues, A.; Morante, J. R.; Janotti, A; Wei, S. -H. & Scheer, R. (2005). Vibrational and crystalline properties of polymorphic $CuInC_2$ (C=Se,S) chalcogenides, *Phys. Rev. B*, Vol. 71, 054303.

Asokan, S.; Krueger, K. M.; Alkhawaldeh, A.; Carreon, A. R.; Mu, Z.; Colvin, V. L.; Mantzaris, N. V. &Wong, M. S. (2005). The use of heat transfer fluids in the synthesis of high-quality CdSe quantum dots, core/shell quantum dots, and quantum rods, *Nanotechnology Vol.* 16, pp. 2000-2011

Asokan, S.; Krueger, K. M.; Colvin, V. L. & Wong, M. S. (2007) Shape-Controlled Synthesis of CdSe Tetrapods Using Cationic Surfactant Ligands, *Small* Vol. 3, pp. 1164-1169

Bae, W. K.; Char, K. H.; Hur H. & Lee, S. (2008). Single-Step Synthesis of Quantum Dots with Chemical Composition Gradients *Chem. Mater.* Vol. 20, pp.531-539

Bailey, R. E. & Nie, S. M. (2003). Alloyed Semiconductor Quantum Dots: Tuning the Optical Properties without Changing the Particle Size, *J. Am. Chem. Soc.* Vol. 125, pp. 7100-7106

Bawendi, M.; Steigerwald, M. & Brus, L. (1990). The Quantum Mechanics of Larger Semiconductor Clusters, *Annu. Rev. Phys. Chem.* Vol. 41, pp. 477-496

Binsma, J.; Giling, L.J.; Bloem, J. (1982). Luminescence of $CuInS_2$: I. The broad band emission and its dependence on the defect chemistry, *J. Lumin.* , Vol. 27 , pp. 35-53.

Comas, F.; Trallero-Giner, C.; Studart, N. & Marques, G. E. (2002). Interface optical phonons in spheroidal dots: Raman selection rules *Phys. Rev.* B Vol. 65, pp. 073303-073305

Dalpian, Gustavo M. & Chelikowsky, J.R. (2006). Self-Purification in Semiconductor Nanocrystals, *Phys. Rev. Lett.*, Vol. 96, pp. 226802

Dean, P.J. (1973). in *Progress in Solid State Chemistry* McCaldin, J.O., Somorjai, G.(Eds), Pergamon: New York, Vol. 8 p. 1-126.

van Driel, A.F.,; Nikolaev, S.; Vergeer, P.; Lodahl, P.; Vanmaekelbergh, D.& Vos, W.L. (2007). Statistical analysis of time-resolved emission from ensembles of semiconductor quantum dots: Interpretation of exponential decay models, *Phys. Rev. B.*, Vol. 75, 035329.

Erwin, S. C.; Zu, L. J.; Haftel, M. I.; Efros, A. L.; Kennedy, T. A. & Norris, D. J. (2005) Doped Nanocrystals, *Nature* Vol. 436, pp. 91-94

Hamanaka,Y.; Kuzuya,T.; Sofue,T.; Kino, T.; Ito, K. & Sugiyama, K. (2008). Defect-induced photoluminescence and third-order nonlinear optical response of chemically synthesized chalcopyrite CuInS2 nanoparticles, *Chem. Phys. Lett.* Vol. 466, pp.176-180

Hankare, P. P.; Delekar, S. D.; Chate, P. A.; Sabane, S. D.; Garadkar, K. M. & Bhuse, V. M. (2005). A novel route to synthesize $Cd_{1-x}Pb_xSe$ thin films from solution phase, *Semicond. Sci. Technol.* Vol. 20, pp. 257-264

Hansen P. C. (1994). Regularization Tools: A Matlab package for analysis and solution of discrete ill-posed problems, *Numerical Algorithms*, Vol.6, pp. 1–35

Hsu, Y. & Lu, S. (2008) Dopant-Induced Formation of Branched CdS Nanocrystals, *Small*, Vol. 4, pp. 951-955

Hofhuis, J.; Schoonman, J.; Goossens, A. (2008). *J. Phys. Chem. C*, Vol. 112, pp.15052-15059.

Jing, P.; Zheng, J.; Ikezawa, M.; Liu, X.; Lv, S.; Kong, X.; Zhao, J. & Masumoto, Y. (2009). Temperature-Dependent Photoluminescence of CdSe-Core CdS/CdZnS/ZnS-Multishell Quantum Dots, *J. Phys. Chem.* C Vol. 113, pp. 13545-13550

Kapitonov, A.M.; Stupak, A.P.; Gaponenko, S.V.; Petrov, E.P.; Rogach, A.L.; Eychmuller, A. (1999). Luminescence Properties of Thiol-Stabilized CdTe Nanocrystals, *J. Phys. Chem. B* , Vol.103, pp. 10109-10113.

Krustok, J.; Raudoja, J.; Schon, J.H.; Yakushev, M.; Collan, H. (2000). The role of deep donor-deep acceptor complexes in CIS-related compounds, *Thin Solid Films,* Vol. 361–362, pp. 406-410

Kudera, S.; Carbone, L.; Manna, L. & Parak, W. J., (2008). Growth mechanism, shape and composition control of semiconductor nanocrystals, In: *Semiconductor Nanocrystal*

Quantum Dots, Rogash, A.(Eds.), pp. 1-34, Springer Verlag, ISBN 978-3-211-75235-7, Wien

Lange, H.; Artemyev, M.; Woggon, U. & Thomsen, C. (2009). Geometry dependence of the phonon modes in CdSe nanorods, *Nanotechnology* Vol. 20, pp. 045705-045709

Li, J & Wang, L. W. (2003). High Energy Excitations in CdSe Quantum Rods, *Nano Letters* Vol. 3, pp. 101-105

Li, J. J.; Wang, Y. A.; Guo, W. Z.; Keay, J. C.; Mishima, T. D.; Johnson, M. B.; Peng & X. G. (2003). Large-Scale Synthesis of Nearly Monodisperse CdSe/CdS Core/Shell Nanocrystals Using Air-Stable Reagents via Successive Ion Layer Adsorption and Reaction, *J. Am. Chem. Soc.*, Vol. 125, pp.12567- 12575

Manna, L.; Scher, E. C. & Alivisatos, A. P. (2000). Synthesis of Soluble and Processable Rod-, Arrow-, Teardrop-, and Tetrapod-Shaped CdSe Nanocrystals, *J. Am. Chem. Soc.* Vol. 122, pp.12700- 12706

Markovich, G.; Collier, C.; Henrichs, S.; Remacle, F.; Levine, R. & Heath, J. R. (1999) Architectonic Quantum Dot Solids, *Acc. Chem. Res.* Vol. 32, pp. 415-423

Murray, C. B.; Norris, D. J. & Bawendi, M. G. (1993). Synthesis and characterization of nearly monodisperse CdE (E = sulfur, selenium, tellurium) semiconductor nanocrystallites *J. Am. Chem. Soc.* Vol. 115, pp. 8706- 8715

Ryu, E.; Kim, S.; Jang, E.; Jun, S.; Jang, H.; Kim, B. & Kim, S.-W. (2009). Step-Wise Synthesis of InP/ZnS Core−Shell Quantum Dots and the Role of Zinc Acetate, *Chem. Mater.* Vol. 21, pp.573- 575

Saunders, A. E.; Ghezelbash, A.; Sood, P. & Korgel, B. A. (2008). Synthesis of High Aspect Ratio Quantum-Size CdS Nanorods and Their Surface-Dependent Photoluminescence, *Langmuir* Vol. 24, pp. 9043-9049

Schlegel, G.; Bohnenberger, J.; Potapova, I. & Mews, A. (2002). Fluorescence Decay Time of Single Semiconductor Nanocrystals, *Phys. Rev. Lett.* Vol.88, 137401.

Uehara, M.; Watanabe K.; Tajiri Y.; Nakamura, H. & Maeda, H. (2008). Synthesis of $CuInS_2$ fluorescent nanocrystals and enhancement of fluorescence by controlling crystal defect, *J. Chem. Phys.*, Vol. 129, 134709

Wang, X.; Qu, L.; Zhang, J.; Peng, X.; Xiao, M. (2003). Surface-Related Emission in Highly Luminescent CdSe Quantum Dots, *Nano Letters,* Vol.3 , pp.1103-1106

Xie, R.; Kolb, U.; Li, J.; Basche, T. & Mews, A. (2005). Synthesis and Characterization of Highly Luminescent CdSe−Core CdS/$Zn_{0.5}Cd_{0.5}S$/ZnS Multishell Nanocrystals, *J. Am. Chem. Soc.* Vol. 127, pp. 7480-7488

Xie, R.; Rutherford, M. & Peng, X. (2009). Formation of high-quality I-III-VI semiconductor nanocrystals by tuning relative reactivity of cationic precursors *Chem. Mater.* , Vol. 22, pp. 5691-5697

Xu, S.; Ziegler, J. & Nann, T. (2008). Rapid synthesis of highly luminescent InP and InP/ZnS nanocrystals, *J. Mater. Chem.* Vol. 18, pp.2653-2656.

Zhang, J. & Yu, W. W. (2006). Formation of CdTe nanostructures with dot, rod, and tetrapod shapes, *Appl. Phys. Lett.*, Vol. 89, pp. 123108-123111

Zhong, H.; Zhou, Y.; Mingfu, Y.; He, Y.; Ye, J.; He, C.; Yang, C. & Li, Y. (2008). Controlled Synthesis and Optical Properties of Colloidal Ternary Chalcogenide $CuInS_2$ Nanocrystals, *Chem. Mater.*, Vol. 20, pp.6434-6443.

Zenkevich, E.; Cichos, F.; Shulga, A.; Petrov, E.P.; Blaudek, T. & von Borczyskowski, C. (2005). Nanoassemblies Designed from Semiconductor Quantum Dots and Molecular Arrays, *J. Phys. Chem. B* 109, pp. 8679-8685.

15

Functional Organic Nanocrystals

Koichi Baba[1], Hitoshi Kasai[2,3], Kohji Nishida[1] and Hachiro Nakanishi[2]
[1]*Osaka University*
[2]*Tohoku University*
[3]*PRESTO, Japan Science and Technology Agency*
Japan

1. Introduction

The topic in this chapter concerns organic nanocrystals. Numerous scientific researches of nanocrystals are focusing on inorganic compounds such as a metal and a semiconductor. The research related to organic nanocrystals is relatively new compared to that of inorganic nanocrystals. Nearly two decades ago, our research team had developed a novel preparation method of organic nanocrystals, what we call the reprecipitation method. Since then, we have been leading the research area of organic nanocrystals. The reprecipitation method has bottom-up type with a wet process procedure to prepare organic nanocrystals, and organic nanocrystals are obtained as their water dispersion. Numerous numbers of organic compounds are allowed to be a candidate for a targeting of organic nanocrystals by means of using the reprecipitation method. Functional organic nanocrystals have been attracted in a scientific and technological area of materials. In this chapter, first we mention the research back ground of organic nanocrystals, and then introduce the preparation method of organic nanocrystals, i.e. the reprecipitation method, and their improved ones. Then, several kinds of functional organic nanocrystals relating to optoelectrical materials, nanomedicine, and biophotonics are mentioned. Recent topics of organic nanocrystals achieved by several groups are briefly introduced. In this chapter, the topics are proceeding as follows:

1. Back ground of the research of organic nanocrystals
2. Preparation method of organic nanocrystals
3. Several functional organic nanocrystals
4. Recent topics of organic nanocrystals achieved by several groups

2. Back ground of the research of organic nanocrystals

Fabrication of nanocrystals and their nanostructure has been investigated extensively for metals (Ung et al., 1997) and semiconductors (Murray et al., 1993), and many characteristic properties have already been shown. For example, a novel and unique optical property called the quantum size effect (Nakamura, 1992) was found for such inorganic nanocrystals in a size of a few nanometres (e.g. nonlinear optics, semiconductor lasing, and quantum dot applications).

For organic nanocrystals, however, nanocrystals fabrication has so far rarely been investigated, owing to their thermal instability, except for the thermally stable phthalocyanine and a kind of low molecular weight aromatic compounds which were nanocrystallized by an evaporation method in an inert gas (Granqvist et al., 1976) or by a crystal growth technique in a porous polymer (Matsui et al., 1995). Recently, we have proposed a simple and convenient method for the fabrication of organic nanocrystals, what we call the reprecipitation method (Kasai et al., 1992). In brief, the hydrophobic solution in which hydrophobic compounds dissolved is injected into water, which results in a reprecipitation owing to its insolubility in water to give nanocrystal dispersion. By this method, many kinds of organic compounds such as polydiacetylene (Katagi et al., 1996), phthalocyanine (Yanagawa et al., 1997), perylene (Kasai et al., 1996) and so on can be nanocrystallized successfully. Interestingly, size-dependent optical properties are observed for these organic nanocrystals even in the 100 nm size range, an order of magnitude larger than the scale where the inorganic nanocrystals usually show size-dependent optical properties. These aqueous organic nanocrystals dispersions may have great potential for use in fundamental and applied optics, e.g. as nonlinear optical (NLO) materials having large $\chi^{(3)}$ without optical loss (Kasai et al., 1995) and having easy process ability to other material forms.

Until now, not only the investigation of fundamental optelectrical properties of organic nanocrystals, but also their interesting physiochemical properties and their peculiar advantages compared to inorganic nanocrystals have been extended to nanomedicine and biophotonics. First we introduce the fabrication technique of organic nanocrystals; the reprecipitation method.

3. Preparation method of organic nanocrystals

3.1 The reprecipitation method

The reprecipitation method, which is a solvent displacement method, was first reported by our research groups (Fig. 1a) (Kasai et al., 1992). It provides a very simple and versatile way to prepare organic nanocrystals dispersions. The method involves a rapid mixing of a small amount of concentrated stock solution of the target compound dissolved in a good solvent with excess of a poor solvent. The great disparity between the solubilities of the target compound in the good and poor solvents, and the good compatibility of the two solvents are essential for this method. The rapid mixing of the stock solution and the poor solvent changes the micro-environment of the target compound molecules. The molecules are exposed to the poor solvent surroundings in a very short time, inducing the nucleation and growth of the molecules to nanocrystals. The size and morphology of nanocrystals can be controlled (Fig. 1b and 1c) by tuning a concentration, temperature, stirring speed, and adding surfactant. Then the precipitation creates nanocrystals dispersed in water. Many kinds of organic compound such as π-conjugate compounds, NLO dyes (Kaneko et al, 2005), fluorescent dyes (Fig. 1d), fullerene (Tan et al., 2008), and several polymers (Ishizaka et al., 2006; Zhao et al., 2009) are candidate for nanocrystals (nanoparticles for polymer). However some kinds of rigid compounds such as phthalocyanine having π-π stack structure are quite hard to be solved in usual organic solvent, thus for the nanocrystal preparation of these attractive functional compound, an advanced preparation technique is required. A supercritical fluid is a candidate for dissolving this problem.

3.2 Supercritical fluid crystallization method

The properties of phthalocyanine are very interesting in both optics and electronics in the fields of nonlinear optics, organic photoconductors, and pigments. Phthalocyanine is known

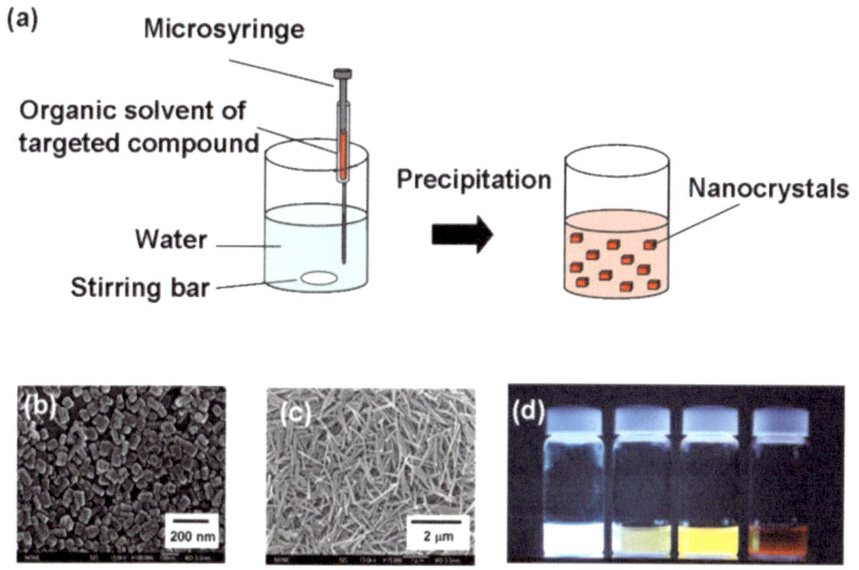

Fig. 1. (a) Scheme of the reprecipitation method. (b,c) Size controlled organic nanocrystals in same compound. (d) Fluorescence images of organic nanocrystals in water.

to have large $\chi^{(3)}$ and ultrafast optical response in the crystalline state (Matsuda et al., 1990).. However, it is almost impossible to make large single crystals of phthalocyanine, because its solubility is very poor in almost all kinds of solvent. Thus, one of the best ways for the application of phthalocyanine crystals would undoubtedly be to make devices with constructed by nanocrystals. Even this, commercial phthalocyanine nanocrystals have been prepared and refined only by the acid-pasting method, and these phthalocyanine nanocrystals have been used as materials for organic photoconductor. However, this method may give rise to environmental problems in the near future due to the use of much sulphuric acid.

On the other hand, supercritical fluid (SCF) has been expected to have a high solubility against insoluble organic compounds, since it shows high solubility under high temperatures and high pressure. Thus, we expected the SCF approach to be an alternative method for making phthalocyanine nanocrystals, instead of the acid pasting method. On the basis of this idea, the use of SCF was examined for the purpose of improving the reprecipitation method. The new method is called supercritical fluid crystallization (SCFC) method (Fig. 2) (Komai et al., 1998). The preparation of phthalocyanine nanocrystals using the SCFC method was used in order to demonstrate this usefulness. The size and modification control of titanyl phthalocyanine crystal have been attempted using the reprecipitation method from SCFC. The fabrication r-form of titanly phthalocyanine nanocrystals was achieved, which is a hopefully candidate for materials for organic photoconductor. Recently, further application using SCFC method has been developed as new one including several organic pigments and inorganic semiconductors. As shown in this section, the applying the additional outsources energy like SCF to the reprecipitation method is quite useful for extending the preparation possibility of organic nanocrystals. As a further extension, the approach using a microwave irradiation was also demonstrated as a useful procedure for preparing organic nanocrystals.

3.3 Microwave-irradiation reprecipitation method

Recently, as a novel fabrication technique of functional materials, a microwave irradiation process is much noted in the field of organic and inorganic synthesis and nanoparticles. For example, high critical temperature superconductors, mono-dispersed nickel, gold, silver nanoparticles, and organic-metal complexes were successfully prepared by using this technique.

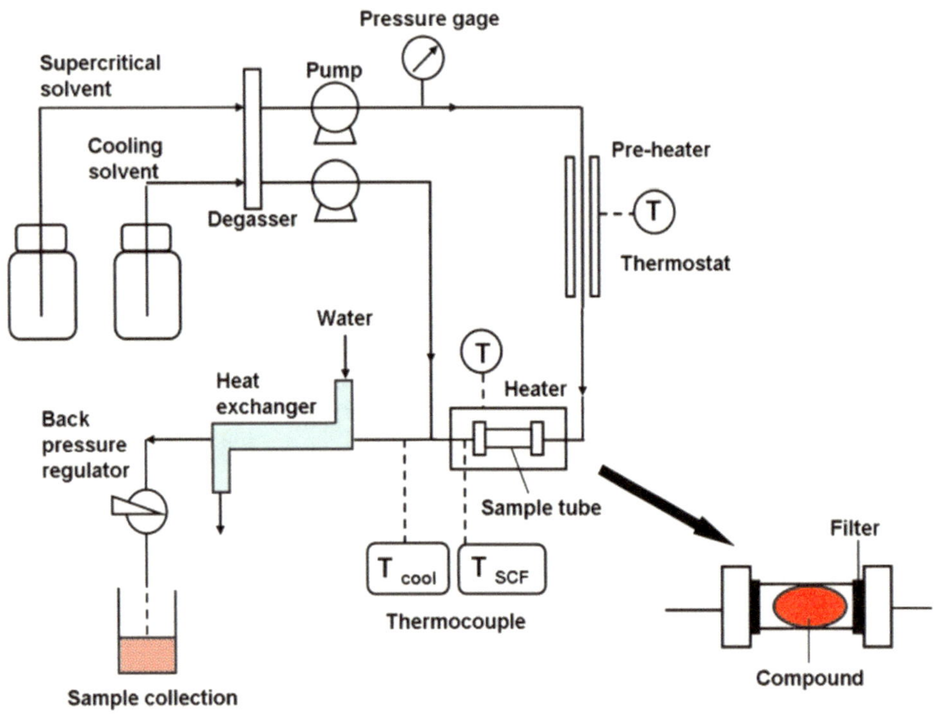

Fig. 2. Scheme of the supercritical fluid crystallization method.

We focused on the unique aspect of microwave at 2.45 GHz. One of the aspects of microwave (2.45 GHz) is that the microwave is well absorbed by water molecule, resulting in effective homogeneous and rapid heating of water is realized. Since nanocrystals are prepared in water environment by the reprecipitation method, we have attempted to apply the microwave-irradiation process to the reprecipitation method for the purpose of fabricating well-defined organic nanocrystals (Fig. 3). Upon microwave-heating was applied to the reprecipitation method, homogeneous and rapid heating increased the homogeneous crystal growth and then the crystallization was completed without particle aggregation. This homogeneous heating resulted in making fine narrow-sized homogeneous nanocrystals dispersed in water (Baba et al., 2000). If the conventional conductive heating such as hot plate heating was used, this gradient heating induced large-sized and random-size distributed crystal growth in aqueous dispersion system.

We demonstrated that the narrow size distributed nanocrystals of TPB, perylene (Fig. 4) (Baba et al., 2000), and DCHD were successfully fabricated within a short time by applying microwave heating just after our conventional reprecipitation procedure. Furthermore, there

was finding that the size and morphology of DCHD nanocrystals and nanofiber were controlled well under using the open- and closed-type vessel (Fig. 5) (Baba et al., 2007a).
The next stage of our interest was how to mass produce these attractive organic nanocrystals for the sake of industrial approach.

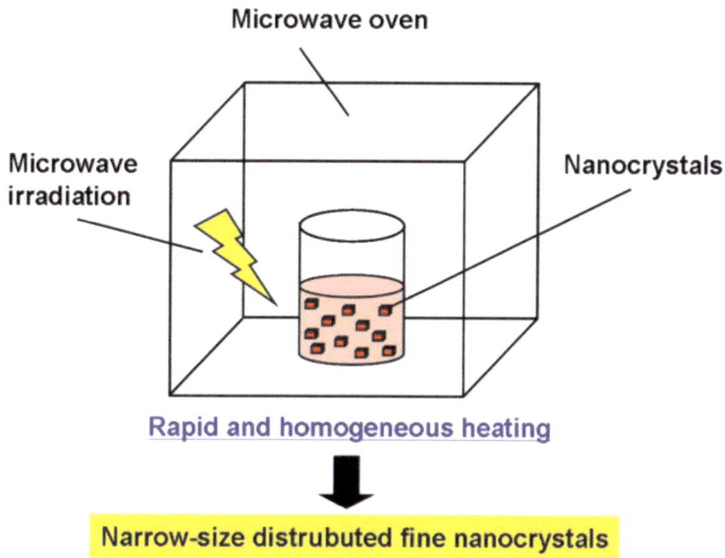

Fig. 3. Scheme of the microwave-irradiation reprecipitation method.

Fig. 4. Perylene nanocrystals prepared by the microwave-irradiation reprecipitation method. (Reprinted from (Baba, et al., 2000): Japanese Journal of Applied Physics. Vol. 39 (2000) pp. L1256-L1258, Part 2, No. 12A, DOI: 10.1143/JJAP.39.L1256. Copyright 2000 The Japan Society of Applied Physics)

3.4 Mass production of organic nanocrystals

The reprecipitation method is an easy and convenient technique to fabricate organic nanocrystals in general. However, only limited amount of nanocrystals dispersion (about several-tens ml) was prepared at once, i.e. the laboratory scale. To overcome this problem, we introduced a pump as an injection apparatus of the reprecipitation method for mass-production of pigment nanocrystals with controlled size and morphology (Ujiiye-Ishii et al., 2006).

For mass-production, most efficient conditions of the reprecipitation method have injecting highly concentrated solution with high injection rate. The former condition was achieved by using 1-methyl-2-pyrrolidone (NMP) and the latter was realized by using a high flow-rate pump. Using the injection pump-assisted reprecipitation method, the amount of obtained quinacridone nanocrystals in dispersion was 700 mg per minute at maximum, and then 3.5 g of the nanocrystals were obtained in one experiment. This value was about 17,500 times

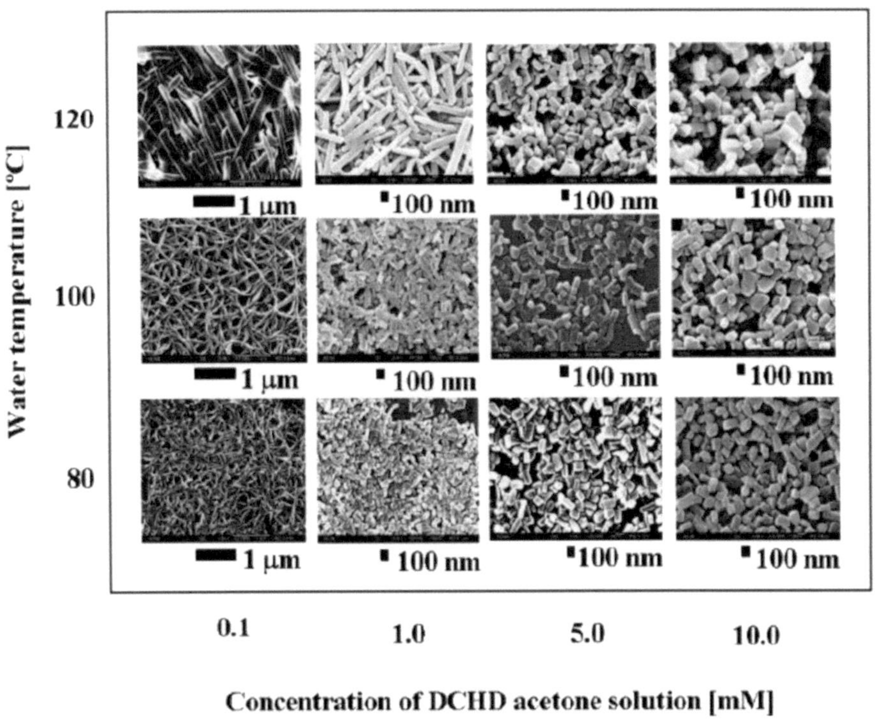

Fig. 5. Poly(DCHD) nanocrystals prepared by the closed-type microwave-irradiation reprecipitation method. (Reprinted from (Baba, et al., 2007a): Japanese Journal of Applied Physics. Vol. 46 (2007) pp. 7558-7561, No. 11, DOI: 10.1143/JJAP.46.7558. Copyright 2007 The Japan Society of Applied Physics)

larger than that obtained by the conventional reprecipitation method with manual injection. The size of quinacridone nanocrystals fabricated was 25–80 nm, which was confirmed by observation using a scanning electron microscope. From the powder X-ray diffraction data, only α-type nanocrystals were obtained. This mass production procedure of pigment nanocrystals is useful for an application of a color filter.

4. Several functional organic nanocrystals

4.1 Color filter using organic pigment nanocrystals

Organic pigments have been widely used in printing and coating industries as the coloring agents of synthetic fibers and plastics. Recent years, some organic pigments have an attention as organic photoconductors such as organic EL materials and electrophotographic photoreceptors. In addition, organic pigments are used for ink-jet inks to produce photographic-quality pictures and for color filters that are essential materials for the full-colorization of digital cameras and liquid crystal displays (LCDs). To increase the performance of the color filter, it is absolutely essential to cancel out the light scattering by reducing the particle size of organic pigments. Yet dyes can be an attractive alternative to overcome this limitation due to the reduced light scattering because they are dissolved in the media and exist in molecule form, however, in order for the dyes to be applied successfully to LCDs manufacturing process, their low thermal stability needs to be improved. Therefore, as the way to reduce the particle size, the breakdown method using beads and/or inorganic salts are commonly used. However, an energy efficiency, inclusion of contaminants such as bead materials (e.g., inorganic salts and zirconia), and a difficulty to reduce the particle size less than 50 nm are causing industrial problems. Therefore, the solution of these problems has been strongly desired. To overcome these problems, mass production procedure of organic pigment nanocrystals by the newly developed reprecipitation method was investigated, and the possibility of their application to color filter for LCDs was evaluated.

To evaluate the prepared organic pigment nanocrystals as a functional material for color filters, it is necessary to significantly increase their preparation amount. For this realization, a highly concentrated pigment solution and an improvement in the injection rate of the solution were necessary. Based on the literature (Ujiiye-Ishii et al., 2006), the system using a pulsation-free pump was built. In this newly developed reprecipitation method, it increased the preparation amount of nanocrystals by 104–106 times (Miyashita et al., 2008).

The organic pigment nanocrystals prepared by the reprecipitation method were finer in mono-dispersed property, compared to the conventional products. The average size of pigment nanocrystals prepared by the reprecipitation method was 25 nm and that of the pigments prepared by the breakdown improved milling method was 39 nm, and the former was in finer size dispersion (i.e., mono-dispersion). The relationship between the particle formation mechanism and particle size in the reprecipitation method is also under study. To clarify the reduction of the light scattering that is affected by the reduced size of pigment particles, the light scattering intensity was measured using a goniometer. In LCDs, if the light scattering occurs by pigment particles, the reduction of the display contrast and/or the luminance ratios occurs. Thus this becomes a serious problem when the display is in ON/OFF in LCDs.

It is apparent that the use of pigment nanocrystals prepared by the reprecipitation method can significantly increase the contrast ratio of LCDs, which is one of the most important performances. As a result, a new production process of organic pigment nanocrystals reduced the light scattering by fine size and well dispersed organic pigment nanocrystals prepared by the reprecipitation method.

It was found that organic pigment nanocrystals with the particle size about 25 nm were successfully achieved by the reprecipitation method. Because the fine size-controlled organic pigment nanocrystals lead to very little light scattering, they are considered to be useful for improving the performance of color filters that are an essential component of LCD.

4.2 Organic nanocrystals for organic field effect transistor

The functionality of organic pigment is attracted in electro devices such as organic electro luminescence devices and organic field effect transistor (OFET). Especially, the research of OFET is hot topic toward a future of optical device. Several kinds of electron-hole transfer organic compounds are well investigated. For OFET, carrier mobility is important factor for high quality devices. A π-conjugate polymer, polydiacetylene (PDA), is one of the promising compounds having high carrier mobility. PDA is a unique material, which combine the properties of an essentially one-dimensional, fully conjugated polymer with the ability of single crystalline monomers to polymerize without disruption of the crystallinity (Takahashi, et al., 2002). PDA has the optical properties of large non-linear optical susceptibility and ultra-fast response time as well as a PDA been predicted to present very high charge carrier mobility, up to 10^3–10^5 cm^2/V s. However the real mobility of PDA crystals measured using time-of-flight technique was found to be 1–10 cm^2/V s, and that is comparable to those measured on polyacene molecular crystals such as anthracene or naphthalene. The use of PDA films as an active layer of OFET is very promising task, however the pure electrical conductivity of these films have restricted their application. Thus to improve the conductivity is attracted work. Several attempts to improve the electrical conductivity have been reported; chemical doping into the stage of monomer crystals and in the course of solid-state polymerization, chemical doping into soluble polydiacetylenes, chemical doping into a Langmuir–Blodgett film of PDA, ion implantation to PDA bulk crystals, and using a scanning tunneling microscopy probe tip to fabricate linear PDA nanowires. Recently, the maximum conductivity for chemically doped PDA crystals has been reported to be as high as 10^{2} S/cm. As a widely held opinion, one of the big issues in increasing the conductivity of PDA is that its rigid crystalline lattice prevents the dopant from penetrating into PDA bulk crystals. This was especially the case in 1,6-di-*N*-carbazolyl-2,4-hexadiyne (DCHD).

To overcome the difficulty of chemical doping to PDA, we tried chemically doping into the nanocrystals of PDA, because PDA nanocrystals are known to have a softened crystal lattice (Baba, et al., 2006) and a large surface area, compared with bulk crystals, and demonstrated almost the highest conductivity of chemically doped PDA ever reported (Baba, et al., 2008). Figure 6 shows that the single layered poly-DCHD nanocrystals on slide glass and their specimen for measuring their conductivity. The success in chemical doping was achieved because poly-DCHD nanocrystals have a large surface area and softened crystal lattices, which allows chemical dopant species to diffuse easily into poly-DCHD nanocrystals in good contrast to the case of bulk crystals. Actually, nanostructure showed unique conformation change during solid state polymerization. Nanostructure have showed crack-less action during polymerization. This also showed that the nanocrystal has softened crystal lattice and has flexible stricture (this is mentioned in the next section). Thus, the PDA nanocrystals can accept the dopant. Aiming at the fully doped single crystalline nanocrystals of PDA, the effects of crystal size, morphology, and several kinds of chemical dopant species, which would increase the conductivity of PDA, are now under investigation.

As future approach, achieving the one dimensional array of the PDA nanofibers layered on the surface of devices is promising approach for high conductivity. We previously reported the one dimensional array of nanofibers. Furthermore, interesting finding concerning morphology changes during solid-state polymerization of PDA nanofibers were reported by our group, which related to topochemical study. This was because the softened crystal lattice of nanocrystals.

4.3 Solid state polymerization of polydiacetylene: topochemical polymerization

The research of the topochemical polymerization has been reported extensively since the 1960s, and PDA is the most attracted one. PDA also has been attracted because this π-conjugated polymer system has received much attention as conductive and non-linear optical materials. However, the accumulation of strain in bulk crystal during the solid-state

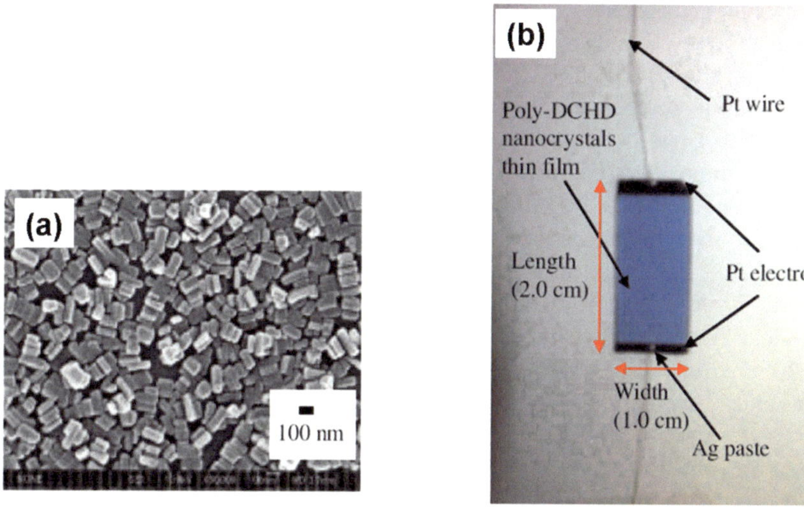

Fig. 6. (a) Single layered poly(DCHD) nanocrystals on quartz slide glass. (b) Image of the specimen for conductivity measurement of single layered poly(DCHD) nanocrystals. (Reprinted from (Baba, et al., 2008): Japanese Journal of Applied Physics., Vol. 47 (2008), pp. 3769-3771, No. 5, DOI: 10.1143/JJAP.47.3769. Copyright 2008 The Japan Society of Applied Physics)

polymerization results in crushing crystal structure, and the rate of polymeric conversion does not occur with high efficiency. We found that the polymerization of nanocrystals achieved high quality polymeric crystal because the strain accumulated in nanocrystals was easily released (Baba, et al, 2006). Nanofibers prepared by the reprecipitation method had an average length of 10 μm and width of 50 nm. We found interesting behavior of morphological change of nanofibers during the stage of the solid-state polymerization. DCHD monomer nanofibers, which were fabricated by the reprecipitation method, were worm-like before solid-state polymerization. Interestingly, worm-like nanofibers converted into straight-like ones through the solid-state polymerization. This phenomenon indicated that the lattice in nanofibers was generally soft, compared with that of bulk crystals. The crystalline nanofibers are useful for the estimation of a nonlinear optical property and a conductivity of a single polymer chain.

4.4 Linear and nonlinear optical property of organic nanocrystals

PDA crystals have attractive nonlinear optical properties. Our interest is the nonlinearity occurred in nanocrystals. The fabrication of polydiacetylene nanocrystals with varied size was achieved using the reprecipitation method. The size effect on the absorption spectra, which was a characteristic to organic nanocrystals, has been experimentally verified. An important point is that size-dependent blue-shift appears at a size an order of magnitude larger than that of semiconductors and metals. That is due to the lattice vibration must become highly frequent with increasing temperature and also with decreasing crystal size. Actually, these thermal effects and sizes were clearly related to phonons in the nanocrystals lattice. The obtained $\chi^{(3)}$ was the order of 10^{-9} esu (Nakanishi & Katagi, 1998). Thus, we believe PDA nanocrystals is promising NLO materials. These attractive data was obtained in two-dimensional structured thin films of polymer matrix. Further our interest was constructing the three-dimensional (3-D) structure of nonlinear materials such as photonic crystals.

Photonic crystals have been a subject of considerable interest for more than a decade. A great deal of effort has been put into preparing highly ordered photonic crystals for the visible spectral range. Initial studies were directed toward the linear optical properties of these crystals, but more recent studies have focused on nonlinear processes in photonic crystals, and dramatic enhancements of these properties have been reported (Markowicz, et al., 2004).

Our interest is to study the optical properties of highly ordered 3-D arrays of polydiacetylene nanocrystals, since 3-D arrays formed by mono-dispersed particles in the range of 10 nm to 1 μm, referred to as colloidal crystals, show some special optical properties and potential applications, particularly in photonic devices. However, polydiacetylene nanoparticles are usually polyhedral and multi-dispersed. Their morphology prevents them from forming large-scale densely packed 3-D structures because the particles that form colloidal crystals are required to have a size variation of less than 5%. Polystyrene is one of the important materials in the fabrication of colloidal crystals, since uniform spherical polystyrene particles were easily synthesized by soap-free emulsion polymerization. When polydiacetylene nanocrystals are enclosed within uniform spherical polystyrene particles, the resulting composite particles are used to construct 3-D arrays of nonlinear materials. We reported the fabrication and characterization of spherical mono-dispersed composite particles consisting of 1,6-di(*N*-carbazolyl)-2,4-hexadiyne (DCHD) nanocrystals cores and polystyrene shells. We succeeded in preparing mono-dispersed composite particles consisting of poly-DCHD cores and polystyrene shells (Wei, et al., 2003). Their 3-D structure is attractive for investigating their nonlinear optical properties.

Recently, an interest of photonic crystals has also been directed toward studying the effect of disorder in photonic crystals. Prasad's group reported on a type of disorder in photonic crystals in which only a random variation of refractive index is introduced, without creating any structural or positional disorder, which was referred to be such disordered photonic crystals as photonic crystal alloys (Tiryaki, et al., 2004). The vertical deposition colloidal crystal method to produce three-dimensional photonic crystal alloys of polystyrene and poly(methyl methacrylate), 300 nm spheres in varying compositions were used. This photonic crystal alloys provided a convenient medium to compositionally tune the bandgap and the wavelength of a phase-matched nonlinear optical response. Prasad's group is saying that the use of photonically active components (such as dye-doped spheres) allows one to conveniently create microcavities and study cavity electrodynamics.

4.5 Nanomedicine: Drug delivery system using drug nanocrystals

Numerous commonly used pharmaceutical drugs are hydrophobic in nature. Thus, special formulations are required to make their aqueous dispersion for these drugs delivery.

Usually, surfactants or other nanoparticle-based delivery vehicles are used. Once systemic administration of these drugs, if the drug target cancer organ, such drug-doped carriers are preferentially taken up by tumour tissues based on the "enhanced permeability and retention effect". The carriers include liposomes, polymeric micelles, oil dispersions (micelles), ceramic, hydrophilic drug–polymer complexes and polymeric nanoparticles. However many surfactants themselves tend to increase the systemic toxicity of the drug formulation. Therefore, there is increasing interest in the development of novel drug formulation and delivery methods without any external agents such as surfactants or other carrier vehicles. One method proposed for dispersion of hydrophobic compound in water, the reprecipitation method was used as demonstration.

Prasad's group have prepared the nanocrystal formulation of a hydrophobic drug for photodynamic therapy (PDT) and compared its efficacy with the conventional surfactant-based formulation (Baba, et al., 2007b). PDT is a promising new modality for the treatment of a variety of cancers and some dermatological and ophthalmic diseases. The main advantage of PDT is the capability to localize the treatment, by using selective light exposure to the tumor site. Typical PDT treatment involves systemic administration of a photosensitizer drug, then localized light exposure at the tumor site using near-infrared light. After being excited with light, these photosensitizer molecules can transfer their excited-state energy to molecular oxygen in the surroundings, forming reactive oxygen species such as singlet oxygen. The locally generated reactive oxygen species destruct various cellular compartments, resulting in irreversible damage to tumor cells. The major disadvantage for successful clinical PDT is the poor water solubility of photosensitizing drugs. Thus, making their stable formulation for systemic administration is highly challenging. To overcome this difficulty, a stable dispersion of these drugs into aqueous systems is achieved using delivery vehicles. However, allergic reactions and their sustained *in vivo* toxicities are caused by the carrier vehicles. Therefore, the ideal formulation for safe and efficient PDT should involve the minimal number of additional agent such as carrier. To overcome these problems, nanocrystals, which were monodispersed with diameter 100 nm (ζ-potential: −40 mV), were prepared by the reprecipitation method (Baba, et al., 2007b).

There was interesting finding that though the fluorescence and photodynamic activity of the drug nanocrystals were initially quenched in aqueous media. However, both recovered under *in vitro* and *in vivo* conditions (Baba, et al., 2007b). This recovery of drug activity and fluorescence was attributed to the interaction of nanocrystals with blood serum or other intracellular components (e.g., serum albumin), resulting in conversion of the drug nanocrystals into the molecular form. Efficacy of the nanocrystals formulation *in vitro* and *in vivo* was found to be comparable with that of the same drug formulated in the conventional delivery vehicle.

These results have not only mentioned the potential of using pure drug nanocrystals for PDT, but also this approach eliminates the need of any external agents such as surfactants or other carrier vehicles for drug delivery. Further studies are expected to increase the efficacy of nanocrystals by controlling their size that may affect long-term *in vivo* circulation and accumulation in the tumor tissue. Potentially, this method of drug formulation can be applied not only for PDT drugs, but also for delivery of other therapeutic drugs including imaging agents such as fluorescent dyes for biophotonics (e.g., bioimaging).

4.6 Bioimaging using fluorescent dye nanocrystals

Fluorescence microscopy is one of the most versatile imaging techniques in biomedical research. That allows the non-invasive imaging of cells and tissues with molecular specificity. Such imaging requires a fluorescent dye to enter cells and tissues prior to visualization. Currently used dyes include members of the coumarin, rhodamine, fluorescein, and carbocyanine families. One of them, 3,3′-Dioctadecyloxacarbocyanine perchlorate [DiO; DiOC18(3)] is the long-chain dialkylcarbocyanine dye, and most commonly used for the visualization of anterograde and retrograde neuronal tracers in living cells. This lipophilic carbocyanine is also employed in many other applications, including cytotoxicity assays, the labeling of lipoproteins, and the tracking of cell migration and lipid diffusion in membranes through fluorescence recovery after photobleaching. The hydrophobic nature of these dyes means that, in many cases, organic solvents [e.g., dimethylformamide, dimethylsulfoxide (DMSO)] and surfactants are required for successful cell imaging. However, unfortunately, organic solvents and surfactants themselves tend to increase cytotoxicity *in vitro* and *in vivo*. On the other hand, the direct applications of micron-sized crystals have also been investigated, although such crystals are probably not small enough to allow the diffusion of dyes into cells, i. e. not efficient imaging approach.

In addition to DiO dyes, many other fluorescent dyes are hydrophobic in nature—including perylene, a widely studied hydrophobic material that functions with a high quantum yield in organic electroluminescence devices. Although it is potentially useful for bioimaging, the preparation of aqueous dispersions of perylene dyes requires special formulation techniques, similar to those for DiO. The reprecipitation method was used to solvent-free and non-invasive bioimaging. We reported an organic solvent-free bioimaging method employing fluorescent dye nanocrystals prepared using the reprecipitation method and their application to *in vitro* fluorescence confocal imaging (Baba et al., 2009, 2010).

These nanocrystal formulations allowed efficient fluorescence confocal imaging of living cells *in vitro*, with performances almost identical to those obtained by treatment using a conventional organic solvent. This nanocrystal formulation approach to *in vitro* bioimaging in cultures eliminates the need for external solubilizing agents (e.g., organic solvent or surfactant), which usually tend to increase cytotoxicity. We expect that this method can be used with a wide range of hydrophobic organic fluorophores. As an advanced functional bioimaging tool, our team recently have developed thermoresponsive fluorescent nanocrystals, which will be useful for temperature-dependent cell imaging as well as photodynamic cancer therapy.

4.7 Thermoresponsive behaviour of fluorescent organic nanocrystals

After the first introduction of a temperature-responsive polymer, poly(*N*-isopropylacrylamide) (PNIPAM) (Heskinsa & Guillet, 1968), numerous studies on this polymer have been reported. The fundamental findings have been applied in various fields such as drug delivery, regenerative medicine, (Nishida, et al., 2004) and analytical chemistry. The property of this polymer is that when the temperature of the aqueous solution where the polymer is dissolved increases above the lower critical solution temperature of the cloud point, the polymer shows phase separation. The hydrophobic groups in the polymer form insoluble aggregates, turning the solution cloudy (Chen & Hoffman, 1995). Using this characteristic of PNIPAM, fluorescent dyes covalently linked to PNIPAM showed on-off fluorescent switching properties above and below the critical temperature. On the other hand, we demonstrated the thermoresponsive behaviour of fluorescent organic nanocrystals using PNIPAM in an aqueous system for the first time (Baba, et al., 2011). For the preparation of thermoresponsive fluorescent organic nanocrystals, fluorescent dyes of perylene, quinacridone, and zinc phthalocyanine were used. As a result, an on-off switching of fluorescence intensity was observed below and above around cloud point (~35°C). This thermoresponsive behaviour of fluorescent organic nanocrystals in the aqueous system was stably repeatable. As an extension of this research, this research finding indicates that the fluorescence properties of organic nanocrystals can be controlled using the phase-transitional functional polymer system, where the key factors for stimulation are not only limited to temperature but also light and pH. This fundamental finding of tuneable fluorescence optical property of organic nanocrystals will have several applications such as thermosensitive fluorescence bioimaging of living cells/tissues using PNIPAM-based fluorescent dye nanocrystals; the on–off switching logic gate for optical devices; and medical treatment by photodynamic therapy using PNIPAM-based on-off switching photosensitizing nanocrystals, where the drug efficacy can be controlled by temperature on demand. We believe this finding will be applied in multidisciplinary fields such as nanomedicine, nanobiology, tissue engineering, regenetic medicine, and applied optical physics.

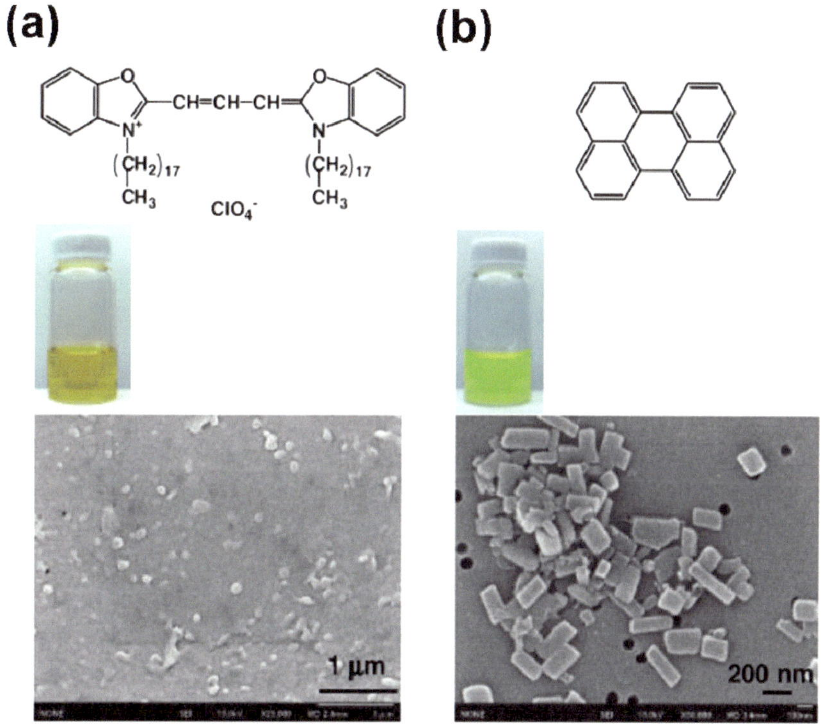

Fig. 7. Chemical structure, photographs of dispersion, SEM images of (a) DiO and (b) perylene. (Reprinted from (Baba, et al., 2009): Japanese Journal of Applied Physics. Vol. 48 (2009), 117002 (4 pages), DOI: 10.1143/JJAP.48.117002. Copyright 2009 The Japan Society of Applied Physics)

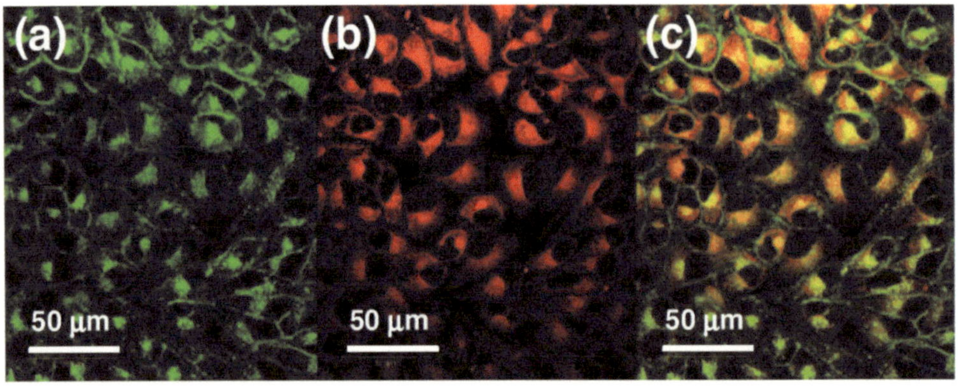

Fig. 8. Confocal fluorescence images of culture cells *in vitro*. Dyes used for the staining are (a) DiO and (b) perylene. (c) Double staining image that is overlapped pictures of (a) and (b). (Reprinted from (Baba, et al., 2009): Japanese Journal of Applied Physics. Vol. 48 (2009), 117002 (4 pages), DOI: 10.1143/JJAP.48.117002. Copyright 2009 The Japan Society of Applied Physics)

5. Recent topics of organic nanocrystals achieved by several groups

Now a day, with using the reprecipitation method, a series of functional organic nanocrystals/nanoparticles were successfully fabricated by several research groups. For example, nanocrystals from perylene and a perylene derivative were prepared and the spectroscopies of single nanoparticles were studied. Nanoparticles were prepared from β-carotene and observed the influence of both supramolecular structure and particle size on the absorption spectra. The size dependence of the luminescence and the enhanced emission of the nanoparticles prepared with this method were studied in Yao's group (Fu & Yao, 2001) and Park's group (An, et al., 2002). One of the interesting feature occurred in

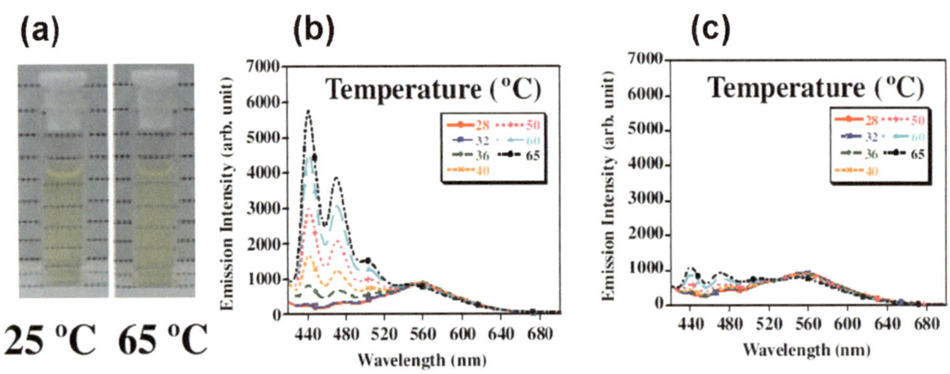

Fig. 9. (a) Photographs of perylene nanocrystals dispersion with PNIPAM at 25 (left) and 65 °C (right). Fluorescence spectra of the nanocrystals (b) with and (c) without PNIPAM. (Reprinted from (Baba, et al., 2011): Japanese Journal of Applied Physics. Vol. 50 (2011), 010202 (3 pages), DOI: 10.1143/JJAP.50.010202. Copyright 2011 The Japan Society of Applied Physics)

nanocrystals, recently Park' group reported a phenomenon named aggregation induced enhanced emission (AIEE). Here, enhanced emission instead of fluorescence quenching was observed in the solid state for some specific fluorophores. In general, the fluorescence efficiency of organic chromophores decreases in the solid state, as a result of concentration quenching, even though they show high fluorescence efficiency in solution. Park's group reported AIEE from organic nanoparticle. First, Park's group synthesized the organicmolecule 1-cyano-trans-1,2-bis-(40-methylbiphenyl)ethylene (CN-MBE), and then prepared the corresponding nanoparticles. It was interesting that although the fluorescent from CN-MBE solution was quite weak, the nanoparticles emitted a quite strong photoluminescence. It was considered that the aggregation induced the planarization of the CN-MBE molecules in the nanoparticles, which in turn resulted in the strong intermolecular interactions causing a specific aggregation that was good for efficient emission. This phenomenon was specific to nanoparticles and not to bulky crystals.

Not only the above mentioned optical properties, but also some other optoelectronic properties of organic nanomaterials have been reported in recently. For example, the electronic conductance through organic nanowires was investigated. The Tetracyanoquinodimethane (TCNQ) charge transfer complexes nanowires and field emission properties of tris(8-hydroxyquinolino)-aluminum (Alq 3) nanowires were reported. The field-effect transistors (FETs) based on single-crystalline organic one-dimensional nanomaterials were fabricated, and also lasers have proved to be another promising application for organic one-dimensional nanomaterials. The optical waveguiding properties of organic nanowires were reported. The details are well reviewed in the literature written by Yao's group (Zhao, et al., 2008).

6. Conclusion

In this chapter, first we explained the preparation method of organic nanocrystals and their improved ones. The reprecipitation method is very useful technique for fabricating several kinds of functional organic nanocrystals. Supercritical fluid crystallization method, microwave-irradiation method, and mass production method extend the potential of the reprecipitation method. Then, several kinds of functional organic nanocrystals including recent topics achieved by several research groups were introduced. The research trend of functional organic nanocrystals had begun with a series of optoelectronic materials, and recently several functional organic nanocrystals were revealed to be useful for nanomedicine such as drug delivery system using pure nanocrystal form of drugs as well as organic solvent free confocal fluorescence imaging of living cells *in vitro*. The unique finding of thermoresponsive behaviour of fluorescent organic nanocrystals will be quite useful for bioimaging, regenetic medicine, applied optics, etc. Toward the future works of functional organic nanocrystals, their basic research and application will much more extend to the fields such as optelectronic materials, applied optics as well as nanomedicine, regenetic medicine, and biophotonics.

7. References

An, B.-K.; Kwon, S.-K.; Jung, S.-D. & Park, S. Y. (2002). Enhanced emission and its switching in fluorescent organic nanoparticles. *Journal of the American Chemical Society*. Vol. 124, No. 48, (November 8, 2002), pp. 14410-14415, ISSN 0002-7863

Baba, K.; Kasai, H.; Okada, S.; Oikawa, H. & Nakanishi, H. (2000). Novel fabrication process of organic microcrystals using microwave-irradiation. *Japanese Journal of Applied Physics*, Vol. 39, No. 12A, (October 16, 2000), pp. L1256-L1258, ISSN 0021-4922

Baba, K.; Kasai, H.; Okada, S.; Nakanishi, H. & Oikawa, H. (2006). Fabrication of diacetylene nanofibers and their dynamic behavior in the course of solid-state polymerization. *Molecular Crystals and Liquid Crystals*, Vol. 445, pp. 161-166, ISSN: 1542-1406

Baba, K.; Kasai, H.; Masuhara, A.; Okada, S.; Oikawa, H. & Nakanishi, H. (2007a). Diacetylene nanowire crystals prepared by reprecipitation/microwave-irradiation method. *Japanese Journal of Applied Physics*, Vol. 46, No. 11, (November 6, 2007), pp. 7558-7561, ISSN 0021-4922

Baba, K.; Pudavar, H. E.; Roy, I.; Ochulchanskyy, T. Y.; Chen, Y.; Pandey, R. K. & Prasad, P. N. (2007b). New method for delivering a hydrophobic drug for photodynamic therapy using pure nanocrystal form of the drug. *Molecular Pharmaceutics*, Vol. 4, No. 2, (February 1, 2007), pp. 289-297, ISSN 1543-8384

Baba, K.; Kasai, H.; Shinohara, Y.; Okada, S.; Oikawa, H. Matsuda, H. & Nakanishi, H. (2008). Chemical doping into nanocrystals of poly(diacetylene). *Japanese Journal of Applied Physics*, Vol. 47, No. 5, (May 16, 2008), pp. 3769-3771

Baba, K.; Kasai, H.; Masuhara, A.; Oikawa, H. & Nakanishi, H. (2009). Organic solvent-free fluorescence confocal imaging of living cells using pure nanocrystal forms of fluorescent dyes. *Japanese Journal of Applied Physics*, Vol. 48, No. 11 (November 20, 2009), pp. 117002/1-117002/4

Baba, K.; Kasai, H.; Nishida, K. & Nakanishi, H. (2010). Organic Nanocrystals for Nanomedicine and Biophotonics, In: *Nanocrystals*, M. Yoshitake, (Ed.), pp. 311-326, Sciyo, ISBN 978-953-307-126-8, Vukovar, Croatia

Baba, K.; Kasai, H.; Nishida, K. & Nakanishi, H. (2011). Poly(*N*-isopropylacrylamide)-based thermoresponsive behavior of fluorescent organic nanocrystals. *Japanese Journal of Applied Physics*, Vol. 50, (January 5, 2011), pp. 010202/1-010202-3

Chen, G. & Hoffman, A. S. (1995). Graft copolymers that exhibit temperature-induced phase transitions over a wide range of pH. *Nature*, Vol. 373, No. 6509, (05 January 1995), pp. 49-52, ISSN 0028-0836

Fu, H. & J. Yao, J. (2001). Size effects on the optical properties of organic nanoparticles. *Journal of the American Chemical Society*, Vol. 123, No. 7, (January 23, 2001), pp. 1434-1439, ISSN 0002-7863

Granqvist, C. G. & Buhrman, R. A. (1976). Ultrafine metal particles. *Journal of Applied Physics*, Vol. 47, No. 5, (May, 1976), pp. 2200-2219

Heskinsa, M. & Guillet, J. E. (1968). Solution properties of poly(N-isopropylacrylamide). *Journal of Macromolecular Science, Chemistry*, Vol. 2, Issue 8, pp. 1441-1455, ISSN 0022-233X

Ishizaka, T.; Kasai, H.; Oikawa, H. & Nakanishi, H. (2006). Unique luminescence properties of Eu3+-doped polyimide. *Journal of Photochemistry and Photobiology, A: Chemistry*, Vol. 183, Issue 3, (25 October 2006) pp. 280-284, ISSN: 1010-6030

Kaneko, Yuji; Shimada, Satoru; Fukuda, Takashi; Kimura, Tatsumi; Yokoi, Hiroyuki; Matsuda, Hiro; Onodera, Tsunenobu; Kasai, Hitoshi; Okada, Shuji; Oikawa, Hidetoshi & Nakanishi, H. (2005). A novel method for fixing the anisotropic orientation of dispersed organic nanocrystals in a magnetic field. *Advanced Materials*, Vol. 17, Issue 2, pp. 160-163, ISSN: 0935-9648

Kasai, H.; Nalwa, H. S.; Oikawa, H.; Okada, S.; Matsuda, H.; Minami, N.; Kakuta, A.; Ono, K.; Mukho, A. & Nakanishi, H. (1992). A novel preparation method of organic microcrystals. *Japanese Journal of Applied Physics*, Vol. 31, No. 8A, (May 29, 1992), pp. L1132-L1134, ISSN 0021-4922

Kasai, H.; Kanbara, H.; Iida, R.; Okada, S.; Matsuda, H.; Oikawa, H. & Nakanishi, H. (1995). Optical Kerr shutter response of organic microcrystals. *Japanese Journal of Applied Physics*, Vol. 34, No. 9B, (August 11, 1995), pp. L1208-L1210, ISSN 0021-4922

Kasai, H.; Kamatani, H.; Okada, S.; Oikawa, H.; Matsuda, H. & Nakanishi, H. (1996). Size-dependent colors and luminescences of organic microcrystals. *Japanese Journal of Applied Physics*, Vol. 35, No. 2B, (January 8, 1996), pp. L221- L223, ISSN 0021-4922

Katagi, H.; Kasai, H.; Okada, S.; Oikawa, H.; Komatsu, K.; Matsuda, H.; Liu, Z. F. & Nakanishi, H. (1996). Size control of polydiacetylene microcrystals. *Japanese Journal of Applied Physics*, Vol. 35, 10B, (August 28, 1996), pp. L1364-L1366, ISSN 0021-4922

Komai, Y.; Kasai, H.; Hirakoso, H.; Hakuta, Y.; Okada, S.; Oikawa, H.; Adschiri, T.; Inomata, H.; Arai, K. & Nakanishi, H. (1998). Size and form control of titanylphthalocyanine microcrystals by supercritical fluid crystallization method. *Molecular Crystals and Liquid Crystals*, Vol. 322, (01 November 1998), pp. 167-172, ISSN 1058-725X

Markowicz, P. P.; Tiryaki, H.; Pudavar, H. & Prasad, P. N. (2004). Dramatic enhancement of third-harmonic generation in three-dimensional photonic crystals. *Physical review letters*, Vol. 92, No. 8, (26 February, 2004), pp. 083903, ISSN 0031-9007

Matsuda, H., Okada, S., Masaki, A., Nakanishi, H., Suda, Y., Shigehara, K. & Yamada, A. (1990). Molecular structural view on the large third order nonlinearity of phthalocyanine derivatives, *Proceedings of SPIE*, Vol. 1337, pp. 105-113, San Diego, CA, USA, (1 December, 1990)

Matsui, A.; Mizuno, K.; Nishi, O.; Matsushima, Y. & Shimizu, M. (1995). Densities of states and bandwidths of excitons in anthracene microcrystallites embedded in PMMA. *Chemical Physics*, Vol. 194, No. 1, (May 1, 1995), pp. 167-174, ISSN 0301-0104

Miyashita, Y.; Baba, K.; Kasai, H.; Nakanishi, H. & Miyashita, T. (2008). A new production process of organic pigment nanocrystals. *Molecular Crystals and Liquid Crystals*, Vol. 492, pp. 268-274, ISSN: 1542-1406

Murray, C. B.; Morris, D. J. & Bawendi M. G. (1993). Synthesis and characterization of nearly monodisperse CdE (E = sulfur, selenium, tellurium) semiconductor nanocrystallites. *Journal of the American Chemical Society*, Vol. 115, No. 19, (March 22, 1993), pp. 8706-8715, ISSN 0002-7863

Nakamura, A.; Tokizaki, T.; Akiyama, H. & Kataoka, T. (1992). Quantum size effects and optical nonlinearity of confined excitons in semiconducting microcrystallites. *Journal of Luminescence*, Vol. 53, No. 1-6, (July 1992), pp. 105-109, ISSN 0022-2313

Nakanishi, H. & Katagi, H. (1998). Microcrystals of polydiacetylene derivatives and their linear and nonlinear optical properties. *Supramolecular Science*, Vol. 5, Issue 3-4, pp. 289-295

Nishida, K.; Yamato, M.; Hayashida, Y.; Watanabe, K.; Yamanoto, K.; Adachi, E.; Nagai, S.; Kikuchi, A.; Maeda, N.; Watanabe, H.; Okano, T. & Tano, Y. (2004). Corneal reconstruction with tissue-engineered cell sheets composed of autologous oral mucosal epithelium. *The New England Journal of Medicine,* Vol. 351, No. 12, (September 16, 2004), pp. 1187-1196, ISSN 0028-4793

Takahashi, S.; Miura, H.; Kasai, H.; Okada, S.; Nakanishi, H. & Oikawa, H. (2002). Single-crystal-to-single-crystal transformation of diolefin derivatives in nanocrystals. *Journal of the American Chemical Society,* Vol. 124, No. 37, (August 24, 2002), pp. (10944-10945), ISSN 0002-7863

Tan, Z.; Masuhara, A.; Kasai, H.; Nakanishi, H. & Oikawa, H. (2008). Multibranched C_{60} micro/nanocrystals fabricated by reprecipitation method. *Japanese Journal of Applied Physics,* Vol. 47, Issue 2, Pt. 2, pp. 1426-1428

Tiryaki, H.; Baba, K.; Markowicz, P. P. & Prasad, P. N. (2004). Linear and nonlinear optical studies in photonic crystal alloys. *Optics Letters,* Vol. 29, No. 19, (October 1, 2004), pp. 2276-2278, ISSN 0146-9592

Ujiiye-Ishii, K.; Baba, K.; Wei, Z.; Kasai, H.; Nakanishi, H.; Okada, S. & Oikawa, H. (2006). Mass-production of pigment nanocrystals by the reprecipitation method and their encapsulation. *Molecular Crystals and Liquid Crystals,* Vol. 445, pp. 177-183, ISSN 1542-1406

Ung, T.; Giersig, M.; Dunstan, D. & Mulvaney, P. (1997). Spectroelectrochemistry of colloidal silver. *Langmuir,* Vol. 13, No. 6, (March 19, 1997), pp. 1773-1782, ISSN 0743-7463

Wei, Z.; Miura, H.; Masuhara, A.; Kasai, H.; Okada, S. & Nakanishi, H. (2003). Monodispersed poly diacetylene-polystyrene composite particles. *Japanese Journal of Applied Physics,* Vol. 42, No. 10A, (October 1, 2003), PP. L1213-L1215

Yanagawa, T.; Kurokawa, Y.; Kasai, H. & Nakanishi, H. (1997). Degenerate four-wave mixing using an optical parametric oscillator as an incoherent light source. *Optics Communications,* Vol. 137, Issue 1-3, (15 April 1997), pp. 103-106, ISSN 0030-4018

Zhao, G.; Ishizaka, T.; Kasai, H.; Hasegawa, M.; Furukawa, T.; Nakanishi, H. & Oikawa, H. (2009). Ultralow-dielectric-constant films prepared from hollow polyimide nanoparticles possessing controllable core sizes. *Chemistry of Materials,* Vol. 21, (29 December, 2008), pp. 419-424, ISSN 0897-4756

Zhao, Y. S.; Fu, H.; Peng, A.; Ma, Y.; Xiao, D. & Yao, J. (2008). Low-dimensional nanomaterials based on small organic molecules: preparation and optoelectronic properties. *Advanced Materials,* Vol. 20, No. 15, (4 JUL 2008) pp. 2859-2876, ISSN 0935-9648

16

Structure of Nanocrystals in Finemets with Different Silicon Content and Stress-Induced Magnetic Anisotropy

Nikolay V. Ershov[1] and Yuri P. Chernenkov[2*]
[1]*Institute of Metal Physics, Ural Branch of Russian Academy of Sciences*
[2]*B.P.Konstantinov Petersburg Nuclear Physics Institute, Russian Academy of Sciences*
Russian Federation

1. Introduction

Nanocrystalline Fe-Si-B-Nb-Cu alloys (Finemets) obtained by crystallization of amorphous ribbons quenched from the melt exhibit high soft magnetic properties: the permeability greater than 10^5, coercive force of about 0.002 Oe and the saturation magnetization exceeding 10 kGs (Yoshizawa et al., 1988). In addition, their permeability can be purposefully controlled by inducing a magnetic anisotropy during annealing in a magnetic field (Herzer, 1992; Yoshizawa & Yamauchi, 1989) or in the field of tensile stress (Glazer et al., 1991; Herzer, 1994; Hofmann & Kronmüller, 1996). Magnetic anisotropy energy exceeding 5000 J/m^3 is attained by annealing under a tensile stress of 400-600 MPa. For the first time the effect of tensile load applied along the ribbon at nanocrystallization annealing was investigated in (Glazer et al., 1991) and it was shown that after such treatment in a sample of the alloy $Fe_{73.5}Si_{13.5}B_9Nb_3Cu_1$ a state with the magnetic anisotropy of the "easy-plane" type is formed. The plane of the magnetic anisotropy is oriented perpendicularly to the direction of stretching. For the first time it was assumed that the transverse magnetic anisotropy is due to residual elastic strain in the lattice of nanocrystals having a negative constant of the magnetostriction.
It is known that the type of the magnetic anisotropy induced in the $Fe_{87-X}Si_XB_9Nb_3Cu_1$ alloy during tensile stress annealing (TSA) depends on the silicon content X (Serikov et al., 2006). If $X \leq 8$, in specimens after TSA the magnetic anisotropy of easy axis type is induced, whose axis is oriented along the tape (i.e. in the direction of load application during TSA). In the process of reversal magnetization along the tape axis, the permeability and the residual magnetization increase, coercive force decreases, the hysteresis loop becomes rectangular. If the concentration of silicon is equal to or greater than 11 at.%, then a transverse magnetic anisotropy of easy-plane type is formed after TSA. In this case, the permeability is constant over a wide range of magnetic fields (up to 125 Oe), the magnetic hysteresis loop becomes inclined, and the magnetization is oriented predominantly in a plane perpendicular to the direction of load application during the annealing.

* Vladimir I. Fedorov[2], Vera A. Lukshina[1], Nadezda M. Kleinerman[1], Vadim V. Serikov[1], Anatoly P. Potapov[1] and Nikita K. Yurchenko[2]

If the formation of a state with magnetic anisotropy after annealing and cooling under a tensile load is associated with residual elastic stresses (Glazer et al. 1991; Herzer, 1994), – a result of the so-called Villari effect, then the type of magnetic anisotropy can be explained by the nature of magnetoelastic interactions in the nanoparticles (Serikov et al., 2006; Filippov, 2006) mainly consisting of Fe-Si crystals having a bcc lattice. This argument is supported by the fact that, in bulk crystalline samples of the $Fe_{1-x}Si_x$ alloy, with increasing concentration of silicon the magnetostriction constant changes its sign from positive to negative at x ~ 0.12 (Bozorth, 1993; Bertotti & Fiorillo, 1994c), and that the silicon concentration in nanoparticles is a few percent higher then the average Si concentration *X* in the alloy (Serikov et al., 2006). For example, if *X* is equal to 8 at.% then silicon in nanocrystals is contained in an amount of about 12 at.%.

In the $Fe_{73.5}Si_{15.5}B_7Nb_3Cu_1$ alloy after TSA, residual deformations of a bcc nanocrystal lattice were detected by X-ray diffraction analysis judging from the shifts of one Bragg reflection (Ohnuma et al., 2003, 2005). In these papers it has been shown that the crystal lattice is stretched in the direction of load application and is compressed in the transverse direction, the strain and the magnetic anisotropy energy being proportional to the load.

Detailed studies of the residual strains have shown that deformations of the bcc lattice of Fe-Si nanocrystals are not isotropic: extension and compression along the <100> directions are maximum, and in the <111> – are minimum (Chernenkov et al., 2010). Therefore, the formation of a state with transverse magnetic anisotropy in the nanocrystals of the $Fe_{73.5}Si_{13.5}B_9Nb_3Cu_1$ alloy, which is characterized by the orientation of magnetic moments transverse to the direction of stretching, is due to the negative magnetoelastic coupling that is caused by residual anisotropic lattice deformations of the Fe-Si nanocrystals with a large fraction of the Fe_3Si phase. The research presented was undertaken to ascertain in detail the atomic structure of nanocrystals of Fe-Si-B-Nb-Cu alloys, depending on the silicon content and the thermal treatment of the samples.

2. Nanocrystalline Fe-Si-B-Nb-Cu alloys and their magnetic properties

The investigations were carried out using $Fe_{87-X}Si_XB_9Nb_3Cu_1$ alloys (*X* = 0, 4, 6, 8, 9.5, 11, 13.5) on samples in the form of ribbons 20 mcm thick and 1 mm wide obtained in the initial amorphous state by melt-spinning technique, where a molten metal alloy is ejected through an orifice onto a rotating copper wheel. The nanocrystallization annealing (NCA) and the stress annealing (TSA) under a tensile stress of 440 MPa were carried out in air at a temperature of 520°C for 120 minutes (under these conditions, no secondary recrystallization still occurs and coercive force shows up no sharp changes in the annealing-temperature range of 510-570°C).

As a result of the heat treatments, samples from the alloys in three different states: initial (or immediately after quenching on the wheel), nanocrystalline (after NCA) and subjected to TSA were obtained. The magnetic state of the samples after NCA and TSA was controlled by the shape of hysteresis loops measured by the ballistic method using an F-190 microwebermeter (Serikov et al., 2006). The hysteresis loops of the samples after NCA and TSA treatments measured upon reversal magnetization along the ribbon are compared in Fig. 1. It is seen that with increasing silicon content *X* from 0 to 13.5 the samples subjected to NCA show a decrease in the saturation magnetization B_s from 14.5 to 12.5 kGs; the coercive force H_c decrease from 1.2 Oe at *X* = 0 to 0.006 Oe at *X* = 13.5 (the data for all compositions are given in Table 1). After TSA, the coercive force decreases only for the alloy without silicon (X = 0). The greatest changes after TSA are observed in the shape of the hysteresis loops, and the character of changes strongly depends on the silicon content.

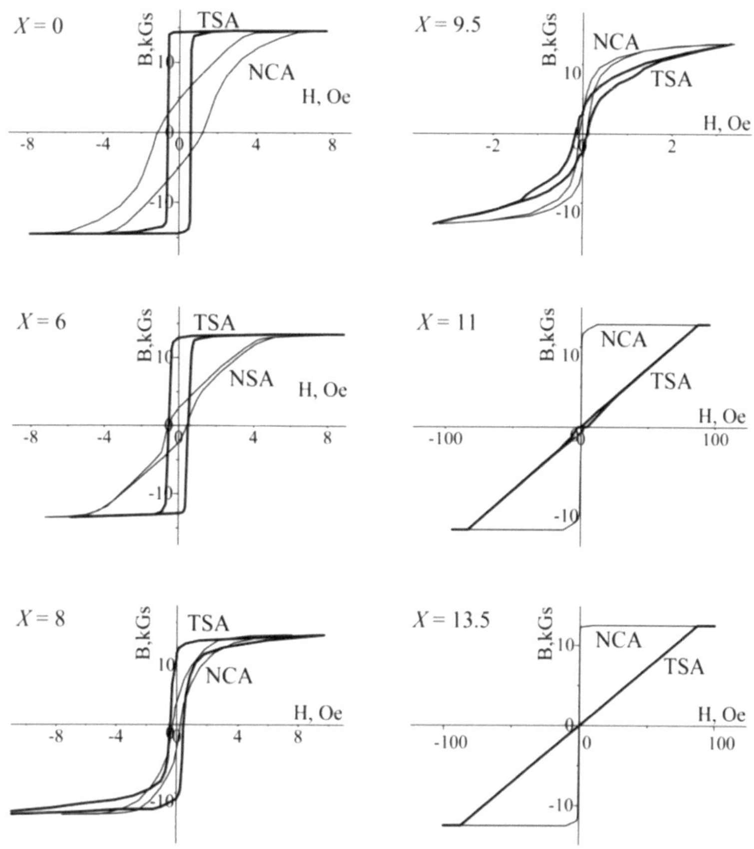

Fig. 1. Hysteresis loops measured for nanocrystalline $Fe_{87-X}Si_XB_9Nb_3Cu_1$ alloy after the nanocrystallization annealing (NCA) and tensile stress annealing (TSA).

For samples after TSA with $X < 8$, the hysteresis loop shape transforms from inclined into rectangular, with a high magnitude of the remanent magnetization. This indicates the inducing the longitudinal magnetic anisotropy (Fig. 1, X = 0 and 6). For samples with X = 8, the loop after the TSA is still steeper than after NCA, and the difference in the slope of the loops is much less than at a lower silicon content. When X = 9.5, the loops of samples after NCA and TSA are also close to each other, but the loop of sample after TSA is lower than for sample after the NCA, i.e., remanent magnetization decreases as a result of TSA. For samples with $X > 9.5$ after TSA treatment, the loops become inclined. This indicates the appearance of a transverse magnetic anisotropy. The sample with X = 13.5 after TSA has a constant magnetic permeability in the range of magnetic fields from -85 to 85 Oe, and its hysteresis loop has an inclined shape.

The magnetic properties of the nanocrystalline $Fe_{87-X}Si_XB_9Nb_3Cu_1$ alloys are controlled by the alloy composition, atomic structure and annealing conditions (Yoshizawa, 2006; Potapov & Filippov, 2009).

Parameters	X					
	0.0	6.0	8.0	9.5	11.0	13.5
B_s, kGs	14.5	13.45	13.2	13.0	12.8	12.5
H_c, Oe	1.2	0.5	0.25	0.075	0.049	0.006

Table 1. Saturation magnetization, B_s, and coercive force, H_c, depending on silicon content, X, in the nanocrystalline alloy $Fe_{87-X}Si_XB_9Nb_3Cu_1$.

3. Nanocrystal structure

3.1 X-ray diffraction analysis

The X-ray diffraction patterns of $Fe_{87-X}Si_XB_9Nb_3Cu_1$ alloys were measured in transmission geometry using a four-circle diffractometer and monochromated Mo K_α radiation (λ = 0.71 Å). The experimental scheme is shown in Fig. 2.

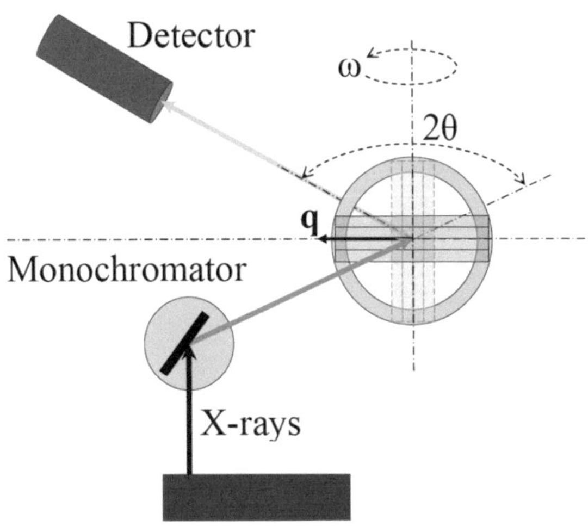

Fig. 2. Scheme of X-ray diffraction experiment. Upon θ-2θ scanning (ω = θ), the scattering vector **q** lies in the plane of the sample, which has the shape of a rectangular plate. The longitudinal scan corresponds to the orientation of the rectangular plate (solid lines), when the scattering vector **q** is parallel to the ribbons. In the transverse scan (orientation of the plate is shown by the dashed line) the scattering vector is perpendicular to the ribbon axis.

The samples in the form of a rectangular plate were prepared from ribbon fragments, which were glued to the thin narrow ring-like holder parallel to each other in several overlapping layers of thickness about 40 mcm. During the θ-2θ scanning, when ω angle is equal to θ, the scattering vector **q** was in the sample plane. The ω angle is equal to zero, when a sample plate is perpendicular to the direction of the incident X-ray beam. By rotating the sample around the horizontal axis by 90°, the transition from the longitudinal scan, when the scattering vector is directed along the axis of the ribbons, to the transverse scan was carried out. In the case of the samples annealed under stress at the longitudinal scan the scattering vector was parallel to the direction of the tensile load application and perpendicular to it in the transverse scan. For each sample, the diffraction patterns were measured for two orientations, i.e. in the form of longitudinal and transverse scans.

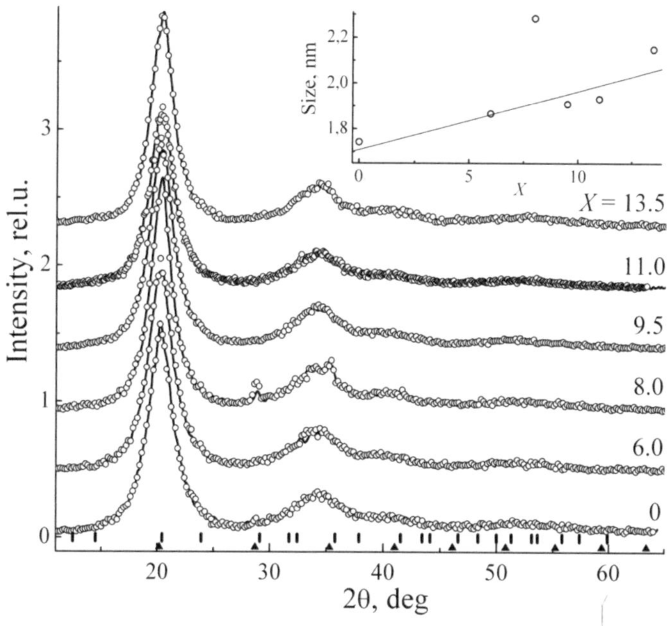

Fig. 3. Diffraction patterns of $Fe_{87-X}Cu_1Nb_3Si_XB_9$ alloys in the initial state measured in the longitudinal (solid line) and transverse (circles) scans. The calculated positions of the peaks for the bcc structure of α-Fe(Si) - ▲, and for Fe_3Si phase - ❙. The inset shows the dependence of the average size of the regions with a short-range order on X.

The X-ray diffraction patterns of $Fe_{87-X}Cu_1Nb_3Si_XB_9$ (X = 0, 6, 8, 9.5, 11, 13.5) alloy samples in the initial state are shown in Fig. 3. Within the accuracy of measurements the longitudinal and transverse scans for the same sample are identical. All these diffraction patterns have a shape typical of the quasi-amorphous material (Yoshizawa et al., 1988; Noskova et al., 1992). However, the position of the first peak ($2\theta \approx 20°$) coinciding with the Bragg reflection (110) for the bcc structure of iron gives reason to suppose that in the initial state in the alloys there are the regions with bcc short-range order in the arrangement of atoms. Since the iron atoms are a major component of the alloy and the shortest distance between two atoms in the α-Fe is 2.48 Å, then the position of the main peak in the diffraction pattern of disordered (amorphous) state can be estimated as $2\theta_{amorphous} = 16.46°$ (Fischer et al., 2006).
Here and below, to describe the shape of reflections, the pseudo-Voigt function defined as a linear combination of Lorentzian and Gaussian with the same halfwidth (FWHM) was selected. After correction for instrumental resolution, integral width of the reflexion was computed, and the average diameter of the bcc grains was calculated by Scherrer formula (Warren, 1969). The dependence of the average sizes of the regions with the bcc order in the atomic arrangement on the silicon concentration is shown in the inset in Fig. 3. Therefore, the structure of the alloy in the initial state can be defined as a fine grained, highly defective bcc structure with the grain sizes about 2 nm. This statement is supported by the fact that the broad diffuse maxima are located close to the calculated positions for the bcc reflections, which are shown in Fig. 3 by triangles. The average size of the ordered regions that contribute to the main peak in the diffraction patterns slightly depends on the concentration of silicon X. One point at X = 8 drops out of a linear dependence, which is apparently due to

a significant number of relatively large, micron-sized grains of iron that give the narrow Bragg peaks (200) and (211) and narrow and intense contribution to the main peak (110) in the diffraction pattern. Previously, a minor amount of such iron grains was observed in the nanocrystalline samples with $X = 0$ and 8 by TEM (Lukshina et al., 2002).

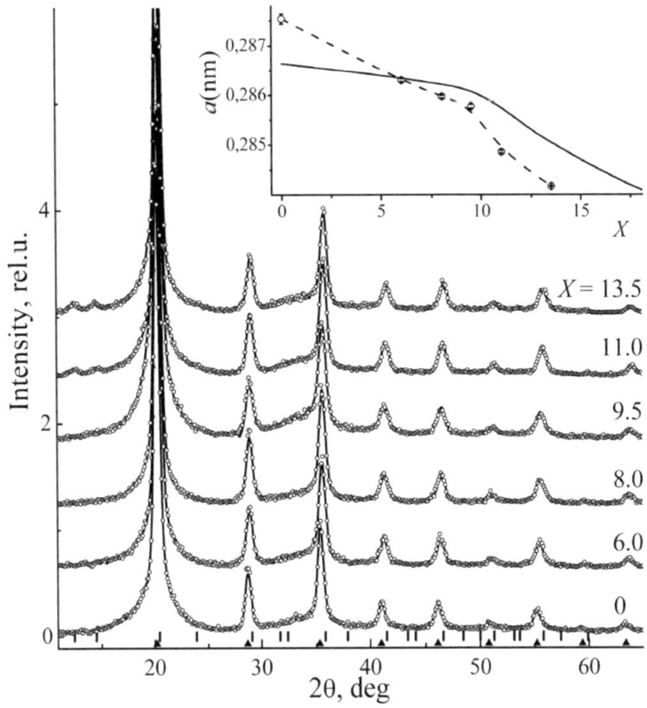

Fig. 4. The same as in Fig. 3, but for samples after the nanocrystallization annealing. The inset shows the dependence of the lattice parameter on the silicon content in the crystalline alloy Fe-Si (solid line) (Bertotti & Fiorillo, 1994a), and in the Fe(Si) nanocrystals (circles fitted by the dashed line) according to our data.

In the diffraction patterns of the $Fe_{87-X}Si_XB_9Nb_3Cu_1$ alloy samples subjected to NCA shown in Fig. 4, there are peaks with the Miller indices (110), (200), (211), (220), (310), (222), (321), and (330) allowed for the bcc lattice of α-Fe(Si). Whereas the intensity of these reflections is sufficient for quantitative analysis, the intensity of other reflections is too small for it. The unit cell parameter of the ordered Fe_3Si phase (structure of $D0_3$ type) is about twice as high as that of the bcc Fe(Si) phase. Hereafter we will use the peak indexing for the bcc lattice. Therefore, when the Fe_3Si phase appears in a sufficient quantity, the superstructure peaks with half-integer indices, e.g. (½ ½ ½), (1½ ½ ½), (1½ 1½ ½), etc., as well as the peaks with an odd sum of integer indices ($h + k + l$), for example (100), (111), (210), etc. should appear. With the available resolution of the diffractometer, the bcc peaks (hkl) of α-Fe (Si) phase in the diffraction patterns coincide with the $D0_3$-peaks ($2h$ $2k$ $2l$) of the Fe_3Si phase.

After annealing at temperature 520°C (NCA treatment), the nanocrystals of α-Fe(FeSi) have appeared in all the alloys, independently of the silicon concentration. The average size of the nanocrystals is 10-12 nm.

The intensity of Bragg reflections at higher angles is rather weak. With increasing silicon content (X) in the samples, and, as a consequence, in the nanocrystals, the Bragg peaks are slightly shifted to the higher angles. It means that the bcc unit cell parameter decreases as in the bulk Fe-Si single crystals (Bozorth, 1993; Bertotti & Fiorillo, 1994a). The dependencies of the unit cell parameter on X in nancrystals and in bulk samples shown in the inset in Fig. 4 are very similar. A sharper decline in the curve $a(X)$ for nancrystals is probably due to the higher concentration of silicon in nanocrystals than that ($C_{Si} = X/100$) in bulk Fe-Si crystal.
A large value of $a(X)$ for alloy with $X = 0$ can be explained by the presence in nancrystals of Nb atoms (atomic radii 0.146 nm) substituting for the Fe atoms (atomic radii 0.126 nm) in the bcc lattice of iron. With increasing X, the Nb atoms are replaced by Si atoms (atomic radii 0.118 nm), and the unit cell parameter rapidly decreases. A kink in the curve $a(X)$ for Fe-Si crystals at $C_{Si} = 0.10$ is caused by the formation of the Fe_3Si phase with the $D0_3$ structure. The $D0_3$ unit cell is composed of eight bcc-unit cells, and its parameter (0.5652 nm) is less then the doubled unit cell parameter of bcc-Fe (0.5732 nm). With increasing X, the volume fraction of $D0_3$ increases, and the unit cell parameter of the Fe-Si alloy decreases faster ($C_{Si} >$ 0.10) then in the α-FeSi phase ($C_{Si} < 0.10$). The similar behaviour of $a(X)$ is observed in the nanocrystals of the $Fe_{87-X}Si_XB_9Nb_3Cu_1$ alloys.
The appearance of the Fe_3Si phase in the $Fe_{87-X}Si_XB_9Nb_3Cu_1$ alloys with $X = 11$ and 13.5 is evidenced by two superstructure peaks (½ ½ ½) and (100) at angles $2\theta \approx 12$ and 14° in the diffraction patterns. Other peaks of the Fe_3Si phase are not distinguishable because they overlap with peaks of α-FeSi phase or their intensity is rather low.

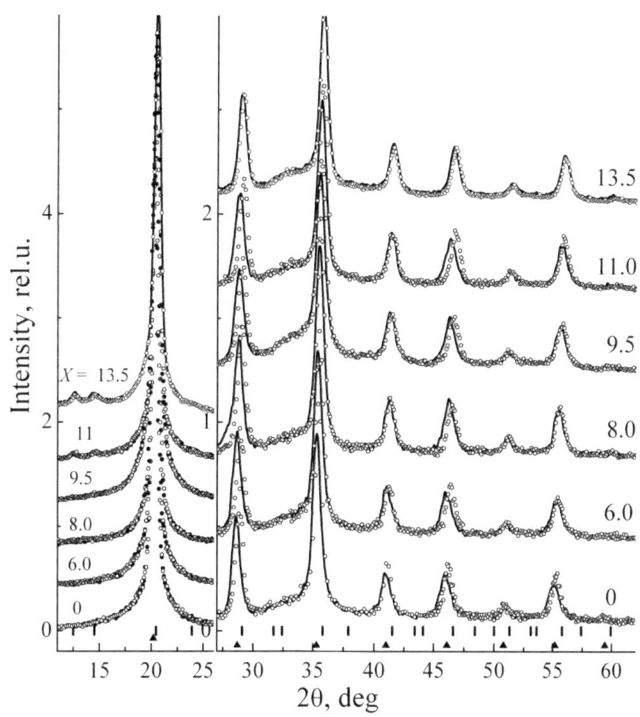

Fig. 5. The same as in Fig. 3 and 4, but for samples after the tensile stress annealing. The scale of intensity in the right part of the figure is two times as large as in the left one.

The diffraction patterns measured in the longitudinal and transverse scans coincide (Fig. 4). The non-linear background is most probably due to the amorphous matrix surrounding the nanocrystallites in the $Fe_{87-X}Si_XB_9Nb_3Cu_1$ alloy. The phase analysis of the alloy with X = 13.5 has shown that contributions of α-FeSi (bcc), Fe_3Si ($D0_3$), and amorphous phase (matrix) are comparable (Chernenkov et al., 2010). The number of separate peaks of the Fe_3Si phase is not enough to obtain the reliable value of the Fe_3Si fraction, but it is predominant. With decreasing X, the Fe_3Si fraction is reduced, and only α-Fe(Si) nanocrystals and amorphous matrix are retained.

In contrast to the alloys in the initial state and alloys subjected to the nanocrystallization annealing, a shift of peaks is observed in the diffraction patterns from alloys subjected to the tensile stress annealing (Fig. 5).

The diffraction peak profiles measured in the longitudinal and transverse scans for the $Fe_{87-X}Si_XB_9Nb_3Cu_1$ alloys with X = 6 and 13.5 are shown in Fig. 6 and 7. The peaks in the longitudinal scans are shifted to the lower angles. This means that the inerplanar distances in the nanocrystals increase along the direction of the stress application during TSA. The peaks in the transverse scans are shifted in opposite direction, and consequently the inerplanar distances decrease.

The relative shifts of the peaks (Fig. 6 and 7) are different, and correspond to the anisotropic residual deformation of the nanocrystals. The largest shifts are observed for reflections (200) and (310), whereas no shifts are seen for the reflections (222) within the limits of the experimental resolution. The character of shifting the reflections in the diffraction patterns of the $Fe_{87-X}Si_XB_9Nb_3Cu_1$ alloys with other values of X is similar to that displayed for X = 6 and 13.5 in Fig. 6 and 7.

The anisotropy of the distortion of the nanocrystals can be characterized by the relative change (Δd/d) of the interplanar spacing d in their lattice. The positions of all the peaks (*hkl*) in the diffraction pattern have been determined by the least-square refinement. The peak profile was described by a pseudo-Voigt function, a linear combination of a Lorentzian and a Gaussian of the same full width at half maximum. Each peak was considered to be a sum of two pseudo-Voigt functions corresponding to the contribution of two lines $K_{\alpha 1}$ and $K_{\alpha 2}$ in the spectrum of the MoK_α. The value of Δd was determined as a difference between the d values calculated from the refined positions of the peaks (*hkl*) in the longitudinal (transverse) scan for the sample subjected to TSA and in the scan for the sample subjected to NCA. As a result of this procedure, relative values of the residual strains of the nanocrystals, Δd/d, were obtained for different directions [*hkl*] in the lattice, parallel and perpendicular to the direction of the load application during the TSA treatment, independently accounting for the residual lattice extensions and compressions, respectively. Their dependences on the angle Φ between the direction of the [*hkl*] vector and the nearest [111] axis are shown in Fig. 8. The residual extensions and compressions change with the silicon content (Fig. 9). The general trend in the Δd/d dependence on X, i.e. the decrease of Δd/d, is in accordance with the enhancement of hardness of the Fe-Si alloys with increasing C_{Si} (Bertotti & Fiorillo, 1994b). Some softening, i.e. an increase of Δd/d, is observed at X = 11, where a sharp decrease of the $a(X)$ takes place (Fig. 4). At lower X, the values of the relative residual strains, Δd/d, in the longitudinal and transverse directions, in the form of extension and compression of the lattice, respectively, are approximately equal, but at X = 11 and 13.5, the residual distortions of nanocrystals in the direction of the tensile stress application are about twice as high as in the transverse one (Fig. 9).

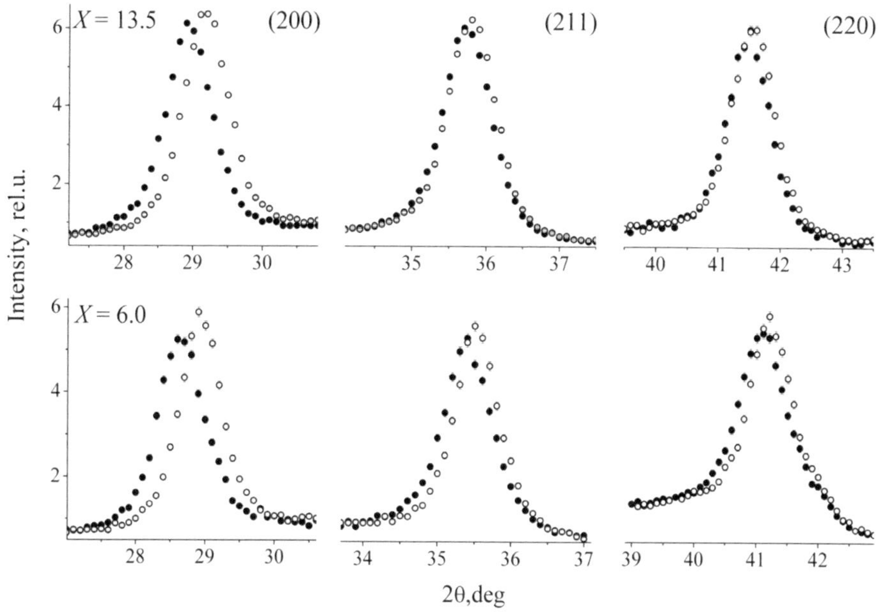

Fig. 6. Profiles of the diffraction peaks (200), (221) and (220) of the $Fe_{87-X}Si_XB_9Nb_3Cu_1$ alloy samples at X = 13.5 and 6 after TSA measured by scanning along (•) and perpendicular (○) to the ribbon or the direction of the load application during TSA.

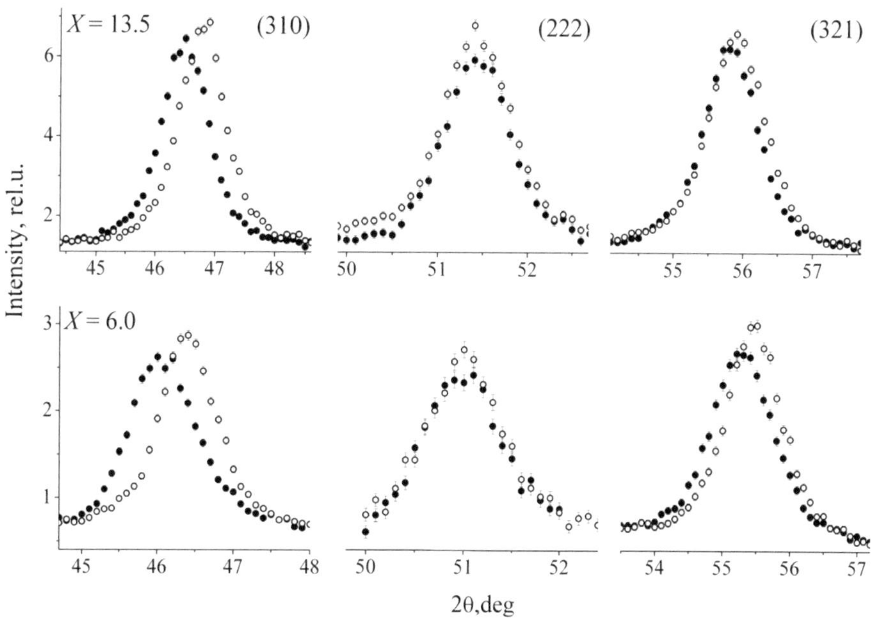

Fig. 7. Profiles of the picks (310), (222) and (321) of the same samples and at the same conditions as in Fig. 6.

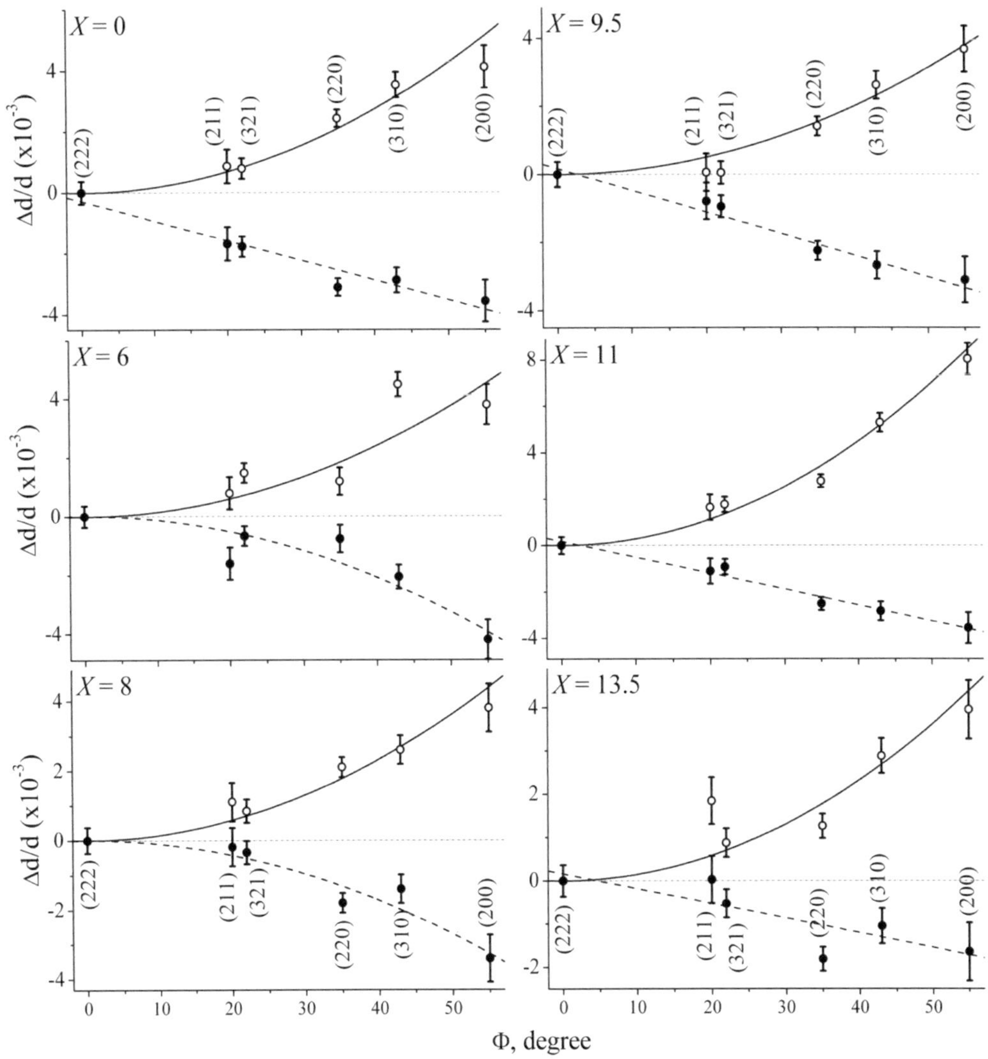

Fig. 8. Relative changes in the interplanar spacing Δd/d: longitudinal strains (○), and transverse reductions (•) in the alloys with different silicon content *X* depending on the angle Φ between the direction [*hkl*] and the closest axis <111>. Lines show the variants of fitting by linear or quadratic functions of the angle.

The distortions of the nanocrystal structure in the $Fe_{87-X}Si_XB_9Nb_3Cu_1$ alloy sample subjected to the TSA treatment observed experimentally can be interpreted as follows. In the case of measurements in the transmission geometry, the sample is adjusted so that, during θ-2θ scanning, the scattering vector **q** is always in the sample plane. In this case, all the nanocrystals, for which the normals to the (*hkl*) planes are parallel to the scattering vector $\mathbf{q}_{hkl}$ within the limits of the instrumental resolution, will contribute to the (*hkl*) reflection in the X-ray diffraction pattern and the position of the maximum in θ angles θ_{hkl} will indicate the mean distance between the (*hkl*) planes in the direction of scanning.

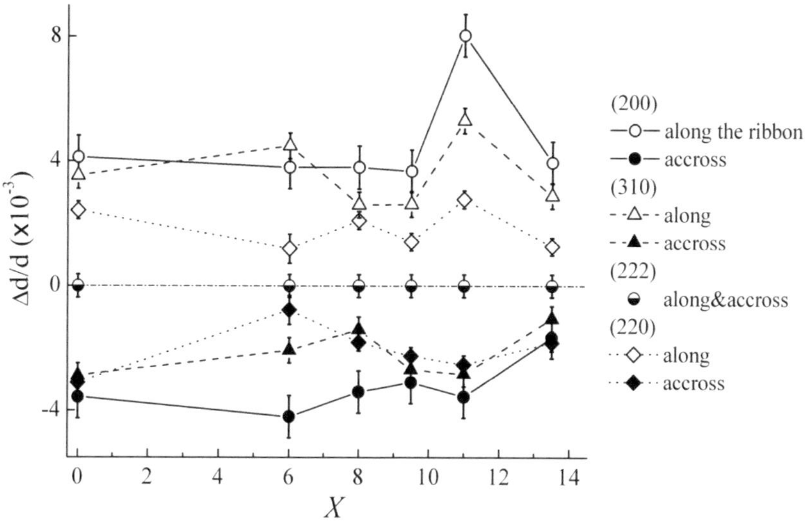

Fig. 9. Dependence of relative extension (along the ribbon) and compression (across the ribbon) of the nanocrystal lattice for the interplanar spacings (200), (220), (310) and (222) on the silicon content *X*.

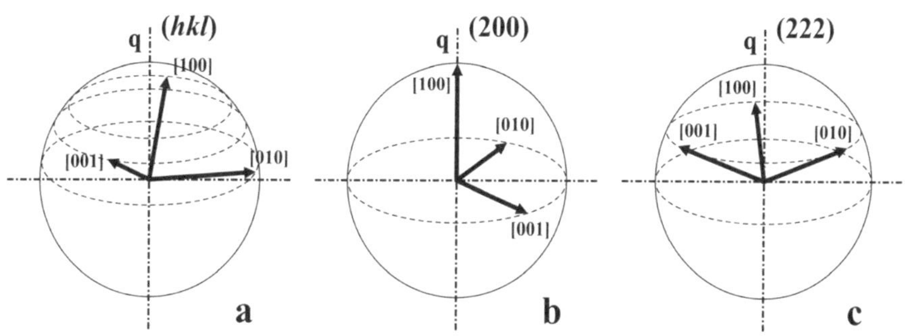

Fig. 10. Orientations of the axes of the nanocrystals contributing to the (*hkl*) reflection in the X-ray diffraction pattern measured during (a) the θ–2θ scanning along the scattering vector **q** and (b, c) special cases for the (b) (200) and (c) (222) reflections.

The ends of the <100> crystallographic axes of the set of such nanocrystals on the sphere of a unit radius will circumscribe circles, belonging to the planes normal to the scattering vector **q** (Fig. 10a). Thus, when scanning along the ribbon, we can trace, from the positions of the (*hkl*) maxima, changes in the interplanar distances of the nanocrystals, in which corresponding (*hkl*) planes are normal to the direction of applying the tensile load **P** during the TSA treatment (in this case, $\mathbf{q}_{hkl} \parallel \mathbf{P}$). In the case of the transverse scan, the positions of the (*hkl*) peaks also characterize the distances between the (*hkl*) planes with the normals oriented crosswise the ribbon or crosswise **P** ($\mathbf{q}_{hkl} \perp \mathbf{P}$). In the case of the (200) reflection, for which the orientation of the axes is shown in Fig. 10b, the angle Φ between the normal to the (200) planes parallel to the scattering vector $\mathbf{q}_{200}$ and the [111] direction in the nanocrystals is

~55°. The largest distortion of the nanocrystalline lattice is exactly observed along the <100> axis (Figs. 8 and 9; Table 2). Thus, after the TSA treatment, the deformation of the initial cubic unit cell has a tetragonal character: an extension along [100] and contraction along [010] and [001]; the opposite pattern is observed for $\mathbf{q} \perp \mathbf{P}$. The contraction can be explained by displacements of atoms in the bcc lattice during the TSA treatment. If the atoms can be considered as hard spheres, they touch each other in the bcc cell in the directions of the cube diagonals. Their close contact is retained with the extension of the unit cell along [100] if eight atoms in the cube vertices are shifted to the central atom along the diagonals of the <110> cube faces.

hkl	Φ, degree	X					
		0	6	8	9.5	11	13.5
along the ribbon - stretching							
200	55.0	0.41(7)	0.38(7)	0.38(7)	0.37(7)	0. 80(7)	0.40(7)
211	20.0	0.08(5)	0.08(5)	0.11(6)	0.00(5)	0. 16(5)	0.18(5)
220	35.0	0.24(3)	0.12(5)	0.21(3)	0.14(3)	0. 28(3)	0.13(3)
310	43.0	0.35(4)	0.45(4)	0.26(4)	0.26(4)	0. 53(4)	0.29(4)
222	0.0	0.00(4)	0.00(4)	0.00(4)	0.00(4)	0. 00(4)	0.00(4)
321	22.0	0.08(3)	0.15(3)	0.08(3)	0.05(3)	0. 18(3)	0.09(3)
across the ribbon – compressing							
200	55.0	-0.36(7)	-0.42(7)	-0.34(7)	-0.31(7)	-0.36(7)	-0.16(7)
211	20.0	-0.17(6)	-0.16(5)	-0.02(5)	-0.08(5)	-0.11(5)	0.00(5)
220	35.0	-0.31(3)	-0.08(5)	-0.18(3)	-0.22(3)	-0.25(3)	-0.18(3)
310	43.0	-0.29(4)	-0.21(4)	-0.14(4)	-0.27(4)	-0.28(4)	-0.11(4)
222	0.0	0.00(4)	0.00(4)	0.00(4)	0.00(4)	0.00(4)	0.00(4)
321	22.0	-0.18(3)	-0.07(3)	-0.04(3)	-0.09(3)	-0.09(3)	-0.05(3)

Table 2. The dependence of the relative changes in the interplanar spacing (in the percentage of Δd/d) on the angle (Φ) between the direction [*hkl*] and the closest axis <111>.

The specific case of the orientation of the nanocrystalline axes for the (222) reflection is shown in Fig. 10c. The <100> axes of the nanocrystals that contribute to the (222) reflection are on the cone surface with the apex angle ~55 × 2° = 110°, and the angle Φ is zero. It can be assumed that the absence of the deformation or its minimum value in the [111] direction is provided by a strong interaction of the nearest neighboring atoms arranged along the cube diagonals in the bcc lattice. The interaction hinders an increase in the distances between nearest neighboring atoms during extension. In the case of the tetragonal distortions, the distance between the nearest neighboring atoms is unchanged and only the directions of the bonds between the nearest neighboring atoms are slightly changed.

The residual structure distortions should be discussed using the elastic deformation tensor. However, it is hardly possible to do this, since the available structural information is very limited and the chemical composition of the nanocrystals is likely inhomogeneous: along with Fe and Si, the composition can also include other atoms. Moreover, the deformed nanocrystals are in the rigid amorphous matrix, in which, after the TSA treatment, residual stresses are likely exist as well.

3.2 NGR-spectroscopy

The local atomic structure, phase composition and orientation of magnetization as a function of silicon content – X were investigated in the alloys after the NCA and TSA by nuclear gamma resonance (NGR, Mössbauer spectroscopy). The Mössbauer spectra were measured on a YaGRS-4M spectrometer in the regime of constant velocity at 20°C with a ^{57}Co in a chromium matrix used as the source. From the Mössbauer spectra, the distribution functions of hyperfine fields (HFFs) - P (H) were constructed.

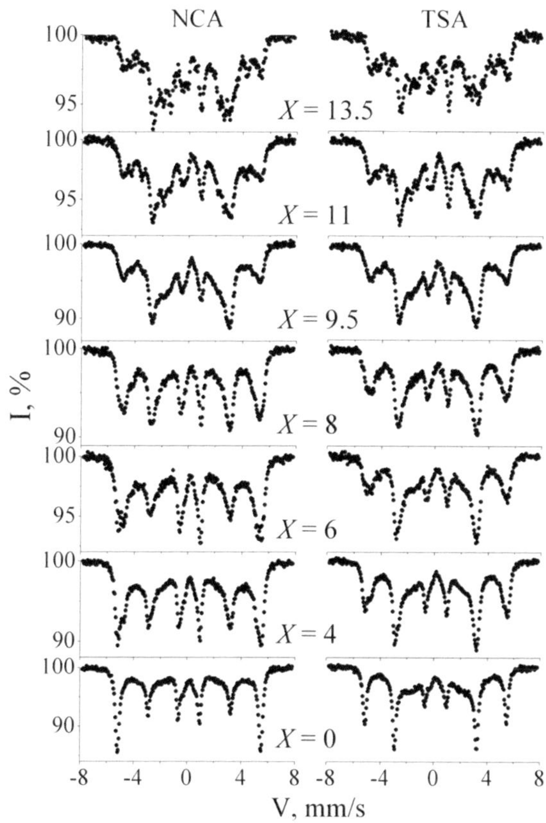

Fig. 11. Mössbauer spectra of the alloys with different values of X after NCA at 520°C for 120 minutes and TSA at 520°C for 120 minutes under a tensile stress of 440 MPa.

Mössbauer spectra of the alloys with different silicon content after the nanocrystallizing annealing and after the annealing under tensile load are shown in Fig. 11. Whereas, the distribution function P (H), in Fig. 12. The intensity of contributions at fields below 100 kOe is quite small; therefore, this range is not shown here. When analyzing distributions of HFFs, the Mössbauer data on the structure of FeSi alloys (Litvinov et al., 1982; Randrianantoandro et al., 1999; Stearns, 1963) were used.

Based on the Mössbauer spectra, the coefficients that characterize the deviation of the magnetic moments from the sample plane (Wertheim, 1964) were determined as a ratio of intensities of the second (fifth) and first (sixth) lines of the Mössbauer sextet. If the magnetic moment is perpendicular to the sample plane, i.e., lies along the direction of incidence of γ

quanta, then $A_2/A_1 = 0$; in the case of an isotropic distribution of the magnetic moment, $A_2/A_1 = 0.67$; if the magnetic moments are located in the sample plane, $A_2/A_1 = 1.33$.

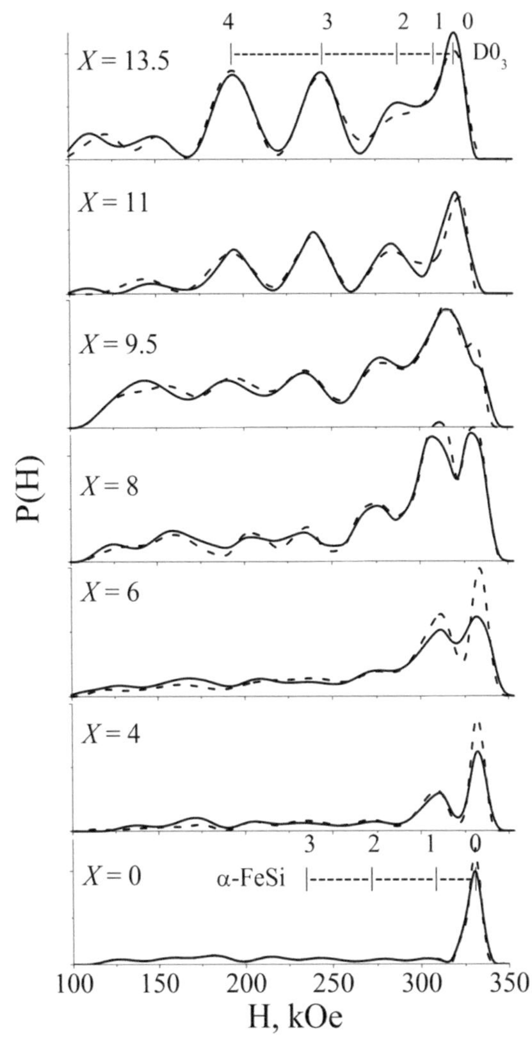

Fig. 12. P(H), distribution functions of HFFs of the nanocrystalline alloys $Fe_{87-X}Si_XB_9Nb_3Cu_1$ with different values of X after NCA (dashed curves) and TSA (solid). With the rulers marked α-FeSi and $D0_3$ the line positions are shown corresponding to surroundings with 0, 1, 2, and 3 silicon atoms in the first coordination sphere for the disordered state and to surroundings with 0, 1, 2, 3 and 4 silicon atoms for the ordered state, respectively.

Fig. 13 displays the concentration dependence of the coefficient A_2/A_1 for the alloys studied, which is compared with the magnitudes of the constant of induced anisotropy, K_{TSA}, found from the hysteresis loops measured on the samples subjected to TSA (Serikov et al., 2006). It is seen that in all the curves there is a feature near the Si content $X \sim 8$. In the case of heat treatment without a tensile stress, the magnetic moments are off the sample plane at small silicon concentrations, which appears to indicate a strong influence of borides on the iron grains (Kleinerman et al., 2004); a further growth of silicon content leads to the orientation of magnetic moments in the sample plane. After TSA the magnetic moments lie in the sample plane already at small silicon concentrations, but are oriented along the ribbon axis; later, they change the orientation to the transverse and then become somewhat off the sample plane.

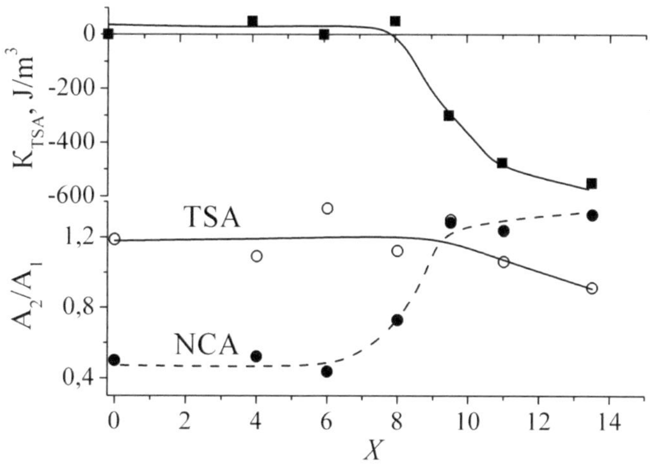

Fig. 13. Variation of the anisotropy constant K_{TSA} and of the ratio A_2/A_1 as functions of the silicon content X for the alloys subjected to NCA (•) and TSA (○).

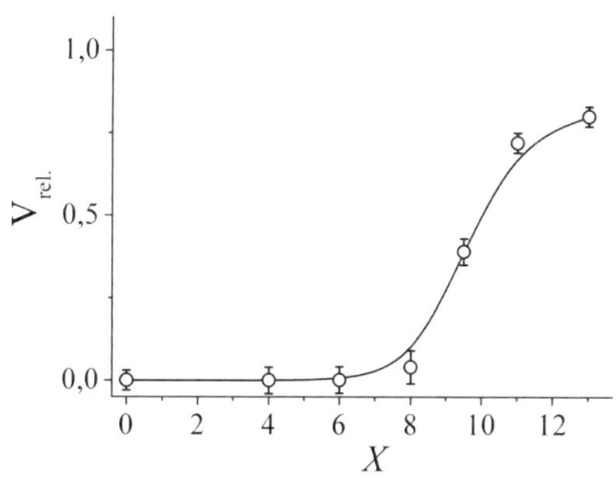

Fig. 14. The relative volume fraction of the ordered phase Fe_3Si - V_{rel} in the nanocrystalline $Fe_{87-X}Si_XB_9Nb_3Cu_1$ alloy depending on the average silicon concentration - X.

The qualitative analysis of the concentration dependence of the distributions P(H) shown in Fig. 12 gives the following results. For $X = 0$ the distribution function over the local surroundings of iron atoms has a peak at H = 330 kOe slightly broadened, but coinciding with the line of 8:0 (8 Fe atoms and 0 Si atoms in the first coordination shell of the Fe atom) for pure iron. The addition of silicon in the alloy ($X = 4$ and 6) leads to a slight growth of the field at the 8:0 peak maximum of the function P(H). Apparently, the effect of Si atoms in the remote (third or fourth) coordination shells of Fe atoms is thus manifested in the change of the hyperfine field at the iron nuclei. In the alloy with $X = 9.5$ there is a significant fraction of the structural component of 8:0 type with a field close in value to 330 kOe, which almost completely disappears from the P(H) at $X = 11$. When X is in the range between 6 and 9.5, there appears a first sign of the local configuration 4:4, i.e., of four Si atoms in the first coordination shell of iron, typical of the $D0_3$ ordering. Because of the six silicon atoms in the second coordination shell of an iron atom that has only iron in the first shell - a configuration characteristic of the $D0_3$ phase, there takes place the decrease of the hyperfine field of the maximum of P(H) that corresponds to coordination 8:0. When $X = 13.5$, more than 80% of the nanocrystal volume is occupied by clusters of the ordered $D0_3$ phase. The concentration dependence of the relative volume fraction of the $D0_3$ phase V_{rel} shown in Fig. 14 was constructed out using the structural and magnetic parameters from (Yelsukov et al., 1989).

An observed fact that after TSA the peaks of 8:0 and 7:1 coordinations are broader than those after the heat treatment without stresses can be explained by the effect of residual strains ($X = 6$). For alloys with $X \geq 8$, the distribution functions P(H) are identical in the main, but the presence of surroundings with two (6:2) and three (5:3) silicon atoms in the first coordination shell of iron atom are indicative of the silicon-concentration heterogeneity of nanocrystals and the low degree of the $D0_3$ type order in them.

4. Residual distortions and magnetic anisotropy

Thus, it is shown that in the samples subjected to TSA, there is a significant residual distortion of the nanocrystal lattice, the anisotropic character of which does not change when the concentration of silicon X varies in the range from 0 to 13.5. The relative lattice distortion along the <100> axis reaches almost 0.01, which is more than two orders of magnitude greater than the saturation magnetostriction in crystalline α-FeSi alloys (Bozorth, 1993). Most likely, the residual strain in the α-Fe(Si) nanocrystals is retained owing to rigidity of the surrounding amorphous matrix (Ohnuma et al., 2005).

The formation of the magnetic anisotropy after annealing and cooling under tensile load (TSA treatment) can be attributed to the residual elastic stresses in the alloy nanoparticles (Glaser et al., 1991; Herzer, 1994). Then, the change of the magnetic-anisotropy type from longitudinal to transverse is explained, respectively, by changing the sign of the magnetoelastic coupling in the nanocrystals from positive to negative with increasing the silicon content. The magnetoelastic Villari effect consisting in the change of magnetization in magnetic materials under the influence of mechanical stretching is the inverse phenomenon of the magnetostriction. If at the positive Villari effect, the magnetization along the direction of elongation increases, then at the negative magnetoelastic effects, on the contrary, an elastic elongation leads to a decrease of magnetization. In the bulk α-Fe(Si) crystals with the positive magnetostriction (and the positive Villari effect), the magnetization is predominantly oriented along one of the easy axes <100> forming the smallest angle with the direction of elongation. Upon saturation of the effect, which corresponds to the strain $\sim 20\text{-}25 \times 10^{-6}$, almost 100% of the magnetic moments are directed along this axis. If the magnetoelastic coupling takes up a negative value, as in the ordered Fe_3Si alloy, the elongation would result in a deviation of the magnetic moments of individual Fe atoms from the longitudinal direction toward the transverse direction. They will be mainly oriented along one of the easy axes <100> that is perpendicular to the direction of the elongation or has the smallest angle with the plane transverse to it. The critical point of silicon concentration C_{Si}, in which the magnetostriction constant changes its sign (see, for example, λ_{100} in Fig. 15), is the value of $C_{Si} \approx 0.12$ (Bertotti & Fiorillo, 1994c).

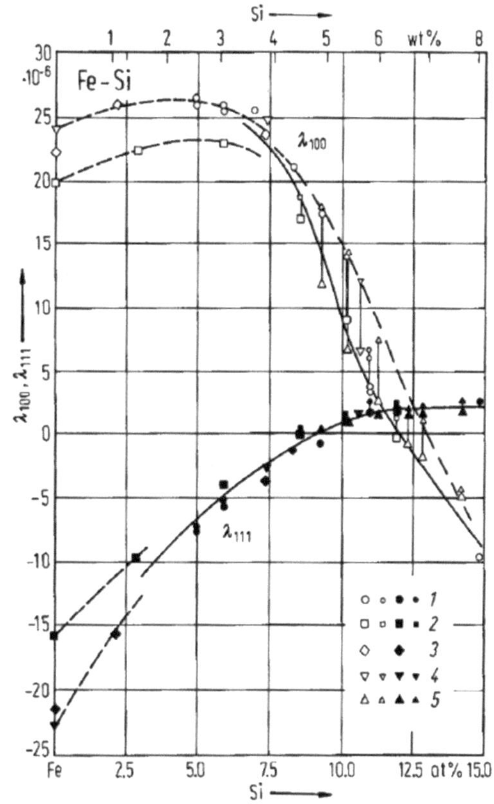

Fig. 15. Magnetostriction constants as functions of Si concentrations, determined by strain-gauge measurements on single crystals (room temperature). Large symbols: slowly cooled specimens. Small symbols: specimens quenched from temperature of the order of 1000°C. Image was taken from the database (Bertotti & Fiorillo, 1994c), Fig. 42.

The dependence of the magnetostriction constant λ_{100} on the C_{Si} largely reflects an increase in the volume fraction of the ordered Fe_3Si phase with increasing concentration of silicon (Hilfrich et al., 1994), which, as known (Bertotti & Fiorillo, 1994c), has a negative value of the magnetostriction constant ($\lambda_{100} \approx -20 \times 10^{-6}$, $\lambda_{111} = -5 \times 10^{-6}$). With the negative magnetoelastic coupling, which is typical of the ordered Fe_3Si phase, the relative compression of the lattice should cause an increase in the magnetization along the direction of reduction.
The microstructure of the nanocrystalline FeSiBNbCu alloy can be represented as a huge number of the isotropically oriented α-Fe (Si) nanocrystals of about 10 nm in average grain diameter with the bcc lattice, and the clusters of nonmagnetic fcc Cu(Fe) grains about 5 nm placed in the residual amorphous matrix phase Fe(Nb)-B (Yoshizawa, 2006). The concentration of silicon in the nanoparticles is a few percent higher than its average concentration in the alloy (Serikov et al., 2006). For example, at an average value of $X = 13.5$,

the silicon content in the nanocrystals reaches 18 at.%. Therefore, if we assume that the magnetoelastic properties of the crystalline Fe-Si alloys can be transferred to the nanocrystalline ones, then the critical point at which the sign of the magnetoelastic coupling changes will be the average silicon concentration $X \approx 9$ in the $Fe_{87-X}Si_XB_9Nb_3Cu_1$ alloy.

At this point it should be noted the role of Fe_3Si phase in the formation of the transverse magnetic anisotropy. The iron-silicon alloy in the ordered Fe_3Si state ($D0_3$ structure) is characterized by a negative constant of the magnetoelastic coupling. The transverse orientation of the magnetization with respect to the extension direction is characteristic of it, the so-called the transverse Villari effect; and negative magnetostriction – in the magnetic field, the length of the sample in the direction of the field application decreases. Therefore, the transverse magnetic anisotropy induced in the nanocrystalline alloy ribbons after the SA treatment should be attributed to the presence of a substantial fraction of Fe_3Si phase in the nanocrystals. The presence of Fe_3Si phase in the nanocrystals grown during NCA or TSA was supported by the following experimental observations. Firstly, as well as in the crystalline α-FeSi alloys the line $a(X)$ (inset in Fig. 4) describing the dependence of the nanocrystal lattice parameter on the silicon concentration has an inflection point at $X \approx 9.5$ associated with the $D0_3$ ordering. Secondly, in the low-angle part of the diffraction patterns of the $Fe_{87-X}Si_XB_9Nb_3Cu_1$ alloy samples subjected to nanocrystallization (by NCA or TSA) with $X = 11$ and 13.5, there are the superstructure peaks (½ ½ ½) and (100). And finally, the distribution functions P(H) obtained from the Mössbauer spectra has a weak manifestation of the contribution from the coordination of 4:4 (four Si and four Fe atoms in the first shell of an iron atom) at $X = 8$, and a further significant increase of it with $X \geq 9$. The relative volume fraction of Fe_3Si phase obtained from these data reaches 80% at $X = 13.5$. The observed changes in the phase composition of the nanocrystals depending on the silicon concentration correlate with the magnetic properties (a form of the magnetic hysteresis loops and the magnetic anisotropy constant). When $X \geq 9.5$ the constant K_{TSA} is negative and grows in magnitude. Most clearly, the transverse magnetic anisotropy is pronounced in the form of the loops for samples with $X = 11$ and 13.5. For the same values of X, the magnetization comes from the sample plane as observed after the TSA treatment.

Fig. 16a represents schematically one of the 48 equivalent spherical triangles. The direction of tensile stress application relative to the nanocrystalline axes is shown by the radius-vector for three most interesting cases, namely, along [010], [110], and [111]. We remind that the nanocrystals have random orientation in the Fe-Si-Nb-Cu-B alloy and, hence, a random one relative to the ribbon axis. The distortions of the nanocrystal lattice are different for the radius-vectors within this spherical triangle. The nearly tetragonal distortions correspond to the case where the radius-vector is directed along an easy-magnetization <100> axis of a nanocrystal, for example, [010] in Fig. 16a. After TSA in the nanocrystals of the $Fe_{87-X}Si_XB_9Nb_3Cu_1$ alloys with the positive magnetoelastic coupling constant (longitudinal Villari effect) at $X < 9$, the magnetization will be oriented along one of the easy axes <100> that makes the smallest angle with the axis of residual strain or with the direction of load application **P**. In this case the maximum deviation from the longitudinal direction can be ~ 55° for those nanocrystals that have [111] axis parallel to the direction of tensile load application **P**. But for this direction the residual distortion is minimum, if any is. If the nanocrystal axis [110] || **P**, then the deviation of magnetization from the longitudinal axis of the ribbon will be 45°. The value of the residual strain along the [110] - $\Delta d_{110}/d_{110}$ is not more then ½ of $\Delta d_{100}/d_{100}$. Since the greatest value of the relative lattice distortion was observed in nanocrystals oriented with one of the axes <100> along **P**, then the magnetization in these nanocrystals is oriented parallel to **P**, making the largest contribution to the formation of the longitudinal magnetic anisotropy. Since the density of the magnetic anisotropy energy is proportional to the interplanar spacing strain $\Delta d/d$, then the nanocrystals with tetragonal deformation of lattice will give the largest contribution to the energy of the longitudinal magnetic anisotropy, when one of the easy magnetization axes <100> is parallel to the ribbon axis and **P**.

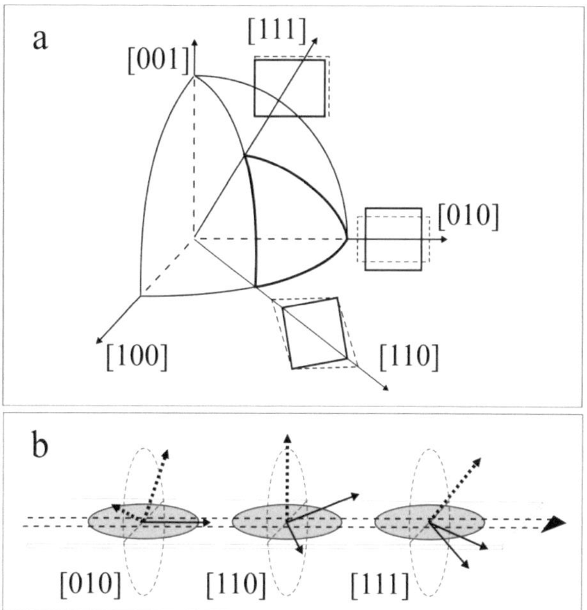

Fig. 16. a - scheme of the distortions of the nanocrystal lattice (dashed lines) for different directions of tensile load application parallel to the radius vectors [111], [010] and [110]. The distortions are shown in projection on (001) plane in the case of extension along [010] and [110], and on the plane (1-1 0) in the case of extension along [111].
b – the corresponding variants of orientations of the crystallographic axes <100> (straight lines with arrows) and magnetization (dashed lines with arrows) in the nanocrystals with the transverse Villari effect relative to the ribbon axis or the direction of load application - double dotted line with an arrow.

For $X > 9$, the magnetoelastic coupling in nanocrystals becomes negative (transverse Villari effect). The tetragonal distortions correspond to the case when the radius vector is directed along the easy axis <100> of the nanocrystal in Fig. 16b - [010]. Owing to the transverse Villari effect, the magnetization of these nanocrystals is to be directed with equal probability along either [100] or [001]. The residual distortion along [110] results in the nanocrystal magnetization along [001] as shown in Fig. 16b. The magnetization of the nanocrystals with the axis of <111> type parallel to the ribbon axis (i.e. the tensile stress direction) is to be directed along one of the easy axes <100>. Their deviation from the ribbon axis is about 55°. Thus, these nanocrystals also give contribution to the transverse magnetization (Fig. 16b). The radius-vectors <111> and <100> correspond to two limiting cases: the magnetization direction of a nanocrystal makes an angle of 55° and 90° with the ribbon axis, and of 35° and

0° with the transverse plane, respectively. The magnetization directions of all the nanocrystals must lie within ±35° to the plane normal to **P**. Taking into account the random orientation of nanocrystals and the anisotropy of lattice strains in the nanocrystals, we can estimate the ratio of the magnetization components that are perpendicular and parallel to the ribbon axis as at least 3 to 1. The anisotropy of the residual strains facilitates an increase in the transversal component nature of the magnetic anisotropy induced during TSA as the greatest contribution to the energy of the transverse magnetic anisotropy is provided by the nanocrystals in which one of the easy axes is oriented along the ribbon and stretched along the same axis. At the same time, the anisotropy of the ribbon shape is an important factor. Therefore, the magnetization vector in the nanocrystals predominantly oriented along the easy magnetization axis that is closer to the ribbon plane.

5. Conclusion

Summing up the results of X-ray diffraction and NGR-spectroscopy studies on the structure of the nanocrystals in soft magnetic alloys $Fe_{87-X}Si_XB_9Nb_3Cu_1$ (X = 0 - 13.5) in the initial state, i.e. after quenching from the melt, after the nanocrystallization annealing and the tensile stress annealing, we can conclude that all the alloys in the initial state independently of the silicon concentration (X) are in a state with an ultra fine grain (average grain size of about 2 nm). Their structure is isotropic, it does not depend on the direction of observation (i.e., along or across the ribbon, or direction of the ribbon output upon quenching) and silicon content. After the nanocrystallization annealing, the intensity distribution in the diffraction patterns remains isotropic, but with increasing crystallite size up to 10-12 nm. A complete set of reflections, corresponding the bcc lattice of the alloy α-Fe (Si) becomes allowed. When the content of silicon X = 11 and 13.5, in the diffraction patterns there appear superstructure peaks of the $D0_3$ phase - (½ ½ ½) and (100). After TSA along the direction of the tensile load application, there is an increase in the interplanar distances and the decrease in the transverse direction. The strain is anisotropic: within the accuracy of the experiment, no distortions in the <111> directions are observed, and the distortions in the <100> directions are maximum and reach 1%. It is shown that the type (longitudinal or transverse) of the magnetic anisotropy depends on the sign (positive or negative) of the magnetoelastic coupling in nanocrystals, which, in turn, is determined by the contribution of the ordered phase Fe_3Si, characterized by a negative constant of magnetoelastic coupling.
For nanocrystalline alloy $Fe_{87-X}Si_XB_9Nb_3Cu_1$, a phenomenological model of the mechanism of inducing the longitudinal magnetic anisotropy at $X < 9$ and the transverse magnetic anisotropy at $X > 9$ due to the anisotropic distortions of nanocrystals was proposed.

6. Acknowledgments

This study was supported by the Russian Foundation for Basic Research (project no. 10-02-00435) and by the Presidium of Russian Academy of Sciences (project no. 09-П-2-1035).

7. References

Bozorth, R.M. (1993). *Ferromagnetism*. IEEE Press, ISBN 0780310322, New York, USA

Bertotti, G., Fiorillo, F. (1994a). 7.1.2.2.1 Phase diagrams, lattice parameters and density, thermal expansion, In: *Landolt-Bornstein. Numerical data and functional relationships in science and technology. New series. Group III: Solid state physics. V. 19. Magnetic properies of metals. Subv. i1. Magnetic alloys for technical applications. Soft magnetic alloys, invar and elinvar alloys*, Wijn, H.P.J. (Ed.), pp. 35-44, Springer-Verlag, ISBN 3540555900/9783540555902/3-540-55590-0, Berlin, Germany, Retrieved from: http://www.springermaterials.com

Bertotti, G., Fiorillo, F. (1994b). 7.1.2.2.2 Elastic constants, mechanical properties, Ibid, Wijn, H.P.J. (Ed.), pp. 45-48, Springer-Verlag, ISBN 3540555900/9783540555902/3-540-55590-0, Berlin, Germany, Retrieved from: http://www.springermaterials.com

Bertotti, G., Fiorillo, F. (1994c). 7.1.2.3.3 Magnetostriction constants, Ibid, Wijn, H.P.J. (Ed.), pp. 55-58, Springer-Verlag, ISBN 3540555900/9783540555902/3-540-55590-0, Berlin, Germany, Retrieved from: http://www.springermaterials.com

Chernenkov, Yu.P., Ershov, N.V., Fedorov, V.I., Lukshina, V.A., & Potapov, A.P. (2010). X-Ray diffraction studies of the structure of nanocrystals in $Fe_{73.5}Si_{13.5}B_9Nb_3Cu_1$ soft magnetic alloys before and after thermomechanical treatment. *Phys. Solid State*, Vol. 52, No. 3, (March, 2010), pp. 554-560, ISSN 1063-7834

Elsukov, E.P., Barinov, V.A., & Konygin, G.N. (1989). The structural and magnetic parameters of the ordered Fe-Si alloys. *Metallofizika*, Vol. 11, No. 4, (April, 1989), pp. 52-59, ISSN 1024-1809 (in Russian)

Filippov, B.N. (2006). Theoretical and technological principles of production of soft magnetic materials with new level of functional properties, *Proceeding of the Result Conference of Projects Leaders "New Materials and Structures"*, p 81, Moscow, Russia, November 30 - December 1, 2006 (in Russian)

Fischer, H., Barnes, A., & Salmon, P. (2006). Neutron and x-ray diffraction studies of liquids and glasses. *Rep. Prog. Phys.*, Vol. 69, No. 1, (January, 2006), pp. 233-299, ISSN 0034-4885

Glazer, A.A., Kleinerman, N.M., Lukshina, V.A., Potapov, A.P., & Serikov, V.V. Thermomechanical treatment of the nanocrystalline alloy $Fe_{73.5}Cu_1Nb_3Si_{13.5}B_9$. *Fiz. Met. Metalloved.*, No. 12, (January, 1991), pp. 56-61, ISSN 1812-7339 (in Russian)

Herzer, G. (1992). Nanocrystalline soft magnetic materials. *J. Magn. Magn. Mater.*, Vol. 112, No.1-3, (July, 1992), pp. 258-262, ISSN 0304-8853

Herzer, G. (1994). Creep induced magnetic anisotropy in nanocrystalline Fe-Cu-Nb-Si-B alloys. *IEEE Transactions on Magnetics*, Vol. 30, No. 6, (June, 1994), pp. 4800-4802, ISSN 0018-9464

Hilfrich, K., Koelker, W., Petry, W., Schärpf, O., & Nembach, E. (1994). The states of order and the phase diagram of $Fe_{1-x}Si_x$, $0.06 \leq x \leq 0.20$, investigated by neutron scattering. *Acta Metal. Mater.* Vol. 42, No. 3, (March 1994), pp. 743-748, ISSN 0956-7151

Hofmann, B., & Kronmüller, H. (1996) Stress-induced magnetic anisotropy in nanocrystalline FeCuNbSiB alloy. *J. Magn. Magn. Mater,* Vol. 152, No. 1-2, (January, 1996), pp. 91-98, ISSN 0304-8853

Kleinerman, N.M., Serikov, V.V., Lukshina, V.A., Potapov, A.P., & Volkova, E.G. (2004). Induced magnetic anisotropy and the structure of nanocrystalline Fe-Co-Cu-Nb-Si-B alloys with different content of Co: II structure of alloys with an induced magnetic anisotropy. *Phys. Met. Metallogr.*, Vol. 107, No. 5, (May, 2004), pp. 449-456, ISSN 0031-918X

Litvinov, V.S., Karakishev, S.D. & Ovchinnikov, V.V. (1982). *Nuclear gamma-resonance spectroscopy of alloys*, Metallurgiya, ISBN 5-1205068, Moscow, USSR (in Russian).

Lukshina, V.A., Dmitrieva, N.V., Noskova, N.I., Volkova, E.G., Kleinerman, N.M., Serikov, V.V., & Potapov, A.P. (2002). Nanocrystalline alloy $Fe_{73.5}Cu_1Nb_3Si_{13.5}B_9$: structure and magnetic properties. II. Thermal stability of induced magnetic anisotropy. *Phys. Met. Metallogr.*, Vol. 93, No. 6, (June, 2002), pp. 536-543, ISSN 0031-918X

Noskova, N.I., Serikov, V.V., Glazer, A.A., Kleinerman, N.M., &Potapov, A.P. (1992). Electron-microscopic investigation structure of $Fe_{73.5}Cu_1Nb_3Si_{13.5}B_9$ alloy in nanicrystalline state. *Phys. Met. Metallogr.* Vol. 74, No. 1. 1, (January, 1992), pp. 80-86, ISSN 0031-918X

Ohnuma, M., Hono, K., Yanai, T., Fukunaga, H., & Yoshizawa, Y. (2003). Direct evidence for structural origin of stress-induced magnetic anisotropy in Fe–Si–B–Nb–Cu nanocrystalline alloys. *Appl. Phys. Lett.*, Vol. 83, No. 14, (September, 2003), pp. 2859-2861, ISSN 0003-6951

Ohnuma, M., Hono, K., Yanai, T., Nakano, M., Fukunaga, H., & Yoshizawa, Y. (2005). Origin of the magnetic anisotropy induced by stress annealing in Fe-based nanocrystalline alloy. *Appl. Phys. Lett.*, Vol. 86, No. 15, (April, 2005), 152513- 152511, ISSN 0003-6951

Potapov, A.P., & Filippov, B.N. (2009). Magnetic properties of soft magnetic nanocrystalline materials. *Proceeding of 3th All-Russian Conference on Nanomaterials, NANO-2009*, pp. 41-43, ISBN 978-5-93667-123-5, Ekaterinburg, Russia, April 20-24, 2009 (in Russian)

Randrianantoandro, N., Gaffet, E., Mira, J. & Greneche, J.-M. (1999). Magnetic hyperfine temperature dependence in Fe-Si crystalline alloys. *Solid State Commun.*, Vol. 111, No. 6, (June, 1999), pp. 323-327, ISSN 0038-1098

Serikov, V.V., Kleinerman, N.M., Volkova, E.G., Lukshina, V.A., Potapov, A.P., & Svalov, A.V. (2006). Structure and magnetic properties of nanocrystalline FeCuNbSiB alloys after a thermomechanical treatment. *Phys. Met. Metallogr.* Vol. 102, No. 3, (March, 2006), pp. 268-273, ISSN 0031-918X

Stearns, M.B. (1963). Internal magnetic fields, isomer shifts, and relative abundance of the various Fe sites in Fe-Si alloys. Phys.Rev., Vol. 129, No. 3, (March, 1963), pp. 1134-1144, ISSN 1098-0121

Warren, B.E. (1969). *X-ray diffraction*. Addison–Wesley, ISBN 0-486-66317-5, New York, USA

Wertheim, G.K. (1964). *Mössbauer effect: Principles and applications*. Academic Press, ISBN 147-3662-1139, New York, USA

Yoshizawa, Y., Oguma, S., & Yamauchi, K. J. (1988). New Fe-based soft magnetic alloys composed of ultrafine grain structure. *J. Appl. Phys.*, Vol. 64, No. 10, (November, 1988), pp. 6044-6046, ISSN 0021-8979

Yoshizawa, Y. & Yamauchi, K. (1989). Effects of magnetic field annealing on magnetic properties in ultrafine crystalline Fe-Cu-Nb-Si-B alloys. *IEEE Trans. Magn.*, Vol. 25, No. 5, (September, 1989), pp. 3324-3326, ISSN 0018-9464

Yoshizawa, Y. (2006). Nanocrystalline Soft Magnetic Materials and Their Applications, In: *Handbook of advanced magnetic materials. Vol. 4: Properties and applications*, Liu, Yi, Sellmyer, D. J., Shindopp D., pp. 124-158, Shpringer, ISBN 1-4020-7983-4, New York, USA

17

Energy Transfer from Silicon Nanocrystals to Er^{3+} Ions Embedded in Silicon Oxide Matrix

Kantisara Pita and Quang Vinh Vu
Photonic Research Centre, School of Electrical and Electronics Engineering, Nanyang Technological University, 50 Nanyang Avenue
CINTRA CNRS/NTU/THALES, UMI 3288, Research Techno Plaza, 50 Nanyang Drive, Border X Block, Level 6
Singapore

1. Introduction

Silicon (Si) based light emitting devices have drawn much attention for the integration of electronic and photonics. Si nanostructures (amorphous clusters or crystals) have been recognized as good candidates for effective light emitting devices (Bulutay, 2007; Seino et al., 2009; Takagahara & Takeda, 2007; Wolkin et al., 1999). However, photons emitted by Si nanostructures can be reabsorbed by Si waveguides due to the higher photon energy compared to bulk Si bandgap. To overcome this problem, Erbium (Er) doped Si nanostructures embedded in SiO_2 matrix has been extensively studied (Fujii et al., 2004; Heitmann et al., 2003; Kik & Polman, 2001; Polman & Veggel, 2004; Savchyn et al., 2007, 2008). Si nanostructures can be excited optically or electrically, then transfer the energy to Er^{3+} ions which then decay radiatively giving emission peaked at 1.53μm, which coincides with the telecommunication wavelength. Hence, light sources made by this material system (Er^{3+} ions doped Si nanostructures embedded in SiO_2 matrix) in the integrated Si platforms can be used directly with telecommunication devices.

To date, however, one of the main challenges of this material system is the low energy transfer efficiency from Si nanostructures to Er^{3+} ions, which we are going to address in this chapter. To increase the energy transfer efficiency, the mechanism of the energy transfer must be well understood. Prior to 2007, many research works reported that the main energy transfer mechanism is from Si nanocrystals to Er^{3+} ions (Fujii et al., 2004; Heitmann et al., 2003; Kik & Polman, 2001; Polman & Veggel, 2004). However, recently, in 2007 and 2008, one leading group in the field suggested that defect mediated energy transfer was the dominant mechanism (Savchyn et al., 2007, 2008). In their work (Savchyn et al., 2007, 2008), the origin of defects that transfer energy to Er^{3+} ions was not discussed. The controversial in the energy transfer mechanism has created much difficulty in improving Er^{3+} emission efficiency. The approach to improve Er^{3+} emission efficiency will clearly very much depend on the dominant mechanism of the energy transfer to Er^{3+} ions, whether the defects or the Si nanocrystals are the dominant factor for the energy transfer.

In this chapter, we present our work on different energy transitions from defects and from Si nanocrystals and their energy transfer efficiency to Er^{3+} ions. Bulk Silicon Rich Oxide

(SRO) and alternate layers of SRO and SiO_2 films (SRO/SiO_2 superlattice) with and without Er^{3+} ions incorporation were developed by co-sputtering method. Through the photoluminescence (PL) study of the samples annealed at temperature ranging from 500°C to 1050°C, with and without passivation (further annealing in a forming gas of 5% H_2 and 95% N_2 mixture), 4 different energy transitions and their energy transfer to Er^{3+} ions have been identified and investigated. Optimum Er^{3+} ions concentration for maximized Er^{3+} emission in superlattice structure has also been found. Through the lifetime measurement, transfer rate and transfer efficiency from exciton recombination in Si nanocrystals to Er^{3+} ions in our superlattice structure have also been found.

2. Preparation of Si nanostructures embedded in SiO_2 matrix with/without Er^{3+} ions incorporation by co-sputtering deposition method

In this section, the preparation of Si nanostructures embedded in SiO_2 matrix is described. The substrates were 4" Si wafers which were cleaned from dust particles by blowing with a nitrogen gun before loaded into a sputtering deposition chamber. Co-sputtering method was chosen as deposition method due to its flexibility to incorporate different species into the films and the well-controlled deposition rate.

Before depositing the films, the chamber was pumped down to a pressure of ~5×10^{-6} torr. The substrate holder was then heated to temperature of 250°C for 1 hour to outgas the substrates. All depositions were carried out at the base pressure of ~5×10^{-7} torr and substrate temperature of 100°C using an Argon gas with a pressure of ~6.3×10^{-3} torr. To ensure uniformity, the substrate was rotated at 20rpm. Prior to each deposition, the targets were pre-sputtered at low power (20W) for 15 minutes to ensure each target was free of surface contamination.

SiO_2 layers were deposited by sputtering a SiO_2 target. To deposit Silicon Rich Oxide (SRO) layers, a SiO_2 target and a Si target were co-sputtered simultaneously. The concentration of Si in SRO layers can be controlled by controlling the power of each target. To incorporate Er^{3+} ions into the films, an Er_2O_3 target was co-sputtered simultaneously. Generally, the Er^{3+} ions are incorporated using the ion implantation technique (Castagna et al., 2003, 2006; Heitmann et al., 2003; Iacona et al., 2006). The use of the co-sputtering method to incorporate Er^{3+} ions has two major advantages: (i) the ability to control the locations of Er^{3+} ions and (ii) no extra defects are introduced which has to be cured by a post annealing process. The ability to control location of Er^{3+} ions is important because the energy transfer to Er^{3+} ions can only be achieved when the donors and acceptors are close to each other (Heitmann et al., 2003). In our process, the deposition rates for SiO_2 and SRO layers are 4.5nm/minute and 8nm/minute, respectively. These slow deposition rates have enabled us to accurately control the layer thickness of few nm.

To study the mechanism of different energy transitions and their energy transfer to Er^{3+} ions, bulk SRO and alternate layers of SRO and SiO_2 films (SRO/SiO_2 superlattice) with and without Er^{3+} ions incorporation were deposited and annealed at different temperatures. The bulk SRO samples with and without Er^{3+} ions incorporation are, respectively, referred to as SRO and Er^{3+}:SRO samples, and the SRO/SiO_2 superlattice samples with and without Er^{3+} ions incorporation are, respectively, referred to as SRO/SiO_2 and Er^{3+}:SRO/SiO_2 samples. Figure 1 shows the cross section of bulk SRO and Er^{3+}:SRO structure. The Si concentration in SRO layer was 52 at%. Figure 2 shows the cross section of SRO/SiO_2 and Er^{3+}:SRO/SiO_2 superlattice structure.

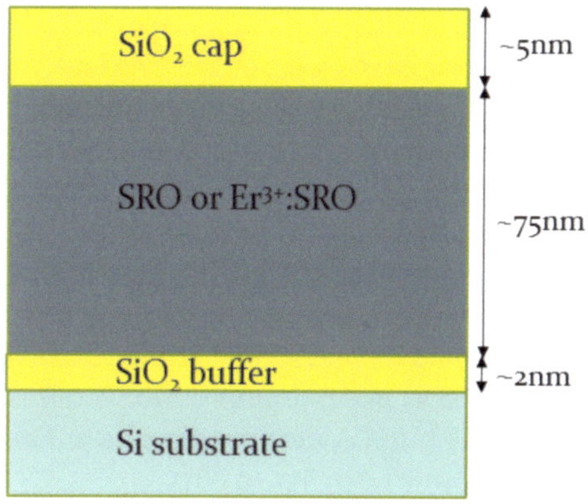

Fig. 1. Cross section of the bulk SRO and Er^{3+}:SRO samples deposited by the co-sputtering process

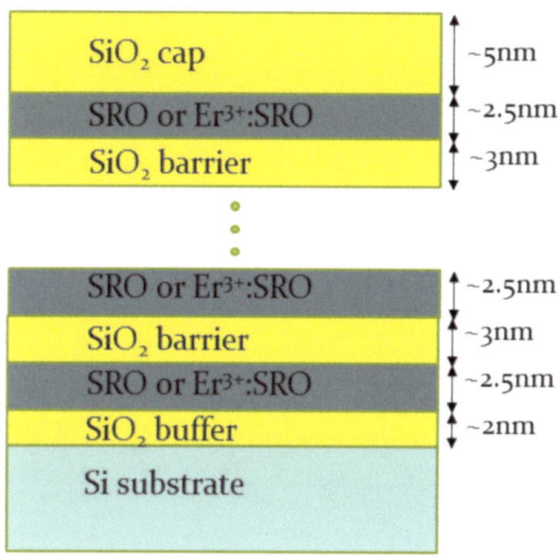

Fig. 2. Cross section of the SRO/SiO_2 and Er^{3+}:SRO/SiO_2 superlattice samples deposited by the co-sputtering process

Similar to bulk SRO, the Si concentration in each SRO layer of SRO/SiO_2 superlattice was 52 at%. The number of SRO layers in these sample was 30, making total thickness of SRO layers 75nm, which is the same as that of the bulk samples.

All bulk and superlattice samples were annealed using a rapid thermal processor (RTP) at various temperatures. Prior to each annealing process, the RTP chamber was vacuum pumped and then filled with nitrogen in several cycles to minimize the presence of oxygen in annealing chamber. All annealing processes were done in nitrogen environment. The annealing temperature ranged from 500°C to 1050°C. The annealing time was fixed at 1 minute.

After the annealing process, some of the samples were passivated. The passivation process was done by annealing the samples in a forming gas (H_2 5% N_2 95%) at 500°C. The samples annealed by the RTP only are referred to as "as-annealed" samples and the samples annealed by the RTP and passivated are referred to as "passivated" samples. The purpose of the passivation is to remove the dangling bond defects surrounding Si nanostructures (Savchyn et al., 2008; Wilkinson & Elliman, 2003). In our study, we have varied the passivation time and found that after 30 minutes of the passivation, the emission intensity of exciton recombinations in Si nanocrystals as well as the exciton lifetime have reached the optimum values and saturated, showing that most, if not all, dangling bond defects have been effectively removed. To ensure that we remove most, if not all, dangling bond defects, we passivated our samples for 1 hour.

In our work, we also varied the concentrations of Er^{3+} ions to obtain the optimum Er^{3+} ion concentration, the energy transfer rate and the transfer efficiency to Er^{3+} ions for our materials system. The samples were prepared using similar structure as shown in Figure 2. The samples were annealed at 1050°C in N_2 ambience for 1 minute and then passivated at 500°C in forming gas for 1 hour. Er^{3+} ion concentration was varied from 0.02 at% to 0.4 at% by varying the Er_2O_3 deposition power from 5W to 35W.

To study different energy transitions from defects and from Si nanocrystals and their energy transfer efficiency to Er^{3+} ions, photoluminescence (PL) was performed on the as-annealed and passivated SRO, Er^{3+}:SRO, SRO/SiO_2, and Er^{3+}:SRO/SiO_2 samples. The PL spectra of all samples were taken using a Fluorolog-3 system. Excitation is fixed at 290nm by using a Xenon lamp and a monocromator. The spectra are corrected to the respond of the detector. All PL measurements were done at room temperature. Two detection ranges were investigated: (i) visible to near Infra-Red (IR) (400nm-850nm) to study emission from different kinds of defects and from exciton recombinations in Si nanocrystals and (ii) IR range (1470nm-1700nm) to detect the Er^{3+} emission peaked at 1535nm. The lifetime was measured using the same Fluorolog-3 system with a Xenon flash lamp. The transmission electron microscopy (TEM) was used to characterize the Si nanostructure in our samples. The concentrations of excess Si and Er^{3+} ions in our samples was estimated by the deposition rate of each target at each deposition power.

3. Si nanocrystals, different kinds of defects in bulk & superlattice structures and their energy transfer to Er^{3+} ions

3.1 TEM images of samples

Figure 3 shows TEM images of the bulk SRO samples annealed at 850°C and 950°C and the SRO/SiO_2 superlattice sample annealed at 950°C respectively.

As shown in Figure 3 (a), at the annealing temperature of 850°C, there are amorphous Si nanostructures, shown in the TEM image as dark coloured patches, in the bulk SRO sample. For our SRO/SiO_2 superlattice samples, it has been established through a TEM image that at the annealing temperature of 850°C, similar to our bulk samples, there are only amorphous Si nanostructures in the SRO/SiO_2 superlattice samples developed by our group (Silalahi et al., 2010). Up to 850°C annealing temperature, there is no crystallization of Si can be observed in both our bulk SRO and SRO/SiO_2 samples. On the other hand, at the annealing temperature of 950°C, as shown in Figures 3 (b) and (c) respectively, the fringes of Si nanocrystals can be clearly observed in both the bulk SRO sample and the SRO/SiO_2 superlattice sample. These results shows that in our samples, the nanocrystals have been formed at the annealing temperature of 950°C while at the annealing temperature of 850°C, the Si nanostructures are still in an amorphous state.

From Figure 3 (b), the diameter of Si nanocrystals in the bulk SRO sample ranges from 2.7nm to 4.2nm with an average of 3.3nm. In the SRO/SiO_2 superlattice sample, from Figure

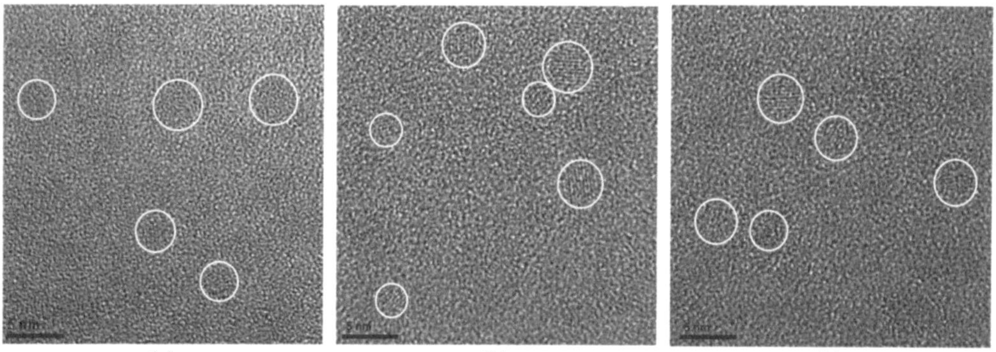

(a) (b) (c)

Fig. 3. TEM images of (a) the bulk SRO sample annealed at 850°C, (b) the bulk SRO sample annealed at 950°C, and (c) the SRO/SiO_2 superlattice sample annealed at 950°C

3 (c), the diameter of Si nanocrystals ranges from 3.3nm to 3.9nm with an average of 3.6nm. The TEM images show that the size distribution of Si nanocrystals in the bulk sample is larger than that in the superlattice sample. Indeed, the aim of using the superlattice structure is to have better control on the size of Si nanocrystals.

3.2 Visible to near IR PL (400nm-850nm) of the bulk SRO and Er^{3+}:SRO samples

Visible to near IR PL spectra of the as-annealed and passivated bulk SRO and Er^{3+}:SRO samples annealed at different temperature are shown in Figure 4 (a) and Figure 4 (b) respectively.

In Figures 4(a) and 4(b), we observe that there are two PL emission bands peaked at ~570nm and ~800nm. First of all, let us consider the emission peaked at ~800nm.

At the annealing temperature lower than 950°C, the emission at ~800nm is not observed in any of our samples. At the annealing temperature of 950°C, the emission at ~800nm is only detected from the passivated SRO and Er^{3+}:SRO samples. For the as-annealed bulk SRO and Er^{3+}:SRO samples, this emission is very low, almost not observable. At the annealing temperature of 1050°C, all of the as-annealed and passivated bulk SRO and Er^{3+}:SRO samples have emission peaked at ~800nm. It has been established that, for our samples, Si nanocrystals have been formed when the samples were annealed at 950°C (see Figure 3 (b) in section 3.1 above). The emission peaked at ~800nm can only be observed when Si nanocrystals are present in the samples and is attributed to the exciton recombinations in Si nanocrystals (Bulutay, 2007; Charvet et al., 1999; Kik & Polman, 2001; Seino et al., 2009; Silalahi et al., 2009, 2010; Wilkinson & Elliman, 2003). The increase of the emission intensity at the higher annealing temperature (in our case, 1050°C compared to 950°C) has also been reported and attributed to the better crystallization of Si nanocrystals (Kahler & Hofmeister, 2002; Kapaklis et al., 2005).

At the annealing temperature of 1050°C, the intensity of the ~800nm emission from the passivated samples are more than 4 times higher than that from the as-annealed samples. The increase of this emission upon passivation is due to the removal of dangling bond defects, which introduce non-radiative paths that quenches the radiative excitation recombinations in Si nanocrystals (Savchyn et al., 2008; Wilkinson & Elliman, 2003). For the 950°C as-annealed SRO and Er^{3+}:SRO samples, the emission from the exciton recombinations in Si nanocrystals is effectively quenched by the dangling bond defects.

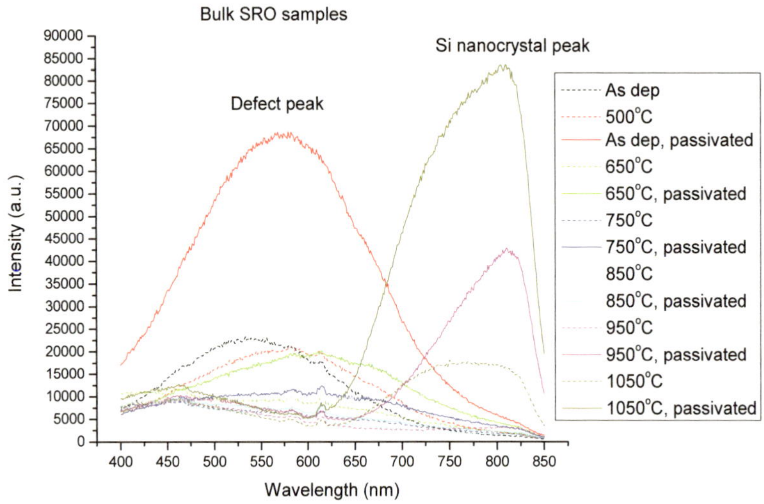

(a)

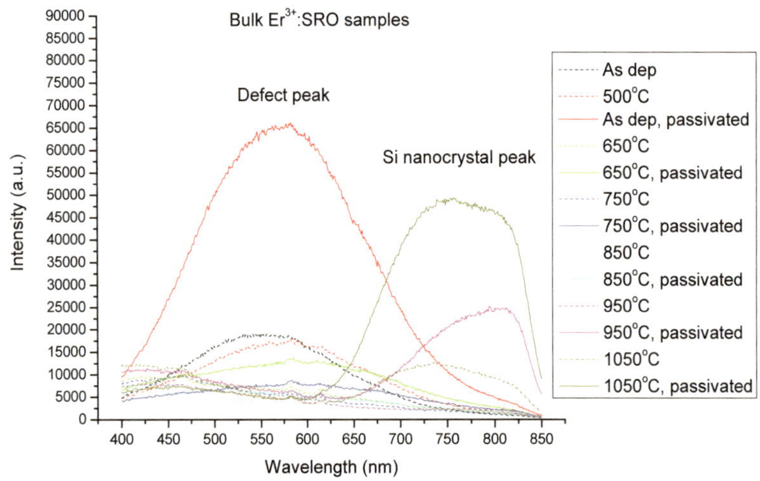

(b)

Fig. 4. Visible to near IR PL spectra of (a) the bulk SRO samples and (b) the bulk Er^{3+}:SRO samples at different annealing temperature under the excitation of 290nm

Upon passivation, where most dangling bond defects have been removed, the emission from exciton recombinations in Si nanocrystals can then be clearly observed.

From Figures 4 (a) and (b), we notice that the emission intensity at ~ 800nm of the bulk Er^{3+}:SRO samples is lower than that of the bulk SRO samples undergoing the same heat treatment and passivation. Furthermore, we also observe that the Er^{3+}:SRO samples have emission peaked at 1535nm due to the transition from excited state $^4I_{13/2}$ to ground state $^4I_{15/2}$ in the Er^{3+} ions (the emission spectra of the Er^{3+} ions will be presented in Figure 8 (a) in

section 3.4 below). The decrease of the emission intensity due to the exciton recombinations in Si nanocrystals and the presence of Er^{3+} emission, upon Er^{3+} ions incorporation, is an indication of an energy transfer from the exciton recombinations in Si nanocrystals to Er^{3+} ions. When Er^{3+} ions are near a Si nanocrystal, excitons in the Si nanocrystal can recombine non-radiatively, transfer their energies to the Er^{3+} ions and excite them. The excited Er^{3+} ions can then decay radiatively from $^4I_{13/2}$ to $^4I_{15/2}$, giving emission peak at 1535nm. This energy transfer mechanism has been reported earlier (Fujii et al., 2004; Heitmann et al., 2003; Kik & Polman, 2001; Polman & Veggel, 2004). The efficiency of energy transfer from exciton recombinations in Si nanocrystals to Er^{3+} ions will be discussed in more detail in section 3.4.
Now, let us consider the emission at ~570nm. As shown in Figure 4 (a) and (b), for both bulk SRO and Er^{3+}:SRO, the as-deposited samples and the samples annealed at low temperature (500°C - 650°C) show a dominant emission peaked at ~570nm. To investigate the evolution of this emission, the peak intensity at 570nm of the as-annealed and passivated SRO and Er^{3+}:SRO samples was plotted as a function of the annealing temperature, as shown in Figure 5.

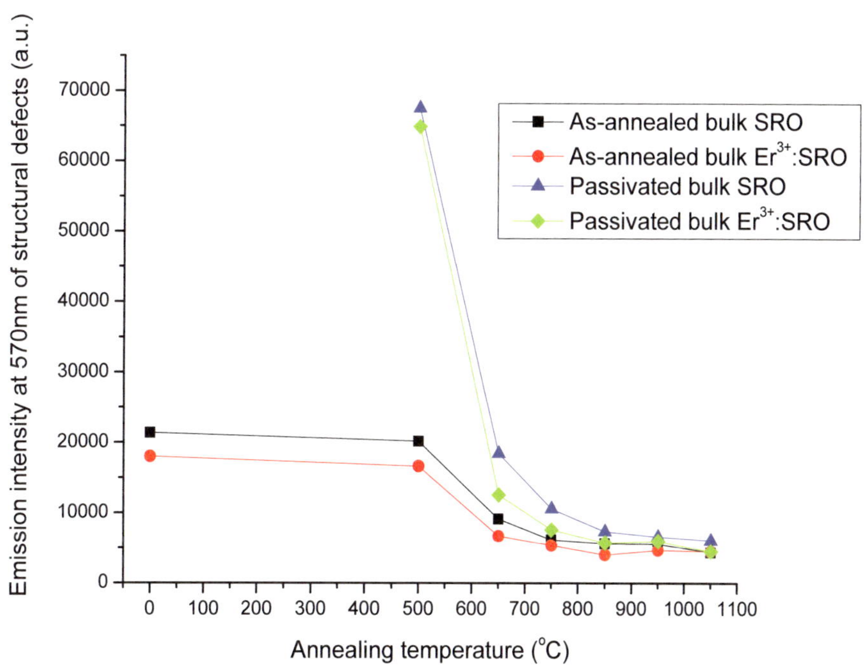

Fig. 5. Evolution of the emission peak at 570nm of the as-annealed and passivated SRO and Er^{3+}:SRO samples as a function of the annealing temperature

From Figure 5, we notice that for the as-annealed SRO and Er^{3+}:SRO samples, the emissions peaked at ~570nm are almost constant from the as-deposited samples up to the post-deposition annealing temperature of 500°C. The emissions then decrease beyond 500°C annealing temperature. We also observe that the emissions from the passivated SRO and Er^{3+}:SRO samples annealed at 500°C are significantly higher than those as-annealed samples and then reduce to the level comparable to the as-annealed samples beyond 500°C annealing temperature. The reduction of the emissions from all the samples at the annealing temperature higher than 500°C (i.e. the emissions decrease with the formation of Silicon amorphous nanostructures and nanocrystals) indicates that these emissions must be due to the defects rather than the Silicon amorphous nanostructures or nanocrystals. Furthermore, the reason that the emissions from the passivated SRO and Er^{3+}:SRO samples are significantly higher than those from the as-annealed SRO and Er^{3+}:SRO samples is because the passivation processes have removed the dangling bond defects which can serve as the non-radiative recombination path. Consequently, this emission which we have attributed to the defects must be due to the structural defects, associated with the presence of rich Silicon atoms and cannot be removed by the passivation process.

In Figure 5, we also see that emissions from Er^{3+}:SRO samples for both the as-annealed and passivated samples are consistently lower than those from just SRO samples. This is due to the energy transfer from those defects to excite the Er^{3+} ions which in turn give emission peaked at ~1535nm. We have indeed obtained the Er^{3+} emissions from our Er^{3+}:SRO samples annealed at 500°C, which are shown in Figure 8 (a). The efficiency of these defects on energy transfer to Er^{3+} ions will be discussed in detail in section 3.4.

3.3 Visible to near IR PL (400nm-850nm) of the SRO/SiO_2 superlattice and Er^{3+}:SRO/SiO_2 superlattice samples

Visible to near IR PL spectra of the as-annealed and passivated SRO/SiO_2 and Er^{3+}:SRO/SiO_2 superlattice samples annealed at different temperature are shown in Figure 6 (a) and Figure 6 (b) respectively.

As shown in Figures 6 (a) and (b), at the annealing temperature of 950°C and 1050°C, all SRO/SiO_2 and Er^{3+}:SRO/SiO_2 samples show the emission from the exciton recombinations in Si nanocrystals at ~800nm. Recall that, in section 3.2 above, the emission from the exciton recombinations in Si nanocrystals is not observed in the as-annealed bulk SRO and Er^{3+}:SRO samples annealed at 950°C. The appearance of this emission in the as-annealed SRO/SiO_2 and Er^{3+}:SRO/SiO_2 samples annealed at 950°C suggests that the superlattice structure has induced better crystallization. Similar to the bulk SRO and Er^{3+}:SRO samples, the emission intensity at ~800nm of the SRO/SiO_2 and Er^{3+}:SRO/SiO_2 samples annealed at 1050°C is higher than that of the samples annealed at 950°C. Furthermore, the intensity of this emission increases upon passivation. Again, as discussed above, the higher emission intensity at the annealing temperature of 1050°C is due to better crystallization of the Si nanocrystals and the increase of intensity upon passivation is the result of dangling bond defects removal. Also similar to the bulk SRO and Er^{3+}:SRO samples, the decreases of the exciton recombination emission intensity with the presence of Er^{3+} ions in the Er^{3+}:SRO/SiO_2 superlattice samples indicates energy transfer from the exciton recombinations in Si nanocrystals to Er^{3+} ions.

We observe that the as-deposited SRO/SiO_2 and Er^{3+}:SRO/SiO_2 samples and those annealed at low temperature (500°C-650°C) show an emission peaked at ~500nm. The peak intensity at 500nm of this emission was plotted as a function of the annealing temperature and is shown in Figure 7.

As shown in Figure 7, the emission of the as-deposited samples is just slightly higher than that of the as-annealed samples annealed at 500°C. Above 500°C, the emissions of the as-annealed and passivated SRO/SiO_2 and Er^{3+}:SRO/SiO_2 samples decrease. The decrease of this emission with the increasing annealing temperature indicates that this emission is originated from defects. Furthermore, all the passivated SRO/SiO_2 and Er^{3+}:SRO/SiO_2 samples show a significantly higher emission intensity than that of the as-annealed samples. As in the case of the bulk samples discussed above, this simply indicates that the dangling bonds, which can serve as non-radiative recombination path, have been removed by the

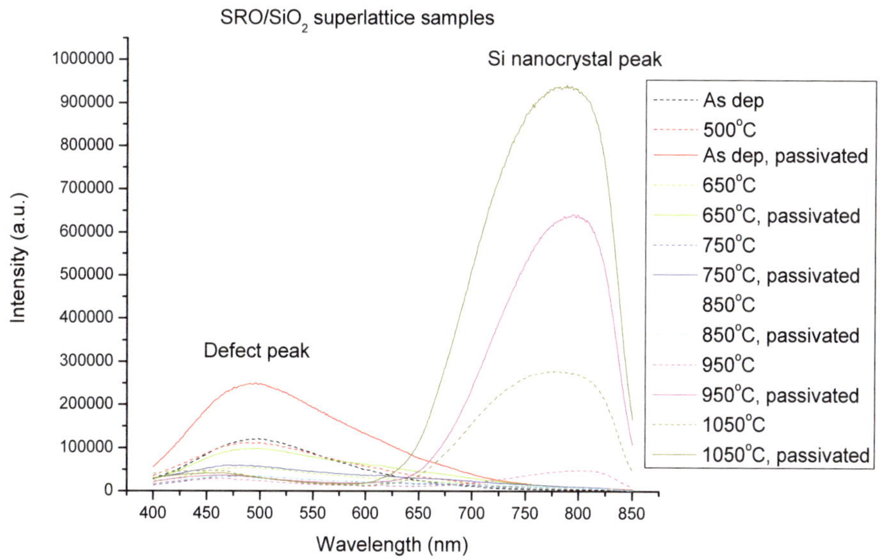

(a)

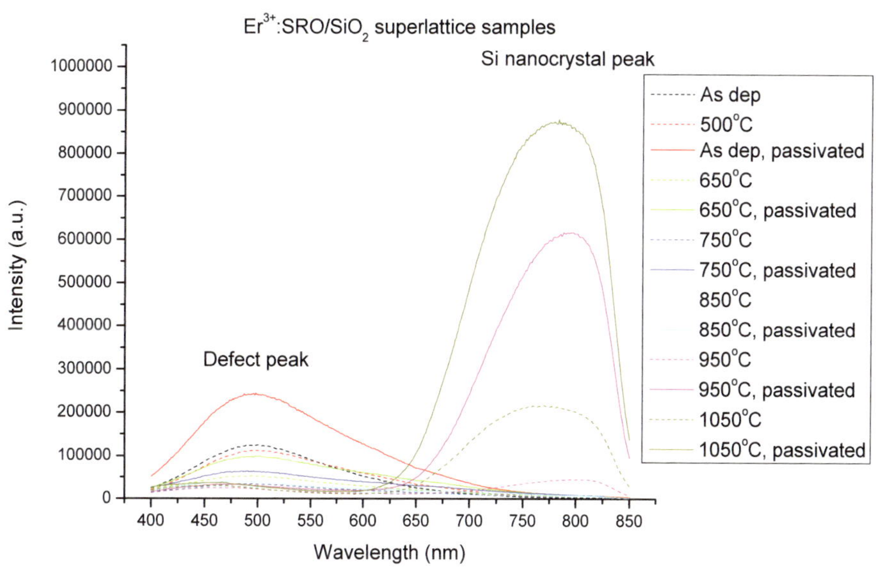

(b)

Fig. 6. Visible to near IR PL spectra of (a) the SRO/SiO_2 superlattice samples and (b) the Er^{3+}:SRO/SiO_2 superlattice samples at different annealing temperature under the excitation of 290nm

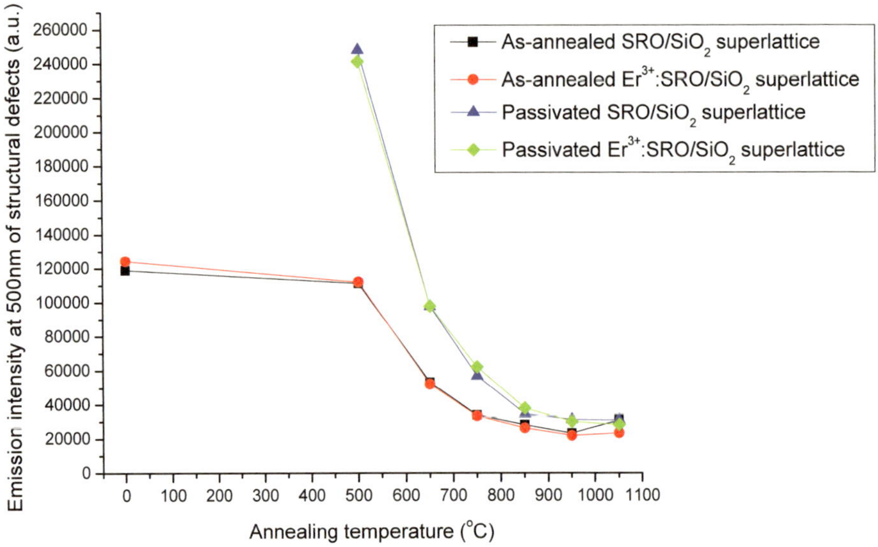

Fig. 7. Evolution of the emission peak at 500nm of the as-annealed and passivated SRO/SiO_2 and $Er^{3+}:SRO/SiO_2$ superlattice samples as a function of the annealing temperature

passivation. Since this defect related emission is only observed in the superlattice samples, the defects are most likely formed by structural defects that only occur in superlattice structure. The stress between SRO (or Er^{3+}:SRO) layers and SiO_2 layers may presumably induce such defects. In contrast to the defect emission of the bulk samples presented in Section 3.2, we notice here that the emission intensity of the SRO/SiO_2 samples are practically the same as that of the $Er^{3+}:SRO/SiO_2$ samples under the same annealing conditions. This suggests that these defects are not transferring their energy to Er^{3+} ions. Accordingly, as shown later in section 3.4, we indeed could not observe the emission of Er^{3+} ions due to the energy transfer from these defects.

We have observed from the PL spectra above that there is no emission that can be attributed to dangling bond defects. An emission due to dangling bond defects is expected to decrease with the passivation. For all of our samples, all detected PL emissions increases with the passivation. In the literature, no emission linked to dangling bond defects was reported either in visible or in IR range (Polman & Veggel, 2004; Savchyn et al., 2008; Silalahi et al., 2010; Wilkinson & Elliman, 2003). Up to this point, we conclude that dangling bond defects can serve as non-radiative de-excitation paths competing with and quenching other radiative de-excitation processes. However, in our study, from the emission intensity of Er^{3+} ions of the as-annealed and passivated $Er^{3+}:SRO/SiO_2$ superlattice samples presented in section 3.4 below, we observe that dangling bond defects may transfer their energy to Er^{3+} ions. This energy transfer mechanism and its efficiency will be discussed in detail in section 3.4.

3.4 IR (1470nm-1700nm) PL of the bulk Er^{3+}:SRO and $Er^{3+}:SRO/SiO_2$ superlattice samples

In section 3.2 and 3.3, four energy transitions have been identified: i) the emission of exciton recombinations in Si nanocrystals, ii) the non-radiative dangling bond de-excitation, iii) the emission from the structural defects in bulk structure, and iv) the emission from the structural defects in superlattice structure. In this section, the IR spectra at 1470nm-1700nm of the bulk Er^{3+}:SRO and Er^{3+}:SRO/SiO_2 superlattice samples will be presented and discussed to provide deeper understanding on the efficiency of those energy transitions in transferring energy to Er^{3+} ions.

Figure 8 (a) shows IR PL spectra from 1470nm to 1700nm of the bulk Er^{3+}:SRO samples. The emission peak at 1535nm is the well-known radiative transition from excited level $^4I_{13/2}$ to ground state $^4I_{15/2}$ of Er^{3+} ions, which has been reported earlier (Castagna et al., 2003, 2006; Heitmann et al., 2003; Iacona et al., 2006; Polman & Veggel, 2004; Savchyn et al., 2007). In this chapter, the emission peak at 1535nm will be referred to as "Er^{3+} peak". Figure 8 (b) shows the evolution of Er^{3+} peak intensity at 1535nm as a function of the annealing temperature of the as-annealed and passivated bulk Er^{3+}:SRO and Er^{3+}:SRO/SiO_2 superlattice samples. The excitation was fixed at 290nm.

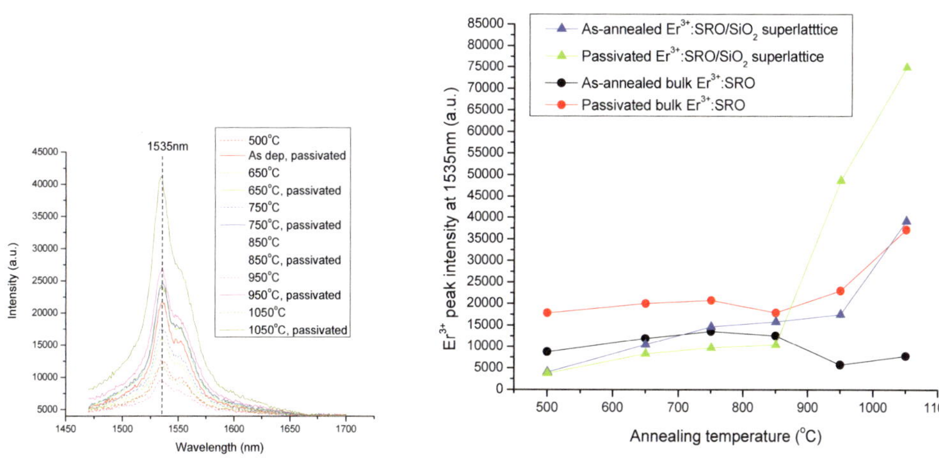

(a) (b)

Fig. 8. (a) IR PL spectra at 1470nm-1700nm of the bulk Er^{3+}:SRO samples (b) Evolution of the Er^{3+} peak intensity at 1535nm as a function of the annealing temperature of the as-annealed and passivated bulk Er^{3+}:SRO and Er^{3+}:SRO/SiO_2 superlattice samples. The excitation was fixed at 290nm

Let us firstly analyse the Er^{3+} emission intensity of the as-annealed and passivated Er^{3+}:SRO/SiO_2 superlattice samples at the annealing temperature from 500°C to 850°C (respectively, blue and green curves in Figure 8 (b)). At these annealing temperatures, the Si nanocrystals have not been formed (Silalahi et al., 2010) and consequently the emission from the exciton recombinations in Si nanocrystals at ~800nm is absent (see Figures 6 (b)). The excitation leading to the emission of Er^{3+} ions is from the energy transfer from defects. As discussed in section 3.3 above, the structural defects in superlattice structure, which gives emission at ~500nm, cannot transfer energy to Er^{3+} ions. Hence, the excitation of Er^{3+} ions is likely from the dangling bond defects. Indeed, upon passivation, which removes the dangling bond defects, the Er^{3+} peak intensity reduces. The decrease of Er^{3+} peak intensity after passivation has also been reported in similar material system (Silalahi et al., 2010).

At the annealing temperature of 950°C and above, Si nanocrystals have been formed in the Er^{3+}:SRO/SiO_2 superlattice samples (see Figure 3 (c)). Before the passivation, the exciton recombinations in Si nanocrystals are quenched by the dangling bond defects (see Figure 6 (b)). The energy transfer to Er^{3+} ions can be from both the exciton recombinations in Si nanocrystals and the dangling bond defects. When the dangling bond defects are removed by the passivation, the Er^{3+} peak intensity is about two times of that of the as-annealed samples (see, respectively, the green and blue curves in Figure 8 (b)). The exciton recombinations in the Si nanocrystals are no longer quenched by the dangling bond defects and, of course, the energy transfer from the dangling bond defects to Er^{3+} ions have also been mostly suppressed, leaving only energy transfer from the exciton recombinations in Si nanocrystals to Er^{3+} ions. The dramatic increase in the Er^{3+} peak intensity shows that the exciton recombinations in Si nanocrystals are much more efficient than the dangling bond defects in transferring their energy to Er^{3+} ions.

Next, let us consider the Er^{3+} emission intensity of the as-annealed and passivated bulk Er^{3+}:SRO samples (respectively, black and red curves in Figure 8 (b)). Again, at the annealing temperature equal or lower than 850°C, Si nanocrystals have not been formed (see Figure 3 (a)) and so the excitation of Er^{3+} ions must consequently be due to the energy transfer from the defects. Before the passivation, where both the structural defects and the dangling bond defects are present, the energy transfer to Er^{3+} ions can be from both of these defects. After the passivation, the dangling bond defects have been removed, and therefore the Er^{3+} ion excitation is dominated by the energy transfer from the structural defects. It is observed that after the passivation, the Er^{3+} peak intensity is ~2 times of that of the as-annealed samples (see, respectively, red and black curves in Figure 8 (b)). The increase of the Er^{3+} peak intensity upon passivation shows that the structural defects are more efficient than the dangling bond defects in transferring energy to Er^{3+} ions.

At the annealing temperature higher than 850°C, the structural defects in the bulk Er^{3+}:SRO samples have been suppressed (see Figure 5) and the Si nanocrystals have been formed (see Figure 3 (b)). Upon passivation, as the dangling bond defects have been removed, the energy transfer to Er^{3+} ions is dominated by the exciton recombinations in Si nanocrystals. The Er^{3+} peak intensity of the passivated Er^{3+}:SRO samples becomes ~4 times of that of the as-annealed samples (see, respectively, red and black curves in Figure 8 (b)). It is important to note that the Er^{3+} peak intensity of the passivated bulk Er^{3+}:SRO samples annealed at temperature higher than 850°C, where the energy transfer to Er^{3+} ions is dominated by the exciton recombinations in Si nanocrystals, is up to ~2 times higher than that of the passivated bulk Er^{3+}:SRO samples annealed at temperature below 850°C, where the energy transfer to Er^{3+} ions is dominated by the structural defects in bulk structure. This implies that the exciton recombinations in Si nanocrystals are more efficient than the bulk structural defects in transferring their energy to Er^{3+} ions.

Comparing the passivated bulk Er^{3+}:SRO and the passivated Er^{3+}:SRO/SiO_2 superlattice samples at the annealing temperature higher than 850°C (respectively, red and green curves in Figure 8 (b)), where the energy transfer is dominated by the exciton recombinations in Si nanocrystals, the Er^{3+} peak intensity of the superlattice samples is more than 2 times higher than that of the bulk samples. One possible explanation is a better size control of Si nanocrystals in superlattice structure compared to the bulk structure (Gourbilleau et al., 2001; Zacharias et al., 2002). As the energy can be transferred from smaller nanocrystals to larger nanocrystals (Iwayama et al., 1999), a better size control may then suppress this energy transfer and hence remove a competitive recombination path of energy transfer to Er^{3+} ions, leading to a more efficient energy transfer to Er^{3+} ions.

Based on the above discussion, we summarize schematically the energy transitions of the exciton recombinations in Si nanocrystal, the structural defects in the bulk Er^{3+}:SRO samples, the structural defects in the Er^{3+}:SRO/SiO_2 superlattice samples, the dangling bond defects and their energy transfer to Er^{3+} ions in Figure 9.

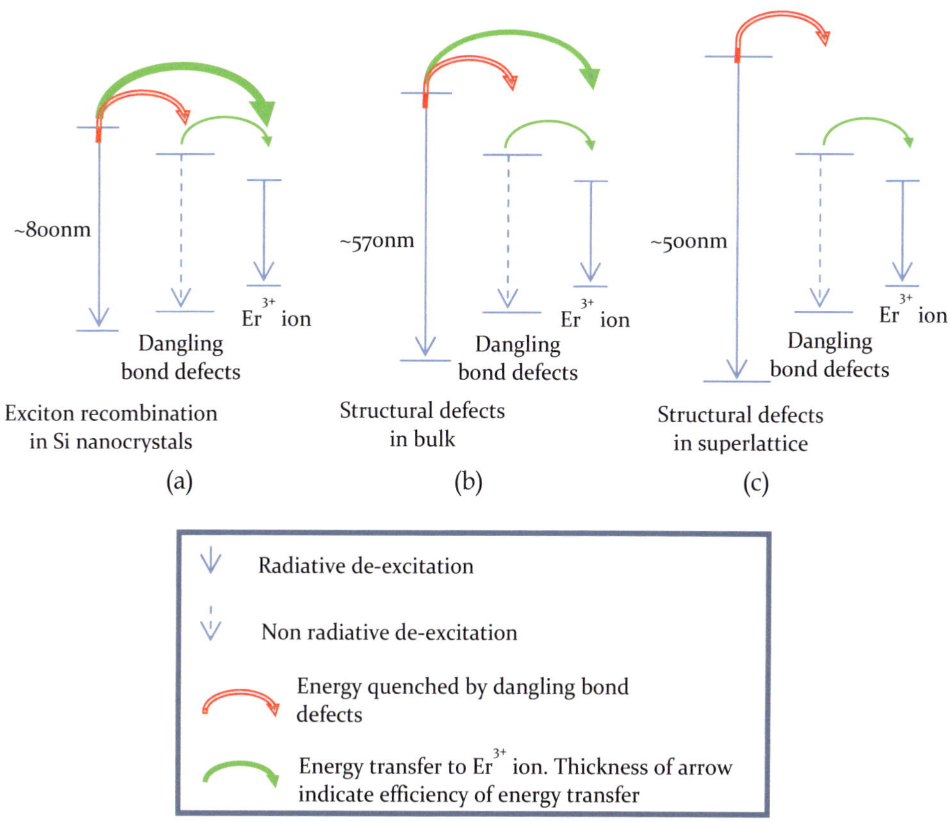

Fig. 9. Different energy transitions and their energy transfer to Er^{3+} ions

Figure 9 (a) shows the exciton radiative recombination, which gives emission at ~800nm, and the energy transfer from the exciton recombination in Si nanocrystals to Er^{3+} ions. The dangling bond defects not only create non-radiative de-excitation paths that quench the exciton recombination in Si nanocrystals, but also transfer their energy to Er^{3+} ions with efficiency much lower than that of the exciton recombination in Si nanocrystals. Figure 9 (b) shows the radiative de-excitation of the structural defects in the bulk Er^{3+}:SRO samples, which gives emission at ~570nm, and their energy transfer to Er^{3+} ions. Note that the energy transfer efficiency from these defects to Er^{3+} ions is lower than that of the exciton recombination in Si nanocrystals. The dangling bond defects create non-radiative de-excitation paths and transfer their energy to Er^{3+} ions with efficiency lower than that of these structural defects. In Figure 9 (c), the structural defects in the Er^{3+}:SRO/SiO_2 superlattice samples show the radiative de-excitation responsible for emission at ~500nm. These defects do not transfer energy to Er^{3+} ions. Similar to the transitions described in Figures 9 (a) and (b), the dangling bond defects create non-radiative de-excitation paths and transfer their energy to Er^{3+} ions. When most defects have been reduced by the high temperature annealing and the passivation, the exciton recombinations in Si nanocrystals in the Er^{3+}:SRO/SiO_2 superlattice samples are more efficient than that in the bulk Er^{3+}:SRO samples in transferring their energy to Er^{3+} ions.

4. Er^{3+} ion concentration effect on Er^{3+} emission in superlattice structure

It has been shown in previous section that the exciton recombinations process in Si nanocrystals in our passivated superlattice samples is the most efficient in energy transfer to Er^{3+} ions. Therefore, in this section, to study the effect of Er^{3+} ion concentration on Er^{3+} emission, we have used Er^{3+}:SRO/SiO_2 superlattice samples annealed at 1050°C and passivated.

4.1 Optimization of Er^{3+} ion concentration

The optimization of Er^{3+} ion concentration was carried out by varying the Er_2O_3 deposition power from 5W to 35W, resulting Er^{3+} ion concentration from 0.02 at% to 0.4 at%. Figure 10 shows the variation of the exciton recombination peak intensities at ~800nm and the Er^{3+} peak intensities at 1535nm with the Er^{3+} ion concentration and the ratio of Er^{3+} ions to the number of Silicon nanocrystals. The ratio of Er^{3+} ions to the number of Silicon nanocrystals is obtained using the concentration of Er^{3+} ions and concentration of Si nanocrystals assuming Si nanocrystals average size of 3.6nm (see section 3.1).

As shown in Figure 10, as the Er^{3+} ion concentration increases, the peak intensity of the exciton recombinations in Si nanocrystals decreases continuously. With increasing number of Er^{3+} ions in the samples, the mean distance between Er^{3+} ions and Si nanocrystals decreases, and so stronger coupling between the two species is expected. The shorter distance to the Er^{3+} ions together with the increasing Er^{3+} ion concentrations make the excitons in Si nanocrystals more likely to recombine non-radiatively and transfer their energy to Er^{3+} ions (Heitmann et al., 2003). Hence, the emission from the exciton recombination Si nanocrystals is quenched with increasing Er^{3+} ion concentration. We would expect that when all the excitons in Si nanocrystals transfer their energy to Er^{3+} ions, we could no longer detect the emission from exciton recombination in Si nanocrystals. For our samples, this occurs when the concentration of Er^{3+} ions reach 0.27 at%, which corresponds to Er^{3+} ions to the number of Silicon nanocrystals ratio of 13. Although, at this Er^{3+} ion concentrations, all the excitons in Si nanocrystals have apparently transferred their energy to Er^{3+} ions, it does not mean that it would give the maximum optimum Er^{3+} emission intensity as we will discuss below.

In Figure 10, we observe that the Er^{3+} peak intensity increases with increasing Er^{3+} ion concentration from 0.02 at% to 0.03 at%. Beyond 0.03 at% of Er^{3+} ion concentration, the Er^{3+} peak intensity decreases. This concentration quenching of Er^{3+} emission can be explained by the energy transfer between neighbouring Er^{3+} ions (Polman & Veggel, 2004). With increasing number of Er^{3+} ions, the mean distance between Er^{3+} ions is reduced. One Er^{3+} ion can de-excite non-radiatively and excite a neighbouring Er^{3+} ion. This process will continue until the energy arrives at a non-radiative defect site in the samples where the energy is given-up non-radiatively (Polman & Veggel, 2004).

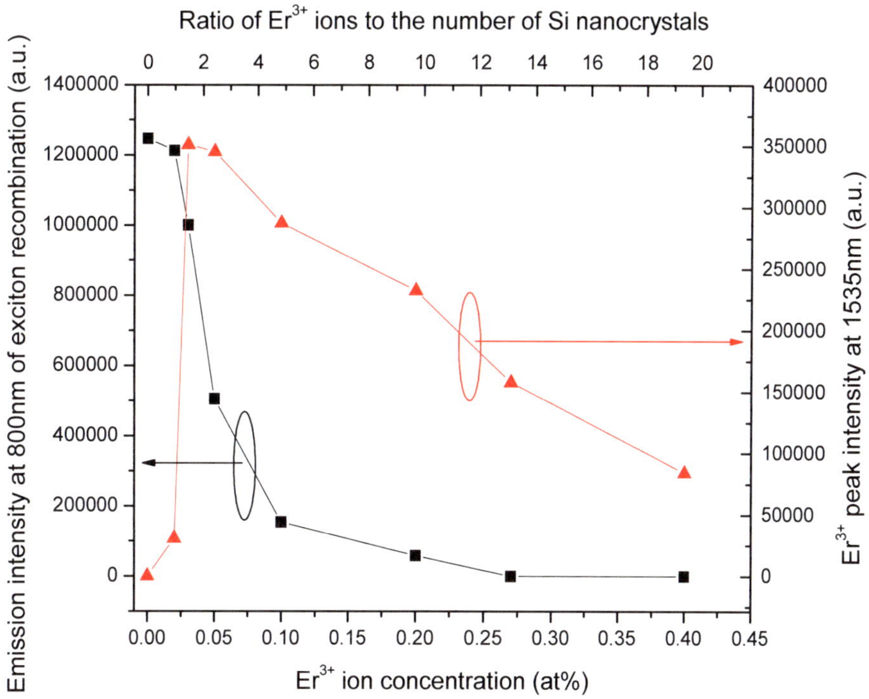

Fig. 10. Variation of the emission intensity at 800nm of the exciton recombinations in Si nanocrystals and the Er^{3+} peak intensity at 1535nm as a function of the Er^{3+} ion concentration and the ratio of Er^{3+} ions to the number of Silicon nanocrystals

As shown in Figure 10, the optimum Er^{3+} peak intensity is obtained at the Er^{3+} ion concentration of 0.03%, which corresponds to the ratio of Er^{3+} ions to the number of Silicon nanocrystals of 1.6. These optimum values of Er^{3+} ion concentration and Er^{3+} ions to Si nanocrystal ratio is comparable to optimum values suggested by Polman & Veggel of 0.02-0.04 at% and 1-2 Er^{3+} ions per Si nanocrystal (Polman & Veggel, 2004). It is noteworthy that the optimum values suggested by Polman & Veggel were derived from the excitation rates of Er^{3+} ions and excitons in Si nanocrystals, both calculated using rise time and decay time of the respective species. In our work, we obtain the optimum Er^{3+} ions to Si nanocrystal ratio from the Er^{3+} emission intensity behaviour.

We see that at the optimum Er^{3+} ion concentration of 0.03 at%, the emission from exciton recombination in Si nanocrystals can still be observed. In this case, the energy transfer from the exciton recombinations in Si nanocrystal to Er^{3+} ions is not fully utilized. One way to fully utilize the Si nanocrystals is to deposit Er^{3+} ions at and near the Er^{3+}:SRO/SiO_2 interface. It is understood that in our superlattice structure, the Si nanocrystals formed may have diameter larger than the thickness of deposited Er^{3+}:SRO layers (see Figure 3 (c)). Hence, when Er^{3+} ions is only deposited in SRO layers, certain areas of the nanocrystals that are in SiO_2 layers will not have any Er^{3+} ions in their vicinity. By depositing Er^{3+} ions at and near the Er^{3+}:SRO/SiO_2 interface, the chance of having Er^{3+} ions near to Si nanocrystals will be improved without the need of increasing Er^{3+} ion concentration in SRO layers.

4.2 Determination of energy transfer rate from the exciton recombinations in Si nanocrystals to Er^{3+} ions

From the previous discussion, when the Er^{3+} ion concentration increases, the coupling between Si nanocrystals and Er^{3+} ions becomes stronger. With stronger coupling, we can expect the energy transfer rate and energy transfer efficiency to increase. Through lifetime measurement, transfer rate as well as transfer efficiency can be deduced. The lifetime measurements were carried out by measuring the emission intensity as a function of time at room temperature. Figure 11 shows the normalized emission decay intensity of the exciton recombinations in Si nanocrystals at 800nm as a function of time of the SRO/SiO_2 and Er^{3+}:SRO/SiO_2 superlattice samples with different Er^{3+} ion concentration.

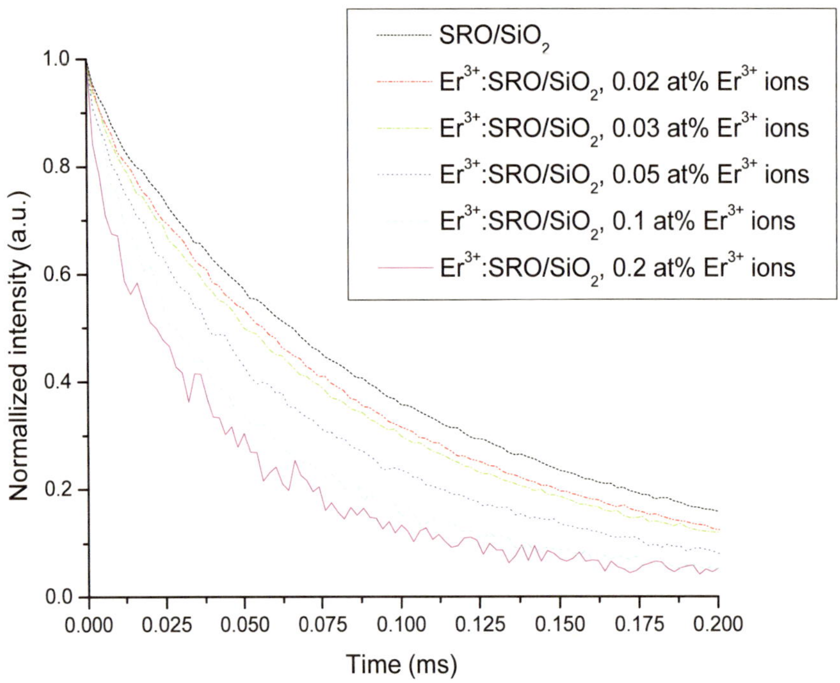

Fig. 11. Normalized room temperature luminescence decay intensity of the exciton recombinations in Si nanocrystals at 800nm of the SRO/SiO_2 and Er^{3+}:SRO/SiO_2 superlattice samples with different Er^{3+} ion concentration

For the samples with Er^{3+} ion concentration larger than 0.2 at%, the emission intensities of exciton recombinations in Si nanocrystals are too low to be measured. The decay curves are well described by stretched exponential decay model (Williams & Watts, 1970):

$$I_{exciton}(t) = I_0 e^{-\left(\frac{t}{\tau_{exciton}}\right)^{\beta}} \quad (1)$$

where $I_{exciton}(t)$ is the intensity of the exciton recombinations in Si nanocrystals measured at time t, I_0 is the intensity of the exciton recombination at time t=0, $\tau_{exciton}$ is the lifetime of the

excitons in Si nanocrystals and β is the stretching exponential coefficient. The value of β ranges from 0 to 1. In the ideal case where species in the material system (Si nanocrystals and Er^{3+} ions) are not interacting with each other, the value of β is equal to 1. In a material system with interacting species, the value of β is less than 1 (Brongersma et al., 1998). Applying equation (1) to the decay curves in Figure 11, we obtained the values of the $\tau_{exciton}$ and β. For our samples, the exciton lifetime in Si nanocrystals decreases from 99.5μs for the SRO/SiO_2 sample to 87μs for the Er^{3+}:SRO/SiO_2 sample with 0.02 at% Er^{3+} ions and then gradually down to 37μs when Er^{3+} ion concentration increases from 0.02 at% to 0.2 at%. The exciton lifetime of each sample is presented in Figure 13 (a) below.
The stretching exponential coefficient β was found to decrease consistently from 0.87 to 0.73 for the SRO/SiO_2 sample, and the Er^{3+}:SRO/SiO_2 samples with Er^{3+} ion concentration increasing from 0.02 at% to 0.2 at%. The decreasing values of β with increasing Er^{3+} ion concentration is consistent with our expectation that the interactions between Si nanocrystals and Er^{3+} ions becomes stronger with increasing Er^{3+} ion concentrations.
Figure 12 shows a simplified energy band diagram of one Si nanocrystal with possible recombination paths.

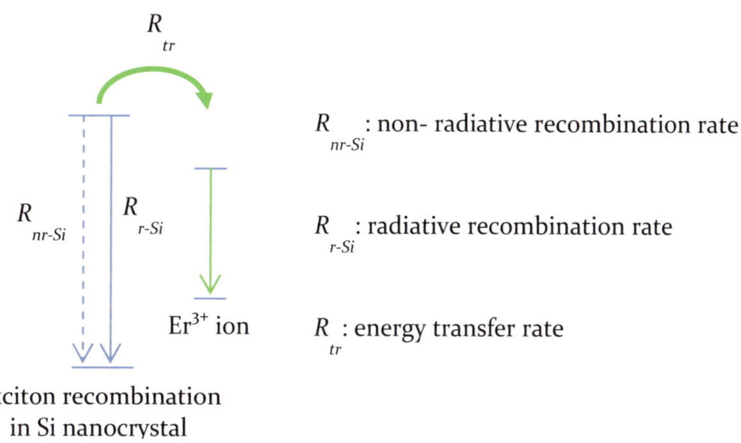

Fig. 12. Possible recombination paths of one Si nanocrystal

Even with our optimum annealing treatment (1050°C annealing in N_2 and passivated for 1 hour), which reduces majority of the defects, there may be still some defects left in the samples that creates non-radiative recombination paths, therefore we incorporate the non-radiative transitions in the exciton recombinations in Si nanocrystals. In the absence of Er^{3+} ions (i.e. for the SRO/SiO_2 samples), the excitons in Si nanocrystals can de-excite through radiative emission or non-radiative recombination. The exciton lifetime, $\tau_{exciton}$, can then be expressed as (Heitmann et al., 2003):

$$\frac{1}{\tau_{exciton}} = R_{nr-Si} + R_{r-Si} \tag{2}$$

where R_{nr-Si} is the non-radiative recombination rate, and R_{r-Si} is the radiative recombination rate of excitons. In the presence of Er^{3+} ions (i.e. for the Er^{3+}:SRO/SiO_2 samples), the exciton lifetime, $\tau_{exciton}$, becomes (Heitmann et al., 2003):

$$\frac{1}{\tau_{exciton}} = R_{nr-Si} + R_{r-Si} + R_{tr} \quad (3)$$

where R_{tr} is the energy transfer rate from an exciton to Er^{3+} ions. The transfer rate, R_{tr}, can then be calculated by subtracting the values of $(1/\tau_{exciton})$ in equation (3) to that in equation (2).

Using the values of R_{tr} and $(1/\tau_{exciton})$ in equation (3), the transfer efficiency E_T of our samples can be calculated according to (Heitmann et al., 2003; Savchyn et al., 2007):

$$E_T = \frac{R_{tr}}{R_{nr-Si} + R_{r-Si} + R_{tr}} \quad (4)$$

Figure 13 (a) shows the exciton lifetime, $\tau_{exciton}$, and the transfer rate, R_{tr}, and Figure 13 (b) shows the transfer efficiency, E_T, of our samples.

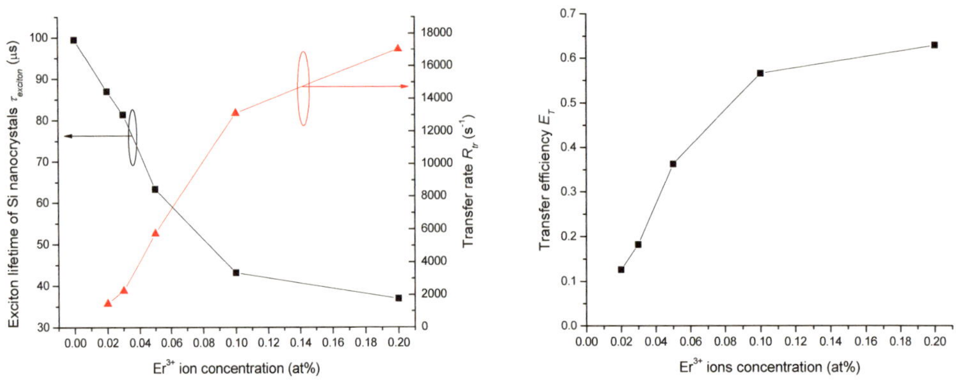

(a) (b)

Fig. 13. (a) Exciton lifetime in Si nanocrystals, $\tau_{exciton}$, transfer rate, R_{tr}, and (b) transfer efficiency, E_T, from the exciton recombinations in Si nanocrystals to Er^{3+} ions as a function of the Er^{3+} ion concentration

As mentioned earlier, when more Er^{3+} ions are incorporated into the samples, the energy transfer rate and transfer efficiency is expected to increase. Indeed, the transfer rate increases from $1.5\times10^{-3}s^{-1}$ to $1.7\times10^{-4}s^{-1}$ (see Figure 13(a)) and transfer efficiency increases from 12% to 63% (see Figure 13 (b)) when Er^{3+} ion concentration increases from 0.02 at% to 0.2 at%. The transfer rate of our material system ranges from 10^3s^{-1} to 10^4s^{-1}, which is lower than reported values of 10^4s^{-1} to 10^5s^{-1} by other groups (Fujii et al., 2004; Savchyn et al., 2007). At this moment, this difference in the energy transfer rate is not obvious to us. We would like to point it out, however, that we have used the superlattice structure, while Fujii et al. and Savchyn et al. were adopting a bulk structure.

As shown in Figure 13 (b), at the optimized Er^{3+} ion concentration of 0.03 at%, the transfer efficiency of our sample is only 18%. Savchyn et al. reported a transfer efficiency of ~60%-75% at Er^{3+} ion concentration of 0.63 at% (Savchyn et al., 2007) and Kik et al reported a transfer efficiency of ~55% at Er^{3+} ion concentration of 1.8 at% (Kik et al., 2000). It is important to note that the transfer efficiency of our sample can reach 63% at the Er^{3+} ion concentration of 0.2 at%. However, at the Er^{3+} ion concentration of 0.2 at%, Er^{3+} concentration quenching has already taken place and the resulting Er^{3+} emission is not optimized. It is also noteworthy that at our optimized Er^{3+} ion concentration of 0.02 at% and optimized heat treatment (1050°C annealing and passivated), Er^{3+} emission from the superlattice sample is more than 2 times higher than that from the bulk sample (see, respectively, green and red curves in Figure 8(b)). The superlattice structure is showing great potential and optimization can be done to enhance Er^{3+} emission from this material structure.

5. Conclusion

In this chapter, through the PL study of the as-annealed and passivated bulk SRO, Er^{3+}:SRO, SRO/SiO_2, and Er^{3+}:SRO/SiO_2 superlattice samples, 4 different energy transitions and their energy transfer to Er^{3+} ions have been identified and investigated:

1. Radiative energy transition which corresponds to the exciton recombinations in Si nanocrystals. In our material system, the energy transfer due to the exciton recombinations is the most efficient energy transfer both in the bulk Er^{3+}:SRO and Er^{3+}:SRO/SiO_2 superlattice samples.
2. Radiative energy transitions associated with structural defects in bulk samples. The efficiency of the energy transfer due to this type of transitions is lower than that due to the exciton recombinations in Si nanocrystals.
3. Radiative energy transitions associated with structural defects in superlattice samples. These defect transitions give a different PL spectrum than those associated with the defects in the bulk samples and show no energy transfer to Er^{3+} ions.
4. Non-radiative energy transitions due to the dangling bond de-excitation, which quenches other radiative transitions. The non-radiative dangling bond de-excitations have some energy transfer to Er^{3+} ions. The efficiency of this energy transfer is less efficient compared to that due to the exciton recombinations in Si nanocrystals and the transition associated with structural defects in the bulk Er^{3+}:SRO samples. The dangling bonds can effectively be removed by annealing in forming gas environment.

In our study, we have found that the exciton recombinations in Si nanocrystals in our Er^{3+}:SRO/SiO_2 superlattice samples are the most efficient in transferring the energy to Er^{3+} ions. Er^{3+} emission by the exciton recombinations in Si nanocrystals energy transfer from the superlattice samples is more than 2 times higher than that from the bulk samples; while the Er^{3+} emission by the exciton recombination in Si nanocrystals energy transfer in the bulk samples is up to ~2 times higher than Er^{3+} emission by the energy transfer from structural defects in the bulk samples.

Using the superlattice structure at the optimized annealing condition (1050°C RTP followed by passivation), Er^{3+} emissions as a function of Er^{3+} ion concentrations have also been investigated. The optimum concentration of Er^{3+} ions in this material system is 0.03 at%, corresponding to a ratio of Er^{3+} ions to the number of Silicon nanocrystals of 1.6 to 1, achieved at the Er_2O_3 deposition power of 10W.

Through the lifetime study, transfer rate and transfer efficiency by exciton recombinations in Si nanocrystals in the superlattice structure have been determined. The transfer rate ranges from $10^3 s^{-1}$ to $10^4 s^{-1}$. The transfer efficiency can reach 63%, but at optimized Er^{3+} ion concentration of 0.03 at%, the transfer efficiency is 18%. One way to improve the transfer rate and transfer efficiency is to optimize the Si nanocrystal by incorporate Er^{3+} ions at and near the Er:SRO/SiO_2 interface.

6. References

Brongersma, M. L.; Polman, A.; Min, K. S.; Boer, E.; Tambo, T. & Atwater, H. A. (1998). Tuning the emission wavelength of Si nanocrystals in SiO2 by oxidation. *Applied Physics Letters*, Vol. 72, No. 20, 2577-2579, 00036951 (ISSN)

Bulutay, C. (2007). Electronic structure and optical properties of silicon nanocrystals along their aggregation stages. *Physica E: Low-Dimensional Systems and Nanostructures*, Vol. 38, No. 1-2, 112-117, 13869477 (ISSN)

Castagna, M. E.; Coffa, S.; Monaco, M.; Caristia, L.; Messina, A.; Mangano, R. & Bongiorno, C. (2003). Si-based materials and devices for light emission in silicon. *Physica E: Low-Dimensional Systems and Nanostructures*, Vol. 16, No. 3-4, 547-553, 13869477 (ISSN)

Castagna, M. E.; Muscara, A.; Leonardi, S.; Coffa, S.; Caristia, L.; Tringali, C. & Lorenti, S. (2006). Si-based erbium-doped light-emitting devices. *Journal of Luminescence*, Vol. 121, No. 2 SPEC. ISS., 187-192, 00222313 (ISSN)

Charvet, S.; Madelon, R.; Gourbilleau, F. & Rizk, R. (1999). Spectroscopic ellipsometry analyses of sputtered Si/SiO2 nanostructures. *Journal of Applied Physics*, Vol. 85, No. 8 I, 4032-4039, 00218979 (ISSN)

Fujii, M.; Imakita, K.; Watanabe, K. & Hayashi, S. (2004). Coexistence of two different energy transfer processes in SiO2 films containing Si nanocrystals and Er. *Journal of Applied Physics*, Vol. 95, No. 1, 272-280, 00218979 (ISSN)

Gourbilleau, F.; Portier, X.; Ternon, C.; Voivenel, P.; Madelon, R. & Rizk, R. (2001). Si-rich/SiO2 nanostructured multilayers by reactive magnetron sputtering. *Applied Physics Letters*, Vol. 78, No. 20, 3058-3060, 00036951 (ISSN)

Heitmann, J.; Schmidt, M.; Zacharias, M.; Timoshenko, V. Y.; Lisachenko, M. G. & Kashkarov, P. K. (2003). Fabrication and photoluminescence properties of erbium doped size-controlled silicon nanocrystals. *Materials Science and Engineering B: Solid-State Materials for Advanced Technology*, Vol. 105, No. 1-3, 214-220, 09215107 (ISSN)

Iacona, F.; Irrera, A.; Franzo, G.; Pacifici, D.; Crupi, I.; Miritello, M.; Presti, C. D. & Priolo, F. (2006). Silicon-based light-emitting devices: Properties and applications of crystalline, amorphous and er-doped nanoclusters. *IEEE Journal on Selected Topics in Quantum Electronics*, Vol. 12, No. 6, 1596-1605, 1077260X (ISSN)

Kahler, U. & Hofmeister, H. (2002). Size evolution and photoluminescence of silicon nanocrystallites in evaporated SiOx thin films upon thermal processing. *Applied Physics A: Materials Science and Processing*, Vol. 74, No. 1, 13-17, 09478396 (ISSN)

Kapaklis, V.; Politis, C.; Poulopoulos, P. & Schweiss, P. (2005). Photoluminescent Si nanoparticles embedded in silicon oxide matrix. *Materials Science and Engineering B: Solid-State Materials for Advanced Technology*, Vol. 124-125, No. SUPPL., 475-478, 09215107 (ISSN)

Kik, P. G.; Brongersma, M. L. & Polman, A. (2000). Strong exciton-erbium coupling in Si nanocrystal-doped SiO2. *Applied Physics Letters*, Vol. 76, No. 17, 2325-2327, 00036951 (ISSN)

Kik, P. G. & Polman, A. (2001). Exciton-erbium energy transfer in Si nanocrystal-doped SiO2. *Materials Science and Engineering B: Solid-State Materials for Advanced Technology*, Vol. 81, No. 1-3, 3-8, 09215107 (ISSN)

Polman, A. & van Veggel, F. C. J. M. (2004). Broadband sensitizers for erbium-doped planar optical amplifiers: Review. *Journal of the Optical Society of America B: Optical Physics*, Vol. 21, No. 5, 871-892, 07403224 (ISSN)

Savchyn, O.; Ruhge, F. R.; Kik, P. G.; Todi, R. M.; Coffey, K. R.; Nukala, H. & Heinrich, H. (2007). Luminescence-center-mediated excitation as the dominant Er sensitization mechanism in Er-doped silicon-rich Si O2 films. *Physical Review B - Condensed Matter and Materials Physics*, Vol. 76, No. 19, 10980121 (ISSN)

Savchyn, O.; Kik, P. G.; Todi, R. M. & Coffey, K. R. (2008). Effect of hydrogen passivation on luminescence-center-mediated Er excitation in Si-rich SiO2 with and without Si nanocrystals. *Physical Review B - Condensed Matter and Materials Physics*, Vol. 77, No. 20, 10980121 (ISSN)

Seino, K.; Bechstedt, F. & Kroll, P. (2009). Influence of SiO2 matrix on electronic and optical properties of Si nanocrystals. *Nanotechnology*, Vol. 20, No. 13, 09574484 (ISSN)

Shimizu-Iwayama, T.; Hole, D. E. & Townsend, P. D. (1999). Optical properties of interacting Si nanoclusters in SiO2 fabricated by ion implantation and annealing. *Nuclear Instruments and Methods in Physics Research, Section B: Beam Interactions with Materials and Atoms*, Vol. 147, No. 1-4, 350-355, 0168583X (ISSN)

Silalahi, S. T. H.; Yang, H. Y.; Pita, K. & Mingbin, Y. (2009). Rapid thermal annealing of sputtered silicon-rich oxide/SiO2 superlattice structure. *Electrochemical and Solid-State Letters*, Vol. 12, No. 4, 10990062 (ISSN)

Silalahi, S. T. H.; Vu, Q. V.; Yang, H. Y.; Pita, K. & Mingbin, Y. (2010). Size control of Si nanocrystals by two-step rapid thermal annealing of sputtered Si-rich oxide/SiO2 superlattice. *Applied Physics A: Materials Science and Processing*, Vol. 98, No. 4, 867-871, 09478396 (ISSN)

Silalahi, S. T. H.; Chen, R.; Vu, Q. V.; Pita, K.; Sun, H. D. & Mingbin, Y. (2010). The Effects of Rapid Annealing and Passivation of Co-sputtered Erbium doped Si-rich Oxide/SiO2 Superlattice Structures. *Photonics Global Conference 2010*, Singapore, 14-16 December 2010

Takagahara, T. & Takeda, K. (1992). Theory of the quantum confinement effect on excitons in quantum dots of indirect-gap materials. *Physical Review B*, Vol. 46, No. 23, 15578-15581, 01631829 (ISSN)

Wilkinson, A. R. & Elliman, R. G. (2003). Kinetics of H2 passivation of Si nanocrystals in SiO2. *Physical Review B - Condensed Matter and Materials Physics*, Vol. 68, No. 15, 1553021-1553028, 01631829 (ISSN)

Williams, G. & Watts, D. C. (1970). Non-symmetrical dielectric relaxation behaviour arising from a simple empirical decay function. *Transactions of the Faraday Society*, Vol. 66, No., 80-85, 0014-7672

Wolkin, M. V.; Jorne, J.; Fauchet, P. M.; Allan, G. & Delerue, C. (1999). Electronic states and luminescence in porous silicon quantum dots: The role of oxygen. *Physical Review Letters*, Vol. 82, No. 1, 197-200, 00319007 (ISSN)

Zacharias, M.; Heitmann, J.; Scholz, R.; Kahler, U.; Schmidt, M. & Blasing, J. (2002). Size-controlled highly luminescent silicon nanocrystals: A SiO/SiO2 superlattice approach. *Applied Physics Letters*, Vol. 80, No. 4, 661, 00036951 (ISSN)

18

Nonlinear Optoelectronic Devices Based on Nanocrystalline Silicon Films: Acoustoelectrical Switchers for Optical Modes, Nonlinear Optical Switchers and Lasers

Dmitry E. Milovzorov
FLUENS TECHNOLOGY GROUP, LTD, Moscow
Russia

1. Introduction

Since 2000, the nanostructured materials have received a great application in optoelectronic devices development. Nanocrystalline silicon films are used now as for performing the new memory devices as to make thin film transistors for displays. Optical switches are used already two decades in optical communication systems. The switching times were varied from μs to ns. However, the possible increasing the rate of switching can be results in new optical computer creation the main principal of which are based on nonlinear switches. Accordingly, the high degree of integration of elements is, also, very attractive. Because, the miniaturization of switch is significant.

This investigation is devoted to possible nonlinear switch design the main principle of which is based on the control of the nonlinear-optical response enhancement for nanostructured silicon films by applying external electrical field. First, it is discused a suitable nonlinear optical system based on various types of silicon films: amorphous, nanocrystalline and polycrystalline (D. Milovzorov et al., 1998, 1999, 2001). Also, each silicon film differs from the other in crystal phase fraction, size distribution of nanocrystals, their crystal orientation, by various types of chemical bonding and defects, their densities, and density and sizes of nanoscale holes. The next type of proposed devices is acoustoelectrical switches for optical modes inside the laser cavity. Acoustoelectric effect was detected in various materials such as n-type germanium (Weinreich et al., 1959), GaAs/$LiNbO_3$ structure (Rotter et al., 1998). In present work the acoustoelectrical effect is studied in nanostructural silicon thin films by applied external electric field due to their specific structural and chemical properies. The third type of optoelectronic devices is lasers based on silicon films such as Raman laser and laser with nonlinear two-photon resonant pumping. The new electronic devices based on natural luminescent and nonlinear properties of oxidized nanocrystalline silicon film can be easily resolved numerous problems of communications and computing by using combination of new spectral lines in IR, visible and UV ranges. Theoretical approach assists to find the main mechanisms of evolution of inverted population and appearance of laser generation modes.

2. Acoustoelectric effects by Raman scattering

2.1 Raman scattering from silicon films

The Raman spectra for poly-Si films consist of a narrow line near 520 cm^{-1} arising from a crystalline phase and a broad line around 480 cm^{-1} from an amorphous phase. The ρ value was estimated from the ratio of the Raman integrated intensity for the crystalline component to the total intensity, using the ratio of the integrated Raman cross-section for the crystalline phase to that for an amorphous phase. The random silicon network Gauss-distributed in their bond lengths with various deviations, and phonon wave numbers is spread. The regular bonding network has the Lorentz shape of its spectral line. Our possible interpretation of a-Si Raman spectra decomposition is a possible recognition the role of point defects and inserted impurities which cause the changes in structural properties of film. It is assumed, that the a-Si film with high density of defects such as silicon vacancies causes the spectral peak around 465 cm^{-1} , but after annealing there is a spectral shift in wavenumber 465 $cm^{-1} \rightarrow 475$ cm^{-1} . By hydrogen dilution of gas mixture by PECVD of a-Si the Raman peak position is changed from 475 cm^{-1} to the 480 cm^{-1} (D.V.Tsu et al., 2001). It is supposed, that the Raman data of hydrogenised amorphous silicon film result in the spectral peak around 45-447 cm^{-1} (D. Han et al., 2000) corresponds the LO mode, but the 480 cm^{-1} for TO mode. For the higher structural relaxed silicon thin film by high level of hydrogen dilution the TO mode reflects in 490 cm^{-1} value of peak position. The spectral peak width changed from the value of 40 cm^{-1} for a-Si to the 70 cm^{-1} for a-Si:H.

There are many interpretations of Raman spectral data concerning the optical response from crystalline silicon phase around 520 cm^{-1}, various kinds of intermediate, µc-Si or nc-Si, or grain boundary response around 494-501 cm^{-1}, and amorphous silicon phase at the wave number value of 480 cm^{-1} , (D. Han et al., 2003). Figure 1 shows the various types of decompositions of Raman spectrum for microcrystalline silicon film with thickness around 100 nm. This film was deposited at 380 ^{o}C. By using this procedure we use three or four peaks approximations. Figure 1 a illustrates the three peak decomposition with c-Si related spectral line around 520 cm^{-1}, nanocrystalline or intermediate spectral line in range 500-510 cm^{-1} and for amorphous silicon is 480 cm^{-1}. The next Fig. 1 b shows the Raman data with noise signal for comparison and fourth spectral component (~490 cm^{-1}) is related to hydrogenated amorphous silicon network. According to the Matsuda work (Matsuda, 1999, 2004) it is possible to deposit the µc-Si:H film with low content of P_b defects at relatively low temperatures 200-300°C. By these temperature conditions the hydrogen diffusion is effective compare to the diffusion rates of SiH_3 and SiH_2 radicals. The hydrogen terminates all the dangling bonds. The defects levels spectral intensity of Raman spectra becomes lower. The spectral characteristics of amorphous-related peaks are changed by the hydrogen saturation of dangling bonds.

2.2 Raman spectroscopy from thin crystalline silicon films on glass substrate

For additional confirmation the nature of optical response from silicon thin film we tried to carry on the analysis the Stocks components of Raman scattering spectral lines and physics of laser light absorption by surface layer. Absorption of radiation by c-Si film on glass substrate can be written as: $I = I_0 \exp(-\alpha Z_{c-Si}) \exp(-\beta Z_{glass})$.

The spectral intensity of radiation by Raman Scattering can be distinguished as following:

$$I^{Scattered} = \int_{w1}^{w2} I_{c-Si}^{Scattered}(\omega)d\omega + \int_{w3}^{w4} I_{glass}^{Scattered}(\omega)d\omega. \quad (1)$$

The energy of radiation can be expressed as addition of energies of all dipoles:

$$\int I_{c-Si}^{Scattered}(\omega)d\omega \cong \eta Z_{c-Si} d\sqrt{I_0}.$$

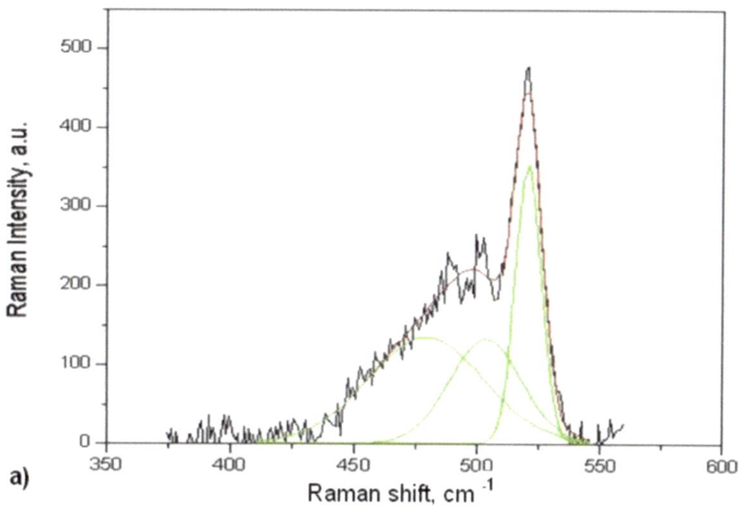

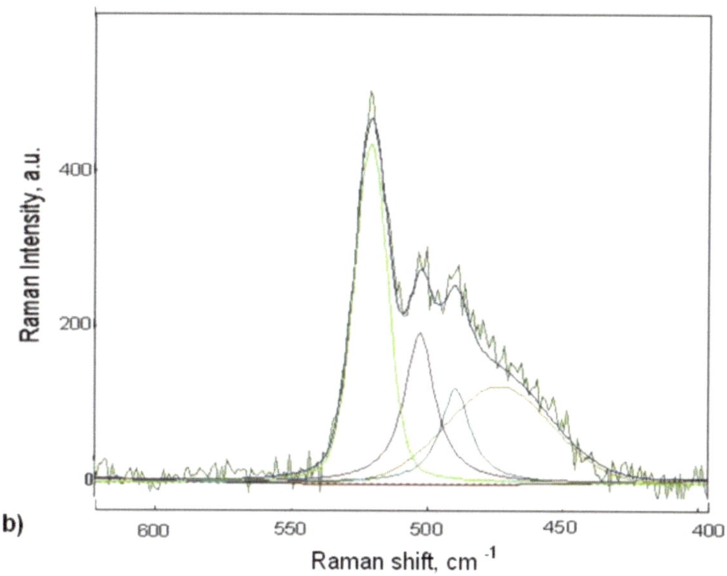

Fig. 1. Raman spectra deconvolution by using a) 3 peaks approximation: c-Si, for 520 cm^{-1}; a-Si for 480 cm^{-1} and intermediate or nanocrystalline for 500-510 cm^{-1}; b) 4 peaks approximation: for c-Si (520 cm^{-1}); a-Si (480 cm^{-1}); nanocrystalline (500-510 cm^{-1}) and hydrogenated a-Si:H for 490cm^{-1}.

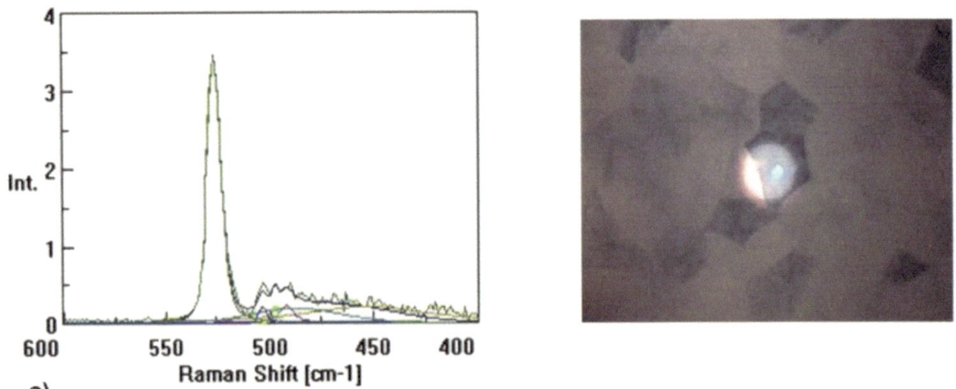

a)

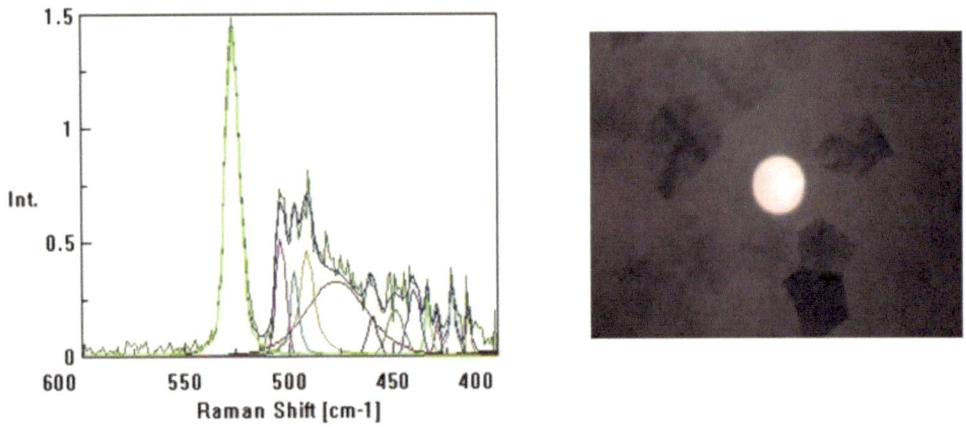

b)

Fig. 2. Raman spectra and micro photo of Secco etched silicon film with thickness of film is around 25 nm -a); Raman spectra micro photo of Secco etched silicon film with thickness of film is around 17 nm- b).

We assume that the D is a laser beam diameter and the density of dipoles can be written as following:

$$N_{Si} = \frac{1}{4}\pi D^2 Z_{c-Si} \tag{2}$$

The thin film thickness can be estimate by using Raman scattering signal from standard film sample and suggested that silicon film is fully crystallized: $Z_{1;c-Si} = Z_{c-Si}\rho$.
For the real thickness of fully crystallized silicon (see Fig. 2) film Z=50 nm, we can estimate the thickness of etched silicon film as ratio between peaks' squares under crystalline phase around 520 cm^{-1} and amorphous phase around 480 cm^{-1} multiplied on 50 nm. For the thin film on Fig. 3 it is seen that the thickness value is around 25 nm. By Secco etching of Si film the rate of etching is around 3-4 nm/s. According to our estimations, the thickness of thin silicon film shown on Fig. 2.4 is equal to 17 nm.

2.3 Raman scattering from multicomponent silicon films

Figure 3 shows the Raman spectra porous silicon film on the Corning 7059 (Fig.3 a) glass substrate, Raman map of film's area 9μm x 9μm (see Fig.3b), the scheme of silicon film structure (Fig. 3c) and its atomic-force microscopic (AFM) photo on the area 1μm² (Fig. 3d). It is seen, that crystal phase volume fraction can be estimated by using the Raman spectroscopy data according to following formula: $\rho = \frac{I_C + I_{nc}}{I_C + I_{nc} + I_A}$, where the values I_C, I_A, I_{nc} are intensities of spectral signal on 520 cm^{-1}, 480 cm^{-1}, and 500-510 cm^{-1}, respectively. The ρ value is more than 70% for the silicon films that were shown on Fig.3. Figure 4 illustrates the optical and structural properties of silicon films with nanocrystals included into silicon oxide material and their photo detected by means of AFM. It is clear, that the ρ value is lower for such type of silicon films than mentioned above. It is in the range of 30-50%. However, it is seen the rectangular shape of large (around several hundreds nm) crystals which are made due to low temperatures of deposition and high degree of hydrogen dilution. Also, it is important to note about low pressure and low flow rates of gases.

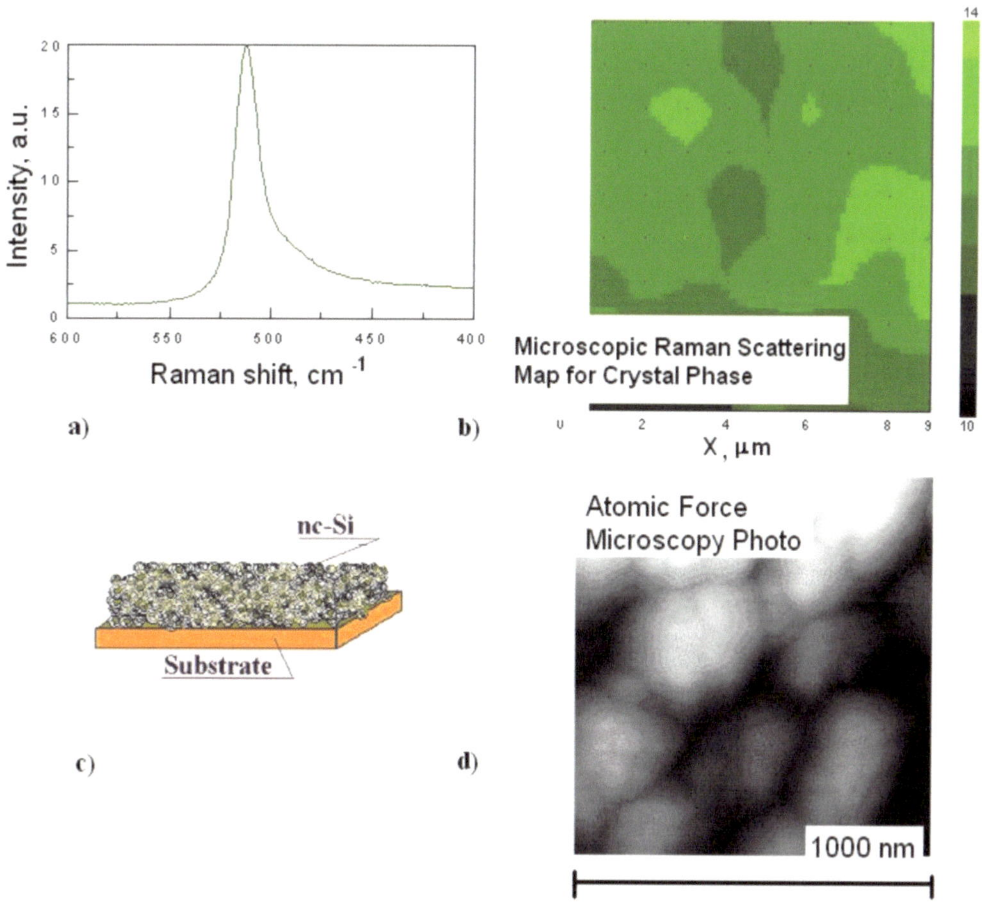

Fig. 3. Raman spectrum -a); and Raman map of crystallized silicon in film on glass substrate -b); scheme of material -c); and its atomic-force microscopic (AFM) photo -d).

The silicon bonding is estimated by using Raman scattering measurements from the ratio:

$$\rho = \frac{\int_0^\infty I_{Si}(\omega)d\omega}{\int_0^\infty I_{Raman}(\omega)d\omega} \tag{3}$$

The ρ value was estimated from the ratio of the Raman integrated intensity for the silicon-related components to the total intensity, using the ratio of the integrated Raman cross-section for them. The local factor is given by

$$L(\omega,\rho) = \frac{\rho}{4\pi}\frac{\left(\varepsilon_{Si}(\omega) - \varepsilon_{SiO2}(\omega)\right)}{1+(\varepsilon_{Si}(\omega) - \varepsilon_{SiO2}(\omega))(\Lambda - \beta\rho)} \tag{4}$$

where the $\varepsilon_{Si}(\omega)$ and ε_{SiO2} are the dielectric functions of silicon and SiO_2, Λ is the depolarization factor (for sphere is 1/3), β is the Lorentz constant (β=1/3 for a homogeneous spherical surrounding).The natural Raman spectral width for Gaussian distributed crystallites in their sizes in film depends on the dispersion of phonon wave vector: $\Delta q=4\pi\sigma/(\delta^2-\sigma^2)$, where the value of σ is equal to max(ε,θ) , ε is a dispersion of spatial crystallites distribution ($\varepsilon=\int f(\delta)\delta^2 d\delta$, f(δ) is Gaussian distribution function),θ* is random shape distortion of crystallites from spherical shape δ*=δθ, where δ* is random Gaussian distributed value of size of crystallite with random spherical shape, <δ*>=δ<θ>. The value of θ* is equal to δ(θ-1).

2.4 Laser picosecond spectroscopy

A mode-locked YAG:Nd^{3+} laser radiation with wavelength 532 nm was used as an optical pump of media, but the second-harmonic radiation (λ=1064 nm) was used for probing of sample's surface. The pulse duration was 120 ps. The pulse repetition rate was 100 MHz and frequency of Q-switched modulation of second-harmonic radiation was 6.2 MHz. The correlation function of reflected signal intensity of probe laser beam was detected by means of pump-probe laser scheme

$$G(t) = < I(t)I(t+\tau) > = \frac{1}{T}\int_0^T I(t)I(t+\tau)dt; \tag{5}$$

where *G (t)* is averaged correlation function, *T* is the time of detection. It is estimated, that $\Delta\omega$ is a width of level that equals to 12 μeV for silicon film with <δ>=8.5 nm, and 16 μeV for silicon film with <δ>=14.7 nm (see Fig. 5 a, upper and down graphs). The most important fact is appearance of quantum beats, which reflect the optical response from two-neighbor level. We suppose that the different fractions of oxygen incorporation in silicon film cause the appearance of neighbor levels inside the band gap with the spectral width (Milovzorov, 2001):

$$\Delta_d = \frac{20\varepsilon_x h^3}{m^2 e^2} N_{Si-O} \tag{6}$$

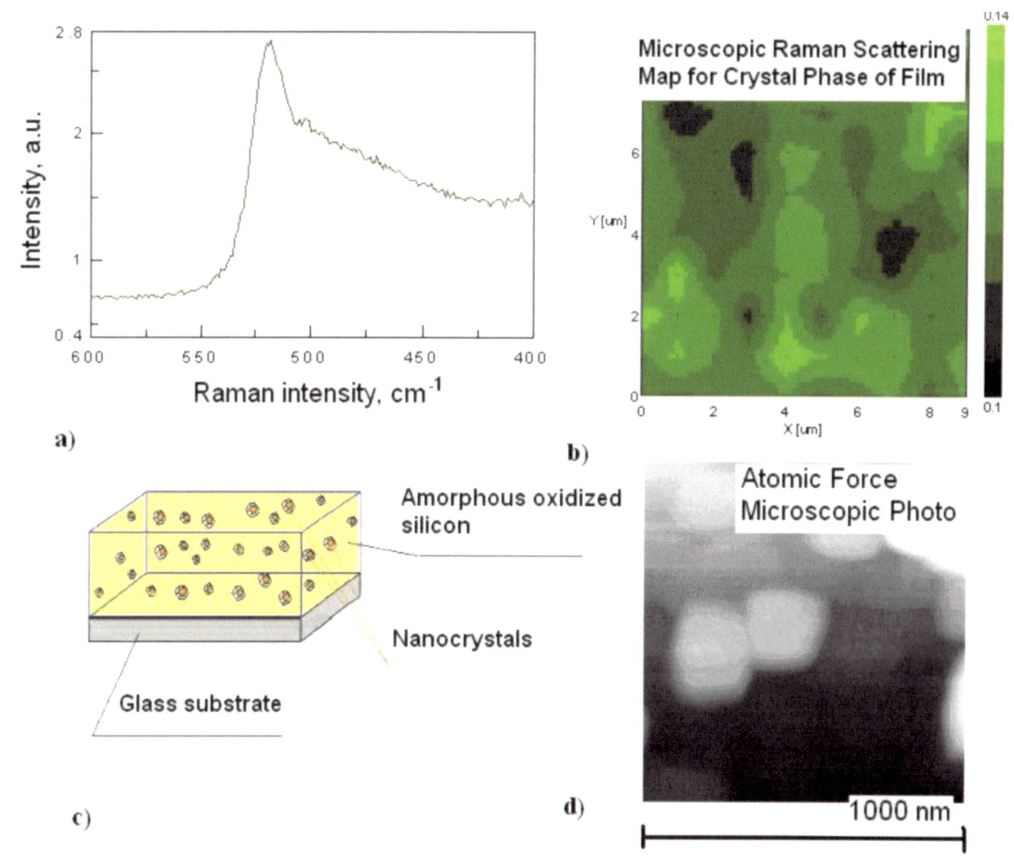

Fig. 4. Silicon nanocrystals included into silicon oxide material and their photo by means of AFM.

the presence of complex SiO configuration. Figure 5 shows the energy diagram for silicon films with nanocrystals' defects' levels, and time resolved spectroscopic data for reflected laser radiation with wavelength 532 nm for three nanocrystalline films and oxidized silicon crystal (111) surface. The wave fuction of excited state is a quantum intereferential picture of two closed located energetic levels;

$$|\psi\rangle = a_1|\phi_1(0)\rangle \exp\left(-\left(iE_{1g}t\right)/\hbar - \Delta\omega_1 t\right) + a_2|\phi_2(0)\rangle \exp\left(-\left(iE_{2g}t\right)/\hbar - \Delta\omega_2 t\right); \quad (7)$$

where $|a_1|^2$ and $|a_2|^2$ are prioportional to probabilities for both excited states 1 and 2. Such quantum interference of states results in quantum beats.

The Raman response (relative value in percents of all spectral area) from crystal phase of silicon around 520 cm^{-1} becomes down to three times lower by applied electric field. Width of level is $\Delta\omega$=7.2 μeV and the gap between levels is Δ=12 μeV for silicon film with $\langle\delta\rangle$=9.7 nm were measured experimentally from the data of laser picosecond spectroscopy (see Fig. 5).The oxygen atoms and dimers are incorporated in the silicon grain boundary and have weak covalent bonds with silicon. But the activation energy for molecular diffusion is low, 0.3 eV, in contrast with the activation energy value for atomic diffusion (1.3 eV).

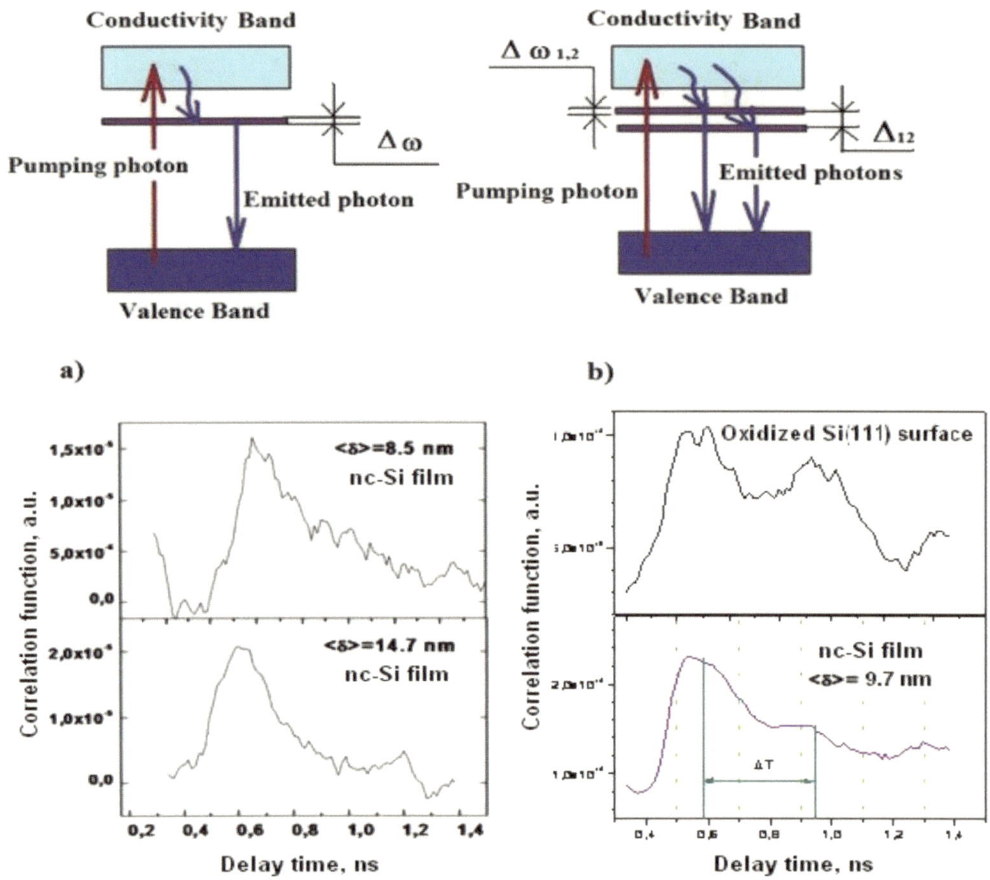

Fig. 5. Energy diagrams for different defect levels inside band gap of silicon: a) one defect level (upper graph); b) two levels which are closed to each other (upper graph); a) time evolution of excited carriers from silicon film by appearance of one defect level (down graph) near the bottom of conduction band (CB); b) quantum beats (down graph) of intensity of radiation by quantum interference of two defect levels which are closed to each other which are placed near the bottom of CB (Milovzorov, 2001).

2.5 Field effect in Raman spectra and acoustoelectric switching the optical mode

By applied electrical field the dipole moment value for Si-O-Si bridge is defined by displacement evolution:

$$\mu = \mu^0 + \mu_x(t); \tag{8}$$

The induced polarization of nanocrystalline silicon film can be explained by using the Raman scattering measurements by applied external electric field (see Fig.7). Dipole moment can be written in form (Kahan, 2000)

$$d = \frac{er}{2} \frac{(E_2 - E_1)^2}{(E_2 - E_1)^2 + 4J^2(r)} [n_2(t) - n_1(t)]; \tag{9}$$

where n_2 *(t)* and n_1 *(t)* are levels' populations, *J* is resonance integral, E_1 and E_2 are energies of electrons. Reflected intensity harmonic oscillations can be explained by using time dependent dipole moment of coupled atoms of Si-Si bonding. The Fermi Golden Rule for Raman scattering can be written according to (Yu & Cardona, 1996):

$$\Gamma_{Raman} = \frac{2\pi}{\hbar}\sum_{k,k'}\left|\sum_i \frac{\langle f|H_{el-phonon}|i\rangle\langle i|H_{el-photon}|0\rangle}{E_{i0}-\hbar\omega}\right|^2 \delta(E_C(k)-E_V(k)-\hbar\omega \pm E_p); \qquad (10)$$

where $H_{electron\text{-}phonon}$ and $H_{electron\text{-}photon}$ are Hamiltonians terms for electron-phonon and electron-photon interactions.

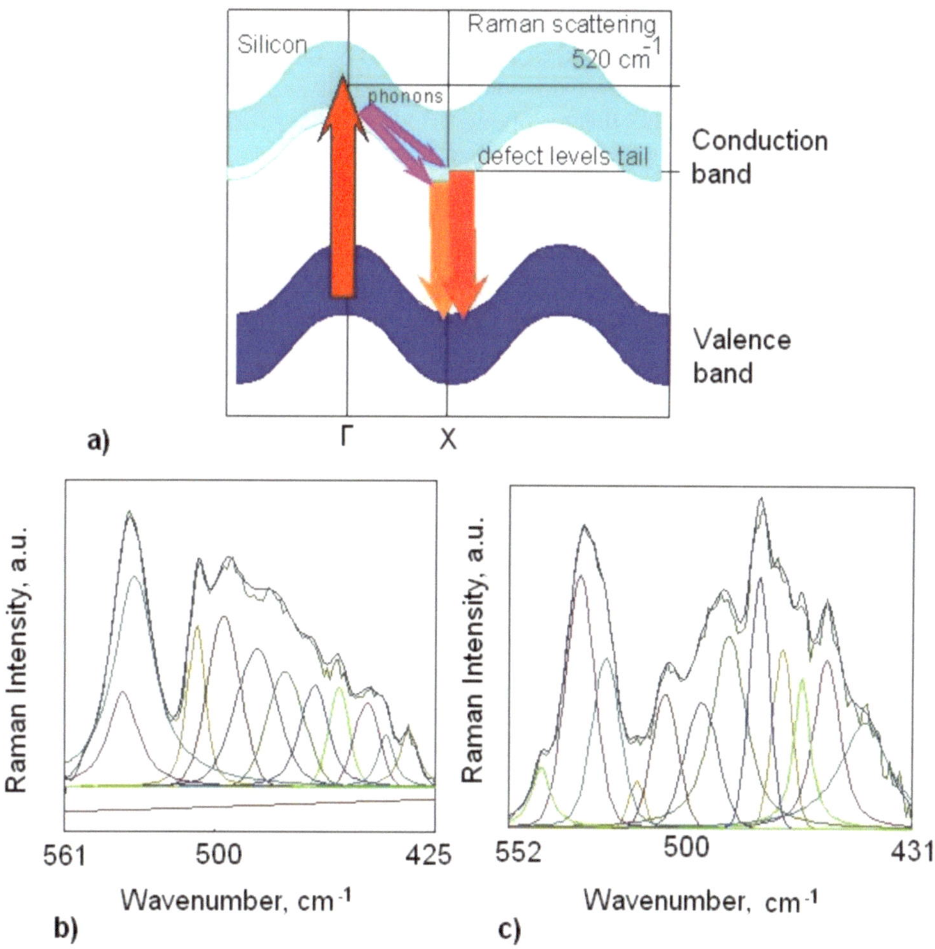

Fig. 7. Model of switching of carriers' population for silicon nanocrystals by applied electric field can be explained by using the energetic diagram of silicon film (a) and Raman scattering spectra (blue arrows) for radiation (hw_{pump}-hw_{phonon}=$hw_{emitted}$) with wavelength 488 nm (red arrow)) for silicon nanocrystalline films without (b) and with (c) applied electric field (E=30 μv/nm).

2.6 Field effect in Raman spectra for silicon films deposited on buffer layer

For the samples of silicon films deposited on buffer layer of cerium dioxide that was deposited at high temperature by magnetron sputtering of target the thickness of buffer layer was 90 nm. This layer was often used by thin silicon crystallized film preparation to stabilize the crystal silicon-amorphous glass disorder. This is because the crystal dielectric cerium dioxide with energy band gap 5.5 eV has the same crystal lattice constant: 5.41 Å is for CeO_2 and 5.43 Å for Si. It is seen in Figs. 8, 9 the Raman spectral data for various external applied electrical field values from silicon film deposited on buffer cerium dioxide layer at low deposition temperature with crystal fraction less than 30%. Figure 8 illustrates the changes in phonon modes which were occur by applying the external electric field. It is seen, that the crystal-related spectral peak is decomposed into two component (LO and TO phonon modes) with their energy positions at 515 cm^{-1} and 524 cm^{-1} without field (E=0), and 514 cm^{-1} and 524 cm^{-1} by E≠0. Figure 9 illustrates the lowering in the magnitude of LO phonon modes by increasing the external electric field. In this case the dipoles in polarized buffer layer generate the electric field which is sufficient for stabilizing the nanocrystals inside the silicon film. Because, there is phonon TO-LO modes splitting which is observed in Fig.8.

2.7 Raman silicon laser development

According to the work of (Klein & Cook, 1967) the solution of transformed Maxwell's equations can be written as combination of plane waves. For the plane wave (with frequency ω) incident on the film and interacted with phonons wavelength Λ the intensity of radiation is represented by ($\Lambda \le 0.01\omega$):

$I = I_0 W^2 \frac{\sin\left(\frac{X\pi L}{\Lambda \cos\theta}\right)}{X}$; where $X^2 = W^2 + (\sin\theta_B - \sin\theta)^2$; $W = \frac{\Lambda}{2\lambda}\frac{\Delta\varepsilon}{\varepsilon}$. I_0 is intensity of the incident laser radiation, $\Delta\varepsilon$ is a perturbation of dielectric function by phonons or electric fields. The spectral width of output radiation depends on the geometrical ratio between thinckness of film d (see Fig. 10) and substrate ℓ, and wavelength of radiation. By applying the external electric field the dielectric function is varied and the spectral width is, also, changed. The cavity with length d is adjusted according to following requirements:

$$\Delta\nu = \frac{c}{2nd}; d = m\lambda.$$

The quality factor of cavity Q characterizes the degree of spectral sharpness of output radiation (Verdeyen, 1995):

$$Q = \frac{2\pi nd}{\lambda_0}\frac{(r_1 r_2)^{1/4}}{1-(r_1 r_2)^{1/2}}. \tag{11}$$

The edge reflectivities r_1, r_2 are sufficient (~0.5), and their values depend on Si/SiO_2 interface. By the addition the CeO_2 buffer layer with strong polarization property with thickness more than 50 nm the efficient transfusion of laser energy into TM or TE mode by applying the external field can be easily achived. It is reflected in spectral pictures that there is a splitting of TO and LO phonon modes by assisted electric field.

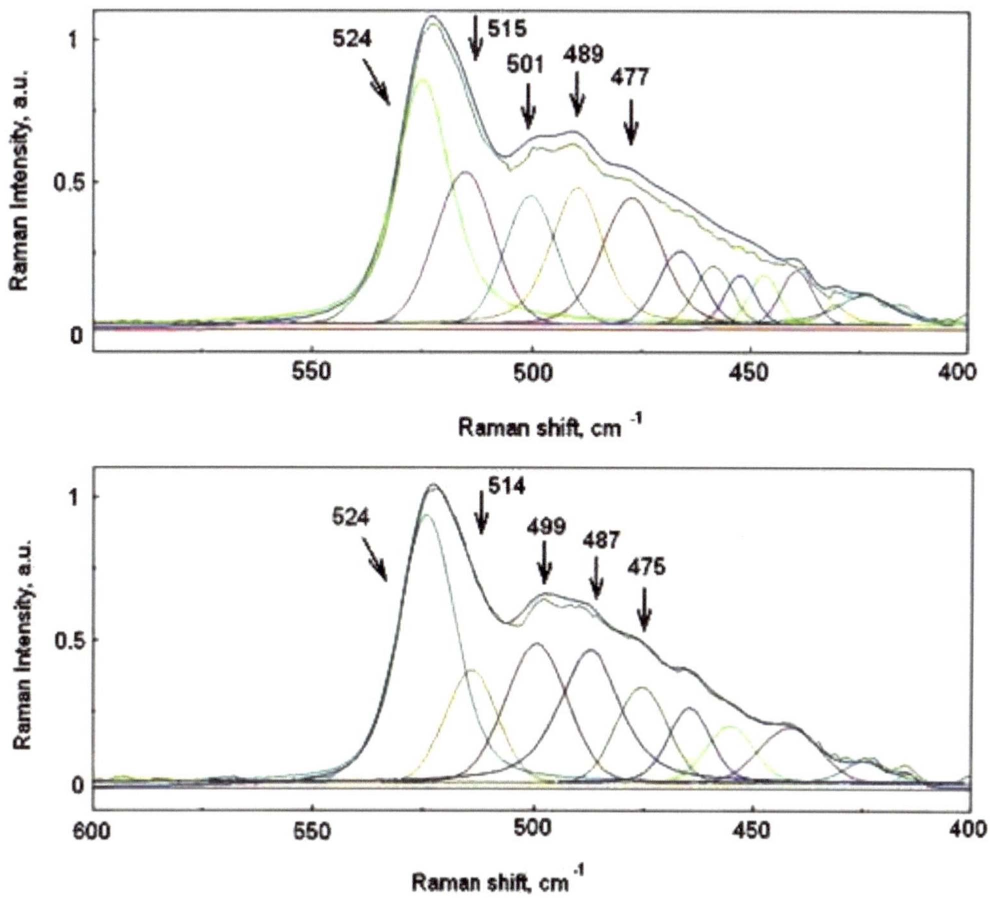

Fig. 8. Raman spectra from silicon film deposited on buffer layer without (upper graph) and by applied external electric field (down graph).

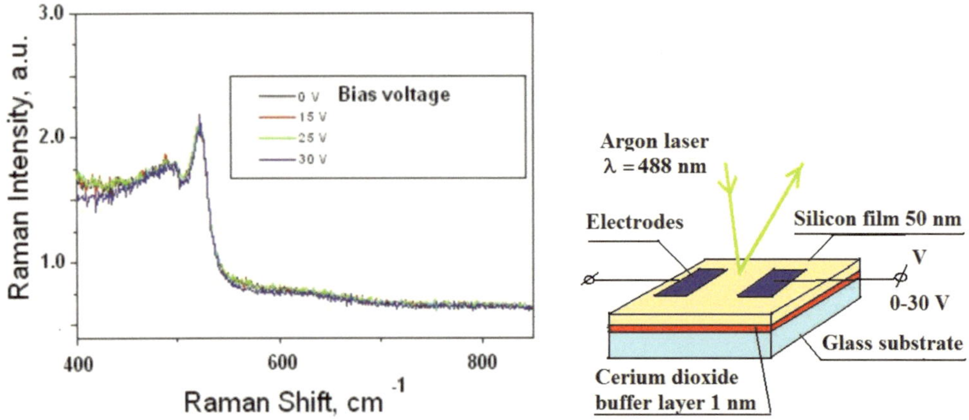

Fig. 9. Raman spectra (left graph) from nanocrystalline silicon film deposited at low temperature on buffer layer of cerium dioxide with thickness 90 nm: a) 0V/cm; b) 30V/cm; c) 50V/cm; d) 60 V/cm. Scheme of acoustoelectric switcher for laser optical modes (right graph).

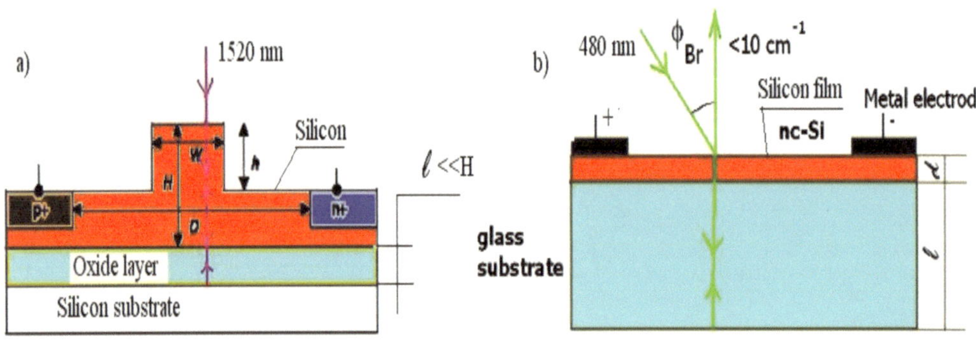

Fig. 10. Typical scheme of ordinary Raman silicon laser (a) and laser cavity for Raman scattered radiation with two separate TM and TE modes (b) based on thin film and glass substrate.

3. Nonlinear optical switchers and lasers

3.1 Chemical bonds as quantum objects

3.1.1 FT-IR spectral components and SHG response

Figure 11 shows the SHG spectra from two silicon films (prepared by using PECVD and gas mixture of silane and hydrogen) with nanocrystals Gauss distributed in their sizes and crystalline volume fraction 60%. There is a slight difference in standard deviation of Gauss distributions functions for nanocrystals' sizes in the area of 1 mm^2 (laser beam diameter on the film). It is clear, that by the same physical properties, chemical contamination, and structural properties the SHG signals is significant varied. This is because of dipole appearance inside silicon film. There is great variations in chemical bonding of silicon and oxygen (or hydrogen and fluorine), kinds of point defects and their spatial distribution which contribute in local field factor. The bond length distortion causes the surface strain. This is because of there is an oxygen atoms incorporation and structural relaxation of surface Si atoms (see Figs.12, 13 a, b). Shklyaev and coworkers (Shklyaev et al., 1999) proposed the model of oxygen nucleation on the silicon surface. There are several critical conditions in oxygen interaction with silicon. The kinetic of Si surface oxidation was studied by SHG method. However, there are no evident information about earlier lateral growth and oxygen adatoms. Figure 12 illustrates the spectral peaks around 1100 cm^{-1} (upper graph) corresponded to the oxygen triplet states $(^3P_{2-0})$ in poly-Si films FTIR spectra. It is supposed that oxygen molecules are inserted in interface amorphous media between silicon nanocrystals and cause the appearance of three spectral peaks around 1100 cm^{-1}. It is assumed that the contents of crystabalite structure Si-O-O-Si is low, but the Si-O-O-O-S is sufficient. There are no triplet states related spectral peaks for buried silicon oxide, by the annealing to the temperature 1100°C. The quartz Si-O-Si bridging structure produces one peak at 1100 cm^{-1} with great magnitude. The spectral splitting of 0.008 eV is because there are two unpaired electrons at two p orbital. However, the lateral position is unpredictable due to the great value of the surface diffusion. It is assumed, that the defects are localized in $Si\text{-}SiO_2$ interface with complex composition SiO_x. The increase in oxygen concentration per silicon atom x results in increasing of SiO bond length. For the sample with smaller grain size the hydrogen termination of dangling bonds is greater, and density of oxygen is low. But, the film with large grains contains great amount of oxygen in Si-O-Si and SiO_2 compositions. Also, it is assumed that the exponential electron density decay is due to the exciton state decay. The short relaxation time can be described as possible relaxation process through the oxygen incorporation related defect states, which density of states is following: $n(E)=(n_{Si}(E)+xn_o(E))/(1+x)$ [22] (Lanoo & Allan, 1978; Delerue et al., 1993). For low graph in Fig.12 the silicon film has crystalline volume fraction 48%, and nonlinear responses in various ranges of SHG spectra are following: I_{shg}=1.65 at λ_{2w}=370 nm; I_{shg}=0.9 at λ_{2w}=535 nm; I_{shg}=0.5 at λ_{2w}=420 nm.

3.1.2 Nonlinear optical signal by chemical bonding variations

The radiation with determined wavelengths for second-harmonic generation was detected for a films deposited on glass, silicon (100) and quartz substrates. The film thickness was more than 300 nm. Figure 13 shows the absorbances of different silicon films prepared by using gas mixture of silane diluted by hydrogen and silicon tetrafluoride diluted by helium as 5:95. It is seen, that there is great variations in chemical bonding properties of films. It is supposed that the spectral lines at 1107 cm^{-1} (see Fig. 13 b) is related to the asymmetric stretching mode for Si-O-Si of interstitial oxygen atom; the spectral line at 913 cm^{-1} (see Fig. 13 c) is for SiH_3 bending mode; the spectral peak at 1174 cm^{-1} (see Fig.13 b) is an assymetric stretching mode of O-Si-O TO (Wang et al. , 1999), and line at 1060 cm^{-1} (see Fig. 13 b) is for stretching mode of O-Si-O bonding (TO mode in phase motion of oxygen atoms) (Wang et al., 1999); spectral line at 956 cm^{-1} (see Fig. 13 d) as scissor mode of $Si\text{-}H_2$ (Chabal et al., 2002), spectral lines at 975 cm^{-1} and 979 cm^{-1} (see Figs. 13c, 13d) are related to silicon fluorine bonding (Milovzorov et al., 1998, Milovzorov, 2010) or divalent silicon specy interacted with two oxygen atoms (Chabal et al., 2002); spectral lines at 902 cm^{-1} (see Fig. 13c) and 901 cm^{-1} (see Fig.13d) are related to silicon hydrogen bonding $Si\text{-}H_2$ scissor mode (Chabal et al., 2002); also, spectral line at 913 cm^{-1} corresponds to the $Si\text{-}F_5$ molecular-like assembly (King et al., 1986); and spectral line at 934 cm^{-1} (see Fig. 13b) and 933 cm-1 (see Fig. 13d) are related to stretching mode Si-F of $Si_9H_{15}F$ configuration (Chatterjee et al., 2001) ; spectral lines at 838 cm^{-1} (see Fig. 13d) and 841 cm^{-1} (see Fig.13c) are related to A-center response (Corbett et al., 1961, 1964). In addition, it is assumed that the spectral lines at 858 cm^{-1} (see Fig. 13d) and at

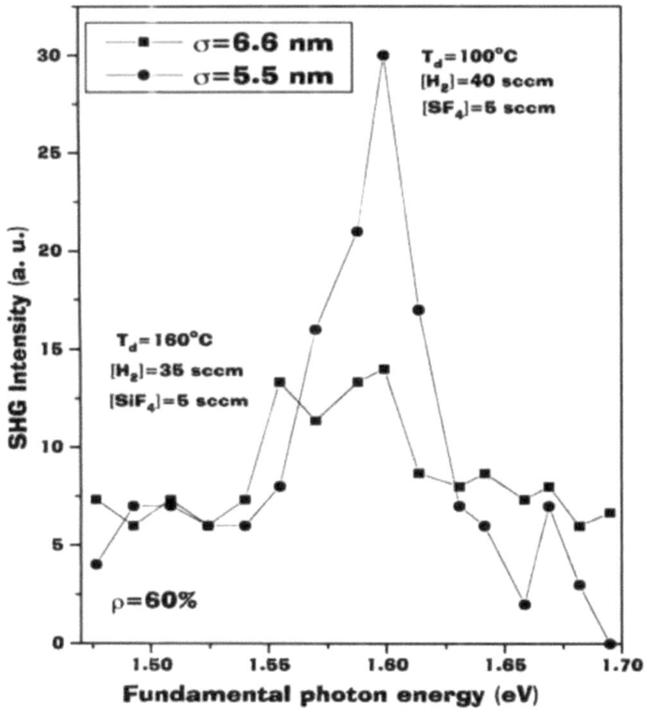

Fig. 11. SHG spectra from silicon films with nanocrystals Gauss distributed in their sizes and crystalline volume fraction 60%.

859 cm^{-1} (see Fig. 13c) is for Si-F stretching mode or Si=O stretching mode of sylynene and electron configuration for the triplet state.
Variation in SHG signal for different nanocrystalline films with their chemical bonding properties at determined wavelengths is presented in Table № 1. In addition, there is SHG responses from the films with silicon substrate, crystalline volume fractions obtained by using Raman data, average grain sizes of nanocrystals estimated by means of X-ray diffraction, and photoluminescent signal. It is supposed that the SHG spectrum is produced by surface component of dipole moment because there is a violence of symmetry by surface and great contribution into the surface area of large size nanocrystals (S/V ratio). The great enhance is due to the appearance of surface's nonzero dipole moment:

$$P^S(2\text{w}) = \chi^{(2)}{}_{surface} : E(\text{w})E(\text{w}) + \chi^{(2)}{}_{Si-O;F} : E(\text{w})E(\text{w}) + \chi^{(2)}{}_{distortion} : E(\text{w})E(\text{w})\,. \qquad (12)$$

There is anisotropy of nonlinear optical properties by the dominant crystal orientation (111) for the films deposited at low temperatures. Such kind of anisotropy can be easily recognized by means of p- or s-polarized laser radiation and by the detection of *s*-polarized, for χ_{xxx} or χ_{zzz} components or p -polarized, for χ_{zzz}, χ_{zxx} and χ_{xxz} components of output radiation. The nonlinear second-order susceptibility depends on the surface hydrogen coverage θ_H according to the formula (Durr & Hofer, 2006):

$$\chi^{(2)}{}_{s,i,j,k}(\theta) = \chi_{zzz} + (1-\theta_H)\chi_{xxz} \qquad (13)$$

where χ_{zzz} and χ_{xxz} are the nonzero tensor components of susceptibility, $0 \le \theta_H \le 1$ ML.

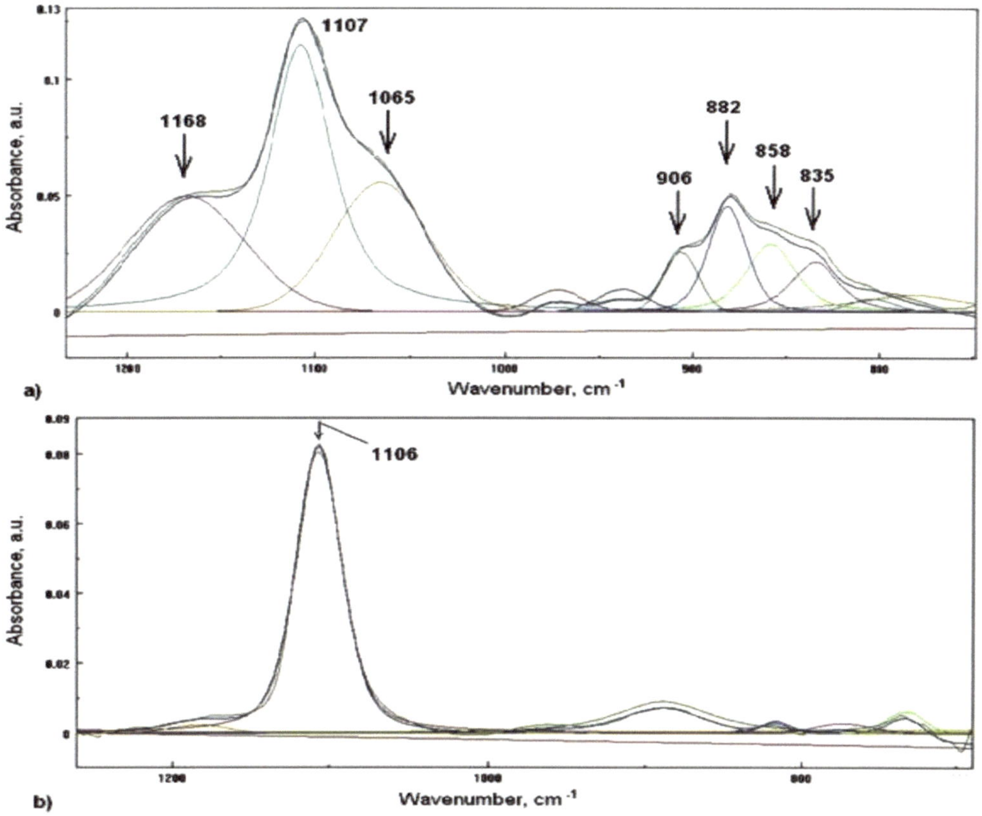

Fig. 12. FT-IR spectra from oxidized nanocrystalline silicon films (a) and (b) with significant SHG response.

3.2 Dipole moments and SHG spectra

The nature of defects is complicated because the types of oxygen (or fluorine) and hydrogen incorporation in nc-Si films are varied. There are different kinds of defects with significant physical properties such as paramagnetic centers or charge traps. The optical luminescent properties depend on defect density and their spatial distribution. Optical absorption measurements help to study the oxygen and hydrogen incorporation, to characterize the types of defects, to evaluate the density of bonds. Dipole moment of silicon dangling bond can be estimated by using combination of *Si (1S)-Si$_{db}$ (3p)* and *Si (3p)-Si$_{db}$ (3p)* atomic orbital:

$$\mu = ed\int_0^{V_{2\xi}} \Psi_{Si1S}\Psi_{Si3p}dv;$$
$$\mu = ed\int_0^{V_{2\xi}} \Psi_{Si3p}\Psi_{Si3p}dv; \qquad (14)$$

where the value $V_{2\xi}$ is the volume of integration. The overlap integrals are shown in the Table № 2 to calculate the dipole moment values.

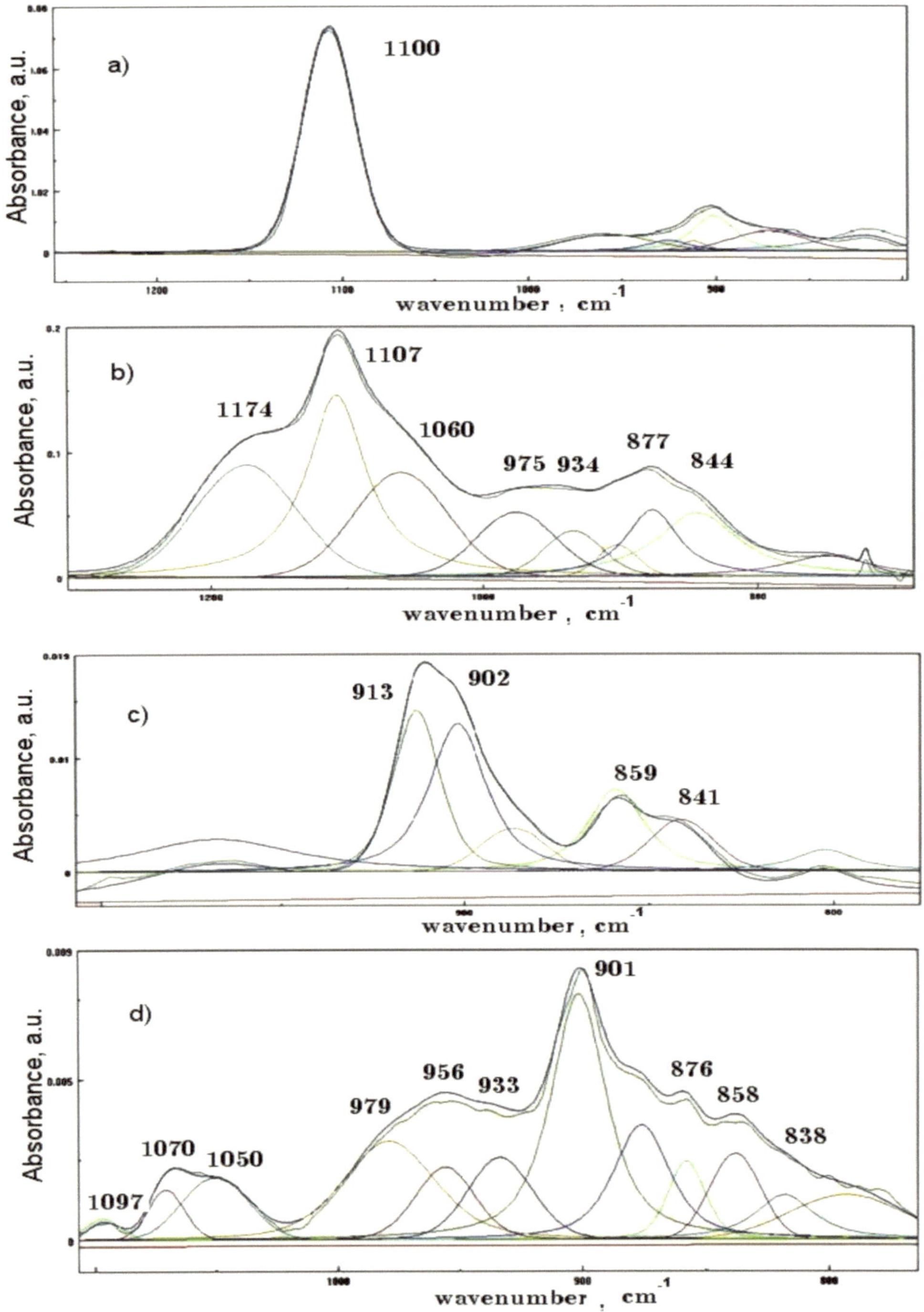

Fig. 13. FT-IR spectra for nanocrystalline silicon films with high SHG response in UV and visible range; silicon films are varied in chemical bondings: a) SD56, b) SD 55, c) SD 58, d) SD53. The SHG signals for all films are in Table № 1.

Sample, Chemical bonding	λ=370 nm	λ=420 nm	λ=450 nm	λ=472 nm	λ=535 nm	Silicon substrate λ=420 nm	Silicon substrate λ=450 nm	Silicon substrate λ=535 nm	nc-Si Size,nm	ρ,%	PL
SD52, Si-O-Si bridge	1.13	0.052	1.06	0.7	17	1.5	1.2	17	10	70	PL
SD53, Si-F	1.06	10	1.5	0.7	10	0.7	2.0	20	11	70	PL
SD55, Si-O	1.65	0.02	0.73	0.3	13	0.12	0.6	20	-	30	PL
SD56, Si-O-Si	0.75	0.0 4	2.0	0.7	10	1.0	24	-	18	70	PL
SD58, Si-F	0.47	0.9	1.8	0.7	10	0.2	1.0	15	-	30	PL

Table 1. SHG signals detected on fixed wavelengths from the 5 different silicon films with various fractions of nanocrystals, deposited on glass and silicon (100) substrates.

d, length for strain or broken bond, Å	Overlap integral value for silicon *3p-3p* orbital, δ_2 10^{-6}	Overlap integral value for silicon *3p-* oxygen *2p* orbital, δ_3 10^{-6}
3.65	6.835 10^4	3.1 10^{-4}
2.21	8.37 10^5	2.8
2.35	6.75 10^5	1.2
3.83	4.84 10^4	9.55 10^{-5}

Table 2. The overlap integral values calculated for silicon bonds and silicon-oxygen bonduing at fixed bond lengths.

For the silicon orbital *(1s)*-oxygen atom orbital *(2p)* and silicon orbital *(3p)*-oxygen orbital *(2p)* interactions we can evaluate dipole moment of Si-O bond by using formula:

$$\mu_{2pO-3pSi} = ed\int_0^{r2\xi} \Psi_{O2p}\Psi_{Si3p}dv;$$
$$\mu_{2pO-1SSi} = ed\int_0^{r2\xi} \Psi_{O2p}\Psi_{Si1S}dv \qquad (15)$$

where d is bond length, $dv = 4\pi r^2 dr d\phi d\theta$, $0 < r \le 2\xi$, ξ is a largest covalent radius. By the $r = 1.6\mathring{A}$, $r/a_0 = 3$ the value $\mu_{2pO-3pSi} = 1.12D$.

Figure 14 illustrates the nonlinear spectra because the SHG response is optically induced in oxidized nanocrystalline silicon films. There are various phenomena which can be responsible for increasing the SHG. The first possible explanation of nonlinear optical properties is appearance of dipoles inside the film such as Si-O bonding with significant dipole moment. The second is a perturbation of electron density by electromagnetic field by

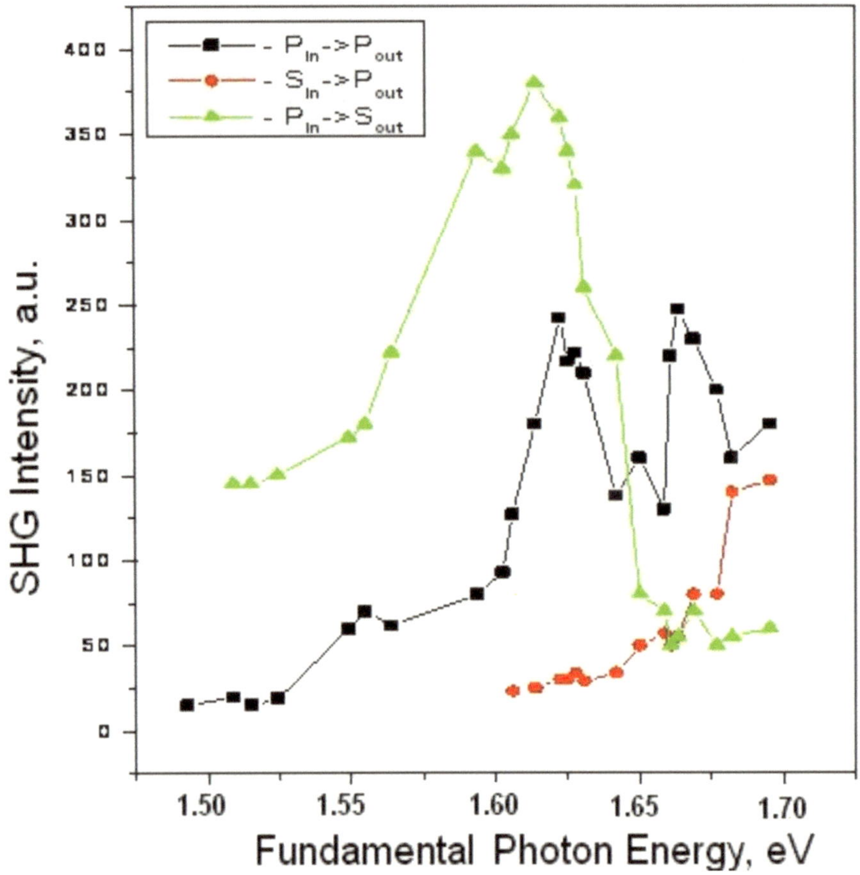

Fig. 14. Nonlinear SHG spectra from the surface of silicon films with 10 nm-sized nanocrystals (111) and crystalline volume fraction at 70%.

applied external electric field or laser field. The third is a quadruple moments from the complex point defects, such as vacancy of oxygen atoms in Si/SiO_2 interface area, dangling bonds which migrate in films due to the weakening of silicon bonds. It is supposed, the value of dipole moment by the angle between the two Si-O bonds is around 109°. The second-order terms of dipole values for polarized media are given by

$$\mu^{(2)} = e\int_0^{VR} \Psi_{Si1S}\Psi_{Si3p}dv(\int_0^{VR} \Psi_{Si1S}x^2\Psi_{Si3p}dv - \left(\int_0^{VR} \Psi_{Si1S}x\Psi_{Si3p}dv\right)^2) / \int_0^{VR} \Psi_{Si1S}x\Psi_{Si3p}dv \tag{16}$$

where a sphere of integration is V_R, x is bond length. It is assumed that x value is a sum of two covalent radii for both atoms, ξ is a largest covalent radius for two atoms, R is a distance by the atomic orbital for each center has spatial the limit of its extension, for the values of distance $R = 4\xi$ we suppose the averaging procedure is possible. This dipole moment is because of the dangling bond causes the breaking in crystal structure's symmetry. The concentration of dangling bonds in silicon film deposited on glass substrate is varied from 10^{16} cm^{-3} to 10^{20} cm^{-3}.

The next two types of dipole moments are due to the oxygen atom incorporation with silicon can be evaluated for determination the role of oxidized silicon in yield of second-harmonic generation:

$$\mu^{(2)} = e\int_0^{VR} \Psi_{Si1S}\Psi_{O2p}dv(\int_0^{VR} \Psi_{Si1S}x^2\Psi_{O2p}dv - \left(\int_0^{VR} \Psi_{Si1S}x\Psi_{O2p}dv\right)^2)/\int_0^{VR} \Psi_{Si1S}x\Psi_{O2p}dv;$$
$$\mu^{(2)} = e\int_0^{VR} \Psi_{Si3p}\Psi_{O2p}dv(\int_0^{VR} \Psi_{Si3p}x^2\Psi_{O2p}dv - \left(\int_0^{VR} \Psi_{Si3p}x\Psi_{O2p}dv\right)^2)/\int_0^{VR} \Psi_{Si3p}x\Psi_{O2p}dv; \quad (17)$$

The first integral in equation describes the charge spatial distribution of two atomic orbital that are located on different centers overlap. The second term in such multiplication in brackets describes the spatial distortion of dipole bond length by various conditions: such as external electromagnetic field or the other coupled $Si-O$ dipole. These changes cause the perturbations in wave functions, too. However, we estimate such dipole moment distortion by using non perturbative wave functions and stressed bond with its length more than ordinary. Such dipole moment deviation causes the electron density distortion and significant nonlinear optical signal. In addition, the induced by external field dipole moment for a pair of silicon atoms can be explained as following:

$$\mu_{Si-Si}(t) = ed\left(\int_0^{r2\xi} \Psi_{3pSi}(t)\Psi^*(t)_{3pSi}dv + \int_0^{r2\xi} \Psi_{1sSi}(t)\Psi^*(t)_{1sSi}dv + \int_0^{r2\xi} \Psi_{3pSi}(t)\Psi^*(t)_{1sSi}dv + \int_0^{r2\xi} \Psi_{1sSi}(t)\Psi^*(t)_{3pSi}dv\right);$$
$$\Psi_{3pSi} = \Psi_{3pI}\exp\left(-iE_{3pI}t/\hbar\right) + \Psi_{3pII}\exp\left(-iE_{3pII}t/\hbar\right); \quad (18)$$
$$\Psi_{1sSi} = \Psi_{1sI}\exp\left(-iE_{1sI}t/\hbar\right) + \Psi_{1sII}\exp\left(-iE_{1sII}t/\hbar\right).$$

The wavefunctions according to the P. Feuer theoretical work (Feuer, 1952) corresponding to the energies E_I and E_{II} by external applied electromagnetic field can be written as

$$\Psi_I = \frac{1}{\sqrt{2}}\sum_m \left(J_{m+l1}(-q_1)a_1(x-x_m)\right) + \left(J_{m+l2}(-q_2)a_2(x-x_m)\right);$$
$$\Psi_{II} = \frac{1}{\sqrt{2}}\sum_m \left(J_{m+l1}(-q_1)a_1(x-x_m)\right) - \left(J_{m+l2}(-q_2)a_2(x-x_m)\right).$$

Here, the J_m is a Bessel function and Taylor expansion of pertubing potential is $V(x) = eFx_m + eF(x-x_m)$.

3.3 χ^2 succeptibility components from polarized light experiments

For spectral measurements, an optical parametric oscillator/amplifier pumped by the third harmonics (0.355 nm) of a Q-switched Nd:YAG laser (Spectra-Physics, MOPO 730) at a 10 Hz repetition rate was used. The p-polarized laser radiation with spectral width of 0.3 cm^{-1} was focused on the sample to a spot with diameter of 0.5 mm. The energy of primary laser radiation was near 3 mJ. Resonance spectra consist of two sharp peaks. The peak at 1.6 eV is caused by SHG response due to $E_0' = \Gamma_{25}$-Γ_{15} transition. We suppose that the second peak can be recognized as SHG response due to $E_1 = L_{2'}$-L_1 transition in silicon nanocrystallites.
It is observed the sharp spectral peaks around 3.26 eV related to SiO bonding in silicon network with energy position E_c*-0.14 eV* (see Figs. 15-16). It is assumed, that the second

sharp peak is related to vacancy of oxygen switching by A-defects (with energy position *E_c-0.17* eV for bulk Si) appearance in silicon films. It can be observed by means of spectroscopic methods (such as FT-IR, EPR, Laser time-resolved spectroscopy) only in fluorinated silicon films. A-defects were detected at first time in silicon irradiated by high energy electron beam by (Watkins & Corbett, 1961). For silicon film synthesized by PECVD method there is H_2SiF_6 acid and ortho-silicon acid which generate nonlinear response due to their great dipole moment. Such incorporation is changed according to fluorine dilution. The *SiO* bonding decreases, but the angle of *Si-O-Si* bridge is changed, there is a *Si-Si* atomic orbital splitting. The two new levels are closed to each other with difference around several μeV (see Fig. 5b).
Figure 16 illustrates the SHG spectra for $P_{in}\rightarrow S_{out}$ polarization scheme for horizontal (left graph) and vertical (right graph) spatial positions of sample with nanocrystalline silicon film with average grain size 97 Å. SHG spectra contains isotropic and anisotropic component of susceptibility tensor $\chi^{(2)}_{zzz}$, $\chi^{(2)}_{zxx}$,$\chi^{(2)}_{xxz}$, and $\chi^{(2)}_{xxx}$.). Figure 17 illustrates the second-harmonic generation by optical pumping of laser light with energy of pumping photon 1.63 eV (black arrow) along with resonant tunneling of carriers through the thin layer of amorphous oxidized silicon. The pumping photons energies $h\omega$ are 1.63 eV, but the emitted second-harmonic photons with 2hω energies (3.26 eV) have the: $2h\omega_{emitted}=h\omega_{pumping}+h\omega_{pumping}$. The red arrow corresponds the pumping photons with summ energies more than 3.4 eV (for direct transitions in bulk Si (111)), and the blue arrows illustrate the photoluminescent radiation, caused by optical transitions through the excited levels which are situated inside band gap of silicon, too. It is necessary to note, that the resonant tunneling processes can be realized from the level with lower energy to the next energy level in closed silicon nanocrystal (see Fig. 17) or other defect levels inside oxidized silicon layer. It is clear, that by the resonance tunneling the population of lower levels decreases but the population of upper levels increases.

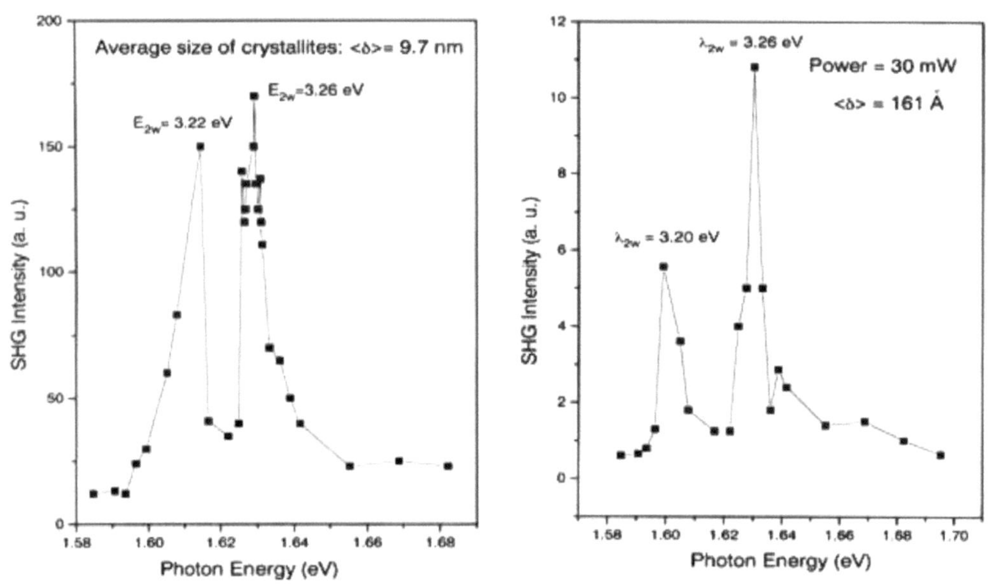

Fig. 15. SHG spectra ($P_{in}\rightarrow P_{out}$) for $\chi^{(2)}_{xxx}$ component of susceptibility tensor for nanocrystalline silicon films with average grain sizes 97 Å (left graph) and 161 Å (right graph).

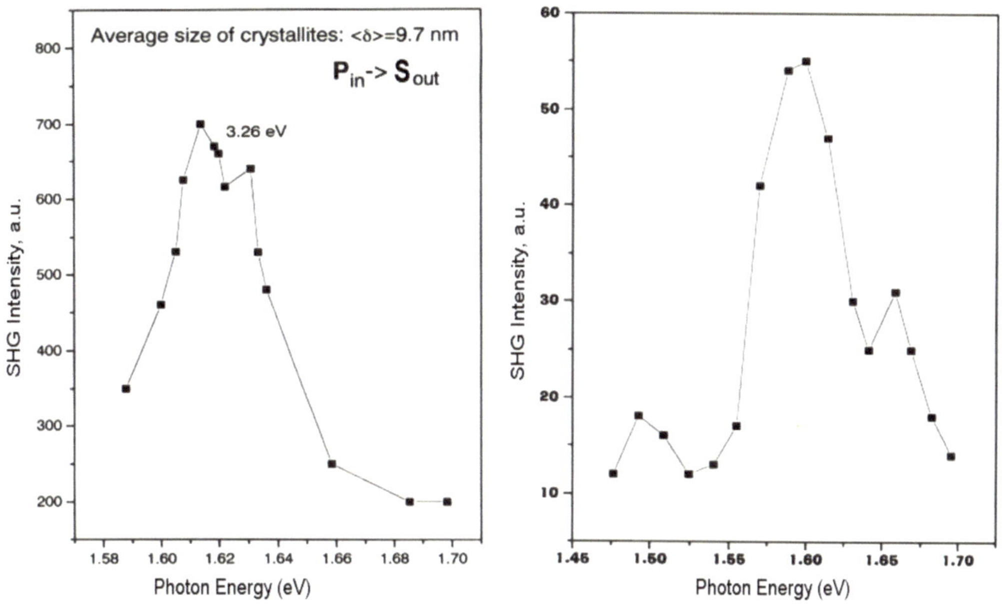

Fig. 16. SHG spectra for $P_{in} \rightarrow S_{out}$ polarization scheme for horizontal (left graph) and vertical (right graph) spatial positions of sample.

The nonresonant SHG intensity as a function of the average grain size in poly-Si films, with crystalline volume fraction 70%, is presented in Fig. 18, where

$$K = \frac{I_{exp}}{I_{exp}^{min}} \frac{|L(2\omega,\rho_{min})L^2(\omega,\rho_{min})|}{|L(2\omega,\rho)L^2(\omega,\rho)|} \quad (19)$$

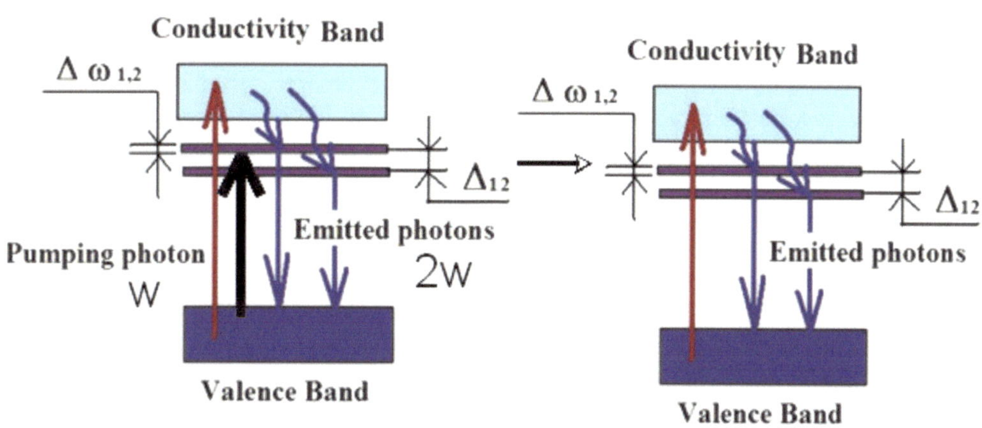

Fig. 17. The energy diagram and resonant SHG scheme by optical pumping of laser light with determined energy of pumping photon 1.63 eV (black arrow). The nonresonant excitation of carriers was illustrated by red arrow.

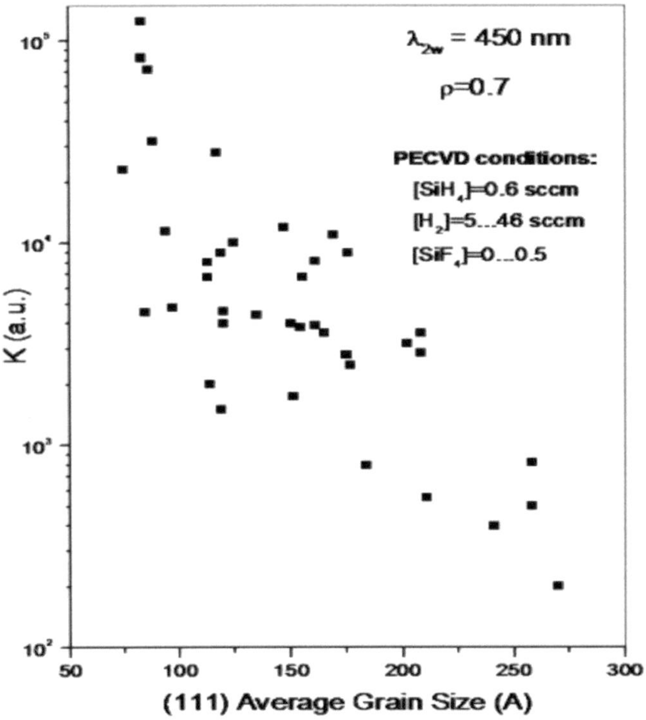

Fig. 18. SHG signals ($P_{in} \rightarrow S_{out}$) as a function of average grain sizes of nanocrystals in various silicon films deposited on glass substrates by detection the SHG radiation with wavelength 450 nm.

is normalized SHG signal, $L(2\omega,\rho)$ and $L(\omega,\rho)$ are the factors of the local field:

$$L(\omega,\rho) = \frac{\rho}{4\pi} \frac{(\varepsilon_c(\omega) - \varepsilon_a(\omega))}{1 + (\varepsilon_c(\omega) - \varepsilon_a(\omega))(\Lambda - \beta\rho)} \tag{20}$$

where ε_c and ε_a are dielectric functions of crystalline and amorphous silicon, respectively. For sphere depolarization factor Λ is equal to 1/3, β is the Lorentz constant (for homogeneous spherical surrounding β =1/3). For calculations the following estimated values were λ_w =900 nm, λ_{2w} =450 nm, $\varepsilon_c'(w)$ =13, $\varepsilon_c''(w)$ = 0.03, $\varepsilon_a'(w)$ = 12, $\varepsilon_a''(w)$ = 0.23, $\varepsilon_c'(2w)$ = 18, $\varepsilon_c''(2w)$ = 0.5, $\varepsilon_a'(2w)$ = 18, $\varepsilon_a''(2w)$ = 7.5 . It is shown on Fig. 18 that decrease in the average size of crystallites from 25 to 7.5 mm results in increasing of SHG intensity more than two orders of magnitude. Each point corresponds to one silicon films with its unique deposition conditions.

3.4 Thin nanocrystalline silicon film for nonlinear devices development

Nonlinear properties such as second-harmonic generation (SHG) were studied by means of spectroscopic technique: laser source with monocromator coupled with photo multiplied tube. Optical parametrical generator with amplifier (MOPO 730, Spectra Physics) was used to probe silicon films. The wavelength was changed in range of 450-550 nm. The diameter of laser beam on silicon film was 1 mm^2.

Figure 14 and Fig. 15 illustrate the SHG spectra from nanocrystalline silicon films with different average grain sizes: 9.7 nm and 16.1 nm (Milovzorov et al, 2001). It is seen, that there are two spectral peaks in SHG spectra for silicon films by laser radiation $P_{in} \rightarrow P_{out}$ polarization scheme. However, in contrast to the $P_{in} \rightarrow S_{out}$ experiments (see Fig.14) with size dependent integral SHG signal there is no strict size dependence for scheme of radiation $P_{in} \rightarrow P_{out}$ (see Fig. 16) due to the negligible integrated signal.

3.5 Quantum description of microholes for SHG

Nonlinear optical semiconductor–based devices are successfully implemented in various fields of electronics as quantum electronics for SHG as fiber optics for the new Raman lasers design. Some of the structures (Pellegrini et al., 1999) such as ZnSe with microhole and Bragg reflectors Si_3N_4/SiO_2 generate the SHG in blue-green spectral range (460-500 nm). These conditions for SHG can be estimated as holes layer thickness L_{cav}=360 nm and length of coherence l_c=1500 nm. For silicon nanostructured films with microholes operating at 1500 nm the optical signal strictly depends on the hole geometry. SHG spectral peaks in all silicon photonic crystals can be detected at 40° and 45° for angles on incidence in the range of 700-950 nm (Dolgova et al., 2002). For nanosized holes included in nanostructured films such as shown on Fig.3 it is supposed that the inner space of holes were covered by hydrogen and oxygen atoms incorporated into silicon. These atoms terminated dangling bonds of silicon by the deposition and cancelled the further growth of crystal structure. The surface diffusion coefficient for hydrogen atom is more than for oxygen which has high electronegativity (3.44 for O and 2.1 for H). Because, the hydrogen atoms (and dangling bonds, respectively) can easily move by laser field interaction and significant dipole moments will appear. Such quantum nanoholes with defects and impurities generate additional local surface levels in electronic structure (or polaritons). The simplest possible description is coupling of exciton and photon that produces the resulted frequency value for polariton. This polariton generates SHG radiation. The frequency eigenvalues are given by (Deveaud, 2007)

$$\omega = \frac{\omega_{ex} + \omega_{ph} - i(\gamma_{ex} - \gamma_{ph})}{2} \pm \frac{1}{2}\sqrt{\Omega_R^{\ 2} + (\omega_{ex} - \omega_{ph} - i(\gamma_{ex} - \gamma_{ph}))^2}; \qquad (21)$$

where ω_{ex} and ω_{ph} are exciton and nanocavity (or nano-hole) photon mode frequency, Ω_R is the Rabi frequency, γ_{ex} and γ_{ph} are, respectively, damping rates. According to the proposed model of quantum description of SHG caused by nano-sized cavities we assume that by the polariton-photon scattering SHG results in $2\omega_{laser}=\omega_{Si\text{-}O}+\ \omega_{ex}+\omega_S$;where the frequency is written as $\omega_S = \frac{\mu E}{\hbar}$; and μ=dQ is dipole moment of nanosized hole, d is the size of hole, Q is charge. Figure 15 shows SHG spectra nanocrystalline silicon films with grain sizes 9.7 nm and 16 nm. The spectral peak around 3.26 eV corresponds to the Si-O oscillated bonds. The second spectral peak has random energetic position. It is assumed that its position depends on the charges distributed on film surface, particularly in man-sized holes. Accordingly, there is no observed microholes on the surface of film by using Raman mapping data (Fig.3). But for atomic-force microscopic photo it is seen the random nanozised holes. Xu proposed the method of fabrication of two-dimensional arrays of nano-holes on the silicon crystalline surface (Xu, 2004). The sharp spectral peak (see Fig. 19) reflects the high efficiency of such structures.

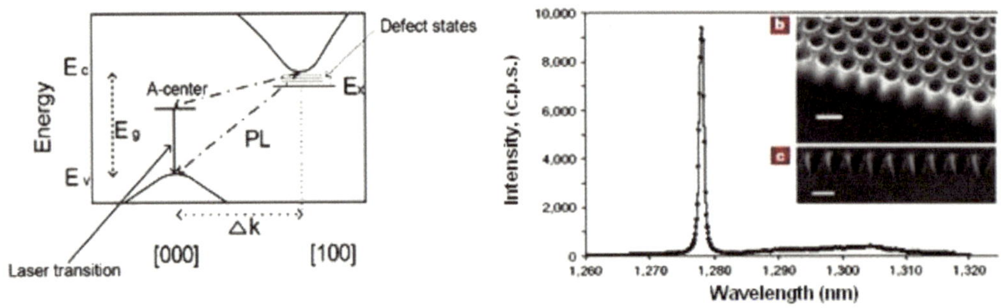

Fig. 19. Silicon energy diagram with laser transition and PL (left graph) and optical spectrum (right graph) with inserted photo of nano-holes (Xu, 2004).

3.6 Dipole moment of strained Si-O-Si

We can estimate the value of dipole moment as following:

$$\mu^{\text{Si-O-Si}}_{\text{2pO-3pSi}} = 2\sqrt{2\pi}e\int_0^{\xi}\int_0^{2\pi}\int_0^{\pi}\Psi_{3\text{px}\text{Si}}\Psi_{3\text{py}\text{Si}}r\Psi_{2\text{px-O}}\Psi_{2\text{py-O}}r^2 dr d\varphi d\theta\,; \tag{22}$$

Usually, for Si-O-Si bridge the angle between two bonds is approximately in the range of 90-180°. The dipole moment can be estimated as $\mu_{Si-O-Si} \approx 1.4\mu_{Si-O}$ for the angle $\alpha \rightarrow 90°$ between two Si-O bonds in Si-O-Si configuration. Such increasing of dipole value can be resulted by stress of film surface or even structural point defects in grain boundary of silicon nanocrystals. The stress by applied electrical field the dipole moment value for Si-O-Si bridge is defined by displacement evolution: $\mu = \mu^0 + \mu_x(t) \approx ex(t)$. We can evaluate the dipole moment by zero displacement $\mu^0{}_{Si-O-Si} = 0.17D$. It is supposed, that the field-stimulated dipole moment can be varied in range of 0-1.4 D. The nonlinear susceptibility tensor is given by $\chi_{\alpha\beta\gamma} = \partial^2 P_\alpha / \partial E_\beta \partial E_\gamma$. The second-harmonic generation is forbidden for centro-symmetric crystal such as bulk silicon because the sum dipole moment is zero, but SHG is possible due to the surface breaking symmetry and quadruple terms contributions. The opposite situation is for nanostructured oxidized silicon film, the surface area for a great amount of nanocrystals is significant, and the breaking symmetry is permanent and lateral isotropic. We can use the following expressions for simplest formula of tensor such as $\alpha = N_{Si-O}/N$ and $\beta = N_{db}/N$:

$$\chi_{nc-Si}{}^{(2)} = \begin{pmatrix} \alpha d^o{}_{11} + \beta d_{11} & -\alpha d^o{}_{11} - \beta d_{11} & \beta d_{13} & \alpha d_{14}{}^o & \beta d_{15} & 0 \\ 0 & 0 & 0 & \beta d_{15} & -\alpha d^o{}_{14} & \alpha d^o{}_{11} - \beta d_{11} \\ \beta d_{31} & \beta d_{31} & \beta d_{31} & 0 & \beta d_{35} & 0 \end{pmatrix}; \tag{23}$$

There are many atoms of surface which are involved into complicated dipole moment of nanocrystals with different shapes. By decreasing the size of nanocrystal it can be observed the increasing in quantity of surface atoms and Si-O bonds which results in an appearance of charges on nanocrystal surface. Most of them are oriented according to the dominant crystal orientation (111) and placed in silicon atoms' sites. Because, there is a spectral peak (see Fig. 15) with energy position around 3.26 eV. The second spectral peaks in the Fig. 15 are related to the defect level because of there are unpaired silicon orbitals.

The spectral peak at 3.26 eV is an optical response from nanocrystal grain boundary that contains oxygen atoms incorporated in silicon as dipoles inside film. Figure 20 shows the resonant SHG response due to the $\chi^{(2)}_{zxx}$ and $\chi^{(2)}_{zyy}$ component of susceptibility tensor for silicon nanocrystals with orientation (111) and average grain size 9.7 nm. Time resolved picosecond spectroscopy was used to investigate the electronic structure of defect levels of silicon (see Fig. 5). There is a possibility to develop optooelectronic device based on polarized nanocrystalline silicon film for switching the spectral TE and TM modes of the radfiation.

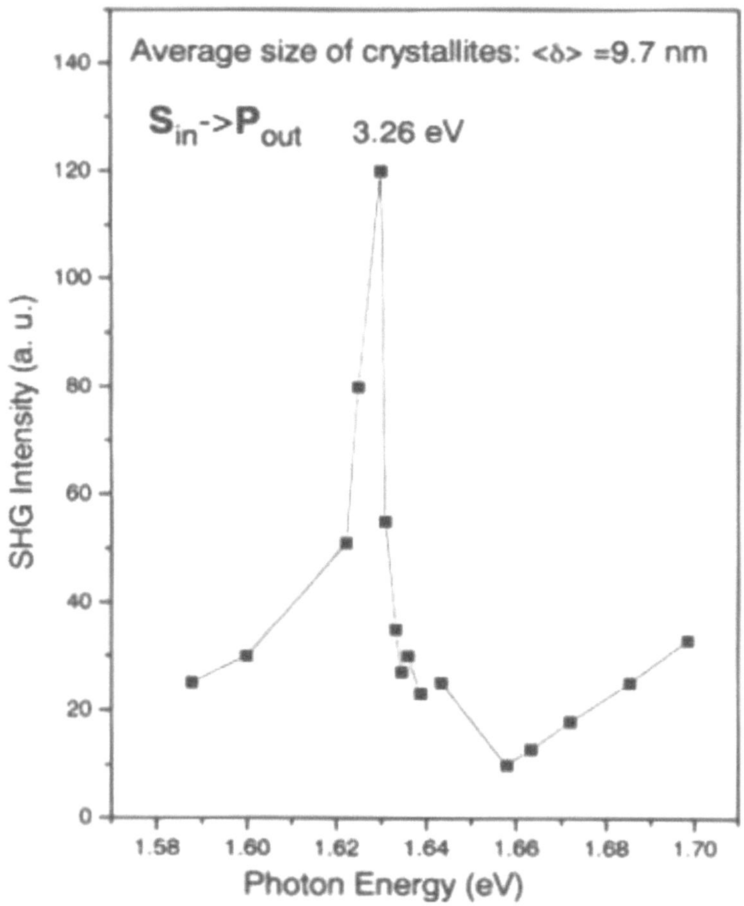

Fig. 20. Resonant SHG spectra for $\chi^{(2)}_{zxx}$ and $\chi^{(2)}_{zyy}$ component of susceptibility tensor for silicon nanocrystals with orientation (111) and average grain size 9.7 nm

3.7 Nonlinear switch parameters

The nonlinear optical semiconductor-based device is proposed. Figure 21 illustrates the scheme of second-harmonic switching. Adjusted two pairs of metal electrodes create the electric lateral field. The incident radiation has its polarization that can be transformed along the two X and Y-axis. To obtain significant SHG response it is needed to orient dipoles in silicon film. Usually, for these purposes the gold electrodes are used. Also, the rectangular geometry with sufficient size for laser beam (from μm to mm) placing on film is popular. The film thickness is varied from the several hundredths of nm to μm. Metal electrodes are deposited by thermal evaporating by using electromagnetic inductor-based evaporator. For reorientation dipoles we need to use electrical field values sufficient for sure dipole manipulation, from one side, and comparable with laser field value, from another side. For the Ar laser radiation with power 2 mW it is used the electric field 30 μV/nm. By this way, to correct the changes in dipoles' orientations by laser field is necessary for significant SHG signal detection. The polarization of film can be written as following:

$$\begin{bmatrix} P_x \\ P_y \\ P_z \end{bmatrix}^{SHG} = \varepsilon_0 \chi_{nc-Si}^{(2)} \begin{bmatrix} E^2_x \\ E^2_y \\ E^2_z \\ 2E_yE_z \\ 2E_xE_z \\ 2E_xE_y \end{bmatrix}; \quad (24)$$

The terms of tensor (from eq. 23) which describe silicon atoms with Si-O bonding are $\chi_{xxx}^{2w} = \alpha d_{11}^{0} + \beta d_{11}$; and $\chi_{yxz}^{2w} = \alpha d_{14}^{0}$. The lowering of oxygen concentration ($\alpha \rightarrow 0$) in silicon film causes the elimination of related component and the view of tensor shows pure silicon surface. In addition, the $P^{2w}{}_y$ component will be equal to zero. The output radiation will be polarized, too. The oxygen incorporation with silicon atoms causes the changes in $\chi^{2w}{}_{xxx}$ and isotropic term ($\chi_{yxz}{}^{2w}$) will appear. By the $E_x=0$ the polarization components of reflected radiation from silicon film will be following

$$\begin{aligned} P_x^{2w} &= \chi_{xyy}{}^{2w} E_y{}^2 + \chi_{xzz}{}^{2w} E_z{}^2 + 2\chi_{xyz}{}^{2w} E_y E_z; \\ P_y^{2w} &= 2\chi_{yyz} E_y E_z; \\ P_z^{2w} &= \chi_{zyy}{}^{2w} (E_y{}^2 + E_z{}^2). \end{aligned} \quad (25)$$

The components along the axis Y and Z of polarization don't depend on the Si-O bonding. However, the nonzero $P_x{}^{2w}$ component contains the terms in which the *Si-O* dipoles are contributed. By the polarized incident radiation $E_y=0$ we will have the

$$\begin{aligned} P_x^{2w} &= \chi^{2w}{}_{xxx} E^2{}_x + \chi_{xzz}{}^{2w} E_z{}^2 + 2\chi_{xxz}{}^{2w} E_x E_z; \\ P_y^{2w} &= 2\chi_{yxz}{}^{2w} E_x E_z; \\ P_z^{2w} &= \chi_{zxx}{}^{2w} (E_x{}^2 + E_z{}^2) + 2\chi_{zxz}{}^{2w} E_x E_z. \end{aligned} \quad (26)$$

The term that describes the oxidation of silicon with Si-O bonding is in the $\chi_{xxx}{}^{2w} = \alpha d_{11}^{0} + \beta d_{11}$; and $\chi_{yxz}{}^{2w} = \alpha d_{14}^{0}$. The $P^{2w}{}_y$ component will be equal to zero.

4.1 Light emitting devices by optical pumping

4.1.1 Light emitting devices based on photoluminescentce in visible range from oxidized nanocrystalline silicon

Controlled PL from silicon nanocrystals in vertical optical cavity (Inokuma et al., 2003) with distributed Bragg reflector (DBR) was performed with deepness 800 nm corresponds to maximum yield of PL wavelength. DBR (see Fig. 22) was made by using ten periodic layers

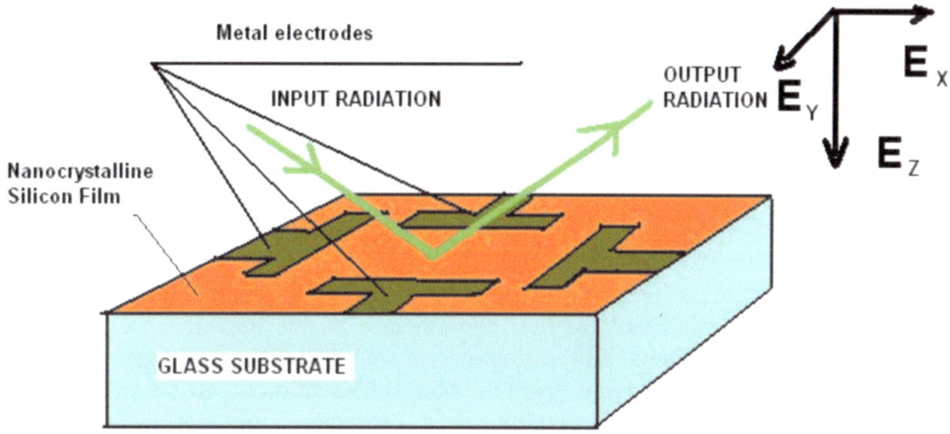

Fig. 21. Scheme of nonlinear optical semiconductor-based device.

of $SiO_{1.3}$ and $SiO_{1.8}$ with thickness 200 nm. Next, active layer with $SiO_{1.6}$ has its thickness of 800 nm. This layer was served, also, as optical cavity. It was completed by using annealing to 1100°C of silicon oxide film to produce the nanocrystals. Light emitting device can surely emitt at the wavelength 800 nm.

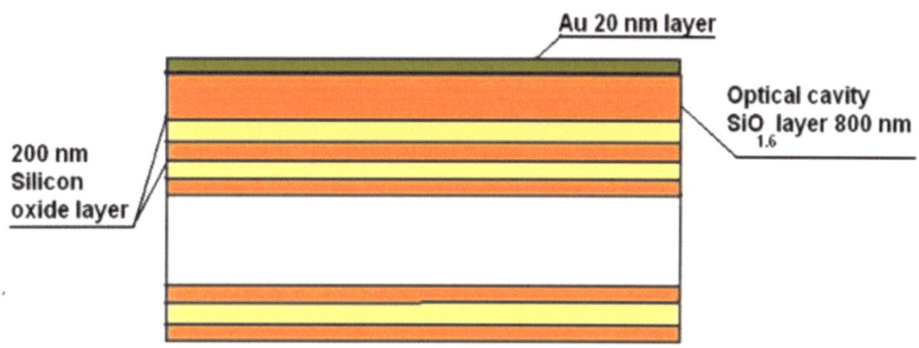

Fig. 22. Scheme of optical cavity with Au 20 nm top reflector layer and DBR (Inokuma et al., 2003).

4.1.2 Frequency-doubled resonant optically pumped vertical external-cavity laser

The process of resonance two-photon absorption can be described by means of cross-section: $W = \sigma^{(2w)} I^2$; where I is a fundamental radiation intensity. The cross-section of resonant two-photon absorption $\sigma^{(2\omega)}$ can be written as

$$\sigma^{(2\omega)} = \frac{512\pi^5}{h^3c^2}\frac{\omega}{1+4\omega^2}\left|\sum_n \left[\frac{<g|\mu|n.><n|\mu|f>}{\omega_{ng}-\omega}\right]\right|^2. \qquad (27)$$

By resonant two-photon absorption, usually, $\sigma^{(2\omega)}$ is approximately the same orders of magnitude as this value for one photon absorption. The energy of two-photon radiation is $E_{2\omega} = 3.26eV, (\lambda = 380nm)$ eV, that corresponds to the energy of defect level (VO) inside band gap of silicon by direct transitions. The cavity with length ℓ (see Fig. 23) is adjusted according to following requirements: $\Delta\nu = \frac{c}{2nl}; l = m\lambda$. Active layer that is made from nanocrystalline oxidized silicon has the thickness d =380 nm. Therefore, the conditions for lasing are the following: $\exp(\gamma d)[1-L]=1;\ L = 1 - r_1 r_2 \exp(-\alpha(l+d))$. The resulted SHG radiation due to the two-photon resonant absorption and appearance of inverted population on the excited level. Gold electrodes (as it is shown in Fig. 23) are nesessary for fine tuning the electronic structure of silicon film to eliminate the detuning between pump laser two photons and defect level position. Al mirror layer has the thickness 10 nm. By the diffraction on the silicon nanocrystals there are two tied cavities by one transition frequency. Each optical path differs from others. The estimations of frequency beats no more than 10^{12} *Hz* which comparable with LO phonons. From the Maxwell equation for electromagnetic media the following equations for fields, polarizations and inverted population in two interacted cavities are given by

$$\ddot{\vec{E}}_1 + \gamma \dot{\vec{E}}_1 + \omega_1^2 \vec{E}_1 = -\gamma \ddot{\vec{P}}_1 + m_1 \vec{E}_2;$$

$$\ddot{\vec{E}}_2 + \gamma \dot{\vec{E}}_2 + \omega_2^2 \vec{E}_2 = -\gamma \ddot{\vec{P}}_2 + m_2 \vec{E}_1;$$

$$\ddot{\vec{P}}_1 + \Delta_2 \dot{\vec{P}}_1 + \Omega^2 \vec{P}_1 = \frac{-4\pi\Omega}{h} L_{coeff} \frac{|\mu_1|^2}{3} N \vec{E}_1;$$

$$\ddot{\vec{P}}_2 + \Delta_2 \dot{\vec{P}}_2 + \Omega^2 \vec{P}_2 = \frac{-4\pi\Omega}{h} L_{coeff} \frac{|\mu_2|^2}{3} N \vec{E}_2;$$

$$\dot{N} + \Delta_1 (N - N_s) = \frac{4\pi}{\Omega} \left(\dot{\vec{P}}_1 + \dot{\vec{P}}_2 \right) \left(\vec{E}_1 + \vec{E}_2 \right).$$

Here, m_1, m_2 characterize the degree of interaction between the cavities, μ_1, μ_2 are matrix elements of dipole moments, L_{coeff} is Lorentz coefficient, $\gamma, \Delta_1, \Delta_2$ are the time duration of field relaxation inside cavity, broadenings of spectral lines, respectively. The solutions of system of equations can be written as harmonic functions by assumption of low changed magnitudes of field, polarization and phase values. By the way of simplification such complicated system of equations it is possible to present as the master equations for magnitudes of fields, phase and inverted population and solve them by using computer simulations:

$$\begin{aligned}
\dot{E}_1' &= E_1' \frac{1}{2}[DgN - \gamma] + \frac{m_1}{2\omega} E_2' \sin\Phi; \\
\dot{E}_2' &= E_2' \frac{1}{2}[DgN - \gamma] - \frac{m_2}{2\omega} E_1' \sin\Phi; \\
\dot{\Phi} &= \frac{\omega_2^2 - \omega_1^2}{2\omega} + \frac{1}{2\omega}\left[m_1 \frac{E_2}{E_1} - m_2 \frac{E_1}{E_2} \right] \cos\Phi; \\
\dot{N} + \Delta_1 (N - N_S) &= -\frac{2\pi DNg\varepsilon}{h\Omega} \left((E_1')^2 + (E_2')^2 + 2E_1' E_2' \cos\Phi \right);
\end{aligned} \tag{28}$$

where

$$D = \frac{4\pi\Omega}{3h\Delta_2\varepsilon} L_{coeff} |\mu|^2; \; \Phi = \phi_2 - \phi_1; \; g = \frac{\Delta_2^{\,2}}{4\left((\omega - \Omega)^2 + 1\right)}.$$

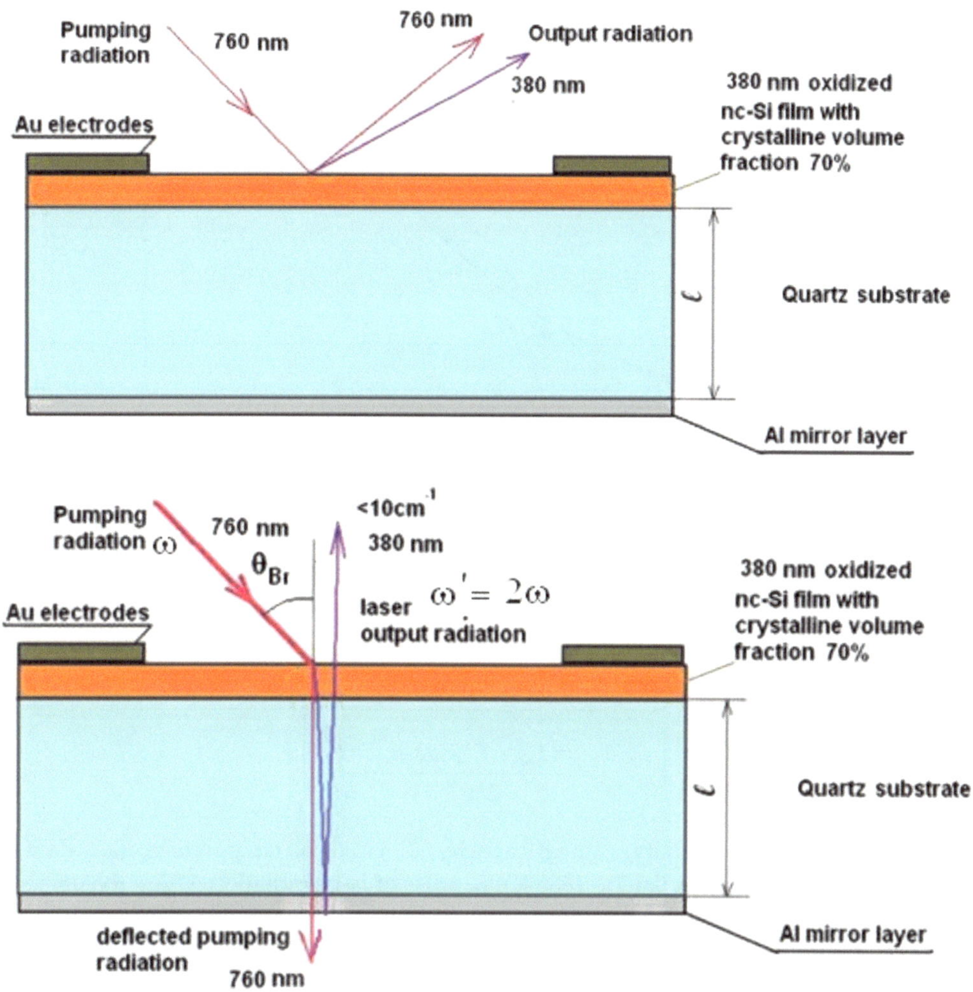

Fig. 23. Schemes of SHG switcher (upper picture) for TE mode and optical cavity and active layer serving as UV laser (down picture) with radiation at wavelength 380 nm for TM mode.

4.1.3 Damage threshold intensity

Figure 24 illustrates the SHG signal as a function of laser power with wavelength of radiation 680 nm. The maximal value of laser power reflects the damage threshold of silicon film. The intensity of SHG has the following view: $I_{SHG} = \eta_{SHG} I^2$ where $\eta_{SHG} \propto |L(2\omega,\rho)L^2(\omega,\rho)|^2 |\chi_{SHG}|^2$; η_{SHG} = 0.05x10-18 cm2/W, for silicon films. By the power

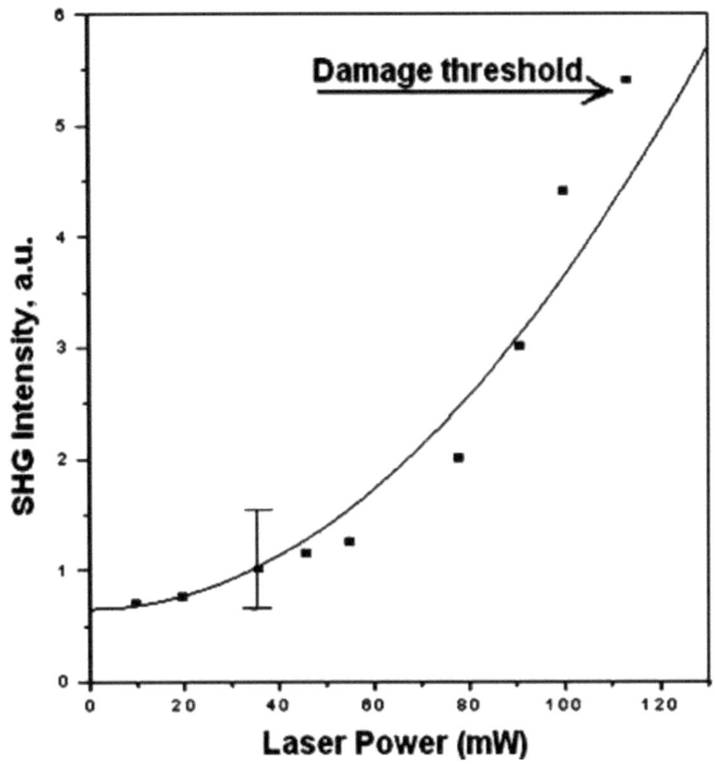

Fig. 24. SHG as a function of primary laser power for the radiation with wavelength 680 nm.

density of primary radiation 375 MW/cm2 the intensity of SHG signal can be estimated as 12.5 mW/cm2 poly-Si films (Milovzorov & Suzuki, 1999). The value of damage threshold intensity for p-polarized light can be represented in the form:

$$I_{threshold,p} = \frac{\pi(T_{melt} - T_0)(\pi\kappa c_p\rho)^{1/2}}{2c\tau_p^{1/2}} \frac{|\varepsilon'|^{3/2}}{\varepsilon''} \tag{29}$$

where κ is thermal conductivity, c_p heat capacity, T_0 is initial temperature, T_{melt} damage threshold temperature (for Si Tmelt =1685 K), τ_p pulse of laser radiation. Using the values of T_{melt}-T_0 = 1412 K, $\tau_p \approx$ 4 ns (by λ_w = 680 nm), κ = 1.56x10^7 erg/cm s deg, c_p=0.69x10^7 erg/g deg, ρ = 2.32 g/cm^3, we can estimate the damage threshold intensity $I_{threshold,p} \approx$ 4x10^8 W/cm^2 from (Eq.22).

4.2 Theoretical model of silicon nanocrystals by laser pumping

By this way the model of photon-assisted electron transport is proposed here. Such approach can be described in second-quantized form of operators of creating and annihilation: $H_{QE} = H + H_L + H_R$

$$H = \sum_{k=1}^{N}[E_{1k}c_{1k}^{+}c_{1k} + E_{2k}c_{2k}^{+}c_{2k} + \sum_{i,j=1}^{2}(V_{ikjk+1}c_{ik}^{+}c_{jk+1} + h.c.) + (V_{12k}c_{1k}^{+}c_{2k} + h.c.)] \tag{30}$$

where E_{1k}, E_{2k} are the energies of levels, V_{ikjk+1} is matrix element of tunneling, H_R and H_L are the terms which describe the right and left branches from the system, V_{12k} is the matrix element that describes the interaction of system with laser radiation. The matrix element of tunneling depends on the time as $V_{ikjk+1} = V^0_{ikjk+1} \exp(-S)$, where, according to the perturbation theory, $U = U_0 + \delta U$, $\delta U(t) = eV_{ext} + eE_l x \cos(\omega_l t)$. Here $\delta U << U_0$. The master equation for the creation-annihilation operators can be written in matrix form: $i\dot{C} = HC$, where the C is vector. Using block matrix 2x2 (the diagonal elements are the elements of energies of levels in two-level system, and others are tunneling matrix elements) we can write the matrix of Hamiltonian as

$$H = \begin{Vmatrix} A_{11} & A_{12} & 0 \\ A_{21} & A_{22} & A_{23} \\ 0 & A_{32} & A_{33} \end{Vmatrix} \tag{31}$$

where diagonal block matrix $A_{jj} = \begin{Vmatrix} E_1^{'} & 0 \\ 0 & E_2^{'} \end{Vmatrix}$ describe the energy of single quantum well interacted with electromagnetic field and eigen values of this "dressed-atom" model are following

$$E_{1,2}^{'} = \frac{E_1 + E_2}{2} \pm \sqrt{\frac{(E_2 - E_1)^2}{4} - V_{12}V_{21}} \tag{32}$$

The non-diagonal block matrix $A_{ijk-1} = \begin{Vmatrix} V'_{ikik-1} & 0 \\ 0 & V'_{jkjk-1} \end{Vmatrix}$,

$$,V'_{ikjk-1} = \frac{V_{ikik-1} + V_{jkjk-1}}{2} \pm \sqrt{\frac{(V_{ikik-1} - V_{jkjk-1})^2}{4} - V_{ikjk-1}V_{jkik-1}}\,.$$

The solution of differential equation

$$i\dot{C}_k = \hat{H} C_k \tag{33}$$

and $C_k(t) = C_k(0)\exp\left(-i\int_0^t \hat{A}_{22}\, d\tau\right)$. The $C_{k+1}(t) = C_{k+1}(0)\exp\left(-i\int_0^t \hat{A}_{33}\, d\tau\right)$ is obtained according to the same procedure. By this way the resulted matrix has the eigen values which correspond to the energies for interacted terms with indexes k and k+1. Matrix tunneling element is possible to combine of two multiplied term. Each of them explains the order 1 and 2 of perturbation energy for tunneling process through the potential barrier between quantum wells.

$$V_{ikjk+1} = V^0_{ikjk+1} \exp\left(-\frac{2}{\hbar}\int_0^d \sqrt{2m(U_0 - E)}dx\right)\exp\left(\frac{1}{\hbar}\int_0^d \frac{m\delta U dx}{\sqrt{2m(U_0 - E)}}\right). \tag{34}$$

It can be used as $V_{ikjk+1} = V^0_{ikjk+1} F(t)$; where $F(t) = \exp\left(Z\cos(w_l t)\right)$ and parameter of field-system interaction:

$$Z = \frac{1}{\hbar\sqrt{2}} \frac{\sqrt{m}d^2 eE_l}{\sqrt{U_0 - E}}. \tag{35}$$

$$n_{2k}(t) = c_{2k}{}^{+}(t)c_{2k}(t); \tag{36}$$

$$c_{2k}(t) = c^0{}_{2k} \exp\left(-i\int_0^t \left[E_{2k} + \frac{\Delta_{12}(t')}{hw}\right]dt'\right) \exp\left(i\int_0^t \frac{\Delta_{2k2k-1}(t)}{E_{2k-1} + \frac{\Delta_{12}(t)}{hw}} dt'\right); \tag{37}$$

For upper level population it can be transformed into following expression:

$$n_{2k}(t) \cong n^0{}_{2k}(1 + \frac{8\pi\Delta_{2k2k-1}}{h\nu(E_{2k} + \frac{\Delta_{12}(t')}{hw})} \sum_m J_m(Z)J_m(-Z)K_{2m}(-2Z_2)); \tag{38}$$

The saturated gain (Verdeyen, 1995) of a laser oscillator is stated as

$G^2 = \frac{1}{R_1 R_2}\left[1 - n_2 c\sigma h\nu \frac{(1 - R_1 R_2)}{P_{out}}\right]$. It is assumed, that for the scheme which is shown in Fig 23 the laser technical features

are following: reflection of Al mirror and silicon film is $R_1 = 0.99, R_2 = 0.57$,respectively, frequency of laser generation by two-photon pumping mode: $\nu = 7.510^{14} Hz$. The value σ is stimulated emission cross section that can be evaluated as 10^{12} cm^2, but output laser power $P_{out} \approx 17mW$. Accordingly, $G^2 << 1$, because of there is no saturation of upper level population. The losses of radiation inside cavity are managed by bias voltage applied to electrodes as shown in Fig.23. The quality factor of the cavity should be more than 10^3. In addition, the unsaturated gain and inverted population are linear related as $G = \sigma\left(n_{2k} - n_{1k}\right)$. Evolution of upper-level population of two-level system by laser pumping is illustrated in Fig. 25. For high carrier diffusion values and small laser field (see Fig. 25 a) and b)) the saturation mode of upper-level populations can be surely observed. By changing in photon energy of laser radiation and electron level detuning, the difference between the electron energy and energy position of level, the Z parameter there is a several modes in population dynamics correspond to the ordinary laser active media modes.

5. Conclusions

Optical spectral characteristics for nanocrystalline silicon films are investigated for their application in optoelectronic device manufacturing. Different optoelectronic devices based on nanocrystalline silicon films such as acoustoelectrical optical mode switcher for laser radiation, nonlinear optical switcher, and frequency-doubled resonant optically pumped vertical external-cavity laser is proposed. There is a significant yield from the using nanocrystalline silicon films in device implementation for optoelectronic. Their application is wide spread and became broader than before by each year.Frequency-doubled, optically pumped vertical external-cavity surface-emitting lasers (VECSEL) have numerous applications as for waveguiding and flat panel displays. In recent years the fiber laser have grew in power as on their fundamental frequencies as in SHG yields.

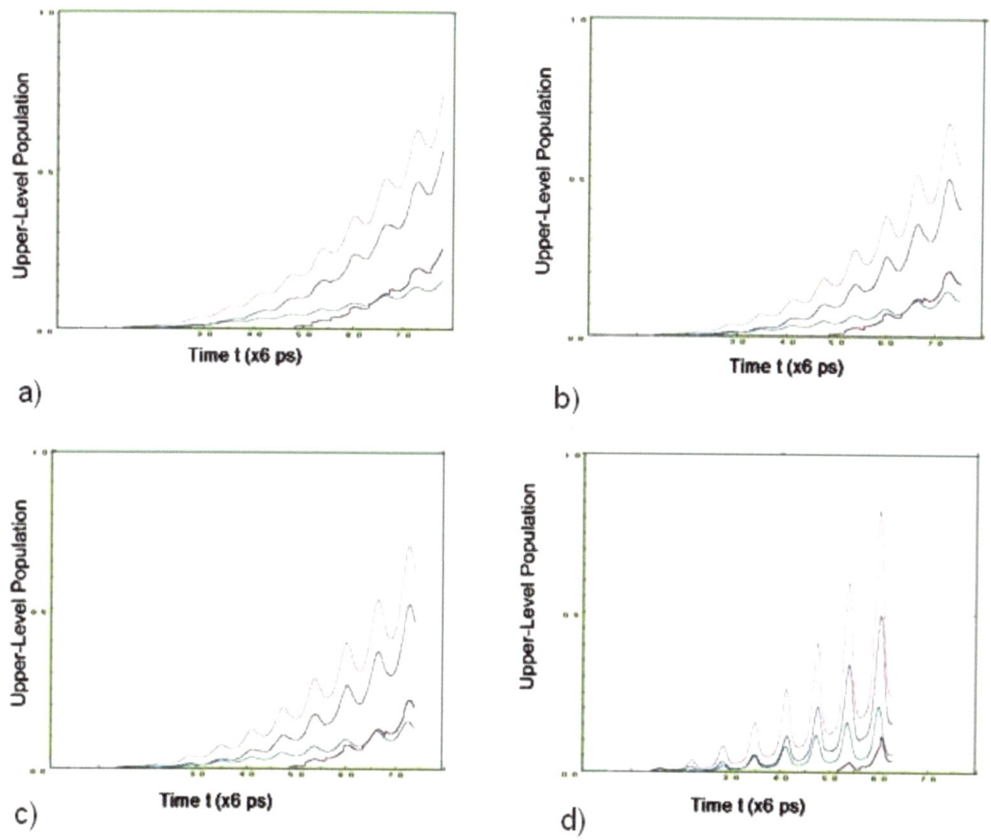

Fig. 25. Evolution of upper-level population for system with index k=38 by the parameters: a) Z=0.3; Ω=90 GHz; b) Z=0.5, Ω=150 GHz; c) Z=0.6, Ω=180GHz;d) Z=1, Ω=0.6 THz.

The nonlinear optical processing the signal often depends on the variation in refraction of index which changes the speed of travelling through the fibre and these results in an error signal appearance. It is dramatic by transmission the signal up to 1000 km and more. There are problems which signal degradation: amplification of sponteneous emission, four-wave mixing, self-phase mixing and cross-phase mixing. All of them results in degrading the signal to noise ratio (SNR). For 10 Gbit/s this ratio 100:1, or 20 dB (Livas, 2002). Using two mode laser with two interacted cavities can be easily used to prihibit high losses and prevent the signal changes due to refraction index deviation. Multiple optical beams in optical network pass through a large number of wavelength switchers which consist of multiplexer, demultiplexer, optical cross connects, optical amplifiers. Using the Raman amplifiers based on stimulated Raman scattering effect is, also, very effective because there is a possibility in using any type of transmission fiber.
Polarization is caused by the interaction of radiation with the molecules of fiber. The degree of polarization depends on the temperature and stress conditions, such as stretching and

bending modes of silica molecules of the fiber. If polarization of fiber is changing the polarization dispersion is increases and polarization-dependent losses increase, too. Fiber amplifier (FA) cuts the determined frequency from the others as filter by its amplifying. By using the FA device to input signal the SNR is affected, by using it to output signal the amplitude is dramatically increasing. To prevent the losses by modal dispersion is possible to change mode and use another single mode fiber. The laser systems with interacted cavities can be used to change the frequency modes from the one to another by efficient transfusion of laser energy. Accordingly, by using conversation law for both photons of modes which are closed each to other it is possible to determine the signal from new frequency mode. Therefore, the further investigations open great possibilities to make new optoelectronic devices fore speed data transforming and data recognition in optical networking

6. References:

Chabal, Y., Ragnhavachari, K., Zhang, X., Garfunkel, E., Silanone on Si (100): intermediate for initial silicon oxidation, Physical *Review B*, 66 (2002) 161315-1-4. ISSN 0163-1829.

Chatterjee, A., Iwasaki, T., Ebina, T., Structural and energetic changes of Si (100) surface with fluorine in presence of water - a density functional study, *International Journal of Molecular Sciences*, 2 (2001) 40-56, ISSN 1422-0067.

Corbett, J., Watkins, G., Chrenko, R., McDonalds, R., Defects in Irradiated Silicon: Infrared Absorption of the Si- A center, *Physical Review*, v.121, n4, (1961) 1015-1022 ISSN 0163-1829.; Corbett, J., Watkins, G., McDonalds, R., New oxygen infrared bands in annealed irradiated silicon, *Physical Review*, v.135, N5A, (1964) A1381-A1385. ISSN 0163-1829.

Delerue C., Allan, G., Lanoo, M., Theoretical aspects of the luminescence of porous silicon, Physical Review B, V.48, N.15, 15 October 1993, 11024-11036. ISSN 0163-1829.

Deveaud, B., *The physics of semiconductor microcavities*, Wiley-VCH Verlag GmbH & Co., Wiendheim, 2007, p. 7.

Dolgova, T., Maidykovski, M., Martemyanov, M., Fedyanin, A., Aktsipetrov, O., Marowsky, G., Yakovlev, V., Mattei, M., Giant microcavity enhancement of second-harmonic generation in all-silicon photonic crystals, *Applied Physics Letters*, 81, 2725-2727 (2002) ISSN 0003-6951.

Durr, M., Hofer, U., Dissociative adsorption of molecular hydrogen on silicon surface, Surface Science Reports, 61 (2006) pp.465-526.

Feuer, P., Electronic States in Crystals under Large Over-All Perturbations, *Physics Review*, v.88, n.1, (1952) 92-100. ISSN 0163-1829, ISSN 0163-1829.

Han, D., Yue, G., Lorentzen, J., Lin, J., Habuchi, H., Wang, Q., Optical and electronic properties of microcrystalline silicon as a function of microcrystallinity, *Journmal of Applied Physics*, 87, 1882 (2000)ISSN0021-8979; Han, D., Lorentzen, J., Weinberg-Wolf, J., MCNeil, L., Raman study of thin films of amorphous-to-microcrystalline silicon prepared by hot-wire chemical vapor deposition, *Journal of Applied Physics*, 94, 2930 (2003) ISSN0021-8979.

Inokuma, T., Kurata, Y., and Hasegawa, S., Controlled photoluminescence from silicon nanocrystals in a vertical optical cavity with distributed Bragg reflectors, *Physica Status Solidi (c)*, 0 , N4(2003) 1258-1261.ISSN 1610-1634.

Kahan, V., High frequency hopping transport in solids and dielectric transparency in compensated semiconductors, *JETP*, 117, (2000) 452-456.

King, R., Mastrykov, V., Schaefter, H., The electron affinities of the silicon fluorides SiF_n (n=1-5), *Journal of Chemical Physics*, 105, (1996) 6880-6886. ISSN 0021-9606.

Klein, W., Cook, B., Unified approach to ultrasonic light difraction, *IEEE Transaction*, SU-14, N3 (1967) 123134.ISSN 0018-9383.

Lanoo, M., Allan, G., A cluster plus effective tight-binding study of SiO_x systems, *Solid State Communications*, 28 (1978) 733-739.ISSN 0038-1098.

Livas, J., Conditioning squeezes the best out of signals, *Lightwave Europe*, October 2002, 18-20 ISSN 1447-5069.

Matsuda, A., Growth mechanism of microcrystalline silicon obtained from reactive plasmas, *Thin Solid Films*, 337 (1999) 1-6 ISSN 0040-6090; Matsuda, A., Microcrystalline silicon. Growth and device application, *Journal of Non-Crystalline Solids*, 338-340 (2004) 1-12.ISSN 0022-3093.

Milovzorov, D., Inokuma, T., Kurata, Y., Hasegawa, S., Relationship between structural and optical properties in polycrystalline silicon films prepared at low temperature by PECVD, *Journal of Electrochemical Society*, 145, (1998) 3615-3620; Milovzorov , D., and Suzuki, T., Size-dependent second harmonic generation of nanocrystals prepared by plasma-enhanced chemical vapor deposition, *Applied Physics Letters*, 75, 4103-4105 (1999) ISSN 0003-6951.;. Milovzorov, D., Ali, A., Inokuma, T., Kurata, Y., Suzuki, T., and Hasegawa, S., Optical properties of silicon nanocrystallites in polycrystalline silicon films prepared at low temperature by plasma-enhanced chemical vapor deposition, *Thin Solid Films*, 382, 53-55 (2001)ISSN0040-6090; Milovzorov, D., Electronic structure of nanocrystalline silicon and oxidized silicon surfaces, *Electrochemical and Solid State Letters*, 4(7), 2001, G61-63 ISSN0013-4651; Milovzorov, D., Point defects in amorphous and nanocrystalline fluorinated silicon films, *Journal of Materials Science and Engineering with Advanced Technology*, v.2, n.1, (2010) 41-59.ISSN0976-1446.

Pellegrini, V., Colombelli, R., Carusotto, I., Beltram, F., Rubini, S.,.Lantier,R., Franciosi, A., Vinegoni, C., Pavesi, L., Resonant second harmonic generation in ZnSe bulk microcavity, *Applied Physics Letters*, 74 , 1945-1947 (1999). ISSN 0003-6951.

Rotter, M., Wixfirth, A., Rulle, W., Bernklau, D., Riechert, H., Giant acoustoelectric effect in GaAs/liNbO3 hybrids, *Applied Physics Letters*, 73 (1998) 2128-2130. ISSN 0003-6951.

Shklyaev, A., Aono, M., Suzuki, T., Thermaly enhanced second-harmonic generation, *Surface Science*, 423, 61 (1999) ISSN0039-6028.

Tsu, D., Chao, B., Ovchinsky, S., Jones, S., Heterogeneity in hydrogenated silicon: Evidence for intermediately ordered chainlike objects, *Physical Review B*, 63, 125338 (2001). ISSN 0163-1829.

Verdeyen, J., *Laser electronics*, Prentice Hall, London, 1995, p. 150. ISBN 0-13-101668-7.

Wang, J., Zou, B., El-Sayed, M., Comparison between the polarized Fourier-transform infrared spectra of aged porous silicon and amorphous silicon dioxide films on Si(100) surface, *Journal of Molecular Structure,* 508 (1999) 87-96. ISSN 0022-2860.

Weinreich, G., Sanders, T., White, H., Acoustoelectric effect in n-type Germanium, *Physical Review,* 114 (1959) 33-44. ISSN 0163-1829.

Xu, J., Div Engineering, Directly Pumped Crystalline Silicon Laser an impossible possibility, Brown University, Providence, RI 02912, 213-215,ISBN:1-4244-0096-1,INSPEC Accession Number:9274227.

Yu P., Cardona M., *Fundamentals of semiconductors, Physics and Materials Properties,* Springer, Berlin, 1996, p.207.ISBN3-540-61461-3.